Gmelin Handbook of Inorganic and Organometallic Chemistry

8th Edition

Gmelin Handbook of Inorganic and Organometallic Chemistry

8th Edition

Gmelin Handbuch der Anorganischen Chemie

Achte, völlig neu bearbeitete Auflage

PREPARED AND ISSUED BY

Gmelin-Institut für Anorganische Chemie
der Max-Planck-Gesellschaft
zur Förderung der Wissenschaften

Director: Ekkehard Fluck

FOUNDED BY Leopold Gmelin

8TH EDITION 8th Edition begun under the auspices of the Deutsche Chemische Gesellschaft by R. J. Meyer

CONTINUED BY E. H. E. Pietsch and A. Kotowski, and by Margot Becke-Goehring

Springer-Verlag
Berlin · Heidelberg · New York · London · Paris · Tokyo ·
Hong Kong · Barcelona · Budapest 1993

Gmelin-Institut für Anorganische Chemie
der Max-Planck-Gesellschaft zur Förderung der Wissenschaften

GMELIN HANDBOOK

Dr. J. von Jouanne

Dr. L. Berg, Dr. H. Bergmann, Dr. J. Faust, J. Füssel, Dr. H. Katscher, Dr. R. Keim, Dipl.-Phys. D. Koschel, Dr. A. Kubny, Dr. P. Merlet, Dr. M. Mirbach, Prof. Dr. W. Petz, Dr. F. A. Schröder, Dr. A. Slawisch, Dr. W. Töpper

Dr. R. Albrecht, Dr. G. Bär, D. Barthel, Dr. N. Baumann, Dr. K. Behrends, Dr. W. Behrendt, D. Benzaid, Dr. R. Bohrer, K. D. Bonn, Dipl.-Chem. U. Boßlet, Dr. U. Busch, A.-K. Castro, Dipl.-Ing. V. A. Chavizon, E. Cloos, A. Dittmar, Dipl.-Geol. R. Ditz, R. Dowideit, Dr. H.-J. Fachmann, B. Fischer, Dr. D. Fischer, Dipl.-Ing. N. Gagel, Dr. K. Greiner, Dipl.-Bibl. W. Grieser, Dr. R. Haubold, Dipl.-Min. H. Hein, H.-P. Hente, H. W. Herold, U. Hettwer, G. Hoell, Dr. G. Hönes, Dr. W. Hoffmann, G. Horndasch, Dr. W. Huisl, Dr. M. Irmler, B. Jaeger, Dr. R. Jotter, Dipl.-Chem. P. Kämpf, Dr. B. Kalbskopf, H.-G. Karrenberg, Dipl.-Phys. H. Keller-Rudek, Dipl.-Chem. C. Koeppel, Dr. M. Körfer, R. Kolb, Dr. M. Kotowski, E. Kranz, E. Krawczyk, Dipl.-Chem. I. Kreuzbichler, Dr. V. Kruppa, Dr. W. Kurtz, M. Langer, Dr. B. Ledüc, H. Mathis, E. Meinhard, M. Meßer, C. Metz, K. Meyer, E. Mlitzke, Dipl.-Chem. B. Mohsin, Dr. U. Neu-Becker, K. Nöring, Dipl.-Min. U. Nohl, Dr. U. Ohms-Bredemann, Dr. H. Pscheidl, Dipl.-Phys. H.-J. Richter-Ditten, E. Rudolph, G. Rudolph, Dr. B. Sarbas, Dr. H. Schäfer, Dr. R. Schemm, Dr. D. Schiöberg, V. Schlicht, Dipl.-Chem. D. Schneider, E. Schneider, A. Schwärzel, Dr. B. Schwager, R. Simeone, Dr. F. Stein, Dr. C. Strametz, Dr. G. Swoboda, Dr. D. Tille, A. Tuttas, Dipl.-Phys. J. Wagner, R. Wagner, M. Walter, Dr. E. Warkentin, Dr. C. Weber, Dr. A. Wietelmann, Dr. M. Winter, Dr. B. Wöbke, K. Wolff

GMELIN ONLINE

Dr. R. Deplanque

Dr. P. Kuhn, Dr. G. Olbrich

Dr. R. Baier, Dr. B. Becker, Dipl.-Chem. E. Best, Dr. H.-U. Böhmer, Dipl.-Phys. R. Bost, Dr. A. Brandl, Dr. R. Braun, Dr. T. Buck, Dipl.-Chem. R. Durban, R. Hanz, Dipl.-Phys. C. Heinrich-Sterzel, Dr. S. Kalwellis-Mohn, Dr. A. Kirchhoff, Dipl.-Chem. H. Köttelwesch, Dr. M. Kunz, Dipl.-Chem. R. Maass, Dr. K. Müller, Dr. A. Nebel, Dipl.-Chem. R. Nohl, Dr. M. Nohlen, H. Reinhardt, Dr. B. Rempfer, Dipl.-Ing. H. Vanecek

The following Gmelin Formula Index volumes have been published up to now:

Formula Index

Volume 1	Ac–Au
Volume 2	B–Br_2
Volume 3	Br_3–C_3
Volume 4	C_4–C_7
Volume 5	C_8–C_{12}
Volume 6	C_{13}–C_{23}
Volume 7	C_{24}–Ca
Volume 8	Cb–Cl
Volume 9	Cm–Fr
Volume 10	Ga–I
Volume 11	In–Ns
Volume 12	O–Zr
	Elements 104–132

Formula Index 1st Supplement

Volume 1	Ac–Au
Volume 2	B–$B_{1.9}$
Volume 3	B_2–B_{100}
Volume 4	Ba–C_7
Volume 5	C_8–C_{17}
Volume 6	C_{18}–C_x
Volume 7	Ca–I
Volume 8	In–Zr
	Elements 104–120

Formula Index 2nd Supplement

Volume 1	Ac–$B_{1.9}$
Volume 2	B_2–Br_x
Volume 3	C–$C_{6.9}$
Volume 4	C_7–$C_{11.4}$
Volume 5	C_{12}–$C_{16.5}$
Volume 6	C_{17}–$C5_{22.5}$
Volume 7	C_{23}–$C_{32.5}$
Volume 8	C_{33}–Cf
Volume 9	Cl–Ho
Volume 10	I–Zr

Formula Index 3rd Supplement

Volume 1	Ag–B_5
Volume 2	B_6–$C_{7.5}$
Volume 3	C_8–C_{14} (present volume)

Gmelin Handbook of Inorganic and Organometallic Chemistry

8th Edition

INDEX

Formula Index

3rd Supplement Volume 3

B_8–C_{14}

AUTHORS Rainer Bohrer, Bernd Kalbskopf, Uwe Nohl, Hans-Jürgen Richter-Ditten, Paul Kämpf

CHIEF EDITORS Uwe Nohl, Gottfried Olbrich

Springer-Verlag
Berlin · Heidelberg · New York · London · Paris · Tokyo · Hong Kong · Barcelona · Budapest 1993

THE VOLUMES OF THE GMELIN HANDBOOK ARE EVALUATED FROM 1988 THROUGH 1992

Library of Congress Catalog Card Number: Agr 25-1383

ISBN 3-540-93678-5 Springer-Verlag, Berlin · Heidelberg · New York · London · Paris · Tokyo
ISBN 0-387-93678-5 Springer-Verlag, New York · Heidelberg · Berlin · London · Paris · Tokyo

Printed in Germany

Typesetting, printing, and bookbinding: Universitätsdruckerei H. Stürtz AG, Würzburg

Preface

The Gmelin Formula Index and the First and Second Supplement covered the volumes of the Eighth Edition of the Gmelin Handbook which appeared up to the end of 1987.

This Third Supplement extends the Gmelin Formula Index and includes the compounds from the volumes until 1992. The publication of the Third Supplement enables to locate all compounds described in the Gmelin Handbook of Inorganic and Organometallic Chemistry since 1924. The basic structure of the Formula Index remains the same as the previous editions.

Computer methods were employed during the preparation and the publication of the Third Supplement. Data acquisition, sorting, and data handling were performed using a suite of computer programs, developed originally by B. Roth, now at Chemplex GmbH. The SGML application for the final data processing for printing was developed in the computer department of the Gmelin Institute and at Universitätsdruckerei H. Stürtz AG, Würzburg.

Frankfurt am Main,
October 1993

U. Nohl, G. Olbrich

Instructions for Users of the Formula Index

First Column (Empirical Formula)

The empirical formulae are arranged in alphabetical order of the element symbols and by increasing values of the subscripts. Any indefinite subscripts are placed at the end of the respective sorting section. Ions always appear after the neutral species, positive ions preceding negative ones.

H_2O is included in empirical formulae only if it is an integral part of a complex, as indicated in the second column. For compounds which are described as solvates only both empirical formulae are given, with and without the solvent molecules. Multicomponent systems (solid solutions, melts, etc.) are listed under the empirical formulae of their respective components. However, solutions are found only under the solute, and polymers of the type $(AB)_n$ are sorted under AB.

Second Column (Linearized Formula)

The second column contains a linearized formula to indicate the constitution and configuration of a compound as close as possible. The formula given corresponds to that given in the handbook, except in cases where additional structural features can be described in more detail. For elements the names are included.

Entries with the same composition but with different structural formulae are arranged in the following order: elements or compounds, isotopic species, polymers, hydrates, and multicomponent systems.

For multicomponent systems the components are arranged in the sequence: inorganic components-organic components-water. The inorganic components are sorted alphabetically, the organic components according to the number of carbon atoms. If a component is a single element it is always represented by the unsubscripted atomic symbol. The term "system" is used in a restricted sense in this index; it represents mixtures described by phase diagrams or sometimes by, e.g., eutectic points.

Elements and compounds whose treatment in the handbook requires a larger amount of space are further characterized by topics like physical properties, preparation, electrochemical behaviour, etc.

Third Column (Volume and Page Numbers)

This column contains the volume descriptor and the page numbers, both separated by a hyphen. The volume descriptors consist of the atomic symbol of the element which is treated in a given volume, followed by an abbreviated form of the type of volume, including the part or section. The following abbreviations are used for the type of volume:

MVol.	Main Volume
SVol.	Supplement Volume
Org. Comp.	Organic Compounds
PFHOrg.	Perfluorohalogenoorganic Compounds of Main Group Elements
SVol.GD	Gmelin-Durrer, Metallurgy of Iron
Biol.Med.Ph.	Bor in Biologie, Medizin und Pharmazie

Volume descriptors like "3rd Suppl. Vol. 4" are abbreviated as "SVol. 3/4". For instance, the entry "B: B Comp.SVol. 3/4-345" indicates that the information can be found on page 345 of the boron volume "Boron Compounds 3rd Supplement Volume 4".

$C_8 - C_{14}$

$C_8CaF_8N_2O_4$	$Ca[OC(O)\text{-}CF_2CF_2\text{-}CN]_2$	F: PFHOrg.SVol.6-100/1, 132
$C_8ClCoH_{16}N_8Pd$	$[(NC)_3Pd\text{-}CN\text{-}Co(NH_2CH_2CH_2NH_2)_2Cl] \cdot n\ H_2O$	Pd: SVol.B2-287
$C_8ClCrH_{18}N_8OPd$	$[Cr(NH_2CH_2CH_2NH_2)_2(H_2O)Cl][Pd(CN)_4] \cdot H_2O$	Pd: SVol.B2-279
$C_8ClF_2H_9MoNO_2P$	$(C_5H_5)Mo(CO)_2(Cl)(F_2P\text{-}NHCH_3)$	Mo: Org.Comp.7-57, 92/3
$C_8ClF_3H_5NOS$	$O{=}S{=}NCCl(CF_3)C_6H_5$	S: S-N Comp.6-109, 112
$C_8ClF_3N_2$	$5\text{-}Cl\text{-}1,3\text{-}(NC)_2\text{-}C_6F_3$	F: PFHOrg.SVol.6-107, 152
$C_8ClF_4GeH_{15}$	$Ge(C_2H_5)_3CFClCF_3$	Ge: Org.Comp.2-129, 138
$C_8ClF_4H_8MoNO_2P_2$	$(C_5H_5)Mo(CO)_2(Cl)[F_2P\text{-}N(CH_3)\text{-}PF_2]$	Mo: Org.Comp.7-56/7, 96
–	$[(C_5H_5)Mo(CO)_2(F_2P\text{-}N(CH_3)\text{-}PF_2)]Cl$	Mo: Org.Comp.7-249, 276
$C_8ClF_4O_5Re$	$(CO)_5Re\text{-}CF{=}C(Cl)CF_3$	Re: Org.Comp.2-138
–	$(CO)_5Re\text{-}CF_2C(Cl){=}CF_2$	Re: Org.Comp.2-136/7
$C_8ClF_5H_7N_5$	$[C_6F_5\text{-}NH\text{-}C({=}NH)\text{-}NH\text{-}C({=}NH)\text{-}NH_3]Cl$	F: PFHOrg.SVol.5-3, 36, 73
$C_8ClF_5N_2$	$C_8F_5ClN_2$	F: PFHOrg.SVol.4-292, 310/1
$C_8ClF_6H_{12}MoN_4OP$	$[Mo(CN\text{-}CH_3)_4(O)(Cl)][PF_6]$	Mo: Org.Comp.5-39, 40, 45
C_8ClF_6NS	$CF_2ClC_7F_4NS$	F: PFHOrg.SVol.4-284, 298
C_8ClF_7HNS	$CF_2Cl\text{-}C({=}S)NH\text{-}C_6F_5$	F: PFHOrg.SVol.5-25, 59, 93
C_8ClF_8N	$C_6F_5\text{-}N{=}CCl\text{-}CF_3$	F: PFHOrg.SVol.6-192/3, 206, 219
$C_8ClF_{12}NO$	$NC\text{-}(CF_2)_6\text{-}CCl{=}O$	F: PFHOrg.SVol.6-101, 140
$C_8ClF_{16}N$	$1\text{-}CF_3CFClCF_2\text{-}NC_5F_{10}$	F: PFHOrg.SVol.4-122, 134
–	$1\text{-}ClCF_2CF_2CF_2\text{-}4\text{-}CF_3\text{-}NC_4F_7$	F: PFHOrg.SVol.4-27/8, 50
–	$1\text{-}ClCF_2CF_2CF_2\text{-}NC_5F_{10}$	F: PFHOrg.SVol.4-122, 134
–	$1\text{-}ClCF_2CF_2\text{-}2\text{-}CF_3\text{-}NC_5F_9$	F: PFHOrg.SVol.4-122, 135
–	$1\text{-}ClCF_2CF_2\text{-}3\text{-}CF_3\text{-}NC_5F_9$	F: PFHOrg.SVol.4-122, 136
–	$1\text{-}ClCF_2CF_2\text{-}3\text{-}C_2F_5\text{-}NC_4F_7$	F: PFHOrg.SVol.4-27/8, 51
–	$1\text{-}ClCF_2CF_2\text{-}4\text{-}CF_3\text{-}NC_5F_9$	F: PFHOrg.SVol.4-122/3, 137
–	$1\text{-}ClCF_2CF_2\text{-}NC_6F_{12}$	F: PFHOrg.SVol.4-274/5
$C_8ClFeH_5O_3$	$[(C_5H_5)Fe(CO)_3]Cl$	Fe: Org.Comp.B15-38, 46
$C_8ClFeH_5O_4$	$[(HO)C_5H_4Fe(CO)_3]Cl$	Fe: Org.Comp.B15-50/1
$C_8ClFeH_7O_7$	$[(CH_2CHCHCHCH_2)Fe(CO)_3][ClO_4]$	Fe: Org.Comp.B15-13, 18
C_8ClFeH_8NO	$C_5H_5Fe(CNCH_3)(CO)Cl$	Fe: Org.Comp.B15-287, 293/4
$C_8ClGaH_{16}O_4$	$[Ga(1,4\text{-}O_2C_4H_8)_2]Cl$	Ga: SVol.D1-142/3
$C_8ClGeH_{11}O_2$	$Ge(CH_3)_3C({=}CHCH{=}C(COCl)O)$	Ge: Org.Comp.2-97

C_8ClGeH_{15}	$CH_3CHClCH_2(CH_3)Ge(-CH_2CH{=}CHCH_2-)$	Ge: Org.Comp.3-266
$C_8ClGeH_{15}S$	$Ge(CH_3)_3C(SCH_3){=}CClCH{=}CH_2$	Ge: Org.Comp.2-17
C_8ClGeH_{19}	$(C_2H_5)_3Ge-CHClCH_3$	Ge: Org.Comp.2-127
–	$(C_2H_5)_3Ge-CH_2CH_2Cl$	Ge: Org.Comp.2-124
$C_8ClH_4O_6Re$	$(CO)_5ReC(O)CH_2CH_2Cl$	Re: Org.Comp.2-118
$C_8ClH_4O_7Re$	$(CO)_5ReC(O)OCH_2CH_2Cl$	Re: Org.Comp.2-116
$C_8ClH_6N_2O_3Re$	$(N{\equiv}C-CH_2CH_2CH_2-C{\equiv}N)Re(CO)_3Cl$	Re: Org.Comp.1-159
$C_8ClH_7MnN_3S^+$	$Mn[S-C(NH_2){=}N-N{=}CH-C_6H_4-4-Cl]^+$	Mn: MVol.D6-338
$C_8ClH_7N_2O_3PS$	$(C_6H_5O)P(O)-CH_2-N(CN)-S(O)Cl$	S: S-N Comp.8-259, 268
$C_8ClH_8MoO_5^-$	$[P(C_6H_5)_4][(CH_2CHCH_2)MoCl(CO)_3-OOCCH_3]$	Mo: Org.Comp.5-292, 293, 295
$C_8ClH_8NO_2S$	$C_6H_5C(OCH_3){=}N-S(O)Cl$	S: S-N Comp.8-272/3
$C_8ClH_8NO_3S^-$	$CH_3CCl(C_6H_5)-N(O)-SO_2^-$, radical anion	S: S-N Comp.8-326/8
$C_8ClH_8O_5Re$	$(CO)_3ReCl[CH_3C(O)CH_2C(O)CH_3]$	Re: Org.Comp.1-197
$C_8ClH_9NO_4Sb$	$3-Cl-4-CH_3CONH-C_6H_3Sb(O)(OH)_2$	Sb: Org.Comp.5-295
$C_8ClH_9N_2O_4S_3$	$ClCH_2S(O)_2N{=}S{=}NS(O)_2-C_6H_4-CH_3-4$	S: S-N Comp.7-100/1
$C_8ClH_{10}MoNO$	$C_5H_5Mo(NO)(Cl)C_3H_5$	Mo: Org.Comp.6-172
$C_8ClH_{10}NOS$	$ClS(O)-N(CH_3)CH_2C_6H_5$	S: S-N Comp.8-259, 267
$C_8ClH_{10}NO_2Sn$	$(CH_3)_2Sn(Cl)-OC(O)-2-NC_5H_4$	Sn: Org.Comp.17-86
–	$(CH_3)_2Sn(Cl)-OC(O)-4-NC_5H_4$	Sn: Org.Comp.17-86
$C_8ClH_{10}NO_3ReS_2^-$	$[(CO)_3ReCl(-S-C(N(C_2H_5)_2){=}S-)]^-$	Re: Org.Comp.1-121/2
$C_8ClH_{10}NO_4S_2$	$(CH_3O)_2S{=}N-S(O)_2-C_6H_4-4-Cl$	S: S-N Comp.8-156/7
$C_8ClH_{10}O_4ReS$	$(CO)_4Re(Cl)S(C_2H_5)_2$	Re: Org.Comp.1-459
$C_8ClH_{10}O_4ReSe$	$(CO)_4Re(Cl)Se(C_2H_5)_2$	Re: Org.Comp.1-462
$C_8ClH_{10}O_4ReTe$	$(CO)_4Re(Cl)Te(C_2H_5)_2$	Re: Org.Comp.1-462
$C_8ClH_{11}MoN_2OS_2$	$C_5H_5Mo(NO)(Cl)S_2CN(CH_3)_2$	Mo: Org.Comp.6-49
$C_8ClH_{11}NO_6PS_2$	$4-Cl-C_6H_4-SO_2-NH-S(O)-P(O)(OCH_3)_2$	S: S-N Comp.8-372
$C_8ClH_{11}O_4PReS$	$(CO)_4Re(Cl)-P(C_2H_5)_2-SH$	Re: Org.Comp.1-448
–	$(CO)_4Re(Cl)-S-PH(C_2H_5)_2$	Re: Org.Comp.1-459
$C_8ClH_{12}I_3MoN_4O$	$[Mo(CN-CH_3)_4(O)(Cl)][I_3]$	Mo: Org.Comp.5-39, 40, 44/5
$C_8ClH_{12}MoN_4O^+$	$[Mo(CN-CH_3)_4(O)(Cl)][I_3]$	Mo: Org.Comp.5-39, 40, 44/5
–	$[Mo(CN-CH_3)_4(O)(Cl)][PF_6]$	Mo: Org.Comp.5-39, 40, 45
$C_8ClH_{12}ORe$	$(CH_3C{\equiv}CCH_3)_2Re(O)Cl$	Re: Org.Comp.2-359, 365
$C_8ClH_{12}O_3ReS_2$	$(CO)_3ReCl(CH_3SCH_2CH_2CH_2SCH_3)$	Re: Org.Comp.1-198, 211/3
$C_8ClH_{12}O_3ReSe_2$	$(CO)_3ReCl(CH_3SeCH_2CH_2CH_2SeCH_3)$	Re: Org.Comp.1-200, 211/3
$C_8ClH_{14}InN_2$	$(CH_3)_2InCl \cdot NC_5H_4-2-NHCH_3$	In: Org.Comp.1-118, 121
$C_8ClH_{15}MoN_2O_2$	$[CH_2C(CH_3)CH_2]MoCl(CO)_2(NH_2CH_2CH_2NH_2)$	Mo: Org.Comp.5-250, 259
$C_8ClH_{15}O_2Sn$	$(C_2H_5)_2Sn(Cl)OOCCH{=}CHCH_3$	Sn: Org.Comp.17-95
$C_8ClH_{15}O_5Th$	$Th[OC(CH_3)_3]Cl(CH_3COO)_2$	Th: SVol.C7-25/9
$C_8ClH_{16}NOS$	$ClS(O)-N(CH_3)CH_2C_6H_{11}-c$	S: S-N Comp.8-259, 267
$C_8ClH_{17}O_2Sn$	$(C_3H_7)_2Sn(Cl)OOCCH_3$	Sn: Org.Comp.17-96
$C_8ClH_{18}IIn^-$	$[(C_4H_9)_2In(Cl)(I)]^-$	In: Org.Comp.1-349/50, 356
$C_8ClH_{18}In$	$(t-C_4H_9)_2InCl$	In: Org.Comp.1-118, 119, 126
$C_8ClH_{18}NOS$	$(i-C_4H_9)_2N-S(O)Cl$	S: S-N Comp.8-259, 264
–	$(s-C_4H_9)_2N-S(O)Cl$	S: S-N Comp.8-259, 264
$C_8ClH_{19}OSn$	$(C_2H_5)_2Sn(Cl)-O-C_4H_9-s$	Sn: Org.Comp.17-94
–	$(C_2H_5)_2Sn(Cl)-O-C_4H_9-t$	Sn: Org.Comp.17-94
–	$(n-C_4H_9)_2Sn(Cl)OH$	Sn: Org.Comp.17-99, 107
–	$(t-C_4H_9)_2Sn(Cl)OH$	Sn: Org.Comp.17-114, 115/6
$C_8ClH_{19}O_2Sn$	$C_2H_5-SnCl(O-C_3H_7-i)_2$	Sn: Org.Comp.17-149

$C_8ClH_{19}O_2Sn$ .... $(n-C_4H_9)_2Sn(Cl)-O-OH$ .... Sn: Org.Comp.17-106, 110
– .... $n-C_8H_{17}-SnCl(OH)_2$ .... Sn: Org.Comp.17-155
$C_8ClH_{22}InSi_2$ .... $[(CH_3)_3Si-CH_2]_2InCl$ .... In: Org.Comp.1-118, 120
$C_8ClH_{22}NOsSi_2$ .... $[(CH_3)_3SiCH_2]_2Os(N)Cl$ .... Os: Org.Comp.A1-17/8
$C_8ClH_{22}N_3O_4S$ .... $[((CH_3)_2N)_2S-N(C_2H_5)_2][ClO_4]$ .... S: S-N Comp.8-230, 239
$C_8ClH_{24}NSi_2Sn$ .... $(CH_3)_3SnN(Si(CH_3)_3)Si(CH_3)_2Cl$ .... Sn: Org.Comp.18-80/1
$C_8Cl_2F_4GeH_{10}O_2$ $Ge(CH_3)_3CH(C(=O)OC(CF_2Cl)_2)$ .... Ge: Org.Comp.2-93
$C_8Cl_2F_4H_{24}MnN_4O_8S_4$
$[N(CH_3)_4]_2[MnCl_2(N(SO_2F)_2)_2]$ .... Mn: MVol.D7-221
$C_8Cl_2F_4N_2$ .... $4,5-Cl_2-2,7-N_2C_8F_4$ .... F: PFHOrg.SVol.4-292, 294, 308
– .... $4-(CCl_2=N)-C_6F_4-CN$ .... F: PFHOrg.SVol.6-192, 205
$C_8Cl_2F_7HN_4$ .... $(n-C_3F_7)C_5Cl_2N_4H$ .... F: PFHOrg.SVol.4-286/7
$C_8Cl_2F_7N$ .... $4-CF_3-C_6F_4-N=CCl_2$ .... F: PFHOrg.SVol.6-192, 206
$C_8Cl_2F_{10}N_2$ .... $2-(ClN=)-3-Cl-4-(i-C_3F_7)-NC_5F_3$ .... F: PFHOrg.SVol.4-149, 160
– .... $2-(ClN=)-5-Cl-4-(i-C_3F_7)-NC_5F_3$ .... F: PFHOrg.SVol.4-149, 159
– .... $2-Cl_2N-4-(i-C_3F_7)-NC_5F_3$ .... F: PFHOrg.SVol.4-149, 159
$C_8Cl_2FeH_5O_2^+$ .... $[C_5H_5(CO)_2Fe=CCl_2]^+$ .... Fe: Org.Comp.B16a-135, 148/9
$C_8Cl_2FeH_6O_3$ .... $[(C_5H_5)Fe(CO)_3][HCl_2]$ .... Fe: Org.Comp.B15-37/8, 46
$C_8Cl_2GaH_{16}O_4^+$ .... $[(C_8H_{16}O_4)GaCl_2][GaCl_4]$ .... Ga: SVol.D1-150/1
$C_8Cl_2GaH_{20}O_2^{2-}$ $[GaCl_2 \cdot 2\ O(C_2H_5)_2]^-$ .... Ga: SVol.D1-99/100
$C_8Cl_2GaH_{20}O_4^+$ .... $[GaCl_2(CH_3-O-CH_2CH_2-O-CH_3)_2][GaCl_4]$ .... Ga: SVol.D1-108/10
$C_8Cl_2GeH_{16}$ .... $(C_2H_5)_2Ge(-CCl_2CH_2CH_2CH_2-)$ .... Ge: Org.Comp.3-244
$C_8Cl_2GeH_{19}OP$ .... $Ge(C_2H_5)_3CH_2CH_2P(O)Cl_2$ .... Ge: Org.Comp.2-126
$C_8Cl_2GeH_{20}Si$ .... $Ge(CH_3)_2(CH_2Cl)CH_2CH_2CH_2Si(CH_3)_2Cl$ .... Ge: Org.Comp.3-194
$C_8Cl_2GeH_{22}Si_2$ .... $Ge(CH_3)_2(CH_2Si(CH_3)_2Cl)_2$ .... Ge: Org.Comp.3-148
$C_8Cl_2H_4N_3O_3Re$ .... $(CO)_2ReCl_2(NC_5H_4-3-CN)(NO)$ .... Re: Org.Comp.1-87
– .... $(CO)_2ReCl_2(NC_5H_4-4-CN)(NO)$ .... Re: Org.Comp.1-88
$C_8Cl_2H_7NOS$ .... $Cl_2S=N-C(O)-C_6H_4-4-CH_3$ .... S: S-N Comp.8-122/4
$C_8Cl_2H_7NO_2S$ .... $4-Cl-C_6H_4-C(OCH_3)=N-S(O)Cl$ .... S: S-N Comp.8-273
– .... $Cl_2S=N-C(O)-C_6H_4-4-OCH_3$ .... S: S-N Comp.8-122/4
$C_8Cl_2H_7N_2O_3Re$ .... $(CO)_2ReCl_2(NC_5H_4-4-CH_3)(NO)$ .... Re: Org.Comp.1-88
$C_8Cl_2H_8MnN_3S_2$ .... $MnCl_2[S-C(SCH_3)=NN=CH-2-C_5H_4N] \cdot 0.5\ H_2O$
Mn: MVol.D6-361, 363
$C_8Cl_2H_8MnN_4O_2S_2$
$[Mn(1,3-N_2C_4H_4(=S-2)(=O-4))_2Cl_2]$ .... Mn: MVol.D7-78/9
$C_8Cl_2H_8MnN_4S_2$ .... $[Mn(1,3-N_2C_4H_4(=S-2))_2Cl_2]$ .... Mn: MVol.D7-78/9
$C_8Cl_2H_8NO_4Sb$ .... $3,5-Cl_2-4-CH_3CONH-C_6H_2Sb(O)(OH)_2$ .... Sb: Org.Comp.5-299
$C_8Cl_2H_{10}MnN_2O_3S$
$Mn(4-NH_2C_6H_4SO_2NH-COCH_3)Cl_2$ .... Mn: MVol.D7-116/23
$C_8Cl_2H_{11}MnNOP$ .... $Mn(NO)[P(C_6H_5)(CH_3)_2]Cl_2$ .... Mn: MVol.D8-46, 61
$C_8Cl_2H_{11}MnOPS_{0.5}$
$Mn[P(C_6H_5)(CH_3)_2](SO_2)_{0.5}Cl_2$ .... Mn: MVol.D8-46
$C_8Cl_2H_{11}MnO_2P$ .... $Mn[P(C_6H_5)(CH_3)_2](O_2)Cl_2$ .... Mn: MVol.D8-56
$C_8Cl_2H_{11}MnP$ .... $Mn[P(C_6H_5)(CH_3)_2]Cl_2$ .... Mn: MVol.D8-38
$C_8Cl_2H_{11}Sb$ .... $(CH_3)_2(C_6H_5)SbCl_2$ .... Sb: Org.Comp.5-55
$C_8Cl_2H_{11}Ti$ .... $[((CH_3)_2CHC_5H_4)TiCl_2]_n$ .... Ti: Org.Comp.5-36
$C_8Cl_2H_{12}In_2$ .... $[(CH_2=CH)_2InCl]_2$ .... In: Org.Comp.1-129

$C_8Cl_2H_{12}MnN_4O_6S_2$	$[Mn(NCCH_3)_4(SO_3Cl)_2]$	Mn:MVol.D7-10
$C_8Cl_2H_{12}MnN_4O_8$	$Mn(NCCH_3)_4(ClO_4)_2$	Mn:MVol.D7-9
$C_8Cl_2H_{12}MnN_4S_2$	$[Mn((2-NH_2)(4-CH_3)C_3HNS-3,1)_2Cl_2]$	Mn:MVol.D7-228/30
$C_8Cl_2H_{12}MoO_2$	$(C_5H_5)MoO(O-C_3H_7-n)Cl_2$	Mo:Org.Comp.6-9, 11
–	$(C_5H_5)MoO(O-C_3H_7-i)Cl_2$	Mo:Org.Comp.6-10, 11
$C_8Cl_2H_{12}N_4OS$	$[CH_3-3-(1,3-N_2C_3H_3)-1-S(O)-1-(1,3-N_2C_3H_3)-3-CH_3]Cl_2$	S: S-N Comp.8-365
$C_8Cl_2H_{13}MnN_3O_2S$	$[Mn(2-CH_3CONH-4-CH_3CO-C_2N_2S-4,3,1-(CH_3)_2-5,5)(H_2O)_2]Cl_2$	Mn:MVol.D7-237/9
$C_8Cl_2H_{14}MnN_4O_2$	$[MnCl_2((CH_3)_2C=N-NHC(O)C(O)NH-N=C(CH_3)_2)] \cdot 2\ H_2O$	Mn:MVol.D6-313, 315
–	$[MnCl_2((CH_3)_2C=N-NHC(O)C(O)NH-N=C(CH_3)_2)]_n$	Mn:MVol.D6-313, 315
$C_8Cl_2H_{15}O_2Sn$	$C_4H_9SnCl_2(OC(CH_3)=C(CH_3)O)$, radical	Sn: Org.Comp.17-179, 183
$C_8Cl_2H_{16}MnO_{10}S_2$	$[Mn(C_4H_8OS-1,4)_2(H_2O)_2](ClO_4)_2$	Mn:MVol.D7-239
$C_8Cl_2H_{16}MnS_2$	$[Mn(SC_4H_8)_2Cl_2]$	Mn:MVol.D7-225/6
$C_8Cl_2H_{16}O_2Sn$	$C_4H_9SnCl_2(OCH(CH_3)COCH_3)$	Sn: Org.Comp.17-178
$C_8Cl_2H_{17}MnN_3O_4S$	$[Mn(2-CH_3CONH-4-CH_3CO-C_2N_2S-4,3,1-(CH_3)_2-5,5)(H_2O)_2]Cl_2$	Mn:MVol.D7-237/9
$C_8Cl_2H_{18}MnN_4O_8$	$Mn[H_2NCH_2CH_2N=CCH_3-CCH_3=NCH_2CH_2NH_2][ClO_4]_2$	Mn:MVol.D6-217/8
$C_8Cl_2H_{18}MnN_6O_2$	$[Mn((CH_3)_2C=N-NHC(O)NH_2)_2Cl_2]$	Mn:MVol.D6-327
$C_8Cl_2H_{18}MnN_6S_2$	$MnCl_2[(CH_3)_2C=N-NHC(=S)NH_2]_2$	Mn:MVol.D6-341
$C_8Cl_2H_{18}NO_2Sn$	$C_4H_9SnCl_2(ON(O)C_4H_9-t)$, radical	Sn: Org.Comp.17-179, 183
$C_8Cl_2H_{18}NO_3PS$	$Cl_2S=N-P(O)(O-C_4H_9-n)_2$	S: S-N Comp.8-93/4
$C_8Cl_2H_{18}NO_4PS$	$(n-C_3H_7O)_2P(O)-N(CH_2CH_2Cl)-S(O)Cl$	S: S-N Comp.8-277/8
$C_8Cl_2H_{18}OSn$	$n-C_4H_9-SnCl_2-O-C_4H_9-t$	Sn: Org.Comp.17-178, 182/3
–	$n-C_8H_{17}-SnCl_2-OH$	Sn: Org.Comp.17-179
$C_8Cl_2H_{18}O_8Sn$	$(C_4H_9)_2Sn(OClO_3)_2$	Sn: Org.Comp.15-336/7
$C_8Cl_2H_{19}InOSi$	$(CH_3)_3Si-CH_2-In(Cl)_2 \cdot OC_4H_8$	In: Org.Comp.1-136, 139
$C_8Cl_2H_{19}NSn$	$n-C_4H_9-SnCl_2-N(C_2H_5)_2$	Sn: Org.Comp.19-157
$C_8Cl_2H_{20}InN$	$t-C_4H_9-CH_2-In(Cl)_2 \cdot N(CH_3)_3$	In: Org.Comp.1-136, 139
$C_8Cl_2H_{20}In_2$	$[(C_2H_5)_2InCl]_2$	In: Org.Comp.1-125
$C_8Cl_2H_{20}MnN_8$	$[Mn(NH_2-N=C(CH_3)C(CH_3)=N-NH_2)_2Cl_2]$	Mn:MVol.D6-264, 265/6
$C_8Cl_2H_{20}MnO_{12}S_2$	$[Mn(C_4H_8OS-1,4)_2(H_2O)_2](ClO_4)_2$	Mn:MVol.D7-239
$C_8Cl_2H_{20}N_2PtS_2$	$[((n-C_3H_7)_2S)PtCl_2(CH_3N=S=NCH_3)]$	S: S-N Comp.7-316/7, 322
$C_8Cl_2H_{21}InOSi$	$(CH_3)_3SiCH_2InCl_2 \cdot O(C_2H_5)_2$	In: Org.Comp.1-136, 139
$C_8Cl_2H_{21}InO_2Si$	$(CH_3)_3SiCH_2InCl_2 \cdot CH_3OCH_2CH_2OCH_3$	In: Org.Comp.1-136, 139
$C_8Cl_2H_{21}N_2PPtS$	$[((C_2H_5)_3P)PtCl_2(CH_3N=S=NCH_3)]$	S: S-N Comp.7-316/7, 320, 322
$C_8Cl_2H_{21}N_2PtSSb$	$[((C_2H_5)_3Sb)PtCl_2(CH_3N=S=NCH_3)]$	S: S-N Comp.7-316/7, 322
$C_8Cl_2H_{22}MnN_4S_2$	$[Mn(SCH_2CH_2NHCH_2CH_2NH_2)_2Cl_2]$	Mn:MVol.D7-28/9
$C_8Cl_2H_{22}Mn_2N_{16}S_4$	$[Mn(NH_2C(S)NC(NH)NH_2)(NH_2C(S)NHC(NH)NH_2)Cl]_2$	Mn:MVol.D7-199/200
$C_8Cl_2H_{22}NOsSi_2^-$	$[N(C_4H_9-n)_4][((CH_3)_3SiCH_2)_2Os(N)(Cl)_2]$	Os: Org.Comp.A1-18/9
$C_8Cl_2H_{24}In_3^-$	$[(CH_3)_2In(Cl)_2(In(CH_3)_3)_2]^-$	In: Org.Comp.1-363

$C_8Cl_2N_6S$	$(NC)_2C{=}CCl{-}N{=}S{=}N{-}CCl{=}C(CN)_2$	S: S-N Comp.7-217
$C_8Cl_3FH_4N_2O_2$	$C_6F{-}(C({=}O)NH_2)_2{-}1,3{-}Cl_3{-}4,5,6$	F: PFHOrg.SVol.5-22, 55
$C_8Cl_3FN_2$	$4,5,6{-}Cl_3{-}1,3{-}(NC)_2{-}C_6F$	F: PFHOrg.SVol.6-107/8, 127, 152
$C_8Cl_3F_3H_2N_2O_3S$	$(HO_3S)(CF_3)C_7Cl_3N_2H$	F: PFHOrg.SVol.4-284
$C_8Cl_3F_3H_5NS$	$Cl_2S{=}N{-}CCl(CF_3){-}C_6H_5$	S: S-N Comp.8-102
$C_8Cl_3F_5HNO$	$CCl_3{-}C({=}O)NH{-}C_6F_5$	F: PFHOrg.SVol.5-24, 59, 70, 93
$C_8Cl_3F_6H_2N_3$	$(H_2N)(CCl_3)C_7F_6N_2$	F: PFHOrg.SVol.4-287, 301
$C_8Cl_3F_6N$	$C_6F_5{-}N{=}CFCCl_3$	F: PFHOrg.SVol.6-192, 206
$C_8Cl_3FeH_8NOSn$	$C_5H_5Fe(CNCH_3)(CO)SnCl_3$	Fe: Org.Comp.B15-288, 295
$C_8Cl_3GaH_4N_2O_4$	$GaCl_3[1{-}(2,5{-}(O)_2NC_4H_2{-}1){-}NC_4H_2(O)_2{-}2,5]$	Ga: SVol.D1-241/2
$C_8Cl_3GaH_8O$	$GaCl_3(1{-}OC_8H_8)$	Ga: SVol.D1-114
–	$GaCl_3[O{=}C(CH_3){-}C_6H_5]$	Ga: SVol.D1-39/42
$C_8Cl_3GaH_8O_2$	$GaCl_3(1,4{-}O_2C_8H_8)$	Ga: SVol.D1-149/50
$C_8Cl_3GaH_{10}O$	$GaCl_3(C_6H_5{-}O{-}C_2H_5)$	Ga: SVol.D1-105, 106/8
$C_8Cl_3GaH_{10}O_2$	$GaCl_3[1,2{-}(CH_3O)_2{-}C_6H_4]$	Ga: SVol.D1-108/10
$C_8Cl_3GaH_{10}O_4$	$GaCl_3[CH_2{=}CH{-}COO{-}(CH_2)_2{-}OOC{-}CH{=}CH_2]$	Ga: SVol.D1-203
$C_8Cl_3GaH_{11}N$	$GaCl_3[N(CH_3)_2{-}C_6H_5]$	Ga: SVol.D1-233, 234
$C_8Cl_3GaH_{16}O$	$GaCl_3[O{=}C(CH_3){-}C_6H_{13}]$	Ga: SVol.D1-36/9
$C_8Cl_3GaH_{16}O_2$	$GaCl_3(OC_4H_8)_2$	Ga: SVol.D1-111
$C_8Cl_3GaH_{16}O_4$	$GaCl_3(1,4{-}O_2C_4H_8)_2$	Ga: SVol.D1-147
–	$GaCl_3[O{=}C(CH_3){-}O{-}C_2H_5]_2$	Ga: SVol.D1-197, 199
$C_8Cl_3GaH_{17}N$	$GaCl_3[N(CH_3)_2{-}C_6H_{11}{-}c]$	Ga: SVol.D1-233, 234
$C_8Cl_3GaH_{18}O$	$GaCl_3(C_4H_9{-}O{-}C_4H_9)$	Ga: SVol.D1-105, 106/8
$C_8Cl_3GaH_{20}O_2$	$GaCl_3(C_2H_5{-}O{-}C_2H_5)_2$	Ga: SVol.D1-100
–	$GaCl_3(HO{-}C_4H_9{-}n)_2$	Ga: SVol.D1-16, 18/9
$C_8Cl_3GeH_{13}Ti$	$Ge(CH_3)_3C_5H_4TiCl_3$	Ge: Org.Comp.2-33
$C_8Cl_3HN_6S$	$(CN)_2C{=}C(Cl)NH{-}S(Cl){=}N{-}CCl{=}C(CN)_2$	S: S-N Comp.8-183
$C_8Cl_3H_5N_2OS$	$O{=}S{=}NC(CCl_3){=}NC_6H_5$	S: S-N Comp.6-180
$C_8Cl_3H_7NRe$	$Re(CNC_6H_4CH_3{-}4)(Cl)_3$	Re: Org.Comp.2-232, 241
$C_8Cl_3H_7O_2Th$	$ThCl_3(C_6H_5CH_2COO)$	Th: SVol.C7-65, 66
$C_8Cl_3H_8NO_2S$	$(CH_3O)_2S{=}N{-}C_6H_2{-}2,4,6{-}Cl_3$	S: S-N Comp.8-166
$C_8Cl_3H_8NO_3S_2$	$4{-}CH_3{-}C_6H_4{-}S(O)_2{-}NH{-}S(O){-}CCl_3$	S: S-N Comp.8-292
$C_8Cl_3H_{12}MoO$	$i{-}C_3H_7C_5H_4Mo(OH)Cl_3 \cdot 0.5\ C_2H_4Cl_2{-}1,2$	Mo: Org.Comp.6-3
$C_8Cl_3H_{14}N_2ORe$	$Re(CNCH(CH_3)_2)_2(Cl)_3O$	Re: Org.Comp.2-257
$C_8Cl_3H_{18}Sb$	$(C_4H_9)_2SbCl_3$	Sb: Org.Comp.5-141
$C_8Cl_3H_{19}MoN_3$	$(Cl)_3Mo(CN{-}CH_3)[N(CH_3)_2{-}CH_2CH_2{-}N(CH_3)_2]$	Mo: Org.Comp.5-2
$C_8Cl_3H_{24}N_3OPSb$	$(CH_3)_2SbCl_3 \cdot OP[N(CH_3)_2]_3$	Sb: Org.Comp.5-142
$C_8Cl_4F_3HN_2$	$CF_3C_7Cl_4N_2H$	F: PFHOrg.SVol.4-284
$C_8Cl_4F_3HN_2O_2S$	$(ClO_2S)(CF_3)C_7Cl_3N_2H$	F: PFHOrg.SVol.4-284, 299, 315
$C_8Cl_4F_3H_3N_2O$	$CF_3{-}C({=}O)NH{-}C_6Cl_4{-}NH_2{-}2$	F: PFHOrg.SVol.5-24, 58, 96
$C_8Cl_4F_4N_2$	$1,3{-}(CCl_2{=}N)_2{-}C_6F_4$	F: PFHOrg.SVol.6-192, 206
–	$1,4{-}(CCl_2{=}N)_2{-}C_6F_4$	F: PFHOrg.SVol.6-192, 206
$C_8Cl_4F_5N$	$C_6F_5{-}N{=}CClCCl_3$	F: PFHOrg.SVol.6-192/3, 206, 219
$C_8Cl_4Fe_2H_5O_3$	$[(C_5H_5)Fe(CO)_3][FeCl_4]$	Fe: Org.Comp.B15-36, 38, 46/7
$C_8Cl_4Ga_2H_{16}O_2$	$Ga_2(OC_4H_8)_2Cl_4$	Ga: SVol.D1-111

$C_8Cl_4Ga_2H_{16}O_4$	$1,4\text{-}O_2C_4H_8\text{-}1\text{-}GaCl_2\text{-}GaCl_2\text{-}1\text{-}(1,4\text{-}O_2C_4H_8)$	Ga: SVol.D1-143/6
$C_8Cl_4H_6N_6O_6Sn$	$[2,4,6\text{-}(O{=})_3\text{-}3,5\text{-}Cl_2\text{-}1,3,5\text{-}N_3C_3\text{-}1]_2Sn(CH_3)_2$	Sn: Org.Comp.19-71
$C_8Cl_4H_8O_8Th$	$Th(OOC\text{-}CH_2Cl)_4$	Th: SVol.C7-57, 61/2
$C_8Cl_4H_9N_2O_2Sb$	$(3\text{-}O_2N\text{-}4\text{-}C_2H_5NHC_6H_3)SbCl_4$	Sb: Org.Comp.5-258
$C_8Cl_4H_{11}OSSb$	$C_6H_5SbCl_4 \cdot OS(CH_3)_2$	Sb: Org.Comp.5-244
$C_8Cl_4H_{12}N_4Ti$	$[(CH_3N{=}CCl)_4Ti]_n$	Ti: Org.Comp.5-324
$C_8Cl_4H_{12}N_6O_2Pd$	$(C_4H_6N_3O)_2[PdCl_4]$	Pd: SVol.B2-121
$C_8Cl_4H_{14}MoN_2$	$MoCl_4 \cdot 2\ n\text{-}C_3H_7CN$	Mo: SVol.B5-374
$C_8Cl_4H_{14}N_4Pd$	$(CH_3C_3H_2N_2H_2)_2[PdCl_4]$	Pd: SVol.B2-119
$C_8Cl_4H_{16}MnO_8P_2$	$[Mn(O_2PCl_2)_2(OC(CH_3)(OC_2H_5))_2]$	Mn: MVol.D8-186/7
$C_8Cl_4H_{16}MoO_2$	$MoCl_4 \cdot 2\ C_4H_8O$	Mo: SVol.B5-372
$C_8Cl_4H_{16}MoO_2S_2$	$MoCl_4 \cdot 2\ C_4H_8OS$	Mo: SVol.B5-324
$C_8Cl_4H_{18}N_2O_2S_2Th$	$ThCl_4 \cdot 2\ (CH_3)_2SO \cdot 2\ CH_3CN$	Th: SVol.D4-204/5
$C_8Cl_4H_{20}NSb$	$[N(CH_3)_4][(CH_2)_4SbCl_4]$	Sb: Org.Comp.5-170
$C_8Cl_4H_{20}Sb_2$	$[(CH_3)_2SbCl_2]_2(CH_2)_4$	Sb: Org.Comp.5-123/4
$C_8Cl_4H_{24}N_2Pd$	$[(CH_3)_4N]_2[PdCl_4]$	Pd: SVol.B2-119
$C_8Cl_4H_{24}O_4S_4Th$	$ThCl_4 \cdot 4\ (CH_3)_2SO$	Th: SVol.D4-204
$C_8Cl_4H_{28}N_4Pd_2$	$[Pd(NH_2C_2H_5)_4][PdCl_4]$	Pd: SVol.B2-144
$C_8Cl_4H_{32}N_{16}O_8Th$	$[Th(O{=}C(NH_2)_2)_8]Cl_4$	
	Chemistry in solution	Th: SVol.D4-161, 205
	Photoemission spectra	Th: SVol.A4-131
$C_8Cl_4H_{32}N_{16}S_8Th$	$ThCl_4 \cdot 8\ S{=}C(NH_2)_2$	Th: SVol.D4-147
$C_8Cl_4O_4Th$	$Th[1,2\text{-}(OOC)_2\text{-}C_6Cl_4]$	Th: SVol.C7-144, 145
$C_8Cl_5GaH_6O$	$GaCl_3[O{=}C(CH_3)\text{-}C_6H_3(Cl)_2\text{-}3,4]$	Ga: SVol.D1-39/42
$C_8Cl_5H_7O_2SSb^-$	$[3\text{-}HOOC\text{-}CH_2\text{-}S\text{-}C_6H_4\text{-}SbCl_5]^-$	Sb: Org.Comp.5-253
–	$[4\text{-}HOOC\text{-}CH_2\text{-}S\text{-}C_6H_4\text{-}SbCl_5]^-$	Sb: Org.Comp.5-253
$C_8Cl_5H_7O_2Sb^-$	$[3\text{-}CH_3\text{-}OOC\text{-}C_6H_4\text{-}SbCl_5]^-$	Sb: Org.Comp.5-257
–	$[4\text{-}CH_3\text{-}OOC\text{-}C_6H_4\text{-}SbCl_5]^-$	Sb: Org.Comp.5-257
–	$[4\text{-}HOOC\text{-}CH_2\text{-}C_6H_4\text{-}SbCl_5]^-$	Sb: Org.Comp.5-257
$C_8Cl_5H_7O_3Sb^-$	$[(4\text{-}HO_2CCH_2OC_6H_4)SbCl_5]^-$	Sb: Org.Comp.5-252
$C_8Cl_5H_7O_4SSb^-$	$[(4\text{-}HO_2CCH_2SO_2C_6H_4)SbCl_5]^-$	Sb: Org.Comp.5-253
$C_8Cl_5H_8NOSSb^-$	$[(4\text{-}H_2NC(O)CH_2SC_6H_4)SbCl_5]^-$	Sb: Org.Comp.5-253
$C_8Cl_5H_9OSSb^-$	$[(4\text{-}HOCH_2CH_2SC_6H_4)SbCl_5]^-$	Sb: Org.Comp.5-252
$C_8Cl_5H_9OSb^-$	$[(4\text{-}C_2H_5OC_6H_4)SbCl_5]^-$	Sb: Org.Comp.5-252
$C_8Cl_6FH_{24}NSb_2$	$[N(CH_3)_4][Cl_3(CH_3)_2SbFSb(CH_3)_2Cl_3]$	Sb: Org.Comp.5-141/2, 145
$C_8Cl_6Ga_2H_4N_2O_4$	$Ga_2Cl_6[1\text{-}(2,5\text{-}(O)_2NC_4H_2\text{-}1)\text{-}NC_4H_2(O)_2\text{-}2,5]$	Ga: SVol.D1-241/2
$C_8Cl_6Ga_2H_8O_2$	$Cl_6Ga_2(1,4\text{-}O_2C_8H_8)$	Ga: SVol.D1-149/50
$C_8Cl_6Ga_2H_{10}O_2$	$Ga_2Cl_6[1,2\text{-}(CH_3O)_2\text{-}C_6H_4]$	Ga: SVol.D1-108/10
$C_8Cl_6Ga_2H_{10}O_4$	$Ga_2Cl_6[CH_2{=}CH\text{-}COO\text{-}(CH_2)_2\text{-}OOC\text{-}CH{=}CH_2]$	Ga: SVol.D1-203
$C_8Cl_6Ga_2H_{16}O_4$	$[(C_8H_{16}O_4)GaCl_2][GaCl_4]$	Ga: SVol.D1-150/1
$C_8Cl_6Ga_2H_{20}O_4$	$[GaCl_2(CH_3\text{-}O\text{-}CH_2CH_2\text{-}O\text{-}CH_3)_2][GaCl_4]$	Ga: SVol.D1-108/10
$C_8Cl_6H_{10}NOSSb$	$[OSN(CH_3)CH_2C_6H_5]SbCl_6$	S: S-N Comp.6-289, 296/7
$C_8Cl_6H_{16}N_4S$	$[(CH_3)_2N\text{-}CCl{=}N\text{-}S(Cl)N{=}C(Cl)N(CH_3)_2]Cl \cdot C_2H_4Cl_2$	S: S-N Comp.8-189/91
$C_8Cl_6H_{18}N_2O_4Pd_3S_2$	$[Pd_2(S(CH_3)CH_2CH(COOH)NH_2)_2Cl_2][PdCl_4]$	Pd: SVol.B2-144
$C_8Cl_6H_{20}N_2O_2Pd_2$	$[(N(CH_3)_2C(O)CH_3)_2H_2][Pd_2Cl_6]$	Pd: SVol.B2-163
$C_8Cl_6H_{24}N_2Po$	$[N(CH_3)_4]_2[PoCl_6]$	Po: SVol.1-300

$C_8Cl_6H_{24}N_2Th$ .... $[(CH_3)_4N]_2ThCl_6$
Crystallographic properties .... Th: SVol.A4-7/8
$C_8Cl_8GeH_{12}$ ..... $Ge(CHClCH_2Cl)_4$ .... Ge: Org.Comp.1-88
$C_8Cl_8H_4O_8Th$ .... $Th(OOC-CHCl_2)_4$ .... Th: SVol.C7-57, 61/2
– .... $Th(OOC-CHCl_2)_4 \cdot H_2O$ .... Th: SVol.C7-58, 62
$C_8Cl_8H_{12}N_4Ti_2$ ... $[Cl_3Ti(CNCH_3)(CCl{=}NCH_3)]_2$ .... Ti: Org.Comp.5-19
$C_8Cl_8H_{12}Pb$ ...... $Pb(CCl_2-CH_3)_4$ .... Pb: Org.Comp.3-67
$C_8Cl_8H_{20}PbTi_2$ ... $Pb(C_2H_5)_4 \cdot 2\ TiCl_4$ .... Pb: Org.Comp.2-270
$C_8Cl_{12}H_8Mo_6N_4$ .. $Mo_6Cl_{12} \cdot 2\ N(CHCH)_2N$ .... Mo: SVol.B5-267
$C_8Cl_{12}H_{14}Mo_6N_2$. $Mo_6Cl_{12} \cdot 2\ n\text{-}C_3H_7CN$ .... Mo: SVol.B5-266
$C_8Cl_{12}O_8Th$ ...... $Th(OOC-CCl_3)_4$ .... Th: SVol.C7-58, 61, 62/3
– .... $Th(OOC-CCl_3)_4 \cdot H_2O$ .... Th: SVol.C7-59, 63
– .... $Th(OOC-CCl_3)_4 \cdot 2\ H_2O$ .... Th: SVol.C7-59, 63
$C_8CoH_{12}N_4O_2PdS_4$
$Pd(SCN)_2Co(NCS)_2(C_2H_5OH)_2$ .... Pd: SVol.B2-311
$C_8CoH_{26}IMoN_6O_8$ $[Co(C_2H_8N_2)_3][MoO_3(C_2O_4)(H_2O)]I \cdot 2\ H_2O$ ... Mo: SVol.B3b-174
$C_8Co_2H_{36}Mo_5N_8O_{25}P_2$
$[(C_2H_8N_2)_2Co(H_2O)OPO_3]_2Mo_5O_{15}$ .... Mo: SVol.B3b-63/4
$C_8CrH_{22}NO_4OsSi_2^-$
$[N(C_4H_9\text{-}n)_4][((CH_3)_3SiCH_2)_2Os(N)(CrO_4)]$ .... Os: Org.Comp.A1-2, 20
$C_8CsF_{12}GaO_8$ .... $Cs[Ga(OC(O)-CF_3)_4]$ .... Ga: SVol.D1-156, 157
$C_8CsF_{14}N$ .... $[NCF_2CF_2C(i\text{-}C_3F_7)CFCF_2]Cs$ .... F: PFHOrg.SVol.4-118, 140
$C_8CsF_{16}NO$ ...... $Cs[i\text{-}C_3F_7\text{-}NC(=O)\text{-}C_4F_9\text{-}t]$ .... F: PFHOrg.SVol.5-24, 58, 70, 92
$C_8Cs_4H_8O_{16}Th$ ... $Cs_4[Th(HCOO)_8]$ .... Th: SVol.C7-44/5
$C_8Cs_4N_8S_8Th$ .... $Cs_4[Th(NCS)_8]$ .... Th: SVol.C7-22, 23, 24
– .... $Cs_4[Th(NCS)_8] \cdot 2\ H_2O$ .... Th: SVol.C7-22, 23/4
$C_8CuF_6H_{12}MoN_4$ .. $Cu[MoF_6] \cdot 4\ CH_3CN$ .... Mo: SVol.B5-176, 185
$C_8CuF_{24}N_8O_4S_4^{2+}$ $[Cu(NS-ON(CF_3)_2)_4]^{2+}$ .... S: S-N Comp.5-254
$C_8FH_{16}NO_2S$ ..... 3,3,5,5-$(CH_3)_4$-4-FS(O)-1,4-ONC_4H_4 .... S: S-N Comp.8-255/6
$C_8F_2FeH_5O_2^+$ .... $[C_5H_5(CO)_2Fe{=}CF_2]^+$ .... Fe: Org.Comp.B16a-135, 148
$C_8F_2GeH_{22}Si_2$ .... $Ge(CH_3)_2(CH_2Si(CH_3)_2F)_2$ .... Ge: Org.Comp.3-147/8
$C_8F_2H_{16}N_2O_2S$ ... $F_2S(4\text{-}1,4\text{-}ONC_4H_8)_2$ .... S: S-N Comp.8-397/400
$C_8F_2H_{18}N_2OS$ .... $F_2S(N(C_2H_5)_2)\text{-}4\text{-}1,4\text{-}ONC_4H_8$ .... S: S-N Comp.8-397/400
$C_8F_2H_{18}N_4PS_2^+$ .. $[F_2P(N{=}S{=}N\text{-}C_4H_9\text{-}t)_2]^+$ .... S: S-N Comp.7-105
$C_8F_2H_{18}O_6S_2Sn$ .. $(C_4H_9)_2Sn(OSO_2F)_2$ .... Sn: Org.Comp.15-348
$C_8F_2H_{20}N_2S$ ..... $F_2S(N(C_2H_5)_2)_2$ .... S: S-N Comp.8-397/400
$C_8F_2H_{22}N_2SSi_2$ ... $F_2S{=}N\text{-}CH_2CH_2\text{-}N(Si(CH_3)_3)_2$ .... S: S-N Comp.8-42/3
$C_8F_2N_3O_5Re$ ..... $(CO)_5ReC_3F_2N_3$ .... Re: Org.Comp.2-145
$C_8F_2N_8$ .... 4,5-$(N_3)_2$-1,2-$(NC)_2$-C_6F_2 .... F: PFHOrg.SVol.6-107, 127
$C_8F_3FeH_5O_4S_2$ ... $(C_5H_5)Fe(CS)(CO)OS(O)_2CF_3$ .... Fe: Org.Comp.B15-268/9
$C_8F_3FeH_{10}P$ ..... $C_5H_5FeCH_2CHCH_2PF_3$ .... Fe: Org.Comp.B17-141
$C_8F_3GeH_{15}$ ...... $Ge(C_2H_5)_3CF{=}CF_2$ .... Ge: Org.Comp.2-188
$C_8F_3GeH_{15}O$ ..... $Ge(C_2H_5)_3COCF_3$ .... Ge: Org.Comp.2-137, 139
$C_8F_3GeH_{19}Sn$ .... $(CH_3)_3Ge\text{-}CF(CHF_2)\text{-}Sn(CH_3)_3$ .... Ge: Org.Comp.1-169
– .... $(CH_3)_3Ge\text{-}CHFCF_2\text{-}Sn(CH_3)_3$ .... Ge: Org.Comp.1-168/9
$C_8F_3H_2O_5Re$ ..... $(CO)_5ReCH{=}CHCF_3$ .... Re: Org.Comp.2-127/8
$C_8F_3H_3NO_8ReS$ .. $[(CO)_5ReNCCH_3][O_3SCF_3]$ .... Re: Org.Comp.2-152
$C_8F_3H_4N_5$ .... $CF_3\text{-}C(NH_2){=}C(CN)\text{-}C(NH_2){=}C(CN)_2$ .... F: PFHOrg.SVol.5-2, 34
$C_8F_3H_4O_2STh^{3+}$ .. $Th[SC_4H_3C(O)CHC(O)CF_3]^{3+}$ .... Th: SVol.D1-128

$C_8F_3H_5NO_4Re$	$(C_5H_5)Re(CO)(NO)-OC(=O)-CF_3$	Re: Org.Comp.3-149
$C_8F_3H_6O_4PoS^+$	$[Po(OH)_2(CF_3C(O)CHC(O)-2-C_4H_3S)]^+$	Po: SVol.1-354/5
$C_8F_3H_8MoO_2P$	$(C_5H_5)Mo(CO)_2(PF_3)CH_3$	Mo: Org.Comp.8-77, 82
$C_8F_3H_{10}NS$	$F_3S-N(C_2H_5)-C_6H_5$	S: S-N Comp.8-391/2
$C_8F_3H_{14}NS$	$F_3S-3-NC_8H_{14}-[3.2.2]$	S: S-N Comp.8-390
$C_8F_3H_{15}N_4SiSn$	$(CH_3)_3SnN(C_3N_3F_2)Si(CH_3)_2F$	Sn: Org.Comp.18-71, 75
$C_8F_3H_{18}NS$	$F_3SN(C_4H_9-n)_2$	S: S-N Comp.8-382
$C_8F_3H_{30}N_{15}O_9Th$	$(C(NH_2)_3)_5[ThF_3(CO_3)_3]$	Th: SVol.C7-14/6
$C_8F_3N_5$	$2-N_3-1,4-(NC)_2-C_6F_3$	F: PFHOrg.SVol.6-107
–	$4-N_3-1,2-(NC)_2-C_6F_3$	F: PFHOrg.SVol.6-107, 127, 151
$C_8F_4GeH_{16}$	$Ge(C_2H_5)_3CHFCF_3$	Ge: Org.Comp.2-127, 138
$C_8F_4HNO_2$	$(O)_2C_8F_4NH$	F: PFHOrg.SVol.4-283
$C_8F_4HNO_3$	$(O)_2C_8F_4NHO$	F: PFHOrg.SVol.4-291/2, 305
$C_8F_4H_2N_2O$	$(O)(NH)C_8F_4NH$	F: PFHOrg.SVol.4-285
$C_8F_4H_3O_5Re$	$(CO)_5ReCF_2CF_2CH_3$	Re: Org.Comp.2-136
$C_8F_4H_8MoNO_2P_2^+$	$[(C_5H_5)Mo(CO)_2(F_2P-N(CH_3)-PF_2)]^+$	Mo: Org.Comp.7-249, 276
$C_8F_4H_{11}MoN$	$MoF_4 \cdot C_6H_5-N(CH_3)_2$	Mo: SVol.B5-92
$C_8F_4H_{18}O_4P_2Sn$	$(C_4H_9)_2Sn(OP(O)F_2)_2$	Sn: Org.Comp.15-356
$C_8F_4H_{24}Mo_2N_2O_4$	$[N(CH_3)_4]_2[Mo_2O_4F_4]$	Mo: SVol.B5-217/20
$C_8F_4N_2$	$1,2-(NC)_2-C_6F_4$	F: PFHOrg.SVol.6-107, 115, 126, 151/2
–	$1,3-(NC)_2-C_6F_4$	F: PFHOrg.SVol.6-107, 127, 152
–	$1,4-(NC)_2-C_6F_4$	F: PFHOrg.SVol.6-107, 115, 127, 150/2
$C_8F_4N_2O_2$	$1,3-(NCO)_2-C_6F_4$	F: PFHOrg.SVol.6-167, 181
$C_8F_5GeH_{15}$	$Ge(C_2H_5)_3C_2F_5$	Ge: Org.Comp.2-128, 138
$C_8F_5H_6NOS$	$(CH_3)_2N-S(O)-C_6F_5$	S: S-N Comp.8-297/8
$C_8F_5H_6N_2S_2^+$	$[(CH_3)_2SN=S=N-C_6F_5]^+$	S: S-N Comp.7-23
$C_8F_5H_6N_5$	$C_6F_5-NH-C(=NH)-NH-C(=NH)-NH_2$	F: PFHOrg.SVol.5-3, 35
C_8F_5NO	$C_6F_5-C(=O)-CN$	F: PFHOrg.SVol.6-108, 127
$C_8F_5N_3O_2$	$NNC(O)N(C_6F_5)C(O)$	F: PFHOrg.SVol.4-44/5, 70
$C_8F_5O_5Re$	$(CO)_5ReCF=C(F)CF_3$	Re: Org.Comp.2-138
$C_8F_5O_6Re$	$(CO)_5ReC(O)C_2F_5$	Re: Org.Comp.2-122/3
$C_8F_5O_7Re$	$(CO)_5ReOC(O)C_2F_5$	Re: Org.Comp.2-23
$C_8F_6FeH_5O_2PS$	$[(C_5H_5)Fe(CS)(CO)_2][PF_6]$	Fe: Org.Comp.B15-273/6
$C_8F_6FeH_5O_3P$	$[(C_5H_5)Fe(CO)_3][PF_6]$	Fe: Org.Comp.B15-36/46
$C_8F_6FeH_5O_3Sb$	$[(C_5H_5)Fe(CO)_3][SbF_6]$	Fe: Org.Comp.B15-36, 38, 46
$C_8F_6FeH_7O_2P$	$[C_5H_5(CO)_2Fe=CH_2][PF_6]$	Fe: Org.Comp.B16a-87, 93/4
$C_8F_6FeH_7O_3P$	$[(CH_2CHCHCHCH_2)Fe(CO)_3][PF_6]$	Fe: Org.Comp.B15-13, 18, 28
$C_8F_6FeH_{11}P_2^+$	$[(C_5H_5)Fe(CH_2=CHCH_3)(PF_3)_2]^+$	Fe: Org.Comp.B16b-5
$C_8F_6GeH_6$	$Ge(CH_3)_2(CCCF_3)_2$	Ge: Org.Comp.3-154, 161
$C_8F_6HN_3O_5$	$(O_2N)_2CF-C(=NH)-O-C_6F_5$	F: PFHOrg.SVol.5-178, 188
$C_8F_6H_5NS$	$F_2S=NCF(C_6H_5)CF_3$	S: S-N Comp.8-43
$C_8F_6H_6O_4Sn$	$(CH_2=CH)_2Sn(OOCCF_3)_2$	Sn: Org.Comp.16-90/1, 94/5
$C_8F_6H_8N_2O_2PRe$	$[(C_5H_5)Re(CO)(NO)(NC-CH_3)][PF_6]$	Re: Org.Comp.3-153, 155
$C_8F_6H_8O_4PRe$	$[(C_2H_4)_2Re(CO)_4][PF_6]$	Re: Org.Comp.2-363, 371
$C_8F_6H_9NO_5PRe$	$[(CO)_5ReNH_2C_3H_7-i][PF_6]$	Re: Org.Comp.2-151
$C_8F_6H_{11}MoN_2O_3P$	$[C_5H_5Mo(NO)_2O=C(CH_3)_2][PF_6]$	Mo: Org.Comp.6-61

$C_8F_6H_{11}N_5O_2$ .... $NHC(O)NHC(CF_3)_2NC(NH_2) \cdot HC(O)N(CH_3)_2$ .. F: PFHOrg.SVol.4-224/5, 239
$C_8F_6H_{14}MoN_2O_5P_2$
$[C_5H_5Mo(NO)_2P(OCH_3)_3][PF_6]$ .... Mo: Org.Comp.6-61
$C_8F_6H_{16}N_2Sn$ .... $(CH_3)_3SnNHC(CF_3)_2N(CH_3)_2$ .... Sn: Org.Comp.18-16/7
$C_8F_6H_{18}N_2O_5S_2Si_2Sn$
$(CH_3)_2Sn[-N(SO_2-CF_3)-Si(CH_3)_2-O-Si(CH_3)_2$
$-N(SO_2-CF_3)-]$ .... Sn: Org.Comp.19-79, 80
$C_8F_6H_{18}N_4OS_2$ ... $[((CH_3)_2N)_3S][S(O)F{=}N-C_2F_5]$ .... S: S-N Comp.8-230, 234, 247
$C_8F_6H_{18}O_7S_2Sb_2$ $CF_3S(O)_2-O-Sb(CH_3)_3-O-Sb(CH_3)_3-O-S(O)_2CF_3$
Sb: Org.Comp.5-106
– .... $CF_3S(O)_2-O-Sb(CH_3)_3-O-Sb(CH_3)_3-O$
$-S(O)_2CF_3 \cdot 2\ H_2O$ .... Sb: Org.Comp.5-106
$C_8F_6H_{24}P_2Si$ ..... $[(CH_3)_4P]_2[SiF_6]$ .... Si: SVol.B7-298
$C_8F_6H_{24}Sb_2Si$ .... $[(CH_3)_4Sb]_2[SiF_6]$ .... Si: SVol.B7-298
$C_8F_6N_2$ .... $C_8F_6N_2$ .... F: PFHOrg.SVol.4-292, 310/1
$C_8F_7GeH_9$ .... $(CH_3)_3Ge-C[{=}C(C_2F_5)-CF_2-]$ .... Ge: Org.Comp.2-29
– .... $(CH_3)_3Ge-C{\equiv}C-C_3F_7$-i .... Ge: Org.Comp.2-57/8
$C_8F_7H_2NO$ .... $C_6F_5-CF_2-C({=}O)NH_2$ .... F: PFHOrg.SVol.5-18, 52
C_8F_7N .... $C_6F_5-CF_2-CN$ .... F: PFHOrg.SVol.6-108, 127
C_8F_7NS .... $CF_3C_7F_4NS$ .... F: PFHOrg.SVol.4-284, 298
$C_8F_7O_5Re$ .... $(CO)_5ReC_3F_7$ .... Re: Org.Comp.2-136
C_8F_8HNO .... $CF_3-C({=}O)NH-C_6F_5$ .... F: PFHOrg.SVol.5-24, 92/3
C_8F_8HNS .... $CF_3-C({=}S)NH-C_6F_5$ .... F: PFHOrg.SVol.5-25, 59, 93
$C_8F_8H_{10}N_2S$ ..... $C_5H_{10}N-1-S(F){=}N-C_3F_7$-i .... S: S-N Comp.8-177/8
$C_8F_8H_{18}N_4P_2S_2$ .. $[F_2P(N{=}S{=}N-C_4H_9-t)_2][PF_6]$ .... S: S-N Comp.7-105
$C_8F_9H_2N_3$ .... $(H_2N)(CF_3)C_7F_6N_2$ .... F: PFHOrg.SVol.4-287, 301
$C_8F_9H_5O_6Sn$ ..... $C_2H_5Sn(OOCCF_3)_3$ .... Sn: Org.Comp.17-27
$C_8F_9H_9N_2O_2S_2$ ... n-$C_4F_9-S(O)_2N{=}S{=}N-C_4H_9$-t .... S: S-N Comp.7-53
$C_8F_9H_{12}NOSSi$ ... $(CH_3)_3Si-N(CH_3)-S(O)-C_4F_9$-n .... S: S-N Comp.8-296/7
$C_8F_9H_{12}NOSSn$ ... $(CH_3)_3SnN(CH_3)S({=}O)C_4F_9$ .... Sn: Org.Comp.18-57/8, 72
C_8F_9N .... 2-NC_8F_9 .... F: PFHOrg.SVol.4-287/8, 301
– .... 4-$(CF_2{=}CCF_3)-NC_5F_4$ .... F: PFHOrg.SVol.4-117/8, 128
– .... $C_6F_5-N{=}CF-CF_3$ .... F: PFHOrg.SVol.6-192, 206
$C_8F_{10}H_2N_2$ ...... $NC(NH_2)CFC(i-C_3F_7)CFCF$ .... F: PFHOrg.SVol.4-149, 159, 163
$C_8F_{10}H_3N_5O_2$ .... $NHNC(NHC(O)C_2F_5)NC(NHC(O)C_2F_5)$ .... F: PFHOrg.SVol.4-44/5, 69
$C_8F_{10}N_2$ .... 1,2-$(NC)_2-C_6F_{10}$-c .... F: PFHOrg.SVol.6-105, 125
– .... 1,3-$(NC)_2-C_6F_{10}$-c .... F: PFHOrg.SVol.6-105, 125
– .... 1,4-$(NC)_2-C_6F_{10}$-c .... F: PFHOrg.SVol.6-105, 125
$C_8F_{10}N_4$ .... 2-(1,3-$N_2C_4F_5$-2)-1,3-$N_2C_4F_5$ .... F: PFHOrg.SVol.4-193, 206
– .... 2-(N_3)-4-(i-C_3F_7)-NC_5F_3 .... F: PFHOrg.SVol.4-145, 163
$C_8F_{11}H_4N_3$ ...... $CF_3-C(NH_2){=}C(CN)-C(CF_3)(C_2F_5)-NH_2$ .... F: PFHOrg.SVol.5-2, 34
$C_8F_{11}N$ .... 4-(i-C_3F_7)-NC_5F_4 .... F: PFHOrg.SVol.4-117, 125, 139
– .... 5-C_2F_5-3-CF_3-NC_5F_3 .... F: PFHOrg.SVol.4-115, 124
$C_8F_{12}GaO_8^-$ ..... $Cs[Ga(OC(O)-CF_3)_4]$ .... Ga: SVol.D1-156, 157
$C_8F_{12}Ge$ .... $Ge(CFCF_2)_4$ .... Ge: Org.Comp.1-96
$C_8F_{12}HN$ .... 1,2,3,4-$(CF_3)_4$-[2.1.0]-5-NC_4H .... F: PFHOrg.SVol.4-280, 295
– .... 2,3,4,5-$(CF_3)_4$-NC_4H .... F: PFHOrg.SVol.4-25/6, 48

Formula	Compound	Reference
$C_8F_{12}HN_3$	1,5,6,7-$(CF_3)_4$-[3.2.0]-2,3,4-N_3C_4H	F: PFHOrg.SVol.4-281, 296
–	$CF_3-C(N_2)C(CF_3)=C(CF_3)-C(CF_3)=NH$	F: PFHOrg.SVol.5-28
$C_8F_{12}H_2N_2O_2$	$[NH(CF_2)_3C(O)]_2$	F: PFHOrg.SVol.4-278/9
$C_8F_{12}H_2N_8$	$(HNN_3C)(CF_2)_6(CN_3NH)$	F: PFHOrg.SVol.4-45, 71
$C_8F_{12}H_3I_2N_3O_2$	$ICF_2CF_2OCF_2C(=NH)N=C(NH_2)CF_2OCF_2CF_2I$	F: PFHOrg.SVol.5-5, 37
$C_8F_{12}H_4N_4O_2$	$OC(CF_3)_2OC(CF_3)_2NC(NHC(NH)NH_2)$	F: PFHOrg.SVol.4-176, 180
$C_8F_{12}H_6N_2Sn$	$(CF_3)_2C=N-Sn(CH_3)_2-N=C(CF_3)_2$	Sn: Org.Comp.19-72
$C_8F_{12}H_{12}HgMo_2N_4$	$Hg[MoF_6]_2 \cdot 4\ CH_3CN$	Mo: SVol.B5-176, 185
$C_8F_{12}HgN_2$	$Hg[C(CF_3)_2CN]_2$	F: PFHOrg.SVol.6-100, 132
$C_8F_{12}MnNOS_4^{2-}$	$[Mn(SC(CF_3)=C(CF_3)S)_2NO]^{2-}$	Mn: MVol.D7-44/5
$C_8F_{12}N_2$	2,6-$(C_2F_5)_2$-1,4-$N_2C_4F_2$	F: PFHOrg.SVol.4-198/9
–	4-(s-C_4F_9)-1,2-$N_2C_4F_3$	F: PFHOrg.SVol.4-185/6
–	4-(s-C_4F_9)-1,3-$N_2C_4F_3$	F: PFHOrg.SVol.4-192, 205
–	5-(s-C_4F_9)-1,3-$N_2C_4F_3$	F: PFHOrg.SVol.4-192, 205
–	$NC-(CF_2)_6-CN$	F: PFHOrg.SVol.6-102, 133
$C_8F_{12}N_4O_2S_2$	$[(CF_3)_2C=NS-NCO]_2$	F: PFHOrg.SVol.6-49
$C_8F_{12}O_4S_2Sn$	$(CF_2=CF)_2Sn(OS(O)CF=CF_2)_2$	Sn: Org.Comp.16-94
$C_8F_{12}O_8Th$	$Th(OOC-CF_3)_4$	Th: SVol.C7-55/6, 61
–	$Th(OOC-CF_3)_4 \cdot H_2O$	Th: SVol.C7-56, 61
$C_8F_{12}Pb$	$Pb(CF=CF_2)_4$	Pb: Org.Comp.3-91
$C_8F_{13}H_2NO_3$	$CF_2=CF-O-CF_2CF(CF_3)-O-CF_2CF_2-C(=O)NH_2$	F: PFHOrg.SVol.5-20/1, 54
$C_8F_{13}N$	2-C_4F_3-(C_3F_7-i)-3-(=N-CF_3)-1	F: PFHOrg.SVol.6-191, 204
–	2-C_4F_4-(C_3F_7-i)-3-(N=CF_2)-1	F: PFHOrg.SVol.6-191, 204
–	3-(i-C_3F_7)-NC_5F_6	F: PFHOrg.SVol.4-117, 126
–	4-(i-C_3F_7)-NC_5F_6	F: PFHOrg.SVol.4-117/8, 126, 128, 139
$C_8F_{13}NO$	ON(c-C_4F_7)$CF_2CFCFCF_2$	F: PFHOrg.SVol.4-175, 179
$C_8F_{13}NO_2$	$CF_2=CF-O-CF_2CF(CF_3)-O-CF_2CF_2-CN$	F: PFHOrg.SVol.6-99, 119, 143, 152
$C_8F_{13}N_3O$	$NC(CF(CF_3)OC_3F_7)NCFNCF$	F: PFHOrg.SVol.4-227, 241
$C_8F_{14}HN$	$C_2F_5-C(CF_3)=C(CF_3)-C(CF_3)=NH$	F: PFHOrg.SVol.5-28
$C_8F_{14}H_2N_2O_2$	n-C_3F_7-C(=O)NH-NHC(=O)-C_3F_7-n	F: PFHOrg.SVol.5-207, 212, 218
$C_8F_{14}H_3N_3$	$C_3F_7-C(=NH)-N=C(NH_2)-C_3F_7$	F: PFHOrg.SVol.5-5, 36, 75
$C_8F_{14}H_4N_2$	$CF_3-C(NH_2)=C(CF_3)-C(CF_3)(C_2F_5)-NH_2$	F: PFHOrg.SVol.5-2, 34
$C_8F_{14}N^-$	$[NCF_2CF_2C$(i-C_3F_7)$CFCF_2]^-$	F: PFHOrg.SVol.4-118, 128
$C_8F_{14}N_2$	$N(C_4F_5)CF_2N(CF_3)CF_2CF_2$	F: PFHOrg.SVol.4-36/7, 62
$C_8F_{14}N_6$	$(CF_3-N=C=NCF=N-CF_3)_2$	F: PFHOrg.SVol.6-191, 204
$C_8F_{15}HN_2O$	$CF_3-C(=O)NH-C(CF_3)_2-N=C(CF_3)_2$	F: PFHOrg.SVol.5-23/4, 58
$C_8F_{15}HN_2O_3$	$ONC(C(CF_3)_2N(OH)CF_3)OC(CF_3)_2$	F: PFHOrg.SVol.4-31, 55
$C_8F_{15}H_2NO$	n-C_7F_{15}-C(=O)NH_2	F: PFHOrg.SVol.5-20, 53, 90, 96
$C_8F_{15}H_2NO_2$	t-C_4F_9-NH-$C(CF_3)_2$-C(=O)OH	F: PFHOrg.SVol.5-15, 46
–	n-C_7F_{15}-C(=O)NH-OH	F: PFHOrg.SVol.5-23, 56, 91
$C_8F_{15}H_2NO_4$	$F(CF_2CF_2-O-)_3CF_2-C(=O)NH_2$	F: PFHOrg.SVol.5-20, 53
$C_8F_{15}H_3N_2$	n-C_7F_{15}-C(=NH)-NH_2	F: PFHOrg.SVol.5-3, 72
$C_8F_{15}N$	n-C_7F_{15}-CN	F: PFHOrg.SVol.6-97/8, 116, 132, 137/9

$C_8F_{15}NO$ .... $t-C_4F_9-C(=O)N=C(CF_3)_2$ .... F: PFHOrg.SVol.6-189/90, 202, 216
– .... $n-C_7F_{15}-NCO$ .... F: PFHOrg.SVol.6-167, 172, 174
– .... $O=CF-N(C_4F_9-t)-C(CF_3)=CF_2$ .... F: PFHOrg.SVol.6-229/30, 239/40, 243
– .... $[-N(C_4F_9-t)-C(=O)-C(CF_3)_2-]$ .... F: PFHOrg.SVol.4-2/3, 6, 11/3
$C_8F_{15}NO_2$ .... $C(CF_3)_2OC(O)N(t-C_4F_9)$ .... F: PFHOrg.SVol.4-21, 23, 24
$C_8F_{16}NO^+$ .... $[i-C_3F_7-NC(=O)-C_4F_9-t]^+$ .... F: PFHOrg.SVol.5-24, 58, 70, 92
$C_8F_{16}N_2$ .... $n-C_3F_7-CF=NN=CF-C_3F_7-n$ .... F: PFHOrg.SVol.5-208, 219
$C_8F_{16}N_4$ .... $(CF_3)C_3F_6N_2-C_3F_4N_2(CF_3)$ .... F: PFHOrg.SVol.4-36/7
– .... $n-C_3F_7-N=C(N_3)-C_4F_9-t$ .... F: PFHOrg.SVol.5-204, 216
– .... $N_2C_3F_3(CF=NCF_3)(=NCF_3)(CF_3)_2$ .... F: PFHOrg.SVol.4-36/7
– .... $N_3C_5F_7(=NCF_3)(CF_3)_2$ .... F: PFHOrg.SVol.4-282, 296
– .... $N_4C_5F_7(CF_3)_3$ .... F: PFHOrg.SVol.4-36/7, 63
$C_8F_{17}N$ .... $1,3-(C_2F_5)_2-NC_4F_7$ .... F: PFHOrg.SVol.4-27/8, 51
– .... $1-C_2F_5-2-CF_3-NC_5F_9$ .... F: PFHOrg.SVol.4-122, 135
– .... $1-C_2F_5-3-CF_3-NC_5F_9$ .... F: PFHOrg.SVol.4-122, 136
– .... $1-C_2F_5-4-CF_3-NC_5F_9$ .... F: PFHOrg.SVol.4-122/3, 137
– .... $1-C_2F_5-NC_6F_{12}$ .... F: PFHOrg.SVol.4-274/5
– .... $n-C_3F_7-N=CF-C_4F_9-t$ .... F: PFHOrg.SVol.6-190, 201
$C_8F_{17}NO$ .... $CF_2CF_2ON(CF_2CF_2C(CF_3)_3)$ .... F: PFHOrg.SVol.4-21, 22, 24
$C_8F_{17}NO_2S$ ...... $2-(C_2F_5N=)-1,3,2-O_2SC_2-4,4,5,5-(CF_3)_4$ ...... S: S-N Comp.8-164/5
$C_8F_{18}N_2O_2$ ...... $CF_3-N(O-C_3F_7-i)C(F)=N-O-C_3F_7-i$ .... F: PFHOrg.SVol.5-115, 126
$C_8F_{18}N_2S$ .... $(CF_3)_2S=N-C(CF_3)_2-N=C(CF_3)_2$ .... S: S-N Comp.8-144/5
– .... $t-C_4F_9-N=S=N-C_4F_9-t$ .... F: PFHOrg.SVol.6-52
$C_8F_{18}N_2S_2$ ...... $t-C_4F_9SN=S=N-C_4F_9-t$ .... S: S-N Comp.7-23/4
$C_8F_{18}N_4O_3$ ...... $[(CF_3)_2N-O-C(=N-CF_3)]_2O$ .... F: PFHOrg.SVol.6-191, 204
$C_8F_{21}N_3O_2S$ ..... $(CF_3)_2NO-S(CF_3)=N-C(CF_3)_2-ON(CF_3)_2$ ...... S: S-N Comp.8-154
$C_8F_{22}HgN_4O_2$ .... $Hg[N(CF_3)CF_2-O-N(CF_3)_2]_2$ .... F: PFHOrg.SVol.5-14, 44, 69
$C_8F_{22}N_2S_2$ ...... $1,3-[i-C_3F_7-N=]_2-1,1,2,2,3,3,4,4-F_8-1,3-S_2C_2$. F: PFHOrg.SVol.6-55, 72
$C_8FeH_5IO_3$ ...... $[(C_5H_5)Fe(CO)_3]I$ .... Fe: Org.Comp.B15-38, 46/7
$C_8FeH_5IO_4$ ...... $[(HO)C_5H_4Fe(CO)_3]I$ .... Fe: Org.Comp.B15-50/1
$C_8FeH_5KO_2S_2$ .... $K[(C_5H_5)Fe(CO)_2-CS_2]$ .... Fe: Org.Comp.B14-162/3
$C_8FeH_5KO_4$ ...... $K[(C_5H_5)Fe(CO)_2-CO_2]$ .... Fe: Org.Comp.B14-161/2
$C_8FeH_5LiO_4$ ..... $Li[(C_5H_5)Fe(CO)_2-CO_2]$ .... Fe: Org.Comp.B14-161/2
– .... $Li[(C_5H_5)Fe(CO)_2-{}^{13}CO_2]$ .... Fe: Org.Comp.B14-161
$C_8FeH_5NaO_2S_2$ ... $Na[(C_5H_5)Fe(CO)_2-CS_2]$ .... Fe: Org.Comp.B14-162/3
$C_8FeH_5NaO_4$ ..... $Na[(C_5H_5)Fe(CO)_2-CO_2]$ .... Fe: Org.Comp.B14-161/2
$C_8FeH_5O_2S^+$ ..... $[(C_5H_5)Fe(CS)(CO)_2]^+$ .... Fe: Org.Comp.B15-273/6
$C_8FeH_5O_2S_2^-$ .... $[(C_5H_5)Fe(CO)_2-CS_2]^-$ .... Fe: Org.Comp.B14-162/3
$C_8FeH_5O_3^+$ ...... $[(C_5H_5)Fe(CO)_3]^+$
Chemical reactions .... Fe: Org.Comp.B15-40/7
Physical properties .... Fe: Org.Comp.B15-38/40
Preparation and formation .... Fe: Org.Comp.B15-36/7
Structure .... Fe: Org.Comp.B15-39

$C_8FeH_5O_4^+$ ...... $[(HO)C_5H_4Fe(CO)_3]^+$ .... Fe: Org.Comp.B15-50/1
$C_8FeH_5O_4^-$ ...... $[C_5H_5Fe(CO)_2CO_2]^-$ .... Fe: Org.Comp.B14-161/2

$C_8FeH_6O_2$ .... $(C_6H_6)Fe(CO)_2$ .... Fe: Org.Comp.B18-21, 40
$C_8FeH_6O_2^+$ ...... $[(C_6H_6)Fe(CO)_2]^+$ .... Fe: Org.Comp.B18-40
$C_8FeH_6O_2S_2$ ..... $(C_5H_5)Fe(CO)_2C(S)SH$ .... Fe: Org.Comp.B13-56/8
$C_8FeH_6O_3$ .... $(C_5H_5)Fe(CO)_2CHO$ .... Fe: Org.Comp.B13-13, 27/8
$C_8FeH_6O_4$ .... $(C_5H_5)Fe(CO)_2COOH$ .... Fe: Org.Comp.B13-23, 43/4
$C_8FeH_7IO_2$ ...... $(C_5H_4I)Fe(CO)_2CH_3$ .... Fe: Org.Comp.B14-58, 64/5
$C_8FeH_7KO_2$ ...... $K[(C_5H_4CH_3)Fe(CO)_2]$ .... Fe: Org.Comp.B14-117
$C_8FeH_7NO_3$ ...... $(C_5H_5)Fe(CO)_2CONH_2$ .... Fe: Org.Comp.B13-9, 24, 45
$C_8FeH_7NaO_2$ ..... $Na[(C_5H_4CH_3)Fe(CO)_2]$ .... Fe: Org.Comp.B14-117
$C_8FeH_7O_2^+$ ...... $[(C_5H_5)(CO)_2Fe{=}CH_2]^+$ .... Fe: Org.Comp.B16a-77/9, 87, 93/4
$C_8FeH_7O_2^-$ ...... $[(C_5H_4CH_3)Fe(CO)_2]^-$ .... Fe: Org.Comp.B14-117
$C_8FeH_7O_3^+$ ...... $[(CH_2C(CH_2)CHCH_2)Fe(CO)_3]^+$ .... Fe: Org.Comp.B15-31/2
– .... $[(CH_2CHCHCHCH_2)Fe(CO)_3]^+$ .... Fe: Org.Comp.B15-2, 12/3, 15/6, 18, 28
$C_8FeH_8^+$ .... $[(C_8H_8)Fe]^+$ .... Fe: Org.Comp.B18-2
C_8FeH_8HgINO.... $C_5H_5Fe(CNCH_3)(CO)HgI$ .... Fe: Org.Comp.B15-289, 295
C_8FeH_8INO ...... $(C_5H_5)Fe(CO)(CN{-}CH_3)I$ .... Fe: Org.Comp.B15-287, 293/4
C_8FeH_8INS ...... $(C_5H_5)Fe(CS)(CN{-}CH_3)I$ .... Fe: Org.Comp.B15-286
– .... $(C_5H_5)Fe(CS)(NC{-}CH_3)I$ .... Fe: Org.Comp.B15-261, 262
$C_8FeH_8I_3NOSn$ ... $C_5H_5Fe(CNCH_3)(CO)SnI_3$ .... Fe: Org.Comp.B15-289, 295
$C_8FeH_8NO^-$ ...... $[(C_5H_5)Fe(CO)CNCH_3]^-$ .... Fe: Org.Comp.B15-293
$C_8FeH_8NO_2^+$ .... $[C_5H_5(CO)_2Fe{=}CH(NH_2)]^+$ .... Fe: Org.Comp.B16a-114, 124/5
$C_8FeH_8N_2O_3$ ..... $(C_5H_5)Fe(CO)_2CONHNH_2$ .... Fe: Org.Comp.B13-25, 46
$C_8FeH_8N_2O_4$ ..... $C_5H_5Fe(CNCH_3)(CO)NO_3$ .... Fe: Org.Comp.B15-287, 294
$C_8FeH_8O_2^+$ ...... $[C_5H_5Fe(CO)_2CH_3]^+$, radical cation .... Fe: Org.Comp.B14-164/5
$C_8FeH_8O_2^-$ ...... $[C_5H_5Fe(CO)_2CH_3]^-$, radical anion .... Fe: Org.Comp.B14-165
C_8FeH_9NO ...... $C_5H_5Fe(CNCH_3)(CO)H$ .... Fe: Org.Comp.B15-287, 293
$C_8FeH_9NO_4Sn$.... $(CH_3)_3SnNC{-}Fe(CO)_4$ .... Sn: Org.Comp.18-128
$C_8FeH_{10}^+$ .... $[(C_8H_{10})Fe]^+$ .... Fe: Org.Comp.B18-2
$C_8FeH_{11}NOSi$ .... $C_5H_5Fe(CNCH_3)(CO)SiH_3$ .... Fe: Org.Comp.B15-288, 295
$C_8Fe_3HNO_9S$..... $(CO)_8(H)Fe_3S(NO)$ .... Fe: Org.Comp.C6a-55
$C_8Fe_3HN_2O_9^-$ .... $[(CO)_8Fe_3(NH)(NO)]^-$ .... Fe: Org.Comp.C6a-75
$C_8Fe_3O_8^+$ .... $[Fe_3(CO)_8]^+$ .... Fe: Org.Comp.C6a-4
$C_8GaH_4O_4^+$ ..... $[Ga(1,2{-}(OC(O))_2{-}C_6H_4)]^+$ .... Ga: SVol.D1-169
$C_8GaH_5O_5$....... $Ga(OH)[1,2{-}(OC(O))_2{-}C_6H_4]$ .... Ga: SVol.D1-169/70
$C_8GaH_6O_3^+$ ..... $[Ga(OC(O){-}CH(O){-}C_6H_5)]^+$ .... Ga: SVol.D1-174
$C_8GaH_7O_3^{2+}$..... $[Ga(OC(O){-}CH(OH){-}C_6H_5)]^{2+}$ .... Ga: SVol.D1-174
$C_8GaH_8NO_2^{2+}$ ... $[Ga(NC_5H{-}2,6{-}(CH_3)_2{-}3{-}O{-}4{-}CH{=}O)]^{2+}$ .... Ga: SVol.D1-255, 256
$C_8GaH_8NO_3^{2+}$ ... $[Ga(NC_5H{-}2{-}CH_3{-}3{-}O{-}4{-}CHO{-}6{-}CH_2OH)]^{2+}$ Ga: SVol.D1-255, 256
$C_8GaH_8O_8^-$...... $[Ga(OC(O){-}CH(CH_3){-}C(O)O)_2]^-$ .... Ga: SVol.D1-166/7
$C_8GaH_8O_{10}^-$ ..... $[Ga(OC(O){-}CH(OH)CH_2{-}C(O)O)_2]^-$ .... Ga: SVol.D1-183/4
$C_8GaH_8O_{12}^-$ ..... $[Ga(OC(O){-}CH(OH)CH(OH){-}C(O)O)_2]^-$ .... Ga: SVol.D1-184/5
$C_8GaH_{10}NO_2$..... $Ga(NH_2{-}CH_3)[OC(O){-}C_6H_5]$ .... Ga: SVol.D1-217
$C_8GaH_{10}NO_2^{2+}$ .. $[Ga(NC_5H_2{-}1{-}C_2H_5{-}2{-}CH_3{-}3,4{-}(O)_2)]^{2+}$ .... Ga: SVol.D1-258
$C_8GaH_{10}O_{10}^+$.... $[Ga(OC(O){-}CH(OH)CH_2{-}C(O)OH)_2]^+$ .... Ga: SVol.D1-183/4
$C_8GaH_{16}O_4^+$..... $[Ga(1,4{-}O_2C_4H_8)_2]Br$ .... Ga: SVol.D1-142/3
– .... $[Ga(1,4{-}O_2C_4H_8)_2]Cl$ .... Ga: SVol.D1-142/3
$C_8GaH_{24}O_4^{3+}$.... $[Ga(HO{-}C_2H_5)_4]^{3+}$ .... Ga: SVol.D1-19/20

$C_8GaH_{24}O_8{}^{3+}$ .... $[Ga(HO-CH_2CH_2-OH)_4]^{3+}$ .... Ga: SVol.D1-23
$C_8GaH_{26}O_9{}^{3+}$ .... $[(H_2O)Ga(HO-CH_2CH_2-OH)_4]^{3+}$ .... Ga: SVol.D1-23
$C_8GaH_{28}O_6{}^{3+}$ .... $[Ga(HO-C_2H_5)_4(H_2O)_2]^{3+}$ .... Ga: SVol.D1-16, 18
$C_8Ga_2H_{16}I_4O_4$ .... $Ga_2(1,4-O_2C_4H_8)_2I_4$ .... Ga: SVol.D1-143/6
C_8GeH_4 .... $Ge(CCH)_4$ .... Ge: Org.Comp.1-103/5
$C_8GeH_8O_2$ .... $Ge(CH_3)_2(CCCHO)_2$ .... Ge: Org.Comp.3-154, 161
$C_8GeH_9O_5Re$ .... $(CO)_5ReGe(CH_3)_3$ .... Re: Org.Comp.2-7, 13/4
$C_8GeH_{10}N_2$ ...... cis-$(CH_3)_2Ge(-CH(CN)CH=CHCH(CN)-)$ ...... Ge: Org.Comp.3-261
C_8GeH_{11} .... $Ge(CH_3)_2C_6H_5$, radical .... Ge: Org.Comp.3-361, 363
C_8GeH_{12} .... $(CH_3)_3Ge-C{\equiv}C-C{\equiv}C-CH_3$ .... Ge: Org.Comp.2-59, 65
– .... $(C_2H_5)_2Ge(C{\equiv}CH)_2$ .... Ge: Org.Comp.3-170, 173
– .... $Ge(-CH_2CH=CHCH_2-)_2$ .... Ge: Org.Comp.3-343, 346
– .... $Ge(CH=CH_2)_4$ .... Ge: Org.Comp.1-96/9
$C_8GeH_{12}O_2$ ...... $(CH_3)_2Ge(C{\equiv}C-CH_2OH)_2$ .... Ge: Org.Comp.3-154
– .... $(CH_3)_3Ge-C[=CHCH=C(CH=O)-O-]$ .... Ge: Org.Comp.2-96, 105
$C_8GeH_{12}O_3$ ...... $Ge(CH_3)_3C(=CHCH=C(COOH)O)$ .... Ge: Org.Comp.2-96/7, 105
C_8GeH_{13} .... $Ge(CH_3)_3C_5H_4$, radical .... Ge: Org.Comp.3-351
$C_8GeH_{13}In$ ...... $In[C_5H_4-Ge(CH_3)_3]$ .... In: Org.Comp.1-379, 380
$C_8GeH_{13}Li$ ...... $Ge(CH_3)_3C_5H_4Li$ .... Ge: Org.Comp.2-46
$C_8GeH_{13}N$ .... 2-$(CH_3)_3Ge-NC_5H_4$ .... Ge: Org.Comp.2-101/2, 106
– .... 3-$(CH_3)_3Ge-NC_5H_4$ .... Ge: Org.Comp.2-102, 106
– .... 4-$(CH_3)_3Ge-NC_5H_4$ .... Ge: Org.Comp.2-102, 106
$C_8GeH_{13}NO$ ..... 3-$(CH_3)_3Ge-NC_5H_4(O)$-1 .... Ge: Org.Comp.2-102, 106
– .... 4-$(CH_3)_3Ge-NC_5H_4(O)$-1 .... Ge: Org.Comp.2-102/3, 106/7
$C_8GeH_{13}NO_2$ ..... $Ge(CH_3)_3C(=CHCH=C(CONH_2)O)$ .... Ge: Org.Comp.2-97
C_8GeH_{14} .... $(CH_3)_2Ge(-CH_2CH=CHCH=CHCH_2-)$ .... Ge: Org.Comp.3-309, 313
– .... $(CH_3)_2Ge[-CH=C(CH_3)C(=CH_2)CH_2-]$ .... Ge: Org.Comp.3-265
– .... $(CH_3)_2Ge[-CH=C(CH_3)C(CH_3)=CH-]$ .... Ge: Org.Comp.3-276/7, 282/3
– .... $(CH_3)_2Ge[-CH_2C(=CH_2)C(=CH_2)CH_2-]$ .... Ge: Org.Comp.3-247, 253
– .... $(CH_3)_3Ge-C_5H_5$ .... Ge: Org.Comp.2-32, 45/6
– .... $C_2H_5-Ge(CH=CH_2)_3$ .... Ge: Org.Comp.3-50
– .... $HC{\equiv}C-Ge(C_2H_5)(-CH_2CH_2CH_2CH_2-)$ .... Ge: Org.Comp.3-248
– .... $CH_2=CHCH_2-Ge(CH_3)(-CH_2CH=CHCH_2-)$ ..... Ge: Org.Comp.3-266
$C_8GeH_{14}O$ .... $Ge(CH_3)_3C(=CHCH=C(CH_3)O)$ .... Ge: Org.Comp.2-96
C_8GeH_{15} .... $Ge(CH_3)_3C_5H_6$, radical .... Ge: Org.Comp.3-351
C_8GeH_{16} .... $(-CH_2CH_2CH_2CH_2-)Ge(-CH_2CH_2CH_2CH_2-)$ ... Ge: Org.Comp.3-343, 346
– .... $CH_2=CH-Ge(CH_3)[-CH(CH_3)CH_2CH_2CH_2-]$ .... Ge: Org.Comp.3-250
– .... $(CH_3)_2Ge(CH_2CH=CH_2)_2$ .... Ge: Org.Comp.3-153, 161
– .... $(CH_3)_2Ge[-CH(CH_3)CH=CHCH(CH_3)-]$ .... Ge: Org.Comp.3-260
– .... $(CH_3)_2Ge[-CH(C_2H_5)-CH=CHCH_2-]$ .... Ge: Org.Comp.3-259
– .... $(CH_3)_2Ge[-CH_2C(CH_3)=C(CH_3)CH_2-]$ .... Ge: Org.Comp.3-259, 270/1
– .... $(CH_3)_2Ge[C(CH_3)=CH_2]_2$ .... Ge: Org.Comp.3-153
– .... $(CH_3)_3Ge-C(=CHCH_2CH_2CH_2-)$ .... Ge: Org.Comp.2-31, 44
– .... $(CH_3)_3Ge-CH(-CH=CHCH_2CH_2-)$ .... Ge: Org.Comp.2-31,44
– .... $(CH_3)_3Ge-CH(-CH_2CH=CHCH_2-)$ .... Ge: Org.Comp.2-31/2
– .... $(CH_3)_3Ge-CH_2CH=CHCH=CH_2$ .... Ge: Org.Comp.2-19
– .... $(CH_3)_3Ge-C{\equiv}C-C_3H_7$-i .... Ge: Org.Comp.2-57
– .... $(C_2H_5)_2Ge(-CH=CHCH_2CH_2-)$ .... Ge: Org.Comp.3-262
– .... $(C_2H_5)_2Ge(-CH_2CH=CHCH_2-)$ .... Ge: Org.Comp.3-257, 270
– .... $(C_2H_5)_2Ge(CH=CH_2)_2$ .... Ge: Org.Comp.3-169, 172/3

C_8GeH_{16}	$(C_2H_5)_3Ge-C{\equiv}CH$	Ge: Org.Comp.2-236, 245/6
$C_8GeH_{16}N_2O$	$Ge(CH_3)_3C(C(OC_2H_5)=CHNHN=)$	Ge: Org.Comp.2-101
$C_8GeH_{16}O$	1,3,3,5-$(CH_3)_4$-[3.1.0]-6,3-$OGeC_4H_4$	Ge: Org.Comp.3-320
–	3,3-$(C_2H_5)_2$-[3.1.0]-6,3-$OGeC_4H_6$	Ge: Org.Comp.3-320
–	$(CH_3)_2Ge[-CH=C(CH_3)C(OH)(CH_3)-CH_2-]$	Ge: Org.Comp.3-265
–	$(CH_3)_2Ge[-CH_2C(OH)(CH_3)-C(=CH_2)CH_2-]$	Ge: Org.Comp.3-248, 253/4
–	$(CH_3)_3Ge-C(=O)-CH=C(CH_3)_2$	Ge: Org.Comp.2-16
–	$(CH_3)_3Ge-C(=O)-CH_2C(CH_3)=CH_2$	Ge: Org.Comp.2-16
–	$(CH_3)_3Ge-C(OCH_3)=CHCH=CH_2$	Ge: Org.Comp.2-16/7
–	$(CH_3)_3Ge-C{\equiv}C-C(CH_3)_2OH$	Ge: Org.Comp.2-58
–	$(C_2H_5)_3Ge-CH=C=O$	Ge: Org.Comp.2-198, 200/1
$C_8GeH_{16}O_2$	$(CH_3)_3Ge-CH_2-OOC-C(CH_3)=CH_2$	Ge: Org.Comp.1-147
–	$(CH_3)_3Ge-CH_2$-c-C_3H_4-COOH	Ge: Org.Comp.1-155, 161
$C_8GeH_{17}N$	$Ge(C_2H_5)_3CH_2CN$	Ge: Org.Comp.2-132, 138/9
$C_8GeH_{17}NO_2$	$Ge(CH_3)_3CH_2CON(CH_3)COCH_3$	Ge: Org.Comp.1-171
$C_8GeH_{17}N_3O$	$(CH_3)_2Ge[-(CH_2)_2-C(=NNHC(O)NH_2)-(CH_2)_2-]$	Ge: Org.Comp.3-297
C_8GeH_{18}	$(CH_3)_2Ge[-CH(CH_3)CH_2CH_2CH_2CH_2-]$	Ge: Org.Comp.3-290
–	$(CH_3)_2Ge[-(CH_2)_6-]$	Ge: Org.Comp.3-309
–	$(CH_3)_2Ge[-CH_2CH(CH_3)CH_2CH_2CH_2-]$	Ge: Org.Comp.3-290, 297
–	$(CD_3)_2Ge[-CH_2CH(CH_3)CH_2CH_2CH_2-]$	Ge: Org.Comp.3-297
–	$(CH_3)_2Ge[-CH_2CH_2CH(CH_3)CH_2CH_2-]$	Ge: Org.Comp.3-290, 297
–	$(CD_3)_2Ge[-CH_2CH_2CH(CH_3)CH_2CH_2-]$	Ge: Org.Comp.3-297
–	$(CH_3)_3Ge-CH(CH_3)CH=CHCH_3$	Ge: Org.Comp.2-15
–	$(CH_3)_3Ge-CH_2C(CH_3)=CHCH_3$	Ge: Org.Comp.2-15
–	$(CH_3)_3Ge-CH_2CH=C(CH_3)_2$	Ge: Org.Comp.2-14/5
–	$(CH_3)_3Ge-CH_2CH=CH-C_2H_5$	Ge: Org.Comp.2-18
–	$(CH_3)_3Ge-C_5H_9$-c	Ge: Org.Comp.1-195, 199
–	$(C_2H_5)_2Ge(-CH_2CH_2CH_2CH_2-)$	Ge: Org.Comp.3-242, 252
–	$(C_2H_5)_2Ge[-CH_2CH(CH_3)CH_2-]$	Ge: Org.Comp.3-236
–	$(C_2H_5)_3Ge-CH=CH_2$	Ge: Org.Comp.2-169, 182/3
–	$C_2H_5-Ge(CH_3)[-CH(CH_3)CH_2CH_2CH_2-]$	Ge: Org.Comp.3-249, 254
$C_8GeH_{18}O$	$(CH_3)_2Ge[-CH_2C(OH)(CH_3)CH(CH_3)CH_2-]$	Ge: Org.Comp.3-247, 251/2
–	$(CH_3)_3Ge-CH[-CH_2CH_2CH(CH_3)-O-]$	Ge: Org.Comp.2-95
–	$(C_2H_5)_2Ge[-CH_2CH(OH)CH_2CH_2-]$	Ge: Org.Comp.3-243, 252
–	$(C_2H_5)_3Ge-C(=O)CH_3$	Ge: Org.Comp.2-137, 139
–	$(C_2H_5)_3Ge-CH_2-CH=O$	Ge: Org.Comp.2-130, 138
$C_8GeH_{18}OS$	$Ge(CH_3)_3C_3H_6SCOCH_3$	Ge: Org.Comp.1-181
$C_8GeH_{18}OSi$	$Ge(CH_3)_3C(Si(CH_3)_3)=C=O$	Ge: Org.Comp.2-7
$C_8GeH_{18}OSn$	$Ge(CH_3)_3C(Sn(CH_3)_3)=C=O$	Ge: Org.Comp.2-7
$C_8GeH_{18}O_2$	$(CH_3)_3Ge-CH_2CH_2CH_2-OOC-CH_3$	Ge: Org.Comp.1-181, 185
–	$(CH_3)_3Ge-CH_2CH_2-COO-C_2H_5$	Ge: Org.Comp.1-184/5
–	$(C_2H_5)_3Ge-CH_2-COOH$	Ge: Org.Comp.2-130, 138
$C_8GeH_{18}S_2$	$Ge(CH_3)_3C(CH_3)(SCH_2)_2CH_2$	Ge: Org.Comp.2-100, 106
$C_8GeH_{18}Si$	$Ge(CH_3)_3C{\equiv}CSi(CH_3)_3$	Ge: Org.Comp.2-54
$C_8GeH_{18}Sn$	$Ge(CH_3)_3C{\equiv}CSn(CH_3)_3$	Ge: Org.Comp.2-54
C_8GeH_{19}	$(C_2H_5)_3Ge-CHCH_3$, radical	Ge: Org.Comp.3-349
–	$(C_2H_5)_3Ge-CH_2CH_2$, radical	Ge: Org.Comp.3-349
$C_8GeH_{19}NO$	$Ge(CH_3)_3CONHC_4H_9$-t	Ge: Org.Comp.1-145
C_8GeH_{20}	$(CH_3)_2Ge(C_2H_5)-C_4H_9$-n	Ge: Org.Comp.3-197, 206
–	$(CH_3)_2Ge(C_3H_7-n)_2$	Ge: Org.Comp.3-149, 160

C_8GeH_{20} .... $(CH_3)_2Ge(C_3H_7\text{-}i)_2$ .... Ge: Org.Comp.3-149
– .... $(CH_3)_3Ge\text{-}C_5H_{11}\text{-}n$ .... Ge: Org.Comp.1-192/3
– .... $(C_2H_5)_2Ge(CH_3)\text{-}C_3H_7\text{-}n$ .... Ge: Org.Comp.3-209, 212
– .... $Ge(C_2H_5)_4$
Chemical reactions .... Ge: Org.Comp.1-59/62
Molecule .... Ge: Org.Comp.1-50/3
Pharmacology .... Ge: Org.Comp.1-64/5
Physical poperties .... Ge: Org.Comp.1-53/9
Preparation and formation .... Ge: Org.Comp.1-47/50
Solubility .... Ge: Org.Comp.1-62
Spectra .... Ge: Org.Comp.1-50/3
Uses .... Ge: Org.Comp.1-64/5

$C_8GeH_{20}Hg$ .... $Ge(CH_3)_3CH_2HgC_4H_9\text{-}i$ .... Ge: Org.Comp.1-142
$C_8GeH_{20}O$ .... $(CH_3)_3Ge\text{-}CH(CH_3)CH_2CH(OH)CH_3$ .... Ge: Org.Comp.1-188
– .... $(C_2H_5)_3Ge\text{-}CH_2CH_2OH$ .... Ge: Org.Comp.2-124
– .... $(C_2H_5)_3Ge\text{-}CH_2\text{-}O\text{-}CH_3$ .... Ge: Org.Comp.2-116
$C_8GeH_{20}O_2Si$ .... $(CH_3)_3Ge\text{-}CH(COOH)\text{-}Si(CH_3)_3$ .... Ge: Org.Comp.1-172
– .... $(CH_3)_3Ge\text{-}CH_2\text{-}COO\text{-}Si(CH_3)_3$ .... Ge: Org.Comp.1-171
$C_8GeH_{20}O_3S$ .... $(CH_3)_3Ge\text{-}CH(C_3H_7\text{-}n)\text{-}OS(O)_2CH_3$ .... Ge: Org.Comp.1-189
– .... $(C_2H_5)_3Ge\text{-}CH_2\text{-}S(O)_2\text{-}OCH_3$ .... Ge: Org.Comp.2-116
$C_8GeH_{20}O_4$ .... $Ge(CH_2OCH_3)_4$ .... Ge: Org.Comp.1-88, 92
$C_8GeH_{20}Si$ .... $(CH_3)_2Ge[\text{-}CH_2Si(CH_3)_2CH_2CH_2CH_2\text{-}]$ .... Ge: Org.Comp.3-294
– .... $(CH_3)_3Ge\text{-}C(=CH_2)\text{-}Si(CH_3)_3$ .... Ge: Org.Comp.2-5, 21
– .... $(CH_3)_3Ge\text{-}CH=CHSi(CH_3)_3$ .... Ge: Org.Comp.2-5, 21
$C_8GeH_{21}N$ .... $(CH_3)_3Ge\text{-}CH_2N(C_2H_5)_2$ .... Ge: Org.Comp.1-139, 148
$C_8GeH_{21}NO_2S$ .... $(CH_3)_3Ge\text{-}CH_2\text{-}S(O)_2\text{-}N(C_2H_5)_2$ .... Ge: Org.Comp.1-138
$C_8GeH_{21}O_3P$ .... $Ge(CH_3)_3CH_2P(O)(OC_2H_5)_2$ .... Ge: Org.Comp.1-140
$C_8GeH_{22}HgSi$ .... $Ge(CH_3)_3CH_2HgCH_2Si(CH_3)_3$ .... Ge: Org.Comp.1-142
$C_8GeH_{22}OSi$ .... $(CH_3)_3Ge\text{-}CD(CH_3)\text{-}Si(CH_3)_2\text{-}OCH_3$ .... Ge: Org.Comp.1-168
– .... $(CH_3)_3Ge\text{-}CH_2CH_2\text{-}O\text{-}Si(CH_3)_3$ .... Ge: Org.Comp.1-167, 176
$C_8GeH_{22}OSi_2$ .... $(CH_3)_2Ge(\text{-}CH_2Si(CH_3)_2OSi(CH_3)_2CH_2\text{-})$ .... Ge: Org.Comp.3-295, 298
$C_8GeH_{22}Si$ .... $Ge(CH_3)_3CH_2CH_2Si(CH_3)_3$ .... Ge: Org.Comp.1-168
$C_8HK_5O_{21}Th_2$ .... $K_5Th_2(OH)(C_2O_4)_2(CO_3)_4 \cdot 2\ H_2O$ .... Th: SVol.C7-100, 102
$C_8H_2MoO_6$ .... $(H\text{-}CC\text{-}H)Mo(CO)_6$ .... Mo: Org.Comp.5-169
$C_8H_3MoO_5^-$ .... $[(C_6H_5)_3P=N=P(C_6H_5)_3][CH_3\text{-}CC\text{-}Mo(CO)_5]$ ... Mo: Org.Comp.5-133/4
$C_8H_3O_5Re$ .... $(CO)_5ReC{\equiv}CCH_3$ .... Re: Org.Comp.2-130
$C_8H_3O_6Re$ .... $(CO)_5ReC(O)CH=CH_2$ .... Re: Org.Comp.2-120
$C_8H_4K_2O_9Th$ .... $K_2ThO(C_4H_2O_4)_2 \cdot 2\ H_2O$ .... Th: SVol.C7-108/10
$C_8H_4MoO_4$ .... $(c\text{-}C_4H_4)Mo(CO)_4$ .... Mo: Org.Comp.5-312, 313, 314
$C_8H_4MoO_5$ .... $(CH_3\text{-}CC\text{-}H)Mo(CO)_5$ .... Mo: Org.Comp.5-168/9
$C_8H_4N_2O_5Re^+$ .... $[(CO)_5ReN(NHCHCHCH)]^+$ .... Re: Org.Comp.2-151
$C_8H_4O_4Th^{2+}$ .... $[Th(1,2\text{-}(OOC)_2C_6H_4)]^{2+}$ .... Th: SVol.D1-74/6, 79
$C_8H_4O_7Re^+$ .... $[(CO)_5Re=C(OCH_2)_2]^+$ .... Re: Org.Comp.2-162
$C_8H_4O_8Th$ .... $Th(C_4H_2O_4)_2 \cdot 2\ H_2O$ .... Th: SVol.C7-108/10
– .... $Th(C_4H_2O_4)_2 \cdot 7\ H_2O\ =\ Th(OH)_2(C_4H_3O_4)_2 \cdot 5\ H_2O$
Th: SVol.C7-108/10
$C_8H_4O_{12}Th$ .... $Th(OOCC(OH)C(OH)COO)_2$ .... Th: SVol.C7-108/10
C_8H_4Pb .... $Pb(CC\text{-}H)_4$ .... Pb: Org.Comp.3-92, 94
$C_8H_5MnN_2O_4S_2^+$ $[Mn(5\text{-}(S)\text{-}2\text{-}(O_3SC_6H_4\text{-}4)\text{-}C_2HN_2O\text{-}4,3,1)(H_2O)_3]^+$
Mn: MVol.D7-63/4

$C_8H_5MoNO_5$	$(CO)_5Mo(CN-C_2H_5)$	Mo:Org.Comp.5-6, 7
$C_8H_5MoO_5^-$	$[(C_5H_5)Mo(CO)_2(-O-C(=O)-O-)]^-$	Mo:Org.Comp.7-190, 202
$C_8H_5N_3OS_2$	$5-(O=S=N)-3-(C_6H_5)C_2N_2S-4,2,1$	S: S-N Comp.6-170, 191
$C_8H_5N_3O_2S$	$2-(O=S=N)-5-(C_6H_5)C_2N_2O-4,3,1$	S: S-N Comp.6-170
$C_8H_5OReS_2$	$(C_5H_5)Re(CO)(CS)_2$	Re: Org.Comp.3-144
$C_8H_5O_5Re$	$(CO)_4Re[-O=C(CH_3)-CH=CH-]$	Re: Org.Comp.1-417/8
–	$(CO)_5Re-CH_2CH=CH_2$	Re: Org.Comp.2-108
$C_8H_5O_6Re$	$(CO)_5Re-C(=O)-C_2H_5$	Re: Org.Comp.2-118
–	$(CO)_5Re-CH_2-C(=O)CH_3$	Re: Org.Comp.2-108
$C_8H_5O_7Re$	$(CO)_5ReC(O)OC_2H_5$	Re: Org.Comp.2-116
$C_8H_6LiO_6Re$	$Li[(CO)_4Re(C(=O)CH_3)_2]$	Re: Org.Comp.1-425
$C_8H_6MnN_4O_4S_2$	$Mn(1,3-N_2C_4H_3(=S-2)(=O)_2-4,6)_2$	Mn: MVol.D7-75/8
$C_8H_6MoN_2O_4$	$(CO)_4Mo(CN-CH_3)_2$	Mo:Org.Comp.5-25, 29
$C_8H_6MoO_4$	$(CH_2=CHCH=CH_2)Mo(CO)_4$	Mo:Org.Comp.5-311
–	$[(CH_2)_3C]Mo(CO)_4$	Mo:Org.Comp.5-312
$C_8H_6MoO_5$	$(CH_3-CH=CH_2)Mo(CO)_5$	Mo:Org.Comp.5-168/9
$C_8H_6MoO_6$	$CH_3O-C(CH_3)=Mo(CO)_5$	Mo:Org.Comp.5-101, 102, 109
$C_8H_6NO_6Re$	$(CO)_5Re-C(=O)-N(CH_3)_2$	Re: Org.Comp.2-117
–	$(CO)_5Re-C(=O)-NH-C_2H_5$	Re: Org.Comp.2-117
$C_8H_6Na_2O_{10}Th$	$Na_2Th(OH)_2(C_4H_2O_4)_2 \cdot 4\ H_2O$	Th: SVol.C7-108/10
$C_8H_6O_5Re^+$	$[(CO)_5Re(CH_2=CHCH_3)]^+$	Re: Org.Comp.2-350, 352/3
$C_8H_6O_6Re$	$(CO)_4Re[CH_3C(O)C(O)CH_3]$, radical	Re: Org.Comp.1-371
$C_8H_6O_6Re^+$	$[(CO)_5ReOC(CH_3)_2]^+$	Re: Org.Comp.2-157
$C_8H_6O_6Re^-$	$[(CO)_4Re(C(=O)CH_3)_2]^-$	Re: Org.Comp.1-425
$C_8H_7I_3MoO_2$	$(CH_3-C_5H_4)Mo(CO)_2(I)_3$	Mo:Org.Comp.7-4
$C_8H_7I_3NRe$	$Re(CNC_6H_4CH_3-4)(I)_3$	Re: Org.Comp.2-232, 241
$C_8H_7MnN_4O_2S^+$	$Mn[S-C(NH_2)=N-N=CH-C_6H_4-4-NO_2]^+$	Mn: MVol.D6-338
$C_8H_7MnO_2S^+$	$Mn(OC(=O)CH_2SC_6H_5)^+$	Mn: MVol.D7-82/3
$C_8H_7MnO_2Se^+$	$Mn(OC(=O)CH_2SeC_6H_5)^+$	Mn: MVol.D7-246
$C_8H_7MnO_3S^+$	$Mn(O=S(C_6H_5)CH_2C(=O)O)^+$	Mn: MVol.D7-107
$C_8H_7MoNO_2^+$	$[C_5H_5Mo(CO)(NO)H-CC-H]^+$	Mo:Org.Comp.6-304
$C_8H_7MoNO_3$	$(CH_3-C_5H_4)Mo(CO)_2(NO)$	Mo:Org.Comp.7-6, 9, 21
$C_8H_7MoNO_5$	$(CH_3)_2N-CH=Mo(CO)_5$	Mo:Org.Comp.5-101
$C_8H_7NNaO_3Re$	$[2-CH_3C_3H_4Re(CO)_3CN]Na$	Re: Org.Comp.2-376
$C_8H_7NNaO_5Re$	$Na[(CO)_4Re(C(O)CH_3)C(=NH)CH_3]$	Re: Org.Comp.1-424/5
$C_8H_7NO_2S$	$O=S=NC(O)C_6H_4CH_3-4$	S: S-N Comp.6-171/9, 190/2, 198, 203/4
$C_8H_7NO_2Si$	$H_3SiN[-C(O)-(1,3-C_6H_4)-C(O)-]$	Si: SVol.B4-311
$C_8H_7NO_3Re^-$	$[2-CH_3C_3H_4Re(CO)_3CN]^-$	Re: Org.Comp.2-376
$C_8H_7NO_3S$	$O=S=N-C(=O)-C_6H_4-OCH_3-4$	S: S-N Comp.6-171/9
–	$O=S=N-C_6H_4-C(=O)OCH_3-4$	S: S-N Comp.6-197
$C_8H_7NO_5Re^-$	$[(CO)_4Re(C(O)CH_3)C(=NH)CH_3]^-$	Re: Org.Comp.1-424/5
$C_8H_7N_3OSSe$	$2-(O=S=N-N=)-3-(CH_3)C_7H_4NSe-3,1$	S: S-N Comp.6-69/71
$C_8H_7N_3OS_2$	$2-(O=S=N-N=)-3-(CH_3)C_7H_4NS-3,1$	S: S-N Comp.6-69/71
$C_8H_7N_5OS$	$5-(O=S=N)-1-(C_6H_5CH_2)CN_4-1,2,3,4$	S: S-N Comp.6-170/1
$C_8H_7O_3Re$	$(CH_2CHCHCHCH_2)Re(CO)_3$	Re: Org.Comp.3-4
–	$(C_5H_5)Re(CO)_2(O=CH_2)$	Re: Org.Comp.3-186
$C_8H_7O_3Th^{3+}$	$[Th(2-CH_3O-C_6H_4-COO)]^{3+}$	Th: SVol.D1-104/10
–	$[Th(C_6H_5-CH(OH)-COO)]^{3+}$	Th: SVol.D1-29, 64, 72
$C_8H_7O_4Re$	$(CH_2CHCH-CH_3)Re(CO)_4$	Re: Org.Comp.2-388, 392/3

$C_8H_7O_4Re$ .... $[CH_2C(CH_3)CH_2]Re(CO)_4$ .... Re: Org.Comp.2-388, 393
$C_8H_7O_6Re$ .... $(CO)_4Re[C(=O)CH_3]=C(OH)CH_3$ .... Re: Org.Comp.1-388, 403/4
$C_8H_8I_2NO_4Sb$ .... 3,5-I_2-4-$CH_3CONHC_6H_2Sb(O)(OH)_2$ .... Sb: Org.Comp.5-300
$C_8H_8I_4O_8Th$ ...... $Th(CH_2ICOO)_4$ .... Th: SVol.C7-61, 63
$C_8H_8LiMoNO_3$ .... $C_5H_5Mo(CO)(NO)=C(CH_3)OLi$ .... Mo: Org.Comp.6-264, 268
$C_8H_8MnNOS^+$ .... $[Mn(SCH_2C(=O)NHC_6H_5)]^+$ .... Mn: MVol.D7-54
$C_8H_8MnNO_2S^+$ ... $Mn(OC(=O)CH_2SC_6H_4NH_2\text{-}4)^+$ .... Mn: MVol.D7-82/3
$C_8H_8MnNO_3{}^+$ .... $Mn[2\text{-}(OOC\text{-}CH_2CH_2\text{-}N=CH)C_4H_3O]^+$ .... Mn: MVol.D6-75/6
$C_8H_8MnNO_3S^+$ ... $Mn((OOC\text{-}4)((2\text{-}C_4H_3O)\text{-}2)C_3H_5NS\text{-}3,1)^+$ ..... Mn: MVol.D7-230/2
$C_8H_8MnNO_6P$ .... $[Mn(O_3POCH_2\text{-}3\text{-}C_5HN\text{-}CHO\text{-}4\text{-}OH\text{-}5\text{-}CH_3\text{-}6)]$
Mn: MVol.D8-152/3
$C_8H_8MnN_4O_2S$ ... $[Mn(SC(NHNH_2)NNHC(O)C_6H_4O)(H_2O)_3] \cdot H_2O$
Mn: MVol.D7-207/9
$C_8H_8MnO_8S_4Sb_2$. $Mn[Sb(SCH_2C(=O)O)_2]_2$ .... Mn: MVol.D7-52
$C_8H_8MoN_2O_2$ .... $(C_5H_5)Mo(CO)(NO)(CN\text{-}CH_3)$ .... Mo: Org.Comp.6-262
– .... $(C_5H_5)Mo(CO)_2(N_2)CH_3$ .... Mo: Org.Comp.8-77, 78/9
– .... $(C_5H_5)Mo(CO)_2\text{-}N=NCH_3$ .... Mo: Org.Comp.7-6, 14
$C_8H_8MoN_2O_3$ .... $[(C_5H_5)Mo(CO)_2(NH_3)\text{-}N=C=O]$ .... Mo: Org.Comp.7-57/9
$C_8H_8MoO_2$ ...... $(C_5H_5)Mo(CO)_2CH_3$ .... Mo: Org.Comp.8-7/8
– .... $(C_5H_5)Mo(^{12}CO)(^{13}CO)CH_3$ .... Mo: Org.Comp.8-7
– .... $(C_5H_5)Mo(^{13}CO)_2CH_3$ .... Mo: Org.Comp.8-7
$C_8H_8MoO_2{}^-$ ...... $[(C_5H_5)Mo(CO)_2CH_3]^-$, radical anion .... Mo: Org.Comp.8-8
– .... $[(C_5H_5)Mo(CO)_2CD_3]^-$, radical anion .... Mo: Org.Comp.8-8
$C_8H_8MoO_4$ ...... $(CH_2=CH_2)_2Mo(CO)_4$ .... Mo: Org.Comp.5-194
$C_8H_8NO_3Re$ ..... $C_3H_5Re(CO)_3CNCH_3$ .... Re: Org.Comp.2-375, 377
$C_8H_8NO_4Re$ ..... $(C_5H_5)Re(CO)(NO)\text{-}C(=O)OCH_3$ .... Re: Org.Comp.3-158/60, 163
$C_8H_8NO_4ReS_2$ .... $(CO)_4Re[\text{-}S=C(NHC_3H_7\text{-}i)\text{-}S\text{-}]$ .... Re: Org.Comp.1-360/1
$C_8H_8NO_4SSb$ .... 4,5-$(\text{-}SCH_2C(=O)NH\text{-})C_6H_3Sb(O)(OH)_2$ .... Sb: Org.Comp.5-308
$C_8H_8NO_5Re$ ..... cis-$(CO)_4Re[C(O)CH_3]C(CH_3)=NH_2$ .... Re: Org.Comp.1-390
$C_8H_8NO_5Sb$ ...... 3,4-$(\text{-}NHC(OH)=CHO\text{-})C_6H_3Sb(O)(OH)_2$ .... Sb: Org.Comp.5-308
$C_8H_8NO_6Sb$ ...... 3-O_2N-4-$CH_3C(O)\text{-}C_6H_3Sb(O)(OH)_2$ .... Sb: Org.Comp.5-298/9
$C_8H_8N_2O_2Re^+$ ... $[(C_5H_5)Re(CO)(NO)(NC\text{-}CH_3)][BF_4]$ .... Re: Org.Comp.3-154, 155
– .... $[(C_5H_5)Re(CO)(NO)(NC\text{-}CH_3)][PF_6]$ .... Re: Org.Comp.3-153, 155
$C_8H_8N_2O_2S$ ...... $O=S=N\text{-}NHC(=O)\text{-}CH_2C_6H_5$ .... S: S-N Comp.6-61, 62
– .... $O=S=N\text{-}NHC(=O)\text{-}C_6H_4\text{-}CH_3\text{-}4$ .... S: S-N Comp.6-61, 63
$C_8H_8N_2O_2Th^{2+}$ ... $Th[2,5\text{-}(O)C_4H_4N]_2{}^{2+}$ .... Th: SVol.D1-129
$C_8H_8N_4Si$ .... $SiN_4C_8H_8$ .... Si: SVol.B4-228
$C_8H_8N_{10}O_{13}Th$ ... $Th(NO_3)_4 \cdot [3\text{-}NH_2\text{-}4\text{-}(NC_5H_4\text{-}3\text{-}NHN)$
$\text{-}1,2\text{-}N_2C_3H=O\text{-}5] \cdot 4\ H_2O$ .... Th: SVol.D4-137
$C_8H_8Na_4O_{16}Th$ ... $Na_4[Th(HCOO)_8]$ .... Th: SVol.C7-44/5
$C_8H_8O_4Re^+$ ..... $[(C_2H_4)_2Re(CO)_4]^+$ .... Re: Org.Comp.2-363, 371
$C_8H_8O_8S_2Th$ ..... $Th(OOCCH_2SCH_2COO)_2$ .... Th: SVol.D1-74/6, 80
$C_8H_8O_8S_4Th^{4-}$ ... $[Th(OOCCH_2S)_4]^{4-}$ .... Th: SVol.D1-66, 72/3
$C_8H_8O_8Th$ .... $Th[OOC\text{-}CH_2CH_2\text{-}COO]_2$ .... Th: SVol.C7-104
$C_8H_8O_{10}Th$ ...... $Th(OH)_2[OOC\text{-}CH=CH\text{-}COOH]_2 \cdot 5\ H_2O$
$= Th[OOC\text{-}CH=CH\text{-}COO]_2 \cdot 7\ H_2O$ .... Th: SVol.C7-108/10
– .... $Th[OOC\text{-}CH(OH)CH_2\text{-}COO]_2 \cdot n\ H_2O$ .... Th: SVol.C7-105/6
– .... $Th[OOC\text{-}CH_2\text{-}O\text{-}CH_2\text{-}COO]_2$ .... Th: SVol.D1-74/6, 79/80
– .... $[Th(OOC\text{-}CH_2\text{-}O\text{-}CH_2\text{-}COO)_2]_n$ .... Th: SVol.C7-115/6
$C_8H_8O_{12}Th$ ...... $Th[(CHOH)_2(COO)_2]_2$ .... Th: SVol.C7-106/7

$C_8H_8O_{16}Rb_4Th$	$Rb_4[Th(HCOO)_8]$	Th: SVol.C7-44/5
$C_8H_8O_{16}Th^{4-}$	$[Th(HCOO)_8]^{4-}$	Th: SVol.C7-44
$C_8H_9INO_4ReSi$	$(CO)_4Re(CNSi(CH_3)_3)I$	Re: Org.Comp.2-249, 251
$C_8H_9INO_4Sb$	$3\text{-}I\text{-}4\text{-}CH_3CONH\text{-}C_6H_3Sb(O)(OH)_2$	Sb: Org.Comp.5-296
$C_8H_9InS_2$	$CH_3\text{-}In[1,2\text{-}(S)_2\text{-}C_6H_3\text{-}4\text{-}CH_3]$	In: Org.Comp.1-246, 247, 250/1
–	$C_6H_5\text{-}In[\text{-}S\text{-}CH_2CH_2\text{-}S\text{-}]$	In: Org.Comp.1-246, 248
$C_8H_9InS_3$	$(C_6H_5)In[\text{-}S\text{-}CH_2\text{-}S\text{-}CH_2\text{-}S\text{-}]$	In: Org.Comp.1-246, 248
$C_8H_9KO_{13}Th$	$KTh(OH)[(CHOH)_2(COO)_2]_2 \cdot 7\ H_2O$	Th: SVol.C7-106/7
$C_8H_9MnN_2O_2^+$	$Mn[O\text{-}C_6H_3(OCH_3\text{-}2)\text{-}4\text{-}CH{=}NNH_2]^+$	Mn: MVol.D6-240
$C_8H_9MnN_4S^+$	$Mn[S\text{-}C(NH_2){=}N\text{-}N{=}CH\text{-}C_6H_4\text{-}4\text{-}NH_2]^+$	Mn: MVol.D6-338
C_8H_9NOS	$O{=}S{=}NCH_2\text{-}C_6H_4\text{-}CH_3$	S: S-N Comp.6-109
–	$O{=}S{=}N\text{-}CH_2CH_2\text{-}C_6H_5$	S: S-N Comp.6-109
–	$O{=}S{=}N\text{-}CH(CH_3)\text{-}C_6H_5$	S: S-N Comp.6-109
–	$O{=}S{=}N\text{-}C_6H_3(CH_3)_2\text{-}2,6$	S: S-N Comp.6-156/7, 202/5, 236
$C_8H_9NOS_2$	$O{=}S{=}N\text{-}S\text{-}C_6H_3(CH_3)_2\text{-}2,4$	S: S-N Comp.6-205
–	$O{=}S{=}N\text{-}S\text{-}C_6H_3(CH_3)_2\text{-}3,5$	S: S-N Comp.6-192/4, 247
$C_8H_9NO_5Re^+$	$[(CO)_5ReNH_2C_3H_7\text{-}i]^+$	Re: Org.Comp.2-151
$C_8H_9N_2O_2ReS_2$	$(C_5H_5)Re(CO)(NO)\text{-}SC({=}S)\text{-}NHCH_3$	Re: Org.Comp.3-150
$C_8H_9N_2O_3Re$	$(C_5H_5)Re(CO)(NO)\text{-}C({=}O)NHCH_3$	Re: Org.Comp.3-161
$C_8H_9N_4O_4Re$	$(C_5H_5)Re(CO)(NO)\text{-}C({=}O)NHNH\text{-}C({=}O)NH_2$	Re: Org.Comp.3-159, 161
$C_8H_9NaO_{11}Th$	$NaTh(OH)[OOCCH(OH)CH_2COO]_2 \cdot 6\ H_2O$	Th: SVol.C7-105/6
$C_8H_9O_4ReS_2Sn$	$(CO)_4Re[\text{-}S{=}C(Sn(CH_3)_3)\text{-}S\text{-}]$	Re: Org.Comp.1-359
$C_8H_9O_4Sb$	$(4\text{-}CH_3COC_6H_4)Sb(O)(OH)_2$	Sb: Org.Comp.5-293
$C_8H_9O_5PbRe$	$(CO)_5RePb(CH_3)_3$	Re: Org.Comp.2-11/2
$C_8H_9O_5ReSi$	$(CO)_5ReSi(CH_3)_3$	Re: Org.Comp.2-4/5, 13
$C_8H_9O_5ReSn$	$(CO)_5ReSn(CH_3)_3$	Re: Org.Comp.2-9/10, 14
$C_8H_9O_5SSb$	$3\text{-}HOOC\text{-}CH_2\text{-}S\text{-}C_6H_4\text{-}Sb(O)(OH)_2$	Sb: Org.Comp.5-288
–	$4\text{-}HOOC\text{-}CH_2\text{-}S\text{-}C_6H_4\text{-}Sb(O)(OH)_2$	Sb: Org.Comp.5-288
$C_8H_9O_5Sb$	$2\text{-}CH_3\text{-}OOC\text{-}C_6H_4\text{-}Sb(O)(OH)_2$	Sb: Org.Comp.5-294
–	$3\text{-}CH_3\text{-}OOC\text{-}C_6H_4\text{-}Sb(O)(OH)_2$	Sb: Org.Comp.5-294
–	$4\text{-}CH_3\text{-}OOC\text{-}C_6H_4\text{-}Sb(O)(OH)_2 \cdot H_2O$	Sb: Org.Comp.5-294, 303
–	$3\text{-}HO\text{-}4\text{-}CH_3C(O)\text{-}C_6H_3\text{-}Sb(O)(OH)_2$	Sb: Org.Comp.5-296
–	$4\text{-}HOOC\text{-}CH_2\text{-}C_6H_4\text{-}Sb(O)(OH)_2$	Sb: Org.Comp.5-295
$C_8H_9O_6Sb$	$(4\text{-}HO_2CCH_2OC_6H_4)Sb(O)(OH)_2$	Sb: Org.Comp.5-287
$C_8H_9O_7ReSSn$	$(CO)_5ReSn(OS(O)CH_3)(CH_3)_2$	Re: Org.Comp.2-9
$C_8H_9O_7SSb$	$(4\text{-}HO_2CCH_2SO_2C_6H_4)Sb(O)(OH)_2$	Sb: Org.Comp.5-288
$C_8H_9O_8Th^+$	$Th[H(OOCC_2H_4COO)_2]^+$	Th: SVol.D1-74/6, 77
$C_8H_{10}IMoNO$	$(C_5H_5)Mo(NO)(I)(CH_2CHCH_2)$	Mo: Org.Comp.6-172, 178
–	$(C_5H_5)Mo(NO)(I)(CH_2CDCH_2)$	Mo: Org.Comp.6-174, 179
–	$(C_5H_5)Mo(NO)(I)(CHDCHCH_2)$	Mo: Org.Comp.6-174, 179
$C_8H_{10}IMoNOS_3$	$C_5H_5Mo(NO)(I)S_2CSC_2H_5$	Mo: Org.Comp.6-49
$C_8H_{10}IMoNO_2S_2$	$C_5H_5Mo(NO)(I)S_2COC_2H_5$	Mo: Org.Comp.6-49
$C_8H_{10}IMoNO_3$	$C_5H_5Mo(NO)(I)O_2CC_2H_5$	Mo: Org.Comp.6-46
$C_8H_{10}INO_3ReS_2^-$	$[(CO)_3ReI(\text{-}S\text{-}C(N(C_2H_5)_2){=}S\text{-})]^-$	Re: Org.Comp.1-122
$C_8H_{10}InNO_3$	$(CH_3)_2In[O\text{-}C_6H_4\text{-}2\text{-}NO_2]$	In: Org.Comp.1-190, 192
$C_8H_{10}K_2O_{14}Th$	$K_2Th(OH)_2[(CHOH)_2(COO)_2]_2 \cdot 8\ H_2O$	Th: SVol.C7-106/7
$C_8H_{10}MnN_2O_3S^{2+}$	$[Mn(4\text{-}NH_2C_6H_4SO_2NH\text{-}COCH_3)]^{2+}$	Mn: MVol.D7-116/23

$C_8H_{10}MnN_4O_2$ . . . $[Mn(-O-2-C_6H_4-CH{=}N-N{=}C(NH_2)NH-)(H_2O)] \cdot 6\ H_2O$ Mn:MVol.D6-336/7
$C_8H_{10}MnN_4O_3S$ . . $[Mn(S{=}C(NHNH_2)NHNHC({=}O)C_6H_4O)(OH)(H_2O)_2] \cdot 3\ H_2O$ . . . Mn:MVol.D7-207/9
$C_8H_{10}MnN_4S_2$ . . . $Mn((2-S)(1-CH_3)C_3H_2N_2-1,3)_2$ . . . Mn:MVol.D7-58/9
$C_8H_{10}Mn_2N_2O_6P_2$ $Mn_2[(O_3PCH_2NH)_2C_6H_4] \cdot 2\ H_2O$ . . . Mn:MVol.D8-130/1
$C_8H_{10}MoN_2O_3$ . . . $(C_5H_5)Mo(CO)_2(NH_3)-C({=}O)NH_2$ . . . Mo:Org.Comp.8-46, 69
$C_8H_{10}NOS^+$ . . . $[OSN(CH_3)CH_2C_6H_5]^+$ . . . S: S-N Comp.6-289, 296/7
$C_8H_{10}NO_3Re$ . . . $(C_5H_5)Re(CO)(NO)-CH_2OCH_3$ . . . Re: Org.Comp.3-159/60, 162
$C_8H_{10}NO_4SSb$ . . . $(4-H_2NC(O)CH_2SC_6H_4)Sb(O)(OH)_2$ . . . Sb: Org.Comp.5-288
$C_8H_{10}NO_4Sb$ . . . $2-CH_3C({=}O)NH-C_6H_4-Sb({=}O)(OH)_2$ . . . Sb: Org.Comp.5-289
– . . . $3-CH_3C({=}O)NH-C_6H_4-Sb({=}O)(OH)_2$ . . . Sb: Org.Comp.5-289
– . . . $4-CH_3C({=}O)NH-C_6H_4-Sb({=}O)(OH)_2$ . . . Sb: Org.Comp.5-289, 302
– . . . $3-H_2N-4-CH_3C({=}O)-C_6H_3-Sb({=}O)(OH)_2$ . . . Sb: Org.Comp.5-297
$C_8H_{10}NO_5Sb$ . . . $4-HO-3-[CH_3C({=}O)NH]-C_6H_3-Sb({=}O)(OH)_2$ . . . Sb: Org.Comp.5-296
– . . . $4-[CH_3-OC({=}O)-NH]-C_6H_4-Sb({=}O)(OH)_2$ . . . Sb: Org.Comp.5-289
– . . . $4-[H_2N-C({=}O)-CH_2-O]-C_6H_4-Sb({=}O)(OH)_2$ . . . Sb: Org.Comp.5-287
$C_8H_{10}N_2OS$ . . . $O{=}S{=}N-C_6H_4-N(CH_3)_2-4$ . . . S: S-N Comp.6-139/40, 205
– . . . $O{=}S{=}N-N(C_6H_5)-C_2H_5$ . . . S: S-N Comp.6-63/8
– . . . $O{=}S{=}N-NH-CH_2CH_2-C_6H_5$ . . . S: S-N Comp.6-54
$C_8H_{10}N_2OS_2$ . . . $O{=}S{=}NSN(CH_3)CH_2C_6H_5$ . . . S: S-N Comp.6-42/6
$C_8H_{10}N_2O_2S_2$ . . . $4-CH_3-C_6H_4-S(O)_2N{=}S{=}NCH_3$ . . . S: S-N Comp.7-61
$C_8H_{10}N_2O_2S_3$ . . . $4-CH_3-C_6H_4-S(O)_2N{=}S{=}NSCH_3$ . . . S: S-N Comp.7-65
$C_8H_{10}N_2O_3S^-$ . . . $[c-C_6H_{10}{=}C(CN)-N(O)-SO_2]^-$, radical anion . . . S: S-N Comp.8-326/8
$C_8H_{10}N_2O_6S_2$ . . . $(CH_3O)_2S{=}N-S(O)_2-C_6H_4-3-NO_2$ . . . S: S-N Comp.8-156/7
$C_8H_{10}N_2O_8Th$ . . . $Th[HN(CH_2-COO)_2]_2$ . . . Th: SVol.D1-97/8
– . . . $Th[HN(CH_2-COO)_2]_2 \cdot 4\ H_2O$ . . . Th: SVol.C7-119/21
– . . . $Th[OOC-CH_2-CH(NH_2)-COO]_2$ . . . Th: SVol.D1-82/3, 87
$C_8H_{10}N_2O_{12}Th$ . . . $Th(NO_3)_2[OOC-CH_2-C({=}O)-CH_3]_2$ . . . Th: SVol.D4-203
$C_8H_{10}N_2S$ . . . $C_6H_5-N{=}S{=}N-C_2H_5$ . . . S: S-N Comp.7-218
$C_8H_{10}N_2S_2$ . . . $S{=}S{=}NC_6H_4N(CH_3)_2-4$ . . . S: S-N Comp.6-321
$C_8H_{10}N_6O_4PdS$ . . . $Pd(SO_4)[1,4-(NH_2NH)_2-C_8H_4N_2-2,3]$ . . . Pd: SVol.B2-238
$C_8H_{10}N_{10}O_{16}Th$ . . $Th(NO_3)_4 \cdot 2\ [4-(O{=}N)-5-CH_3-1,2-N_2C_3H_2({=}O)-3]$ Th: SVol.D4-136, 168
$C_8H_{10}Na_2O_{12}Th$ . . $Na_2[Th(OH)_2(OOCCH(OH)CH_2COO)_2] \cdot 4\ H_2O$ Th: SVol.C7-105/6
$C_8H_{10}O_4PReS_2$ . . . $(CO)_4Re[-S{=}P(C_2H_5)_2-S-]$ . . . Re: Org.Comp.1-362/3, 366/7
$C_8H_{10}O_4Sn$ . . . $(CH_2{=}CH-CH_2)_2Sn[-O-C({=}O)-C({=}O)-O-]$ . . . Sn: Org.Comp.16-93
– . . . $C_6H_5-Sn(OH)_2-O-C({=}O)-CH_3$ . . . Sn: Org.Comp.17-75/6
$C_8H_{10}O_6Th$ . . . $Th(OH)_3(2-HO-3-CH_3C_6H_3COO) \cdot 2\ H_2O$ . . . Th: SVol.C7-134, 135
$C_8H_{10}O_{14}Po^{2-}$ . . . $[Po(OH)_2(OOC-CH(OH)CH(OH)-COO)_2]^{2-}$ . . . Po: SVol.1-354/5
$C_8H_{10}Ti$ . . . $[(C_5H_5)Ti(C_3H_5)]_n$ . . . Ti: Org.Comp.5-328
$C_8H_{11}IMoN_2OS_2$ . . $C_5H_5Mo(NO)(I)S_2CN(CH_3)_2$ . . . Mo:Org.Comp.6-50
$C_8H_{11}IO_4PReS$ . . . $(CO)_4Re(I)-P(C_2H_5)_2-SH$ . . . Re: Org.Comp.1-453
– . . . $(CO)_4Re(I)-S-PH(C_2H_5)_2$ . . . Re: Org.Comp.1-461
$C_8H_{11}I_2MnNOP$ . . . $Mn(NO)[P(C_6H_5)(CH_3)_2]I_2$ . . . Mn:MVol.D8-46, 61
$C_8H_{11}I_2MnOP$ . . . $[Mn((CH_3)_2(C_6H_5)P{=}O)I_2]$ . . . Mn:MVol.D8-101/2
$C_8H_{11}I_2MnO_2P$ . . . $Mn[P(C_6H_5)(CH_3)_2](O_2)I_2$ . . . Mn:MVol.D8-56
$C_8H_{11}I_2MnP$ . . . $Mn(I)_2(SO_2)_n-P(C_6H_5)(CH_3)_2$ . . . Mn:MVol.D8-46, 57/8
– . . . $Mn[P(C_6H_5)(CH_3)_2]I_2$ . . . Mn:MVol.D8-38
$C_8H_{11}I_2Sb$ . . . $(CH_3)_2(C_6H_5)SbI_2$ . . . Sb: Org.Comp.5-68

$C_8H_{11}InN_2O$	$(CH_3)_2In-ON=CH-2-NC_5H_4$	In: Org.Comp.1-211, 214, 217
$C_8H_{11}InO$	$(CH_3)_2In-O-C_6H_5$	In: Org.Comp.1-181, 182
$C_8H_{11}InO_3S$	$(CH_3)_2In-OS(O)O-C_6H_5$	In: Org.Comp.1-212, 213
$C_8H_{11}MnN_2O_3S^+$	$[Mn((OOCCH_2CH_2)(O=)C_5H_7N_2S)]^+$	Mn: MVol.D7-227
$C_8H_{11}MnN_2O_5P$	$[Mn(O_3POCH_2-3-C_5HN-CH_2NH_2-4-OH-5-CH_3-6)]$	Mn: MVol.D8-152/3
$C_8H_{11}MnN_2O_7S_2^+$	$[Mn((5-S=)(2-(4-C_6H_4SO_3))C_2HN_2O-4,3,1)(H_2O)_3]^+$	Mn: MVol.D7-63/4
$C_8H_{11}MoN_2O_3^+$	$[C_5H_5Mo(NO)_2O=C(CH_3)_2]^+$	Mo: Org.Comp.6-61
$C_8H_{11}NOS$	$O=S=N-C_6H_{10}-2-C \equiv CH$	S: S-N Comp.6-108, 111
$C_8H_{11}NOS_2$	$(CH_3)_2N-S(O)-S-C_6H_5$	S: S-N Comp.8-332/3
$C_8H_{11}NO_2S$	$(CH_3)_2N-S(O)O-C_6H_5$	S: S-N Comp.8-306/7, 311
–	$(CH_3O)_2S=N-C_6H_5$	S: S-N Comp.8-166
$C_8H_{11}NO_3PTh^{3+}$	$[Th(C_6H_5-CH_2-NHCH_2-PO_3H)]^{3+}$	Th: SVol.D1-132
$C_8H_{11}NO_4S_2$	$(CH_3O)_2S=N-S(O)_2-C_6H_5$	S: S-N Comp.8-156/7
$C_8H_{11}N_2O_4Sb$	$3-H_2NC(=O)CH_2NH-C_6H_4-Sb(=O)(OH)_2$	Sb: Org.Comp.5-289
–	$4-H_2NC(=O)CH_2NH-C_6H_4-Sb(=O)(OH)_2$	Sb: Org.Comp.5-289
$C_8H_{11}N_2O_5Sb$	$4-C_2H_5NH-3-O_2N-C_6H_3Sb(O)(OH)_2$	Sb: Org.Comp.5-297
$C_8H_{11}N_2S_2^+$	$[(CH_3)_2SN=S=N-C_6H_5]^+$	S: S-N Comp.7-23
$C_8H_{11}O_3Sb$	$2,4-(CH_3)_2-C_6H_3-Sb(=O)(OH)_2$	Sb: Org.Comp.5-299
–	$4-C_2H_5-C_6H_4-Sb(=O)(OH)_2$	Sb: Org.Comp.5-293
$C_8H_{11}O_4SSb$	$(4-HOCH_2CH_2SC_6H_4)Sb(O)(OH)_2$	Sb: Org.Comp.5-288
$C_8H_{11}O_4Sb$	$2-(C_2H_5-O)-C_6H_4-Sb(=O)(OH)_2$	Sb: Org.Comp.5-287
–	$4-(C_2H_5-O)-C_6H_4-Sb(=O)(OH)_2$	Sb: Org.Comp.5-287
$C_8H_{12}HgMnN_4O_4S_4$	$Mn(O=S(CH_3)_2)_2Hg(SCN)_2(OCN)_2$	Mn: MVol.D7-98
$C_8H_{12}IMoN_5O$	$(I)Mo(CN-CH_3)_4(NO)$	Mo: Org.Comp.5-39, 40
$C_8H_{12}IORe$	$(CH_3-C \equiv C-CH_3)_2Re(O)I$	Re: Org.Comp.2-359, 365/8
$C_8H_{12}IO_3ReS_2$	$(CO)_3ReI(CH_3SCH_2CH_2CH_2SCH_3)$	Re: Org.Comp.1-209, 211/3
$C_8H_{12}IO_3ReSe_2$	$(CO)_3ReI(CH_3SeCH_2CH_2CH_2SeCH_3)$	Re: Org.Comp.1-210, 211/3
$C_8H_{12}InNO$	$1-(CH_3)_2In-NC_4H_3-2-C(=O)CH_3$	In: Org.Comp.1-276/7, 278
$C_8H_{12}InN_4^-$	$[(CH_3)_2In(1,2-N_2C_3H_3)_2]^-$	In: Org.Comp.1-366
$C_8H_{12}InN_4Na$	$Na[(CH_3)_2In(1,2-N_2C_3H_3)_2]$	In: Org.Comp.1-366
$C_8H_{12}In_2O_4$	$(CH_3)_2In[1,2-(O)_2-C_4-3,4-(O)_2]In(CH_3)_2$	In: Org.Comp.1-207/9
–	$(CH_3)_2In[1,2-(O)_2-C_4-3,4-(O)_2]In(CH_3)_2 \cdot 4\ H_2O$	In: Org.Comp.1-208/9
$C_8H_{12}MnN_2O_4S_2$	$Mn(OOC-4-C_3H_6NS-3,1)_2$	Mn: MVol.D7-230/2
$C_8H_{12}MnN_2O_{10}P_2^{4-}$	$[Mn(O_3PCH_2N(CH_2COO)CH_2CH_2N(CH_2COO)CH_2PO_3)]^{4-}$	Mn: MVol.D8-131/2
$C_8H_{12}MnN_4^{2+}$	$[Mn(NC-CH_3)_4]^{2+}$	Mn: MVol.D7-9
$C_8H_{12}MnN_4O_2$	$Mn[(CH_3)_2C=N-N=C(O-)C(O-)=N-N=C(CH_3)_2]$	Mn: MVol.D6-313/4
$C_8H_{12}MnN_4S_8Zn$	$MnZn(SC(=S)NHCH_2CH_2NHC(=S)S)_2$	Mn: MVol.D7-180/4
$C_8H_{12}MnN_6O_2S_2$	$Mn[N(CN)_2]_2 \cdot 2\ O=S(CH_3)_2$	Mn: MVol.D7-98
$C_8H_{12}MnN_6O_4S_2$	$Mn[OOCC(CH_3)=N-NHC(=S)NH_2]_2$	Mn: MVol.D6-344
$C_8H_{12}MnN_6S_4$	$[Mn((2-S=)C_3H_6N_2-1,3)_2(NCS)_2]$	Mn: MVol.D7-58/9
$C_8H_{12}MnN_7O_2S_2$	$[Mn(SC(NH_2)=NNH_2)_2(OOCC_5H_4N)]$	Mn: MVol.D7-206/7
$C_8H_{12}MnO_3S$	$[Mn(SC_6H_4O-2)(OHCH_3)_2]$	Mn: MVol.D7-30/1
$C_8H_{12}MnO_4S_2$	$[Mn((OC(=O)CH_2SCH_2)_2(CH_2CH_2))(H_2O)_2]$	Mn: MVol.D7-86/8
$C_8H_{12}MnO_4S_3$	$Mn[(OC(=O)-CH_2-S-CH_2CH_2)_2S]$	Mn: MVol.D7-86/8

$C_8H_{12}MnO_4S_3$ .... $Mn[(OC(=O)-CH_2-S-CH_2CH_2)_2S] \cdot 2\ H_2O$ Mn:MVol.D7-86/8

$C_8H_{12}Mn_5N_2O_{16}P_4$
$Mn_5[(O_3PCH_2)_2NCH_2COO]_2 \cdot 5\ H_2O$ Mn:MVol.D8-128/9

$C_8H_{12}MoO_{16}^{2-}$ ... $[(MoO_4 \cdot 2\ HOOC(CHOH)_2COOH)]^{2-}$ Mo:SVol.B3b-174/5

$C_8H_{12}NO_3Sb$ ..... $(4-(CH_3)_2NC_6H_4)Sb(O)(OH)_2$ .... Sb: Org.Comp.5-288

$C_8H_{12}NO_4Sb$ ..... $(4-HOCH_2CH_2NHC_6H_4)Sb(O)(OH)_2$ .... Sb: Org.Comp.5-289

$C_8H_{12}N_2O_2S$ ..... $(CH_3)_2N-S(O)O-CH_2-2-NC_5H_4$ .... S: S-N Comp.8-306/7, 311

– .... $C_6H_5-NH_2-CH_2CH_2-NH-SO_2$ .... S: S-N Comp.8-300

$C_8H_{12}N_4NiO_2PdS_4$ $Pd(SCN)_2Ni(NCS)_2(C_2H_5OH)_2$ .... Pd: SVol.B2-311

$C_8H_{12}N_4OS^{2+}$ .... $[CH_3-3-(1,3-N_2C_3H_3)-1-S(O)-1-(1,3-N_2C_3H_3)-3-CH_3]^{2+}$
S: S-N Comp.8-365

$C_8H_{12}N_4O_2S_4Sn$ .. $[2-(S=)-4-(O=)-1,3-SNC_3H_2-3-NH-]_2Sn(CH_3)_2$ Sn: Org.Comp.19-62

$C_8H_{12}N_4S$ .... $(CH_3)_2C(CN)N=S=N-C(CN)(CH_3)_2$ .... S: S-N Comp.7-195

$C_8H_{12}N_4SSn$ ..... $(CH_3)_3SnC_5H_3N_4S$ .... Sn: Org.Comp.18-84, 93, 102

$C_8H_{12}N_6O_{12}Th$ ... $[C(NH_2)_3]_2[Th(C_2O_4)_3] \cdot 6\ H_2O$ .... Th: SVol.C7-86, 92

– .... $[C(NH_2)_3]_2[Th(C_2O_4)_3] \cdot 8\ H_2O$ .... Th: SVol.C7-86

– .... $[C(NH_2)_3]_2[Th(C_2O_4)_3] \cdot x\ H_2O$ .... Th: SVol.C7-86

$C_8H_{12}N_6PdS_4$ .... $Pd(SCN)_2[1,3-N_2C_3H_6(=S)-2]_2$ .... Pd: SVol.B2-309

$C_8H_{12}Ni_2O_{14}S_8Th$ $Th(H_2O)_6[OC(S)C(S)O]_4Ni_2 \cdot 6.5\ H_2O$ .... Th: SVol.C7-159

$C_8H_{12}O_4PReSe$ ... cis-$(CO)_4Re(SeCH_3)P(CH_3)_3$ .... Re: Org.Comp.1-439/40

$C_8H_{12}O_4S_2Sn$ .... $(CH_2=CH)_2Sn(OS(O)CH=CH_2)_2$ .... Sn: Org.Comp.16-90, 92

$C_8H_{12}O_4Sn$ ...... $(CH_2=CH)_2Sn(OOC-CH_3)_2$ .... Sn: Org.Comp.16-89, 91

– .... $Sn[CH_2CH(CH_3)COO]_2$ .... Sn: Org.Comp.16-221

$C_8H_{12}O_5PRe$ ..... $(CO)_4Re[P(CH_3)_2OCH_3]CH_3$ .... Re: Org.Comp.1-466

$C_8H_{12}O_8Th$ ...... $Th(CH_3COO)_4$ .... Th: SVol.A4-130, 131, 182
Th: SVol.C7-45/6, 47/8
Th: SVol.D1-66/8
Th: SVol.D4-130, 131, 156, 187

– .... $Th(CH_3COO)_4$ solutions
$Th(CH_3COO)_4-H_2NC_2H_4NH_2$ .... Th: SVol.D1-18/9
$Th(CH_3COO)_4-H_2O$ .... Th: SVol.D1-18

$C_8H_{12}O_{10}Th$ ..... $Th(OH)_2(HOOCC_2H_4COO)_2$ .... Th: SVol.C7-104

$C_8H_{12}O_{12}Th$ ..... $Th(HO-CH_2-COO)_4$ .... Th: SVol.D1-66/71

– .... $Th(HO-CH_2-COO)_4 \cdot 2\ H_2O$ .... Th: SVol.C7-52/3

$C_8H_{12}Pb$ .... $Pb(CH=CH_2)_4$ .... Pb: Org.Comp.3-76/89

$C_8H_{13}InSi$ .... $In[C_5H_4-Si(CH_3)_3]$ .... In: Org.Comp.1-379, 380, 383

$C_8H_{13}MnN_3O_2S^{2+}$ $[Mn((2-CH_3CONH)(4-CH_3CO)C_2N_2S-4,3,1-(CH_3)_2-5,5)(H_2O)_2]^{2+}$ Mn:MVol.D7-237/9

$C_8H_{13}MnO_2S_2^{+}$ .. $[Mn(1,2-S_2C_3H_5-3-CH_2CH_2-CH_2CH_2-COO)]^{+}$ Mn:MVol.D7-226

$C_8H_{13}MnO_4S_3^{+}$ .. $[Mn(OOC-CH_2S-CH_2CH_2-S-CH_2CH_2-SCH_2-COOH)]^{+}$
Mn:MVol.D7-86/8

$C_8H_{13}NO_4PReS_2$ .. $(CO)_4Re(NH_3)SP(S)(C_2H_5)_2$ .... Re: Org.Comp.1-430

$C_8H_{13}N_2O_3ReS_2$ .. $(CH_3)_2NH-Re(CO)_3[-S-C(N(CH_3)_2)=S-]$ ...... Re: Org.Comp.1-128

$C_8H_{13}N_3O_{12}Th$ ... $[Th(OOCCH_2CH(NH_2)COO)(HOOCCH_2CH(NH_2)COO)(H_2O)]NO_3 \cdot 3\ H_2O$ .... Th: SVol.C7-107/8

$C_8H_{14}IMoN_3O$ .... $C_5H_5Mo(NO)(I)N(CH_3)N(CH_3)_2$ .... Mo: Org.Comp.6-53, 54

$C_8H_{14}I_2MoNOP$ ... $C_5H_5Mo(NO)(P(CH_3)_3)I_2$ .... Mo: Org.Comp.6-34

$C_8H_{14}I_2MoNO_4P$ .. $C_5H_5Mo(NO)(P(OCH_3)_3)I_2$ .... Mo: Org.Comp.6-35

$C_8H_{14}MnN_2O_3S$ .. $[Mn(-O-C(CH_3)=CH-C(CH_3)=N-N=C(OC_2H_5)S-)(H_2O)]$ Mn:MVol.D6-361

$C_8H_{14}MnN_2O_{10}P_2^{2-}$ $[MnH_2(O_3PCH_2N(CH_2COO)CH_2CH_2N(CH_2COO)CH_2PO_3)]^{2-}$ Mn:MVol.D8-131/2

$C_8H_{14}MnN_2S_4$ .... $Mn(SC(=S)NHCH_2CH_2CH_2CH_2CH_2CH_2NHC(=S)S)$ Mn:MVol.D7-180/4

$C_8H_{14}MnN_2S_4^-$ ... $[Mn(SCH_2CH_2S)_2(NCCH_3)_2]^-$ Mn:MVol.D7-33/8

$C_8H_{14}MnN_4O_5S$ .. $Mn[NH_2-NH-C(S)-NHNH-C(=O)-C_6H_4-O](OH)(H_2O)_2 \cdot 3\ H_2O$ Mn:MVol.D7-207/9

– ... $Mn[NH_2-NH-C(S)-NNH-C(=O)-C_6H_4-O](H_2O)_3 \cdot H_2O$ Mn:MVol.D7-207/9

$C_8H_{14}MoN_2OS_4$ .. $(HC{\equiv}CH)Mo(O)[S_2C-N(CH_3)_2]_2$ Mo:Org.Comp.5-139/40, 142

$C_8H_{14}MoN_2O_5P^+$ $[C_5H_5Mo(NO)_2P(OCH_3)_3]^+$ Mo:Org.Comp.6-61

$C_8H_{14}N_4O_6Th^{2+}$ .. $[Th(H_2NC(O)CH_2CH(NH_2)COO)_2]^{2+}$ Th: SVol.D1-82/3, 87

$C_8H_{14}N_4PdS_2$ .... $Pd(SCN)_2[1,2-(NH_2)_2C_6H_{10}]$ Pd: SVol.B2-307

$C_8H_{14}N_6Sn$ ...... $[1,2,4-N_3C_2H_2-1-]_2Sn(C_2H_5)_2$ Sn: Org.Comp.19-82

$C_8H_{14}O_4Th^{2+}$ .... $[Th(OOC-C_3H_7-n)_2]^{2+}$ Th: SVol.D1-68

– ... $[Th(OOC-C_3H_7-i)_2]^{2+}$ Th: SVol.D1-69

$C_8H_{14}O_6Th^{2+}$ .... $[Th(OOC-C(CH_3)_2-OH)_2]^{2+}$ Th: SVol.D1-72

– ... $[Th(OOC-CH_2CH_2CH_2-OH)_2]^{2+}$ Th: SVol.D1-72

– ... $[Th(OOC-CH_2-CH(OH)-CH_3)_2]^{2+}$ Th: SVol.D1-72

$C_8H_{15}IMoN_3O^+$ ... $[C_5H_5Mo(NO)(N(CH_3)HN(CH_3)_2)I]^+$ Mo:Org.Comp.6-39

$C_8H_{15}I_2MoN_3O$ ... $C_5H_5Mo(NO)(N(CH_3)HN(CH_3)_2)I_2$ Mo:Org.Comp.6-37

$C_8H_{15}InN_2$ .... $1-(C_2H_5)_2In-1,2-N_2C_3H_2-3-CH_3$ In: Org.Comp.1-276/7, 280

$C_8H_{15}InO$ .... $(CH_2=CH)_3In \cdot O(CH_3)_2$ In: Org.Comp.1-84, 85

$C_8H_{15}MnNO_6P_2^{2-}$ $[Mn((O_3PCH_2)_2NC_6H_{11}-c)]^{2-}$ Mn:MVol.D8-129/30

$C_8H_{15}MnN_2O_2PS_2$ $Mn[P(C_2H_5)_3](O_2)(NCS)_2$ Mn:MVol.D8-56

$C_8H_{15}MnN_2PS_2$ ... $Mn[P(C_2H_5)_3](NCS)_2$ Mn:MVol.D8-38

$C_8H_{15}NO_5S^-$ ..... $[CH_3C(OC(O)CH_3)(C_4H_9-t)-N(O)-SO_2]^-$, radical anion S: S-N Comp.8-326/8

$C_8H_{15}N_4O_3Re^{2+}$ .. $[ON-Re(CO)_2(1,4,7-N_3C_6H_{15})]^{2+}$ Re: Org.Comp.1-62

$C_8H_{15}O_2Th^{3+}$ .... $[Th(OOC-C(CH_3)_2-C_4H_9-n)]^{3+}$ Th: SVol.D1-70

– ... $[Th(OOC-C_7H_{15}-n)]^{3+}$ Th: SVol.D1-70

$C_8H_{15}O_6Sb$ ...... $(CH_3)_2Sb(O_2CCH_3)_3$ Sb: Org.Comp.5-180/1

$C_8H_{16}I_2MnS_2$ .... $[Mn(SC_4H_8)_2I_2]$ Mn:MVol.D7-225/6

$C_8H_{16}In_2N_2^+$ ..... $[1,4-((CH_3)_2In)_2-1,4-N_2C_4H_4]^+$, radical cation In: Org.Comp.1-294

$C_8H_{16}In_2N_4NiO_4$ .. $Ni[(CH_3)_2In(-O-NCHCHN-O-)_2In(CH_3)_2]$ In: Org.Comp.1-228/30

$C_8H_{16}MnNO_6P_2^-$ $[Mn(H(O_3PCH_2)_2NC_6H_{11}-c)]^-$ Mn:MVol.D8-129/30

$C_8H_{16}MnN_2S_4$ .... $Mn(SC(=S)NHC_3H_7-i)_2$ Mn:MVol.D7-136/8

$C_8H_{16}MnN_4O_4$ ... $[Mn((CH_3)_2C=N-N=C(O-)C(O-)=N-N=C(CH_3)_2)(H_2O)_2]$ Mn:MVol.D6-313/4

$C_8H_{16}MnN_6O_4S_4$ .. $[Mn(NH(C(NH_2)=S)_2)_2(OC(CH_3)=O)_2] \cdot C_2H_5OH$ Mn:MVol.D7-199/200

$C_8H_{16}MnN_8OS_6$ .. $[Mn(NH(C(NH_2)=S)_2)_2(NCS)_2] \cdot C_2H_5OH$ Mn:MVol.D7-199/200

$C_8H_{16}MnO_2S_2^{2+}$ .. $[Mn(C_4H_8OS-1,4)_2(H_2O)_2]^{2+}$ Mn:MVol.D7-239

$C_8H_{16}MnO_6S_2$ .... $[Mn((OC(=O)CH_2SCH_2)_2(CH_2CH_2))(H_2O)_2]$ Mn:MVol.D7-86/8

$C_8H_{16}MnO_{12}S_4$ ... $Mn[O_3S-CH_2CH_2CH_2CH_2-SO_3]_2$ Mn:MVol.D7-114/5

– ... $Mn[O_3S-CH_2CH_2CH_2CH_2-SO_3]_2 \cdot 4\ H_2O$ Mn:MVol.D7-114/5

$C_8H_{16}Mn_2S_8^{2-}$ ... $[Mn_2(SCH_2CH_2S)_4]^{2-}$ Mn:MVol.D7-33/8, 42/3

$C_8H_{16}N_2OS$	C_4H_8N-1-S(O)-1-NC_4H_8	S: S-N Comp.8-346, 359
$C_8H_{16}N_2O_2S$	O=S=NNHC(O)C_7H_{15}	S: S-N Comp.6-61, 62
$C_8H_{16}N_2O_3S$	1,4-ONC_4H_8-4-S(O)-4-C_4H_8ON-1,4	S: S-N Comp.8-347, 359/60
$C_8H_{16}N_2O_4PReS_2$	[cis-$(CO)_4Re(NH_3)_2$][$S_2P(C_2H_5)_2$]	Re: Org.Comp.1-480
$C_8H_{16}N_2O_4Th^{2+}$	$[Th(C_2H_5CH(NH_2)COO)_2]^{2+}$	Th: SVol.D1-82/3, 85
$C_8H_{16}N_2O_6Th^{2+}$	$[Th(CH_3CH(OH)CH(NH_2)COO)_2]^{2+}$	Th: SVol.D1-82/3, 86
$C_8H_{16}N_4O_2Pd$	$Pd(NCO)_2[(CH_3)_2NCH_2CH_2N(CH_3)_2]$	Pd: SVol.B2-288
$C_8H_{16}N_4O_{16}Th$	$Th(NO_3)_4 \cdot [-CH_2CH_2-O-]_4$	Th: SVol.D4-158
–	$[NH_4]_4[Th(C_2O_4)_4]$	Th: SVol.C7-94
–	$[NH_4]_4[Th(C_2O_4)_4] \cdot 3\ H_2O$	Th: SVol.C7-87, 94
–	$[NH_4]_4[Th(C_2O_4)_4] \cdot 6.5\ H_2O$	Th: SVol.C7-87, 94
–	$[NH_4]_4[Th(C_2O_4)_4] \cdot x\ H_2O$	Th: SVol.C7-88
$C_8H_{16}N_6O_4Th^{2+}$	$[Th(H_2NC(NH)N(CH_3)CH_2COO)_2]^{2+}$	Th: SVol.D1-82/3, 87
$C_8H_{16}N_6Pd$	$[C_2H_5NH_3]_2[Pd(CN)_4]$	Pd: SVol.B2-270
$C_8H_{16}N_8Pd_2S_4$	$[Pd(NH_2CH_2CH_2NH_2)_2][Pd(SCN)_4]$	Pd: SVol.B2-300
$C_8H_{16}N_{12}O_4S_4Th$	$Th(SCN)_4 \cdot 4\ O{=}C(NH_2)_2 \cdot 7\ H_2O$	Th: SVol.D4-153
$C_8H_{16}N_{12}Pd_2S_8$	$[Pd(NH_2C({=}S)NH_2)_4][Pd(SCN)_4]$	Pd: SVol.B2-300
$C_8H_{16}N_{12}S_8Th$	$[NH_4]_4[Th(NCS)_8]$	Th: SVol.C7-22
–	$[NH_4]_4[Th(NCS)_8] \cdot 2\ H_2O$	Th: SVol.C7-22, 23/4
$C_8H_{16}O_2Sn$	$(CH_2{=}CHCH_2)_2Sn(OCH_3)_2$	Sn: Org.Comp.16-89, 93
$C_8H_{16}O_6Th$	$Th(OH)_2$(n-$C_3H_7COO)_2$	Th: SVol.C7-71/2
$C_8H_{16}O_{14}Th$	$[Th(O(CH_2COO)_2)_2(H_2O)_4] \cdot 6\ H_2O$	Th: SVol.C7-115/6
$C_8H_{16}Pb$	$Pb(-CH_2CH_2CH_2CH_2-)_2$	Pb: Org.Comp.3-235/6, 238
$C_8H_{17}IN_4O_2Re^+$	[(CO)Re(1,4,7-c-$N_3C_6H_{15}$)(NO)$CH_2I]^+$	Re: Org.Comp.1-52/3
$C_8H_{17}InN_2S_4$	C_2H_5-In[S-C(=S)-$N(CH_3)_2]_2$	In: Org.Comp.1-246, 247
$C_8H_{17}MnNO_6P_2$	[Mn($H_2(O_3PCH_2)_2NC_6H_{11}$-c)]	Mn: MVol.D8-129/30
$C_8H_{17}MnN_3O_4S^{2+}$	[Mn((2-CH_3CONH)(4-CH_3CO)C_2N_2S-4,3,1-$(CH_3)_2$-5,5)$(H_2O)_2]^{2+}$	Mn: MVol.D7-237/9
$C_8H_{17}MnO_4P$	$[Mn(O_3POC_8H_{17})]$	Mn: MVol.D8-146/7
$C_8H_{17}NOS$	O=S=N-$C(CH_3)_2$-C_5H_{11}-n	S: S-N Comp.6-105
–	O=S=N-C_8H_{17}-n	S: S-N Comp.6-105
$C_8H_{17}NOSSn$	$(C_2H_5)_2Sn(NCS)OC_3H_7$-i	Sn: Org.Comp.17-137
$C_8H_{17}NO_2S$	$(CH_3)_2N$-S(O)O-C_6H_{11}-c	S: S-N Comp.8-306/7, 309
$C_8H_{17}NO_2S_2$	1,4-ONC_4H_8-4-S(O)-S-C_4H_9-t	S: S-N Comp.8-333
$C_8H_{17}NO_3Sn$	C_2H_5-C(=O)O-Sn$(CH_3)_2$-NH-C(=O)-C_2H_5	Sn: Org.Comp.19-129, 130
–	C_2H_5-Sn$(-O-CH_2CH_2-)_3N$	Sn: Org.Comp.17-27, 28/9
$C_8H_{17}N_3O_4S_2$	CH_3S-C(CH_3)=N-OC(O)N(CH_3)-S(O)-N(CH_3)O-C_2H_5	S: S-N Comp.8-351
$C_8H_{17}N_3O_5S_3$	CH_3S-C(CH_3)=N-OC(O)N(CH_3)-S(O)-N(C_2H_5)-S$(O)_2CH_3$	S: S-N Comp.8-355
$C_8H_{18}IInN_3^-$	$[(C_4H_9)_2In(N_3)(I)]^-$	In: Org.Comp.1-363
$C_8H_{18}I_4In_2$	[n-C_4H_9-In$(I)_2]_2$	In: Org.Comp.1-162
$C_8H_{18}InN$	CH_3-In[$-CH_2CH_2CH_2$-N(CH_3)-$CH_2CH_2CH_2$-]	In: Org.Comp.1-101, 109/10
$C_8H_{18}InN_3$	$[(C_4H_9)_2In(N_3)]_x$	In: Org.Comp.1-176/7, 179
$C_8H_{18}In_2N_2O_2$	$(CH_3)_2In$-N(CH_3)-C(=O)-C(=O)-N(CH_3)-In$(CH_3)_2$	In: Org.Comp.1-290/1
–	$(CH_3)_2In$-O-C(=N-CH_3)-C(=O)-N(CH_3)-In$(CH_3)_2$	In: Org.Comp.1-290/1
–	CH_3C(=O)-N[In$(CH_3)_2$]-N[In$(CH_3)_2$]-C(=O)CH_3	In: Org.Comp.1-289

$C_8H_{18}In_2N_2S_2$	$(CH_3)_2In-N(CH_3)-C(=S)-C(=S)-N(CH_3)-In(CH_3)_2$	In: Org.Comp.1-293/4
$C_8H_{18}In_2O_4$	$[(CH_3)_2InOC(O)-CH_3]_2$	In: Org.Comp.1-197/8
$C_8H_{18}KN_2PS$	$K[(t-C_4H_9)_2PN=S=N]$	S: S-N Comp.7-103
$C_8H_{18}KN_2S$	$K[(t-C_4H_9-N)_2S]$, radical	S: S-N Comp.7-334/6
$C_8H_{18}LiN_2S$	$Li[(t-C_4H_9-N)_2S]$, radical	S: S-N Comp.7-334/6
$C_8H_{18}MnNO_5S_2^+$	$[Mn(O=S(C_3H_7)CH_2CH_2S(C_3H_7)=O)NO_3]^+$	Mn: MVol.D7-108/9
$C_8H_{18}MnN_2O_6P_2^{2-}$	$[Mn(O_3PC(CH_3)_2NHCH_2CH_2NHC(CH_3)_2PO_3)]^{2-}$	Mn: MVol.D8-130
$C_8H_{18}MnN_2O_8S_2$	$[Mn(O=S(C_3H_7)CH_2CH_2S(C_3H_7)=O)NO_3]NO_3$	Mn: MVol.D7-108/9
$C_8H_{18}MnN_6O_4S_3$	$[Mn((CH_3)_2C=N-NHC(=S)NH_2)_2(SO_4)]$	Mn: MVol.D6-341
$C_8H_{18}MnN_6O_6S$	$[Mn((CH_3)_2C=N-NHC(O)NH_2)_2SO_4]$	Mn: MVol.D6-327
$C_8H_{18}MnN_8S_2$	$Mn(NH=C(SC_2H_5)NC(=NH)NH_2)_2$	Mn: MVol.D7-199/200
$C_8H_{18}MnO_6S_2$	$Mn(O_3S-C_4H_9)_2$	Mn: MVol.D7-114/5
–	$Mn(O_3S-C_4H_9)_2 \cdot 2\ H_2O$	Mn: MVol.D7-114/5
$C_8H_{18}MnO_8P_2^{2-}$	$[Mn(O_3POC_4H_9)_2]^{2-}$	Mn: MVol.D8-146/7
$C_8H_{18}Mn_2N_2O_6P_2$	$Mn_2[O_3PC(CH_3)_2NHCH_2CH_2NHC(CH_3)_2PO_3] \cdot 2\ H_2O$	Mn: MVol.D8-130/1
$C_8H_{18}NO_2S$	$t-C_4H_9-N-S(O)O-C_4H_9-t$, radical	S: S-N Comp.8-325/6
$C_8H_{18}NO_4PS$	$O=S=NP(O)(OC_4H_9-n)_2$	S: S-N Comp.6-76/9
$C_8H_{18}N_2NaS$	$Na[(t-C_4H_9-N)_2S]$, radical	S: S-N Comp.7-334/6
$C_8H_{18}N_2OS_2Sb_2$	$[(CH_3)_3SbNCS]_2O$	Sb: Org.Comp.5-92
$C_8H_{18}N_2O_3Sb_2$	$[(CH_3)_3SbNCO]_2O$	Sb: Org.Comp.5-92
$C_8H_{18}N_2O_6P_2PdS_2$	$Pd(NCS)(SCN)[P(OCH_3)_3]_2$	Pd: SVol.B2-341
–	$Pd(NCS)_2[P(OCH_3)_3]_2$	Pd: SVol.B2-341
–	$Pd(SCN)_2[P(OCH_3)_3]_2$	Pd: SVol.B2-341
$C_8H_{18}N_2O_6Sn$	$(C_4H_9)_2Sn(ONO_2)_2$	Sn: Org.Comp.15-337
$C_8H_{18}N_2PS^-$	$[(t-C_4H_9)_2PN=S=N]^-$	S: S-N Comp.7-103
$C_8H_{18}N_2S$	$n-C_4H_9-N=S=N-C_4H_9-n$	S: S-N Comp.7-195/6
–	$t-C_4H_9-N=S=N-C_4H_9-t$	S: S-N Comp.7-197/210
$C_8H_{18}N_2S^-$	$[(t-C_4H_9-N)_2S]^-$, radical anion	S: S-N Comp.7-334/6
$C_8H_{18}N_2Si$	$1,2-(C_4H_9-t)_2-N_2Si$	Si: SVol.B4-154/5
–	$(NC_4H_8-1)_2SiH_2$	Si: SVol.B4-188
–	$Si(=N-C_4H_9-t)_2$	Si: SVol.B4-153/4
$C_8H_{18}N_2Sn$	$(CH_3)_3SnN=C=NC_4H_9-t$	Sn: Org.Comp.18-107, 110
$C_8H_{18}N_3O_5PS_3$	$CH_3S-C(CH_3)=N-OC(O)-N(CH_3)-S(O)-N(CH_3)-P(S)(OCH_3)_2$	S: S-N Comp.8-356
$C_8H_{18}N_4O_2Re^+$	$[(CO)Re(1,4,7-c-N_3C_6H_{15})(NO)CH_3]^+$	Re: Org.Comp.1-52
$C_8H_{18}N_4S_3$	$t-C_4H_9-N=S=NSN=S=N-C_4H_9-t$	S: S-N Comp.7-50/1
$C_8H_{18}N_4Sn$	$(C_2H_5)_3SnN(N=C(NH_2)N=CH)$	Sn: Org.Comp.18-142, 148
$C_8H_{18}O_3Os$	$(CH_3)_2Os(O)[-O-C(CH_3)_2-C(CH_3)_2-O-]$	Os: Org.Comp.A1-2, 14
$C_8H_{18}O_3SSn$	$(C_4H_9)_2SnOSO_2$	Sn: Org.Comp.15-346
$C_8H_{18}O_3Sn$	$(C_2H_5)_2Sn(OC_2H_5)OOCCH_3$	Sn: Org.Comp.16-181
$C_8H_{18}O_3SnTi$	$(C_4H_9)_2SnOTiO_2$	Sn: Org.Comp.15-368
$C_8H_{18}O_4SSn$	$(C_4H_9)_2SnOSO_3$	Sn: Org.Comp.15-346
$C_8H_{18}O_5Sb_2$	$[(CH_3)_3Sb]_2(C_2O_4)O$	Sb: Org.Comp.5-132
$C_8H_{19}InO$	$(C_2H_5)_2In-O-C_4H_9-t$	In: Org.Comp.1-182, 186
$C_8H_{19}NOSn$	$2,2-(n-C_3H_7)_2-1,3,2-ONSnC_2H_5$	Sn: Org.Comp.19-129, 132
$C_8H_{19}NO_2S$	$(n-C_3H_7)_2N-S(O)O-C_2H_5$	S: S-N Comp.8-316
$C_8H_{19}NO_2Sn$	$(C_2H_5)_2Sn(OC_2H_5)ON=CHCH_3$	Sn: Org.Comp.16-181

$C_8H_{19}NO_4Sn$ ..... $(C_4H_9)_2Sn(OH)ONO_2$ ..... Sn: Org.Comp.16-187, 197
$C_8H_{19}NSn$ ..... $(CH_3)_3SnN(CH_2)_5$ ..... Sn: Org.Comp.18-83/4, 86, 98/9
$C_8H_{19}N_3S_2$ ..... $(C_2H_5)_2NSN{=}S{=}N{-}C_4H_9{-}t$ ..... S: S-N Comp.7-45
$C_8H_{19}O_2Sb$ ..... $(C_4H_9)_2Sb(O)OH$ ..... Sb: Org.Comp.5-205
$C_8H_{19}O_3PSn$ ..... $(C_4H_9)_2SnOP(O)(H)O$ ..... Sn: Org.Comp.15-358
$C_8H_{19}O_3Sb$ ..... $(CH_3)_3Sb(OCH_3)O_2CC_3H_7{-}i$ ..... Sb: Org.Comp.5-45
$C_8H_{19}O_4PSn$ ..... $(C_4H_9)_2SnOP(O)(OH)O$ ..... Sn: Org.Comp.15-359
$C_8H_{19}O_5PTh^{2+}$ ..... $Th(OH)[O_2P(OC_4H_9)_2]^{2+}$ ..... Th: SVol.D1-130
$C_8H_{20}IN_3S$ ..... $[((CH_3)_2N)_2S{-}1{-}NC_4H_8]I$ ..... S: S-N Comp.8-230, 239
$C_8H_{20}I_2In_2$ ..... $[(C_2H_5)_2InI]_2$ ..... In: Org.Comp.1-159
$C_8H_{20}In^-$ ..... $[In(C_2H_5)_4]^-$ ..... In: Org.Comp.1-339
$C_8H_{20}InK$ ..... $K[In(C_2H_5)_4]$ ..... In: Org.Comp.1-339
$C_8H_{20}InN$ ..... $(CH_3)_2In{-}N(C_3H_7{-}i)_2$ ..... In: Org.Comp.1-253/6, 260/1
– ..... $(C_2H_5)_2In{-}N(C_2H_5)_2$ ..... In: Org.Comp.1-252/3, 257
$C_8H_{20}InNO$ ..... $(C_2H_5)_2In[{-}O{-}CH_2CH_2{-}N(CH_3)_2{-}]$ ..... In: Org.Comp.1-190, 192
$C_8H_{20}InNa$ ..... $Na[In(C_2H_5)_4]$ ..... In: Org.Comp.1-339
$C_8H_{20}InP$ ..... $(CH_3)_2In{-}P(C_3H_7{-}i)_2$ ..... In: Org.Comp.1-312, 313, 317
– ..... $(C_2H_5)_2In{-}P(C_2H_5)_2$ ..... In: Org.Comp.1-312, 314
$C_8H_{20}In_2N_6$ ..... $[(C_2H_5)_2In(N_3)]_2$ ..... In: Org.Comp.1-178/9
$C_8H_{20}Li_2N_4Si$ ..... $Li_2[(C_2H_5N)_2Si({-}NCH_3{-}CH_2{-}CH_2{-}NCH_3{-})]$ ..... Si: SVol.B4-224
$C_8H_{20}MnN_2O_6P_2$ ..... $[MnH_2(O_3PC(CH_3)_2NHCH_2CH_2NHC(CH_3)_2PO_3)]$ Mn: MVol.D8-130
$C_8H_{20}MnN_2O_{10}S_4$ $[Mn(O{=}S(CH_3)CH_2CH_2S(CH_3){=}O)_2(NO_3)_2]$ ..... Mn: MVol.D7-108/9
$C_8H_{20}MnO_2P_2S_2$ ..... $[Mn(OP(S)(C_2H_5)_2)_2]_n$ ..... Mn: MVol.D8-188
$C_8H_{20}MnO_2S_6^-$ ..... $[Mn(SCH_2CH_2S)_2(OS(CH_3)_2)_2]^-$ ..... Mn: MVol.D7-33/8
$C_8H_{20}MnO_4P_2$ ..... $Mn[O_2P(C_2H_5)_2]_2$ ..... Mn: MVol.D8-105/8
$C_8H_{20}MnO_4P_2S_4$ ..... $[Mn(S_2P(OC_2H_5)_2)_2]$ ..... Mn: MVol.D8-193/5
$C_8H_{20}MnO_4S_2^{2+}$ ..... $[Mn(C_4H_8OS{-}1,4)_2(H_2O)_2]^{2+}$ ..... Mn: MVol.D7-239
$C_8H_{20}MnO_6P_2$ ..... $[Mn(OP({=}O)(CH_3)OC_3H_7{-}i)_2]$ ..... Mn: MVol.D8-139
$C_8H_{20}MnO_6P_2S_2$ ..... $[Mn(OP(S)(OC_2H_5)_2)_2]$ ..... Mn: MVol.D8-193
$C_8H_{20}MnO_8P_2$ ..... $Mn[O_2P(O{-}C_2H_5)_2]_2$ ..... Mn: MVol.D8-156/7
– ..... $[Mn(O_2P(O{-}C_2H_5)_2)_2]_n$ ..... Mn: MVol.D8-155
$C_8H_{20}MnP_2S_4$ ..... $[Mn(S_2P(C_2H_5)_2)_2]_n$ ..... Mn: MVol.D8-189/91
$C_8H_{20}Mo_4O_{21}^{2-}$ ..... $[Mo_4O_{13}(C_4H_{10}O_4)_2]^{2-}$ ..... Mo: SVol.B3b-170/1
$C_8H_{20}NOs^-$ ..... $[N(C_4H_9{-}n)_4][(C_2H_5)_4Os(N)]$ ..... Os: Org.Comp.A1-2, 29
$C_8H_{20}N_2OS$ ..... $((C_2H_5)_2N)_2SO$ ..... S: S-N Comp.8-342/4
$C_8H_{20}N_2O_2Sn$ ..... $2{-}(n{-}C_4H_9){-}2{-}(H_2N{-}CH_2CH_2{-}O){-}1,3,2{-}ONSnC_2H_5$ Sn: Org.Comp.19-159
– ..... $(C_2H_5)_3Sn{-}N(C_2H_5){-}NO_2$ ..... Sn: Org.Comp.18-141
$C_8H_{20}N_2O_5S_2V$ ..... $((C_2H_5)_2NS(O)O)_2VO$ ..... S: S-N Comp.8-323
$C_8H_{20}N_3S^+$ ..... $[((CH_3)_2N)_2S{-}1{-}NC_4H_8]^+$ ..... S: S-N Comp.8-230, 239
$C_8H_{20}N_4Si$ ..... $Si({-}NCH_3{-}CH_2{-}CH_2{-}NCH_3{-})_2$ ..... Si: SVol.B4-225
$C_8H_{20}N_{16}O_{20}Th$ ..... $Th(NO_3)_4 \cdot 4\ H_2N{-}C({=}O){-}NH{-}C({=}O){-}NH_2$ ..... Th: SVol.D4-134
$C_8H_{20}OOs$ ..... $(C_2H_5)_4Os{=}O$ ..... Os: Org.Comp.A1-2, 29
$C_8H_{20}O_2Sn$ ..... $(n{-}C_4H_9)_2Sn(OH)_2$ ..... Sn: Org.Comp.15-1
– ..... $(t{-}C_4H_9)_2Sn(OH)_2$ ..... Sn: Org.Comp.15-373, 376
$C_8H_{20}O_2Ti$ ..... $(CH_3)_2Ti(O{-}C_3H_7{-}i)_2$ ..... Ti: Org.Comp.5-8, 12
$C_8H_{20}O_3Sn$ ..... $C_2H_5Sn(O{-}C_2H_5)_3$ ..... Sn: Org.Comp.17-25, 28
$C_8H_{20}O_4S_4Th$ ..... $Th[S{-}CH_2CH_2{-}OH]_4$ ..... Th: SVol.D1-130
$C_8H_{20}O_4Th$ ..... $Th(O{-}C_2H_5)_4$ ..... Th: SVol.C7-25/9

$C_8H_{20}O_5W$ .... $WO(O-C_2H_5)_4$ .... W: SVol.A5a-136
$C_8H_{20}Pb$ .... $Pb(C_2H_5)_4$
Analysis .... Pb: Org.Comp.2-68/92
Chemical reactions
Oxidation .... Pb: Org.Comp.2-130/8
Photolysis .... Pb: Org.Comp.2-121/6
Thermal decomposition .... Pb: Org.Comp.2-113/21
with H .... Pb: Org.Comp.2-139/40
with halogens and interhalogen compounds
Pb: Org.Comp.2-139/43
with metal compounds .... Pb: Org.Comp.2-178/91
with metals .... Pb: Org.Comp.2-176/8
with nonmetal compounds .... Pb: Org.Comp.2-143/76
Coordination compounds .... Pb: Org.Comp.2-270/1
Electrochemical behavior .... Pb: Org.Comp.2-191/3
Environmental behavior .... Pb: Org.Comp.2-262/70
Extraction .... Pb: Org.Comp.2-194
Formation and preparation .... Pb: Org.Comp.2-1/68
Molecule .... Pb: Org.Comp.2-93/105
Physical properties .... Pb: Org.Comp.2-105/13
Physiological properties .... Pb: Org.Comp.2-198/227
Radiochemical studies .... Pb: Org.Comp.2-227/8
Removal from solutions .... Pb: Org.Comp.2-195
Solubility .... Pb: Org.Comp.2-193/8
Spectra
ESR spectra .... Pb: Org.Comp.2-128/9
Mass spectra .... Pb: Org.Comp.2-126/8
Nuclear magnetic resonance spectra .. Pb: Org.Comp.2-96/9
Photoelectron spectra .... Pb: Org.Comp.2-93/5
UV spectra .... Pb: Org.Comp.2-104/5
Vibrational spectra .... Pb: Org.Comp.2-99/103
Thermodynamic data of formation .... Pb: Org.Comp.2-47/8
Toxicity .... Pb: Org.Comp.2-198/227
Uses .... Pb: Org.Comp.2-228/62
– .... $Pb(C_2H_5)_4$ solutions
$Pb(C_2H_5)_4-H_2O$ .... Pb: Org.Comp.2-193/5
$Pb(C_2H_5)_4$-organic solvents .... Pb: Org.Comp.2-194/5
$Pb(C_2H_5)_4$-silicon oil E301 .... Pb: Org.Comp.2-195
– .... $Pb(C_2H_5)_4$ systems
$Pb(C_2H_5)_4-Si(C_2H_5)_4$ .... Pb: Org.Comp.2-195
$Pb(C_2H_5)_4-Sn(CH_3)_4$ .... Pb: Org.Comp.2-195
$Pb(C_2H_5)_4-Sn(C_2H_5)_4$ .... Pb: Org.Comp.2-195
$C_8H_{20}Si$ .... $Si(C_2H_5)_4$ systems
$Si(C_2H_5)_4-Pb(C_2H_5)_4$ .... Pb: Org.Comp.2-195
$C_8H_{20}Sn$ .... $Sn(C_2H_5)_4$ systems
$Sn(C_2H_5)_4-Pb(C_2H_5)_4$ .... Pb: Org.Comp.2-195
$C_8H_{21}I_2InN_2$ ..... $C_2H_5-In(I)_2 \cdot (CH_3)_2N-CH_2CH_2-N(CH_3)_2$ ..... In: Org.Comp.1-163, 165/6
$C_8H_{21}InNP$ ...... $(CH_3)_2In-N=P(C_2H_5)_3$ .... In: Org.Comp.1-283, 286

$C_8H_{21}InOSn$	$(CH_3)_2In{-}O{-}Sn(C_2H_5)_3$	In: Org.Comp.1-187/90
$C_8H_{21}NSn$	$(C_2H_5)_3SnN(CH_3)_2$	Sn: Org.Comp.18-131/2
$C_8H_{21}N_4OPS$	$[(CH_3)_2N]_2P(=O)N=S=N{-}C_4H_9{-}t$	S: S-N Comp.7-104
$C_8H_{21}N_4PSn$	$(CH_3)_2Sn(N_3){-}N=P(C_2H_5)_3$	Sn: Org.Comp.19-127
$C_8H_{21}N_5Si$	$(CH_3)_2NSi(NHCH_2CH_2)_3N$	Si: SVol.B4-224
$C_8H_{22}IInSi_2$	$[(CH_3)_3Si{-}CH_2]_2InI$	In: Org.Comp.1-156, 157
$C_8H_{22}InLiO$	$[Li(C_2H_5{-}O{-}C_2H_5)][In(CH_3)_4]$	In: Org.Comp.1-340
$C_8H_{22}InNP_2$	$[{-}In(CH_3)_2{-}CH_2{-}P(CH_3)_2NP(CH_3)_2{-}CH_2{-}]$	In: Org.Comp.1-367/8
$C_8H_{22}In_2Sn$	$(CH_3)_2In{-}CH_2{-}Sn(CH_3)_2{-}CH_2{-}In(CH_3)_2$	In: Org.Comp.1-101, 104
$C_8H_{22}MnN_6O_2S_2{}^{2+}$	$[Mn((CH_3)_2C=N{-}NHC(=S)NH_2)_2(H_2O)_2]^{2+}$	Mn: MVol.D6-341
$C_8H_{22}MnN_8O_4S_4$	$[Mn(S=C(NH_2)_2)_4(OC(CH_3)=O)_2]$	Mn: MVol.D7-187/9
–	$[Mn(S=C(NH_2)_2)_4(OC(CH_3)=O)_2] \cdot H_2O$	Mn: MVol.D7-189
$C_8H_{22}NO_4OsSSi_2{}^-$	$[N(C_4H_9{-}n)_4][((CH_3)_3SiCH_2)_2Os(N)(SO_4)]$	Os: Org.Comp.A1-2, 19
$C_8H_{22}N_2OSiSn$	$(CH_3)_3SnNHC(CH_3)=NOSi(CH_3)_3$	Sn: Org.Comp.18-15/6, 18
$C_8H_{22}N_2O_6S_2Sn$	$(C_4H_9)_2Sn(OSO_2NH_2)_2$	Sn: Org.Comp.15-348
$C_8H_{22}N_2Sn$	$(CH_3)_2N{-}Sn(C_2H_5)_2{-}N(CH_3)_2$	Sn: Org.Comp.19-81
–	$C_2H_5{-}N(CH_3){-}Sn(CH_3)_2{-}N(CH_3){-}C_2H_5$	Sn: Org.Comp.19-67/70
$C_8H_{22}N_3S^+$	$[((CH_3)_2N)_2S{-}N(C_2H_5)_2]^+$	S: S-N Comp.8-230, 239
$C_8H_{22}N_4S_2Si$	$(C_2H_5)_2NSi(CH_3)_2N=S=NSN(CH_3)_2$	S: S-N Comp.7-148
$C_8H_{22}N_4Si$	$(C_2H_5NH)_2Si({-}NCH_3{-}CH_2{-}CH_2{-}NCH_3{-})$	Si: SVol.B4-223/4
$C_8H_{22}N_7O_3PPdS_4$	$[Pd(NCS)(SC(NH_2)_2)_3][OP(OC_2H_5)_2]$	Pd: SVol.B2-341
$C_8H_{22}N_{10}O_{13}Th$	$[C(NH_2)_3]_3[NH_4][Th(C_2O_4)(CO_3)_3] \cdot n\ H_2O$ ($n = 1$ to 3.5)	Th: SVol.C7-99, 101
$C_8H_{22}O_4P_2Sn$	$(C_4H_9)_2Sn(OP(O)H_2)_2$	Sn: Org.Comp.15-354, 356
$C_8H_{23}InN_2$	$(CH_3)_3In \cdot (CH_3)_2N{-}CH_2{-}N(CH_3)_2$	In: Org.Comp.1-27, 34
$C_8H_{23}N_3Sn$	$[(CH_3)_2N]_3Sn{-}C_2H_5$	Sn: Org.Comp.19-111, 113
$C_8H_{24}I_2MnO_4S_4$	$[Mn(O=S(CH_3)_2)_4I_2]$	Mn: MVol.D7-97/8
$C_8H_{24}InNSi_2$	$(CH_3)_2In{-}N[Si(CH_3)_3]_2$	In: Org.Comp.1-253/5, 256, 262
$C_8H_{24}InNaO_2Si_2$	$Na[(CH_3)_2In(OSi(CH_3)_3)_2]$	In: Org.Comp.1-364
$C_8H_{24}InO_2Si_2{}^-$	$[(CH_3)_2In(OSi(CH_3)_3)_2]^-$	In: Org.Comp.1-364
$C_8H_{24}InSb$	$[Sb(CH_3)_4][In(CH_3)_4]$	In: Org.Comp.1-337, 339, 342/3
$C_8H_{24}In_2N_2$	$[(CH_3)_2In{-}N(CH_3)_2]_2$	In: Org.Comp.1-254, 258
$C_8H_{24}In_2O_4P_2$	$[(CH_3)_2In{-}OP(O)(CH_3)_2]_2$	In: Org.Comp.1-211, 218
$C_8H_{24}Li_2Re$	$Li_2[(CH_3)_8Re]$	Re: Org.Comp.1-2
$C_8H_{24}Mn_2O_{10}Si_4$	$[Mn(O{-}Si(OH)(C_2H_5){-}OSi(OH)(C_2H_5){-}O)]_2$	Mn: MVol.D8-23/4
$C_8H_{24}MoN_2S_4$	$[(CH_3)_4N]_2[MoS_4]$	Mo: SVol.B7-278, 281, 285, 288, 294, 296/7
–	$[(C_2H_5)_2NH_2]_2[MoS_4]$	Mo: SVol.B7-278, 280/1, 287
$C_8H_{24}Mo_2N_2S_{11}$	$[(C_2H_5)_2NH_2]_2[(S_2)_2SMo(S)MoS(S_2)_2]$	Mo: SVol.B7-318/9
$C_8H_{24}Mo_3N_2O_{19}W_3$	$[(CH_3)_4N]_2[Mo_3W_3O_{19}] \cdot y$ Solv	Mo: SVol.B3b-29
$C_8H_{24}N_2OSSi_2$	$(CH_3)_2N{-}S(=O){-}N[Si(CH_3)_3]_2$	S: S-N Comp.8-357
–	$(CH_3)_2N{-}S[O{-}Si(CH_3)_3]=N{-}Si(CH_3)_3$	S: S-N Comp.8-202
$C_8H_{24}N_4O_4S_4Si$	$Si(NSO(CH_3)_2)_4$	Si: SVol.B4-147

$C_8H_{24}N_4O_{14.67}P_{2.67}Th$		
	$Th(NO_3)_4 \cdot 2.67\ (CH_3)_3PO$	
	Crystallographic properties	Th: SVol.A4-7/8
$C_8H_{24}N_4O_{16}S_4Th$	$Th(NO_3)_4 \cdot 4\ (CH_3)_2SO$	Th: SVol.D4-205
$C_8H_{24}N_4Si$	$Si(NH-C_2H_5)_4$	Si: SVol.B4-201
–	$Si[N(CH_3)_2]_4$	Si: SVol.B4-209/14
$C_8H_{24}N_6O_{12}Pd$	$[N(CH_3)_4]_2[Pd(NO_3)_4]$	Pd: SVol.B2-42
$C_8H_{24}N_{12}O_{12}Th$	$[C(NH_2)_3]_4[Th(CO_3)_4]$	Th: SVol.C7-4, 8
–	$[C(NH_2)_3]_4[Th(CO_3)_4] \cdot 6\ H_2O$	Th: SVol.C7-4, 8
$C_8H_{24}O_7S_2Sb_2$	$[(CH_3)_3SbOSO_2CH_3]_2O$	Sb: Org.Comp.5-106
$C_8H_{24}Re^{2-}$	$[(CH_3)_8Re]^{2-}$	Re: Org.Comp.1-2
$C_8H_{26}Mo_5N_3NaO_{21}P_2$		
	$Na[(CH_3)_4N][(H_3NC_2H_4PO_3)_2Mo_5O_{15}] \cdot 5\ H_2O$	Mo: SVol.B3b-124
$C_8H_{28}N_8O_{12}Th$	$Th(NO_3)_4 \cdot 4\ (CH_3)_2NH \cdot 8\ H_2O$	
	Photoemission spectra	Th: SVol.A4-131
$C_8H_{28}N_8Si$	$Si(NHN(CH_3)_2)_4$	Si: SVol.B4-259
$C_8H_{30}Mo_3N_2O_{19}V_3$		
	$[(CH_3)_4N]_2H_6Mo_3V_3O_{19} \cdot 5\ H_2O$	Mo: SVol.B3b-23/4
$C_8H_{30}N_{12}O_{15}Th$	$[NH_4]_3[C(NH_2)_3]_3[Th(CO_3)_5] \cdot 3\ H_2O$	Th: SVol.C7-6, 10
$C_8H_{32}I_4N_{16}O_8Th$	$ThI_4 \cdot 8\ O{=}C(NH_2)_2$	Th: SVol.D4-161, 205
	Photoemission spectra	Th: SVol.A4-131
$C_8H_{32}N_{20}O_{20}Th$	$[Th((NH_2)_2CO)_8][NO_3]_4$	
	Photoemission spectra	Th: SVol.A4-131
$C_8H_{33}N_{15}O_{12}Th$	$(C(NH_2)_3)_5[Th(CO_3)_3(OH)_3] \cdot 5\ H_2O$	Th: SVol.C7-14
$C_8K_2MnO_4S_4$	$K_2[Mn(1,2\text{-}S_2C_4(=O)_2\text{-}3,4)_2] \cdot 4\ H_2O$	Mn: MVol.D7-40
$C_8K_4N_8S_8Th$	$K_4[Th(NCS)_8] \cdot 1.5\ H_2O$	Th: SVol.C7-21, 23
–	$K_4[Th(NCS)_8] \cdot 3.5\ H_2O$	Th: SVol.C7-21, 23
$C_8K_4O_{16}Th$	$K_4[Th(C_2O_4)_4]$	Th: SVol.C7-87, 92, 94
–	$K_4[Th(C_2O_4)_4] \cdot 3.38\ H_2O$	Th: SVol.C7-87
–	$K_4[Th(C_2O_4)_4] \cdot 4\ H_2O$	Th: SVol.C7-87, 92/4
–	$K_4[Th(C_2O_4)_4] \cdot 4.18\ H_2O$	Th: SVol.C7-87
–	$K_4[Th(C_2O_4)_4] \cdot 4.21\ H_2O$	Th: SVol.C7-87
–	$K_4[Th(C_2O_4)_4] \cdot 4.42\ H_2O$	Th: SVol.C7-87
$C_8MnN_4S_4^{2-}$	$[Mn(SC(CN){=}C(CN)S)_2]^{2-}$	Mn: MVol.D7-38, 39
$C_8MnN_5OS_4^{2-}$	$[Mn(SC(CN){=}C(CN)S)_2NO]^{2-}$	Mn: MVol.D7-44
C_8MnN_6	$Mn[C(CN)_3]_2$	Mn: MVol.D7-14/9
–	$Mn[C(CN)_3]_2 \cdot 0.25\ H_2O$	Mn: MVol.D7-14/9
$C_8MnO_4S_4^{2-}$	$[Mn(3,4\text{-}S_2C_4(=O)_2\text{-}1,2)_2]^{2-}$	Mn: MVol.D7-40/1
$C_8MnS_8^{2-}$	$[Mn((S)_2C_4(=S)_2)_2]^{2-}$	Mn: MVol.D7-55/6
$C_8Mo_2N_8S_2^{6-}$	$[(CN)_4MoS_2Mo(CN)_4]^{6-}$	Mo: SVol.B3b-189, 196
$C_8N_2O_4ReS_2^-$	$[(CO)_4Re(\text{-}S{=}C(CN)\text{-}C(CN){=}S\text{-})]^-$	Re: Org.Comp.1-369/70
$C_8N_6O_8Th^{2-}$	$Th[2,4,6\text{-}(O)_3\text{-}5\text{-}(ON)\text{-}1,3\text{-}N_2C_4]_2^{2-}$	Th: SVol.D1-120
$C_8N_8Rb_4S_8Th$	$Rb_4[Th(NCS)_8]$	Th: SVol.C7-21, 23, 24
–	$Rb_4[Th(NCS)_8] \cdot 2\ H_2O$	Th: SVol.C7-21/2, 23/4
–	$Rb_4[Th(NCS)_8] \cdot 3\ H_2O$	Th: SVol.C7-22
$C_8N_8S_8Th^{4-}$	$[Th(NCS)_8]^{4-}$	Th: SVol.C7-24
$C_8N_8Th_2$	$Th[Th(CN)_8]$	Th: SVol.C7-16
$C_8Na_4O_{16}Th$	$Na_4[Th(C_2O_4)_4] \cdot 5.5\ H_2O$	Th: SVol.C7-86, 92
–	$Na_4[Th(C_2O_4)_4] \cdot 6\ H_2O$	Th: SVol.C7-92

$C_8Na_8O_{20}Th$ ..... $Na_8Th(C_2O_4)_2(CO_3)_4$ ..... Th: SVol.C7-98, 100
– ..... $Na_8Th(C_2O_4)_2(CO_3)_4 \cdot n\ H_2O$ (n = 9 to 11) .. Th: SVol.C7-98, 100
C_8O_8Th ..... $Th[C_4O_4]_2$ ..... Th: SVol.D1-129
$C_8O_{16}Th^{4-}$ ..... $[Th(C_2O_4)_4]^{4-}$ ..... Th: SVol.C7-91/3
Th: SVol.D1-74/7

$C_{8.4}Cl_{1.4}H_{17.4}O_{3.6}Th$
CH_3-COO-$ThCl_{1.4}(O$-C_4H_9-$t)_{1.6}$ ..... Th: SVol.C7-25/9, 49

$C_9ClF_2H_9N_2O_3S_2$ CH_3-C(O)-NH-S($CClF_2$)=N-$S(O)_2$-C_6H_5 ..... S: S-N Comp.8-194
– ..... C_6H_5-C(O)-NH-S($CClF_2$)=N-$S(O)_2$-CH_3 ..... S: S-N Comp.8-194
$C_9ClF_3H_{16}N_2O_2S$ CF_3SCl-(4-1,4-$ONC_4H_8)_2$ ..... S: S-N Comp.8-401, 403
$C_9ClF_3H_{18}O_3SSn$ $(C_4H_9)_2Sn(Cl)OSO_2CF_3$ ..... Sn: Org.Comp.17-106, 110
$C_9ClF_3H_{20}N_4S$.... CF_3SCl-$((NC_2H_5)_2)_2$ ..... S: S-N Comp.8-401, 402
$C_9ClF_5H_{15}NSn$ ... $(C_2H_5)_3SnN=C(CF_3)CF_2Cl$ ..... Sn: Org.Comp.18-151
C_9ClF_6N ..... (3-Cl)C_9F_6N ..... F: PFHOrg.SVol.4-289, 302
$C_9ClF_6O_5Re$ ..... $(CO)_5ReC_4$(Cl-2)F_6-c ..... Re: Org.Comp.2-141
$C_9ClF_{10}N$..... NCFCFC(c-C_4F_6Cl)CFCF ..... F: PFHOrg.SVol.4-121, 133
$C_9ClF_{14}NO_3$ ..... CF_3-CF(CN)-O-$CF_2CF_2CF_2$-O-CF(CF_3)-CCl=O
F: PFHOrg.SVol.6-101, 121, 140
$C_9ClF_{16}N$..... N(c-C_4F_6Cl)$CF_2CF_2CF_2CF_2CF_2$ ..... F: PFHOrg.SVol.4-123, 134
$C_9ClF_{17}H_2N_2$..... n-C_8F_{17}-C(=NCl)-NH_2 ..... F: PFHOrg.SVol.6-1, 15
$C_9ClF_{18}N$..... 1-(CF_2Cl-CF_2CF_2)-2-CF_3-NC_5F_9 ..... F: PFHOrg.SVol.4-122, 135
– ..... 1-(CF_2Cl-CF_2CF_2)-3-CF_3-NC_5F_9 ..... F: PFHOrg.SVol.4-122, 137
– ..... 1-(CF_2Cl-CF_2CF_2)-4-CF_3-NC_5F_9 ..... F: PFHOrg.SVol.4-122/3, 138
– ..... 1-(CF_2Cl-CF_2CF_2)-3-C_2F_5-NC_4F_7 ..... F: PFHOrg.SVol.4-27/8, 52
– ..... 1-(CF_2Cl-CF_2CF_2)-NC_6F_{12} ..... F: PFHOrg.SVol.4-274/5
$C_9ClF_{18}NO_3S$ .... n-C_8F_{17}-C(=NCl)-$OS(O)_2F$ ..... F: PFHOrg.SVol.6-10, 43
$C_9ClFeH_6O_3^+$ .... $[(1\text{-}ClC_6H_6)Fe(CO)_3]^+$ ..... Fe: Org.Comp.B15-96/7, 98, 135
– ..... $[(2\text{-}ClC_6H_6)Fe(CO)_3]^+$ ..... Fe: Org.Comp.B15-96/7, 102, 135
– ..... $[(3\text{-}ClC_6H_6)Fe(CO)_3]^+$ ..... Fe: Org.Comp.B15-96/7, 105, 135
$C_9ClFeH_7O_3$ ..... $[CH_3C_5H_4Fe(CO)_3]Cl$ ..... Fe: Org.Comp.B15-51
$C_9ClFeH_7O_7$ ..... $[(c\text{-}C_6H_7)Fe(CO)_3][ClO_4]$ ..... Fe: Org.Comp.B15-54/70
$C_9ClFeH_8NO_2$ .... $[(C_5H_5)Fe(CO)_2(CH_2{=}C{=}NH)]Cl$ ..... Fe: Org.Comp.B17-125/6
– ..... $[(C_5H_5)Fe(CO)_2(CH_2{=}C{=}ND)]Cl$ ..... Fe: Org.Comp.B17-126
– ..... $[(C_5H_5)Fe(CO)_2(CH_2{=}C{=}NH)]Cl \cdot H_2O$ ..... Fe: Org.Comp.B17-125
– ..... $[(C_5H_5)Fe(CO)_2(CN\text{-}CH_3)]Cl$ ..... Fe: Org.Comp.B15-311
$C_9ClFeH_8NO_6$ .... $[(C_5H_5)Fe(CO)_2(CH_2{=}C{=}NH)][ClO_4]$ ..... Fe: Org.Comp.B17-125
C_9ClFeH_9O ..... $(C_5H_5)Fe(CO)(CH_2CHCHCl)$ ..... Fe: Org.Comp.B17-147/8
$C_9ClFeH_9O_3$ ..... $[C_5H_5(CO)_2Fe{=}C(CH_3)OH]Cl$ ..... Fe: Org.Comp.B16a-108, 118
$C_9ClFeH_9O_6$ ..... $[(C_5H_5)Fe(CO)_2(CH_2{=}CH_2)][ClO_4]$ ..... Fe: Org.Comp.B17-3
$C_9ClFeH_9O_7$ ..... $[(CH_2(CH)_4CH_3)Fe(CO)_3][ClO_4]$ ..... Fe: Org.Comp.B15-13, 18
$C_9ClFeH_{11}HgN_2$ .. $C_5H_5Fe(CNCH_3)_2HgCl$ ..... Fe: Org.Comp.B15-320
$C_9ClFeH_{11}N_2$ .... $C_5H_5Fe(CNCH_3)_2Cl$ ..... Fe: Org.Comp.B15-320
$C_9ClFe_3GeN_3O_{12}$ $[Ge(Fe(CO)_3NO)_3Cl]$ ..... Fe: Org.Comp.C6a-300
$C_9ClFe_3N_3O_{12}Sn$ $[Sn(Fe(CO)_3NO)_3Cl]$ ..... Fe: Org.Comp.C6a-300/1
$C_9ClGaH_4NO_4S^+$ $[Ga(1\text{-}NC_9H_4\text{-}5\text{-}SO_3\text{-}7\text{-}Cl\text{-}8\text{-}O)]^+$ ..... Ga: SVol.D1-286/90

C_9ClGeH_{13} ...... CH_2Cl-$Ge(CH_3)_2$-C_6H_5 .... Ge: Org.Comp.3-201
– .... $Ge(CH_3)_3$-C_6H_4Cl-2 .... Ge: Org.Comp.2-70, 82
– .... $Ge(CH_3)_3$-C_6H_4Cl-3 .... Ge: Org.Comp.2-70, 82
– .... $Ge(CH_3)_3$-C_6H_4Cl-4 .... Ge: Org.Comp.2-70, 82
C_9ClGeH_{15} ...... $(CH_3)_2Ge[-CH=C(CH_3)-CCl=C(CH_3)-CH_2-]$ ... Ge: Org.Comp.3-305
– .... $(C_2H_5)_2Ge[-CH=CH-CCl=CH-CH_2-]$ .... Ge: Org.Comp.3-301, 307
$C_9ClGeH_{15}O_2$ .... $Ge(CH_3)_3C(=C(OC_2H_5)CHClC(=O))$ .... Ge: Org.Comp.2-30
$C_9ClGeH_{17}S$ ..... $Ge(CH_3)_3C(SC_2H_5)=CClCH=CH_2$ .... Ge: Org.Comp.2-17
C_9ClGeH_{19} ...... $(C_2H_5)_3Ge-C(CH_2Cl)=CH_2$ .... Ge: Org.Comp.2-191, 200
– .... $(C_2H_5)_3Ge-CH=CH-CH_2Cl$ .... Ge: Org.Comp.2-203
$C_9ClGeH_{19}O$ ..... $Ge(C_2H_5)_3CH_2CH_2COCl$ .... Ge: Org.Comp.2-150
C_9ClGeH_{21} ...... $Ge(C_2H_5)_3CH_2CH_2CH_2Cl$ .... Ge: Org.Comp.2-143, 153
$C_9ClGeH_{21}O_2Si$ .. $Ge(CH_3)_2(CH_2Cl)CH_2CH_2COOSi(CH_3)_3$ .... Ge: Org.Comp.3-194
$C_9ClGe_2H_{12}O_5Re$ $(CO)_5ReGe(CH_3)_2Ge(CH_3)_2Cl$ .... Re: Org.Comp.2-6/7
$C_9ClH_5NO_4Re$ .... cis-$(CO)_4Re(Cl)NC_5H_5$ .... Re: Org.Comp.1-432, 436
$C_9ClH_6MnNO_3$ ... $Mn[-2-O-C_6H_3(Cl-5)-CH=NCH_2COO-]$ .... Mn: MVol.D6-59
$C_9ClH_8MoNO_2$ ... $(C_5H_5)Mo(CO)_2(Cl)(CN-CH_3)$ .... Mo: Org.Comp.8-9, 11
$C_9ClH_8N_9O_{13}Th$ .. $Th(NO_3)_4$ · [4-(4-Cl-C_6H_4-N=N)-5-NH_2
-1,2-$N_2C_3H_2(=O)$-3] Th: SVol.D4-137
$C_9ClH_8O_7Re$ ..... $(CO)_3ReCl[=C(-OCH_2CH_2O-)]_2$ .... Re: Org.Comp.1-123
$C_9ClH_9N_3O_7Re$ ... $[(CO)_3Re(NC-CH_3)_3][ClO_4]$ .... Re: Org.Comp.1-308/9
$C_9ClH_{10}MoO_5^-$ ... $[P(C_6H_5)_4][(CH_2C(CH_3)CH_2)MoCl(CO)_3-OC(O)CH_3]$
Mo: Org.Comp.5-292, 294, 295
$C_9ClH_{10}NO_3S^-$ ... $[CH_3CCl(CH_2C_6H_5)-N(O)-SO_2]^-$, radical anion
S: S-N Comp.8-326/8
$C_9ClH_{11}MoN_2O_2$ .. $(CH_2CHCH_2)Mo(Cl)(CO)_2(CN-CH_3)_2$ .... Mo: Org.Comp.5-296/9
– .... $(CH_2CHCH_2)Mo(Cl)(CO)_2(NC-CH_3)_2$ .... Mo: Org.Comp.5-235, 236, 240/1
$C_9ClH_{11}NO_5Sb$ ... $(4-ClCH_2CH_2O_2CNHC_6H_4)Sb(O)(OH)_2$ .... Sb: Org.Comp.5-290
$C_9ClH_{11}O_2Sn$ .... 1-$[(CH_3)_2SnCl-O]-C_7H_5(=O-2)$ .... Sn: Org.Comp.17-82
– .... $(CH_3)_2SnCl-OC(=O)-C_6H_5$ .... Sn: Org.Comp.17-85, 90
$C_9ClH_{12}MoNO$ ... $C_5H_5Mo(NO)(Cl)C_3H_4CH_3$ .... Mo: Org.Comp.6-174/5
$C_9ClH_{12}NO_8PRe$ .. $[(CO)_4Re(P(CH_3)_3)(NCCH_3)][ClO_4]$ .... Re: Org.Comp.1-482
$C_9ClH_{12}O_5Re$ .... $(CO)_3ReCl[CH_3C(O)CH_3]_2$ .... Re: Org.Comp.1-285
$C_9ClH_{13}MoN_2O_2$ .. $[(C_5H_5)Mo(CO)_2(NH_2-CH_2CH_2-NH_2)]Cl$ .... Mo: Org.Comp.7-249, 251
$C_9ClH_{13}NO_4Sb$ ... 4-$(C_2H_5O)(NH=)CC_6H_4Sb(O)(OH)_2$ · HCl ..... Sb: Org.Comp.5-294
$C_9ClH_{13}N_2O_2S_2$ .. $(CH_3)_2N-S(Cl)=N-S(O)_2-C_6H_4-4-CH_3$ .... S: S-N Comp.8-186
$C_9ClH_{13}N_2S_2Si$ ... $(CH_3)_3SiN=S=NS-C_6H_4Cl-4$ .... S: S-N Comp.7-145
$C_9ClH_{13}O_5PRe$ ... $(CO)_4Re(Cl)P(CH_3)_2O-C_3H_7$ .... Re: Org.Comp.1-447
$C_9ClH_{14}MnN_4O_3S^+$
$[MnCl(5-HOCH_2-C_5HN(CH_3-2)(OH-3)-4-CH=N$
$-NHC(=S)NH_2)(H_2O)]^+$ .... Mn: MVol.D6-347/8
$C_9ClH_{14}NO_2SSn$ .. $(CH_3)_3Sn-NCl-S(=O)_2-C_6H_5$ .... Sn: Org.Comp.18-77
– .... $(CH_3)_3Sn-NH-S(=O)_2-C_6H_4Cl-4$ .... Sn: Org.Comp.18-19, 21
$C_9ClH_{14}NSn$ ..... $(CH_3)_3SnNHC_6H_4Cl$-4 .... Sn: Org.Comp.18-17
$C_9ClH_{14}O_3ReS_2$ .. $(CO)_3ReCl(C_2H_5SCH_2CH_2SC_2H_5)$ .... Re: Org.Comp.1-198
$C_9ClH_{15}MoO_3$ .... $C_5H_5MoO(OC_2H_5)_2Cl$ .... Mo: Org.Comp.6-10, 11/2
$C_9ClH_{15}NO_7Sb$ ... $[3-(CH_3)_3N-C_6H_4-Sb(=O)(OH)_2][ClO_4]$ .... Sb: Org.Comp.5-289
– .... $[4-(CH_3)_3N-C_6H_4-Sb(=O)(OH)_2][ClO_4]$ .... Sb: Org.Comp.5-289
$C_9ClH_{15}O_4S_2Sn$ .. $CH_2=CHCH_2SnCl(OS(O)CH_2CH=CH_2)_2$ .... Sn: Org.Comp.17-158

$C_9ClH_{15}O_6Th$ .... $ThCl(C_2H_5COO)_3$ .... Th: SVol.C7-68/9
$C_9ClH_{16}N_2O_3Re$ .. cis-$(CO)_3ReCl[(CH_3)_2NCH_2CH_2N(CH_3)_2]$ .... Re: Org.Comp.1-158
$C_9ClH_{18}MnN_3O_4S_6$
$[Mn(SC(=S)N(CH_3)_2)_3]ClO_4$ .... Mn: MVol.D7-162
$C_9ClH_{18}NOS$ .... 1-$ClS(O)$-2,2,6,6-$(CH_3)_4$-NC_5H_6 .... S: S-N Comp.8-259, 266
$C_9ClH_{18}NO_3S$ .... n-C_7H_{15}-$OC(O)$-$N(CH_3)$-$S(O)Cl$ .... S: S-N Comp.8-268/70
$C_9ClH_{18}O_3P_2Re$ .. $(CO)_3ReCl[P(CH_3)_3]_2$ .... Re: Org.Comp.1-240
$C_9ClH_{18}O_3P_2ReS_6$ $(CO)_3ReCl[P(SCH_3)_3]_2$ .... Re: Org.Comp.1-262
$C_9ClH_{18}O_9P_2Re$ .. $(CO)_3ReCl[P(OCH_3)_3]_2$ .... Re: Org.Comp.1-261
$C_9ClH_{19}O_2Sn$ .... $(C_4H_9)_2Sn(Cl)OOCH$ .... Sn: Org.Comp.17-103
$C_9ClH_{21}NO_2Sn$ ... $(C_4H_9)_2Sn(Cl)ON(O)CH_3$, radical .... Sn: Org.Comp.17-106, 110
$C_9ClH_{21}OSn$ .... $(C_4H_9)_2Sn(Cl)OCH_3$ .... Sn: Org.Comp.17-99, 107/8
$C_9ClH_{21}O_3Th$ .... $ThCl(O$-C_3H_7-i$)_3$ .... Th: SVol.C7-25/9
$C_9ClH_{22}OSb$ .... (i-$C_3H_7)_3Sb(Cl)OH$ .... Sb: Org.Comp.5-5/6
$C_9ClH_{23}O_8P_2Sn$ .. $CH_3SnCl(OP(O)(OC_2H_5)_2)_2$ .... Sn: Org.Comp.17-146
$C_9ClH_{24}NSiSn$ .... $(CH_3)_2Sn(Cl)$-$N(C_4H_9$-t$)$-$Si(CH_3)_3$ .... Sn: Org.Comp.19-117/8, 122
$C_9ClH_{25}NPSn$ .... $[(CH_3)_3SnNHP(C_2H_5)_3]Cl$ .... Sn: Org.Comp.18-16, 19
$C_9ClN_2O_5Re$ .... $(CO)_5ReCCl=C(CN)_2$ .... Re: Org.Comp.2-127
C_9ClO_9OsRe .... $(CO)_5ReOs(CO)_4Cl$ .... Re: Org.Comp.2-201, 210
$C_9Cl_2FH_9N_2O_3S_2$ $CH_3C(O)$-NH-$S(CCl_2F)=N$-$S(O)_2$-C_6H_5 .... S: S-N Comp.8-194
– .... C_6H_5-$C(O)$-NH-$S(CCl_2F)=N$-$S(O)_2$-CH_3 .... S: S-N Comp.8-194
$C_9Cl_2F_4H_{15}NSn$ .. $(C_2H_5)_3SnN=C(CF_2Cl)_2$ .... Sn: Org.Comp.18-151
$C_9Cl_2F_5N$ .... 3,4-Cl_2-C_9F_5N .... F: PFHOrg.SVol.4-288, 308/9
$C_9Cl_2F_9H_{10}Sb$ .... $(CF_3)_2(2$-$CF_3C_6H_{10}$-c$)SbCl_2$ .... Sb: Org.Comp.5-55
$C_9Cl_2F_9N$ .... C_9F_9-5-NCl_2 .... F: PFHOrg.SVol.6-6, 25, 37
$C_9Cl_2F_{14}H_2N_4O$ .. $((CF_3)_2C_2FClN)NHC(O)NH((CF_3)_2C_2FClN)$ .... F: PFHOrg.SVol.4-2
$C_9Cl_2GaH_6NO$ .... $GaCl_2(1$-NC_9H_6-8-$O)$ .... Ga: SVol.D1-284
– .... $GaCl_2(1$-NC_9H_6-8-$O) \cdot 2\ H_2O$ .... Ga: SVol.D1-284
$C_9Cl_2GaH_7O_2$ .... $GaCl_2[OC(O)$-$CH=CH$-$C_6H_5]$ .... Ga: SVol.D1-158, 159
$C_9Cl_2GeH_{14}O_2$ ... $Ge(CH_3)_3C(=C(OC_2H_5)C(Cl)_2C(=O))$ .... Ge: Org.Comp.2-30
$C_9Cl_2GeH_{16}$ .... 1,3,3,5-$(CH_3)_4$-6,6-Cl_2-[3.1.0]-3-GeC_5H_4 .... Ge: Org.Comp.3-319
– .... 3,3-$(C_2H_5)_2$-6,6-Cl_2-[3.1.0]-3-GeC_5H_6 .... Ge: Org.Comp.3-318
$C_9Cl_2GeH_{18}$ .... $Ge(C_2H_5)_3$-$C_3H_3Cl_2$ .... Ge: Org.Comp.2-167
$C_9Cl_2GeH_{21}OP$ ... $Ge(C_2H_5)_3CH_2CH_2CH_2P(O)Cl_2$ .... Ge: Org.Comp.2-147
$C_9Cl_2H_4NOTh^{3+}$ .. $Th[5,7$-$(Cl)_2$-8-(O)-1-$NC_9H_4]^{3+}$ .... Th: SVol.D1-105/6, 118
$C_9Cl_2H_7Ti$ .... $[(C_9H_7)TiCl_2]_n$ .... Ti: Org.Comp.5-36
$C_9Cl_2H_8N_2S$ .... $Cl_2S=N$-$C(CN)(CH_3)$-C_6H_5 .... S: S-N Comp.8-104
$C_9Cl_2H_9MnN_3O_2S_2$
$Mn(4$-$NH_2C_6H_4SO_2NH$-2-C_3H_2NS-3,1$)Cl_2$ .... Mn: MVol.D7-116/23
$C_9Cl_2H_9N_2O_3Re$ .. $(CO)_2ReCl_2[NC_5H_3$-3,4-$(CH_3)_2](NO)$ .... Re: Org.Comp.1-88/9
$C_9Cl_2H_{11}In$ .... 2,4,6-$(CH_3)_3$-C_6H_2-$In(Cl)_2$ .... In: Org.Comp.1-136, 138, 144
$C_9Cl_2H_{11}MnN_3O$ .. $MnCl_2[(CH_3)_2C=N$-$NHC(O)$-4-$C_5H_4N]$ .... Mn: MVol.D6-294, 295/6
$C_9Cl_2H_{11}MnN_3OS$ $[Mn(S=C(NHC_6H_5)NHNHCOCH_3)Cl_2]$ .... Mn: MVol.D7-205/6
$C_9Cl_2H_{11}MnN_3S_2$ $[MnCl_2(NC_5H_4$-2-$CH=NN=C(SCH_3)_2)]_n$ .... Mn: MVol.D6-364
– .... $[MnCl_2(NC_5H_4$-2-$CH=NN(CH_3)$-$C(=S)SCH_3)]_n$ .... Mn: MVol.D6-361, 362
$C_9Cl_2H_{11}MnOP$ ... $Mn(CO)[P(C_6H_5)(CH_3)_2]Cl_2$ .... Mn: MVol.D8-47
$C_9Cl_2H_{11}Mo$ .... $C_5H_5Mo(Cl)_2H_2C=CHCH=CH_2$ .... Mo: Org.Comp.6-189, 197/8
$C_9Cl_2H_{11}N_3O_2S_2$ .. $(CH_3)_2N$-$CCl=N$-$S(Cl)=N$-$S(O)_2$-C_6H_5 .... S: S-N Comp.8-187/9
$C_9Cl_2H_{12}MnN_3P$ .. $Mn[P(CH_2CH_2CN)_3]Cl_2$ .... Mn: MVol.D8-69/70
$C_9Cl_2H_{12}N_2OS$ ... $ClS(O)$-$N(C_6H_5)$-$CHCl$-$N(CH_3)_2$ .... S: S-N Comp.8-259, 267

$C_9Cl_2H_{13}NO_2SSn$	$(CH_3)_3SnNClSO_2C_6H_4Cl$-4	Sn: Org.Comp.18-77
$C_9Cl_2H_{14}MnN_4O_3S$	$[MnCl(5\text{-}HOCH_2\text{-}C_5HN(CH_3\text{-}2)(OH\text{-}3)\text{-}4\text{-}CH{=}N$ $\text{-}NHC({=}S)NH_2)(H_2O)][Cl]$	Mn: MVol.D6-347/8
$C_9Cl_2H_{14}MoO_2$	$(C_5H_5)MoCl_2(O)(O\text{-}C_4H_9\text{-}n)$	Mo: Org.Comp.6-10, 11
–	$(C_5H_5)MoCl_2(O)(O\text{-}C_4H_9\text{-}i)$	Mo: Org.Comp.6-10, 11
$C_9Cl_2H_{14}O_4Sn$	$CH_3OOCCH_2CH_2SnCl_2(OC(CH_3){=}CHCOCH_3)$	Sn: Org.Comp.17-179
$C_9Cl_2H_{15}MnP$	$Mn[P(CH_2CH{=}CH_2)_3]Cl_2$	Mn: MVol.D8-38
$C_9Cl_2H_{15}MoO_2$	$C_5H_5Mo(OC_2H_5)_2Cl_2$	Mo: Org.Comp.6-3, 6/7
$C_9Cl_2H_{15}O_2Sb$	$(CH_2)_4Sb(Cl_2)OC(CH_3){=}CHC(O)CH_3$	Sb: Org.Comp.5-188
$C_9Cl_2H_{16}MnN_4O_2$	$[MnCl_2((CH_3)_2C{=}N\text{-}NHC(O)CH_2C(O)NH\text{-}N{=}C(CH_3)_2)]_n$	Mn: MVol.D6-313, 315
$C_9Cl_2H_{17}O_2Sb$	$(C_2H_5)_2Sb(Cl_2)OC(CH_3){=}CHC(O)CH_3$	Sb: Org.Comp.5-187
$C_9Cl_2H_{18}N_2OS$	$(n\text{-}C_4H_9)_2N\text{-}CCl{=}N\text{-}S(O)Cl$	S: S-N Comp.8-273/4
$C_9Cl_2H_{21}MnNOP$	$Mn(NO)[P(C_3H_7)_3]Cl_2$	Mn: MVol.D8-46, 61
$C_9Cl_2H_{21}MnO_2P$	$Mn[P(C_3H_7)_3](O_2)Cl_2$	Mn: MVol.D8-44/5, 53/7
$C_9Cl_2H_{21}MnP$	$Mn[P(C_3H_7)_3]Cl_2$	Mn: MVol.D8-38
$C_9Cl_2H_{21}PSn$	$(t\text{-}C_4H_9)_2P\text{-}SnCl_2\text{-}CH_3$	Sn: Org.Comp.19-226/7
$C_9Cl_2H_{27}MnN_5O_2P_2$	$[Mn((O{=}P(N(CH_3)_2)_2)_2NCH_3)Cl_2]$	Mn: MVol.D8-180
$C_9Cl_3F_3H_7NS$	$Cl_2S{=}N\text{-}CCl(CF_3)\text{-}C_6H_4\text{-}4\text{-}CH_3$	S: S-N Comp.8-102
$C_9Cl_3F_3H_{15}NSn$	$(C_2H_5)_3SnN{=}C(CF_2Cl)CFCl_2$	Sn: Org.Comp.18-151/2
$C_9Cl_3F_6H_4NS$	$Cl_2S{=}N\text{-}CCl(CF_3)\text{-}C_6H_4\text{-}4\text{-}CF_3$	S: S-N Comp.8-102
$C_9Cl_3F_7HN_3$	$(C_3F_7)C_6Cl_3N_3H$	F: PFHOrg.SVol.4-285/6, 300
$C_9Cl_3F_8H_2N_3$	$(H_2N)(CCl_3)C_8F_8N_2$	F: PFHOrg.SVol.4-292, 306
$C_9Cl_3FeH_9O_2PS^+$	$[C_5H_5Fe(CO)_2CH(SCH_3)PCl_3]^+$	Fe: Org.Comp.B14-133, 140
$C_9Cl_3GaH_{10}O$	$GaCl_3(1\text{-}OC_8H_7\text{-}2\text{-}CH_3)$	Ga: SVol.D1-114
–	$GaCl_3(1\text{-}OC_9H_{10})$	Ga: SVol.D1-116/7
$C_9Cl_3GaH_{10}O_2$	$GaCl_3(1,5\text{-}O_2C_9H_{10})$	Ga: SVol.D1-149/50
–	$GaCl_3[O{=}C(CH_3)\text{-}C_6H_4\text{-}OCH_3\text{-}4]$	Ga: SVol.D1-39/40
$C_9Cl_3GaH_{12}O$	$GaCl_3(CH_3\text{-}O\text{-}C_6H_4\text{-}2\text{-}C_2H_5)$	Ga: SVol.D1-105, 106/8
$C_9Cl_3GaH_{20}O$	$GaCl_3(HO\text{-}C_9H_{19}\text{-}n)$	Ga: SVol.D1-16, 18/9
$C_9Cl_3GaH_{21}N$	$GaCl_3[N(C_3H_7\text{-}n)_3]$	Ga: SVol.D1-233, 234
$C_9Cl_3GeH_8MoO_3^-$	$[N(C_2H_5)_4][(C_5H_5)Mo(CO)_2(GeCl_3)\text{-}C({=}O)CH_3]$	Mo: Org.Comp.8-1, 4
$C_9Cl_3GeH_{13}$	$Ge(CH{=}CH_2)_3CHClCH_2CHCl_2$	Ge: Org.Comp.3-51
$C_9Cl_3GeH_{15}O_2$	$Ge(CH_3)_3C(COOC_2H_5){=}CHCCl_3$	Ge: Org.Comp.2-10/1
$C_9Cl_3H_7N_2OS$	$O{=}S{=}NC(CCl_3){=}NC_6H_4CH_3\text{-}4$	S: S-N Comp.6-180
$C_9Cl_3H_8NO_2S$	$2\text{-}(2,4,6\text{-}Cl_3\text{-}C_6H_2\text{-}N{=})\text{-}1,3,2\text{-}O_2SC_3H_6$	S: S-N Comp.8-167
$C_9Cl_3H_9N_2O_3S_2$	$CH_3\text{-}C(O)\text{-}NH\text{-}S(CCl_3){=}N\text{-}S(O)_2\text{-}C_6H_5$	S: S-N Comp.8-195
–	$C_6H_5\text{-}C(O)\text{-}NH\text{-}S(CCl_3){=}N\text{-}S(O)_2\text{-}CH_3$	S: S-N Comp.8-195
$C_9Cl_3H_{10}NOS$	$C_6H_5\text{-}CH_2\text{-}N(CH_3)\text{-}S(O)\text{-}CCl_3$	S: S-N Comp.8-292
$C_9Cl_3H_{12}O_6Th^+$	$[Th(OOCCH_2CH_2Cl)_3]^+$	Th: SVol.D1-71
$C_9Cl_3H_{14}MoO$	$n\text{-}C_4H_9C_5H_4Mo(OH)Cl_3$	Mo: Org.Comp.6-3
$C_9Cl_3H_{14}MoO_4P$	$C_5H_5Mo(CO)(P(OCH_3)_3)Cl_3$	Mo: Org.Comp.6-219
$C_9Cl_3H_{18}N_6S^+$	$[((CH_3)_2N\text{-}CCl{=}N)_3S]^+$	S: S-N Comp.8-230, 244/5, 247
$C_9Cl_3H_{23}InN$	$[N(C_2H_5)_4][CH_3\text{-}In(Cl)_3]$	In: Org.Comp.1-349/50, 352
$C_9Cl_3H_{25}MoN_3$	$MoCl_3(NH(CH_2)_2CH_3)_2 \cdot NH_2(CH_2)_2CH_3$	Mo: SVol.B5-373
$C_9Cl_4Fe_2H_7O_3$	$[(CH_3\text{-}C_5H_4)Fe(CO)_3][FeCl_4]$	Fe: Org.Comp.B15-51
–	$[(c\text{-}C_6H_7)Fe(CO)_3][FeCl_4]$	Fe: Org.Comp.B15-54/70

$C_9Cl_4Ga_2H_{21}NO_3$	$[Ga(N(CH_2CH_2\text{-}OCH_3)_3)][GaCl_4]$	Ga: SVol.D1-233
$C_9Cl_4GeH_{14}O_2$	$Ge(CH_3)_3C(COCCl_3)=C(Cl)OC_2H_5$	Ge: Org.Comp.2-6
$C_9Cl_4GeH_{18}OSi$	$Ge(CH_3)_3C(SiCl_3)=C(Cl)OC_4H_9$	Ge: Org.Comp.2-5
$C_9Cl_4H_8N_2O_3S_2$	$CH_3C(O)\text{-}NH\text{-}S(CCl_3)=N\text{-}S(O)_2\text{-}C_6H_4\text{-}4\text{-}Cl$	S: S-N Comp.8-195
$C_9Cl_4H_9O_4Sb$	$(CH_3)Sb(OCH_3)_2(1\text{-}OC_6Cl_4O\text{-}2)$	Sb: Org.Comp.5-312
$C_9Cl_4H_{13}OSSb$	$(4\text{-}CH_3C_6H_4)SbCl_4 \cdot OS(CH_3)_2$	Sb: Org.Comp.5-255, 260
$C_9Cl_4H_{14}MoO$	$i\text{-}C_3H_7C_5H_4Mo(OH)Cl_3 \cdot 0.5\ C_2H_4Cl_2\text{-}1,2$	Mo: Org.Comp.6-3
$C_9Cl_4H_{21}N_3O_3Th$	$ThCl_4 \cdot 3\ CH_3\text{-}C(=O)\text{-}NH\text{-}CH_3$	Th: SVol.D4-161
$C_9Cl_4H_{22}NSb$	$[N(CH_3)_4][(CH_2)_5SbCl_4]$	Sb: Org.Comp.5-170
$C_9Cl_4H_{22}Sb_2$	$[(CH_3)_2SbCl_2]_2(CH_2)_5$	Sb: Org.Comp.5-124
$C_9Cl_5FeH_7O_3Sn$	$[(c\text{-}C_6H_7)Fe(CO)_3][SnCl_5]$	Fe: Org.Comp.B15-54/70
$C_9Cl_5H_6NSb^-$	$[NC_9H_6\text{-}5\text{-}SbCl_5]^-$	Sb: Org.Comp.5-261/2
–	$[NC_9H_6\text{-}6\text{-}SbCl_5]^-$	Sb: Org.Comp.5-261/2
–	$[NC_9H_6\text{-}8\text{-}SbCl_5]^-$	Sb: Org.Comp.5-261/2
$C_9Cl_5H_7NSb$	$H[NC_9H_6\text{-}5\text{-}SbCl_5]$	Sb: Org.Comp.5-261/2
–	$H[NC_9H_6\text{-}6\text{-}SbCl_5]$	Sb: Org.Comp.5-261/2
–	$H[NC_9H_6\text{-}8\text{-}SbCl_5]$	Sb: Org.Comp.5-261/2
$C_9Cl_5H_7N_2O_2S_2Sb^-$	$[(4\text{-}(2'\text{-}C_3H_2NSNHSO_2)C_6H_4)SbCl_5]^-$	Sb: Org.Comp.5-254
$C_9Cl_5H_9O_2SSb^-$	$[4\text{-}(CH_3\text{-}OC(=O)\text{-}CH_2\text{-}S)\text{-}C_6H_4\text{-}SbCl_5]^-$	Sb: Org.Comp.5-253
$C_9Cl_5H_9O_2Sb^-$	$[3\text{-}(C_2H_5\text{-}OC(=O))\text{-}C_6H_4\text{-}SbCl_5]^-$	Sb: Org.Comp.5-257
–	$[4\text{-}(C_2H_5\text{-}OC(=O))\text{-}C_6H_4\text{-}SbCl_5]^-$	Sb: Org.Comp.5-257
–	$[4\text{-}(HO\text{-}C(=O)\text{-}CH_2CH_2)\text{-}C_6H_4\text{-}SbCl_5]^-$	Sb: Org.Comp.5-257
$C_9Cl_5H_9O_3Sb^-$	$[3\text{-}HO\text{-}4\text{-}(C_2H_5O\text{-}C(=O))\text{-}C_6H_3\text{-}SbCl_5]^-$	Sb: Org.Comp.5-259
–	$[4\text{-}(CH_3O\text{-}C(=O)\text{-}CH_2\text{-}O)\text{-}C_6H_4\text{-}SbCl_5]^-$	Sb: Org.Comp.5-252
$C_9Cl_5H_{12}NSb^-$	$[(4\text{-}(CH_3)_2NCH_2C_6H_4)SbCl_5]^-$	Sb: Org.Comp.5-256
$C_9Cl_5H_{13}NSb$	$3\text{-}(CH_3)_3N\text{-}C_6H_4\text{-}SbCl_5$	Sb: Org.Comp.5-254, 259
–	$4\text{-}(CH_3)_3N\text{-}C_6H_4\text{-}SbCl_5$	Sb: Org.Comp.5-254, 259
–	$H[4\text{-}(CH_3)_2NCH_2\text{-}C_6H_4\text{-}SbCl_5]$	Sb: Org.Comp.5-256
$C_9Cl_5H_{20}NOOs$	$[N(C_2H_5)_4][Os(CO)(Cl)_5]$	Os: Org.Comp.A1-53, 57
$C_9Cl_5H_{24}N_2OOs$	$[N(CH_3)_4]_2[Os(CO)(Cl)_5]$	Os: Org.Comp.A1-50, 51
$C_9Cl_5H_{30}N_3Pd$	$[(CH_3)_3NH]_2[PdCl_4] \cdot (CH_3)_3NHCl$	Pd: SVol.B2-118/9
$C_9Cl_5N_5S$	$NC_5Cl_4\text{-}2\text{-}N=S=N\text{-}CCl=C(CN)_2$	S: S-N Comp.7-219/20
$C_9Cl_6FeH_8NO_2Sb$	$[(C_5H_5)Fe(CO)_2(CH_2=C=NH)][SbCl_6]$	Fe: Org.Comp.B17-126
–	$[(C_5H_5)Fe(CO)_2(CH_2=C=ND)][SbCl_6]$	Fe: Org.Comp.B17-126
$C_9Cl_6FeH_9O_3Sb$	$[(CH_2(CH)_4CH_3)Fe(CO)_3][SbCl_6]$	Fe: Org.Comp.B15-19
$C_9Cl_6Ga_2H_{10}O_2$	$Cl_6Ga_2(1,5\text{-}O_2C_9H_{10})$	Ga: SVol.D1-149/50
$C_9Cl_6H_{13}N_2O_2S_2Sb$	$[(CH_3)_2NS=NS(O)_2C_6H_4\text{-}CH_3\text{-}4][SbCl_6]$	S: S-N Comp.7-332/3
$C_9Cl_6H_{18}N_6S$	$[((CH_3)_2N\text{-}CCl=N)_3S][Cl_3]$	S: S-N Comp.8-230, 244/5, 247
$C_9Cl_6H_{24}N_2Sb_2$	$[N(CH_3)_4][Cl_3(CH_3)_2SbCNSb(CH_3)_2Cl_3]$	Sb: Org.Comp.5-142, 145
$C_9Cl_{10}H_{10}Ti_4$	$[CHTi_2Cl_5]_2 \cdot C_6H_5CH_3$	Ti: Org.Comp.5-293
$C_9Cl_{18}H_{10}Ti_6$	$[CH(TiCl_3)_2]_2 \cdot 2\ TiCl_3 \cdot C_6H_5CH_3$	Ti: Org.Comp.5-293
C_9CoO_9Re	$(CO)_5ReCo(CO)_4$	Re: Org.Comp.2-203, 211/2
$C_9CrH_9NO_5Sn$	$(CH_3)_3SnNC\text{-}Cr(CO)_5$	Sn: Org.Comp.18-127
$C_9CrH_{10}N_2O_5S$	$[Cr(CO)_5(C_2H_5\text{-}N=S=N\text{-}C_2H_5)]$	S: S-N Comp.7-299/302
$C_9CsF_{16}N_3$	$Cs[CF_2NC(i\text{-}C_3F_7)NC(i\text{-}C_3F_7)N]$	F: PFHOrg.SVol.4-227, 241
$C_9FFeH_8NO_5S$	$[(C_5H_5)Fe(CN\text{-}CH_3)(CO)_2][FSO_3]$	Fe: Org.Comp.B15-311
$C_9FFeH_9O_5S$	$[(C_5H_5)(CO)_2Fe=CH\text{-}CH_3][FSO_3]$	Fe: Org.Comp.B16a-87, 95

C_9FGeH_{13}	$Ge(CH_3)_3$-C_6H_4F-2	Ge: Org.Comp.2-69
–	$Ge(CH_3)_3$-C_6H_4F-3	Ge: Org.Comp.2-69
–	$Ge(CH_3)_3$-C_6H_4F-4	Ge: Org.Comp.2-69
$C_9FH_{16}N_2O_3Re$	$(CO)_3ReF[(CH_3)_2NCH_2CH_2N(CH_3)_2]$	Re: Org.Comp.1-157, 176, 177
$C_9FH_{16}N_3O_3S$	(1,4-ONC_4H_8)-4-S(F)=N-C(O)-4-(1,4-ONC_4H_8)	S: S-N Comp.8-179
–	(1,4-ONC_4H_8-4)$_2$S=N-C(O)F	S: S-N Comp.8-207/8, 212
$C_9FH_{18}NOS$	1-FS(O)-3,3,5,5-$(CH_3)_4$-NC_5H_6	S: S-N Comp.8-255/6
$C_9FH_{21}OSn$	$(C_4H_9)_2Sn(F)OCH_3$	Sn: Org.Comp.17-80
$C_9F_2GeH_{18}$	$Ge(C_2H_5)_3CH(CH_2CF_2)$	Ge: Org.Comp.2-167
$C_9F_2H_{18}N_2OS$	F_2S(1-NC_5H_{10})-4-1,4-ONC_4H_8	S: S-N Comp.8-397/400
$C_9F_2H_{18}N_4SiSn$	1,3,5-$N_3C_3F_2$-2-N[$Si(CH_3)_3$]-$Sn(CH_3)_3$	Sn: Org.Comp.18-58, 71
$C_9F_2H_{20}N_2S$	$F_2S(N(C_2H_5)_2)$1-NC_5H_{10}	S: S-N Comp.8-397
$C_9F_2H_{27}N_3SSi$	$[((CH_3)_2N)_3S][Si(CH_3)_3F_2]$	S: S-N Comp.8-222/4, 226/7
$C_9F_3FeH_5O_2$	$(C_5H_5)Fe(CO)_2CF{=}CF_2$	Fe: Org.Comp.B13-85, 114/5
$C_9F_3FeH_5O_3$	$(C_5H_5)Fe(CO)_2COCF_3$	Fe: Org.Comp.B13-9/10, 13/4, 33
$C_9F_3FeH_5O_5S_2$	$[(C_5H_5)Fe(CS)(CO)_2][SO_3CF_3]$	Fe: Org.Comp.B15-273/4
$C_9F_3FeH_5O_6S$	$[(C_5H_5)Fe(CO)_3][SO_3CF_3]$	Fe: Org.Comp.B15-37/8, 40, 42/4, 46
$C_9F_3FeH_8NO_5S$	$[C_5H_5(CO)_2Fe{=}CH(NH_2)][CF_3SO_3]$	Fe: Org.Comp.B16a-114, 124/5
$C_9F_3FeH_{12}P$	$C_5H_5FeCH_2CHCHCH_3PF_3$	Fe: Org.Comp.B17-141
$C_9F_3GeH_{15}$	$Ge(C_2H_5)_3$-C≡C-CF_3	Ge: Org.Comp.2-259, 270
$C_9F_3GeH_{16}I$	$Ge(C_2H_5)_3$-C(I)=CH-CF_3	Ge: Org.Comp.2-207
$C_9F_3GeH_{18}I$	$Ge(C_2H_5)_3CHICH_2CF_3$	Ge: Org.Comp.2-148, 154
$C_9F_3H_4O_6Re$	$(OC)_4Re$[-O=C(CH_3)-CH=C(CF_3)-O-]	Re: Org.Comp.1-356
$C_9F_3H_{10}NS$	F_3S-2-NC_9H_{10}	S: S-N Comp.8-386
$C_9F_3H_{18}N_2S$	t-C_4H_9-N(CF_3)-SN-C_4H_9-t, radical	S: S-N Comp.7-331/2
$C_9F_3H_{18}N_2S_2$	t-C_4H_9-N(S-CF_3)-SN-C_4H_9-t, radical	S: S-N Comp.7-331/2
$C_9F_3H_{18}N_3O_3SSn$	$(C_4H_9)_2Sn(N_3)OSO_2CF_3$	Sn: Org.Comp.17-139
$C_9F_3H_{19}O_4SSn$	$(C_4H_9)_2Sn(OH)OSO_2CF_3$	Sn: Org.Comp.16-187
$C_9F_3N_2O_5Re$	$(CO)_5ReC_4F_3N_2$	Re: Org.Comp.2-145
$C_9F_3N_3$	1,3,5-$(NC)_3$-C_6F_3	F: PFHOrg.SVol.6-107, 152
$C_9F_4HNO_3$	$(O)_3C_9F_4NH$	F: PFHOrg.SVol.4-291/2, 304
$C_9F_4H_{10}N_2SSi$	$(CH_3)_3SiN{=}S{=}N$-C_6H-F_4-2,3,5,6	S: S-N Comp.7-142
$C_9F_5GeH_9$	$Ge(CH_3)_3C_6F_5$	Ge: Org.Comp.2-69/70, 82
$C_9F_5HN_2O_4$	$(O)_2(O_2N)C_9F_5NH$	F: PFHOrg.SVol.4-291/2, 304
$C_9F_5H_9N_2SSi$	$(CH_3)_3SiN{=}S{=}N$-C_6F_5	S: S-N Comp.7-142
$C_9F_5H_9N_2S_2Si$	$(CH_3)_3SiN{=}S{=}NS$-C_6F_5	S: S-N Comp.7-144
$C_9F_5H_{10}NSn$	$(CH_3)_3SnNHC_6F_5$	Sn: Org.Comp.18-15, 17, 21
$C_9F_5NO_2$	5,8-$(O)_2C_9F_5N$	F: PFHOrg.SVol.4-290, 303
$C_9F_5O_5Re$	$(CO)_5Re$-C(CF_3)=C=CF_2	Re: Org.Comp.2-139/40
–	$(CO)_5Re$-CF=CF-CF=CF_2	Re: Org.Comp.2-140
–	$(CO)_5Re$-C_4F_5-c	Re: Org.Comp.2-141
$C_9F_6FeH_7O_3P$	$[(CH_3C_5H_4)Fe(CO)_3][PF_6]$	Fe: Org.Comp.B15-51
–	$[(C_6H_7)Fe(CO)_3][PF_6]$	Fe: Org.Comp.B15-54/70, 72/3, 76, 78/9, 81, 85
–	$[(C_6H_6D)Fe(CO)_3][PF_6]$	Fe: Org.Comp.B15-56
$C_9F_6FeH_7O_4P$	[(HO-2-C_6H_6)$Fe(CO)_3$][PF_6]	Fe: Org.Comp.B15-102, 138

$C_9F_6FeH_8NO_2P$	$[(C_5H_5)Fe(CNCH_3)(CO)_2][PF_6]$	Fe: Org.Comp.B15-312, 315/6
$C_9F_6FeH_9O_2P$	$[(C_5H_5)Fe(CO)_2{=}CH{-}CH_3][PF_6]$	Fe: Org.Comp.B16a-87, 95
–	$[(C_5H_5)Fe(CO)_2(CH_2{=}CH_2)][PF_6]$	Fe: Org.Comp.B17-3/4, 43
$C_9F_6FeH_9O_2PS$	$[C_5H_5(CO)_2Fe{=}CH(SCH_3)][PF_6]$	Fe: Org.Comp.B16a-113, 122/3
$C_9F_6FeH_9O_3P$	$[(CH_2CHCHCHCH{-}CH_3)Fe(CO)_3][PF_6]$	Fe: Org.Comp.B15-15, 18/9, 28/9
–	$[(CH_2CHCHCHCH{-}CHD_2)Fe(CO)_3][PF_6]$	Fe: Org.Comp.B15-11, 29
–	$[(C_5H_5)Fe(CO)_2{=}CH{-}OCH_3][PF_6]$	Fe: Org.Comp.B16a-108, 118
–	$[(C_5H_5)Fe(CO)_2(CH_2{=}CHOH)][PF_6]$	Fe: Org.Comp.B17-20
$C_9F_6FeH_9PS$	$[C_5H_5FeC_4H_4S][PF_6]$	Fe: Org.Comp.B17-213
$C_9F_6FeH_{10}NO_2P$	$[C_5H_5(CO)_2Fe{=}CH(NHCH_3)][PF_6]$	Fe: Org.Comp.B16a-114, 125
$C_9F_6FeH_{13}P_2^+$	$[(C_5H_5)Fe(CH_2{=}C(CH_3)_2)(PF_3)_2]^+$	Fe: Org.Comp.B16b-7, 10
$C_9F_6GeH_{10}O_2$	$Ge(CH_3)_3CH(C({=}O)OC({=}C(CF_3)_2))$	Ge: Org.Comp.2-94
C_9F_6HNO	$2{-}(O)C_9F_6NH$	F: PFHOrg.SVol.4-290, 304
$C_9F_6HO_5Re$	$(CO)_5Re{-}C(CF_3){=}CH{-}CF_3$	Re: Org.Comp.2-139
–	$(CO)_5Re{-}CF{=}CF{-}CF_2{-}CHF_2$	Re: Org.Comp.2-139
$C_9F_6HO_6Re$	$(CO)_4Re[{-}O{=}C(CF_3){-}CH{=}C(CF_3){-}O{-}]$	Re: Org.Comp.1-356
$C_9F_6H_5IMoNOS_2^-$	$[C_5H_5Mo(NO)(I)({-}SC(CF_3){=}C(CF_3)S{-})]^-$	Mo: Org.Comp.6-30
$C_9F_6H_5InS_2$	$(c{-}C_5H_5)In[{-}S{-}C(CF_3){=}C(CF_3){-}S{-}]$	In: Org.Comp.1-249, 251
$C_9F_6H_5MoS_2$	$C_5H_5Mo[{-}SC(CF_3){=}C(CF_3)S{-}]$	Mo: Org.Comp.6-108
$C_9F_6H_5NOS$	$(CF_3)_2S{=}N{-}C(O)C_6H_5$	S: S-N Comp.8-145/6
$C_9F_6H_7NO_5PRe$	$[(CO)_5Re(CN{-}C_3H_7{-}i)][PF_6]$	Re: Org.Comp.2-252
$C_9F_6H_8N_4O_6PRe$	$[(OC)_3(ON)Re{-}OC(O)CH(NH_2)CH_2{-}4{-}(1,3{-}N_2C_3H_3)][PF_6]$	Re: Org.Comp.1-137
$C_9F_6H_9N_3O_3PRe$	$[(CO)_3Re(CN{-}CH_3)_3][PF_6]$	Re: Org.Comp.2-276
–	$[(CO)_3Re(NC{-}CH_3)_3][PF_6]$	Re: Org.Comp.1-308, 312
$C_9F_6H_{10}MoNO_2P$	$[C_5H_5Mo(CO)(NO)C_3H_5][PF_6]$	Mo: Org.Comp.6-346, 351/2
$C_9F_6H_{10}MoNO_2Sb$	$[C_5H_5Mo(CO)(NO)C_3H_5][SbF_6]$	Mo: Org.Comp.6-346
$C_9F_6H_{11}NO_3PRe$	$[(C_5H_5)Re(CO)(NO)(O{=}C(CH_3)_2)][PF_6]$	Re: Org.Comp.3-153, 154/5
$C_9F_6H_{12}NPS$	$[SN][PF_6] \cdot 1,3,5{-}(CH_3)_3C_6H_3$	S: S-N Comp.5-47/9
$C_9F_6H_{12}NSSb$	$[SN][SbF_6] \cdot 1,3,5{-}(CH_3)_3C_6H_3$	S: S-N Comp.5-47/9
$C_9F_6H_{12}O_4Sn$	$(C_2H_5)(C_3H_7)Sn(OOCCF_3)_2$	Sn: Org.Comp.16-211, 214
$C_9F_6H_{13}MoN_2O_2P$	$[(C_5H_5)Mo(CO)_2(NH_2{-}CH_2CH_2{-}NH_2)][PF_6]$	Mo: Org.Comp.7-249, 251
$C_9F_6H_{15}NSn$	$(C_2H_5)_3SnN{=}C(CF_3)_2$	Sn: Org.Comp.18-151
$C_9F_6H_{16}NOP_2Re$	$[(C_5H_5)Re(NO)(P(CH_3)_3){=}CH_2][PF_6]$	Re: Org.Comp.3-109/12
$C_9F_6H_{24}N_6O_6OsS_2$	$[2,6{-}(CH_3)_2{-}NC_5H_3{-}4{-}Os(NH_3)_5][CF_3SO_3]_2$	Os: Org.Comp.A1-11/2
$C_9F_7HN^+$	$[C_9F_7NH]^+$	F: PFHOrg.SVol.4-289, 308/9
$C_9F_7HN_2OS$	$(CF_3C(O)NH)C_7F_4NS$	F: PFHOrg.SVol.4-284, 299
$C_9F_7H_{18}N_3O_2S_2$	$[((CH_3)_2N)_3S][(CF_3)_2CSO_2F]$	S: S-N Comp.8-230, 235, 247
$C_9F_7H_{18}N_3S_2$	$[((CH_3)_2N)_3S][S{-}C_3F_7{-}i]$	S: S-N Comp.8-230, 235
C_9F_7N	NC_9F_7	F: PFHOrg.SVol.4-288, 290, 308/11
$C_9F_7O_5Re$	$(CO)_5ReC(CF_3){=}C(CF_3)F$	Re: Org.Comp.2-138/9
$C_9F_7O_6Re$	$(CO)_5ReC(O)C_3F_7$	Re: Org.Comp.2-123
$C_9F_7O_7Re$	$(CO)_5ReOC(O)C_3F_7{-}n$	Re: Org.Comp.2-23
$C_9F_8H_{10}N_2$	$NCF_2CF_2CF_2CF_2C(N(C_2H_5)_2)$	F: PFHOrg.SVol.4-102
$C_9F_8H_{11}NO_3S$	$1,4{-}ONC_4H_8{-}4{-}S(O)O{-}CH_2CF_2CF_2CF_2CF_2H$	S: S-N Comp.8-318
$C_9F_8H_{13}NO_2S$	$(C_2H_5)_2N{-}S(O)O{-}CH_2CF_2CF_2CF_2CF_2H$	S: S-N Comp.8-315/6

$C_9F_9GeH_9$ . . . . 1-$(CH_3)_3$Ge-2,2,3-$(CF_3)_3$-C_3 . . . . Ge: Org.Comp.2-29
– . . . . 1-$(CH_3)_3$Ge-2-(i-C_3F_7)-C_3F_2 . . . . Ge: Org.Comp.2-29/30
$C_9F_9GeH_9N_2$ . . . . . Ge$(CH_3)_3$C(=C(CF_3)C$(CF_3)_2$N=N) . . . . Ge: Org.Comp.2-101
$C_9F_9H_2N$ . . . . C_9F_9-5-NH_2 . . . . F: PFHOrg.SVol.5-11, 82
$C_9F_9H_5O_9S_3Sn$ . . . $C_6H_5Sn(OSO_2CF_3)_3$ . . . . Sn: Org.Comp.17-65
$C_9F_9H_7O_6Sn$ . . . . . $C_3H_7Sn(OOCCF_3)_3$ . . . . Sn: Org.Comp.17-31
$C_9F_9H_{12}NO_2S$ . . . . $((CH_3)_2C(CF_3)O)_2$S=N-CF_3 . . . . S: S-N Comp.8-158, 161
$C_9F_9H_{18}N_3O_3S_3Si_3$
1,3,5-$(CF_3S(O))_3$-2,2,4,4,6,6-$(CH_3)_6$-N_3Si_3 . . . . S: S-N Comp.8-289/90
$C_9F_9NO_2$ . . . . C_9F_9-NO_2-2 . . . . F: PFHOrg.SVol.5-182
$C_9F_9N_3$ . . . . C_6F_5-N[-N=N-C(CF_3)=CF-] . . . . F: PFHOrg.SVol.4-44, 69
– . . . . C_6F_5-N[-N=N-CF=C(CF_3)-] . . . . F: PFHOrg.SVol.4-44, 69
$C_9F_{10}H_{11}NO_2S$ . . . $HCF_2CF_2CF_2CF_2CH_2O$-S(F_2)-4-1,4-ONC_4H_8 . . S: S-N Comp.8-395/7
$C_9F_{10}H_{13}NOS$ . . . . $HCF_2CF_2CF_2CF_2CH_2O$-S(F_2)-N$(C_2H_5)_2$ . . . . S: S-N Comp.8-395/7
$C_9F_{11}H_2N$ . . . . 2,4,6-$(CF_3)_3$-C_6F_2-NH_2 . . . . F: PFHOrg.SVol.5-10
$C_9F_{11}H_2N_3$ . . . . . . $(H_2N)(CF_3)C_8F_8N_2$ . . . . F: PFHOrg.SVol.4-292, 306
$C_9F_{11}H_{18}N_3S_2$ . . . . $[((CH_3)_2N)_3S][SF_4$-C_3F_7-i] . . . . S: S-N Comp.8-230, 233, 247
$C_9F_{11}N$ . . . . 1-NC_9F_{11} . . . . F: PFHOrg.SVol.4-289, 302
– . . . . 2-NC_9F_{11} . . . . F: PFHOrg.SVol.4-291
– . . . . 4-CF_3-2-NC_8F_8 . . . . F: PFHOrg.SVol.4-287/8, 301
– . . . . C_6F_5-N=C$(CF_3)_2$ . . . . F: PFHOrg.SVol.6-192/3, 207
$C_9F_{11}NO_2$ . . . . 6-NO_2-C_6F_4-(C_3F_7-i)-3 . . . . F: PFHOrg.SVol.5-181, 191
$C_9F_{12}N_2$ . . . . 1,3-$C_5(=N_2)$-5-$(CF_3)_4$-1,2,3,4 . . . . F: PFHOrg.SVol.5-203, 214, 221, 223
$C_9F_{13}HNSb$ . . . . . . $[1-NC_9F_7H][SbF_6]$ . . . . F: PFHOrg.SVol.4-289
– . . . . $[2-NC_9F_7H][SbF_6]$ . . . . F: PFHOrg.SVol.4-290/1
$C_9F_{13}N$ . . . . $C_9F_{13}N$ . . . . F: PFHOrg.SVol.4-291, 305
$C_9F_{14}N_2O$ . . . . NC-CF_2CF_2-CF_2-CF_2CF_2-O-CF(CF_3)-CN . . . . F: PFHOrg.SVol.6-103/4, 133
$C_9F_{14}N_2O_2$ . . . . . . NC-CF_2CF_2-CF_2CF_2-O-CF_2CF_2-O-CF_2-CN . . F: PFHOrg.SVol.6-103/4, 135
$C_9F_{15}HN_2O_2$ . . . . . C$(CF_3)_2$NC(CF_3)C(NC$(CF_3)_2$OH)O . . . . F: PFHOrg.SVol.4-29/30, 53
$C_9F_{15}H_2N$ . . . . c-C_4F_4-CF_3-1-NH_2-1-[=C(CF_3)-C_2F_5]-2 . . . . . F: PFHOrg.SVol.5-6/7, 37, 76/7
$C_9F_{15}H_3N_4O$ . . . . . NHC$(CF_3)_2$NHC$(CF_3)_2$NC(NHC(O)CF_3) . . . . F: PFHOrg.SVol.4-225, 239
$C_9F_{15}H_6NO_2S$ . . . . $((CF_3)_2C(CH_3)O)_2$S=N-CF_3 . . . . S: S-N Comp.8-158, 161
$C_9F_{15}N$ . . . . N$(CF_2)_3$CF$(CF_2)_4$C . . . . F: PFHOrg.SVol.4-289, 303
$C_9F_{15}NO$ . . . . ON(c-C_5F_9)$CF_2CFCFCF_2$ . . . . F: PFHOrg.SVol.4-175, 179
$C_9F_{15}NO_2$ . . . . CF_2=$CFCF_2$-O-CF_2CF(CF_3)-O-CF_2CF_2-CN . . . F: PFHOrg.SVol.6-99, 119
– . . . . CF_2=CF-O-CF_2CF(CF_3)-O-CF_2CF(CF_3)-CN . . . F: PFHOrg.SVol.6-99, 119
$C_9F_{15}NO_3$ . . . . 2,3,3-F_3-OC_2-2-CF_2-O-CF_2CF(CF_3)-O-CF_2CF_2-CN F: PFHOrg.SVol.6-104, 124/5, 143
– . . . . n-C_7F_{15}-C[=N-O-C(=O)-O-] . . . . F: PFHOrg.SVol.4-30/1, 54, 73
$C_9F_{15}N_3$ . . . . 2,4,6-$(C_2F_5)_3$-1,3,5-N_3C_3 . . . . F: PFHOrg.SVol.4-228/9, 253
– . . . . 2,4-(i-$C_3F_7)_2$-1,3,5-N_3C_3F . . . . F: PFHOrg.SVol.4-227, 240
– . . . . 3,5-(i-$C_3F_7)_2$-1,2,4-N_3C_3F . . . . F: PFHOrg.SVol.4-232/3, 245
– . . . . 3,6-(i-$C_3F_7)_2$-1,2,4-N_3C_3F . . . . F: PFHOrg.SVol.4-232/3, 245
$C_9F_{15}N_3O$ . . . . NC-C$(CF_3)_2$-O-N(CF_3)-C$(CF_3)_2$-CN . . . . F: PFHOrg.SVol.6-104, 124
$C_9F_{16}N_2$ . . . . N$(C_5F_7)CF_2N(CF_3)CF_2CF_2$ . . . . F: PFHOrg.SVol.4-36/7, 63

$C_9F_{16}N_2OS_2$ ..... $(C_4F_8S-1-)=N-C(O)-N=(-1-SC_4F_8)$ .... S: S-N Comp.8-152/3
$C_9F_{17}H_2NO_3$ ..... $C_3F_7-O-CF(CF_3)CF_2-O-CF(CF_3)-C(=O)NH_2$ ... F: PFHOrg.SVol.5-20, 87
$C_9F_{17}H_3N_2O_2$ .... $C_3F_7-O-CF(CF_3)CF_2-O-CF(CF_3)C(=NH)-NH_2$ .. F: PFHOrg.SVol.5-3, 72/3
$C_9F_{17}H_3N_2O_3$ .... $C_3F_7-O-CF(CF_3)CF_2-O-CF(CF_3)-C(N=OH)-NH_2$ F: PFHOrg.SVol.5-4, 36, 69
$C_9F_{17}H_4NO_2S$ .... $(n-C_3F_7-CH_2O)_2S=N-CF_3$ .... S: S-N Comp.8-158, 162
$C_9F_{17}N$ .... $1-NC_9F_{17}$ .... F: PFHOrg.SVol.4-290
– .... $4-NC_9F_{17}$ .... F: PFHOrg.SVol.4-293, 306
– .... $1-(c-C_5F_9)-NC_4F_8$ .... F: PFHOrg.SVol.4-28
– .... $n-C_8F_{17}-CN$ .... F: PFHOrg.SVol.6-97/8, 116
– .... $[2.2.1]-C_7F_{11}-N(CF_3)_2$ .... F: PFHOrg.SVol.6-228/9
$C_9F_{17}NO_2$ .... $C_3F_7-O-CF(CF_3)-CF_2-O-CF(CF_3)-CN$ .... F: PFHOrg.SVol.6-98, 117
$C_9F_{17}NO_3$ .... $CF_3[-O-CF_2CF(CF_3)]_2-O-CF_2-CN$. .... F: PFHOrg.SVol.6-98/9, 117
$C_9F_{17}NO_4$ .... $CF_3-O(-CF_2CF_2-O)_3-CF_2-CN$ .... F: PFHOrg.SVol.6-98, 133
$C_9F_{17}N_3O_3$ ...... $N_3-CF_2CF(O-C_3F_7-n)CF_2-O-CF(CF_3)-CF=O$ .. F: PFHOrg.SVol.5-204, 216
$C_9F_{18}N_2$ .... $1-[(CF_3)_2C=N]-2,2,3,3-(CF_3)_4-NC_2$ .... F: PFHOrg.SVol.4-2
– .... $(C_4F_9)_2C=N_2$ .... F: PFHOrg.SVol.5-203
$C_9F_{18}N_2O$ .... $NC(CF_3)(ON(CF_3)_2)CF_2CF_2CF_2C(CF_3)$ .... F: PFHOrg.SVol.4-144, 154
$C_9F_{18}N_4S_3$ ...... $[(CF_3)_2C=NS]_3N$ .... F: PFHOrg.SVol.6-49, 58
$C_9F_{18}N_6$ .... $NNC(N(CF_3)_2)NC(N(CF_3)_2)C(N(CF_3)_2)$ .... F: PFHOrg.SVol.4-232/3, 245
$C_9F_{19}N$ .... $1-(n-C_3F_7)-2-CF_3-NC_5F_9$ .... F: PFHOrg.SVol.4-122, 135
– .... $1-(n-C_3F_7)-3-CF_3-NC_5F_9$ .... F: PFHOrg.SVol.4-122, 136
– .... $1-(n-C_3F_7)-4-CF_3-NC_5F_9$ .... F: PFHOrg.SVol.4-122/3, 138
– .... $1-(n-C_3F_7)-3-C_2F_5-NC_4F_7$ .... F: PFHOrg.SVol.4-27/8, 51
– .... $1-(n-C_3F_7)-NC_6F_{12}$ .... F: PFHOrg.SVol.4-274/5
– .... $CF_3-N=C(C_3F_7-i)-C_4F_9-t$. .... F: PFHOrg.SVol.6-190/1, 203
– .... $CF_3-N=CF-CF(C_2F_5)-C_4F_9-n$. .... F: PFHOrg.SVol.6-190/1, 203
– .... $n-C_4F_9-N=CF-C_4F_9-t$. .... F: PFHOrg.SVol.6-190/1, 202
$C_9F_{19}NO$ .... $N(CF_2CF_2OCF_2CF_3)CF_2CF_2CF_2CF_2CF_2$ .... F: PFHOrg.SVol.4-121, 133
$C_9F_{19}N_3O_2$ ...... $NCF(ON(CF_3)_2)CF_2CF_2CF_2C(ON(CF_3)_2)$. ...... F: PFHOrg.SVol.4-143/4, 154
$C_9F_{19}N_3S$ .... $i-C_3F_7-N=S=NC(CF_3)_2N=C(CF_3)_2$. .... S: S-N Comp.7-266
$C_9F_{21}N$ .... $(n-C_3F_7)_3N$ .... F: PFHOrg.SVol.6-223, 232, 240/2
– .... $(n-C_4F_9)_2N-CF_3$ .... F: PFHOrg.SVol.6-225/6, 240, 242
$C_9F_{21}NO$ .... $(n-C_3F_7)_2N-O-C_3F_7-n$ .... F: PFHOrg.SVol.5-115, 126/7
$C_9F_{21}NO_2$ .... $CF_3-N(CF_2CF_2-O-C_2F_5)_2$. .... F: PFHOrg.SVol.6-226/7, 238
$C_9F_{27}N_4O_4P$ ..... $[(CF_3)_2N-O]_4P-CF_3$ .... F: PFHOrg.SVol.5-117
$C_9FeGeH_{10}O_3$ .... $(CH_3)_2Ge(-CH=CHCH=CH-)Fe(CO)_3$ .... Ge: Org.Comp.3-279, 285
$C_9FeH_5NaO_6$ ..... $Na[C_5H_5Fe(CO)_2C_2O_4]$ .... Fe: Org.Comp.B14-162
$C_9FeH_5O_6^-$ ...... $[C_5H_5Fe(CO)_2C_2O_4]^-$ .... Fe: Org.Comp.B14-162
$C_9FeH_6O_2$ .... $(C_5H_5)Fe(CO)_2C{\equiv}CH$. .... Fe: Org.Comp.B13-154
$C_9FeH_6O_3$ .... $(C_5H_5)Fe(CO)_2CH=C=O$ .... Fe: Org.Comp.B13-112, 139
$C_9FeH_7KO_2S_2$ .... $K[C_5H_4CH_3Fe(CO)_2CS_2]$. .... Fe: Org.Comp.B14-163
$C_9FeH_7LiO_3$ ..... $Li[(CH_3-C(=O)-C_5H_4)Fe(CO)_2]$. .... Fe: Org.Comp.B14-118
– .... $Li[(C_5H_5)Fe(CO)_2-C(O)CH_2]$. .... Fe: Org.Comp.B14-160/1
$C_9FeH_7O_2^+$ ...... $[C_5H_5(CO)_2Fe=C=CH_2]^+$ .... Fe: Org.Comp.B16a-212
$C_9FeH_7O_2S_2^-$ .... $[C_5H_4CH_3Fe(CO)_2CS_2]^-$ .... Fe: Org.Comp.B14-163
$C_9FeH_7O_3^+$ ...... $[(CH_3C_5H_4)Fe(CO)_3]^+$ .... Fe: Org.Comp.B15-50/1

$C_9FeH_7O_3^+$ ...... $[(C_6H_7)Fe(CO)_3]^+$
Calculations .... Fe: Org.Comp.B15-58/9
Chemical behavior .... Fe: Org.Comp.B15-59/86
Physical properties .... Fe: Org.Comp.B15-57
Preparation and formation .... Fe: Org.Comp.B15-54/7
Spectra .... Fe: Org.Comp.B15-57/8

– .... $[(C_6H_6D)Fe(CO)_3]^+$ .... Fe: Org.Comp.B15-56
– .... $[(C_6H_5D_2)Fe(CO)_3]^+$ .... Fe: Org.Comp.B15-56/7
$C_9FeH_7O_3^-$ ...... $[(CH_3-C(=O)-C_5H_4)Fe(CO)_2]^-$ .... Fe: Org.Comp.B14-118
– .... $[(C_5H_5)Fe(CO)_2-C(O)CH_2]^-$ .... Fe: Org.Comp.B14-160/1
$C_9FeH_7O_4^+$ ...... $[(HO-1-C_6H_6)Fe(CO)_3]^+$ .... Fe: Org.Comp.B15-91
– .... $[(HO-2-C_6H_6)Fe(CO)_3]^+$ .... Fe: Org.Comp.B15-102, 138
– .... $[(HO-6-C_6H_6)Fe(CO)_3]^+$ .... Fe: Org.Comp.B15-107, 159
$C_9FeH_7O_4^-$ ...... $[C_5H_5Fe(CO)_2CH_2COO]^-$ .... Fe: Org.Comp.B14-160
$C_9FeH_8INO_2$ ..... $[C_5H_5Fe(CNCH_3)(CO)_2]I$ .... Fe: Org.Comp.B15-312, 315
$C_9FeH_8NOS^+$ .... $[C_5H_5Fe(CS)(CO)NCCH_3]^+$ .... Fe: Org.Comp.B15-271
$C_9FeH_8NO_2^+$ .... $[(C_5H_5)Fe(CO)_2(CH_2=C=NH)]^+$ .... Fe: Org.Comp.B17-125
– .... $[(C_5H_5)Fe(CO)_2(CH_2=C=ND)]^+$ .... Fe: Org.Comp.B17-126
– .... $[(C_5H_5)Fe(CO)_2(CN-CH_3)]^+$ .... Fe: Org.Comp.B15-312/3, 315/7
$C_9FeH_8N_2O$ ...... $C_5H_5Fe(CNCH_3)(CO)CN$ .... Fe: Org.Comp.B15-288, 294/5
$C_9FeH_8O_2$ .... $(C_5H_5)Fe(CO)_2CH=CH_2$ .... Fe: Org.Comp.B13-84, 113/4
$C_9FeH_8O_2S_2$ ..... $(C_5H_5)Fe(CO)_2C(S)SCH_3$ .... Fe: Org.Comp.B13-56, 58
$C_9FeH_8O_3$ .... $(C_5H_5)Fe(^{12}CO)(^{13}CO)-COCH_3$ .... Fe: Org.Comp.B13-30
– .... $(C_5H_5)Fe(CO)_2-COCH_3$ .... Fe: Org.Comp.B13-9/10, 13, 28/32
– .... $(C_5H_5)Fe(CO)_2-^{13}COCH_3$ .... Fe: Org.Comp.B13-13, 32/3
$C_9FeH_8O_3^+$ ...... $[(C_5H_5)Fe(CO)_2C(O)-CH_3]^+$, radical cation.... Fe: Org.Comp.B14-165
$C_9FeH_8O_3S$ ...... $(C_5H_5)Fe(CO)_2C(S)OCH_3$ .... Fe: Org.Comp.B13-55, 57
$C_9FeH_8O_4$ .... $(C_5H_5)Fe(CO)_2-COOCH_3$ .... Fe: Org.Comp.B13-24, 44
– .... $(c-C_6H_7)Fe(CO)_2-COOH$ .... Fe: Org.Comp.B14-80, 81
C_9FeH_9N .... $C_5H_5FeC_4H_4N$ .... Fe: Org.Comp.B17-201, 227/9
$C_9FeH_9NO_2$ ...... $(C_5H_5)Fe(CH_2=CH_2)(CO)NCO$ .... Fe: Org.Comp.B16b-73, 78
$C_9FeH_9NO_2S$ ..... $(C_5H_5)Fe(CO)_2C(S)NHCH_3$ .... Fe: Org.Comp.B13-57/8
$C_9FeH_9NO_3$ ...... $(C_5H_5)Fe(CO)_2CONHCH_3$ .... Fe: Org.Comp.B13-9, 24, 45
$C_9FeH_9NaO_2$ ..... $Na[(c-C_7H_9)Fe(CO)_2]$ .... Fe: Org.Comp.B14-122/3
$C_9FeH_9O_2^+$ ...... $[(C_5H_5)Fe(CO)_2=CH-CH_3]^+$ .... Fe: Org.Comp.B16a-79/82, 87, 95
– .... $[(C_5H_5)Fe(CO)_2(CH_2=CH_2)]^+$ .... Fe: Org.Comp.B17-3, 43
$C_9FeH_9O_2^-$ ...... $[(c-C_7H_9)Fe(CO)_2]^-$ .... Fe: Org.Comp.B14-122/3
$C_9FeH_9O_2S^+$ ..... $[(C_5H_5)Fe(CO)_2=CH-SCH_3]^+$ .... Fe: Org.Comp.B16a-113, 122/3
– .... $[(C_5H_5)Fe(CO)_2=CD-SCH_3]^+$ .... Fe: Org.Comp.B16a-113, 123
$C_9FeH_9O_3^+$ ...... $[(CH_2C(CH_2)CHCH-CH_3)Fe(CO)_3]^+$ .... Fe: Org.Comp.B15-32/3
– .... $[(CH_2CHCHCHCH-CH_3)Fe(CO)_3]^+$ .... Fe: Org.Comp.B15-2/3, 12/9, 28
– .... $[(CH_2CHCHCHCH-CHD_2)Fe(CO)_3]^+$ .... Fe: Org.Comp.B15-2, 11, 29
– .... $[(C_5H_5)Fe(CO)_2=C(CH_3)OH]^+$ .... Fe: Org.Comp.B13-31
Fe: Org.Comp.B16a-108, 118
– .... $[(C_5H_5)Fe(CO)_2=CH-OCH_3]^+$ .... Fe: Org.Comp.B16a-108, 118

$C_9FeH_9O_3^+$ . . . . . . $[(C_5H_5)Fe(CO)_2(CH_2{=}CHOH)]^+$ . . . . . . Fe: Org.Comp.B17-19/20
$C_9FeH_9O_4^+$ . . . . . . $[(CH_3(CH)_5OH)Fe(CO)_3]^+$ . . . . . . Fe: Org.Comp.B15-3/4, 8
C_9FeH_9P . . . . . . $C_5H_5FeC_4H_4P$ . . . . . . Fe: Org.Comp.B17-204, 231
$C_9FeH_9S^+$ . . . . . . $[C_5H_5FeC_4H_4S]^+$ . . . . . . Fe: Org.Comp.B17-213
$C_9FeH_{10}N^+$ . . . . . . $[(C_5H_5)FeC_4H_4(NH)]^+$ . . . . . . Fe: Org.Comp.B16b-64
– . . . . . . $[(C_5H_5)Fe(NC_4H_5)]^+$ . . . . . . Fe: Org.Comp.B17-203
$C_9FeH_{10}NO_2^+$ . . . . $[(C_5H_5)Fe(CO)_2{=}CH\text{-}NHCH_3]^+$ . . . . . . Fe: Org.Comp.B16a-114, 125
$C_9FeH_{10}NO_3^+$ . . . . $[C_5H_5(CO)_2Fe{=}C(OCH_3)NH_2]^+$ . . . . . . Fe: Org.Comp.B16a-137
$C_9FeH_{10}N_2O_3$ . . . . $(C_5H_5)Fe(CO)_2CONHNHCH_3$ . . . . . . Fe: Org.Comp.B13-25, 46
$C_9FeH_{10}N_2O_4$ . . . . $(C_5H_5)Fe(CN\text{-}C_2H_5)(CO)NO_3$ . . . . . . Fe: Org.Comp.B15-286
$C_9FeH_{10}O$ . . . . . . $C_5H_5Fe(CH_2CHCH_2)CO$ . . . . . . Fe: Org.Comp.B17-148, 154
$C_9FeH_{10}O_2$ . . . . . . $(CH_3\text{-}C_5H_4)Fe(CO)_2\text{-}CH_3$ . . . . . . Fe: Org.Comp.B14-53/4, 61/2
– . . . . . . $(C_5H_5)Fe(CH_2{=}CH_2)(CO)\text{-}CHO$ . . . . . . Fe: Org.Comp.B16b-72
$C_9FeH_{10}O_2^+$ . . . . . $[C_5H_4CH_3Fe(CO)_2CH_3]^+$, radical cation . . . . . . Fe: Org.Comp.B14-165
$C_9FeH_{11}^-$ . . . . . . $[C_5H_5Fe(CH_2{=}CHCH{=}CH_2)]^-$ . . . . . . Fe: Org.Comp.B17-186
$C_9FeH_{11}HgIN_2$ . . . $C_5H_5Fe(CNCH_3)_2HgI$ . . . . . . Fe: Org.Comp.B15-320
$C_9FeH_{11}IN_2$ . . . . . . $C_5H_5Fe(CNCH_3)_2I$ . . . . . . Fe: Org.Comp.B15-320, 324/5
$C_9FeH_{11}IOS_2$ . . . . $C_5H_5(CO)(I)Fe{=}C(SCH_3)_2$ . . . . . . Fe: Org.Comp.B16a-172, 173
$C_9FeH_{11}NaO_2$ . . . . $Na[(CH_2{=}C(CH_3)\text{-}CH\text{-}C(CH_3){=}CH_2)Fe(CO)_2]$ . . . Fe: Org.Comp.B14-123
$C_9FeH_{11}O_2^-$ . . . . . $[(CH_2{=}C(CH_3)\text{-}CH\text{-}C(CH_3){=}CH_2)Fe(CO)_2]^-$ . . . . Fe: Org.Comp.B14-123
$C_9FeH_{14}IO_3PS$ . . . $C_5H_5Fe(CS)(P(OCH_3)_3)I$ . . . . . . Fe: Org.Comp.B15-262
$C_9FeH_{14}N_2Si$ . . . . . $C_5H_5Fe(CNCH_3)_2SiH_3$ . . . . . . Fe: Org.Comp.B15-320
$C_9FeH_{14}O_4P$ . . . . . $[C_5H_5Fe(CO)P(OCH_3)_3]$, radical . . . . . . Fe: Org.Comp.B14-125
$C_9FeNO_{10}Re$ . . . . . $[Re(CO)_6][Fe(CO)_3NO]$ . . . . . . Re: Org.Comp.2-219
$C_9Fe_3GeIN_3O_{12}$ . . $[Ge(Fe(CO)_3NO)_3I]$ . . . . . . Fe: Org.Comp.C6a-300
$C_9Fe_3HKO_9S$ . . . . . $K[(CO)_9(H)Fe_3S]$ . . . . . . Fe: Org.Comp.C6a-133
$C_9Fe_3HNaO_9S$ . . . . $Na[(CO)_9(H)Fe_3S]$ . . . . . . Fe: Org.Comp.C6a-133
$C_9Fe_3HO_9S^-$ . . . . . $[(CO)_9(H)Fe_3S]^-$ . . . . . . Fe: Org.Comp.C6a-132/3
$C_9Fe_3HO_{10}^-$ . . . . . $[(CO)_9Fe_3(OH)]^-$ . . . . . . Fe: Org.Comp.C6a-132
$C_9Fe_3H_2N_2O_9$ . . . . $(CO)_9Fe_3(NH)_2$ . . . . . . Fe: Org.Comp.C6a-148, 150/1
$C_9Fe_3H_2O_9S$ . . . . . $(CO)_9(H)_2Fe_3S$ . . . . . . Fe: Org.Comp.C6a-127/8
$C_9Fe_3H_2O_9S_2$ . . . . $(CO)_9(H)_2Fe_3S_2$ . . . . . . Fe: Org.Comp.C6a-128
$C_9Fe_3H_3NO_9Te_2$ . . $(CO)_9Fe_3Te_2(NH_3)$ . . . . . . Fe: Org.Comp.C6a-121
$C_9Fe_3InN_3O_{12}$ . . . . $[In(Fe(CO)_3NO)_3]$ . . . . . . Fe: Org.Comp.C6a-300
$C_9Fe_3K_2O_9S$ . . . . . $K_2[(CO)_9Fe_3S]$ . . . . . . Fe: Org.Comp.C6a-133/4
$C_9Fe_3Na_2O_9S$ . . . . $Na_2[(CO)_9Fe_3S]$ . . . . . . Fe: Org.Comp.C6a-133
$C_9Fe_3O_8S_3$ . . . . . . $(CO)_8Fe_3S_2(CS)$ . . . . . . Fe: Org.Comp.C6a-53
$C_9Fe_3O_9^+$ . . . . . . $[Fe_3(CO)_9]^+$ . . . . . . Fe: Org.Comp.C6a-4
$C_9Fe_3O_9^-$ . . . . . . $[(CO)_9Fe_3]^-$ . . . . . . Fe: Org.Comp.C6a-4
$C_9Fe_3O_9P_2$ . . . . . . $(CO)_9Fe_3P_2$ . . . . . . Fe: Org.Comp.C6a-140
$C_9Fe_3O_9S^{2-}$ . . . . . $[(CO)_9Fe_3S]^{2-}$ . . . . . . Fe: Org.Comp.C6a-133/5
$C_9Fe_3O_9SSe$ . . . . . $(CO)_9Fe_3SSe$ . . . . . . Fe: Org.Comp.C6a-104/5, 107
$C_9Fe_3O_9STe$ . . . . . $(CO)_9Fe_3STe$ . . . . . . Fe: Org.Comp.C6a-104/5, 107
$C_9Fe_3O_9S_2$ . . . . . . $(CO)_9Fe_3S_2$ . . . . . . Fe: Org.Comp.C6a-104/13
$C_9Fe_3O_9S_2^-$ . . . . . $[(CO)_9Fe_3S_2]^-$, radical anion . . . . . . Fe: Org.Comp.C6a-111
$C_9Fe_3O_9Sb_2$ . . . . . $(CO)_9Fe_3Sb_2$ . . . . . . Fe: Org.Comp.C6a-140
$C_9Fe_3O_9SeTe$ . . . . $(CO)_9Fe_3SeTe$ . . . . . . Fe: Org.Comp.C6a-104, 106/7, 117
$C_9Fe_3O_9Se_2$ . . . . . $(CO)_9Fe_3Se_2$ . . . . . . Fe: Org.Comp.C6a-104/7, 113
$C_9Fe_3O_9Se_2^-$ . . . . $[(CO)_9Fe_3Se_2]^-$, radical anion . . . . . . Fe: Org.Comp.C6a-113

$C_9Fe_3O_9Te_2$ ..... $(CO)_9Fe_3Te_2$ ..... Fe: Org.Comp.C6a-104/7, 113/7
$C_9Fe_3O_{10}{}^{2-}$ ..... $[(CO)_9Fe_3(O)]^{2-}$ ..... Fe: Org.Comp.C6a-131/2
$C_9Fe_3O_{10}S_2$ ..... $(CO)_9Fe_3S(SO)$ ..... Fe: Org.Comp.C6a-128/30
$C_9Fe_3O_{10}Se_2$ ..... $(CO)_9Fe_3Se(SeO)$ ..... Fe: Org.Comp.C6a-130
$C_9Fe_3O_{12}S_2{}^{2-}$ ..... $[(CO)_7Fe_3(C\text{-}CO)(SO_2)_2]^{2-}$ ..... Fe: Org.Comp.C6a-39/40
$C_9GaH_4INO_4S^+$ ..... $[Ga(1\text{-}NC_9H_4\text{-}5\text{-}SO_3\text{-}7\text{-}I\text{-}8\text{-}O)]^+$ ..... Ga: SVol.D1-286/90
$C_9GaH_5INO_4S^{2+}$ ..... $[Ga(1\text{-}NC_9H_4\text{-}5\text{-}SO_3\text{-}7\text{-}I\text{-}8\text{-}OH)]^{2+}$ ..... Ga: SVol.D1-286/90
$C_9GaH_5NO_4S^+$ ..... $[Ga(1\text{-}NC_9H_5\text{-}5\text{-}SO_3\text{-}8\text{-}O)]^+$ ..... Ga: SVol.D1-286/90
$C_9GaH_6INO_6S^-$ ..... $[(HO)_2Ga(1\text{-}NC_9H_4\text{-}5\text{-}SO_3\text{-}7\text{-}I\text{-}8\text{-}O)]^-$ ..... Ga: SVol.D1-286/90
$C_9GaH_6K_3O_{12}$ ..... $K_3[Ga(OC(O)\text{-}CH_2\text{-}C(O)O)_3]$ ..... Ga: SVol.D1-166, 167
– ..... $K_3[Ga(OC(O)\text{-}CH_2\text{-}C(O)O)_3] \cdot 2\ H_2O$ ..... Ga: SVol.D1-166, 167/8
– ..... $K_3[Ga(OC(O)\text{-}CH_2\text{-}C(O)O)_3] \cdot 3\ H_2O$ ..... Ga: SVol.D1-166, 167/8
$C_9GaH_6NO^{2+}$ ..... $[Ga(1\text{-}NC_9H_6\text{-}8\text{-}O)]^{2+}$ ..... Ga: SVol.D1-275/8
$C_9GaH_6N_3O_5$ ..... $Ga(NO_2)_2(1\text{-}NC_9H_6\text{-}8\text{-}O)$ ..... Ga: SVol.D1-284
$C_9GaH_6N_3O_7$ ..... $Ga(NO_3)_2(1\text{-}NC_9H_6\text{-}8\text{-}O)$ ..... Ga: SVol.D1-284
$C_9GaH_6Na_3O_{12}$ ..... $Na_3[Ga(OC(O)\text{-}CH_2\text{-}C(O)O)_3] \cdot 3\ H_2O$ ..... Ga: SVol.D1-166, 167/8
$C_9GaH_6O_4{}^+$ ..... $[Ga(OC(O)\text{-}CH(C_6H_5)\text{-}C(O)O)]^+$ ..... Ga: SVol.D1-166/7
$C_9GaH_6O_{12}{}^{3-}$ ..... $[Ga(OC(O)\text{-}CH_2\text{-}C(O)O)_3]^{3-}$ ..... Ga: SVol.D1-166/8
$C_9GaH_6O_{12}Tl_3$ ..... $Tl_3[Ga(OC(O)\text{-}CH_2\text{-}C(O)O)_3]$ ..... Ga: SVol.D1-166, 167
– ..... $Tl_3[Ga(OC(O)\text{-}CH_2\text{-}C(O)O)_3] \cdot H_2O$ ..... Ga: SVol.D1-166, 167/8
$C_9GaH_7O_8$ ..... $Ga[H(1,2,3,4\text{-}(OC(O))_4\text{-}C_5H_6)] \cdot 10\ H_2O$ ..... Ga: SVol.D1-170
$C_9GaH_8NO_3$ ..... $Ga(OH)_2(1\text{-}NC_9H_6\text{-}8\text{-}O)$ ..... Ga: SVol.D1-277
$C_9GaH_9O_4$ ..... $(HO)_2Ga[OC(O)\text{-}CH{=}CH\text{-}C_6H_5]$ ..... Ga: SVol.D1-158/9
$C_9GaH_{10}N_2O_5Re$ ..... $(OC)_4Re[\text{-}1,2\text{-}(1,2\text{-}N_2C_3H_3)\text{-}Ga(CH_3)_2\text{-}O(H)\text{-}]$ Re: Org.Comp.1-354
$C_9GaH_{15}O_5$ ..... $Ga(OH)[OC(O)\text{-}(CH_2)_7\text{-}C(O)O]$ ..... Ga: SVol.D1-168
$C_9GaH_{15}O_6$ ..... $Ga[OC(O)\text{-}C_2H_5]_3$ ..... Ga: SVol.D1-155
$C_9GaH_{15}O_9$ ..... $Ga[OC(O)\text{-}CH(OH)\text{-}CH_3]_3$ ..... Ga: SVol.D1-172/4
– ..... $Ga[OC(O)\text{-}CH(OH)\text{-}CH_3]_3 \cdot 2\ H_2O$ ..... Ga: SVol.D1-173/4
$C_9GaH_{17}O_4$ ..... $(C_2H_5\text{-}O)_2Ga[CH_3\text{-}C(O)CHC(O)\text{-}CH_3]$ ..... Ga: SVol.D1-68/9
$C_9GaH_{21}NO_3{}^+$ ..... $[Ga(N(CH_2CH_2\text{-}OCH_3)_3)][GaCl_4]$ ..... Ga: SVol.D1-233
$C_9GaH_{24}N$ ..... $GaH_3[N(C_3H_7\text{-}n)_3]$ ..... Ga: SVol.D1-233
$C_9GaH_{24}N_9S_3$ ..... $[Ga(NH_2\text{-}CH_2CH_2\text{-}NH_2)_3][NCS]_3$ ..... Ga: SVol.D1-236
$C_9GaH_{24}O_6{}^{3+}$ ..... $[Ga(HO\text{-}CH_2\text{-}CH(OH)\text{-}CH_3)_3]^{3+}$ ..... Ga: SVol.D1-23
$C_9GaH_{26}O_7{}^{3+}$ ..... $[(H_2O)Ga(HO\text{-}CH_2\text{-}CH(OH)\text{-}CH_3)_3]^{3+}$ ..... Ga: SVol.D1-23
$C_9GaH_{28}O_8{}^{3+}$ ..... $[(H_2O)_2Ga(HO\text{-}CH_2\text{-}CH(OH)\text{-}CH_3)_3]^{3+}$ ..... Ga: SVol.D1-23
$C_9Ga_2H_{28}O_8{}^{3+}$ ..... $[(H_2O)_2Ga_2(HO\text{-}CH_2\text{-}CH(OH)\text{-}CH_3)_3]^{3+}$ ..... Ga: SVol.D1-23
$C_9GeH_9MoNO_5$ ..... $(CO)_5Mo[CN\text{-}Ge(CH_3)_3]$ ..... Mo: Org.Comp.5-6, 9, 10
C_9GeH_{12} ..... $(CH_3)_2Ge(\text{-}CH_2C_6H_4\text{-})$ ..... Ge: Org.Comp.3-318
$C_9GeH_{12}O_2$ ..... $Ge(CH_3)_2(C_6H_5)COOH$ ..... Ge: Org.Comp.3-202, 207
C_9GeH_{13} ..... $Ge(CH_3)_2CH_2C_6H_5$, radical ..... Ge: Org.Comp.3-360, 363
$C_9GeH_{13}NO_2$ ..... $Ge(CH_3)_3C_6H_4NO_2\text{-}4$ ..... Ge: Org.Comp.2-72, 84
C_9GeH_{14} ..... $Ge(CH_3)_3\text{-}C{\equiv}C\text{-}CH_2\text{-}C{\equiv}C\text{-}CH_3$ ..... Ge: Org.Comp.2-59
– ..... $Ge(CH_3)_3\text{-}C{\equiv}C\text{-}C{\equiv}C\text{-}C_2H_5$ ..... Ge: Org.Comp.2-59, 65
– ..... $Ge(CH_3)_3\text{-}C_6H_5$ ..... Ge: Org.Comp.2-68/9, 80/2
– ..... $[\text{-}CH_2\text{-}CH{=}CH\text{-}CH_2\text{-}]Ge[\text{-}CH_2\text{-}C(CH_3){=}CH\text{-}CH_2\text{-}]$ Ge: Org.Comp.3-343, 346
$C_9GeH_{14}{}^-$ ..... $[Ge(CH_3)_3C_6H_5]^-$, radical anion ..... Ge: Org.Comp.3-352
$C_9GeH_{14}O$ ..... $Ge(CH_3)_3C_6H_4OH\text{-}4$ ..... Ge: Org.Comp.2-71, 82
$C_9GeH_{14}S$ ..... $Ge(CH_3)_3C_6H_4SH\text{-}4$ ..... Ge: Org.Comp.2-72

C_9GeH_{15} .... $Ge(CH_3)_3C_6H_6$, radical .... Ge: Org.Comp.3-351
$C_9GeH_{15}N$ .... 2-$(CH_3)_3$Ge-4-CH_3-NC_5H_3 .... Ge: Org.Comp.2-102
– .... 2-[$Ge(CH_3)_3CH_2$]-NC_5H_4 .... Ge: Org.Comp.1-158, 163
– .... $Ge(CH_3)_3$-C_6H_4-NH_2-4 .... Ge: Org.Comp.2-72
$C_9GeH_{15}N_3OS$ ... $Ge(CH_3)_3C(=CHCH=C(CH=NNHCSNH_2)O)$ ..... Ge: Org.Comp.2-97, 105
$C_9GeH_{15}N_3O_2$ .... $Ge(CH_3)_3C(=CHCH=C(CH=NNHCONH_2)O)$ ..... Ge: Org.Comp.2-97, 105
C_9GeH_{16} .... 1-C_4H_9-1-CH≡C-GeC_3H_6 .... Ge: Org.Comp.3-235
– .... $(CH_3)_3$Ge-C_5H_4-CH_3 .... Ge: Org.Comp.2-32, 46
– .... $(CH_3)_3$Ge-C_6H_7-c .... Ge: Org.Comp.2-37, 47
– .... C_2H_5-$Ge(CH_3)_2$-C_5H_5-c .... Ge: Org.Comp.3-196, 206
$C_9GeH_{16}N_2$ ...... $Ge(C_2H_5)_3CH(CN)_2$ .... Ge: Org.Comp.2-135
$C_9GeH_{16}O$ .... $Ge(C_2H_5)_3$-C≡C-CHO .... Ge: Org.Comp.2-262
$C_9GeH_{16}O_2$ ...... $Ge(CH_3)_3$-CH=C=CH-COO-C_2H_5 .... Ge: Org.Comp.2-18
– .... $Ge(CH_3)_3$-C≡C-CH_2-COO-C_2H_5 .... Ge: Org.Comp.2-58
$C_9GeH_{16}O_3$ ...... $(CH_3)_2Ge(-CH_2CH(C(=O)OCH_3)C(=O)CH_2CH_2-)$ Ge: Org.Comp.3-291
C_9GeH_{18} .... 1,1-$(C_2H_5)_2$-3-CH_3-GeC_4H_5 .... Ge: Org.Comp.3-258, 270/1
– .... 1-(CH_2=CH-CH_2)-1,2-$(CH_3)_2$-GeC_4H_7 .... Ge: Org.Comp.3-250, 254
– .... 1-(n-C_4H_9)-1-(CH_2=CH)-GeC_3H_6 .... Ge: Org.Comp.3-235
– .... 1-(i-C_4H_9)-1-CH_3-GeC_4H_6 .... Ge: Org.Comp.3-266
– .... 1-(t-C_4H_9)-1-CH_3-GeC_4H_6 .... Ge: Org.Comp.3-266
– .... 3,3-$(C_2H_5)_2$-[3.1.0]-3-GeC_5H_8 .... Ge: Org.Comp.3-318
– .... (-C_5H_{10}-)Ge(-C_4H_8-) .... Ge: Org.Comp.3-345
– .... $Ge(CH_3)_3$-C≡C-C_4H_9-n .... Ge: Org.Comp.2-59
– .... $Ge(CH_3)_3$-C≡C-C_4H_9-t .... Ge: Org.Comp.2-58, 65
– .... $Ge(CH_3)_3$-C_6H_9-c .... Ge: Org.Comp.2-34
– .... $Ge(C_2H_5)_3$-CH=C=CH_2 .... Ge: Org.Comp.2-173, 184
– .... $Ge(C_2H_5)_3$-CH_2C≡CH .... Ge: Org.Comp.2-237/8, 246/7
– .... $Ge(C_2H_5)_3$-C≡CCH_3 .... Ge: Org.Comp.2-237
$C_9GeH_{18}N_2O$ ..... $Ge(C_2H_5)_3C(=N_2)COCH_3$ .... Ge: Org.Comp.2-152
$C_9GeH_{18}N_2O_2$ .... $Ge(C_2H_5)_3C(=N_2)COOCH_3$ .... Ge: Org.Comp.2-137, 139
$C_9GeH_{18}O$ .... 3,3-$(C_2H_5)_2$-1-CH_3-[3.1.0]-3,6-$GeOC_4H_5$ ..... Ge: Org.Comp.3-320
– .... $(CH_3)_2Ge[-CH(CH_3)-CH_2-C(=O)CH_2-CH(CH_3)-]$ Ge: Org.Comp.3-293
– .... $(CH_3)_2Ge[-CH_2CH_2CH_2-C(=O)-CH_2CH_2CH_2-]$ Ge: Org.Comp.3-311
– .... $(C_2H_5)_2Ge[-CH_2CH_2-C(=O)-CH_2CH_2-]$ .... Ge: Org.Comp.3-290, 297
– .... $Ge(CH_3)_3$-$C(OCH_3)$=CH-$C(CH_3)$=CH_2 .... Ge: Org.Comp.2-17
– .... $Ge(CH_3)_3$-C≡C-$C(CH_3)(OH)$-C_2H_5 .... Ge: Org.Comp.2-58
– .... $Ge(CH_3)_3$-C≡C-O-C_4H_9-n .... Ge: Org.Comp.2-53
– .... $Ge(C_2H_5)_3$-$C(CH_3)$=C=O .... Ge: Org.Comp.2-198, 201
– .... $Ge(C_2H_5)_3$-C≡C-CH_2OH .... Ge: Org.Comp.2-260, 270
– .... $Ge(C_2H_5)_3$-C_3H_3(=O-2) .... Ge: Org.Comp.2-167
$C_9GeH_{18}OSi$ ..... 2-$(CH_3)_3$Ge-5-$(CH_3)_2$SiH-OC_4H_2 .... Ge: Org.Comp.2-98
– .... $(CH_3)_3$Ge-C(=O)-C≡C-$Si(CH_3)_3$ .... Ge: Org.Comp.2-56
$C_9GeH_{18}O_4$ ...... $Ge(CH_3)_3CH_2CH(COOCH_3)_2$ .... Ge: Org.Comp.1-184
$C_9GeH_{19}N$ .... $Ge(CH_3)_3$-C≡C-$N(C_2H_5)_2$ .... Ge: Org.Comp.2-53
– .... $Ge(C_2H_5)_3$-CH_2CH_2-CN .... Ge: Org.Comp.2-152
$C_9GeH_{19}NS$ ...... $Ge(CH_3)_3C_3H_6SCH_2CH_2CN$ .... Ge: Org.Comp.1-181
$C_9GeH_{19}O$ .... $(C_2H_5)_2Ge(-CH_2C(OCH_3)CH_2CH_2-)$ .... Ge: Org.Comp.3-243
C_9GeH_{20} .... 1,1,3,5-$(CH_3)_4$-GeC_5H_8 .... Ge: Org.Comp.3-291

C_9GeH_{20}	$1,1-(CH_3)_2-GeC_7H_{14}$	Ge: Org.Comp.3-310, 314
–	$1,1-(C_2H_5)_2-3-CH_3-GeC_4H_7$	Ge: Org.Comp.3-244
–	$1,1-(C_2H_5)_2-GeC_5H_{10}$	Ge: Org.Comp.3-288, 297
–	$1-(n-C_4H_9)-1-C_2H_5-GeC_3H_6$	Ge: Org.Comp.3-235
–	$Ge(CH_3)_3-CH{=}CH-C_4H_9-n$	Ge: Org.Comp.2-19
–	$Ge(CH_3)_3-CH_2CH_2CH_2CH_2-CH{=}CH_2$	Ge: Org.Comp.2-19
–	$Ge(CH_3)_3-C_5H_8-2-CH_3$	Ge: Org.Comp.1-195
–	$Ge(CH_3)_3-C_6H_{11}-c$	Ge: Org.Comp.1-196
–	$Ge(C_2H_5)_3-C(CH_3){=}CH_2$	Ge: Org.Comp.2-170
–	$Ge(C_2H_5)_3-CH{=}CH-CH_3$	Ge: Org.Comp.2-171
–	$Ge(C_2H_5)_3-CH_2CH{=}CH_2$	Ge: Org.Comp.2-171/2, 183/4
$C_9GeH_{20}HgO_3$	$Ge(CH_3)_3CH_2CH(OH)CH_2CH_2HgOOCCH_3$	Ge: Org.Comp.1-189
$C_9GeH_{20}O$	$(C_2H_5)_2Ge[-CH_2-CH(OH)-CH(CH_3)-CH_2-]$	Ge: Org.Comp.3-245, 253
–	$Ge(CH_3)_3-C_6H_{10}-2-OH$	Ge: Org.Comp.1-196
–	$Ge(C_2H_5)_3-CH{=}CH-CH_2OH$	Ge: Org.Comp.2-204
–	$Ge(C_2H_5)_3-CH_2CH_2-CHO$	Ge: Org.Comp.2-149
–	$Ge(C_2H_5)_3-CH_2-COCH_3$	Ge: Org.Comp.2-151, 155
–	$Ge(C_2H_5)_3-C_3H_4-2-OH$	Ge: Org.Comp.2-167
$C_9GeH_{20}O_2$	$Ge(CH_3)_3-CH_2CH_2CH_2CH_2-OC({=}O)-CH_3$	Ge: Org.Comp.1-189
–	$Ge(C_2H_5)_3-CH_2CH_2-COOH$	Ge: Org.Comp.2-149, 154
–	$Ge(C_2H_5)_3-CH_2-COO-CH_3$	Ge: Org.Comp.2-130/1, 138
$C_9GeH_{20}O_3S$	$Ge(C_2H_5)_3CH(CH_2CH_2OS(O)_2)$	Ge: Org.Comp.2-289
$C_9GeH_{20}Si$	$Ge(CH_3)_3C{\equiv}CCH_2Si(CH_3)_3$	Ge: Org.Comp.2-56
C_9GeH_{21}	$Ge(C_3H_7-n)_3$, radical	Ge: Org.Comp.3-359
–	$Ge(C_3H_7-i)_3$, radical	Ge: Org.Comp.3-359
$C_9GeH_{21}NO$	$Ge(CH_3)_3-CH_2-C({=}O)-N(C_2H_5)_2$	Ge: Org.Comp.1-171
–	$Ge(C_2H_5)_3-C({=}O)-N(CH_3)_2$	Ge: Org.Comp.2-119
–	$Ge(C_2H_5)_3-CH_2-C({=}O)-NHCH_3$	Ge: Org.Comp.2-131
$C_9GeH_{21}O_4P$	$Ge(CH_3)_3CH_2COP(O)(OC_2H_5)_2$	Ge: Org.Comp.1-171
C_9GeH_{22}	$C_2H_5-Ge(C_3H_7-n)_2-CH_3$	Ge: Org.Comp.3-213
–	$n-C_3H_7-Ge(C_2H_5)_3$	Ge: Org.Comp.2-110, 113
–	$i-C_3H_7-Ge(C_2H_5)_3$	Ge: Org.Comp.2-110
–	$n-C_4H_9-Ge(CH_3)_2-C_3H_7-n$	Ge: Org.Comp.3-197, 206
–	$n-C_4H_9-Ge(C_2H_5)_2-CH_3$	Ge: Org.Comp.3-209, 212
–	$Ge(CH_3)_3-C_6H_{13}-n$	Ge: Org.Comp.1-192/3
$C_9GeH_{22}O$	$Ge(C_2H_5)_3-CH_2CH_2CH_2-OH$	Ge: Org.Comp.2-143/4, 153
–	$Ge(C_2H_5)_3-CH_2-CH(OH)-CH_3$	Ge: Org.Comp.2-148
–	$Ge(C_2H_5)_3-C(OH)(CH_3)_2$	Ge: Org.Comp.2-129
$C_9GeH_{22}OSi$	$Ge(CH_3)_3-CH{=}C(CH_3)-O-Si(CH_3)_3$	Ge: Org.Comp.2-9
–	$Ge(CH_3)_3-CH[Si(CH_3)_3]-C({=}O)-CH_3$	Ge: Org.Comp.1-183, 185
–	$Ge(CH_3)_3-CH_2-C({=}O)-CH_2-Si(CH_3)_3$	Ge: Org.Comp.1-183
–	$Ge(CH_3)_3-C_3H_4-2-[O-Si(CH_3)_3]$	Ge: Org.Comp.1-194
$C_9GeH_{22}OSn$	$Ge(CH_3)_3C_3H_4-2-[OSn(CH_3)_3]$	Ge: Org.Comp.1-194
$C_9GeH_{22}O_2$	$Ge(C_2H_5)_3CH_2CH(OH)CH_2OH$	Ge: Org.Comp.2-148
$C_9GeH_{22}O_3S$	$Ge(CH_3)_3-C(CH_3)(C_3H_7-n)-O-S({=}O)_2-CH_3$	Ge: Org.Comp.1-189
–	$Ge(C_2H_5)_3-CH(CH_3)-S({=}O)_2-OCH_3$	Ge: Org.Comp.2-127
$C_9GeH_{23}N$	$Ge(CH_3)_3-CH_2CH_2-N(C_2H_5)_2$	Ge: Org.Comp.1-168
–	$Ge(C_2H_5)_3-CH_2CH_2CH_2-NH_2$	Ge: Org.Comp.2-146
$C_9GeH_{23}NOSi$	$Ge(CH_3)_3-CH[Si(CH_3)_3]-C({=}O)-NHCH_3$	Ge: Org.Comp.1-172
–	$Ge(CH_3)_3-CH_2-C({=}O)-N(CH_3)-Si(CH_3)_3$	Ge: Org.Comp.1-171

$C_9GeH_{24}IN$	$[Ge(CH_3)_3CH_2NCH_3(C_2H_5)_2]I$	Ge: Org.Comp.1-139
$C_9GeH_{24}OSi$	$Ge(CH_3)_3C_3H_6OSi(CH_3)_3$	Ge: Org.Comp.1-181, 185
$C_9GeH_{24}Pb$	$Ge(CH_3)_3C_3H_6Pb(CH_3)_3$	Ge: Org.Comp.1-182, 185
$C_9GeH_{24}Si$	$Ge(CH_3)_3C_3H_6Si(CH_3)_3$	Ge: Org.Comp.1-182, 185
$C_9GeH_{24}Sn$	$Ge(CH_3)_3C_3H_6Sn(CH_3)_3$	Ge: Org.Comp.1-182, 185
$C_9GeH_{25}Si_2$	$GeCH_3(CH_2Si(CH_3)_3)_2$, radical	Ge: Org.Comp.3-360, 361, 363
$C_9GeH_{27}InNP$	$(CH_3)_3In \cdot (CH_3)_3Ge{-}N{=}P(CH_3)_3$	In: Org.Comp.1-27, 37
$C_9GeH_{27}PSiSn$	$(CH_3)_3Sn{-}P[Si(CH_3)_3]{-}Ge(CH_3)_3$	Sn: Org.Comp.19-176
$C_9Ge_2H_{13}O_5Re$	$(CO)_5ReGe(CH_3)_2Ge(CH_3)_2H$	Re: Org.Comp.2-7
$C_9Ge_2H_{27}PSn$	$(CH_3)_3Sn{-}P[Ge(CH_3)_3]_2$	Sn: Org.Comp.19-171
$C_9Ge_3H_{27}MnN_3O_3P$	$Mn(NO)_3[P(Ge(CH_3)_3)_3]$	Mn: MVol.D8-73/4
C_9HO_9OsRe	$(CO)_5ReOs(CO)_4H$	Re: Org.Comp.2-201, 210
$C_9H_2Na_{10}O_{23}Th$	$Na_{10}Th(OH)_2(C_2O_4)_3(CO_3)_3$	Th: SVol.C7-100, 102
–	$Na_{10}Th(OH)_2(C_2O_4)_3(CO_3)_3 \cdot n\ H_2O$ (n = 8 to 9)	Th: SVol.C7-100, 102
$C_9H_3K_3MoO_3$	$K_3[(HC{\equiv}C)_3Mo(CO)_3]$	Mo: Org.Comp.5-134
$C_9H_3MoO_3^{3-}$	$K_3[(HC{\equiv}C)_3Mo(CO)_3]$	Mo: Org.Comp.5-134
$C_9H_4N_2O_5Re^+$	$[(CO)_5ReC_4H_4N_2]^+$	Re: Org.Comp.2-152
$C_9H_5IMoN_3OS_2^-$	$[(C_5H_5)Mo(NO)(I)({-}SC(CN){=}C(CN)S{-})]^-$	Mo: Org.Comp.6-30
–	$[(C_5H_5)Mo(NO)(I)(S_2C{=}C(CN)_2)]^-$	Mo: Org.Comp.6-29/30
$C_9H_5INO_4Re$	cis-$(CO)_4Re(I)NC_5H_5$	Re: Org.Comp.1-435, 436
$C_9H_5KMoN_2O_2$	$K[(C_5H_5)Mo(CO)_2(CN)_2]$	Mo: Org.Comp.7-41/2
$C_9H_5KMoN_2O_2S_2$	$K[(C_5H_5)Mo(CO)_2(SCN)_2]$	Mo: Org.Comp.7-41/2
$C_9H_5MnNNaO_3S_2^+$	$[Mn(NC_9H_5(SO_3Na{-}5)(S{-}8))]^+$	Mn: MVol.D7-66/9
$C_9H_5MoNO_2S_2$	$(C_5H_5)Mo(CO)_2[{-}S{-}C(CN){-}S{-}]$	Mo: Org.Comp.7-188, 192
$C_9H_5MoNO_7S^{2-}$	$[MoO_3(OC_9H_5N(SO_3){-}5)]^{2-}$	Mo: SVol.B3b-161
$C_9H_5MoN_2O_2^-$	$[(C_5H_5)Mo(CO)_2(CN)_2]^-$	Mo: Org.Comp.7-41/2
$C_9H_5MoN_2O_2S_2^-$	$[(C_5H_5)Mo(CO)_2(SCN)_2]^-$	Mo: Org.Comp.7-41/2
$C_9H_5NO_4STh^{2+}$	$Th[5{-}(SO_3){-}8{-}(O){-}1{-}NC_9H_5]^{2+}$	Th: SVol.D1-105, 118/9
$C_9H_5N_5OS_4Si$	$Si(NCS)_4 \cdot O{-}NC_5H_5$	Si: SVol.B4-300/1
$C_9H_5N_5S_4Si$	$Si(NCS)_4 \cdot NC_5H_5$	Si: SVol.B4-300/1
$C_9H_5O_6Re$	$(CO)_5Re{-}C({=}O){-}C_3H_5{-}c$	Re: Org.Comp.2-120
–	$(CO)_5Re{-}CH{=}CH{-}C({=}O)CH_3$	Re: Org.Comp.2-127
$C_9H_6K_2O_{12}Th$	$K_2[Th(CH_2(COO)_2)_3]$	Th: SVol.C7-103
$C_9H_6MnNS^+$	$[Mn(NC_9H_6(S{-}8))]^+$	Mn: MVol.D7-67
$C_9H_6MoNO_5^{3-}$	$[MoO_4(OC_9H_6N)]^{3-}$	Mo: SVol.B3b-143/6, 160
$C_9H_6MoN_2O_5$	$(CO)_5Mo[CN{-}N{=}C(CH_3)_2]$	Mo: Org.Comp.5-10, 11
$C_9H_6MoO_5$	$(CH_2{=}CHCH{=}CH_2)Mo(CO)_5$	Mo: Org.Comp.5-168/9
$C_9H_6MoO_6$	$CH_2{=}CH{-}C(OCH_3){=}Mo(CO)_5$	Mo: Org.Comp.5-100, 102
$C_9H_6NOTh^{3+}$	$Th[8{-}(O){-}1{-}NC_9H_6]^{3+}$	Th: SVol.D1-105/6, 118
$C_9H_6O_5Re^+$	$[(CO)_5Re(CH_2{=}CHCH{=}CH_2)]^+$	Re: Org.Comp.2-350, 353
$C_9H_6O_5ReS^+$	$[(CO)_5ReSC_4H_6]^+$	Re: Org.Comp.2-158
$C_9H_6O_8Th$	$Th[1,2,3,4{-}(OOC)_4C_5H_6{-}c]$	Th: SVol.D1-74/6, 80/1
$C_9H_6O_{12}Th^{2-}$	$[Th(OOCCH_2COO)_3]^{2-}$	Th: SVol.D1-74/6, 77
$C_9H_7INO_4Sb$	$(5{-}C_9H_4N(I{-}7)OH{-}8)Sb(O)(OH)_2$	Sb: Org.Comp.5-307
$C_9H_7InO_3$	$8{-}In{-}2{-}OC_9H_7({=}O)_2{-}1,3$	In: Org.Comp.1-370, 376
$C_9H_7MnNO_3$	$Mn[{-}O{-}2{-}C_6H_4{-}CH{=}NCH_2C(O)O{-}]$	Mn: MVol.D6-34/5
$C_9H_7MnN_2OS^+$	$[Mn(1,3{-}N_2C_8H_4({=}S{-}2)(CH_3{-}3)({=}O{-}4))]^+$	Mn: MVol.D7-75/8

$C_9H_7MnN_3O_3$	Mn[-O-2-C$_6$H$_4$-CH=N-N=C(C(O)NH$_2$)O-]	Mn:MVol.D6-287
$C_9H_7MoNO_3$	(CH$_2$=CHC$_5$H$_4$)Mo(CO)$_2$(NO)	Mo:Org.Comp.7-6, 10, 22
$C_9H_7MoNO_4$	(C$_5$H$_5$)Mo(CO)$_2$[-O-C(=O)-CH=NH-]	Mo:Org.Comp.7-208/9, 238
$C_9H_7MoNO_5$	(CH$_3$OC(O)C$_5$H$_4$)Mo(CO)$_2$(NO)	Mo:Org.Comp.7-6, 10
$C_9H_7MoNO_5{}^{2-}$	[MoO$_3$(OH)(OC$_9$H$_6$N)]$^{2-}$	Mo:SVol.B3b-143/6, 160
$C_9H_7MoN_3O_5$	[-N(CH$_3$)-CH=N-N(CH$_3$)-]C=Mo(CO)$_5$	Mo:Org.Comp.5-100, 106
$C_9H_7MoO_2{}^+$	[(C$_5$H$_5$)Mo(CO)$_2$(HC≡CH)]$^+$	Mo:Org.Comp.8-186
$C_9H_7NO_5Re^+$	[(CO)$_5$Re(CN-C$_3$H$_7$-i)]$^+$	Re: Org.Comp.2-252
$C_9H_7N_2O_5Sb$	(5-C$_9$H$_5$N(NO$_2$-7))Sb(O)(OH)$_2$	Sb: Org.Comp.5-308
$C_9H_7N_3O_2S$	2-(O=S=N)-5-(C$_6$H$_4$CH$_3$-4)C$_2$N$_2$O-4,3,1	S: S-N Comp.6-170
$C_9H_7N_4PdS_3{}^-$	[Pd(SCN)$_3$(CH$_3$-C$_5$H$_4$N)]$^-$	Pd: SVol.B2-303
$C_9H_7O_2Th^{3+}$	[Th(C$_6$H$_5$CHCHCOO)]$^{3+}$	Th: SVol.D1-113
$C_9H_7O_6Re$	(CO)$_4$Re[-O=C(CH$_3$)-CH=C(CH$_3$)-O-]	Re: Org.Comp.1-355
–	(CO)$_5$Re-C(=O)-C$_3$H$_7$-i	Re: Org.Comp.2-118
$C_9H_7O_7Re$	(CO)$_5$ReCH$_2$C(O)OC$_2$H$_5$	Re: Org.Comp.2-107
$C_9H_8HgIMoNO_2$	(C$_5$H$_5$)Mo(CO)$_2$(HgI)(CN-CH$_3$)	Mo:Org.Comp.8-9, 11
$C_9H_8IMoNO_2$	[(C$_5$H$_5$)Mo(CO)$_2$(NCCH$_3$)(I)]	Mo:Org.Comp.7-57, 59
$C_9H_8MnN_3O_2S_2{}^+$	[Mn(4-NH$_2$-C$_6$H$_4$-SO$_2$=N-2-C$_3$H$_2$NS-3,1)]$^+$	Mn:MVol.D7-116/23
$C_9H_8MnN_4S_3$	[Mn(S=C(NH$_2$)NHC$_6$H$_5$)(NCS)$_2$] · 2 C$_2$H$_5$OH	Mn:MVol.D7-195/6
$C_9H_8MoNNaO_2$	Na[(C$_5$H$_5$)Mo(CO)$_2$(CN-CH$_3$)]	Mo:Org.Comp.8-1, 2, 6
$C_9H_8MoNO_2{}^-$	Na[(C$_5$H$_5$)Mo(CO)$_2$(CN-CH$_3$)]	Mo:Org.Comp.8-1, 2, 6
$C_9H_8MoNO_5{}^-$	[MoO$_2$(OH)$_2$(OC$_9$H$_6$N)]$^-$	Mo:SVol.B3b-143/6, 160
–	[MoO$_3$(H$_2$O)(OC$_9$H$_6$N)]$^-$	Mo:SVol.B3b-143
–	[MoO$_3$(OH)(HOC$_9$H$_6$N)]$^-$	Mo:SVol.B3b-145, 160
$C_9H_8MoO_2$	(C$_5$H$_5$)Mo(CO)$_2$(C-CH$_3$)	Mo:Org.Comp.8-33, 34/5
$C_9H_8MoO_2S_3$	(C$_5$H$_5$)Mo(CO)$_2$[-S-C(SCH$_3$)-S-]	Mo:Org.Comp.7-189, 197, 204
$C_9H_8MoO_3$	(C$_5$H$_5$)Mo(CO)$_2$-C(=O)CH$_3$	Mo:Org.Comp.8-7
$C_9H_8MoO_3S$	(C$_5$H$_5$)Mo(CO)$_2$[-O-C(CH$_3$)-S-]	Mo:Org.Comp.7-191
$C_9H_8MoO_3S_2$	(C$_5$H$_5$)Mo(CO)$_2$[-S-C(OCH$_3$)-S-]	Mo:Org.Comp.7-189, 197
$C_9H_8MoO_4$	(C$_5$H$_5$)Mo(CO)$_2$[-O-C(CH$_3$)-O-]	Mo:Org.Comp.7-188, 189
$C_9H_8MoO_6$	C$_2$H$_5$-O-C(CH$_3$)=Mo(CO)$_5$	Mo:Org.Comp.5-100, 103, 110
$C_9H_8NO_3Sb$	NC$_9$H$_6$-3-Sb(=O)(OH)$_2$	Sb: Org.Comp.5-307
–	NC$_9$H$_6$-5-Sb(=O)(OH)$_2$	Sb: Org.Comp.5-307
–	NC$_9$H$_6$-6-Sb(=O)(OH)$_2$	Sb: Org.Comp.5-307
–	NC$_9$H$_6$-7-Sb(=O)(OH)$_2$	Sb: Org.Comp.5-307
–	NC$_9$H$_6$-8-Sb(=O)(OH)$_2$	Sb: Org.Comp.5-307
$C_9H_8NO_4Sb$	2-HO-NC$_9$H$_5$-6-Sb(=O)(OH)$_2$	Sb: Org.Comp.5-308
–	8-HO-NC$_9$H$_5$-5-Sb(=O)(OH)$_2$	Sb: Org.Comp.5-307
$C_9H_8NO_5Re$	(CO)$_4$Re(NCCH$_3$)C(O)C$_2$H$_5$	Re: Org.Comp.1-472, 473/4
$C_9H_8NO_5ReSe_2$	(CO)$_4$Re[-Se=C(NC$_4$H$_8$O)-Se-]	Re: Org.Comp.1-365
$C_9H_8NO_6Re$	(CO)$_5$ReC(O)NHC$_3$H$_7$-i	Re: Org.Comp.2-117
$C_9H_8N_3O_5Re$	(CO)$_3$Re[-NH$_2$-CH(CH$_2$-N$_2$C$_3$H$_3$)-C(O)-O-]	Re: Org.Comp.1-118
$C_9H_8N_4O_6Re^+$	[(OC)$_3$(ON)Re-OC(O)CH(NH$_2$)CH$_2$-4-(1,3-N$_2$C$_3$H$_3$)]$^+$	Re: Org.Comp.1-137
$C_9H_8N_{10}O_{15}Th$	Th(NO$_3$)$_4$ · [4-(4-O$_2$N-C$_6$H$_4$-N=N)-5-NH$_2$ -1,2-N$_2$C$_3$H$_2$(=O)-3]	Th: SVol.D4-137
$C_9H_8O_6Re^+$	[(CO)$_5$ReC$_4$H$_8$O]$^+$	Re: Org.Comp.2-158
$C_9H_8O_6Sn$	C$_6$H$_5$Sn(OOCH)$_3$	Sn: Org.Comp.17-64
$C_9H_9LiMoO_3$	Li[(C$_5$H$_5$)Mo(CO)$_2$(-CH(CH$_3$)-O-)]	Mo:Org.Comp.8-112, 126, 164/5

$C_9H_9LiMoO_3$ ..... $Li[(C_5H_5)Mo(CO)_2(-CH(CD_3)-O-)]$ Mo:Org.Comp.8-112, 126
– ..... $Li[(C_5H_5)Mo(CO)_2(-CD(CH_3)-O-)]$ Mo:Org.Comp.8-112, 126
– ..... $Li[(C_5H_5)Mo(CO)_2(CH_3)-CH{=}O]$ Mo:Org.Comp.8-176, 177, 183
– ..... $Li[(C_5H_5)Mo(CO)_2(CH_3)-CD{=}O]$ Mo:Org.Comp.8-176, 177, 183
– ..... $Li[(C_5H_5)Mo(CO)_2(CD_3)-CH{=}O]$ Mo:Org.Comp.8-176, 177, 183
– ..... $Li[(C_5H_5)Mo(CO)_2(H)-C(=O)CH_3]$ Mo:Org.Comp.8-3, 6
– ..... $Li[(C_5H_5)Mo(CO)_2(H)-C(=O)CD_3]$ Mo:Org.Comp.8-3, 6
– ..... $Li[(C_5H_5)Mo(CO)_2(D)-C(=O)CH_3]$ Mo:Org.Comp.8-3, 6
$C_9H_9Li_2O_6Re$ .... $Li_2[(CO)_3Re(C(O)CH_3)_3]$ .... Re: Org.Comp.1-123
$C_9H_9MnNS_4$ ..... $[Mn(SCH{=}CHS)_2(NC_5H_5)]$ Mn:MVol.D7-33/8
$C_9H_9MnNS_4^-$ ..... $[Mn(SCH{=}CHS)_2(NC_5H_5)]^-$ Mn:MVol.D7-33/8, 45/6
$C_9H_9MnNS_4^{2-}$ .... $[Mn(SCH{=}CHS)_2(NC_5H_5)]^{2-}$ Mn:MVol.D7-33/8
$C_9H_9MnN_3O_2S_2^{2+}$ $[Mn(4-NH_2-C_6H_4-SO_2-NH-2-C_3H_2NS-3,1)]^{2+}$ Mn:MVol.D7-116/23
$C_9H_9MnN_3O_3S_2$ .. $[Mn(-5-S-1,3,4-SN_2C_2-2-N{=}CH-C_6H_4-2-O-)(H_2O)_2]$
Mn:MVol.D6-39, 41/2
$C_9H_9MnN_4O_2S_2^+$ $[Mn(4-NH_2-C_6H_4-SO_2{=}N-2-C_2N_2S-4,3,1-CH_3-5)]^+$
Mn:MVol.D7-116/23
$C_9H_9MnO_2S^+$ .... $[Mn(4-CH_3-C_6H_4-S-CH_2-COO)]^+$ Mn:MVol.D7-82/3
– ..... $[Mn(C_6H_5-CH_2-S-CH_2-COO)]^+$ Mn:MVol.D7-82/3
$C_9H_9MnO_3S^+$ .... $[Mn(2-CH_3O-C_6H_4-S-CH_2-COO)]^+$ Mn:MVol.D7-82/3
– ..... $[Mn(3-CH_3O-C_6H_4-S-CH_2-COO)]^+$ Mn:MVol.D7-82/3
– ..... $[Mn(C_6H_5-CH_2-S(=O)-CH_2-COO)]^+$ Mn:MVol.D7-107
$C_9H_9MnO_3Se^+$ ... $[Mn(2-CH_3O-C_6H_4-Se-CH_2-COO)]^+$ Mn:MVol.D7-246
$C_9H_9MoNO_2$ ..... $(C_5H_5)Mo(CO)_2(H)(CN-CH_3)$ Mo:Org.Comp.8-9/10
$C_9H_9MoNO_2S_2$ ... $(C_5H_5)Mo(CO)_2[-S-C(NHCH_3)-S-]$ Mo:Org.Comp.7-188, 193/4, 203
$C_9H_9MoNO_4$ ..... $(C_5H_5)Mo(CO)_2[-O-C(=O)-CH_2-NH_2-]$ Mo:Org.Comp.7-208, 209, 230
$C_9H_9MoNO_5$ ..... $MoO(OH)_3(OC_9H_6N)$ Mo:SVol.B3b-143/6, 160
$C_9H_9MoNO_5Si$.... $(CO)_5Mo[CN-Si(CH_3)_3]$ Mo:Org.Comp.5-6, 8/9
$C_9H_9MoNO_5Sn$ ... $(CH_3)_3Sn-N{\equiv}C-Mo(CO)_5$ Mo:Org.Comp.5-6, 10
Sn: Org.Comp.18-127/8
$C_9H_9MoN_3O_3$ .... $(CO)_3Mo(CN-CH_3)_3$ Mo:Org.Comp.5-34, 35, 37
$C_9H_9MoO_3^-$ ...... $Li[(C_5H_5)Mo(CO)_2(-CH(CH_3)-O-)]$ Mo:Org.Comp.8-112, 126, 164/5
– ..... $Li[(C_5H_5)Mo(CO)_2(-CD(CH_3)-O-)]$ Mo:Org.Comp.8-112, 126
– ..... $Li[(C_5H_5)Mo(CO)_2(-CH(CD_3)-O-)]$ Mo:Org.Comp.8-112, 126
– ..... $[Li \cdot (-CH_2CH_2-O-)_4][(C_5H_5)Mo(CO)_2(-CH(CH_3)-O-)]$
Mo:Org.Comp.8-112, 126
– ..... $Li[(C_5H_5)Mo(CO)_2(CH_3)-CH{=}O]$ Mo:Org.Comp.8-176, 177, 183
– ..... $Li[(C_5H_5)Mo(CO)_2(CH_3)-CD{=}O]$ Mo:Org.Comp.8-176, 177, 183
– ..... $Li[(C_5H_5)Mo(CO)_2(CD_3)-CH{=}O]$ Mo:Org.Comp.8-176, 177, 183
– ..... $Li[(C_5H_5)Mo(CO)_2(H)-C(=O)CH_3]$ Mo:Org.Comp.8-3, 6
– ..... $Li[(C_5H_5)Mo(CO)_2(D)-C(=O)CH_3]$ Mo:Org.Comp.8-3, 6
– ..... $Li[(C_5H_5)Mo(CO)_2(H)-C(=O)CD_3]$ Mo:Org.Comp.8-3, 6
$C_9H_9NO_5S^-$ ...... $[4-C_2H_5OC(O)-C_6H_4-N(O)-SO_2]^-$, radical anion
S: S-N Comp.8-326/8
$C_9H_9NO_5SnW$ .... $(CH_3)_3SnNC-W(CO)_5$ .... Sn: Org.Comp.18-127/8
$C_9H_9NO_6Re$ ..... $(CO)_5ReN(O)C_4H_9$-t, radical .... Re: Org.Comp.2-170
– ..... $(CO)_5ReONC_4H_9$-t, radical .... Re: Org.Comp.2-171
$C_9H_9N_2O_3Sb$..... $(5-C_9H_5N(NH_2-7))Sb(O)(OH)_2$ .... Sb: Org.Comp.5-308

$C_9H_9N_2O_5S_2Sb$	$(4\text{-}(2'\text{-}C_3H_2SNNHSO_2)C_6H_4)Sb(O)(OH)_2$	Sb: Org.Comp.5-288
$C_9H_9N_3O_2S_2$	$2\text{-}(O{=}S{=}N\text{-}N{=})\text{-}3\text{-}(CH_3)\text{-}6\text{-}(OCH_3)C_7H_3NS\text{-}3,1$	S: S-N Comp.6-69/71
$C_9H_9N_3O_3Re^+$	$[(CO)_3Re(CN\text{-}CH_3)_3]^+$	Re: Org.Comp.2-276
–	$[(CO)_3Re(NC\text{-}CH_3)_3]^+$	Re: Org.Comp.1-308/9, 312
$C_9H_9N_4Sn$	$(CH_3)_3SnN{=}C{=}C(CN)C(CN)_2$, radical	Sn: Org.Comp.18-109, 117
$C_9H_9N_6Re^-$	$[Re(CNCH_3)_3(CN)_3]^-$	Re: Org.Comp.2-273
$C_9H_9N_9O_{13}Th$	$Th(NO_3)_4 \cdot [4\text{-}(C_6H_5\text{-}N{=}N)\text{-}5\text{-}NH_2\text{-}1,2\text{-}N_2C_3H_2({=}O)\text{-}3]$	Th: SVol.D4-136
$C_9H_9O_5Re$	$(CO)_5ReC_4H_9\text{-}i$	Re: Org.Comp.2-107
$C_9H_9O_6Re^{2-}$	$[(CO)_3Re(C(O)CH_3)_3]^{2-}$	Re: Org.Comp.1-123/4
$C_9H_{10}IMoNO_4$	$(C_2H_5)_2N\text{-}CMo(CO)_4(I)$	Mo: Org.Comp.5-97/100
$C_9H_{10}MnNOS^+$	$[Mn(SCH_2C({=}O)NHC_6H_4CH_3\text{-}4)]^+$	Mn: MVol.D7-54
$C_9H_{10}MnN_3O_2S^+$	$[Mn(C_2H_5\text{-}OOC\text{-}NC({=}S)NH\text{-}2\text{-}C_5H_4N)]^+$	Mn: MVol.D7-198
$C_9H_{10}MoNO_2{}^+$	$[C_5H_5Mo(CO)(NO)C_3H_5]^+$	Mo: Org.Comp.6-346, 351/2
$C_9H_{10}MoNO_8{}^{3-}$	$[HMoO_4H(3,4\text{-}(O)_2C_6H_3CH_2CH(NH_2)COO)]^{3-}$	Mo: SVol.B3b-150/6
–	$[MoO_4H_2(3,4\text{-}(O)_2C_6H_3CH_2CH(NH_2)COO)]^{3-}$	Mo: SVol.B3b-150/6
$C_9H_{10}MoN_2OS$	$C_5H_5Mo(NO)(NCS)C_3H_5$	Mo: Org.Comp.6-174
$C_9H_{10}MoN_2O_2$	$(C_5H_5)Mo(CO)(NO)\text{-}CN\text{-}C_2H_5$	Mo: Org.Comp.6-262
–	$(C_5H_5)Mo(NO)(NCO)(CH_2CHCH_2)$	Mo: Org.Comp.6-174
$C_9H_{10}MoN_2O_5S$	$[Mo(CO)_5(C_2H_5\text{-}N{=}S{=}N\text{-}C_2H_5)]$	S: S-N Comp.7-300, 302/3
$C_9H_{10}MoO$	$C_5H_5Mo(CO)(HC{\equiv}CH)CH_3$	Mo: Org.Comp.6-312
$C_9H_{10}MoO_2$	$(C_5H_5)Mo(CO)_2(H)(CH_2{=}CH_2)$	Mo: Org.Comp.8-187, 197
–	$(C_5H_5)Mo(CO)_2\text{-}C_2H_5$	Mo: Org.Comp.8-8
$C_9H_{10}MoO_2S$	$(C_5H_5)Mo(CO)_2[\text{-}CH_2\text{-}S(CH_3)\text{-}]$	Mo: Org.Comp.8-128, 165/7
$C_9H_{10}NO_2Th^{3+}$	$[Th(C_6H_5\text{-}CH(NH_2)\text{-}CH_2\text{-}COO)]^{3+}$	Th: SVol.D1-82/3, 85
–	$[Th(C_6H_5\text{-}CH_2\text{-}CH(NH_2)\text{-}COO)]^{3+}$	Th: SVol.D1-82/4
$C_9H_{10}NO_3Re$	$2\text{-}CH_3C_3H_4Re(CO)_3CNCH_3$	Re: Org.Comp.2-376
$C_9H_{10}NO_4ReS_2$	$(CO)_4Re[\text{-}S{=}C(N(C_2H_5)_2)\text{-}S\text{-}]$	Re: Org.Comp.1-361/2
$C_9H_{10}NO_4ReSe_2$	$(CO)_4Re[\text{-}Se{=}C(N(C_2H_5)_2)\text{-}Se\text{-}]$	Re: Org.Comp.1-365
$C_9H_{10}NO_5Re$	$(CO)_4Re[C(O)CH_3]C(CH_3){=}NHCH_3$	Re: Org.Comp.1-390/1
$C_9H_{10}N_2O_5SW$	$[W(CO)_5(C_2H_5\text{-}N{=}S{=}N\text{-}C_2H_5)]$	S: S-N Comp.7-300/5
$C_9H_{10}N_4OS$	$2\text{-}(O{=}S{=}N\text{-}N{=})\text{-}1,3\text{-}(CH_3)_2C_7H_4N_2\text{-}1,3$	S: S-N Comp.6-69/71
$C_9H_{11}IMoN_2O_2$	$(CH_2CHCH_2)Mo(I)(CO)_2(NC\text{-}CH_3)_2$	Mo: Org.Comp.5-235, 236/7, 242/3
$C_9H_{11}I_2In$	$2,4,6\text{-}(CH_3)_3\text{-}C_6H_2\text{-}In(I)_2$	In: Org.Comp.1-163, 165/6
$C_9H_{11}I_2MnOP$	$Mn(CO)[P(C_6H_5)(CH_3)_2]I_2$	Mn: MVol.D8-47
$C_9H_{11}I_2Mo$	$C_5H_5Mo(I)_2H_2C{=}CHCH{=}CH_2$	Mo: Org.Comp.6-189/90
$C_9H_{11}InN_2$	$1\text{-}(CH_3)_2In\text{-}1,3\text{-}N_2C_7H_5$	In: Org.Comp.1-276, 279
–	$2\text{-}(CH_3)_2In\text{-}1,2\text{-}N_2C_7H_5$	In: Org.Comp.1-276, 279
$C_9H_{11}InO_2$	$(CH_3)_2In[\text{-}O\text{-}CH\text{-}C_6H_4\text{-}2\text{-}O\text{-}]$	In: Org.Comp.1-190, 192, 195
–	$(CH_3)_2In[\text{-}O\text{-}C_7H_5\text{-}2\text{-}O\text{-}]$	In: Org.Comp.1-190, 191
$C_9H_{11}InS_2$	$(C_6H_5)In[\text{-}S\text{-}CH_2CH_2CH_2\text{-}S\text{-}]$	In: Org.Comp.1-246, 248
$C_9H_{11}Li_2MoO_2P$	$Li_2[(C_5H_5)Mo(CO)_2P(CH_3)_2]$	Mo: Org.Comp.7-41
$C_9H_{11}MnNO_6P_2{}^{2-}$	$[Mn((O_3P)_2CH\text{-}N(CH_3)\text{-}CH_2\text{-}C_6H_5)]^{2-}$	Mn: MVol.D8-126
–	$[Mn((O_3P\text{-}CH_2)_2N\text{-}CH_2\text{-}C_6H_5)]^{2-}$	Mn: MVol.D8-129/30
$C_9H_{11}MnN_3O_5$	$[Mn(\text{-}O\text{-}2\text{-}C_6H_4\text{-}CH{=}N\text{-}N{=}C(C(O)NH_2)O\text{-})(H_2O)_2]$	Mn: MVol.D6-287
$C_9H_{11}MoNO_2$	$(C_5H_5)Mo(CO)(NO)(CH_2{=}CHCH_3)$	Mo: Org.Comp.6-292
–	$(C_5H_5)Mo(CO)(NO)(CH_2{=}CHCH_2D)$	Mo: Org.Comp.6-293
$C_9H_{11}MoNO_3$	$C_5H_5Mo(CO)(NO)CH_2{=}CHCH_2OH$	Mo: Org.Comp.6-294

$C_9H_{11}MoNO_3S_4$ .. $C_5H_5Mo(NO)(SC(S)OCH_3)S_2COCH_3$ ... Mo:Org.Comp.6-49
$C_9H_{11}MoNO_8{}^{2-}$ ... $[HMoO_4H_2(3,4\text{-}(O)_2C_6H_3CH_2CH(NH_2)COO)]^{2-}$ Mo:SVol.B3b-150/6
– ... $[MoO_4H_3(3,4\text{-}(O)_2C_6H_3CH_2CH(NH_2)COO)]^{2-}$ ... Mo:SVol.B3b-150/6
$C_9H_{11}MoN_2O_2Sb$ $(C_5H_5)Mo(CO)_2(N_2)\text{-}Sb(CH_3)_2$ ... Mo:Org.Comp.7-140
$C_9H_{11}MoO_2P^{2-}$ ... $[(C_5H_5)Mo(CO)_2P(CH_3)_2]^{2-}$ ... Mo:Org.Comp.7-41
$C_9H_{11}MoO_2Sb$ ... $(C_5H_5)Mo(CO)_2Sb(CH_3)_2$ ... Mo:Org.Comp.7-2, 37
$C_9H_{11}MoO_4PS_2$ .. $(C_5H_5)Mo(CO)_2[\text{-}S\text{-}P(OCH_3)_2\text{-}S\text{-}]$ ... Mo:Org.Comp.7-189, 200
$C_9H_{11}MoO_4S_4{}^-$ ... $[(CH_2CHCH_2)Mo(CO)_2(SC(=S)\text{-}OCH_3)_2]^-$ ... Mo:Org.Comp.5-283, 287
$C_9H_{11}NOS$ ... $O=S=N\text{-}CH_2CH_2CH_2\text{-}C_6H_5$ ... S: S-N Comp.6-109
– ... $O=S=N\text{-}C_6H_2(CH_3)_3\text{-}2,4,6$ ... S: S-N Comp.6-157/8, 205, 242/4, 247
$C_9H_{11}NO_3Re^+$ ... $[(C_5H_5)Re(CO)(NO)(O=C(CH_3)_2)][PF_6]$ ... Re: Org.Comp.3-153, 154/5
$C_9H_{11}NO_4SSn$ ... $2\text{-}[(CH_3)_2Sn(OH)]\text{-}1,2\text{-}SNC_7H_4(=O)_3\text{-}1,1,3 \cdot H_2O$ Sn: Org.Comp.19-130, 141
$C_9H_{11}NO_7Th$ ... $Th(OH)_3[2\text{-}(OOCCH_2NH)C_6H_4COOH]$ ... Th: SVol.C7-146/8
$C_9H_{11}NS_2$ ... $S=S=NC_6H_2(CH_3)_3\text{-}2,4,6$ ... S: S-N Comp.6-322
$C_9H_{11}N_2O_6Re$ ... $[\text{-}O\text{-}C(O)\text{-}CH(CH(OH)CH_3)\text{-}NH_2\text{-}]Re(CO)_3(NCCH_3)$ Re: Org.Comp.1-126
$C_9H_{11}N_2O_8Os^-$ ... $[2,4,6\text{-}(CH_3)_3C_6H_2\text{-}N_2][2,4,6\text{-}(CH_3)_3C_6H_2\text{-}Os(O)_2(ONO_2)_2]$ ... Os: Org.Comp.A1-12/3, 38
$C_9H_{11}N_4O_6Sb$ ... $3\text{-}O_2N\text{-}4\text{-}CH_3C(=NNHCONH_2)C_6H_3Sb(O)(OH)_2$ Sb: Org.Comp.5-299
$C_9H_{11}N_6O_3Sb$ ... $4\text{-}[\text{-}NC(NH_2)NC(NH_2)N=]CNHC_6H_4Sb(O)(OH)_2$ Sb: Org.Comp.5-290
$C_9H_{11}O_5ReSi$ ... $(CO)_5ReCH_2Si(CH_3)_3$ ... Re: Org.Comp.2-106
$C_9H_{11}O_5Sb$ ... $3\text{-}[C_2H_5\text{-}OC(=O)]\text{-}C_6H_4\text{-}Sb(=O)(OH)_2$ ... Sb: Org.Comp.5-294
– ... $4\text{-}[C_2H_5\text{-}OC(=O)]\text{-}C_6H_4\text{-}Sb(=O)(OH)_2 \cdot H_2O$ .. Sb: Org.Comp.5-294
$C_9H_{11}O_6Sb$ ... $(4\text{-}CH_3O_2CCH_2OC_6H_4)Sb(O)(OH)_2$ ... Sb: Org.Comp.5-287
$C_9H_{12}IMoNO_3$ ... $(C_5H_5)Mo(NO)(I)O_2C\text{-}C_3H_7\text{-}n$ ... Mo:Org.Comp.6-46
– ... $(C_5H_5)Mo(NO)(I)O_2C\text{-}C_3H_7\text{-}i$ ... Mo:Org.Comp.6-46
$C_9H_{12}I_2MnN_3P$ ... $Mn[P(CH_2CH_2CN)_3]I_2$ ... Mn:MVol.D8-69/70
$C_9H_{12}LiMoO_2P$ ... $Li[(C_5H_5)Mo(CO)_2(HP(CH_3)_2)]$ ... Mo:Org.Comp.7-47/8, 50
$C_9H_{12}MnNO_6P_2{}^-$ $[MnH((O_3P)_2CH\text{-}N(CH_3)\text{-}CH_2\text{-}C_6H_5)]^-$ ... Mn:MVol.D8-126
– ... $[MnH((O_3P\text{-}CH_2)_2N\text{-}CH_2\text{-}C_6H_5)]^-$ ... Mn:MVol.D8-129/30
$C_9H_{12}MoNO_7{}^{3-}$ ... $[HMoO_4H(3,4\text{-}(O)_2C_6H_3CH(O)CH_2NHCH_3)]^{3-}$ .. Mo:SVol.B3b-150/6
– ... $[MoO_4H_2(3,4\text{-}(O)_2C_6H_3CH(O)CH_2NHCH_3)]^{3-}$ ... Mo:SVol.B3b-150/6
$C_9H_{12}MoO_2P^-$ ... $[(C_5H_5)Mo(CO)_2(HP(CH_3)_2)]^-$ ... Mo:Org.Comp.7-47/8, 50
$C_9H_{12}NO_4PRe^+$ .. $[(CO)_4Re(P(CH_3)_3)(NC\text{-}CH_3)]^+$ ... Re: Org.Comp.1-482
$C_9H_{12}NO_4Re$ ... $(CO)_4Re[\text{-}CH_2\text{-}N(C_2H_5)_2\text{-}]$ ... Re: Org.Comp.1-408
$C_9H_{12}NO_4Sb$ ... $4\text{-}[(CH_3)_2N\text{-}C(=O)]\text{-}C_6H_4\text{-}Sb(=O)(OH)_2$ ... Sb: Org.Comp.5-294
– ... $4\text{-}[C_2H_5\text{-}O\text{-}C(=NH)]\text{-}C_6H_4\text{-}Sb(=O)(OH)_2 \cdot HCl$ Sb: Org.Comp.5-294
$C_9H_{12}NO_5Sb$ ... $2\text{-}CH_3CONH\text{-}4\text{-}CH_3O\text{-}C_6H_3\text{-}Sb(=O)(OH)_2$ ... Sb: Org.Comp.5-296
– ... $2\text{-}CH_3CONH\text{-}5\text{-}CH_3O\text{-}C_6H_3\text{-}Sb(=O)(OH)_2$ ... Sb: Org.Comp.5-296
– ... $2\text{-}CH_3CONH\text{-}4\text{-}HOCH_2\text{-}C_6H_3\text{-}Sb(=O)(OH)_2$ ... Sb: Org.Comp.5-298
– ... $4\text{-}CH_3CONH\text{-}3\text{-}HOCH_2\text{-}C_6H_3\text{-}Sb(=O)(OH)_2$ ... Sb: Org.Comp.5-298
– ... $4\text{-}[C_2H_5\text{-}OC(=O)\text{-}NH]\text{-}C_6H_4\text{-}Sb(=O)(OH)_2$ ... Sb: Org.Comp.5-290
– ... $4\text{-}[HO\text{-}CH_2CH_2\text{-}NH\text{-}C(=O)]\text{-}C_6H_4\text{-}Sb(=O)(OH)_2$ Sb: Org.Comp.5-294
– ... $4\text{-}[H_2N\text{-}CH(COOH)\text{-}CH_2]\text{-}C_6H_4\text{-}Sb(=O)(OH)_2$ .. Sb: Org.Comp.5-295
$C_9H_{12}N_2O_5ReSSi^+$ $[(CO)_5ReNSN(CH_3)Si(CH_3)_3]^+$ ... Re: Org.Comp.2-154

$C_9H_{12}N_3O_5Re$	$[-O-C(O)-CH(CH_2CH_2NH_2)-NH_2-]Re(CO)_3(NCCH_3)$	Re: Org.Comp.1-126
$C_9H_{12}N_3O_5Sb$	$3-HO-4-CH_3C(=NNHCONH_2)-C_6H_3Sb(O)(OH)_2$	Sb: Org.Comp.5-296
$C_9H_{12}N_3O_5Th^{3+}$	$[Th(C_9H_{12}N_3O_5)]^{3+}$	Th: SVol.D1-120
$C_9H_{12}N_3O_8Sb$	$(CH_3)_3Sb(OH)OC_6H_2(NO_2)_3-2,4,6$	Sb: Org.Comp.5-39
$C_9H_{12}N_4O_4S_2Si$	$(CH_3)_3SiN=S=NS-C_6H_3(NO_2)_2-2,4$	S: S-N Comp.7-146
$C_9H_{12}N_6O_4S_4Si$	$(CH_3)_3SiN=S=NSN=S=NSC_6H_3(NO_2)_2-2,4$	S: S-N Comp.7-149
$C_9H_{12}O_2Sn$	$(CH_2=CHCH_2)(C_6H_5)Sn(OH)_2$	Sn: Org.Comp.16-213, 217
$C_9H_{12}O_4Sn$	$C_6H_5Sn(OH)_2OOCC_2H_5$	Sn: Org.Comp.17-76
$C_9H_{12}O_5PRe$	$(CO)_4Re[-P(CH_3)_2-O-CH_2-CH_2-CH_2-]$	Re: Org.Comp.1-414
$C_9H_{13}In$	$In(C_5H_4-C_4H_9-t)$	In: Org.Comp.1-379, 380
$C_9H_{13}MnNO_6P_2$	$[Mn(H_2(O_3PCH_2)_2NCH_2C_6H_5)]$	Mn: MVol.D8-129/30
$C_9H_{13}MoNO_7^{2-}$	$[HMoO_4H_2(3,4-(O)_2C_6H_3CH(O)CH_2NHCH_3)]^{2-}$	Mo: SVol.B3b-150/6
–	$[MoO_4H_3(3,4-(O)_2C_6H_3CH(O)CH_2NHCH_3)]^{2-}$	Mo: SVol.B3b-150/6
$C_9H_{13}MoN_2O_2^+$	$[(C_5H_5)Mo(CO)_2(NH_2-CH_2CH_2-NH_2)]^+$	Mo: Org.Comp.7-249, 251
$C_9H_{13}MoN_2O_2S_4^-$	$[(CH_2CHCH_2)Mo(CO)_2(SC(=S)-NHCH_3)_2]^-$	Mo: Org.Comp.5-283, 287
$C_9H_{13}MoOP$	$C_5H_5Mo(CO)(PH_3)C_3H_5$	Mo: Org.Comp.6-331
$C_9H_{13}MoO_2P$	$(C_5H_5)Mo(CO)_2(HP(CH_3)_2)(H)$	Mo: Org.Comp.7-55, 57, 73
$C_9H_{13}NOS_2$	$(CH_3)_2N-S(O)-S-CH_2C_6H_5$	S: S-N Comp.8-332/3
$C_9H_{13}NO_2S$	$(CH_3)_2N-S(O)O-CH_2C_6H_5$	S: S-N Comp.8-306/7, 310
–	$C_6H_5-N(CH_3)-S(O)O-C_2H_5$	S: S-N Comp.8-320
$C_9H_{13}NO_5S^-$	$[c-C_6H_{10}=C(OC(O)CH_3)-N(O)-SO_2]^-$, radical anion	S: S-N Comp.8-326/8
$C_9H_{13}N_2O_2S_2^+$	$[(CH_3)_2NS=NS(O)_2C_6H_4-CH_3-4]^+$	S: S-N Comp.7-332/3
$C_9H_{13}N_2O_4Sb$	$(3-CH_3NHC(O)CH_2NHC_6H_4)Sb(O)(OH)_2$	Sb: Org.Comp.5-289
$C_9H_{13}N_2O_5Re$	$(CO)_4Re(NH_2CH_3)C(O)NH-C_3H_7-i$	Re: Org.Comp.1-470
–	$(CO)_4Re[NH(CH_3)_2]C(O)N(CH_3)_2$	Re: Org.Comp.1-471
$C_9H_{13}N_2O_5Sb$	$2-n-C_3H_7NH-5-O_2N-C_6H_3Sb(O)(OH)_2$	Sb: Org.Comp.5-297
$C_9H_{13}N_2O_7Re$	$[-O-C(O)-CH(CH_3)-NH_2-]Re(CO)_3-NH_2-CH(CH_3)COOH$	Re: Org.Comp.1-126
$C_9H_{13}N_2O_8SSb$	$(CH_3)_3Sb(OH)OSO_2C_6H_3(NO_2)_2-2,4$	Sb: Org.Comp.5-39
$C_9H_{13}N_3O_2S_2Si$	$(CH_3)_3SiN=S=NS-C_6H_4-NO_2-2$	S: S-N Comp.7-145
–	$(CH_3)_3SiN=S=NS-C_6H_4-NO_2-4$	S: S-N Comp.7-145/6
$C_9H_{13}N_3Sn$	$(CH_3)_3SnN(N=NC_6H_4-1,2)$	Sn: Org.Comp.18-83/4, 90
$C_9H_{13}N_4O_4Sb$	$3-H_2N-4-CH_3C(=NNHCONH_2)C_6H_3Sb(O)(OH)_2$	Sb: Org.Comp.5-297
$C_9H_{13}N_5O_2S_4Si$	$(CH_3)_3SiN=S=NSN=S=NS-C_6H_4-NO_2-2$	S: S-N Comp.7-149
$C_9H_{13}O_2ReSi$	$(C_5H_5)Re(H)(CO)_2-SiH(CH_3)_2$	Re: Org.Comp.3-175/6, 178
$C_9H_{14}I_3MoO_4P$	$C_5H_5Mo(CO)(P(OCH_3)_3)I_3$	Mo: Org.Comp.6-219
$C_9H_{14}InN$	$(CH_3)_2In-N(CH_3)-C_6H_5$	In: Org.Comp.1-252/5, 256, 261/2
$C_9H_{14}InNO_2$	$(CH_3)_2In[-OC(CH_3)O-] \cdot NC_5H_5$	In: Org.Comp.1-223/4, 225
$C_9H_{14}LiNSn$	$(CH_3)_3SnN(C_6H_5)Li$	Sn: Org.Comp.18-67, 74
$C_9H_{14}MnN_4O_2$	$Mn[(CH_3)_2C=N-N=C(O-)CH_2C(O-)=N-N=C(CH_3)_2]$	Mn: MVol.D6-313/4
$C_9H_{14}MnN_4O_7S$	$[Mn(S=C(NH_2)NHC_6H_5)(NO_3)_2] \cdot C_2H_5OH$	Mn: MVol.D7-195/6
$C_9H_{14}MnN_4S_8Zn$	$MnZn(SC(=S)NHCH_2CH_2NHC(=S)S)(SC(=S)NHCH_2CH_2CH_2NHC(=S)S)$	Mn: MVol.D7-180/4
$C_9H_{14}MoNO_2P$	$C_5H_5Mo(P(CH_3)_3)(NO)CO$	Mo: Org.Comp.6-237, 248
$C_9H_{14}MoNO_5P$	$C_5H_5Mo(P(OCH_3)_3)(NO)CO$	Mo: Org.Comp.6-243
$C_9H_{14}MoN_2OS_4$	$(HC{\equiv}CH)Mo(CO)[S_2C-N(CH_3)_2]_2$	Mo: Org.Comp.5-145/8

$C_9H_{14}MoN_2O_2$ ... $C_5H_5Mo(NO)_2C_4H_9$-i ... Mo:Org.Comp.6-95, 96
$C_9H_{14}NO_2PRe^+$ .. $[(C_5H_5)Re(CO)(NO)(P(CH_3)_3)][BF_4]$ ... Re: Org.Comp.3-154, 155/6
$C_9H_{14}NO_2ReSi$ ... $(C_5H_5)Re(CO)(NO)$-$Si(CH_3)_3$ ... Re: Org.Comp.3-148, 150
$C_9H_{14}NO_4Sb$ ... (4-$HOCH_2CH_2CH_2NHC_6H_4)Sb(O)(OH)_2$ ... Sb: Org.Comp.5-289
$C_9H_{14}N_2O_3ReS^+$. $[(CO)_3Re(1,4,7\text{-}SN_2C_6H_{14}\text{-}c)]^+$ ... Re: Org.Comp.1-219
$C_9H_{14}N_2SSi$ ... $(CH_3)_3SiN{=}S{=}N\text{-}C_6H_5$ ... S: S-N Comp.7-141
$C_9H_{14}N_2S_2Si$ ... $(CH_3)_3SiN{=}S{=}NS\text{-}C_6H_5$ ... S: S-N Comp.7-144
$C_9H_{14}N_4S_4Sb_2$ ... $[(CH_3)_2Sb(NCS)_2]_2CH_2$ ... Sb: Org.Comp.5-126
$C_9H_{14}O_3Sn$ ... $C_6H_5Sn(OCH_3)_3$ ... Sn: Org.Comp.17-62
$C_9H_{14}O_4PRe^{2+}$ ... $[(C_5H_5)Re(CO)(P(OCH_3)_3)]Br_2$ ... Re: Org.Comp.3-143, 145
$C_9H_{14}O_4Th^{2+}$ ... $[Th(OOCC_7H_{14}COO)]^{2+}$ ... Th: SVol.D1-74/6, 78
$C_9H_{15}IMoNO_3PS_2$ $C_5H_5Mo(NO)(I)S_2P(OC_2H_5)_2$ ... Mo:Org.Comp.6-47/8
$C_9H_{15}I_3In_2$ ... $(CH_2{=}CH\text{-}CH_2)_2In(\text{-}I\text{-})_2InI\text{-}CH_2\text{-}CH{=}CH_2$ ... In: Org.Comp.1-167, 170/1
$C_9H_{15}In$ ... $In(C_3H_5\text{-}c)_3$ ... In: Org.Comp.1-73, 77, 82
– ... c-C_5H_5-$In(C_2H_5)_2$ ... In: Org.Comp.1-101, 105
$C_9H_{15}KMnO_3S_6$ .. $K[Mn(SC({=}S)OC_2H_5)_3]$ ... Mn:MVol.D7-130/3
$C_9H_{15}MnO_3S_6$ ... $Mn(SC({=}S)OC_2H_5)_3$ ... Mn:MVol.D7-130/3
$C_9H_{15}MnO_3S_6^-$ ... $[Mn(SC({=}S)OC_2H_5)_3]^-$ ... Mn:MVol.D7-130/3
$C_9H_{15}MnS_9$ ... $Mn((S_2C)SC_2H_5)_3$ ... Mn:MVol.D7-210
$C_9H_{15}NO_2SSn$ ... $(CH_3)_3SnNHSO_2C_6H_5$ ... Sn: Org.Comp.18-18, 21
$C_9H_{15}NO_3Sb^+$ ... $[3\text{-}(CH_3)_3N\text{-}C_6H_4\text{-}Sb({=}O)(OH)_2]^+$ ... Sb: Org.Comp.5-289
– ... $[4\text{-}(CH_3)_3N\text{-}C_6H_4\text{-}Sb({=}O)(OH)_2]^+$ ... Sb: Org.Comp.5-289
$C_9H_{15}NSn$ ... $(CH_3)_3SnNHC_6H_5$ ... Sn: Org.Comp.18-15, 17, 21
$C_9H_{15}N_3O_3Re^+$ ... $[(CO)_3Re(c\text{-}1,4,7\text{-}N_3C_6H_{15})]^+$ ... Re: Org.Comp.1-218
$C_9H_{15}N_3O_4Sn$ ... $(CH_3)_3SnN(N{=}C(COOCH_3)C(COOCH_3){=}N)$ ... Sn: Org.Comp.18-84, 91
$C_9H_{15}O_6Th^+$ ... $[Th(OOCC_2H_5)_3]^+$ ... Th: SVol.D1-68
$C_9H_{15}O_9Th^+$ ... $[Th(OOCCH(OH)CH_3)_3]^+$ ... Th: SVol.D1-71
$C_9H_{15}PSn$ ... $(CH_3)_3Sn\text{-}PH\text{-}C_6H_5$ ... Sn: Org.Comp.19-163/4
$C_9H_{15}Re$ ... $(CH_2CHCH_2)_3Re$ ... Re: Org.Comp.2-396
$C_9H_{16}InP$ ... $(CH_3)_3In \cdot PH_2\text{-}C_6H_5$ ... In: Org.Comp.1-38
$C_9H_{16}MnN_3O_5S_2$ .. $[Mn((SC({=}S))NC_4H_8O)(NH_2CH_2COO)_2]$ ... Mn:MVol.D7-177/8
$C_9H_{16}NOPRe^+$ ... $[(C_5H_5)Re(NO)(P(CH_3)_3){=}CH_2][PF_6]$ ... Re: Org.Comp.3-109/12
$C_9H_{16}N_2OS_3$ ... $S{=}S{=}NS\text{-}NC_5H_4({=}O)\text{-}4\text{-}(CH_3)_4\text{-}2,2,6,6$ ... S: S-N Comp.6-314/7
$C_9H_{16}OPRe$ ... $(C_5H_5)ReH_2(CO)[P(CH_3)_3]$ ... Re: Org.Comp.3-144
$C_9H_{16}O_7Th$ ... $Th(OH)(C_2H_5COO)_3$ ... Th: SVol.C7-68/9
– ... $Th(OH)(C_2H_5COO)_3 \cdot 4\ H_2O$ ... Th: SVol.C7-68/9
$C_9H_{17}InO_4$ ... $(C_2H_5)_2In[\text{-}OC(OCH_3)CHC(OCH_3)O\text{-}]$ ... In: Org.Comp.1-190, 192
$C_9H_{17}NOPRe$ ... $(C_5H_5)Re(NO)[P(CH_3)_3]\text{-}CH_3$ ... Re: Org.Comp.3-44, 45, 46, 47/8
$C_9H_{17}NO_2Sn$ ... $(C_2H_5)_3SnN(C({=}O))_2CH_2$ ... Sn: Org.Comp.18-143
$C_9H_{17}O_2Th^{3+}$ ... $[Th(OOCC_8H_{17}\text{-}n)]^{3+}$ ... Th: SVol.D1-70
$C_9H_{17}PSn$ ... $1\text{-}(CH_3)_3Sn\text{-}3,4\text{-}(CH_3)_2\text{-}PC_4H_2$ ... Sn: Org.Comp.19-177
$C_9H_{18}IInNSe^-$ ... $[(C_4H_9)_2In(SeCN)(I)]^-$ ... In: Org.Comp.1-363
$C_9H_{18}InN$ ... $(CH_2{=}CH)_3In \cdot N(CH_3)_3$ ... In: Org.Comp.1-84, 85
– ... c-C_5H_5-$In(CH_3)_2 \cdot HN(CH_3)_2$ ... In: Org.Comp.1-106
$C_9H_{18}InNO$ ... $[(C_4H_9)_2In(NCO)]_x$ ... In: Org.Comp.1-176, 178
$C_9H_{18}InNS$ ... $[(C_4H_9)_2In(NCS)]_x$ ... In: Org.Comp.1-176, 178
$C_9H_{18}InNSe$ ... $[(C_4H_9)_2In(NCSe)]_x$ ... In: Org.Comp.1-176, 178
$C_9H_{18}MnN_3O_9P_3^{4-}$
$[Mn((O_3PCH_2)_3\text{-}1,4,7\text{-}N_3C_6H_{12})]^{4-}$ ... Mn:MVol.D8-133/4

$C_9H_{18}MnN_3O_{15}P_5^{8-}$
$[Mn(((O_3PCH_2)_2NCH_2CH_2)_2NCH_2PO_3)]^{8-}$ Mn:MVol.D8-136/7
$C_9H_{18}MnN_3S_6$ .... $[Mn(SC(=S)N(CH_3)_2)_3]$ Mn:MVol.D7-146/60
$C_9H_{18}MnN_3S_6^+$ .. $[Mn(SC(=S)N(CH_3)_2)_3]^+$ Mn:MVol.D7-162
$C_9H_{18}MnN_4O_4$ ... $[Mn((CH_3)_2C{=}N{-}N{=}C(O)CH_2C(O){=}N{-}N{=}C(CH_3)_2)(H_2O)_2]$ Mn:MVol.D6-313/4
$C_9H_{18}MoN_4S_7$ .... $Mo(NS)(S_2CN(CH_3)_2)_3$ .... S: S-N Comp.5-50/2, 56/7
$C_9H_{18}N_2OS$ ...... $O{=}S{=}N{-}NC_5H_6(CH_3)_4$-2,2,6,6 .... S: S-N Comp.6-71/2
$C_9H_{18}N_2O_3SSn$ ... 2,2-$(CH_3)_2$-3-$[NH_2{-}CH_2{-}C(=O)]$ -4-$(CH_3{-}S{-}CH_2CH_2)$-1,3,2-$ONSnC_2H(=O)$-5 . Sn: Org.Comp.19-129, 131, 141
$C_9H_{18}N_2O_4Re^+$ ... $[(CO)_3Re(H_2O)(N(CH_3)_2C_2H_4{-}N(CH_3)_2)]^+$ ..... Re: Org.Comp.1-300, 304/5
$C_9H_{18}N_2S_3$ ...... $S{=}S{=}NS{-}NC_5H_6(CH_3)_4$-2,2,6,6 .... S: S-N Comp.6-314/7
$C_9H_{18}N_2Sn$ ...... 1-$(C_2H_5)_3Sn$-1,2-$N_2C_3H_3$ .... Sn: Org.Comp.18-142, 147
– .... 1-$(C_2H_5)_3Sn$-1,3-$N_2C_3H_3$ .... Sn: Org.Comp.18-142/3, 147, 149
$C_9H_{18}O_5Si_2Sn$ .... $O(Si(CH_3)_2CH_2)_2Sn(OOC)_2CH_2$ .... Sn: Org.Comp.16-220
$C_9H_{19}MnN_3O_9P_3^{3-}$
$[MnH((O_3PCH_2)_3$-1,4,7-$N_3C_6H_{12})]^{3-}$ Mn:MVol.D8-133/4
$C_9H_{19}MnN_3O_{15}P_5^{7-}$
$[MnH(((O_3PCH_2)_2NCH_2CH_2)_2NCH_2PO_3)]^{7-}$ Mn:MVol.D8-136/7
$C_9H_{19}NOS_2$ ...... $C_5H_{10}N$-1-$S(O){-}S{-}C_4H_9$-t .... S: S-N Comp.8-333
$C_9H_{19}NSn$ .... 1-(t-C_4H_9)-2,2-$(CH_3)_2$-1,2-$NSnC_3H_4$ .... Sn: Org.Comp.19-53/4, 58
– .... $(C_2H_5)_2N{-}Sn(CH_3)_2{-}C{\equiv}C{-}CH_3$ .... Sn: Org.Comp.19-52, 57
$C_9H_{19}N_3O_2Sn$ .... $(CH_3)_2Sn[-N(CH_3){-}C(=O){-}CH_2{-}N(CH_3)$ $-CH_2{-}C(=O){-}N(CH_3)-]$ .... Sn: Org.Comp.19-76/7, 79/80
$C_9H_{19}N_3O_3S_2$ .... $CH_3S{-}C(CH_3)_2{-}CH{=}N{-}OC(O){-}N(CH_3){-}S(O){-}N(CH_3)_2$ S: S-N Comp.8-350
$C_9H_{19}N_3Sn$ ...... $(CH_3)_3SnN(N{=}C(C_2H_5)C(C_2H_5){=}N)$ .... Sn: Org.Comp.18-84, 90
$C_9H_{19}O_3P_2ReS$ ... $(CO)_3Re[P(CH_3)_3]_2SH$ .... Re: Org.Comp.1-256
$C_9H_{19}O_3Sb$ ...... $(CH_3)_3Sb(OCH_3)OC(CH_3){=}CHCOCH_3$ .... Sb: Org.Comp.5-46
$C_9H_{19}O_4Sb$ ...... $(CH_3)Sb[-O{-}CH(CH_3){-}CH(CH_3){-}O-]_2$ .... Sb: Org.Comp.5-265
– .... $(CH_3)Sb[-O{-}CH_2CH_2CH_2CH_2{-}O-]_2$ .... Sb: Org.Comp.5-265
$C_9H_{20}InN$ .... 1-$(CH_3)_2In{-}NC_5H_8$-2,6-$(CH_3)_2$ .... In: Org.Comp.1-272, 273, 274/5
– .... 1-$(C_2H_5)_2In{-}NC_5H_{10}$ .... In: Org.Comp.1-274
$C_9H_{20}MnN_3O_9P_3^{2-}$
$[MnH_2((O_3PCH_2)_3$-1,4,7-$N_3C_6H_{12})]^{2-}$ Mn:MVol.D8-133/4
$C_9H_{20}MnN_3O_{15}P_5^{6-}$
$[MnH_2(((O_3PCH_2)_2NCH_2CH_2)_2NCH_2PO_3)]^{6-}$ ... Mn:MVol.D8-136/7
$C_9H_{20}N_3O_5PS_3$ ... $CH_3S{-}C(CH_3){=}N{-}OC(O){-}N(CH_3){-}S(O)$ $-N(C_2H_5){-}P(S)(OCH_3)_2$ .... S: S-N Comp.8-356
$C_9H_{20}O_2Ti$ .... $(CH_3)_2Ti(OCH_2CH(CH_3)CH(C_3H_7)O)$ .... Ti: Org.Comp.5-11
$C_9H_{20}O_3Sn$ ...... $C_4H_9Sn(OCH_2CH_2O)OC_3H_7$-i .... Sn: Org.Comp.17-73
$C_9H_{21}IOSn$ ...... $(C_4H_9)_2Sn(I)OCH_3$ .... Sn: Org.Comp.17-134
$C_9H_{21}I_2MnNOP$ ... $Mn(I)_2(NO){-}P(C_3H_7$-n$)_3$ Mn:MVol.D8-46, 61
$C_9H_{21}I_2MnO_2P$ ... $Mn(I)_2(O_2){-}P(C_3H_7$-n$)_3$ Mn:MVol.D8-55/7
$C_9H_{21}I_2MnP$ ..... $Mn(I)_2(SO_2)_n{-}P(C_3H_7$-n$)_3$ Mn:MVol.D8-46, 57/8
– .... $Mn(I)_2{-}P(C_3H_7$-n$)_3$ Mn:MVol.D8-38
$C_9H_{21}I_3In_2$ .... (n-$C_3H_7)_2In(-I-)_2InI{-}C_3H_7$-n .... In: Org.Comp.1-167, 170

$C_9H_{21}In$ ... $In(C_3H_7\text{-}n)_3$ ... In: Org.Comp.1-73/5, 78
– ... $In(C_3H_7\text{-}i)_3$ ... In: Org.Comp.1-73/5, 78
$C_9H_{21}InN_2$ ... $(CH_3)_3In \cdot N(\text{-}CH_2CH_2\text{-})_3N$ ... In: Org.Comp.1-27, 36, 43
$C_9H_{21}InN_4$ ... $(CH_3)_3In \cdot N_4C_6H_{12}$ ... In: Org.Comp.1-27, 34
$C_9H_{21}MnN_3O_9P_3^-$ $[MnH_3((O_3PCH_2)_3\text{-}1,4,7\text{-}N_3C_6H_{12})]^-$ ... Mn: MVol.D8-133/4
$C_9H_{21}MnN_3O_{15}P_5^{5-}$
$[MnH_3(((O_3PCH_2)_2NCH_2CH_2)_2NCH_2PO_3)]^{5-}$ ... Mn: MVol.D8-136/7
$C_9H_{21}MnN_3S_4$ ... $[Mn((N(CH_3)CH_2CH_2)_3)(S_4)(H_2O)]$ ... Mn: MVol.D7-91/3
$C_9H_{21}NOSn$ ... $2,2\text{-}(n\text{-}C_3H_7)_2\text{-}1,3,2\text{-}ONSnC_3H_7$ ... Sn: Org.Comp.19-129, 132
– ... $(C_2H_5)_3Sn\text{-}N(CH_3)\text{-}COCH_3$ ... Sn: Org.Comp.18-137
$C_9H_{21}NO_2Sn$ ... $2\text{-}(n\text{-}C_4H_9)\text{-}2\text{-}(i\text{-}C_3H_7\text{-}O)\text{-}1,3,2\text{-}ONSnC_2H_5$ ... Sn: Org.Comp.19-158/9
– ... $(C_2H_5)_2Sn(OC_2H_5)\text{-}O\text{-}N{=}C(CH_3)_2$ ... Sn: Org.Comp.16-181
$C_9H_{21}NSn$ ... $1\text{-}(CH_3)_3Sn\text{-}NC_6H_{12}$ ... Sn: Org.Comp.18-84, 86, 98/9
– ... $(CH_3)_3Sn\text{-}N(C_2H_5)\text{-}CH{=}C(CH_3)_2$ ... Sn: Org.Comp.18-47, 49, 53
– ... $(CH_3)_3Sn\text{-}N(C_3H_7\text{-}i)\text{-}CH{=}CHCH_3$ ... Sn: Org.Comp.18-47, 49, 53
$C_9H_{21}N_2S^+$ ... $[t\text{-}C_4H_9\text{-}N(CH_3)S{=}N\text{-}C_4H_9\text{-}t]^+$ ... S: S-N Comp.7-333/4
$C_9H_{21}N_7O_{15}Th$ ... $Th(NO_3)_4 \cdot 3\ CH_3\text{-}C({=}O)\text{-}NH\text{-}CH_3$ ... Th: SVol.D4-134
$C_9H_{22}I_4InN$ ... $[N(C_2H_5)_4][(I)_3In\text{-}CH_2I]$ ... In: Org.Comp.1-364
$C_9H_{22}InN$ ... $(CH_3)_2In\text{-}CH_2CH_2CH_2\text{-}N(C_2H_5)_2$ ... In: Org.Comp.1-101, 106/7
– ... $(C_2H_5)_2In\text{-}CH_2CH_2CH_2\text{-}N(CH_3)_2$ ... In: Org.Comp.1-101, 108
$C_9H_{22}MnN_3O_{15}P_5^{4-}$
$[MnH_4(((O_3PCH_2)_2NCH_2CH_2)_2NCH_2PO_3)]^{4-}$ ... Mn: MVol.D8-136/7
$C_9H_{22}NO_3OsSi_2^-$ $[N(C_4H_9\text{-}n)_4][((CH_3)_3SiCH_2)_2Os(N)(CO_3)]$ ... Os: Org.Comp.A1-2, 19
$C_9H_{22}N_2OSn$ ... $2\text{-}(n\text{-}C_4H_9)\text{-}2\text{-}(i\text{-}C_3H_7\text{-}O)\text{-}1,3,2\text{-}N_2SnC_2H_6$ ... Sn: Org.Comp.19-153/7
$C_9H_{22}N_2O_2Sn$ ... $(C_2H_5)_3SnN(C_3H_7\text{-}i)NO_2$ ... Sn: Org.Comp.18-141
$C_9H_{22}N_{10}O_{14}Th$ ... $[C(NH_2)_3]_3[NH_4][Th(C_2O_4)_2(CO_3)_2] \cdot 3\ H_2O$ ... Th: SVol.C7-99, 101
$C_9H_{22}OSn$ ... $(C_4H_9)_2Sn(H)OCH_3$ ... Sn: Org.Comp.17-78
$C_9H_{23}IInNO$ ... $[(C_2H_5)_2In\text{-}O\text{-}CH_2CH_2\text{-}N(CH_3)_3]I$ ... In: Org.Comp.1-193
$C_9H_{23}I_3InN$ ... $[N(C_2H_5)_4][CH_3\text{-}In(I)_3]$ ... In: Org.Comp.1-349/50, 355
$C_9H_{23}InNO^+$ ... $[(C_2H_5)_2In\text{-}O\text{-}CH_2CH_2\text{-}N(CH_3)_3]^+$ ... In: Org.Comp.1-193
$C_9H_{23}MnN_3OS_4$ ... $[Mn((N(CH_3)CH_2CH_2)_3)(S_4)(H_2O)]$ ... Mn: MVol.D7-91/3
$C_9H_{23}MnN_3O_{15}P_5^{3-}$
$[MnH_5(((O_3PCH_2)_2NCH_2CH_2)_2NCH_2PO_3)]^{3-}$ ... Mn: MVol.D8-136/7
$C_9H_{23}NO_2Sn$ ... $(CH_3)_2Sn(OC_3H_7\text{-}i)OCH_2CH_2N(CH_3)_2$ ... Sn: Org.Comp.16-173
– ... $(CH_3)_2Sn(OC_3H_7\text{-}i)ON(C_2H_5)_2$ ... Sn: Org.Comp.16-173
$C_9H_{23}NS_2Sb_2$ ... $[(CH_3)_2SbS]_2[\text{-}CH_2CH_2N(CH_3)CH_2CH_2\text{-}]$ ... Sb: Org.Comp.5-126
$C_9H_{23}NSn$ ... $(CH_3)_3Sn\text{-}N(C_2H_5)\text{-}C_4H_9\text{-}t$ ... Sn: Org.Comp.18-47, 49, 53
– ... $(CH_3)_3Sn\text{-}N(C_3H_7\text{-}n)_2$ ... Sn: Org.Comp.18-23/4
– ... $(CH_3)_3Sn\text{-}N(C_3H_7\text{-}i)_2$ ... Sn: Org.Comp.18-23/4
$C_9H_{23}O_3Sb$ ... $(C_3H_7)_2Sb(OCH_3)_3$ ... Sb: Org.Comp.5-179
$C_9H_{23}O_4Sb$ ... $(CH_3)Sb(OC_2H_5)_4$ ... Sb: Org.Comp.5-263
$C_9H_{23}PSSn$ ... $(C_2H_5)_3Sn\text{-}P(CH_3)\text{-}S\text{-}C_2H_5$ ... Sn: Org.Comp.19-190/1, 193
$C_9H_{24}InN$ ... $(CH_3)_3In \cdot N(C_2H_5)_3$ ... In: Org.Comp.1-28, 34, 42
$C_9H_{24}InP$ ... $(CH_3)_3In \cdot P(C_2H_5)_3$ ... In: Org.Comp.1-27/8, 33, 39, 45/6
$C_9H_{24}InSb$ ... $(CH_3)_3In \cdot Sb(C_2H_5)_3$ ... In: Org.Comp.1-27, 40
$C_9H_{24}In_2N_2O$ ... $(CH_3)_2In[\text{-}N(CH_3)\text{-}N(CH_3)\text{-}C(CH_3){=}O] \cdot In(CH_3)_3$
In: Org.Comp.1-271
$C_9H_{24}MnN_3O_{15}P_5^{2-}$
$[MnH_6(((O_3PCH_2)_2NCH_2CH_2)_2NCH_2PO_3)]^{2-}$ ... Mn: MVol.D8-136/7

$C_9H_{24}NO_3PSn$	$(C_2H_5)_3SnN(CH_3)P(=O)(OCH_3)_2$	Sn: Org.Comp.18-141
$C_9H_{24}NPSn$	$(CH_3)_3SnN=P(C_2H_5)_3$	Sn: Org.Comp.18-107, 113, 119
$C_9H_{24}N_2OSiSn$	$(CH_3)_3Sn-N(CH_3)C(=N-CH_3)-O-Si(CH_3)_3$	Sn: Org.Comp.18-47/9, 53
–	$(CH_3)_3Sn-NH-C(C_2H_5)=N-O-Si(CH_3)_3$	Sn: Org.Comp.18-15/6, 18
$C_9H_{24}N_6Os^{2+}$	$[2,6-(CH_3)_2-NC_5H_3-4-Os(NH_3)_4(NC-CH_3)]^{2+}$	Os: Org.Comp.A1-12
$C_9H_{24}O_9P_2Sn$	$C_4H_9Sn(OP(O)(OCH_3)_2)_2OCH_3$	Sn: Org.Comp.17-73
$C_9H_{25}InNP$	$(CH_3)_3In \cdot HN=P(C_2H_5)_3$	In: Org.Comp.1-27, 37
$C_9H_{25}InOSi_2$	$[(CH_3)_3Si-CH_2]_2In-OCH_3$	In: Org.Comp.1-183
$C_9H_{25}InSn_2$	$CH_3-In[CH_2-Sn(CH_3)_3]_2$	In: Org.Comp.1-101, 105
$C_9H_{25}NPSn^+$	$[(CH_3)_3SnNHP(C_2H_5)_3]^+$	Sn: Org.Comp.18-16, 19
$C_9H_{25}N_3Os$	$(CH_3)_3Os(N)[(CH_3)_2N-CH_2CH_2-N(CH_3)_2]$	Os: Org.Comp.A1-26
$C_9H_{25}N_3Si$	$SiH(NHC_3H_7)_3$	Si: SVol.B4-190
$C_9H_{25}N_5OOs^{2+}$	$[CH_3-1-NC_5H_4-4-Os(NH_3)_4(O=C(CD_3)_2)]^{2+}$	Os: Org.Comp.A1-11
$C_9H_{26}InNP_2$	$(CH_3)_3In \cdot CH_2=P(CH_3)_2-N=P(CH_3)_3$	In: Org.Comp.1-56
$C_9H_{26}N_2SiSn$	$(CH_3)_2N-Sn(CH_3)_2-CH_2-Si(CH_3)_2-N(CH_3)_2$	Sn: Org.Comp.19-47/8
$C_9H_{26}O_4Sb_2$	$[(CH_3)_2Sb(OCH_3)_2]_2CH_2$	Sb: Org.Comp.5-121
$C_9H_{27}I_5In_2O_4S_4$	$CH_3In_2I_5 \cdot 4\ OS(CH_3)_2$	In: Org.Comp.1-167
$C_9H_{27}InNPSi$	$(CH_3)_3In \cdot (CH_3)_3Si-N=P(CH_3)_3$	In: Org.Comp.1-27, 37
$C_9H_{27}InNPSn$	$(CH_3)_3In \cdot (CH_3)_3Sn-N=P(CH_3)_3$	In: Org.Comp.1-27, 38
$C_9H_{27}InN_2SSi_2$	$(CH_3)_2In[-N(Si(CH_3)_3)-S(CH_3)-N(Si(CH_3)_3)-]$	In: Org.Comp.1-283, 284/5
$C_9H_{27}InN_3P$	$(CH_3)_3In \cdot P[N(CH_3)_2]_3$	In: Org.Comp.1-27, 33, 39
$C_9H_{27}MnN_3O_3PSi_3$	$Mn(NO)_3[P(Si(CH_3)_3)_3]$	Mn: MVol.D8-73/4
$C_9H_{27}MnN_3O_3PSn_3$	$Mn(NO)_3[P(Sn(CH_3)_3)_3]$	Mn: MVol.D8-73/4
$C_9H_{27}MnN_7O_8P_2$	$[Mn((O=P(N(CH_3)_2)_2)_2NCH_3)(NO_3)_2]$	Mn: MVol.D8-180
$C_9H_{27}MnN_9O_{12}P_3^-$	$[Mn(O=P(OCH_3)_3)_3(N_3)_3]^-$	Mn: MVol.D8-160
$C_9H_{27}NO_2Si_3Sn$	$(CH_3)_3SnN(Si(CH_3)_2O)_2Si(CH_3)_2$	Sn: Org.Comp.18-84, 93, 103
$C_9H_{27}NPb_2Sn$	$(CH_3)_3SnN(Pb(CH_3)_3)_2$	Sn: Org.Comp.18-26, 41
$C_9H_{27}NSi_2Sn$	$(CH_3)_3SnN(Si(CH_3)_3)_2$	Sn: Org.Comp.18-23, 26, 41
$C_9H_{27}N_3O_7S_2$	$[HOCH(CH_3)CH_2NH_3]_2[OS(O)-N(CH_2CH(CH_3)OH)-SO_2]$	S: S-N Comp.8-328/9
$C_9H_{27}N_3S_2Si_3$	$(CH_3)_3SiN=S=NSN[Si(CH_3)_3]_2$	S: S-N Comp.7-149
$C_9H_{27}N_4O_{15}P_3Th$	$Th(NO_3)_4 \cdot 3\ (CH_3)_3PO$	Th: SVol.D4-130, 132, 161, 207
$C_9H_{27}N_5OOsS^{2+}$	$[2,6-(CH_3)_2-NC_5H_3-4-Os(NH_3)_4(O=S(CH_3)_2)]^{2+}$	Os: Org.Comp.A1-12
$C_9H_{27}PSi_2Sn$	$(CH_3)_3Sn-P[Si(CH_3)_3]_2$	Sn: Org.Comp.19-171
$C_9H_{27}P_7Si_2Sn$	$3,7-[(CH_3)_3Si]_2-5-(CH_3)_3Sn-[2.2.1.0(2,6)]-P_7$	Sn: Org.Comp.19-179
$C_9H_{28}N_2SSi_2Sn$	$(CH_3)_3SnN(SNHSi(CH_3)_3)Si(CH_3)_3$	Sn: Org.Comp.18-79, 81
$C_9H_{28}N_4Si_2Sn$	$(CH_3)_3Sn-NH-N=N-N[Si(CH_3)_3]_2$	Sn: Org.Comp.18-124, 126
–	$(CH_3)_3Sn-N[Si(CH_3)_3]-N=N-NH-Si(CH_3)_3$	Sn: Org.Comp.18-124, 126
$C_9H_{29}N_3Si_3Sn$	$(CH_3)_3SnN(Si(CH_3)_2NH)_2Si(CH_3)_2$	Sn: Org.Comp.18-83, 95
$C_9H_{33}N_9Si_3$	$[Si(NHCH_3)_2N(CH_3)]_3$	Si: SVol.B4-249/50
$C_9K_6O_{23}Th_2$	$K_6Th_2(C_2O_4)_2(CO_3)_5 \cdot 2\ H_2O$	Th: SVol.C7-99, 101
$C_9Na_{10}O_{23}Th$	$Na_{10}Th(C_2O_4)_2(CO_3)_5 \cdot 10\ H_2O$	Th: SVol.C7-98, 100
–	$Na_{10}Th(C_2O_4)_2(CO_3)_5 \cdot 11\ H_2O$	Th: SVol.C7-98, 100
–	$Na_{10}Th(C_2O_4)_2(CO_3)_5 \cdot 11.5\ H_2O$	Th: SVol.C7-98, 100

$C_9Na_{10}O_{23}Th$ …. $Na_{10}Th(C_2O_4)_2(CO_3)_5 \cdot 16\ H_2O$ …. Th: SVol.C7-98, 100

$C_{9.2}Cl_{1.4}H_{19.8}O_{3.2}Th$
$(t\text{-}C_4H_9\text{-}O)_2ThCl_{1.4}(CH_3COO)_{0.6}$ …. Th: SVol.C7-25/9, 49

$C_{9.5}H_{13.5}InN_{0.5}S_2$ $CH_3\text{-}In[1,2\text{-}(S)_2\text{-}C_6H_3\text{-}4\text{-}CH_3] \cdot 0.5\ N(CH_3)_3$ …. In: Org.Comp.1-251

$C_{10}CdF_{12}H_{15}Mo_2N_5$
$Cd[MoF_6]_2 \cdot 5\ CH_3CN$ …. Mo: SVol.B5-176, 185

$C_{10}ClFGeH_{20}O_2$ …. $Ge(C_2H_5)_3CFClCOOC_2H_5$ …. Ge: Org.Comp.2-135, 139

$C_{10}ClF_3H_{18}O_2Sn$ …. $(C_4H_9)_2Sn(Cl)OOCCF_3$ …. Sn: Org.Comp.17-103

$C_{10}ClF_4FeH_5O_2$ …. $(C_5H_5)Fe(CO)_2CF{=}CClCF_3$ …. Fe: Org.Comp.B13-87, 117

$C_{10}ClF_4H_{17}MoN_2OP_2$
$C_5H_5Mo(CO)(PF_2N(CH_3)_2)_2Cl$ …. Mo: Org.Comp.6-227

$C_{10}ClF_5H_{12}N_2SSn$ $(CH_3)_3SnN{=}C{=}C(CN)C(CF_3)(CF_2Cl)SCH_3$ …. Sn: Org.Comp.18-107, 109

$C_{10}ClF_6FeH_{10}HgO_2P$
$[(C_5H_5)Fe(CO)_2(CH_2{=}CHCH_2HgCl)][PF_6]$ …. Fe: Org.Comp.B17-28

$C_{10}ClF_6H_2NO$ …. $C_{10}F_6\text{-}Cl\text{-}1\text{-}({=}O)\text{-}2\text{-}NH_2\text{-}4$ …. F: PFHOrg.SVol.5-11/2, 42

$C_{10}ClF_6H_6MoO_4^-$ $[As(C_6H_5)_4][(CH_2CHCH_2)Mo(Cl)(CO)_2(CF_3\text{-}C(O)CHC(O)\text{-}CF_3)]$ …. Mo: Org.Comp.5-269/71

$C_{10}ClF_6NO_3$ …. $C_{10}F_6({=}O)\text{-}1\text{-}NO_2\text{-}4\text{-}Cl\text{-}5$ …. F: PFHOrg.SVol.5-182/3

– …. $C_{10}F_6({=}O)\text{-}1\text{-}NO_2\text{-}4\text{-}Cl\text{-}8$ …. F: PFHOrg.SVol.5-182/3

$C_{10}ClF_6O_5Re$ …. $(CO)_5ReC_5(Cl\text{-}2)F_6\text{-}c$ …. Re: Org.Comp.2-141

$C_{10}ClF_7H_4N_2$ …. $[C_{10}F_7\text{-}2\text{-}N_2H_4]Cl$ …. F: PFHOrg.SVol.5-205, 217

$C_{10}ClF_8NO_2$ …. $C_{10}F_8\text{-}NO_2\text{-}1\text{-}Cl\text{-}2$ …. F: PFHOrg.SVol.5-182/3

– …. $C_{10}F_8\text{-}NO_2\text{-}1\text{-}Cl\text{-}3$ …. F: PFHOrg.SVol.5-182/3

– …. $C_{10}F_8\text{-}NO_2\text{-}1\text{-}Cl\text{-}5$ …. F: PFHOrg.SVol.5-182/3

– …. $C_{10}F_8\text{-}NO_2\text{-}1\text{-}Cl\text{-}6$ …. F: PFHOrg.SVol.5-182/3

– …. $C_{10}F_8\text{-}NO_2\text{-}1\text{-}Cl\text{-}7$ …. F: PFHOrg.SVol.5-182/3

– …. $C_{10}F_8\text{-}NO_2\text{-}1\text{-}Cl\text{-}8$ …. F: PFHOrg.SVol.5-182/3

$C_{10}ClF_{15}N_2$ …. $1,4,5,6,7\text{-}(CF_3)_5\text{-}4\text{-}Cl\text{-}[3.2.0]\text{-}2,3\text{-}N_2C_5$ …. F: PFHOrg.SVol.4-281, 296

– …. $CF_3\text{-}C({=}N_2)\text{-}C(CF_3){=}C(CF_3)\text{-}C(CF_3){=}CCl\text{-}CF_3$ F: PFHOrg.SVol.4-314
F: PFHOrg.SVol.5-203, 214

$C_{10}ClFeGeH_{14}NO$ $C_5H_5Fe(CNCH_3)(CO)GeCl(CH_3)_2$ …. Fe: Org.Comp.B15-288, 295

$C_{10}ClFeH_6O_4^+$ …. $[(1\text{-}Cl\text{-}7\text{-}O{=}C_7H_6)Fe(CO)_3]^+$ …. Fe: Org.Comp.B15-212

– …. $[(7\text{-}Cl\text{-}6\text{-}O{=}C_7H_6)Fe(CO)_3]^+$ …. Fe: Org.Comp.B15-213, 227

$C_{10}ClFeH_7Li_2$ …. $(Li\text{-}C_5H_4)Fe(C_5H_3(Li)\text{-}Cl)$ …. Fe: Org.Comp.A10-322/4

$C_{10}ClFeH_8HgI$ …. $(C_5H_5)Fe[C_5H_3(HgCl)\text{-}I]$ …. Fe: Org.Comp.A10-147, 149, 152

– …. $(HgCl\text{-}C_5H_4)Fe(C_5H_4\text{-}I)$ …. Fe: Org.Comp.A10-141, 143, 145

$C_{10}ClFeH_8Li$ …. $(C_5H_5)Fe[C_5H_3(Li)\text{-}Cl]$ …. Fe: Org.Comp.A10-118/22

$C_{10}ClFeH_9O_2$ …. $(C_5H_5)Fe(CO)_2CH_2CH{=}CHCl$ …. Fe: Org.Comp.B13-101, 131

– …. $[(C_5H_5)Fe(CO)_2{=}C{=}CHCH_3]Cl$ …. Fe: Org.Comp.B16a-212

$C_{10}ClFeH_9O_3$ …. $(CH_3\text{-}C_5H_4)Fe(CO)_2\text{-}CH_2C(O)Cl$ …. Fe: Org.Comp.B14-54, 62

– …. $[(c\text{-}C_7H_9)Fe(CO)_3]Cl$ …. Fe: Org.Comp.B15-191

$C_{10}ClFeH_9O_5S_2$ …. $ClO_2S\text{-}C_5H_4FeC_5H_4SO_3H$ …. Fe: Org.Comp.A9-233

$C_{10}ClFeH_9O_7$ …. $[(c\text{-}C_7H_9)Fe(CO)_3][ClO_4]$ …. Fe: Org.Comp.B15-191/4

$C_{10}ClFeH_{10}HgO_2^+$ $[(C_5H_5)Fe(CO)_2(CH_2{=}CHCH_2HgCl)]^+$ …. Fe: Org.Comp.B17-27/8

$C_{10}ClFeH_{10}N$ …. $H_2N\text{-}C_5H_4FeC_5H_4\text{-}Cl$ …. Fe: Org.Comp.A9-3, 4

$C_{10}ClFeH_{10}NO_2$ …. $[(C_5H_5)Fe(CO)_2(CH_3CH{=}C{=}ND)]Cl$ …. Fe: Org.Comp.B17-127

– …. $[(C_5H_5)Fe(CO)_2(CH_3CH{=}C{=}NH)]Cl \cdot H_2O$ …. Fe: Org.Comp.B17-126

$C_{10}ClFeH_{10}NO_3$	$[(C_5H_5)Fe(CO)_2{=}C(-NH-CH_2-CH_2-O-)]Cl$	Fe: Org.Comp.B14-37
$C_{10}ClFeH_{10}NO_6$	$[(C_5H_5)Fe(CO)_2(CH_3CH{=}C{=}NH)][ClO_4] \cdot H_2O$	Fe: Org.Comp.B17-126
$C_{10}ClFeH_{10}O_2{}^+$	$[(C_5H_5)Fe(CO)_2(CH_2{=}CHCH_2Cl)]^+$	Fe: Org.Comp.B17-23
$C_{10}ClFeH_{11}N_2O$	$[(C_5H_5)Fe(CNCH_3)_2CO]Cl$	Fe: Org.Comp.B15-332, 339
$C_{10}ClFeH_{11}O_2$	$[(C_5H_5)Fe(CO)_2(CH_2{=}CHCH_3)]Cl$	Fe: Org.Comp.B17-5
–	$[(C_5H_5)Fe(CO)_2(CH_2{=}CHCH_2D)]Cl$	Fe: Org.Comp.B17-7
$C_{10}ClFeH_{11}O_3$	$[(C_5H_5)Fe(CO)_2(CH_2{=}C(CH_3)OH)]Cl$	Fe: Org.Comp.B17-66
$C_{10}ClFeH_{11}O_6$	$[(C_5H_5)Fe(CO)_2(CH_2{=}CHCH_3)][ClO_4]$	Fe: Org.Comp.B17-5
–	$[(C_5H_5)Fe(CO)_2(CH_2{=}CHCH_2D)][ClO_4]$	Fe: Org.Comp.B17-7
–	$[(C_5H_5)Fe(CO)_2(CH_2{=}CDCH_3)][ClO_4]$	Fe: Org.Comp.B17-8
$C_{10}ClFeH_{14}NO_2$	$[C_5H_5Fe(CO)_2CH_2NH(CH_3)_2]Cl$	Fe: Org.Comp.B14-136, 143
$C_{10}ClGaH_{10}I_3N_2{}^-$	$[H(NC_5H_5)_2][GaClI_3(NC_5H_5)_2]$	Ga: SVol.D1-250
$C_{10}ClGaH_{16}O_4$	$GaCl[CH_3-C({=}O)CH_2C({=}O)-CH_3]_2$ = $[Ga(CH_3-C({=}O)CH_2C({=}O)-CH_3)_2]Cl$	Ga: SVol.D1-53/4
$C_{10}ClGeH_{15}$	$Ge(CH_3)_3-CH_2-C_6H_4Cl-3$	Ge: Org.Comp.1-157, 163
–	$Ge(CH_3)_3-CH_2-C_6H_4Cl-4$	Ge: Org.Comp.1-157, 163
$C_{10}ClGeH_{17}$	$(C_2H_5)_2Ge[-CH{=}CHCCl{=}C(CH_3)CH_2-]$	Ge: Org.Comp.3-303
$C_{10}ClGeH_{18}N$	$[Ge(CH_3)_3CH_2CH_2C_5H_4NH-4]Cl$	Ge: Org.Comp.1-174
$C_{10}ClGeH_{19}O$	$Ge(C_2H_5)_3-C{\equiv}C-CH_2OCH_2Cl$	Ge: Org.Comp.2-260
$C_{10}ClGeH_{21}O$	$Ge(C_2H_5)_3-CH{=}CH-CH_2-O-CH_2Cl$	Ge: Org.Comp.2-204
–	$Ge(C_2H_5)_3-CH_2-CH(CH_3)-C({=}O)Cl$	Ge: Org.Comp.2-150
$C_{10}ClGeH_{21}O_2$	$Ge(C_2H_5)_3-CHCl-COO-C_2H_5$	Ge: Org.Comp.2-133
–	$Ge(C_2H_5)_3-CH_2CH_2CH_2-OC({=}O)Cl$	Ge: Org.Comp.2-145
$C_{10}ClGeH_{23}$	$Ge(C_2H_5)_3-CH_2CH(CH_3)-CH_2Cl$	Ge: Org.Comp.2-147
–	$Ge(C_2H_5)_3-CH_2CH_2CH_2-CH_2Cl$	Ge: Org.Comp.2-158
$C_{10}ClH_7NO_4Re$	$(CO)_4Re(Cl)-NC_5H_4-CH_3-4$	Re: Org.Comp.1-432
$C_{10}ClH_9Mn_2N_2O_8S_2$	$[Mn_2(-O_3S-CH_2N{=}CH-C_6H_2(Cl)(O)-CH{=}NCH_2-SO_3-)(OH)] \cdot 4\ H_2O$	Mn: MVol.D6-211, 213/4
$C_{10}ClH_9MoN_2O_2$	$[(C_5H_5)Mo(CO)_2(Cl)-2-(1,2-N_2C_3H_4)]$	Mo: Org.Comp.7-56/7, 60
–	$[(C_5H_5)Mo(CO)_2(Cl)-3-(1,3-N_2C_3H_4)]$	Mo: Org.Comp.7-56/7, 60
$C_{10}ClH_9MoO_2$	$(C_5H_5)Mo(CO)_2(CH_2CClCH_2)$	Mo: Org.Comp.8-205, 210
$C_{10}ClH_{10}NO_4PRe$	$(CO)_3ReCl[-(1,2-C_5H_4N)-O-P(CH_3)_2-]$	Re: Org.Comp.1-187
$C_{10}ClH_{10}N_2OS^+$	$[ClS(O)-1-NC_5H_5-4-(1-NC_5H_5)]^+$	S: S-N Comp.8-276/7
$C_{10}ClH_{10}Ti$	$(C_5H_5)_2TiCl$	Ti: Org.Comp.5-99/100
$C_{10}ClH_{11}MoN_2O$	$C_5H_5Mo(CO)(CNCH_3)_2Cl \cdot C_6H_5CH_3$	Mo: Org.Comp.6-270
$C_{10}ClH_{11}MoO$	$C_5H_5Mo(CO)(Cl)H_2C{=}CHCH{=}CH_2$	Mo: Org.Comp.6-357
$C_{10}ClH_{11}NO_5Re$	$(CO)_4Re[C(O)CH_3]C(CH_3){=}NH-CH_2CH_2Cl$	Re: Org.Comp.1-393/4
$C_{10}ClH_{12}MoO_4{}^-$	$[N(C_2H_5)_4][(CH_2CHCH_2)Mo(Cl)(CO)_2(CH_3-C(O)CHC(O)-CH_3)]$	Mo: Org.Comp.5-269/71
$C_{10}ClH_{12}NO_3S$	$i-C_3H_7-OC(O)-N(C_6H_5)-S(O)Cl$	S: S-N Comp.8-268/70
–	$C_6H_5-CH_2-OC(O)-N(C_2H_5)-S(O)Cl$	S: S-N Comp.8-268/70
$C_{10}ClH_{12}NO_5PS$	$(C_6H_5O)P(O)-CH_2-N(CH_2C(O)OCH_3)-S(O)Cl$	S: S-N Comp.8-259, 268
$C_{10}ClH_{12}N_3O_3S$	$4-(4-NO_2-C_6H_4-N{=}S(Cl))-1,4-ONC_4H_8$	S: S-N Comp.8-185
$C_{10}ClH_{12}N_3O_5S_2$	$4-(3-NO_2-C_6H_4-S(O)_2-N{=}S(Cl))-1,4-ONC_4H_8$	S: S-N Comp.8-186
–	$4-(4-NO_2-C_6H_4-S(O)_2-N{=}S(Cl))-1,4-ONC_4H_8$	S: S-N Comp.8-186
$C_{10}ClH_{13}MoN_2O_2$	$[CH_2C(CH_3)CH_2]Mo(Cl)(CO)_2(NC-CH_3)_2$	Mo: Org.Comp.5-235, 238, 243/4
$C_{10}ClH_{13}NO_5Sb$	$(4-Cl(CH_2)_3O_2CNHC_6H_4)Sb(O)(OH)_2$	Sb: Org.Comp.5-290
$C_{10}ClH_{13}N_2O_3S_2$	$4-(C_6H_5-S(O)_2-N{=}S(Cl))-1,4-ONC_4H_8$	S: S-N Comp.8-186

$C_{10}ClH_{13}N_2O_4Sn$ $C_6H_5SnCl(ONHCOCH_3)_2$ . . . Sn: Org.Comp.17-161

$C_{10}ClH_{13}N_2S$ . . . 3-Cl-C_6H_4-N=S=N-C_4H_9-t . . . S: S-N Comp.7-218

$C_{10}ClH_{13}O_2Sn$ . . . $(C_6H_5)Sn(Cl)[-OO-CH_2CH_2CH_2CH_2-]$ . . . Sn: Org.Comp.17-142

$C_{10}ClH_{14}HgMoO_5P$

$[(C_5H_5)Mo(CO)_2(P(OCH_3)_3)HgCl]$ . . . Mo: Org.Comp.7-122, 145/6

$C_{10}ClH_{14}MnO_4$ . . . $[Mn(CH_3COCHCOCH_3)_2Cl] \cdot CH_3CN$ . . . Mn: MVol.D7-6

$C_{10}ClH_{14}MoO_5P$ . . $(C_5H_5)Mo(CO)_2(Cl)[P(OCH_3)_3]$ . . . Mo: Org.Comp.7-56/7, 84/5, 111

$C_{10}ClH_{14}NO_2S$ . . . C_6H_5-$N(CH_2CH_2Cl)$-S(O)O-C_2H_5 . . . S: S-N Comp.8-321

$C_{10}ClH_{14}OSSb$ . . . $(CH_3)_3Sb(Cl)SCOC_6H_5$ . . . Sb: Org.Comp.5-14

$C_{10}ClH_{15}MoN_2O_2$ $[(CH_3)_5C_5]Mo(NO)_2Cl$ . . . Mo: Org.Comp.6-56

– . . . $[(C_5H_5)Mo(CO)_2(NH_2$-$CH(CH_3)CH_2$-$NH_2)]Cl$ . . . Mo: Org.Comp.7-249, 251

$C_{10}ClH_{15}O_2Sn$ . . . $C_6H_5(HOCH_2CH_2CH_2CH_2)Sn(Cl)OH$ . . . Sn: Org.Comp.17-142

$C_{10}ClH_{15}O_4PReS_3$ $(CO)_4Re(Cl)P(S$-$C_2H_5)_3$ . . . Re: Org.Comp.1-448

$C_{10}ClH_{15}O_5PRe$ . . $(CO)_4Re(Cl)P(CH_3)_2O$-C_4H_9-n . . . Re: Org.Comp.1-447/8

$C_{10}ClH_{16}MnN_2O_2S_4$

$[Mn((SC(=S))NC_4H_8O)_2Cl]$ . . . Mn: MVol.D7-177

$C_{10}ClH_{16}MnN_3OS_4$

$[Mn((SC(=S))NC_4H_8)_2(NO)Cl]$ . . . Mn: MVol.D7-168

$C_{10}ClH_{16}MnN_3O_3S_4$

$[Mn((SC(=S))NC_4H_8O)_2(NO)Cl]$ . . . Mn: MVol.D7-168

$C_{10}ClH_{16}NO_2SSn$ $(CH_3)_3SnNClSO_2C_6H_4CH_3$-4 . . . Sn: Org.Comp.18-77/8

$C_{10}ClH_{16}O_2Re$ . . . $[CH{\equiv}CC(CH_3)_2OH]_2ReCl$ . . . Re: Org.Comp.2-358, 365

$C_{10}ClH_{18}MnNOS_6$ $[Mn((S_2C)SC_4H_9)_2(NO)Cl]$ . . . Mn: MVol.D7-210

$C_{10}ClH_{18}MnNO_3S_4$

$[Mn(SC(=S)OC_4H_9)_2(NO)Cl]$ . . . Mn: MVol.D7-130/3

$C_{10}ClH_{19}OSn$ . . . $(CH_2{=}CH)_2Sn(Cl)OC_6H_{13}$ . . . Sn: Org.Comp.17-117

$C_{10}ClH_{20}IO_2Sn$ . . . $(C_4H_9)_2Sn(Cl)OOCCH_2I$ . . . Sn: Org.Comp.17-105

$C_{10}ClH_{20}MnN_2S_4$ $[Mn(SC(=S)N(C_2H_5)_2)_2Cl]$ . . . Mn: MVol.D7-161

$C_{10}ClH_{20}MnN_3OS_4$

$Mn[SC(=S)N(C_2H_5)_2]_2(NO)Cl$ . . . Mn: MVol.D7-144

– . . . $Mn[SC(=S)NH$-C_4H_9-$n]_2(NO)Cl$ . . . Mn: MVol.D7-136/8

$C_{10}ClH_{21}OSn$ . . . $(C_2H_5)_2Sn(Cl)O$-C_6H_{11}-c . . . Sn: Org.Comp.17-94

$C_{10}ClH_{21}O_2Sn$ . . . $(C_4H_9)_2Sn(Cl)OOCCH_3$ . . . Sn: Org.Comp.17-103, 109

$C_{10}ClH_{22}In$ . . . $(t$-C_4H_9-$CH_2)_2InCl$ . . . In: Org.Comp.1-118, 119/20

$C_{10}ClH_{23}OSn$ . . . $(C_3H_7)_2Sn(Cl)OC_4H_9$-t . . . Sn: Org.Comp.17-96, 97

$C_{10}ClH_{23}O_2Sn$ . . . $C_4H_9SnCl(OC_3H_7$-$i)_2$ . . . Sn: Org.Comp.17-151

$C_{10}ClH_{24}InN_2$ . . . $[(CH_3)_2N$-$CH_2CH_2CH_2]_2InCl$ . . . In: Org.Comp.1-122

$C_{10}ClH_{24}N_2PSn$ . . $(CH_3)_2Sn(Cl)$-$N(C_4H_9$-$t)$-P=N-C_4H_9-t . . . Sn: Org.Comp.19-119/20, 123

$C_{10}ClH_{24}N_2Re$ . . . $(CH_3)_2Re(=N$-C_4H_9-$t)_2Cl$ . . . Re: Org.Comp.1-14/5

$C_{10}ClH_{24}PSn$ . . . $(CH_3)_2SnCl$-$P(C_4H_9$-$t)_2$ . . . Sn: Org.Comp.19-218/20

$C_{10}ClH_{26}InSi_2$ . . . $[(CH_3)_3Si]_2CH$-InCl-C_3H_7-i . . . In: Org.Comp.1-127/8

$C_{10}ClH_{26}MnNOSi_2$ $[Mn(N(Si(CH_3)_3)_2)Cl(OC_4H_8)]$ . . . Mn: MVol.D8-30

$C_{10}ClH_{27}N_2O_{10}P_3Re$

$(CO)ReCl[P(OCH_3)_3]_3(N_2)$ . . . Re: Org.Comp.1-47

$C_{10}ClH_{29}InPSi_3$ . . . $[(CH_3)_3Si$-$CH_2]InCl$-$P[Si(CH_3)_3]_2$ . . . In: Org.Comp.1-325

$C_{10}Cl_2F_2H_{18}O_2Sn$ $(C_4H_9)_2Sn(Cl)OOCCF_2Cl$ . . . Sn: Org.Comp.17-104

$C_{10}Cl_2F_4GeH_{14}O_2$ $Ge(CH_3)_3C(=C(OC_2H_5)OC(CF_2Cl)_2)$ . . . Ge: Org.Comp.2-94

$C_{10}Cl_2F_5H_8InO_2$ . . C_6F_5-$In(Cl)_2 \cdot$ 1,4-$O_2C_4H_8$ . . . In: Org.Comp.1-136/7, 140, 145

$C_{10}Cl_2F_6H_{27}OOsP_4$	$[(CO)Os(P(CH_3)_3)_3(Cl)_2][PF_6]$	Os: Org.Comp.A1-120
$C_{10}Cl_2F_6N_2$	3-(5-Cl-NC_5F_3-3)-5-Cl-NC_5F_3	F: PFHOrg.SVol.4-170, 172
–	4-(3-Cl-NC_5F_3-4)-3-Cl-NC_5F_3	F: PFHOrg.SVol.4-170/1, 172
$C_{10}Cl_2F_6N_4$	$(C_5F_3N(Cl))NN(C_5F_3N(Cl))$	F: PFHOrg.SVol.4-146, 156
$C_{10}Cl_2F_7N$	1-Cl-$C_{10}F_7$(=NCl)-2	F: PFHOrg.SVol.6-10/1, 34, 37
–	$C_{10}F_7$-2-NCl_2	F: PFHOrg.SVol.6-6, 26, 37
$C_{10}Cl_2FeGeH_{14}$	$C_5H_5Fe(CH_2=CHCH=CH_2)GeCl_2CH_3$	Fe: Org.Comp.B17-186, 190/1
$C_{10}Cl_2FeH_7Li$	$(C_5H_5)Fe[C_5H_2(Li)(Cl)_2]$	Fe: Org.Comp.A10-327
–	$(Cl-C_5H_4)Fe[C_5H_3(Li)-Cl]$	Fe: Org.Comp.A10-322/5
$C_{10}Cl_2FeH_8Hg$	$(C_5H_5)Fe[C_5H_3(Cl)-HgCl]$	Fe: Org.Comp.A10-146, 147, 149
–	$(HgCl-C_5H_4)Fe(C_5H_4-Cl)$	Fe: Org.Comp.A10-141, 142, 143, 145
$C_{10}Cl_2FeH_8Hg_2$	$(C_5H_5)Fe[C_5H_3(HgCl)_2]$	Fe: Org.Comp.A10-147, 148, 151
–	$Fe(C_5H_4-HgCl)_2$	Fe: Org.Comp.A10-132/5
$C_{10}Cl_2FeH_8Hg_2^+$	$[Fe(C_5H_4-HgCl)_2]^+$	Fe: Org.Comp.A10-133
$C_{10}Cl_2FeH_8O_4S_2$	$Fe(C_5H_4SO_2Cl)_2$	Fe: Org.Comp.A9-233, 241/2
$C_{10}Cl_2FeH_8Zn_2$	$Fe(C_5H_4-ZnCl)_2$	Fe: Org.Comp.A10-131/2
$C_{10}Cl_2FeH_{10}HgO_2$	$[(C_5H_5)Fe(CO)_2(CH_2=CHCH_2HgCl)]Cl$	Fe: Org.Comp.B17-27/8
$C_{10}Cl_2GaH_6N_2O_4^+$	$[Ga((NC_5H_2$-2,3-$(O)_2$-5-Cl)H$)_2]^+$	Ga: SVol.D1-255, 256
$C_{10}Cl_2GaH_6O_2$	$GaCl_2[1,2-(O)_2C_{10}H_6]$, radical	Ga: SVol.D1-90/2
$C_{10}Cl_2GaH_{20}O_5^+$	$[(-O-CH_2CH_2-)_5GaCl_2]Cl \cdot H_2O$	Ga: SVol.D1-150, 151
–	$[(-O-CH_2CH_2-)_5GaCl_2][GaCl_4]$	Ga: SVol.D1-150/1
$C_{10}Cl_2GaH_{22}N_2^+$	$[GaCl_2(NC_5H_{11})_2]Cl$	Ga: SVol.D1-243
–	$[GaCl_2(NC_5H_{11})_2][GaCl_4]$	Ga: SVol.D1-243
$C_{10}Cl_2GeH_{14}Hg$	$Ge(CH_3)_3CCl_2HgC_6H_5$	Ge: Org.Comp.1-145
$C_{10}Cl_2GeH_{20}$	$(C_2H_5)_2Ge(-CH_2CH(CHCl_2)CH_2CH_2CH_2-)$	Ge: Org.Comp.3-291
$C_{10}Cl_2GeH_{22}$	$Ge(C_2H_5)_3-CH_2CH(CH_3)CHCl_2$	Ge: Org.Comp.2-148, 153/4
–	$Ge(C_4H_9-n)_2(CH_2Cl)_2$	Ge: Org.Comp.3-178
$C_{10}Cl_2GeH_{24}OSn$	$(C_4H_9)_2Sn(Cl)OGe(CH_3)_2Cl$	Sn: Org.Comp.17-107
$C_{10}Cl_2H_6N_4O_2S_2$	$C_6H_5-S(O)_2NH-S(Cl)=N-CCl=C(CN)_2$	S: S-N Comp.8-184
$C_{10}Cl_2H_6O_4Sn$	$(CH_2=CH)_2SnO_2C_6Cl_2(O)_2$	Sn: Org.Comp.16-90/1
$C_{10}Cl_2H_7MnN_3S$	$[Mn(2-(4-C_3H_2NS-3,1)C_7H_5N_2-1,3)Cl_2]$	Mn: MVol.D7-228/30
$C_{10}Cl_2H_7NO_2S_2$	$Cl_2S=NS(O)_2-1-C_{10}H_7$	S: S-N Comp.8-90
–	$Cl_2S=NS(O)_2-2-C_{10}H_7$	S: S-N Comp.8-90
$C_{10}Cl_2H_7OSb$	$(1-C_{10}H_7)Sb(Cl_2)O$	Sb: Org.Comp.5-273
$C_{10}Cl_2H_{10}MnN_2O_2S_2$	$[Mn(SC_4H_3(CH=NOH-2))_2Cl_2]$	Mn: MVol.D7-222/5
$C_{10}Cl_2H_{10}MnN_4O_2S_2$	$Mn(SC(2-C_4H_3O)=NNH_2)_2Cl_2$	Mn: MVol.D7-218/20
$C_{10}Cl_2H_{10}MoN_2O$	$C_5H_5Mo(NO)(C_5H_5N)Cl_2$	Mo: Org.Comp.6-31
$C_{10}Cl_2H_{10}N_2OS$	$[ClS(O)-1-NC_5H_5-4-(1-NC_5H_5)]Cl$	S: S-N Comp.8-276/7
$C_{10}Cl_2H_{10}Ti$	$[(C_5H_5)_2TiCl_2]_n$	Ti: Org.Comp.5-320
$C_{10}Cl_2H_{12}MnN_4O_2S_2$	$[Mn(SC_4H_3CONHNH_2)_2Cl_2]$	Mn: MVol.D7-222/5
$C_{10}Cl_2H_{12}MnN_4S_2$	$[Mn(1,3-N_2C_4H_3(=S-2)(CH_3-1))_2Cl_2]$	Mn: MVol.D7-78/9

$C_{10}Cl_2H_{12}N_2O_2S$	$Cl_2S{=}N{-}C_6{-}2,3,5,6{-}(CH_3)_4{-}4{-}NO_2$	S: S-N Comp.8-109/15
$C_{10}Cl_2H_{12}N_2O_3S_2$	$4{-}(4{-}Cl{-}C_6H_4{-}S(O)_2{-}N{=}S(Cl)){-}1,4{-}ONC_4H_8$	S: S-N Comp.8-186
$C_{10}Cl_2H_{13}InO_2$	$C_6H_5{-}In(Cl)_2 \cdot 1,4{-}O_2C_4H_8$	In: Org.Comp.1-136, 139
$C_{10}Cl_2H_{13}Mo$	$(C_5H_5)Mo(Cl)_2(CH_2{=}CHCH{=}CH{-}CH_3)$	Mo: Org.Comp.6-190/1
–	$(C_5H_5)Mo(Cl)_2[CH_2{=}C(CH_3){-}CH{=}CH_2]$	Mo: Org.Comp.6-191
$C_{10}Cl_2H_{13}NO_2Sn$	$C_4H_9SnCl_2(OOCC_5H_4N{-}2)$	Sn: Org.Comp.17-179
$C_{10}Cl_2H_{13}Sb$	$(CH_2)_4(C_6H_5)SbCl_2$	Sb: Org.Comp.5-75
$C_{10}Cl_2H_{14}O_4Sn$	$(CH_2{=}CHCH_2)_2Sn(OOCCH_2Cl)_2$	Sn: Org.Comp.16-90, 93
$C_{10}Cl_2H_{15}In$	$[(CH_3)_5C_5]In(Cl)_2$	In: Org.Comp.1-136, 138
$C_{10}Cl_2H_{15}MnNOP$	$Mn(NO)[P(C_6H_5)(C_2H_5)_2]Cl_2$	Mn: MVol.D8-46, 61
$C_{10}Cl_2H_{15}MnO_2P$	$Mn[P(C_6H_5)(C_2H_5)_2](O_2)Cl_2$	Mn: MVol.D8-44/5, 53
$C_{10}Cl_2H_{15}MnP$	$Cl_2Mn[P(CH_3)_2{-}C_6H_5](CH_2{=}CH_2)$	Mn: MVol.D8-46/7
–	$Cl_2Mn[P(C_2H_5)_2{-}C_6H_5]$	Mn: MVol.D8-42/3
$C_{10}Cl_2H_{15}MoN_5$	$Mo(CN{-}CH_3)_5(Cl)_2$	Mo: Org.Comp.5-54
$C_{10}Cl_2H_{15}Sb$	$(C_2H_5)_2(C_6H_5)SbCl_2$	Sb: Org.Comp.5-56
$C_{10}Cl_2H_{16}MnN_2O_2S_4$	$[Mn((SC(=S))NC_4H_8O)_2Cl_2]$	Mn: MVol.D7-180
$C_{10}Cl_2H_{16}MoO_2$	$C_5H_5MoO(OC_5H_{11}{-}i)Cl_2$	Mo: Org.Comp.6-10, 11
$C_{10}Cl_2H_{17}N_2PPtS$	$[(C_6H_5{-}P(CH_3)_2)PtCl_2(CH_3N{=}S{=}NCH_3)]$	S: S-N Comp.7-316/7, 320
$C_{10}Cl_2H_{17}O_2Sb$	$(CH_2)_5Sb(Cl_2)OC(CH_3){=}CHC(O)CH_3$	Sb: Org.Comp.5-188/9
$C_{10}Cl_2H_{18}MnN_4O_2$	$[MnCl_2((CH_3)_2C{=}N{-}NHC(O)CH_2CH_2C(O)NH{-}N{=}C(CH_3)_2)]_n$	Mn: MVol.D6-313, 315
$C_{10}Cl_2H_{19}MnO_7PS$	$MnCl_2 \cdot O{=}P(OCH_3)_2SCH(COOC_2H_5)CH_2COOC_2H_5$	Mn: MVol.D8-193
$C_{10}Cl_2H_{20}MnN_2S_4$	$[Mn(SC(=S)N(C_2H_5)_2)_2Cl_2]$	Mn: MVol.D7-163
$C_{10}Cl_2H_{20}N_3S_5Tc$	$Tc(NS)Cl_2(S_2CN(C_2H_5)_2)_2$	S: S-N Comp.5-60/2
$C_{10}Cl_2H_{20}O_2Sn$	$(C_4H_9)_2Sn(Cl)OOCCH_2Cl$	Sn: Org.Comp.17-104
$C_{10}Cl_2H_{21}MnOP$	$Mn(CO)[P(C_3H_7)_3]Cl_2$	Mn: MVol.D8-47
$C_{10}Cl_2H_{22}MnN_4$	$[MnCl_2(H_2N{-}(CH_2)_3{-}N{=}C(CH_3)C(CH_3){=}N{-}(CH_2)_3{-}NH_2)]$	Mn: MVol.D6-217, 218
$C_{10}Cl_2H_{22}MnN_6O_2$	$[Mn((CH_3)(C_2H_5)C{=}N{-}NHC(O)NH_2)_2Cl_2]$	Mn: MVol.D6-327
$C_{10}Cl_2H_{22}MnN_6S_2$	$MnCl_2[C_2H_5C(CH_3){=}N{-}NHC(=S)NH_2]_2$	Mn: MVol.D6-341
$C_{10}Cl_2H_{22}MoN_4$	$Cl_2Mo(CN{-}CH_3)_2[N(CH_3)_2{-}CH_2CH_2{-}N(CH_3)_2]$	Mo: Org.Comp.5-12, 14, 19/20
$C_{10}Cl_2H_{22}NO_3PS$	$Cl_2S{=}N{-}P(O)(O{-}C_5H_{11}{-}n)_2$	S: S-N Comp.8-93/4
$C_{10}Cl_2H_{22}N_2PtS$	$(CH_2{=}CH_2)PtCl_2(t{-}C_4H_9{-}N{=}S{=}N{-}C_4H_9{-}t)$	S: S-N Comp.7-326/8
$C_{10}Cl_2H_{23}MnO_3P$	$Mn[P(C_2H_5)_3](OC_4H_8)(O_2)Cl_2$	Mn: MVol.D8-51
$C_{10}Cl_2H_{24}OSiSn$	$(C_4H_9)_2Sn(Cl)OSi(CH_3)_2Cl$	Sn: Org.Comp.17-107
$C_{10}Cl_2H_{25}N_2PPtS$	$[((C_2H_5)_3P)PtCl_2(C_2H_5{-}N{=}S{=}N{-}C_2H_5)]$	S: S-N Comp.7-316/7, 319, 321/2
$C_{10}Cl_2H_{26}InN$	$[N(C_2H_5)_4][(CH_3)_2In(Cl)_2]$	In: Org.Comp.1-349/50, 351
$C_{10}Cl_2H_{26}MnN_4S_2$	$[Mn(SCH_2CH_2NHCH_2CH_2CH_2NH_2)_2Cl_2]$	Mn: MVol.D7-28/9
$C_{10}Cl_2H_{27}OOsP_3$	$(CO)(Cl)_2Os[P(CH_3)_3]_3$	Os: Org.Comp.A1-105/6
$C_{10}Cl_2H_{27}OOsP_3{}^+$	$[(CO)Os(P(CH_3)_3)_3(Cl)_2][PF_6]$	Os: Org.Comp.A1-120
$C_{10}Cl_2H_{30}O_{11}S_2Sb_2$	$[((CH_3)_3SbOS(CH_3)_2)_2O][ClO_4]_2$	Sb: Org.Comp.5-101
$C_{10}Cl_2MnO_{10}ReSn$	$(CO)_5ReSnCl_2{-}Mn(CO)_5$	Re: Org.Comp.2-173, 181
$C_{10}Cl_3FeH_7$	$(C_5H_5)Fe[C_5H_2(Cl)_3]$	Fe: Org.Comp.A10-216
–	$(Cl{-}C_5H_4)Fe[C_5H_3(Cl)_2]$	Fe: Org.Comp.A10-216

$C_{10}Cl_3FeH_8$	$[(C_{10}H_8)FeCl]_nCl_{2n}$	Fe: Org.Comp.B18-2
$C_{10}Cl_3GaH_6N_2O_4$	$GaCl_3[1\text{-}(4\text{-}NO_2\text{-}C_6H_4)\text{-}NC_4H_2(=O)_2\text{-}2,5]$	Ga: SVol.D1-240/1
$C_{10}Cl_3GaH_6O_2$	$GaCl_3[1,4\text{-}(O=)_2C_{10}H_6]$	Ga: SVol.D1-95
$C_{10}Cl_3GaH_7NO_2$	$GaCl_3[1\text{-}C_6H_5\text{-}NC_4H_2(=O)_2\text{-}2,5]$	Ga: SVol.D1-240/1
$C_{10}Cl_3GaH_8N_2O_5$	$GaCl_3[2,5\text{-}(O=)_2\text{-}NC_4H_2\text{-}1\text{-}CH_2OCH_2\text{-}1\text{-}NC_4H_2(=O)_2\text{-}2,5]$	Ga: SVol.D1-241/2
$C_{10}Cl_3GaH_{10}IN_2^-$	$[H(NC_5H_5)_2][GaCl_3I(NC_5H_5)_2]$	Ga: SVol.D1-250
$C_{10}Cl_3GaH_{10}N_2$	$[GaCl_2(NC_5H_5)_2]Cl = [GaCl_2(NC_5H_5)_4][GaCl_4]$	Ga: SVol.D1-247
$C_{10}Cl_3GaH_{12}O$	$GaCl_3(1\text{-}OC_{10}H_{12})$	Ga: SVol.D1-142
$C_{10}Cl_3GaH_{12}O_2$	$GaCl_3[1,4\text{-}(O=)_2C_6(CH_3)_4]$	Ga: SVol.D1-93/4
$C_{10}Cl_3GaH_{20}O_5$	$[(C_{10}H_{20}O_5)GaCl_2]Cl \cdot H_2O$	Ga: SVol.D1-150, 151
$C_{10}Cl_3GaH_{22}N_2$	$GaCl_3(NC_5H_{11})_2$	Ga: SVol.D1-242/3
–	$[GaCl_2(NC_5H_{11})_2]Cl$	Ga: SVol.D1-243
$C_{10}Cl_3GaH_{24}O_2$	$GaCl_3(HO\text{-}C_5H_{11}\text{-}i)_2$	Ga: SVol.D1-16, 18/9
$C_{10}Cl_3GeH_{11}MoO_3$	$(C_5H_5)Mo(CO)_2(GeCl_3)=C(CH_3)\text{-}OCH_3$	Mo: Org.Comp.8-16, 19
$C_{10}Cl_3Ge_2H_{15}$	$[GeC_5(CH_3)_5][GeCl_3]$	Ge: Org.Comp.3-383
$C_{10}Cl_3H_7OTh$	$C_{10}H_7\text{-}1\text{-}O\text{-}ThCl_3$	Th: SVol.C7-30/2
–	$C_{10}H_7\text{-}2\text{-}O\text{-}ThCl_3$	Th: SVol.C7-30/2
$C_{10}Cl_3H_8N_3OsS$	$Os(NS)Cl_3(N_2C_{10}H_8)$	S: S-N Comp.5-79, 82/3
$C_{10}Cl_3H_9NOSb$	$(CH_3)Sb(Cl_3)OC_9H_6N$	Sb: Org.Comp.5-270
$C_{10}Cl_3H_9N_2O_2S$	$C_6H_5\text{-}C(O)\text{-}NH\text{-}S(CCl_3)=N\text{-}C(O)CH_3$	S: S-N Comp.8-199
$C_{10}Cl_3H_{10}NO_3S$	$2\text{-}(2,4,6\text{-}Cl_3\text{-}C_6H_2\text{-}N=)\text{-}1,3,6,2\text{-}O_3SC_4H_8$	S: S-N Comp.8-169
$C_{10}Cl_3H_{10}N_3OsS$	$Os(NS)Cl_3(NC_5H_5)_2$	S: S-N Comp.5-79, 82/3
$C_{10}Cl_3H_{10}N_4ReS_2$	$Re(NS)(NSCl)Cl_2(NC_5H_5)_2$	S: S-N Comp.5-51, 66/7
$C_{10}Cl_3H_{11}N_2O_3S_2$	$3\text{-}CH_3\text{-}C_6H_4C(O)\text{-}NH\text{-}S(CCl_3)=N\text{-}S(O)_2CH_3$	S: S-N Comp.8-195
–	$4\text{-}CH_3\text{-}C_6H_4C(O)\text{-}NH\text{-}S(CCl_3)=N\text{-}S(O)_2CH_3$	S: S-N Comp.8-195
–	$CH_3C(O)\text{-}NH\text{-}S(CCl_3)=N\text{-}S(O)_2\text{-}C_6H_4\text{-}4\text{-}CH_3$	S: S-N Comp.8-195
–	$C_6H_5\text{-}CH_2C(O)\text{-}NH\text{-}S(CCl_3)=N\text{-}S(O)_2CH_3$	S: S-N Comp.8-195
$C_{10}Cl_3H_{11}N_2O_4S_2$	$3\text{-}CH_3O\text{-}C_6H_4C(O)\text{-}NH\text{-}S(CCl_3)=N\text{-}S(O)_2CH_3$	S: S-N Comp.8-195
–	$4\text{-}CH_3O\text{-}C_6H_4C(O)\text{-}NH\text{-}S(CCl_3)=N\text{-}S(O)_2CH_3$	S: S-N Comp.8-195
$C_{10}Cl_3H_{13}NO_2Sn$	$C_4H_9SnCl_2[ON(O)C_6H_4Cl\text{-}4]$, radical	Sn: Org.Comp.17-179, 183
$C_{10}Cl_3H_{14}MoO_5PSn$	$(C_5H_5)Mo(CO)_2[P(OCH_3)_3]SnCl_3$	Mo: Org.Comp.7-119, 128
$C_{10}Cl_3H_{16}N_3O_3S$	$(1,4\text{-}ONC_4H_8\text{-}4)_2S=N\text{-}C(O)CCl_3$	S: S-N Comp.8-207/8, 213
$C_{10}Cl_3H_{18}N_2ORe$	$Re(CNC_4H_9\text{-}t)_2(Cl_3)O$	Re: Org.Comp.2-259, 265/6
$C_{10}Cl_3H_{19}O_2Sn$	$(C_4H_9)_2Sn(Cl)OOCCHCl_2$	Sn: Org.Comp.17-104
$C_{10}Cl_3H_{20}N_4S^+$	$[(C_2H_5)_2N\text{-}CCl=N\text{-}S(Cl)\text{-}N=CCl\text{-}N(C_2H_5)_2]^+$	S: S-N Comp.8-192/3
$C_{10}Cl_3H_{27}InSi_3^-$	$[(Cl)_3In\text{-}C(Si(CH_3)_3)_3]^-$	In: Org.Comp.1-353, 358/9
$C_{10}Cl_3H_{28}N_3OPSb$	$(C_2H_5)_2SbCl_3 \cdot OP[N(CH_3)_2]_3$	Sb: Org.Comp.5-143
$C_{10}Cl_3H_{29}OsP_2Si$	$(CH_3)_3Si\text{-}CH_2\text{-}Os(Cl)_3[P(CH_3)_3]_2$	Os: Org.Comp.A1-7/8
$C_{10}Cl_4F_4N_2$	$(3,5\text{-}Cl_2\text{-}4\text{-}C_5F_2N)(4\text{-}C_5F_2N\text{-}Cl_2\text{-}3,5)$	F: PFHOrg.SVol.4-170/2
$C_{10}Cl_4F_7FeIO_2$	$(C_5Cl_4I)Fe(CO)_2C_3F_7\text{-}n$	Fe: Org.Comp.B14-73, 80
$C_{10}Cl_4Fe_2H_9O_2S_2$	$[C_5H_5(CO)_2Fe=C(SCH_2)_2][FeCl_4]$	Fe: Org.Comp.B16a-144, 155
$C_{10}Cl_4Fe_2H_9O_3$	$[(c\text{-}C_7H_9)Fe(CO)_3][FeCl_4]$	Fe: Org.Comp.B15-191/2, 194
$C_{10}Cl_4Fe_3N_3O_{10}P_3$	$(CO)_{10}Fe_3[P_3N_3(Cl)_4]$	Fe: Org.Comp.C6b-43/5
$C_{10}Cl_4Ga_2H_{10}N_2$	$Ga_2Cl_4(NC_5H_5)_2$	Ga: SVol.D1-244/5
$C_{10}Cl_4Ga_2H_{16}O_4$	$[Ga(CH_3\text{-}C(=O)CH_2C(=O)\text{-}CH_3)_2][GaCl_4]$	Ga: SVol.D1-54

$C_{10}Cl_4Ga_2H_{20}O_2$	$Ga_2Cl_4(OC_5H_{10})_2$	Ga: SVol.D1-116/7
$C_{10}Cl_4H_8MoN_2$	$MoCl_4 \cdot C_{10}H_8N_2$	Mo: SVol.B5-324
$C_{10}Cl_4H_{10}MoN_2$	$MoCl_4 \cdot 2\ C_5H_5N$	Mo: SVol.B5-324
$C_{10}Cl_4H_{10}N_2O_2Th$	$ThCl_4 \cdot 2\ (1\text{-}O\text{-}NC_5H_5)$	Th: SVol.D4-162
$C_{10}Cl_4H_{10}OTi_2$	$[(C_5H_5)TiCl_2]_2O$	Ti: Org.Comp.5-48/9
$C_{10}Cl_4H_{12}N_2Pd$	$(C_5H_5NH)_2[PdCl_4]$	Pd: SVol.B2-122/3
$C_{10}Cl_4H_{12}O_4Sn$	$(CH_2{=}CHCH_2)_2Sn(OOCCHCl_2)_2$	Sn: Org.Comp.16-93
$C_{10}Cl_4H_{14}N_4Pd$	$[1\text{-}CH_2{=}CH\text{-}(1,3\text{-}N_2C_3H_4)]_2[PdCl_4]$	Pd: SVol.B2-122
–	$[4,8\text{-}(NH_2)_2\text{-}(4.2.2.2^{2,5})\text{-}3,7\text{-}N_2C_{10}H_{10}][PdCl_4]$	Pd: SVol.B2-124, 125
$C_{10}Cl_4H_{15}Mo$	$(CH_3)_5C_5MoCl_4$	Mo: Org.Comp.6-3
$C_{10}Cl_4H_{18}O_2Sn$	$(C_4H_9)_2Sn(Cl)OOCCCl_3$	Sn: Org.Comp.17-104
$C_{10}Cl_4H_{20}MoO_2$	$MoCl_4 \cdot 2\ C_5H_{10}O$	Mo: SVol.B5-372
$C_{10}Cl_4H_{20}NO_2Re$	$[N(C_2H_5)_4][(CO)_2ReCl_4]$	Re: Org.Comp.1-57
$C_{10}Cl_4H_{20}N_4S$	$(C_2H_5)_2N\text{-}CCl_2\text{-}N{=}S{=}N\text{-}CCl_2\text{-}N(C_2H_5)_2$	S: S-N Comp.7-267
–	$[(C_2H_5)_2N\text{-}CCl{=}N\text{-}S(Cl)\text{-}N{=}CCl\text{-}N(C_2H_5)_2]Cl$	S: S-N Comp.8-192/3
$C_{10}Cl_4H_{24}N_2Pd$	$(C_5H_{10}NH_2)_2[PdCl_4]$	Pd: SVol.B2-123
$C_{10}Cl_4H_{24}Sb_2$	$[(CH_3)_2SbCl_2]_2(CH_2)_6$	Sb: Org.Comp.5-124
$C_{10}Cl_4H_{26}NSb$	$[N(C_2H_5)_4][(CH_3)_2SbCl_4]$	Sb: Org.Comp.5-141
$C_{10}Cl_4H_{30}O_5S_5Th$	$ThCl_4 \cdot 5\ (CH_3)_2SO$	Th: SVol.D4-147, 172
$C_{10}Cl_4O_6Re$	$(CO)_4Re[1,2\text{-}(O)_2C_6Cl_4\text{-}3,4,5,6]$, radical	Re: Org.Comp.1-371
$C_{10}Cl_5FeH_7Si$	$(Cl\text{-}C_5H_4)Fe[C_5H_3(Cl)\text{-}SiCl_3]$	Fe: Org.Comp.A10-300
$C_{10}Cl_5H_7Sb^-$	$[(1\text{-}C_{10}H_7)SbCl_5]^-$	Sb: Org.Comp.5-257
–	$[(2\text{-}C_{10}H_7)SbCl_5]^-$	Sb: Org.Comp.5-257/8
$C_{10}Cl_5H_8N_3O_2SSb^-$	$[(4\text{-}N_2C_4H_3NHSO_2C_6H_4)SbCl_5]^-$	Sb: Org.Comp.5-253
$C_{10}Cl_5H_{11}NSb$	$[NH_4][(1\text{-}C_{10}H_7)SbCl_5]$	Sb: Org.Comp.5-257
–	$[NH_4][(2\text{-}C_{10}H_7)SbCl_5]$	Sb: Org.Comp.5-257
$C_{10}Cl_5H_{11}O_2SSb^-$	$[3\text{-}(C_2H_5\text{-}OC({=}O)\text{-}CH_2\text{-}S)\text{-}C_6H_4\text{-}SbCl_5]^-$	Sb: Org.Comp.5-253
–	$[4\text{-}(C_2H_5\text{-}OC({=}O)\text{-}CH_2\text{-}S)\text{-}C_6H_4\text{-}SbCl_5]^-$	Sb: Org.Comp.5-253
$C_{10}Cl_5H_{11}O_2Sb^-$	$[(4\text{-}C_2H_5O_2CCH_2C_6H_4)SbCl_5]^-$	Sb: Org.Comp.5-257
$C_{10}Cl_5H_{11}O_3Sb^-$	$[(4\text{-}C_2H_5O_2CCH_2OC_6H_4)SbCl_5]^-$	Sb: Org.Comp.5-252
$C_{10}Cl_5H_{17}NSb$	$[N(CH_3)_4][C_6H_5SbCl_5]$	Sb: Org.Comp.5-245
$C_{10}Cl_5H_{35}MoN_5$	$MoCl_5 \cdot 5\ NH(CH_3)_2$	Mo: SVol.B5-373
$C_{10}Cl_6F_8N_4P_2$	$[(4\text{-}C_5F_4N)NPCl_3]_2$	F: PFHOrg.SVol.4-162
$C_{10}Cl_6FeH_8N_3P_3$	$[(Cl)_5N_3P_3\text{-}C_5H_4]Fe(C_5H_4\text{-}Cl)$	Fe: Org.Comp.A10-85
$C_{10}Cl_6FeH_9O_2Sb$	$[(C_5H_5)Fe(CO)_2(CH_2{=}C{=}CH_2)][SbCl_6]$	Fe: Org.Comp.B17-110
$C_{10}Cl_6FeH_{10}NO_2Sb$	$[(C_5H_5)Fe(CO)_2(CH_3CH{=}C{=}NH)][SbCl_6]$	Fe: Org.Comp.B17-126
–	$[(C_5H_5)Fe(CO)_2(CH_3CH{=}C{=}ND)][SbCl_6]$	Fe: Org.Comp.B17-127
$C_{10}Cl_6Fe_3N_4O_{10}P_4$	$(CO)_{10}Fe_3[P_4N_4(Cl)_6]$	Fe: Org.Comp.C6b-45
$C_{10}Cl_6Ga_2H_6O_2$	$Ga_2[1,4\text{-}(O{=})_2C_{10}H_6]Cl_6$	Ga: SVol.D1-95
$C_{10}Cl_6Ga_2H_7NO_2$	$Ga_2Cl_6[1\text{-}C_6H_5\text{-}NC_4H_2({=}O)_2\text{-}2,5]$	Ga: SVol.D1-240/1
$C_{10}Cl_6Ga_2H_{12}O_2$	$Ga_2Cl_6[1,4\text{-}(O{=})_2C_6(CH_3)_4]$	Ga: SVol.D1-93/4
$C_{10}Cl_6Ga_2H_{20}O_5$	$[(C_{10}H_{20}O_5)GaCl_2][GaCl_4]$	Ga: SVol.D1-150/1
$C_{10}Cl_6Ga_2H_{22}N_2$	$[GaCl_2(NC_5H_{11})_2][GaCl_4]$	Ga: SVol.D1-243
$C_{10}Cl_6GeH_{10}$	$Ge(CH_3)_3C_7HCl_6$	Ge: Org.Comp.2-40
$C_{10}Cl_6H_{10}Ti_2$	$[(C_5H_5)TiCl_3]_2$	Ti: Org.Comp.5-60/1
$C_{10}Cl_7H_{26}NTi_2$	$[(C_2H_5)_4N][(CH_3)_2Ti_2Cl_7]$	Ti: Org.Comp.5-4
$C_{10}Cl_8H_{10}O_9STh$	$Th(CHCl_2COO)_4 \cdot (CH_3)_2SO$	Th: SVol.C7-58

$C_{10}Cl_8N_4S$	NC_5Cl_4-2-N=S=N-2-NC_5Cl_4	S: S-N Comp.7-247, 254
–	NC_5Cl_4-4-N=S=N-4-NC_5Cl_4	S: S-N Comp.7-248, 254
$C_{10}Cl_{10}FeH_8N_6P_6$	$Fe[C_5H_4-P_3N_3(Cl)_5]_2$	Fe: Org.Comp.A10-83/4
–	$[(Cl)_5P_3N_3-P_3N_3(Cl)_4-C_5H_4]Fe(C_5H_4-Cl)$	Fe: Org.Comp.A10-84, 85
$C_{10}Cl_{12}FeH_8N_7P_7$	$[(Cl)_7P_4N_4-P_3N_3(Cl)_4-C_5H_4]Fe(C_5H_4-Cl)$	Fe: Org.Comp.A10-84, 85
$C_{10}Cl_{12}H_{10}Mo_6N_2$	$Mo_6Cl_{12} \cdot 2\ C_5H_5N$	Mo: SVol.B5-267
$C_{10}Cl_{12}H_{10}Mo_6N_2O_2$	$Mo_6Cl_{12} \cdot 2\ C_5H_5NO$	Mo: SVol.B5-267
$C_{10}CoF_{12}H_{15}Mo_2N_5$	$Co[MoF_6]_2 \cdot 5\ CH_3CN$	Mo: SVol.B5-176, 185
$C_{10}CoH_{21}N_2O_4Si$	$[SiH_3(N(CH_3)_3)_2][Co(CO)_4] = SiH_3Co(CO)_4 \cdot 2\ N(CH_3)_3$	Si: SVol.B4-324/5
$C_{10}CoO_{10}Re$	$[Re(CO)_6][Co(CO)_4]$	Re: Org.Comp.2-219, 225
$C_{10}Co_2H_{36}N_{12}O_{20}Th$	$[Co(NH_3)_6]_2[Th(C_2O_4)_5] \cdot 3\ H_2O$	Th: SVol.C7-89, 95
–	$[Co(NH_3)_6]_2[Th(C_2O_4)_5] \cdot 3.64\ H_2O$	Th: SVol.C7-89
$C_{10}CrFH_{16}N_6OPd$	$[CrF(C_6H_{10}(NH_2)_2)(H_2O)][Pd(CN)_4]$	Pd: SVol.B2-279
$C_{10}CrFH_{22}N_8OPd$	$[CrF(H_2O)(NH_2CH_2CH_2CH_2NH_2)_2][Pd(CN)_4]$	Pd: SVol.B2-279
$C_{10}CrF_6H_9O_5PSn$	$(CH_3)_3Sn-P(CF_3)_2Cr(CO)_5$	Sn: Org.Comp.19-181, 184
$C_{10}CrF_{12}H_{15}N_6P_2S$	$[Cr(NS)(NCCH_3)_5][PF_6]_2$	S: S-N Comp.5-50, 55
$C_{10}CrGeH_{11}NO_5$	$Ge(CH_3)_3CH_2NCCr(CO)_5$	Ge: Org.Comp.1-140
$C_{10}CrHO_{10}Re$	$(CO)_5ReH-Cr(CO)_5$	Re: Org.Comp.2-178, 182
$C_{10}CrH_6N_6O_6ReS_4$	$[Re(CO)_6][Cr(SCN)_4(NH_3)_2]$	Re: Org.Comp.2-218
$C_{10}CrH_{12}MoN_2O_{12}^{3-}$	$[Cr((OOCCH_2)_2N-CH_2CH_2-N(CH_2COO)_2)MoO_4]^{3-}$	Mo: SVol.B3b-194/5
$C_{10}CrH_{15}N_6S^{2+}$	$[Cr(NS)(NC-CH_3)_5]^{2+}$	S: S-N Comp.5-50, 55
$C_{10}CrH_{15}O_5PSn$	$(CH_3)_3Sn-P(CH_3)_2Cr(CO)_5$	Sn: Org.Comp.19-181/2
$C_{10}CrO_{10}Re^-$	$[(CO)_5ReCr(CO)_5]^-$	Re: Org.Comp.2-190, 204
$C_{10}Cr_2H_{36}N_{12}O_{20}Th$	$[Cr(NH_3)_6]_2[Th(C_2O_4)_5] \cdot 20\ H_2O$	Th: SVol.C7-89, 95
$C_{10}CsF_{14}N$	$Cs[NC-C(CF=C(CF_3)_2)_2]$	F: PFHOrg.SVol.6-100, 120, 131
$C_{10}CuF_{12}H_{15}Mo_2N_5$	$Cu[MoF_6]_2 \cdot 5\ CH_3CN$	Mo: SVol.B5-176, 185
$C_{10}Cu_2FeH_8S_2$	$Cu_2[Fe(C_5H_4S)_2]$	Fe: Org.Comp.A9-204
$C_{10}FFeH_{11}N_2O_4S$	$[(C_5H_5)Fe(CNCH_3)_2CO][SO_3F]$	Fe: Org.Comp.B15-331
$C_{10}FFeH_{13}O_5S_2$	$[C_5H_5Fe(CO)_2CH_2S(CH_3)_2]FSO_3$	Fe: Org.Comp.B14-133/4, 142
$C_{10}FGeH_{15}$	$Ge(CH_3)_3-CH_2-C_6H_4F$-3	Ge: Org.Comp.1-157
–	$Ge(CH_3)_3-CH_2-C_6H_4F$-4	Ge: Org.Comp.1-157
$C_{10}FGeH_{21}O_2$	$Ge(C_2H_5)_3CHFCOOC_2H_5$	Ge: Org.Comp.2-132, 139
$C_{10}FH_{12}N_2O_8Th^-$	$ThF[((OOCCH_2)_2N)_2C_2H_4]^-$	Th: SVol.D1-136
$C_{10}FH_{16}NSn$	$(CH_3)_2N-Sn(CH_3)_2-C_6H_4$-4-F	Sn: Org.Comp.19-47/8, 53
$C_{10}F_2FeH_5N_3O_2$	2-$[(C_5H_5)Fe(CO)_2]$-1,3,5-$N_3C_3(F_2$-4,6)	Fe: Org.Comp.B14-26, 30
$C_{10}F_2GeH_{20}O_2$	$Ge(C_2H_5)_3CF_2COOC_2H_5$	Ge: Org.Comp.2-135, 139
$C_{10}F_2H_{12}N_2O_8Th^{2-}$	$ThF_2[((OOCCH_2)_2N)_2C_2H_4]^{2-}$	Th: SVol.D1-136
$C_{10}F_2H_{20}N_2S$	$F_2S(1-NC_5H_{10})_2$	S: S-N Comp.8-397/400

$C_{10}F_3FeH_5O_2$ .... $(C_5H_5)Fe(CO)_2C{\equiv}CCF_3$ .... Fe: Org.Comp.B13-148, 154, 158
$C_{10}F_3FeH_7O_2$ .... $(C_5H_5)Fe(CO)_2\text{-}C(CF_3){=}CH_2$ .... Fe: Org.Comp.B13-86/7, 117
– .... $(C_5H_5)Fe(CO)_2\text{-}CH{=}CHCF_3$ .... Fe: Org.Comp.B13-86, 116/7
$C_{10}F_3FeH_7O_5S$ ... $[C_5H_5(CO)_2Fe{=}C{=}CH_2][CF_3SO_3]$ .... Fe: Org.Comp.B16a-212
$C_{10}F_3FeH_8NO_4S_2$ $[C_5H_5Fe(CS)(CO)NCCH_3][SO_3CF_3]$ .... Fe: Org.Comp.B15-271
$C_{10}F_3FeH_9O_5S$ ... $[C_5H_5(CO)_2Fe{=}CH(CH_3)][CF_3SO_3]$ .... Fe: Org.Comp.B16a-87, 95
$C_{10}F_3FeH_9O_5S_2$ .. $[C_5H_5(CO)_2Fe{=}CH(SCH_3)][CF_3SO_3]$ .... Fe: Org.Comp.B16a-113, 122/3
– .... $[C_5H_5(CO)_2Fe{=}CD(SCH_3)][CF_3SO_3]$ .... Fe: Org.Comp.B16a-113, 123
$C_{10}F_3FeH_9O_6S$ ... $[(C_5H_5)Fe(CO)_2{=}C(CH_3)OH][CF_3SO_3]$ .... Fe: Org.Comp.B16a-108, 118
– .... $[(C_5H_5)Fe(CO)_2(CH_2{=}CHOH)][CF_3SO_3]$ .... Fe: Org.Comp.B17-20
$C_{10}F_3FeH_{10}NO_6S$ $[C_5H_5(CO)_2Fe{=}C(OCH_3)NH_2][CF_3SO_3]$ .... Fe: Org.Comp.B16a-137
$C_{10}F_3GeH_{13}$ .... $Ge(CH_3)_3\text{-}C_6H_4\text{-}CF_3\text{-}3$ .... Ge: Org.Comp.2-79
– .... $Ge(CH_3)_3\text{-}C_6H_4\text{-}CF_3\text{-}4$ .... Ge: Org.Comp.2-74
$C_{10}F_3H_6O_8ReS_2$ .. $[(CO)_5ReSC_4H_6][O_3SCF_3]$ .... Re: Org.Comp.2-158
$C_{10}F_3H_9NO_7Re$ ... $[(CO)_5ReNH_2C_3H_7\text{-}i][O_2CCF_3]$ .... Re: Org.Comp.2-151
$C_{10}F_3H_{11}MoN_2O_3S_2$
$(C_5H_5)Mo(NO)(OC(O)CF_3)S_2CN(CH_3)_2$ .... Mo: Org.Comp.6-50
$C_{10}F_3H_{11}MoO$ .... $(C_5H_5)Mo(O)(CF_3)(H_3C\text{-}C{\equiv}C\text{-}CH_3)$ .... Mo: Org.Comp.6-151
$C_{10}F_3H_{11}O_3Sn$ ... $(CH_3)_2Sn(OC_6H_5)OOCCF_3$ .... Sn: Org.Comp.16-174, 178
$C_{10}F_3H_{13}MoO_6$ ... $(CH_2CHCH_2)Mo(CO)_2(CH_3O\text{-}CH_2CH_2OH)\text{-}OC(O)CF_3$ Mo: Org.Comp.5-268
$C_{10}F_3H_{16}InO_2$ .... $(CH_3)_2In[\text{-}OC(CF_3)CHC(C_4H_9\text{-}t)O\text{-}]$ .... In: Org.Comp.1-190, 191, 194
$C_{10}F_4GaH_8N_2^-$ ... $[GaF_4(2\text{-}(NC_5H_4\text{-}2)\text{-}NC_5H_4)]^-$ .... Ga: SVol.D1-263
$C_{10}F_4GeH_9MnO_5$ $Ge(CH_3)_3CF_2CF_2Mn(CO)_5$ .... Ge: Org.Comp.1-169, 176
$C_{10}F_4H_{10}MoN_2$ ... $MoF_4 \cdot 2\ C_5H_5N$ .... Mo: SVol.B5-92
$C_{10}F_4H_{12}N_2SSi$ ... $(CH_3)_3SiN{=}S{=}N\text{-}C_6F_4\text{-}CH_3\text{-}4$ .... S: S-N Comp.7-142
$C_{10}F_4MnNO_5$ .... $4\text{-}(CO)_5Mn\text{-}C_5F_4N$ .... F: PFHOrg.SVol.4-110
$C_{10}F_4NO_5Re$ .... $4\text{-}(CO)_5Re\text{-}NC_5F_4$ .... Re: Org.Comp.2-144
$C_{10}F_5FeH_5O_2$ .... $(C_5H_5)Fe(CO)_2CF{=}CFCF_3$ .... Fe: Org.Comp.B13-87, 117
$C_{10}F_5FeH_5O_3$ .... $(C_5H_5)Fe(CO)_2COC_2F_5$ .... Fe: Org.Comp.B13-15, 34
$C_{10}F_5GeH_{15}$ .... $Ge(C_2H_5)_3C({=}C(CF_3)CF_2)$ .... Ge: Org.Comp.2-230
$C_{10}F_5H_6N_5$ .... $NC(NH_2)NC(C_6F_4NH_2)CFC(NH_2)$ .... F: PFHOrg.SVol.4-194, 207
$C_{10}F_5H_9N_2S$ .... $C_6F_5\text{-}N{=}S{=}N\text{-}C_4H_9\text{-}t$ .... S: S-N Comp.7-219
$C_{10}F_5H_{10}MoN_2$ ... $MoF_5 \cdot 2\ C_5H_5N$ .... Mo: SVol.B5-115
$C_{10}F_5H_{17}NSb$ .... $[N(CH_3)_4][C_6H_5SbF_5]$ .... Sb: Org.Comp.5-238
$C_{10}F_6FeH_9O_2P$ ... $[(C_5H_5)Fe(CO)_2(CH_2{=}C{=}CH_2)][PF_6]$ .... Fe: Org.Comp.B17-110, 116
– .... $[(C_5H_5)Fe(CO)_2(HC{\equiv}C\text{-}CH_3)][PF_6]$ .... Fe: Org.Comp.B17-121
$C_{10}F_6FeH_9O_2PS_2$ $[C_5H_5(CO)_2Fe{=}C(SCH_2)_2][PF_6]$ .... Fe: Org.Comp.B16a-143/4, 155
$C_{10}F_6FeH_9O_3P$ ... $[(1\text{-}CH_3\text{-}C_6H_6)Fe(CO)_3][PF_6]$ .... Fe: Org.Comp.B15-98, 135
– .... $[(2\text{-}CH_3\text{-}C_6H_6)Fe(CO)_3][PF_6]$ .... Fe: Org.Comp.B15-103/4, 144, 155/6
– .... $[(3\text{-}CH_3\text{-}C_6H_6)Fe(CO)_3][PF_6]$ .... Fe: Org.Comp.B15-106, 157/8
– .... $[(C_5H_5)Fe(CO)_2(CH_2{=}CHCHO)][PF_6]$ .... Fe: Org.Comp.B17-29
– .... $[(c\text{-}C_7H_9)Fe(CO)_3][PF_6]$ .... Fe: Org.Comp.B15-191
$C_{10}F_6FeH_9O_4P$ ... $[(2\text{-}CH_3O\text{-}C_6H_6)Fe(CO)_3][PF_6]$ .... Fe: Org.Comp.B15-103, 139/54
– .... $[(3\text{-}CH_3O\text{-}C_6H_6)Fe(CO)_3][PF_6]$ .... Fe: Org.Comp.B15-105, 157

$C_{10}F_6FeH_9O_4P$ ... $[(C_5H_5)Fe(CO)_2{=}C(OCH_2)_2][PF_6]$ .... Fe: Org.Comp.B16a-136, 139/40
$C_{10}F_6FeH_{10}NO_2PS$
$[C_5H_5(CO)_2Fe{=}C(-NHCH_2CH_2S-)][PF_6]$ .... Fe: Org.Comp.B16a-146/7
$C_{10}F_6FeH_{10}NO_3P$ $[(C_5H_5)Fe(CO)_2{=}C(-NH-CH_2CH_2-O-)][PF_6]$ ... Fe: Org.Comp.B16a-139, 151
$C_{10}F_6FeH_{11}N_2OP$ $[(C_5H_5)Fe(CO)(CN-CH_3)(NC-CH_3)][PF_6]$ ...... Fe: Org.Comp.B15-299/300
– .... $[(C_5H_5)Fe(CO)(CN-CH_3)_2][PF_6]$ .... Fe: Org.Comp.B15-333, 339/40
$C_{10}F_6FeH_{11}N_2O_2P$ $[C_5H_5(CO)_2Fe{=}C(NHCH_2)_2][PF_6]$ .... Fe: Org.Comp.B16a-147
$C_{10}F_6FeH_{11}N_2PS$ $[(C_5H_5)Fe(CS)(NCCH_3)_2][PF_6]$ .... Fe: Org.Comp.B15-264/5
$C_{10}F_6FeH_{11}OP$ ... $[(C_6H_6)Fe(CO)(CH_2CHCH_2)][PF_6]$ .... Fe: Org.Comp.B18-95/7
$C_{10}F_6FeH_{11}O_2P$ .. $[(C_5H_5)Fe(CO)_2(CH_2{=}CHCH_3)][PF_6]$ .... Fe: Org.Comp.B17-6, 44/5
– .... $[(C_5H_5)Fe(CO)_2(CH_2{=}CHCH_2D)][PF_6]$ .... Fe: Org.Comp.B17-8
$C_{10}F_6FeH_{11}O_2PS$ $[(C_5H_5)Fe(CO)_2{=}C(CH_3)-SCH_3][PF_6]$ .... Fe: Org.Comp.B16a-114, 124
– .... $[(C_5H_5)Fe(CS)(CO)-OC(CH_3)_2][PF_6]$ .... Fe: Org.Comp.B15-271
$C_{10}F_6FeH_{11}O_2PS_2$ $[C_5H_5(CO)_2Fe{=}C(SCH_3)_2][PF_6]$ .... Fe: Org.Comp.B16a-140, 152/4
$C_{10}F_6FeH_{11}O_3P$ .. $[(CH_2C(CH_3)CHC(CH_3)CH_2)Fe(CO)_3][PF_6]$ ..... Fe: Org.Comp.B15-25
– .... $[(CH_2CHCHCHCH-C_2H_5)Fe(CO)_3][PF_6]$ .... Fe: Org.Comp.B15-20
– .... $[(C_5H_5)Fe(CO)_2{=}C(CH_3)OCH_3][PF_6]$ .... Fe: Org.Comp.B16a-108, 118/9
– .... $[(C_5H_5)Fe(CO)_2{=}CH-OC_2H_5][PF_6]$ .... Fe: Org.Comp.B16a-108, 118
– .... $[(C_5H_5)Fe(CO)_2(CH_2{=}CHOCH_3)][PF_6]$ .... Fe: Org.Comp.B17-21, 48
$C_{10}F_6FeH_{11}O_3PS$ $[C_5H_5(CO)_2Fe{=}C(OCH_3)SCH_3][PF_6]$ .... Fe: Org.Comp.B16a-137, 150
$C_{10}F_6FeH_{11}O_3PSi$ $[(1,1-(CH_3)_2SiC_5H_5)Fe(CO)_3][PF_6]$ .... Fe: Org.Comp.B15-134
$C_{10}F_6FeH_{11}O_4P$ .. $[(C_5H_5)Fe(CO)_2{=}C(OCH_3)_2][PF_6]$ .... Fe: Org.Comp.B16a-135/6, 148
– .... $[(C_5H_5)Fe(CO)_2(CH_2{=}C(OH)-OCH_3)][PF_6]$ ..... Fe: Org.Comp.B17-68
– .... $[(C_5H_5)Fe(CO)_2-CH_2-C(OH)-OCH_3][PF_6]$ ..... Fe: Org.Comp.B14-127
$C_{10}F_6FeH_{11}PS$ ... $[(C_5H_5)Fe(SC_4H_3-2-CH_3)][PF_6]$ .... Fe: Org.Comp.B17-214
– .... $[(C_5H_5)Fe(SC_4H_3-3-CH_3)][PF_6]$ .... Fe: Org.Comp.B17-214
$C_{10}F_6FeH_{12}NOP$ .. $[(C_5H_5)Fe(CH_2{=}CH_2)(CNCH_3)(CO)][PF_6]$ ...... Fe: Org.Comp.B17-137/8
$C_{10}F_6FeH_{12}NO_2P$ $[(C_5H_5)(CO)_2Fe{=}CH(N(CH_3)_2)][PF_6]$ .... Fe: Org.Comp.B16a-115
$C_{10}F_6FeH_{12}NP$ ... $[C_5H_5FeC_4H_4NCH_3-1][PF_6]$ .... Fe: Org.Comp.B17-204
$C_{10}F_6FeH_{13}O_2PS$ $[(C_5H_5)Fe(CO)_2-CH_2S(CH_3)_2][PF_6]$ .... Fe: Org.Comp.B14-133/4, 142
$C_{10}F_6FeH_{17}PS_3$ .. $[(C_5H_5)Fe(CS)(S(CH_3)_2)_2][PF_6]$ .... Fe: Org.Comp.B15-264
$C_{10}F_6GeH_{14}O_2$ ... 3-$[(CH_3)_3Ge]$-2-(C_2H_5-O)-4,4-$(CF_3)_2$-OC_3 ... Ge: Org.Comp.2-94
– .... $Ge(CH_3)_3-C(COO-C_2H_5){=}C(CF_3)_2$ .... Ge: Org.Comp.2-10, 22
$C_{10}F_6GeH_{16}$ ..... $Ge(C_2H_5)_3-C(CF_3){=}CH-CF_3$ .... Ge: Org.Comp.2-208
$C_{10}F_6H_3NO$ ...... $C_{10}F_6$-OH-1-NH_2-3 .... F: PFHOrg.SVol.5-11, 41, 82
$C_{10}F_6H_4N_4$ ...... $NC(NH_2)NC(NH_2)CFC(C_6F_5)$ .... F: PFHOrg.SVol.4-194, 207
$C_{10}F_6H_6IMoNO_3$ .. $(C_5H_5)Mo(NO)(I)[-O-C(CF_3){=}CH-C(CF_3)-O-]$ .. Mo: Org.Comp.6-47
$C_{10}F_6H_7NOS$ ..... $(CF_3)_2S{=}N-C(O)CH_2C_6H_5$ .... S: S-N Comp.8-145
$C_{10}F_6H_8NPS$ ..... $[SN][PF_6] \cdot C_{10}H_8$ .... S: S-N Comp.5-47/9
$C_{10}F_6H_8NSSb$ .... $[SN][SbF_6] \cdot C_{10}H_8$ .... S: S-N Comp.5-47/9
$C_{10}F_6H_9MoO_5PSn$ $(CH_3)_3Sn-P(CF_3)_2Mo(CO)_5$ .... Sn: Org.Comp.19-181, 184
$C_{10}F_6H_{12}MoNO_2P$ $[(C_5H_5)Mo(CO)(NO)(CH_2C(CH_3)CH_2)][PF_6]$ .... Mo: Org.Comp.6-348, 352
– .... $[(C_5H_5)Mo(CO)(NO)(CH_2CHCH-CH_3)][PF_6]$ .... Mo: Org.Comp.6-347, 352
$C_{10}F_6H_{12}MoN_2OS_4$
$(CF_3-C{\equiv}C-CF_3)Mo(O)[S_2C-N(CH_3)_2]_2$ .... Mo: Org.Comp.5-139/41, 143

$C_{10}F_6H_{12}N_2SSn$ . . $(CH_3)_3SnN{=}C{=}C(CN)C(CF_3)_2SCH_3$ . . . . Sn: Org.Comp.18-107, 109
$C_{10}F_6H_{12}N_4O_2PRe$ cis-$[(CO)_2Re(CNCH_3)_4][PF_6]$ . . . . Re: Org.Comp.2-283
$C_{10}F_6H_{13}MoN_2OP$ $[C_5H_5Mo(NO)(NCCH_3)C_3H_5][PF_6]$ . . . . Mo: Org.Comp.6-177, 182
$C_{10}F_6H_{13}NO_3PRe$ $[(C_5H_5)Re(CO)(NO)(OC_4H_8)][PF_6]$ . . . . Re: Org.Comp.3-153, 154, 155
$C_{10}F_6H_{15}MoN_2O_2P$
$[(C_5H_5)Mo(CO)_2(NH_2-CH(CH_3)CH_2-NH_2)][PF_6]$ Mo: Org.Comp.7-249, 251
$C_{10}F_6H_{15}MoN_6OP$ $[Mo(CN-CH_3)_5(NO)][PF_6]$ . . . . Mo: Org.Comp.5-52/3
$C_{10}F_6H_{18}O_6S_2Sn$ $(C_4H_9)_2Sn(OSO_2CF_3)_2$ . . . . Sn: Org.Comp.15-348
$C_{10}F_6N_4$ . . . . C_9F_6-2-CN-3-N_3 . . . . F: PFHOrg.SVol.6-106
$C_{10}F_7FeH_4IO_2$ . . . . $(C_5H_4I)Fe(CO)_2C_3F_7$-n . . . . Fe: Org.Comp.B14-58, 65
$C_{10}F_7HN_6$ . . . . $NCFNC(N_3)CFC(NHC_6F_5)$ . . . . F: PFHOrg.SVol.4-188/9, 203
$C_{10}F_7H_2N$ . . . . $C_{10}F_7$-2-NH_2 . . . . F: PFHOrg.SVol.5-11, 41, 82
$C_{10}F_7H_2NO$ . . . . . . $C_{10}F_7(=O)$-1-NH_2-3 . . . . F: PFHOrg.SVol.5-11/2, 42
$C_{10}F_7H_2N_3$ . . . . . . $NCFNC(NH_2)CFC(C_6F_5)$ . . . . F: PFHOrg.SVol.4-194, 207
$C_{10}F_7H_3N_2$ . . . . . . $C_{10}F_7$-2-NH-NH_2 . . . . F: PFHOrg.SVol.5-205, 217, 226
$C_{10}F_7H_9N_2SSi$. . . . $(CH_3)_3SiN{=}S{=}N-C_6F_4-CF_3$-4 . . . . S: S-N Comp.7-142
$C_{10}F_7N$ . . . . C_9F_7-2-CN . . . . F: PFHOrg.SVol.6-106, 126, 146
$C_{10}F_7N_3$ . . . . $C_{10}F_7$-2-N_3. . . . . F: PFHOrg.SVol.5-205, 221
$C_{10}F_8HN_3$ . . . . $(4-C_5F_4N)_2NH$. . . . . F: PFHOrg.SVol.4-149/50, 160
$C_{10}F_8H_3O_5Re$ . . . . $(CO)_5Re(CF_2)_4CH_3$ . . . . Re: Org.Comp.2-137
$C_{10}F_8H_9NO_2SSn$. . $(CH_3)_3SnN(C_6F_5)SO_2CF_3$ . . . . Sn: Org.Comp.18-58, 67, 75
$C_{10}F_8N_2$ . . . . 3-$(NC_5F_4$-3)-NC_5F_4 . . . . F: PFHOrg.SVol.4-170/2
– . . . . 4-$(NC_5F_4$-4)-NC_5F_4 . . . . F: PFHOrg.SVol.4-170, 172
– . . . . 4-C_6F_5-1,3-$N_2C_4F_3$ . . . . F: PFHOrg.SVol.4-193, 206
$C_{10}F_8N_2^-$. . . . . $[(4-C_5F_4N)(4-C_5F_4N)]^-$, radical anion . . . . F: PFHOrg.SVol.4-170, 172
$C_{10}F_8N_2O_6$ . . . . . . 1,7-$O_2C_{10}F_8-(NO_2)_2$-4,8 . . . . F: PFHOrg.SVol.5-183, 192
$C_{10}F_8N_4$ . . . . $(C_5F_4N)NN(C_5F_4N)$ . . . . F: PFHOrg.SVol.4-146, 156
$C_{10}F_9GaH_8O_7$ . . . . $(OC_4H_8)Ga[OC(O)-CF_3]_3$ . . . . Ga: SVol.D1-112
$C_{10}F_9GeH_3$ . . . . . . $Ge(C{\equiv}C-CF_3)_3-CH_3$ . . . . Ge: Org.Comp.3-54, 56
$C_{10}F_9GeH_{15}$ . . . . . $Ge(C_2H_5)_3C(CF_3)_2CF_3$ . . . . Ge: Org.Comp.2-129
$C_{10}F_9HN_2O$ . . . . . . 4-(3-OH-C_5F_4N(F))-C_5F_4N . . . . F: PFHOrg.SVol.4-108
$C_{10}F_9H_9O_6Sn$ . . . . $C_4H_9Sn(OOCCF_3)_3$ . . . . Sn: Org.Comp.17-45
$C_{10}F_9H_{18}I_2N_3S$ . . . $[((CH_3)_2N)_3S][I(CF_2)_4I-F]_n$ . . . . S: S-N Comp.8-230, 237, 248/9
$C_{10}F_9H_{18}N_3S$ . . . . $[((CH_3)_2N)_3S][C_4F_9$-t] . . . . S: S-N Comp.8-230, 235
$C_{10}F_{10}GeH_6$ . . . . . $(2-CF_3-C_3F_2-1)_2Ge(CH_3)_2$ . . . . Ge: Org.Comp.3-153
– . . . . $(C_2F_5-C{\equiv}C)_2Ge(CH_3)_2$ . . . . Ge: Org.Comp.3-154/5
$C_{10}F_{10}H_{13}NOS$ . . . $HCF_2CF_2CF_2CF_2CH_2O-S(F_2)$-1-$NC_5H_{10}$ . . . . . . S: S-N Comp.8-395/7
$C_{10}F_{11}H_2NO$ . . . . . 4-(i-C_3F_7)-C_6F_4-C(=O)NH_2 . . . . F: PFHOrg.SVol.5-22, 55, 85
$C_{10}F_{11}H_2NO_2$ . . . . $C_6F_5-O-C(CF_3)_2-C(=O)NH_2$. . . . . F: PFHOrg.SVol.5-19, 52
$C_{10}F_{11}N$ . . . . 4-i-$C_3F_7-C_6F_4$-CN. . . . . F: PFHOrg.SVol.6-107, 126, 148
$C_{10}F_{11}NO_2$ . . . . . . $C_9F_8-NO_2$-2-CF_3-2 . . . . F: PFHOrg.SVol.5-182, 191/2
$C_{10}F_{12}HNO$ . . . . . . n-C_3F_7-C(=O)NH-C_6F_5 . . . . F: PFHOrg.SVol.5-24, 59
$C_{10}F_{12}H_{10}NNaO_3$ $Na[O-C(CF_3)_2-N{=}C(CF_3)_2] \cdot CH_3-O-CH_2CH_2-O-CH_3$
F: PFHOrg.SVol.6-188/9, 195, 200, 218

$C_{10}F_{12}H_{12}Li_2N_2Sn$	$(CH_3)_2Sn[N(Li)-C(CF_3)_2-CH_3]_2$	Sn:	Org.Comp.19-67/70
$C_{10}F_{12}H_{12}N_4S$	$(CH_3)_2NC(CF_3)_2-N=S=N-C(CF_3)_2N(CH_3)_2$	S:	S-N Comp.7-266
$C_{10}F_{12}H_{15}Mo_2N_5Zn$	$Zn[MoF_6]_2 \cdot 5\ CH_3CN$	Mo:	SVol.B5-176, 185
$C_{10}F_{13}H_2N$	$2,4,5,6-(CF_3)_4-C_6F-NH_2$	F:	PFHOrg.SVol.5-10, 40
$C_{10}F_{13}N$	$4-(2-CF_3-C_4F_6-1)-NC_5F_4$	F:	PFHOrg.SVol.4-121, 133
–	$C_6F_5-N=CF-C_3F_7$	F:	PFHOrg.SVol.6-192/3, 207, 219
$C_{10}F_{13}NO$	$4-[2,2-(CF_3)_2-OC_3F_3-3]-NC_5F_4$	F:	PFHOrg.SVol.4-121, 133
$C_{10}F_{14}^-$	$[NC-C(CF=C(CF_3)_2)_2]^-$	F:	PFHOrg.SVol.6-100, 120, 131
$C_{10}F_{15}HN_2$	$1,2,4,6,7-(CF_3)_5-[4.1.0.0^{2,7}]-3,5-N_2C_5H$	F:	PFHOrg.SVol.4-293, 306, 315
–	$1,3,5,6,7-(CF_3)_5-[3.2.0]-2,4-N_2C_5H$	F:	PFHOrg.SVol.4-280/1, 296
–	$1,4,5,6,7-(CF_3)_5-[3.2.0]-2,3-N_2C_5H$	F:	PFHOrg.SVol.4-280/1, 296
$C_{10}F_{15}HN_2S$	$(CF_3)_5C_5HN_2S$	F:	PFHOrg.SVol.4-281
$C_{10}F_{15}I_2N_3O_2$	$NC(CF_3)NC(CF_2OCF_2CF_2I)NC(CF_2OCF_2CF_2I)$	F:	PFHOrg.SVol.4-230, 243
$C_{10}F_{15}N$	$NC(i-C_3F_7)C(CF_3)CFC(CF_3)CF$	F:	PFHOrg.SVol.4-118/9, 128
$C_{10}F_{16}H_4N_2O_2$	$H_2NC(=O)-(CF_2)_8-C(=O)NH_2$	F:	PFHOrg.SVol.5-21, 90
$C_{10}F_{16}H_6N_4$	$H_2N-C(=NH)-(CF_2)_8-C(=NH)-NH_2$	F:	PFHOrg.SVol.5-3, 35, 73
$C_{10}F_{16}H_6N_4O_2$	$H_2N-C(N=OH)-(CF_2)_8-C(N=OH)-NH_2$	F:	PFHOrg.SVol.5-4, 36
$C_{10}F_{16}N_2$	$2,5-(i-C_3F_7)_2-1,4-N_2C_4F_2$	F:	PFHOrg.SVol.4-198/9
–	$3,5-(i-C_3F_7)_2-1,2-N_2C_4F_2$	F:	PFHOrg.SVol.4-185/6, 213
–	$4,5-(i-C_3F_7)_2-1,2-N_2C_4F_2$	F:	PFHOrg.SVol.4-185/6
–	$4,5-(i-C_3F_7)_2-1,3-N_2C_4F_2$	F:	PFHOrg.SVol.4-191/2
–	$4,6-(i-C_3F_7)_2-1,3-N_2C_4F_2$	F:	PFHOrg.SVol.4-191/2
–	$NC-(CF_2)_8-CN$	F:	PFHOrg.SVol.6-102, 135
$C_{10}F_{16}N_2O_2S_2$	$(C_4F_8S-1-)=N-C(O)C(O)-N=(-1-SC_4F_8)$	S:	S-N Comp.8-153
$C_{10}F_{17}N$	$C_{10}F_{15}-NF_2$	F:	PFHOrg.SVol.6-6, 26
$C_{10}F_{17}NO_3$	$CF_2=CF-[O-CF_2CF(CF_3)]_2-O-CF_2-CN$	F:	PFHOrg.SVol.6-99, 143, 152
–	$CF_3-CF(CN)-[CF_2-O-CF(CF_3)]_2-CF=O$	F:	PFHOrg.SVol.6-101, 121
$C_{10}F_{17}N_3$	$NC(n-C_3F_7)NC(n-C_3F_7)NC(CF_3)$	F:	PFHOrg.SVol.4-228/9, 241
$C_{10}F_{18}H_2N_2O_2$	$n-C_4F_9-C(=O)NH-NHC(=O)-C_4F_9-n$	F:	PFHOrg.SVol.5-207, 212
$C_{10}F_{18}H_3N_3O$	$C_3F_7-C(=NH)-N=C(NH_2)-CF(CF_3)-O-C_3F_7$	F:	PFHOrg.SVol.5-5, 75
–	$C_3F_7-O-CF(CF_3)C(=NH)-N=C(NH_2)-C_3F_7$	F:	PFHOrg.SVol.5-5, 36
$C_{10}F_{18}H_3N_3Sn$	$[(CF_3)_2C=N]_3Sn-CH_3$	Sn:	Org.Comp.19-111, 113
$C_{10}F_{18}N_2$	$1-(c-C_6F_9-1)-3-CF_3-1,3-N_2C_3F_6$	F:	PFHOrg.SVol.4-37, 63
–	$2-(NC_5F_{10}-1)-NC_5F_8$	F:	PFHOrg.SVol.4-171/2
–	$2,5-(i-C_3F_7)_2-1,4-N_2C_4F_4$	F:	PFHOrg.SVol.4-199, 212
–	$4,6-(i-C_3F_7)_2-1,3-N_2C_4F_4$	F:	PFHOrg.SVol.4-192, 206
–	$i-C_3F_7-C(=CF_2)-N=CF-N=CF-C_3F_7-i$	F:	PFHOrg.SVol.6-190/1, 204
$C_{10}F_{19}HN_2O$	$n-C_3F_7-C(=O)NH-C(CF_3)_2-N=C(CF_3)_2$	F:	PFHOrg.SVol.5-23/4, 58
$C_{10}F_{19}N$	$1-CF_3-1-NC_9F_{16}$	F:	PFHOrg.SVol.4-290/1, 303, 312
–	$1-(CF_3-c-C_5F_8)-NC_4F_8$	F:	PFHOrg.SVol.4-28
–	$1-(c-C_5F_9)-2-CF_3-NC_4F_7$	F:	PFHOrg.SVol.4-28
–	$1-(c-C_5F_9-CF_2)-NC_4F_8$	F:	PFHOrg.SVol.4-28
–	$1-(c-C_6F_{11})-NC_4F_8$	F:	PFHOrg.SVol.4-28

$C_{10}F_{19}N$ .... $2\text{-}CF_3\text{-}2\text{-}NC_9F_{16}$ .... F: PFHOrg.SVol.4-291, 305/6, 312/3
– .... $2\text{-}(i\text{-}C_3F_7)\text{-}2,6\text{-}(CF_3)_2\text{-}NC_5F_6$ .... F: PFHOrg.SVol.4-119, 130
– .... $NC_8F_{13}(CF_3)_2$ .... F: PFHOrg.SVol.4-282, 285, 297
– .... $[2.2.1]\text{-}C_7F_{10}(CF_3)\text{-}N(CF_3)_2$ .... F: PFHOrg.SVol.6-228/9
– .... $[2.2.1]\text{-}C_7F_{11}\text{-}N(CF_3)\text{-}C_2F_5$ .... F: PFHOrg.SVol.6-230
$C_{10}F_{20}N_2$ .... $(CF_2)_5NN(CF_2)_5$ .... F: PFHOrg.SVol.4-171
$C_{10}F_{20}N_2S$ .... $1\text{-}((CF_3)_2C{=}N\text{-}C(CF_3)_2\text{-}N{=})\text{-}SC_4F_8$ .... S: S-N Comp.8-151/2
$C_{10}F_{20}N_4$ .... $(N(CF_3)CN(CF_3)CF_2CF_2)_2$ .... F: PFHOrg.SVol.4-35, 61, 74
$C_{10}F_{21}N$ .... $(C_2F_5)_2N\text{-}CF{=}C(CF_3)\text{-}C_3F_7\text{-}n$ .... F: PFHOrg.SVol.6-226, 237/8
– .... $t\text{-}C_4F_9\text{-}CF_2\text{-}N{=}CF\text{-}C_4F_9\text{-}t$ .... F: PFHOrg.SVol.6-190/1, 203
– .... $t\text{-}C_4F_9\text{-}N{=}CF\text{-}CF_2\text{-}C_4F_9\text{-}t$ .... F: PFHOrg.SVol.6-190/1, 203
$C_{10}F_{21}NO$ .... $(CF_3)_2N\text{-}C({=}O)C_7F_{15}\text{-}n$ .... F: PFHOrg.SVol.6-225, 236
$C_{10}F_{23}N$ .... $(CF_3)_2N\text{-}C_8F_{17}\text{-}n$ .... F: PFHOrg.SVol.6-225, 236
$C_{10}F_{23}NO_2$ .... $C_2F_5\text{-}N(CF_2CF_2\text{-}O\text{-}C_2F_5)_2$ .... F: PFHOrg.SVol.6-226/7
$C_{10}F_{23}NO_2S$ .... $(t\text{-}C_4F_9\text{-}O)_2S{=}N\text{-}C_2F_5$ .... F: PFHOrg.SVol.6-52, 65; S: S-N Comp.8-159, 162
$C_{10}F_{24}N_4O_2$ .... $(CF_3)_2N\text{-}O\text{-}C(CF_3)_2\text{-}N{=}N\text{-}C(CF_3)_2\text{-}O\text{-}N(CF_3)_2$ F: PFHOrg.SVol.5-116, 127
$C_{10}F_{24}N_4O_2S$ .... $(CF_3)_2N\text{-}O\text{-}C(CF_3)_2\text{-}N{=}S{=}N\text{-}C(CF_3)_2\text{-}O\text{-}N(CF_3)_2$ F: PFHOrg.SVol.6-52, 66/7; S: S-N Comp.7-265
$C_{10}F_{24}N_4O_3$ .... $(CF_3)_2N\text{-}O\text{-}C(CF_3)_2\text{-}N(O){=}N\text{-}C(CF_3)_2\text{-}O\text{-}N(CF_3)_2$ F: PFHOrg.SVol.5-116, 127, 132
$C_{10}F_{24}N_4O_4S_2$ ... $1,3\text{-}S_2C_2\text{-}[NO(CF_3)_2]_4\text{-}2,2,4,4$ .... F: PFHOrg.SVol.5-117, 128
$C_{10}FeGeH_{12}O_3$ ... $(CH_3)_2Ge(\text{-}CH{=}C(CH_3)CH{=}CH\text{-})Fe(CO)_3$ ...... Ge: Org.Comp.3-279/80, 285
$C_{10}FeH_7K_3$ .... $Fe[C_{10}H_7(K)_3]$ .... Fe: Org.Comp.A10-329
$C_{10}FeH_7Li_3$ .... $(Li\text{-}C_5H_4)Fe[C_5H_3(Li)_2]$ .... Fe: Org.Comp.A10-323, 325
$C_{10}FeH_7Na_3$ .... $Fe[C_{10}H_7(Na)_3]$ .... Fe: Org.Comp.A10-328
$C_{10}FeH_7O_3^+$ .... $[(C_7H_7)Fe(CO)_3]^+$ .... Fe: Org.Comp.B15-217, 230
$C_{10}FeH_7O_4^+$ .... $[(1\text{-}HO\text{-}C_7H_6)Fe(CO)_3]^+$ .... Fe: Org.Comp.B15-217
– .... $[(1\text{-}(O{=}CH)C_6H_6)Fe(CO)_3]^+$ .... Fe: Org.Comp.B15-101
– .... $[(6\text{-}(O{=}CH)C_6H_6)Fe(CO)_3]^+$ .... Fe: Org.Comp.B15-91
– .... $[(6\text{-}(O{=})C_7H_7)Fe(CO)_3]^+$ .... Fe: Org.Comp.B15-209, 223/5
– .... $[(6\text{-}(O{=})C_7H_6D)Fe(CO)_3]^+$ .... Fe: Org.Comp.B15-224
– .... $[(6\text{-}(O{=})C_7H_5D_2)Fe(CO)_3]^+$ .... Fe: Org.Comp.B15-224/5
– .... $[(6\text{-}(O{=})C_7H_4D_3)Fe(CO)_3]^+$ .... Fe: Org.Comp.B15-224/5
– .... $[C_5H_5\text{-}CH_2C(O)Fe(CO)_3]^+$ .... Fe: Org.Comp.B15-246/7
– .... $[C_5H_5\text{-}CHDC(O)Fe(CO)_3]^+$ .... Fe: Org.Comp.B15-246/7
– .... $[C_5H_4D\text{-}CHDC(O)Fe(CO)_3]^+$ .... Fe: Org.Comp.B15-246/7
– .... $[C_5H_3D_2\text{-}CHDC(O)Fe(CO)_3]^+$ .... Fe: Org.Comp.B15-246/7
$C_{10}FeH_7O_5^+$ .... $[(1\text{-}HOOC\text{-}C_6H_6)Fe(CO)_3]^+$ .... Fe: Org.Comp.B15-96/7, 101, 136
– .... $[(2\text{-}HOOC\text{-}C_6H_6)Fe(CO)_3]^+$ .... Fe: Org.Comp.B15-96/7, 104, 136
– .... $[(3\text{-}HOOC\text{-}C_6H_6)Fe(CO)_3]^+$ .... Fe: Org.Comp.B15-96/7, 106, 136
– .... $[(6\text{-}HOOC\text{-}C_6H_6)Fe(CO)_3]^+$ .... Fe: Org.Comp.B15-96/7, 109
– .... $[(CH_3C(O)O\text{-}C_5H_4)Fe(CO)_3]^+$ .... Fe: Org.Comp.B15-51

$C_{10}FeH_8K_2$	$Fe(C_5H_4\text{-}K)_2$	Fe: Org.Comp.A10-131
$C_{10}FeH_8Li_2$	$(C_5H_5)Fe(C_5H_3\text{-}Li_2)$	Fe: Org.Comp.A10-119, 120, 123
–	$Fe(C_5H_4\text{-}Li)_2$	Fe: Org.Comp.A10-95/100
$C_{10}FeH_8Li_2O_3$	$Li_2[(C_5H_4C(O)CHCH_3)Fe(CO)_2]$	Fe: Org.Comp.B14-118
$C_{10}FeH_8NO_2^+$	$[(C_5H_5)Fe(CO)_2(CH_2{=}CHCN)]^+$	Fe: Org.Comp.B17-30
$C_{10}FeH_8N_4^{2+}$	$[Fe(C_5H_4\text{-}N_2)_2]^{2+}$	Fe: Org.Comp.A9-146
$C_{10}FeH_8N_6$	$Fe(C_5H_4\text{-}N_3)_2$	Fe: Org.Comp.A9-144, 147
$C_{10}FeH_8N_6O_4S_2$	$Fe(C_5H_4SO_2N_3)_2$	Fe: Org.Comp.A9-237, 243/4
$C_{10}FeH_8Na_2$	$Fe(C_5H_4\text{-}Na)_2$	Fe: Org.Comp.A10-130
$C_{10}FeH_8Na_2O_4S_2$	$Na_2[(O_2S\text{-}C_5H_4)_2Fe] \cdot 3\ H_2O$	Fe: Org.Comp.A9-230
$C_{10}FeH_8Na_2O_6S_2$	$Na_2[Fe(C_5H_4SO_3)_2]$	Fe: Org.Comp.A9-240
$C_{10}FeH_8Na_2S_2$	$Na_2[Fe(C_5H_4S)_2]$	Fe: Org.Comp.A9-203
$C_{10}FeH_8O_2$	$(C_5H_5)Fe(CO)_2\text{-}CH{=}C{=}CH_2$	Fe: Org.Comp.B13-111, 138
–	$(C_5H_5)Fe(CO)_2\text{-}CH_2C{\equiv}CH$	Fe: Org.Comp.B13-148, 155, 158
–	$(C_5H_5)Fe(CO)_2\text{-}C{\equiv}CCH_3$	Fe: Org.Comp.B13-148, 154, 157/8
$C_{10}FeH_8O_3$	$(C_5H_5)Fe(CO)_2\text{-}C({=}O)\text{-}CH{=}CH_2$	Fe: Org.Comp.B13-10, 18, 37
–	$C_5H_4\text{-}CH_2CH_2C({=}O)\text{-}Fe(CO)_2$	Fe: Org.Comp.B18-18/20, 25
$C_{10}FeH_8O_3^{2-}$	$[(C_5H_4C(O)CHCH_3)Fe(CO)_2]^{2-}$	Fe: Org.Comp.B14-118
$C_{10}FeH_8O_4$	$(C_5H_5)Fe(CO)_2\text{-}CH{=}CHCOOH$	Fe: Org.Comp.B13-90
$C_{10}FeH_8O_4S$	$(C_5H_5)Fe(CO)_2(CH_2{=}C{=}CHSO_2)$	Fe: Org.Comp.B17-114
$C_{10}FeH_8O_4S_2^{2-}$	$[(O_2SC_5H_4)_2Fe]^{2-}$	Fe: Org.Comp.A9-230
$C_{10}FeH_8O_6PbS_2$	$Pb[Fe(C_5H_4SO_3)_2] \cdot 4\ H_2O$	Fe: Org.Comp.A9-240
$C_{10}FeH_8O_6S_2^{2-}$	$[Fe(C_5H_4SO_3)_2]^{2-}$	Fe: Org.Comp.A9-240
$C_{10}FeH_8S_2^{2-}$	$[Fe(C_5H_4S)_2]^{2-}$	Fe: Org.Comp.A9-203/4
$C_{10}FeH_9LiO_3$	$Li[(C_5H_4C(O)C_2H_5)Fe(CO)_2]$	Fe: Org.Comp.B14-118
–	$Li[(C_5H_5)Fe(CO)_2\text{-}C(O)CHCH_3]$	Fe: Org.Comp.B14-161
$C_{10}FeH_9NO_3$	$2\text{-}[(C_5H_5)Fe(CO)_2]\text{-}1,3\text{-}ONC_3H_4$	Fe: Org.Comp.B14-32, 37
–	$(C_5H_5)Fe(CO)_2\text{-}CH{=}CHC(O)NH_2$	Fe: Org.Comp.B13-91, 119
$C_{10}FeH_9O_2^+$	$[(C_5H_5)Fe(CO)_2{=}C{=}CHCH_3]^+$	Fe: Org.Comp.B16a-212
–	$[(C_5H_5)Fe(CO)_2(CH_2{=}C{=}CH_2)]^+$	Fe: Org.Comp.B17-110/1, 116
–	$[(C_5H_5)Fe(CO)_2(HC{\equiv}C\text{-}CH_3)]^+$	Fe: Org.Comp.B17-121
$C_{10}FeH_9O_2S_2^+$	$[C_5H_5(CO)_2Fe{=}C(SCH_2)_2]^+$	Fe: Org.Comp.B16a-143/4, 155
$C_{10}FeH_9O_3^+$	$[(1\text{-}CH_3\text{-}C_6H_6)Fe(CO)_3]^+$	Fe: Org.Comp.B15-98/9, 135
–	$[(1\text{-}CH_3\text{-}C_6H_5D)Fe(CO)_3]^+$	Fe: Org.Comp.B15-135
–	$[(2\text{-}CH_3\text{-}C_6H_6)Fe(CO)_3]^+$	Fe: Org.Comp.B15-97, 103/4, 144, 155/6
–	$[(3\text{-}CH_3\text{-}C_6H_6)Fe(CO)_3]^+$	Fe: Org.Comp.B15-106, 157/8
–	$[(CH_3\text{-}C_6H_6)Fe(CO)_3]^+$	Fe: Org.Comp.B15-96
–	$[(C_5H_5)Fe(CO)_2(CH_2{=}CHCHO)]^+$	Fe: Org.Comp.B17-29/30
–	$[(C_7H_9)Fe(CO)_3]^+$	Fe: Org.Comp.B15-191/201
–	$[(C_7H_8D)Fe(CO)_3]^+$	Fe: Org.Comp.B15-192/6
–	$[(C_7H_7D_2)Fe(CO)_3]^+$	Fe: Org.Comp.B15-192
$C_{10}FeH_9O_3^-$	$[(C_5H_5)Fe(CO)_2C(O)CHCH_3]^-$	Fe: Org.Comp.B14-161
–	$[(C_5H_4\text{-}C(O)C_2H_5)Fe(CO)_2]^-$	Fe: Org.Comp.B14-118
$C_{10}FeH_9O_4^+$	$[(1\text{-}CH_3O\text{-}C_6H_6)Fe(CO)_3]^+$	Fe: Org.Comp.B15-96/7, 98, 135

$C_{10}FeH_9O_4^+$ ..... $[(2-CH_3O-C_6H_6)Fe(CO)_3]^+$ ..... Fe: Org.Comp.B15-97, 103, 139/54
– ..... $[(3-CH_3O-C_6H_6)Fe(CO)_3]^+$ ..... Fe: Org.Comp.B15-105, 157
– ..... $[(CH_3O-C_6H_6)Fe(CO)_3]^+$ ..... Fe: Org.Comp.B15-96
– ..... $[(1-HO-C_7H_8)Fe(CO)_3]^+$ ..... Fe: Org.Comp.B15-205, 222
– ..... $[(1-HO-6-D-C_7H_7)Fe(CO)_3]^+$ ..... Fe: Org.Comp.B15-222
– ..... $[(1-HO-7-D-C_7H_7)Fe(CO)_3]^+$ ..... Fe: Org.Comp.B15-222
– ..... $[(1-HO-2,7-D_2-C_7H_6)Fe(CO)_3]^+$ ..... Fe: Org.Comp.B15-222
– ..... $[(1-HO-6,7-D_2-C_7H_6)Fe(CO)_3]^+$ ..... Fe: Org.Comp.B15-222
– ..... $[(1-HO-7,7-D_2-C_7H_6)Fe(CO)_3]^+$ ..... Fe: Org.Comp.B15-222
– ..... $[(1-HO-2,6,7-D_3-C_7H_5)Fe(CO)_3]^+$ ..... Fe: Org.Comp.B15-222
– ..... $[(1-HO-2,7,7-D_3-C_7H_5)Fe(CO)_3]^+$ ..... Fe: Org.Comp.B15-222
– ..... $[(1-HO-2,6,7,7-D_4-C_7H_4)Fe(CO)_3]^+$ ..... Fe: Org.Comp.B15-222
– ..... $[(C_5H_5)Fe(CO)_2(1,3-O_2C_3H_4)]^+$ ..... Fe: Org.Comp.B16a-136, 149/50
$C_{10}FeH_9O_5^+$ ..... $[(CH_2(CH)_4OC(O)CH_3)Fe(CO)_3]^+$ ..... Fe: Org.Comp.B15-9
$C_{10}FeH_{10}$ ..... $(C_5H_5)FeC_4H_4(CH)$ ..... Fe: Org.Comp.B16b-64
$C_{10}FeH_{10}NO_2^+$ ... $[(C_5H_5)Fe(CO)_2(CH_3CH=C=NH)]^+$ ..... Fe: Org.Comp.B17-126/7
– ..... $[(C_5H_5)Fe(CO)_2(CH_3CH=C=ND)]^+$ ..... Fe: Org.Comp.B17-127
– ..... $[(C_6H_7)Fe(CN-CH_3)(CO)_2]^+$ ..... Fe: Org.Comp.B15-350
$C_{10}FeH_{10}NO_2S^+$ .. $[C_5H_5(CO)_2Fe=C(-NHCH_2CH_2S-)]^+$ ..... Fe: Org.Comp.B16a-146/7
$C_{10}FeH_{10}NO_3^+$ ... $[(C_5H_5)Fe(CO)_2=C(-NH-CH_2CH_2-O-)]^+$ ..... Fe: Org.Comp.B14-37
Fe: Org.Comp.B16a-139, 151
$C_{10}FeH_{10}N_2O$ .... $(C_5H_5)Fe(CN-CH_3)(CO)-CH_2CN$ ..... Fe: Org.Comp.B15-305, 308
– ..... $(C_5H_5)Fe(CN-C_2H_5)(CO)-CN$ ..... Fe: Org.Comp.B15-289, 295
$C_{10}FeH_{10}N_4O_4S_2$ $N_3O_2S-C_5H_4FeC_5H_4SO_2NH_2$ ..... Fe: Org.Comp.A9-236, 242/3
$C_{10}FeH_{10}O_2$ ..... $(C_5H_5)Fe(CO)_2C(CH_3)=CH_2$ ..... Fe: Org.Comp.B13-86, 116
– ..... $(C_5H_5)Fe(CO)_2CH=CHCH_3$ ..... Fe: Org.Comp.B13-85, 116
– ..... $(C_5H_5)Fe(CO)_2CH_2CH=CH_2$ ..... Fe: Org.Comp.B13-83, 100, 128/30
– ..... $(C_5H_5)Fe(CO)_2CH_2CH=CD_2$ ..... Fe: Org.Comp.B13-100
– ..... $(C_5H_5)Fe(CO)_2CD_2CH=CH_2$ ..... Fe: Org.Comp.B13-100
– ..... $(C_5H_5)Fe(CO)_2CD_2CH=CD_2$ ..... Fe: Org.Comp.B13-100, 130
– ..... $(C_5H_5)Fe(CO)_2-c-C_3H_5$ ..... Fe: Org.Comp.B13-196, 234
– ..... $(C_5H_5)Fe(CO)_2-c-C_3H_4D-1$ ..... Fe: Org.Comp.B13-196, 234
– ..... $C_5H_4-CH(CH_3)CH_2-Fe(CO)_2$ ..... Fe: Org.Comp.B18-18, 23/4
– ..... $C_5H_4-CH_2CH(CH_3)-Fe(CO)_2$ ..... Fe: Org.Comp.B18-18, 24
– ..... $[1,4-(CH_3)_2C_6H_4]Fe(CO)_2$ ..... Fe: Org.Comp.B18-21
$C_{10}FeH_{10}O_2S_2$ ... $(C_5H_5)Fe(CO)_2C(S)S-C_2H_5$ ..... Fe: Org.Comp.B13-56
– ..... $(C_5H_4-CH_3)Fe(CO)_2C(S)S-CH_3$ ..... Fe: Org.Comp.B14-55, 63
$C_{10}FeH_{10}O_3$ ..... $(C_5H_5)Fe(CO)-C(=O)O-CH_2CH=CH_2$ ..... Fe: Org.Comp.B17-168
– ..... $(C_5H_5)Fe(CO)_2-C(CH_2OH)=CH_2$ ..... Fe: Org.Comp.B13-87, 117/8
– ..... $(C_5H_5)Fe(CO)_2-C(O)C_2H_5$ ..... Fe: Org.Comp.B13-15, 35
– ..... $(C_5H_5)Fe(CO)_2-C(OCH_3)=CH_2$ ..... Fe: Org.Comp.B13-85, 115
– ..... $(C_5H_4-CH_3)Fe(CO)_2-C(O)CH_3$ ..... Fe: Org.Comp.B14-55
– ..... $[CH_3C(O)-C_5H_4]Fe(CO)_2-CH_3$ ..... Fe: Org.Comp.B14-57, 64
$C_{10}FeH_{10}O_3S$ .... $(C_5H_5)Fe(CO)_2C(S)OC_2H_5$ ..... Fe: Org.Comp.B13-55, 57
$C_{10}FeH_{10}O_4$ ..... $(C_5H_5)Fe(CO)_2-C(O)CH_2OCH_3$ ..... Fe: Org.Comp.B13-14, 33
– ..... $(C_5H_4-CH_3)Fe(CO)_2-CH_2-COOH$ ..... Fe: Org.Comp.B14-54
– ..... $(c-C_7H_9)Fe(CO)_2-COOH$ ..... Fe: Org.Comp.B14-80, 81

$C_{10}FeH_{10}O_4S$ .... $(C_5H_5)Fe(CO)_2(CH_2{=}CHCH_2{-}SO_2)$ .... Fe: Org.Comp.B17-128, 129
– .... $(C_5H_5)Fe(CO)_2{-}CH[-CH_2{-}CH_2{-}S(=O){-}O{-}]$ .... Fe: Org.Comp.B14-31, 36
– .... $(C_5H_5)Fe(CO)_2{-}CD[-CH_2{-}CH_2{-}S(=O){-}O{-}]$ .... Fe: Org.Comp.B14-32
– .... $(C_5H_5)Fe(CO)_2{-}CH[-CH_2{-}S(=O)_2{-}CH_2{-}]$ .... Fe: Org.Comp.B14-3, 11
$C_{10}FeH_{10}O_4S_2$ .... $Fe(C_5H_4SO_2H)_2$ .... Fe: Org.Comp.A9-230
$C_{10}FeH_{10}O_5$ .... $[C_6H_6(OCH_3)]Fe(CO)_2COOH$ .... Fe: Org.Comp.B14-80, 81
$C_{10}FeH_{10}O_6S_2$ .... $Fe(C_5H_4{-}SO_3H)_2$ .... Fe: Org.Comp.A9-231, 239/40
– .... $Fe(C_5H_4{-}SO_3H)_2 \cdot 4\ H_2O$ .... Fe: Org.Comp.A9-239/40
$C_{10}FeH_{10}O_6S_2^+$ .... $[Fe(C_5H_4SO_3H)_2]^+$ .... Fe: Org.Comp.A9-239
$C_{10}FeH_{10}S_2$ .... $Fe(C_5H_4SH)_2$ .... Fe: Org.Comp.A9-199, 202/4
$C_{10}FeH_{10}Se_2$ .... $Fe(C_5H_4SeH)_2$ .... Fe: Org.Comp.A9-270/1
$C_{10}FeH_{11}HgN_3$ .... $C_5H_5Fe(CNCH_3)_2HgCN$ .... Fe: Org.Comp.B15-319
$C_{10}FeH_{11}IN_2O$ .... $[(C_5H_5)Fe(CO)(CN{-}CH_3)(NC{-}CH_3)]I$ .... Fe: Org.Comp.B15-299
– .... $[(C_5H_5)Fe(CO)(CN{-}CH_3)_2]I$ .... Fe: Org.Comp.B15-332, 339
$C_{10}FeH_{11}IO_2$ .... $C_5H_5(CO)(I)Fe{=}C(-OCH_2CH_2CH_2{-})$ .... Fe: Org.Comp.B16a-172
$C_{10}FeH_{11}N$ .... $(CH_3{-}C_5H_4)Fe(NC_4H_4)$ .... Fe: Org.Comp.B17-216
– .... $(C_5H_5)Fe(NC_4H_3{-}2{-}CH_3)$ .... Fe: Org.Comp.B17-201, 229
$C_{10}FeH_{11}NO_2$ .... $(C_5H_5)Fe(CO)(CN{-}CH_3){-}C(O)CH_3$ .... Fe: Org.Comp.B15-305, 309
– .... $(C_5H_5)Fe(CO){-}C(O){-}NH{-}CH_2CH{=}CH_2$ .... Fe: Org.Comp.B17-168
$C_{10}FeH_{11}NO_2S$ .... $(C_5H_5)Fe(CO)_2C(S)NHC_2H_5$ .... Fe: Org.Comp.B13-57/8
$C_{10}FeH_{11}NO_3$ .... $(C_5H_5)Fe(CO)_2{-}C(O)N(CH_3)_2$ .... Fe: Org.Comp.B13-9, 24, 45/6
– .... $(C_5H_5)Fe(CO)_2{-}C(O)NH{-}C_2H_5$ .... Fe: Org.Comp.B13-9, 24
$C_{10}FeH_{11}N_2O^+$ .... $[(C_5H_5)Fe(CO)(CN{-}CH_3)_2]^+$ .... Fe: Org.Comp.B15-331/3, 339/41
– .... $[(C_5H_5)Fe(CO)(CN{-}CH_3){-}NC{-}CH_3]^+$ .... Fe: Org.Comp.B15-299/300
$C_{10}FeH_{11}N_2O_2^+$ .... $[C_5H_5(CO)_2Fe{=}C(NHCH_2)_2]^+$ .... Fe: Org.Comp.B16a-147
$C_{10}FeH_{11}N_2S^+$ .... $[(C_5H_5)Fe(CS)(NCCH_3)_2]^+$ .... Fe: Org.Comp.B15-264/5
$C_{10}FeH_{11}N_3$ .... $C_5H_5Fe(CNCH_3)_2CN$ .... Fe: Org.Comp.B15-319
$C_{10}FeH_{11}N_3O_4$ .... $[(C_5H_5)Fe(CNCH_3)(CO)NCCH_3][NO_3]$ .... Fe: Org.Comp.B15-299
$C_{10}FeH_{11}O^+$ .... $[(C_5H_5)Fe(CO){-}CH_2{=}CHCH{=}CH_2]^+$ .... Fe: Org.Comp.B17-186
– .... $[(C_6H_6)Fe(CO)(CH_2CHCH_2)]^+$ .... Fe: Org.Comp.B18-95/7
$C_{10}FeH_{11}O_2^+$ .... $[(C_5H_5)Fe(CO)_2{=}C(CH_3)_2]^+$ .... Fe: Org.Comp.B16a-85/7, 91, 99
– .... $[(C_5H_5)Fe(CO)_2{=}CH{-}C_2H_5]^+$ .... Fe: Org.Comp.B16a-87, 95
– .... $[(C_5H_5)Fe(CO)_2{=}CD{-}C_2H_5]^+$ .... Fe: Org.Comp.B16a-95
– .... $[(C_5H_5)Fe(CO)_2(CH_2{=}CHCH_3)]^+$ .... Fe: Org.Comp.B17-5
– .... $[(C_5H_5)Fe(CO)_2(CH_2{=}CHCH_2D)]^+$ .... Fe: Org.Comp.B17-7
– .... $[(C_5H_5)Fe(CO)_2(CH_2{=}CDCH_3)]^+$ .... Fe: Org.Comp.B17-8
– .... $[(C_5H_5)Fe(CO)_2(CHD{=}CHCH_3)]^+$ .... Fe: Org.Comp.B17-8
– .... $[(C_5H_5)Fe(CO)_2(CD_2{=}CHCD_3)]^+$ .... Fe: Org.Comp.B17-8
$C_{10}FeH_{11}O_2S^+$ .... $[C_5H_5(CO)_2Fe{=}C(CH_3)SCH_3]^+$ .... Fe: Org.Comp.B16a-114, 124
– .... $[C_5H_5Fe(CS)(CO)OC(CH_3)_2]^+$ .... Fe: Org.Comp.B15-271
$C_{10}FeH_{11}O_2S_2^+$ .... $[C_5H_5(CO)_2Fe{=}C(SCH_3)_2]^+$ .... Fe: Org.Comp.B16a-140, 152/4
$C_{10}FeH_{11}O_3^+$ .... $[(CH_2C(CH_3)CHC(CH_3)CH_2)Fe(CO)_3]^+$ .... Fe: Org.Comp.B15-25, 31
– .... $[(CH_2CHCHCHCH{-}C_2H_5)Fe(CO)_3]^+$ .... Fe: Org.Comp.B15-2/3, 12/3, 15, 19/20
– .... $[(CH_3{-}CHCHCHCHCH{-}CH_3)Fe(CO)_3]^+$ .... Fe: Org.Comp.B15-4/5, 12/7, 21

$C_{10}FeH_{11}O_3^+$	$[(C_5H_5)Fe(CO)_2{=}C(CH_3)OCH_3]^+$	Fe: Org.Comp.B16a-108, 118/9
–	$[(C_5H_5)Fe(CO)_2{=}CH{-}O{-}C_2H_5]^+$	Fe: Org.Comp.B16a-108, 118
–	$[(C_5H_5)Fe(CO)_2(CH_2{=}C(CH_3)OH)]^+$	Fe: Org.Comp.B17-66/7
–	$[(C_5H_5)Fe(CO)_2(CH_2{=}C(CH_3)OD)]^+$	Fe: Org.Comp.B17-67
–	$[(C_5H_5)Fe(CO)_2(CH_2{=}CHCH_2OH)]^+$	Fe: Org.Comp.B17-24
–	$[(C_5H_5)Fe(CO)_2(CH_2{=}CHOCH_3)]^+$	Fe: Org.Comp.B17-21, 48
$C_{10}FeH_{11}O_3S^+$	$[C_5H_5(CO)_2Fe{=}C(OCH_3)SCH_3]^+$	Fe: Org.Comp.B16a-137, 150
$C_{10}FeH_{11}O_3Si^+$	$[(1,1{-}(CH_3)_2SiC_5H_5)Fe(CO)_3]^+$	Fe: Org.Comp.B15-134
$C_{10}FeH_{11}O_4^+$	$[(CH_3{-}CHCHCHCHC(OH){-}CH_3)Fe(CO)_3]^+$	Fe: Org.Comp.B15-5/6, 8, 17, 23, 30
–	$[(CH_3{-}CHCDCHCDC(OH){-}CH_3)Fe(CO)_3]^+$	Fe: Org.Comp.B15-5/6, 8
–	$[(C_5H_5)Fe(CO)_2{=}C(OCH_3)_2]^+$	Fe: Org.Comp.B16a-135/6, 148
–	$[(C_5H_5)Fe(CO)_2(CH_2{=}C(OH){-}OCH_3)]^+$	Fe: Org.Comp.B17-68
–	$[(C_5H_5)Fe(CO)_2(HO{-}CH{=}CH{-}OCH_3)]^+$	Fe: Org.Comp.B17-68/9
–	$[(C_5H_5)Fe(CO)_2{-}CH_2{-}C(OH){-}OCH_3]^+$	Fe: Org.Comp.B14-127
$C_{10}FeH_{11}P$	$(C_5H_5)Fe(PC_4H_3{-}2{-}CH_3)$	Fe: Org.Comp.B17-204
–	$(C_5H_5)Fe(PC_4H_3{-}3{-}CH_3)$	Fe: Org.Comp.B17-205
$C_{10}FeH_{11}S^+$	$[(C_5H_5)Fe(SC_4H_3{-}2{-}CH_3)]^+$	Fe: Org.Comp.B17-214
–	$[(C_5H_5)Fe(SC_4H_3{-}3{-}CH_3)]^+$	Fe: Org.Comp.B17-214
$C_{10}FeH_{12}$	$(C_5H_5)Fe(CH_2CHCHCHCH_2)$	Fe: Org.Comp.B17-253, 254
–	$(C_6H_6)Fe(CH_2CHCHCH_2)$	Fe: Org.Comp.B18-103, 105
$C_{10}FeH_{12}IN$	$[(C_5H_5)Fe(NC_4H_4{-}1{-}CH_3)]I$	Fe: Org.Comp.B17-203
$C_{10}FeH_{12}N^+$	$[(C_5H_5)Fe(NC_4H_4{-}1{-}CH_3)]^+$	Fe: Org.Comp.B17-203/4
$C_{10}FeH_{12}NO^+$	$[(C_5H_5)Fe(CH_2{=}CH_2)(CNCH_3)(CO)]^+$	Fe: Org.Comp.B17-137/8
$C_{10}FeH_{12}NO_2^+$	$[C_5H_5(CO)_2Fe{=}CH(N(CH_3)_2)]^+$	Fe: Org.Comp.B16a-115
$C_{10}FeH_{12}NO_3^+$	$[C_5H_5(CO)_2Fe{=}C(OCH_3)NHCH_3]^+$	Fe: Org.Comp.B16a-137, 151
$C_{10}FeH_{12}N_2$	$Fe(C_5H_4{-}NH_2)_2$	Fe: Org.Comp.A9-2, 3
$C_{10}FeH_{12}N_2O_2$	$C_5H_5Fe(CNCH_3)(CO)C(O)NHCH_3$	Fe: Org.Comp.B15-304
$C_{10}FeH_{12}N_2O_3$	$(C_5H_5)Fe(CO)_2CON(CH_3)NHCH_3$	Fe: Org.Comp.B13-25, 46
$C_{10}FeH_{12}N_2O_4S_2$	$Fe(C_5H_4SO_2NH_2)_2$	Fe: Org.Comp.A9-234
$C_{10}FeH_{12}O$	$(C_5H_5)Fe(CO)(CH_2C(CH_3)CH_2)$	Fe: Org.Comp.B17-148/9
–	$(C_5H_5)Fe(CO)(CH_2CHCH{-}CH_3)$	Fe: Org.Comp.B17-148, 154
$C_{10}FeH_{12}O_2$	$(CH_3{-}C_5H_4)Fe(CO)_2{-}C_2H_5$	Fe: Org.Comp.B14-54
–	$(C_2H_5{-}C_5H_4)Fe(CO)_2{-}CH_3$	Fe: Org.Comp.B14-53, 55, 63
–	$(C_5H_5)Fe(CO)(CH_2CHCH{-}OCH_3)$	Fe: Org.Comp.B17-149
–	$(C_5H_5)Fe(CO){-}CH_2OCH_2CH{=}CH_2$	Fe: Org.Comp.B17-168
–	$(c{-}C_7H_9)Fe(CO)_2{-}CH_3$	Fe: Org.Comp.B14-80, 81
$C_{10}FeH_{12}O_3$	$(C_2H_5O{-}C_5H_4)Fe(CO)_2CH_3$	Fe: Org.Comp.B14-58, 65
$C_{10}FeH_{13}IO_2S$	$[C_5H_5Fe(CO)_2CH_2S(CH_3)_2]I$	Fe: Org.Comp.B14-133/5, 142
$C_{10}FeH_{13}LiO_2Si$	$Li[(C_5H_4Si(CH_3)_3)Fe(CO)_2]$	Fe: Org.Comp.B14-119
$C_{10}FeH_{13}N_2O_2^+$	$[C_5H_5(CO)(CH_3CN)Fe{=}C(OCH_3)NH_2]^+$	Fe: Org.Comp.B16a-66
$C_{10}FeH_{13}N_3O$	$C_5H_5(CO)(CN)Fe{=}C(NHCH_3)_2$	Fe: Org.Comp.B16a-172, 173/4
$C_{10}FeH_{13}NaO_2Si$	$Na[(C_5H_4Si(CH_3)_3)Fe(CO)_2]$	Fe: Org.Comp.B14-119
$C_{10}FeH_{13}O_2S^+$	$[C_5H_5Fe(CO)_2CH_2S(CH_3)_2]^+$	Fe: Org.Comp.B14-133/4, 142
$C_{10}FeH_{13}O_2Si^-$	$[(C_5H_4Si(CH_3)_3)Fe(CO)_2]^-$	Fe: Org.Comp.B14-119
$C_{10}FeH_{14}$	$(CH_2CHCHCHCH_2)_2Fe$	Fe: Org.Comp.B17-247, 248/9

$C_{10}FeH_{14}$	$(C_6H_6)Fe(CH_2{=}CH_2)_2$	Fe: Org.Comp.B18-81, 82, 90/1
$C_{10}FeH_{14}NO_2^+$	$[(C_5H_5)Fe(CO)_2\text{-}CH_2CH_2\text{-}NH_2\text{-}CH_3]^+$	Fe: Org.Comp.B14-149, 155
–	$[(C_5H_5)Fe(CO)_2\text{-}CH_2\text{-}NH(CH_3)_2]^+$	Fe: Org.Comp.B14-136, 143
$C_{10}FeH_{14}O_2$	$[CH_2{=}C(CH_3)\text{-}CH\text{-}C(CH_3){=}CH_2]Fe(CO)_2CH_3$	Fe: Org.Comp.B14-80/1
$C_{10}FeH_{14}O_4PS^+$	$[C_5H_5Fe(CS)(CO)P(OCH_3)_3]^+$	Fe: Org.Comp.B15-272
$C_{10}FeH_{15}NO_2Sn$	$(CH_3)_3SnNC\text{-}Fe(CO)_2(CH_2{=}CHCH{=}CH_2)$	Sn: Org.Comp.18-128
$C_{10}FeH_{15}OP$	$(C_6H_6)Fe(CO)P(CH_3)_3$	Fe: Org.Comp.B18-20/1
$C_{10}FeH_{16}N_2O_6S_2$	$[NH_4]_2[Fe(C_5H_4SO_3)_2] \cdot n\ H_2O$	Fe: Org.Comp.A9-240
$C_{10}FeH_{17}S_3^+$	$[C_5H_5Fe(CS)(S(CH_3)_2)_2]^+$	Fe: Org.Comp.B15-264
$C_{10}Fe_2H_8O_6S_2$	$Fe[Fe(C_5H_4SO_3)_2] \cdot 4\ H_2O$	Fe: Org.Comp.A9-239
$C_{10}Fe_3HNO_{10}$	$(CO)_{10}Fe_3N\text{-}H$	Fe: Org.Comp.C6b-27, 31/2
$C_{10}Fe_3HO_9Sb$	$(CO)_9Fe_3SbC\text{-}H$	Fe: Org.Comp.C6a-266/7, 269
$C_{10}Fe_3H_3NO_8S_2$	$(CO)_8Fe_3S_2(NC\text{-}CH_3)$	Fe: Org.Comp.C6a-42, 47
$C_{10}Fe_3H_3O_9^-$	$[(CO)_9(H)Fe_3(CH_2)]^-$	Fe: Org.Comp.C6a-285/6
$C_{10}Fe_3H_3O_9PS$	$(CO)_9Fe_3SP\text{-}CH_3$	Fe: Org.Comp.C6a-202/3, 204
$C_{10}Fe_3H_3O_9S^-$	$[(CO)_9Fe_3S\text{-}CH_3]^-$	Fe: Org.Comp.C6a-136
$C_{10}Fe_3H_3O_{10}^-$	$[(CO)_9Fe_3(OCH_3)]^-$	Fe: Org.Comp.C6a-132
$C_{10}Fe_3H_4O_9$	$(CO)_9(H)_3Fe_3(C\text{-}H)$	Fe: Org.Comp.C6a-249/52
$C_{10}Fe_3H_4O_9P^-$	$[(CO)_9(H)Fe_3P\text{-}CH_3]^-$	Fe: Org.Comp.C6a-223, 224, 226
$C_{10}Fe_3H_4O_9S$	$(CO)_9(H)Fe_3S\text{-}CH_3$	Fe: Org.Comp.C6a-94
$C_{10}Fe_3H_5O_9P$	$(CO)_9(H)_2Fe_3P\text{-}CH_3$	Fe: Org.Comp.C6a-187, 188, 191
$C_{10}Fe_3H_7NO_8S_2$	$(CO)_8Fe_3S_2[NH(CH_3)_2]$	Fe: Org.Comp.C6a-42, 46/7
$C_{10}Fe_3H_{18}N_3O_7P_3$	$(NO)(CO)_2Fe[\text{-}P(CH_3)_2\text{-}]Fe(CO)_2[\text{-}P(CH_3)_2\text{-}]_2Fe(NO)_2$	Fe: Org.Comp.C6a-5
$C_{10}Fe_3O_{10}^+$	$[Fe_3(CO)_{10}]^+$	Fe: Org.Comp.C6a-4
$C_{10}Fe_3O_{10}^-$	$[(CO)_{10}Fe_3]^-$	Fe: Org.Comp.C6a-4
$C_{10}Fe_3O_{10}S$	$(CO)_{10}Fe_3S$	Fe: Org.Comp.C6b-22/4
$C_{10}Fe_3O_{10}Te_2$	$Fe_3Te_2(CO)_{10}$	Fe: Org.Comp.C6b-25/6
$C_{10}GaH_4O_8S_2^-$	$[Ga(4,5\text{-}(O)_2\text{-}C_{10}H_4\text{-}2,7\text{-}(SO_3)_2)]^-$	Ga: SVol.D1-33/4
$C_{10}GaH_6I_2O_2$	$GaI_2[1,2\text{-}(O)_2C_{10}H_6]$, radical	Ga: SVol.D1-90/2
$C_{10}GaH_8K_3N_6O_4S_3$	$K_3[Ga(NCO)_3(NCS)_3] \cdot OC_4H_8$	Ga: SVol.D1-113
$C_{10}GaH_9N_2O^{2+}$	$[(HO)Ga(2\text{-}(NC_5H_4\text{-}2)\text{-}NC_5H_4)]^{2+}$	Ga: SVol.D1-263
$C_{10}GaH_{10}I_4N_2^-$	$[H(NC_5H_5)_2][GaI_4(NC_5H_5)_2]$	Ga: SVol.D1-250
$C_{10}GaH_{10}N_2O_2^+$	$[(HO)_2Ga(2\text{-}(NC_5H_4\text{-}2)\text{-}NC_5H_4)]^+$	Ga: SVol.D1-263
$C_{10}GaH_{11}N_2O_3$	$(HO)_3Ga[2\text{-}(NC_5H_4\text{-}2)\text{-}NC_5H_4]$	Ga: SVol.D1-263
$C_{10}GaH_{11}O_2^{2+}$	$[Ga(1,2\text{-}(O)_2C_7H_4\text{-}4\text{-}C_3H_7\text{-}i)]^{2+}$	Ga: SVol.D1-50/1
$C_{10}GaH_{11}O_5$	$(C_6H_5\text{-}O)Ga[OC(O)\text{-}CH_3]_2$	Ga: SVol.D1-154
$C_{10}GaH_{12}N_2O_4^-$	$[(HO)_4Ga(2\text{-}(NC_5H_4\text{-}2)\text{-}NC_5H_4)]^-$	Ga: SVol.D1-263
$C_{10}GaH_{13}MoN_2O_2$	$[CH_2C(CH_3)CH_2]Mo(CO)_2[\text{-}N_2C_3H_3\text{-}]Ga\text{-}CH_3$	Mo: Org.Comp.5-235
$C_{10}GaH_{14}O_4^+$	$[Ga(CH_3\text{-}C(O)CHC(O)\text{-}CH_3)_2]^+$	Ga: SVol.D1-54/6
$C_{10}GaH_{16}O_4^+$	$[Ga(CH_3\text{-}C({=}O)CH_2C({=}O)\text{-}CH_3)_2]Br$	Ga: SVol.D1-53/4
–	$[Ga(CH_3\text{-}C({=}O)CH_2C({=}O)\text{-}CH_3)_2]Cl$	Ga: SVol.D1-53/4
–	$[Ga(CH_3\text{-}C({=}O)CH_2C({=}O)\text{-}CH_3)_2][GaBr_4]$	Ga: SVol.D1-54
–	$[Ga(CH_3\text{-}C({=}O)CH_2C({=}O)\text{-}CH_3)_2][GaCl_4]$	Ga: SVol.D1-54
–	$[Ga(CH_3\text{-}C({=}O)CH_2C({=}O)\text{-}CH_3)_2][GaI_4]$	Ga: SVol.D1-54
$C_{10}GaH_{16}O_6^-$	$[(HO)_2Ga(CH_3\text{-}C(O)CHC(O)\text{-}CH_3)_2]^-$	Ga: SVol.D1-56

$C_{10}GaH_{17}O_5$	$Ga(OH)[OC(O)-(CH_2)_8-C(O)O]$	Ga: SVol.D1-168
$C_{10}GaH_{30}O_{10}{}^{3+}$	$[Ga(HO-CH_2CH_2-OH)_5]^{3+}$	Ga: SVol.D1-23
$C_{10}GaH_{32}O_6{}^{3+}$	$[Ga(HO-C_2H_5)_5(H_2O)]^{3+}$	Ga: SVol.D1-16, 18
$C_{10}Ga_2H_{10}I_4N_2$	$Ga_2I_4(NC_5H_5)_2$	Ga: SVol.D1-244/5
$C_{10}Ga_2H_{16}I_4O_4$	$[Ga(CH_3-C(=O)CH_2C(=O)-CH_3)_2][GaI_4]$	Ga: SVol.D1-54
$C_{10}GeH_{10}$	$Ge(C_5H_5)_2$	Ge: Org.Comp.3-372/3, 376/7
$C_{10}GeH_{11}MoNO_5$	$(CO)_5Mo[CN-CH_2-Ge(CH_3)_3]$	Ge: Org.Comp.1-140 Mo: Org.Comp.5-6, 7
$C_{10}GeH_{12}O_2$	$(OC_4H_3-2)_2Ge(CH_3)_2$	Ge: Org.Comp.3-157, 162/3
–	$(OC_4H_3-3)_2Ge(CH_3)_2$	Ge: Org.Comp.3-157
–	$[CH_3-C(=O)-C{\equiv}C]_2Ge(CH_3)_2$	Ge: Org.Comp.3-155, 161
$C_{10}GeH_{12}O_3$	$(CH_3)_2Ge(C_8H_6O_3)$	Ge: Org.Comp.3-322
$C_{10}GeH_{12}S_2$	$Ge(CH_3)_2(C(-SCH=CHCH=))_2$	Ge: Org.Comp.3-157
$C_{10}GeH_{14}$	$1,1-(CH_3)_2-1-GeC_8H_8$	Ge: Org.Comp.3-321, 333
–	$2,2-(CH_3)_2-2-GeC_8H_8$	Ge: Org.Comp.3-321
–	$1,1-(CH_3)_2-2-C_6H_5-GeC_2H_3$	Ge: Org.Comp.3-234
$C_{10}GeH_{14}MoN_2O_2$	$C_5H_5Mo(CO)(NO)CNGe(CH_3)_3$	Mo: Org.Comp.6-263
$C_{10}GeH_{14}N_2$	$Ge(CH_3)_3C(C_6H_5)N_2$	Ge: Org.Comp.1-161
$C_{10}GeH_{14}O$	$Ge(CH_3)_3COC_6H_5$	Ge: Org.Comp.1-160, 164
$C_{10}GeH_{14}O_2$	$Ge(CH_3)_3C_6H_4COOH-4$	Ge: Org.Comp.2-75
$C_{10}GeH_{14}O_2S$	$SC_4H_3-2-Ge(CH_3)_2-C(=CH_2)-C(O)O-CH_3$	Ge: Org.Comp.3-204, 207
–	$SC_4H_3-2-Ge(CH_3)_2-CH=CH-C(O)O-CH_3$	Ge: Org.Comp.3-205, 207
$C_{10}GeH_{14}O_3$	$Ge(CH_3)_3C(=CHCH=C(CH=CHCOOH)O)$	Ge: Org.Comp.2-97, 105
$C_{10}GeH_{15}{}^+$	$[GeC_5(CH_3)_5]^+$	Ge: Org.Comp.3-383/4
$C_{10}GeH_{15}LiS$	$Ge(CH_3)_3CH(Li)SC_6H_5$	Ge: Org.Comp.1-143
$C_{10}GeH_{15}N_3O_2$	$Ge(CH_3)_2(C_6H_5)CH_2N(NO)CONH_2$	Ge: Org.Comp.3-201, 206
$C_{10}GeH_{16}$	$2-(CH_3)_3Ge-[2.2.1]-C_7H_7$	Ge: Org.Comp.2-40
–	$(CH_2=CH-CH_2)_2Ge(-CH_2-CH=CH-CH_2-)$	Ge: Org.Comp.3-257
–	$C_6H_5-Ge(CH_3)_2C_2H_5$	Ge: Org.Comp.3-202
–	$Ge(CH_3)_3-CH_2C_6H_5$	Ge: Org.Comp.1-156, 162
–	$Ge(CH_3)_3-C{\equiv}C-C{\equiv}CC_3H_7-n$	Ge: Org.Comp.2-59
–	$Ge(CH_3)_3-C_6H_4CH_3-2$	Ge: Org.Comp.2-73, 84/5
–	$Ge(CH_3)_3-C_6H_4CH_3-3$	Ge: Org.Comp.2-74, 84/5
–	$Ge(CH_3)_3-C_6H_4CH_3-4$	Ge: Org.Comp.2-74, 84/5
–	$Ge(C_2H_5)_2(C{\equiv}C-CH_3)_2$	Ge: Org.Comp.3-170
–	$Ge(C_2H_5)_3-C{\equiv}C-C{\equiv}CH$	Ge: Org.Comp.2-239
–	$[-CH_2-C(CH_3)=CH-CH_2-]_2Ge$	Ge: Org.Comp.3-343, 346
–	$[-CH_2-CH=CH-CH_2-]Ge[-CH_2-C(CH_3)=C(CH_3)-CH_2-]$	Ge: Org.Comp.3-343, 346
$C_{10}GeH_{16}N_2O$	$Ge(CH_3)_2(C_6H_5)CH_2NHCONH_2$	Ge: Org.Comp.3-201
$C_{10}GeH_{16}O$	$Ge(CH_3)_3-CH(OH)C_6H_5$	Ge: Org.Comp.1-159, 163
–	$Ge(CH_3)_3-CD(OH)C_6H_5$	Ge: Org.Comp.1-159, 163
–	$Ge(CH_3)_3-C_6H_4-OCH_3-2$	Ge: Org.Comp.2-71, 82/3
–	$Ge(CH_3)_3-C_6H_4-OCH_3-3$	Ge: Org.Comp.2-71, 82/3
–	$Ge(CH_3)_3-C_6H_4-OCH_3-4$	Ge: Org.Comp.2-71/2, 82/3
$C_{10}GeH_{16}OS$	$Ge(CH_3)_3-CH_2-S(O)C_6H_5$	Ge: Org.Comp.1-138
–	$Ge(CH_3)_3-C_6H_4-S(O)CH_3-4$	Ge: Org.Comp.2-72
–	$SC_4H_3-2-Ge(CH_3)_2-C(=CH_2)-CH_2-O-CH_3$	Ge: Org.Comp.3-204, 207
–	$SC_4H_3-2-Ge(CH_3)_2-CH=CH-CH_2-O-CH_3$	Ge: Org.Comp.3-204, 207
$C_{10}GeH_{16}O_2$	$Ge(CH_3)_2[C{\equiv}C-CH(OH)CH_3]_2$	Ge: Org.Comp.3-155

$C_{10}GeH_{16}O_2$	OC_4H_7-2-$Ge(CH_3)_2$-2-OC_4H_3	Ge: Org.Comp.3-204
$C_{10}GeH_{16}O_3$	$Ge(CH_3)_3CH{=}CHCH_2CH(C({=}O)OC({=}O)CH_2)$	Ge: Org.Comp.2-11
$C_{10}GeH_{16}O_4$	$(CH_3)_2Ge(-CH(C({=}O)OCH_3)CH{=}CHCH(COOCH_3)-)$	Ge: Org.Comp.3-261
$C_{10}GeH_{16}S$	$Ge(CH_3)_3$-$CH(SH)$-C_6H_5	Ge: Org.Comp.1-159, 163
–	$Ge(CH_3)_3$-CH_2-S-C_6H_5	Ge: Org.Comp.1-138
–	$Ge(CH_3)_3$-C_6H_4-SCH_3-4	Ge: Org.Comp.2-72, 83
$C_{10}GeH_{16}S_4$	$((-SCH_2CH_2S-)CH)_2Ge(-CH_2CH{=}CHCH_2-)$	Ge: Org.Comp.3-268
$C_{10}GeH_{17}N$	$Ge(CH_3)_3CH_2CH_2C_5H_4N$-4	Ge: Org.Comp.1-174
$C_{10}GeH_{17}NO_2$	$Ge(C_2H_5)_3C{\equiv}C$-$CH{=}CHNO_2$	Ge: Org.Comp.2-266, 271
$C_{10}GeH_{18}$	1,1-$(C_2H_5)_2$-GeC_6H_8	Ge: Org.Comp.3-309
–	1-$(CH_2{=}CH$-$CH_2)$-1,3,4-$(CH_3)_3$-GeC_4H_4	Ge: Org.Comp.3-267, 273
–	1-$(CH_3)_3Ge$-2-CH_3-2-(c-C_3H_5)-C_3H	Ge: Org.Comp.2-30
–	1-$(HC{\equiv}C)$-1-(n-C_4H_9)-GeC_4H_8	Ge: Org.Comp.3-249, 252
–	$(CH_3)_3Ge$-C_6H_6-3-CH_3	Ge: Org.Comp.2-37
–	$(CH_3)_3Ge$-C_6H_6-4-CH_3	Ge: Org.Comp.2-37
–	$(CH_3)_3Ge$-C_7H_9	Ge: Org.Comp.1-198 Ge: Org.Comp.2-38, 40
–	CH_3-$Ge(CH_2CH{=}CH_2)_3$	Ge: Org.Comp.3-52, 53
–	CH_3-GeC_9H_{15}	Ge: Org.Comp.3-330, 338
–	$Ge(C_2H_5)_3$-$C{\equiv}C$-$CH{=}CH_2$	Ge: Org.Comp.2-238, 247
$C_{10}GeH_{18}N_2O$	$Ge(C_2H_5)_3C({=}C(CHO)CH{=}NNH)$	Ge: Org.Comp.2-290
$C_{10}GeH_{18}O$	2-$(C_2H_5)_3Ge$-OC_4H_3	Ge: Org.Comp.2-286, 294
–	$(C_2H_5)_3Ge$-$C{\equiv}C$-$C({=}O)CH_3$	Ge: Org.Comp.2-264/5
$C_{10}GeH_{19}NO$	$Ge(CH_3)_3$-$C{\equiv}C$-$CH_2C_4H_8NO$	Ge: Org.Comp.2-56
$C_{10}GeH_{19}N_3S$	$Ge(C_2H_5)_3$-$C{\equiv}C$-$CH{=}NNHCSNH_2$	Ge: Org.Comp.2-262
$C_{10}GeH_{20}$	1-$(CH_2{=}CH$-$CH_2)$-1-(n-C_3H_7)-GeC_4H_8	Ge: Org.Comp.3-249
–	3,3-$(C_2H_5)_2$-1-CH_3-[3.1.0]-3-GeC_5H_7	Ge: Org.Comp.3-319
–	$(CH_3)_3Ge$-CH[-CH=CH-$CH(CH_3)$-CH_2CH_2-]	Ge: Org.Comp.2-34/5, 47
–	$(CH_3)_3Ge$-CH[-CH=CH-CH_2CH_2-$CHCH_3$-]	Ge: Org.Comp.2-36
–	$(CH_3)_3Ge$-CH[-CH=CH-CH_2-$CH(CH_3)$-CH_2-]	Ge: Org.Comp.2-35, 47
–	$(CH_3)_3Ge$-CH[-CH_2-$C(CH_3){=}CH$-CH_2CH_2-]	Ge: Org.Comp.2-36
–	$(CH_3)_3Ge$-CH[-CH_2-$CH{=}C(CH_3)$-CH_2CH_2-]	Ge: Org.Comp.2-37
–	$(CH_3)_3Ge$-C[=CH-$CH_2CH_2CH_2$-CH_2CH_2-]	Ge: Org.Comp.2-37
–	$(CH_3)_3Ge$-C[=CH-CH_2CH_2-$CH(CH_3)$-CH_2-]	Ge: Org.Comp.2-34
–	$(C_2H_5)_2Ge$[-$CH{=}C(CH_3)$-$CH(CH_3)$-CH_2-]	Ge: Org.Comp.3-265
–	$(C_2H_5)_2Ge$[-CH_2-$C(CH_3){=}C(CH_3)$-CH_2-]	Ge: Org.Comp.3-259, 272
–	$(C_2H_5)_3Ge$-C_4H_5	Ge: Org.Comp.2-230
–	$Ge(C_2H_5)_2[C(CH_3){=}CH_2]_2$	Ge: Org.Comp.3-169
–	$Ge(C_2H_5)_2[CH_2CH{=}CH_2]_2$	Ge: Org.Comp.3-169, 173
–	$Ge(C_2H_5)_3$-$CH{=}CHCH{=}CH_2$	Ge: Org.Comp.2-175, 185
–	$Ge(C_2H_5)_3$-$C{\equiv}C$-C_2H_5	Ge: Org.Comp.2-238
–	Ge[-CH_2CH_2-CH_2CH_2-CH_2-$]_2$	Ge: Org.Comp.3-345
–	Ge[-CH_2-$C(CH_3)_2$-CH_2-$]_2$	Ge: Org.Comp.3-342, 346
$C_{10}GeH_{20}N_2O$	$Ge(C_2H_5)_3C({=}N_2)COCH_2CH_3$	Ge: Org.Comp.2-161
$C_{10}GeH_{20}N_2O_2$	$Ge(C_2H_5)_3C({=}N_2)COOC_2H_5$	Ge: Org.Comp.2-137, 139
$C_{10}GeH_{20}O$	1,5-$(CH_3)_2$-3,3-$(C_2H_5)_2$-[3.1.0]-6,3-$OGeC_4H_4$	Ge: Org.Comp.3-321
–	$(C_2H_5)_3Ge$-CH[-CH_2CH_2-C(=O)-]	Ge: Org.Comp.2-168
–	$(C_2H_5)_3Ge$-CH[-CH_2-C(=O)-CH_2-]	Ge: Org.Comp.2-168
–	$Ge(C_2H_5)_3$-$C(C_2H_5){=}C{=}O$	Ge: Org.Comp.2-198

$C_{10}GeH_{20}O$	$Ge(C_2H_5)_3-C{\equiv}C-CH(OH)CH_3$	Ge: Org.Comp.2-263
–	$Ge(C_2H_5)_3-C{\equiv}C-CH_2OCH_3$	Ge: Org.Comp.2-260, 270
–	$Ge(C_2H_5)_3-C{\equiv}C-OC_2H_5$	Ge: Org.Comp.2-252, 257
$C_{10}GeH_{20}OSi$	$Ge(CH_3)_3C(=CHCH=C(Si(CH_3)_3)O)$	Ge: Org.Comp.2-98
$C_{10}GeH_{20}OSn$	$Ge(CH_3)_3C(=CHCH=C(Sn(CH_3)_3)O)$	Ge: Org.Comp.2-98
$C_{10}GeH_{20}O_2$	$(CH_3)_3Ge-CH_2-C_3H_4-2-(COO-C_2H_5)$	Ge: Org.Comp.1-155
–	$Ge(CH_3)_3-C(COO-C_2H_5)=C(CH_3)_2$	Ge: Org.Comp.2-10
–	$Ge(C_2H_5)_3-C{\equiv}C-CH(OH)OCH_3$	Ge: Org.Comp.2-260
$C_{10}GeH_{20}O_4$	$Ge(CH_3)_2(CH_2CH_2COOCH_3)_2$	Ge: Org.Comp.3-149
$C_{10}GeH_{20}S$	$Ge(C_2H_5)_3C{\equiv}CSC_2H_5$	Ge: Org.Comp.2-252, 257/8
$C_{10}GeH_{20}Si$	$Ge(CH=CH_2)_3CH_2Si(CH_3)_3$	Ge: Org.Comp.3-50
$C_{10}GeH_{21}IO$	$Ge(C_2H_5)_3CH=C(I)OC_2H_5$	Ge: Org.Comp.2-190
$C_{10}GeH_{21}N$	$Ge(CH_3)_3-C{\equiv}C-CH_2N(C_2H_5)_2$	Ge: Org.Comp.2-56
–	$Ge(C_2H_5)_3-C(CH_3)_2CN$	Ge: Org.Comp.2-136
–	$Ge(C_2H_5)_3-C{\equiv}C-N(CH_3)_2$	Ge: Org.Comp.2-252
$C_{10}GeH_{21}NO$	$Ge(C_2H_5)_3CH_2CH_2CH_2NCO$	Ge: Org.Comp.2-146, 153
$C_{10}GeH_{21}N_3O$	$(C_2H_5)_2Ge(-CH_2CH_2C(=NNHC(=O)NH_2)CH_2CH_2-)$	Ge: Org.Comp.3-297
$C_{10}GeH_{22}$	$1,1-(C_2H_5)_2-3,4-(CH_3)_2-GeC_4H_6$	Ge: Org.Comp.3-246, 253
–	$1,1-(C_2H_5)_2-3-CH_3-GeC_5H_9$	Ge: Org.Comp.3-290, 297
–	$1,1-(C_2H_5)_2-GeC_6H_{12}$	Ge: Org.Comp.3-309
–	$1-(n-C_4H_9)-1,2-(CH_3)_2-GeC_4H_7$	Ge: Org.Comp.3-249, 254
–	$Ge(CH_3)_3-C(CH_3)_2CH_2C(CH_3)=CH_2$	Ge: Org.Comp.2-16
–	$Ge(CH_3)_3-C(C_3H_7-n)=CH-C_2H_5$	Ge: Org.Comp.2-14
–	$Ge(CH_3)_3-CH(CH_3)CH_2CH_2CH=CHCH_3$	Ge: Org.Comp.2-19
–	$Ge(CH_3)_3-CH(CH_3)CH_2CH_2CH_2CH=CH_2$	Ge: Org.Comp.2-19
–	$Ge(CH_3)_3-CH_2-C_5H_8CH_3-2$	Ge: Org.Comp.1-156
–	$Ge(CH_3)_3-C_6H_{10}-4-CH_3$	Ge: Org.Comp.1-196, 199/200
–	$Ge(C_2H_5)_3-C(CH_3)=CHCH_3$	Ge: Org.Comp.2-171
–	$Ge(C_2H_5)_3-C(C_2H_5)=CH_2$	Ge: Org.Comp.2-170
–	$Ge(C_2H_5)_3-CH_2C(CH_3)=CH_2$	Ge: Org.Comp.2-172
–	$Ge(C_2H_5)_3-CH_2CH=CHCH_3$	Ge: Org.Comp.2-173, 184
–	$Ge(C_2H_5)_3-CD_2CD=CDCHD_2$	Ge: Org.Comp.2-174
–	$Ge(C_2H_5)_3-CH_2CH_2CH=CH_2$	Ge: Org.Comp.2-174
$C_{10}GeH_{22}I_2$	$Ge(C_4H_9-n)_2(CH_2I)_2$	Ge: Org.Comp.3-178
$C_{10}GeH_{22}O$	$1,1-(C_2H_5)_2-3-HO-3,4-(CH_3)_2-GeC_4H_5$	Ge: Org.Comp.3-247, 253
–	$Ge(C_2H_5)_3-C(CH_3)=CH-OCH_3$	Ge: Org.Comp.2-192/3, 200
–	$Ge(C_2H_5)_3-CH(OH)-CH=CHCH_3$	Ge: Org.Comp.2-214
–	$Ge(C_2H_5)_3-CH_2CH_2-C(=O)CH_3$	Ge: Org.Comp.2-160, 166
–	$Ge(C_2H_5)_3-CH_2-C(=O)-C_2H_5$	Ge: Org.Comp.2-160/1
–	$Ge(C_2H_5)_3-C(OCH_3)=CHCH_3$	Ge: Org.Comp.2-207
$C_{10}GeH_{22}O_2$	$Ge(C_2H_5)_3-CH(CH_3)-COO-CH_3$	Ge: Org.Comp.2-133
–	$Ge(C_2H_5)_3-CH_2CH(CH_3)-COOH$	Ge: Org.Comp.2-149
–	$Ge(C_2H_5)_3-CH_2CH_2CH_2-COOH$	Ge: Org.Comp.2-159/60
–	$Ge(C_2H_5)_3-CH_2CH_2CH_2-OC(=O)H$	Ge: Org.Comp.2-144/5
–	$Ge(C_2H_5)_3-CH_2CH_2-OC(=O)-CH_3$	Ge: Org.Comp.2-125
–	$Ge(C_2H_5)_3-CH_2-COO-C_2H_5$	Ge: Org.Comp.2-131, 138
$C_{10}GeH_{22}O_2S$	$Ge(C_2H_5)_3CH_2CH_2SCH_2COOH$	Ge: Org.Comp.2-125
$C_{10}GeH_{22}O_3S$	$Ge(C_2H_5)_3CH(CH_2CH_2CH_2OS(O)_2)$	Ge: Org.Comp.2-290
$C_{10}GeH_{22}S$	$(C_4H_9)_2Ge(-CH_2SCH_2-)$	Ge: Org.Comp.3-236/7

$C_{10}GeH_{22}S_2$	$(C_2H_5)_3Ge\text{-}CH[\text{-}S\text{-}CH_2CH_2CH_2\text{-}S\text{-}]$	Ge: Org.Comp.2-289
–	$(n\text{-}C_4H_9)_2Ge[\text{-}CH_2\text{-}SS\text{-}CH_2\text{-}]$	Ge: Org.Comp.3-251
$C_{10}GeH_{23}NO$	$Ge(C_2H_5)_3CH_2CON(CH_3)_2$	Ge: Org.Comp.2-132
$C_{10}GeH_{23}NO_2S$	$Ge(C_2H_5)_3CH(CH_2CH_2N(CH_3)S(O)_2)$	Ge: Org.Comp.2-290
$C_{10}GeH_{23}O_4P$	$Ge(CH_3)_3CH(CHO)CH_2P(O)(OC_2H_5)_2$	Ge: Org.Comp.1-169
$C_{10}GeH_{24}$	$CH_3\text{-}Ge(C_3H_7)_3$	Ge: Org.Comp.3-1, 6
–	$CH_3\text{-}Ge(C_3H_7\text{-}i)_3$	Ge: Org.Comp.3-14, 16
–	$n\text{-}C_4H_9\text{-}Ge(CH_3)(C_2H_5)\text{-}C_3H_7\text{-}n$	Ge: Org.Comp.3-218
–	$n\text{-}C_4H_9\text{-}Ge(C_2H_5)_3$	Ge: Org.Comp.2-110, 113
–	$i\text{-}C_4H_9\text{-}Ge(C_2H_5)_3$	Ge: Org.Comp.2-110/1
–	$Ge(CH_3)_2(C_4H_9\text{-}n)_2$	Ge: Org.Comp.3-150, 160
–	$Ge(C_2H_5)_2(C_3H_7\text{-}n)_2$	Ge: Org.Comp.3-167, 172
$C_{10}GeH_{24}N_2O$	$Ge(C_2H_5)_3CH_2CH_2CH_2NHCONH_2$	Ge: Org.Comp.2-146
$C_{10}GeH_{24}O$	$Ge(CH_3)_3\text{-}C(CH_3)_2\text{-}CH_2\text{-}C(OH)(CH_3)_2$	Ge: Org.Comp.1-188, 190
–	$Ge(C_2H_5)_3\text{-}CH_2CH_2CH_2\text{-}CH_2OH$	Ge: Org.Comp.2-158
$C_{10}GeH_{24}OSi$	$Ge(C_2H_5)_3COSi(CH_3)_3$	Ge: Org.Comp.2-119
$C_{10}GeH_{24}O_2$	$Ge(CH_3)_2(CH_2CH_2CH_2CH_2OH)_2$	Ge: Org.Comp.3-150
–	$Ge(C_3H_7\text{-}n)_2(CH_2CH_2OH)_2$	Ge: Org.Comp.3-175
$C_{10}GeH_{24}O_3S$	$Ge(C_2H_5)_3C(SO_2OCH_3)(CH_3)_2$	Ge: Org.Comp.2-129
$C_{10}GeH_{25}N$	$Ge(C_2H_5)_3CH_2CH_2N(CH_3)_2$	Ge: Org.Comp.2-125
$C_{10}GeH_{26}OSi$	$Ge(CH_3)_3C_4H_8OSi(CH_3)_3$	Ge: Org.Comp.1-188/90
$C_{10}GeH_{26}O_3SSi$	$Ge(C_2H_5)_3CH_2SO_2OSi(CH_3)_3$	Ge: Org.Comp.2-116/7
$C_{10}GeH_{26}Si$	$Ge(C_2H_5)_3CH_2Si(CH_3)_3$	Ge: Org.Comp.2-117
$C_{10}GeH_{27}PSi$	$Ge(CH_3)_3C[Si(CH_3)_3]P(CH_3)_3$	Ge: Org.Comp.1-147, 151
$C_{10}GeH_{27}PSn$	$(CH_3)_3Sn\text{-}P(C_4H_9\text{-}t)\text{-}Ge(CH_3)_3$	Sn: Org.Comp.19-175
$C_{10}GeH_{28}N_2Sn$	$(CH_3)_3SnN(Ge(CH_3)_3)NHC_4H_9\text{-}t$	Sn: Org.Comp.18-121, 123, 125
$C_{10}GeH_{28}N_5P$	$Ge(CH_3)_3CHNNP[N(CH_3)_2]_3$	Ge: Org.Comp.1-146
$C_{10}GeH_{28}Si_2$	$Ge(CH_3)_2(CH_2Si(CH_3)_3)_2$	Ge: Org.Comp.3-147
$C_{10}GeH_{28}Sn_2$	$Ge(CH_3)_2(CH_2Sn(CH_3)_3)_2$	Ge: Org.Comp.3-148
$C_{10}Ge_2H_{18}O_4Re^-$	$[(CO)_4Re(Ge(CH_3)_3)_2]^-$	Re: Org.Comp.1-343/4
$C_{10}Ge_2H_{27}MnN_3O_3P$	$Mn(NO)_3[P(C_4H_9\text{-}t)(Ge(CH_3)_3)_2]$	Mn: MVol.D8-73/4
$C_{10}Ge_2H_{30}In_2O_2$	$[(CH_3)_2In\text{-}O\text{-}Ge(CH_3)_3]_2$	In: Org.Comp.1-188
$C_{10}HMnO_{10}Re$	$(CO)_5ReMn(CO)_4CHO$	Re: Org.Comp.2-199, 206
$C_{10}HO_{10}ReW$	$(CO)_5ReH\text{-}W(CO)_5$	Re: Org.Comp.2-178/9, 183
$C_{10}H_2O_8Th$	$[Th(1,2,4,5\text{-}C_6H_2(COO)_4)]_n$	Th: SVol.C7-144, 145
$C_{10}H_2O_{20}Th_2$	$H_2[Th_2(C_2O_4)_5] \cdot 9\ H_2O$	Th: SVol.C7-90, 91, 95
$C_{10}H_4MoO_{11}S_2^{4-}$	$[MoO_3(1,8\text{-}(O)_2C_{10}H_4(SO_3)_2\text{-}3,6)]^{4-}$	Mo: SVol.B3b-166
$C_{10}H_4NO_8S_2Th^+$	$Th[1\text{-}(ON)\text{-}2\text{-}(O)\text{-}3,6\text{-}(O_3S)_2C_{10}H_4]^+$	Th: SVol.D1-114
$C_{10}H_4O_8ReW^-$	$[(CO)_4Re(H)(CH{=}CH_2)W(CO)_4]^-$	Re: Org.Comp.1-505
$C_{10}H_4O_8S_2Th$	$Th[1,8\text{-}(O)_2\text{-}3,6\text{-}(O_3S)_2C_{10}H_4]$	Th: SVol.D1-114
$C_{10}H_5K_2MoN_5$	$K_2[C_5H_5Mo(CN)_5]$	Mo: Org.Comp.6-5
$C_{10}H_5MoN_3O_2$	$(C_5H_5)Mo(CO)_2(CN)_3$	Mo: Org.Comp.7-2/3
$C_{10}H_5MoN_5^-$	$[C_5H_5Mo(CN)_5]^-$	Mo: Org.Comp.6-5
$C_{10}H_5NO_5Re^+$	$[(CO)_5ReC_5H_5N]^+$	Re: Org.Comp.2-151/2
$C_{10}H_5O_3Th^{3+}$	$Th[1,2,4\text{-}(O)_3C_{10}H_5]^{3+}$	Th: SVol.D1-116
$C_{10}H_5O_8S_2Th^+$	$Th[1\text{-}(HO)\text{-}8\text{-}(O)\text{-}3,6\text{-}(O_3S)_2C_{10}H_4]^+$	Th: SVol.D1-106/7
$C_{10}H_6MnNO_2S^+$	$[Mn(3\text{-}C_6H_5\text{-}1,2\text{-}SNC_3H\text{-}5\text{-}COO)]^+$	Mn: MVol.D7-228/30
$C_{10}H_6MnN_2O_2S_2$	$[Mn(SC(2\text{-}C_4H_3O)NNCS\text{-}2\text{-}C_4H_3O)(H_2O)_2]$	Mn: MVol.D7-218/20

$C_{10}H_6MnN_2O_3S$ .. $[Mn(SC(2\text{-}C_4H_3O)NNCO\text{-}2\text{-}C_4H_3O)(H_2O)_2]$ Mn:MVol.D7-218/20
$C_{10}H_6MnN_3O_3S^+$ $[Mn(1,3\text{-}N_2C_4H(=S\text{-}2)(C_6H_5\text{-}3)(NO\text{-}5)(=O)_2\text{-}4,6)]^+$ Mn:MVol.D7-75/8
$C_{10}H_6MnN_4O_2$ ... $Mn[OC(CH_3)=C(CN)_2]_2 \cdot 2\ NC_5H_4(CH_3\text{-}2)$ Mn:MVol.D7-14/9
– .... $Mn[OC(CH_3)=C(CN)_2]_2 \cdot 2\ NC_5H_4(CH_3\text{-}4)$ Mn:MVol.D7-14/9
– .... $Mn[OC(CH_3)=C(CN)_2]_2 \cdot 2\ NC_5H_5$ Mn:MVol.D7-14/9
– .... $Mn[OC(CH_3)=C(CN)_2]_2 \cdot 4\ NC_5H_5$ Mn:MVol.D7-14/9
$C_{10}H_6NO_2Th^{3+}$ ... $Th[1\text{-}(O)\text{-}2\text{-}(ON)C_{10}H_6]^{3+}$ .... Th: SVol.D1-113/4
– .... $Th[1\text{-}(ON)\text{-}2\text{-}(O)C_{10}H_6]^{3+}$ .... Th: SVol.D1-114
$C_{10}H_6NO_{14}Th^{3-}$ .. $Th[N(CH_2COO)_3](OOCCOO)_2{}^{3-}$ .... Th: SVol.D1-136
$C_{10}H_6N_2O_2S_2$ .... $1,2\text{-}(O=S=N)_2\text{-}C_{10}H_6$ .... S: S-N Comp.6-166
– .... $1,5\text{-}(O=S=N)_2\text{-}C_{10}H_6$ .... S: S-N Comp.6-166
$C_{10}H_6N_6Sn$ ...... $(CH_3)_2Sn[N=C=C(CN)_2]_2$ .... Sn: Org.Comp.19-72
$C_{10}H_6O_{22}Th_2$ .... $(H_3O)_2[Th_2(C_2O_4)_5] \cdot 5\ H_2O$ .... Th: SVol.C7-89, 91, 95
$C_{10}H_7INO_4Re$ .... $(CO)_4Re(I)NC_5H_4\text{-}4\text{-}CH_3$ .... Re: Org.Comp.1-435/6
$C_{10}H_7MnN_2O_2S^+$ $[Mn(1,3\text{-}N_2C_4H_2(=S\text{-}2)(C_6H_5\text{-}3)(=O)_2\text{-}4,6)]^+$ ... Mn:MVol.D7-75/8
$C_{10}H_7MoNO_6$ .... $(CO)_5Mo(CN\text{-}2\text{-}C_4H_7O)$ Mo:Org.Comp.5-6, 8
$C_{10}H_7NOS$....... $1\text{-}O=S=N\text{-}C_{10}H_7$ .... S: S-N Comp.6-164/5, 191, 197
– .... $2\text{-}O=S=N\text{-}C_{10}H_7$ .... S: S-N Comp.6-165, 191, 197
$C_{10}H_7NO_7Re^+$ ... $[(CO)_5Re(CNCH_2COOC_2H_5)]^+$ .... Re: Org.Comp.2-252, 254
$C_{10}H_7N_5S_4Si$..... $Si(NCS)_4 \cdot NC_5H_4CH_3\text{-}4$ .... Si: SVol.B4-300/1
$C_{10}H_7O_6Re$...... $(CO)_5ReC(O)CH(CH_2)_3$ .... Re: Org.Comp.2-120
$C_{10}H_8KMoNO_3$ ... $K[(C_5H_5)Mo(CO)_2(CN)\text{-}C(=O)CH_3]$.......... Mo:Org.Comp.8-1, 3
– .... $K[(C_5H_5)Mo(CO)_2(CN)\text{-}C(=O)CH_3] \cdot n\ OC_4H_8$ Mo:Org.Comp.8-1, 3
$C_{10}H_8MnN_2O_2S^-$ $[Mn(1\text{-}SC_6H_4COO\text{-}2)(N_2HC_3H_3)]^-$.......... Mn:MVol.D7-53/4
$C_{10}H_8MnN_2O_2S_2$.. $Mn[(ON=CH)\text{-}2\text{-}SC_4H_3]_2$ Mn:MVol.D7-222/5
– .... $[Mn((3\text{-}O)(2\text{-}S=)C_5H_3NH)_2] \cdot H_2O$.......... Mn:MVol.D7-65
$C_{10}H_8MnN_3O_3{}^+$ .. $Mn[H(CH_3\text{-}ONC_3(=O)=N\text{-}N\text{-}C_6H_4\text{-}O)]^+$....... Mn:MVol.D6-254/5
$C_{10}H_8MoNO_2{}^-$.... $[C_5H_5Mo(CO)(CN)(C(O)CCH_3)]^-$............ Mo:Org.Comp.6-280
$C_{10}H_8MoNO_3{}^-$.... $K[(C_5H_5)Mo(CO)_2(CN)\text{-}C(=O)CH_3]$.......... Mo:Org.Comp.8-1, 3
– .... $K[(C_5H_5)Mo(CO)_2(CN)\text{-}C(=O)CH_3] \cdot n\ OC_4H_8$ Mo:Org.Comp.8-1, 3
– .... $[As(C_6H_5)_4][(C_5H_5)Mo(CO)_2(CN)\text{-}C(=O)CH_3]$... Mo:Org.Comp.8-1, 3/4
$C_{10}H_8MoN_2O_4$ ... $1\text{-}(CH_2=CHCH_2)\text{-}1,2\text{-}N_2C_3H_3\text{-}2\text{-}Mo(CO)_4$..... Mo:Org.Comp.5-164, 165
$C_{10}H_8MoN_2O_5$ ... $[\text{-}CH=CH\text{-}N(CH_3)\text{-}N(CH_3)\text{-}]C=Mo(CO)_5$....... Mo:Org.Comp.5-100, 105
– .... $[\text{-}N(CH_3)\text{-}CH=CH\text{-}N(CH_3)\text{-}]C=Mo(CO)_5$....... Mo:Org.Comp.5-100, 106
$C_{10}H_8MoO_2$...... $(c\text{-}C_4H_4)_2Mo(CO)_2$ Mo:Org.Comp.5-350
$C_{10}H_8MoO_3S_2$.... $(C_5H_5)Mo(CO)_2(CS_2)\text{-}C(=O)\text{-}CH_3$ Mo:Org.Comp.8-176, 177
$C_{10}H_8NO_5Sb$..... $4\text{-}NO_2\text{-}C_{10}H_6Sb(O)(OH)_2$................ Sb: Org.Comp.5-300
$C_{10}H_8NO_6Re$..... $(CO)_5ReC(O)N(CH_2)_4$ .... Re: Org.Comp.2-117
$C_{10}H_8N_2OS$...... $O=S=NNHC_{10}H_7$ .... S: S-N Comp.6-57/60
$C_{10}H_8N_2O_{20}Th_2$ .. $2\ Th(C_2O_4)_2 \cdot (NH_4)_2C_2O_4 \cdot 2\ H_2O$ .... Th: SVol.C7-83
– .... $[NH_4]_2[Th_2(C_2O_4)_5] \cdot 4\ H_2O$ .... Th: SVol.C7-90, 95
– .... $[NH_4]_2[Th_2(C_2O_4)_5] \cdot 7\ H_2O$ .... Th: SVol.C7-83, 90, 95
$C_{10}H_8N_6O_{12}Th$ ... $Th(NO_3)_4 \cdot 2\text{-}(NC_5H_4\text{-}2)\text{-}NC_5H_4$........... Th: SVol.D4-159
$C_{10}H_8N_{10}Th^{2+}$ ... $Th[(H_2N)H_2C_5N_4]_2{}^{2+}$ .... Th: SVol.D1-120
$C_{10}H_8O_6Th$ ...... $Th[4\text{-}CH_3\text{-}6,7\text{-}(O\text{-})_2\text{-}1\text{-}OC_9H_3(=O\text{-}2)](OH)_2$ ... Th: SVol.C7-32/5
– .... $Th[4\text{-}CH_3\text{-}7,8\text{-}(O\text{-})_2\text{-}1\text{-}OC_9H_3(=O\text{-}2)](OH)_2 \cdot 4\ H_2O$ Th: SVol.C7-32/5

$C_{10}H_8Ti$	$[(C_5H_4)_2Ti]_n$	Ti: Org.Comp.5-351
$C_{10}H_9MnNO_3$	$Mn[-2-O-C_6H_3(CH_3-5)-CH=N-CH_2-C(O)O-]$	Mn: MVol.D6-59
–	$Mn[-O-2-C_6H_4-CH=N-CH_2CH_2-C(O)O-]$	Mn: MVol.D6-34/5
$C_{10}H_9MnN_4O_2S^+$	$[Mn(4-NH_2-C_6H_4-SO_2=N-2-C_4H_3N_2-1,3)]^+$	Mn: MVol.D7-116/23
$C_{10}H_9MnO_4P$	$[Mn(C_6H_5)P(CH_2COO)_2]$	Mn: MVol.D8-81/2
$C_{10}H_9MoNO_3$	$(CH_2=C(CH_3)C_5H_4)Mo(CO)_2(NO)$	Mo: Org.Comp.7-6, 10
$C_{10}H_9MoNO_4$	$(C_5H_5)Mo(CO)_2[-O-C(=O)-C(CH_3)=NH-]$	Mo: Org.Comp.7-208/9, 238
$C_{10}H_9MoNO_5$	$(CO)_5Mo(CN-C_4H_9-t)$	Mo: Org.Comp.5-6, 7/8, 10
$C_{10}H_9MoNaO_4$	$Na[(C_5H_5)Mo(CO)_2(-CH_2CH_2-O-C(=O)-)] \cdot CH_3O-CH_2CH_2-OCH_3$	Mo: Org.Comp.8-196
$C_{10}H_9MoO_4^-$	$Na[(C_5H_5)Mo(CO)_2(-CH_2CH_2-O-C(=O)-)] \cdot CH_3O-CH_2CH_2-OCH_3$	Mo: Org.Comp.8-196
$C_{10}H_9NO_5Re^+$	$[(CO)_5Re(CNC_4H_9-t)]^+$	Re: Org.Comp.2-252, 253
$C_{10}H_9NO_7S_2Th$	$Th(OH)_3(C_{10}H_6NO_4S_2) \cdot 2\ H_2O$	Th: SVol.C7-134, 136
$C_{10}H_9N_2O_5Re$	$(CO)_4Re(CNC_4H_9-t)NCO$	Re: Org.Comp.2-247
$C_{10}H_9O_3Sb$	$C_{10}H_7-1-Sb(=O)(OH)_2$	Sb: Org.Comp.5-299
–	$C_{10}H_7-2-Sb(=O)(OH)_2$	Sb: Org.Comp.5-299
$C_{10}H_9O_5Re$	$1-[(CO)_5Re-CH_2]-1-CH_3-c-C_3H_4$	Re: Org.Comp.2-108
–	$(CO)_5Re-CH_2CH=C(CH_3)_2$	Re: Org.Comp.2-109
$C_{10}H_9O_7ReS$	$(CO)_5ReS(O)_2C(CH_3)_2CH=CH_2$	Re: Org.Comp.2-28
$C_{10}H_9O_7Sb$	$(C_6H_5)Sb(O)[-OCH(CO_2H)CH(CO_2H)O-] \cdot H_2O$	Sb: Org.Comp.5-311
$C_{10}H_9Ti$	$[(C_{10}H_9)Ti]_n$	Ti: Org.Comp.5-321
$C_{10}H_{10}IIn$	$(c-C_5H_5)_2InI$	In: Org.Comp.1-156, 157
$C_{10}H_{10}I_2MnN_2O_2S_2$	$[Mn(SC_4H_3(CH=NOH-2))_2I_2]$	Mn: MVol.D7-222/5
$C_{10}H_{10}I_2MoN_2O$	$C_5H_5Mo(NO)(C_5H_5N)I_2$	Mo: Org.Comp.6-33/4
$C_{10}H_{10}LiMoNO_4S$	$Li[(C_5H_5)Mo(CO)_2(-S-CH_2-CH(COO)-NH_2-)]$	Mo: Org.Comp.7-208, 209, 234
$C_{10}H_{10}LiO_6Re$	$Li[(CO)_4Re(C(=O)CH_3)C(=O)C_3H_7-i]$	Re: Org.Comp.1-425
$C_{10}H_{10}MnNO_2S^+$	$[Mn(2-C_6H_5-1,3-SNC_3H_5-4-COO)]^+$	Mn: MVol.D7-230/2
$C_{10}H_{10}MnN_2O_4$	$Mn[-O-3-NC_5H(CH_3-2)(CH_2OH-5)-4-CH=N-CH_2COO-]$	Mn: MVol.D6-84
$C_{10}H_{10}MnN_2O_4S_2$	$[Mn(SC(2-C_4H_3O)NNCS-2-C_4H_3O)(H_2O)_2]$	Mn: MVol.D7-218/20
$C_{10}H_{10}MnN_2O_5S$	$[Mn(SC(2-C_4H_3O)NNCO-2-C_4H_3O)(H_2O)_2]$	Mn: MVol.D7-218/20
$C_{10}H_{10}MnN_2O_6S_3$	$[Mn(SC_4H_3(CH=NOH-2))_2SO_4]$	Mn: MVol.D7-222/5
$C_{10}H_{10}MnN_3O_3S^+$	$[Mn(3-(4-NH_2-C_6H_4-SO_2=N)-5-CH_3-1,2-ONC_3H)]^+$	Mn: MVol.D7-116/23
$C_{10}H_{10}MnN_4OS^{2+}$	$Mn[1-NC_8H_4(CH_3-1)(=O-2)=N-NHC(=S)NH_2-3]^{2+}$	Mn: MVol.D6-350
$C_{10}H_{10}MnN_4O_2S_2$	$[Mn(OC_4H_3-2-C(S)=NNH_2)_2(H_2O)_2]$	Mn: MVol.D7-218/20
$C_{10}H_{10}MnN_4O_3S_3$	$Mn(4-NH_2C_6H_4SO_2NH-COCH_3)(NCS)_2$	Mn: MVol.D7-116/23
$C_{10}H_{10}MnN_4O_4S_2$	$Mn(1,3-N_2C_4H_2(=S-2)(CH_3-3)(=O)_2-4,6)_2$	Mn: MVol.D7-75/8
$C_{10}H_{10}MnN_4O_8S_2$	$[Mn(SC_4H_3(CH=NOH-2))_2(NO_3)_2]$	Mn: MVol.D7-222/5
$C_{10}H_{10}MnN_5O^+$	$Mn[3-(NH-C(NH_2)=N-N=)-1-NC_8H_4(CH_3-1)=O-2]^+$	Mn: MVol.D6-336
$C_{10}H_{10}MnO_8S_4^{2-}$	$[Mn((OC(=O)CH_2S)_2CHCH(SCH_2C(=O)O)_2)]^{2-}$	Mn: MVol.D7-89/90
$C_{10}H_{10}Mn_2O_8S_4$	$Mn_2[(OC(=O)CH_2S)_2CHCH(SCH_2C(=O)O)_2]$	Mn: MVol.D7-89/90
–	$Mn_2[(OC(=O)CH_2S)_2CHCH(SCH_2C(=O)O)_2] \cdot 12\ H_2O$	Mn: MVol.D7-89/90
$C_{10}H_{10}MoNO_4S^-$	$[(C_5H_5)Mo(CO)_2(-S-CH_2-CH(COO)-NH_2-)]^-$	Mo: Org.Comp.7-208, 209, 234
$C_{10}H_{10}MoNO_5^+$	$[(C_2H_5)_2N-CMo(CO)_5][BF_4]$	Mo: Org.Comp.5-108, 113

$C_{10}H_{10}MoN_2O_2$ . . . $C_5H_5Mo(NO)_2C_5H_5$ . . . Mo:Org.Comp.6-96
$C_{10}H_{10}MoN_2O_3$ . . . $(C_5H_5)Mo(CO)_2[-C(=O)-NH-C(CH_3)=NH-]$ . . . Mo:Org.Comp.8-112, 140
$C_{10}H_{10}MoN_2O_4$ . . . $(CO)_4Mo(CN-C_2H_5)_2$ . . . Mo:Org.Comp.5-25
$C_{10}H_{10}MoN_2O_4S$ . . $(C_2H_5)_2N-CMo(CO)_4(SCN)$ . . . Mo:Org.Comp.5-97/100
$C_{10}H_{10}MoN_2O_5$ . . . $[-N(CH_3)-CH_2CH_2-N(CH_3)-]C=Mo(CO)_5$ . . . Mo:Org.Comp.5-100, 106, 112/3
$C_{10}H_{10}MoN_{10}O_2$ . . $Mo(H_2NO)(NO)(N_3)_2(C_{10}H_8N_2)$ . . . Mo:SVol.B3b-200
$C_{10}H_{10}MoO_2$ . . . . . $(C_5H_5)Mo(CO)_2(CH_2CHCH_2)$ . . . Mo:Org.Comp.8-205, 208, 239/41
$C_{10}H_{10}MoO_2S_3$ . . . $(C_5H_5)Mo(CO)_2[-S-C(S-C_2H_5)-S-]$ . . . Mo:Org.Comp.7-189, 197/8
$C_{10}H_{10}MoO_3$ . . . . . $(C_5H_5)Mo(CO)_2[CH_2C(CH_3)O]$ . . . Mo:Org.Comp.8-186, 195
– . . . $(C_5H_5)Mo(CO)_2-C(=O)-C_2H_5$ . . . Mo:Org.Comp.8-7
$C_{10}H_{10}MoO_3S_2$ . . . $(C_5H_5)Mo(CO)_2[-S-C(O-C_2H_5)-S-]$ . . . Mo:Org.Comp.7-189, 197
$C_{10}H_{10}Mo_8N_2O_{26}{}^{4-}$
$[Mo_8O_{26}(C_5H_5N)_2]^{4-}$ . . . Mo:SVol.B3b-130
$C_{10}H_{10}NOTi$ . . . . . $[(C_5H_5)_2Ti(NO)]_n$ . . . Ti: Org.Comp.5-331
$C_{10}H_{10}NO_6Re$ . . . . $(CO)_4Re(CN-C_4H_9-t)COOH$ . . . Re: Org.Comp.2-247
– . . . $(CO)_5ReC(O)N(C_2H_5)_2$ . . . Re: Org.Comp.2-117
– . . . $(CO)_5ReC(O)NHC_4H_9-n$ . . . Re: Org.Comp.2-117
– . . . $(CO)_5ReC(O)NHC_4H_9-s$ . . . Re: Org.Comp.2-117
– . . . $(CO)_5ReC(O)NHC_4H_9-t$ . . . Re: Org.Comp.2-117
$C_{10}H_{10}NO_7Re$ . . . . cis-$(CO)_4Re[C(O)CH_3]C(CH_3)=NH-CH_2COOH$ Re: Org.Comp.1-401
$C_{10}H_{10}N_2O_2S_2Ti$ . . $(C_5H_5)_2Ti(N=S=O)_2$ . . . S: S-N Comp.6-254/5
$C_{10}H_{10}N_2O_4PdS$ . . $Pd(SO_4)(C_5H_5N)_2 \cdot H_2O$ . . . Pd: SVol.B2-237/8
$C_{10}H_{10}N_3O_3Sb$ . . . 4-$(-NHCH=CHCH=)CN=NC_6H_4Sb(O)(OH)_2$ . . . Sb: Org.Comp.5-292
$C_{10}H_{10}N_3O_5SSb$ . . $(4-(2'-C_4H_3N_2NHSO_2)C_6H_4)Sb(O)(OH)_2$ . . . Sb: Org.Comp.5-288
$C_{10}H_{10}N_4O_4S_2$ . . . 4-$NO_2-C_6H_4-S(O)_2N=S=NC(CH_3)_2CN$ . . . S: S-N Comp.7-60/1
$C_{10}H_{10}N_6O_2S_2$ . . . $[-NH-2-C_5H_3N-6-NH-S(O)-NH-2-C_5H_3N-6-NH-S(O)-]_n$ S: S-N Comp.8-368, 370, 371
$C_{10}H_{10}N_6O_{12}Th$ . . $Th(NO_3)_4 \cdot 2\ NC_5H_5$ . . . Th: SVol.D4-159
$C_{10}H_{10}N_8O_{10}Th$ . . $Th(NO_3)_3[4-(4-CH_3-C_6H_4-N=N)-5-NH_2$ $-1,2-N_2C_3H(=O)-3] \cdot 4\ H_2O$ . . . Th: SVol.D4-143
$C_{10}H_{10}N_{10}O_{14}Th$ . . $Th(NO_3)_4 \cdot 2\ [2-(NH_2-CO)-1,4-N_2C_4H_3]$ . . . Th: SVol.D4-135
$C_{10}H_{10}OTi$ . . . . $[(C_5H_5)_2TiO]_n$ . . . Ti: Org.Comp.5-332/3
$C_{10}H_{10}O_2Ti$ . . . . . . $[(C_5H_5)_2TiO_2 \cdot 2\ H_2O]_n$ . . . Ti: Org.Comp.5-333
– . . . $[(C_5H_5)_2TiO_2]_n$ . . . Ti: Org.Comp.5-333
$C_{10}H_{10}O_3Ti_2$ . . . . . $[(C_5H_5)_2Ti_2O_3]_n$ . . . Ti: Org.Comp.5-325/6
$C_{10}H_{10}O_5Re^+$ . . . . $[(CO)_5Re(CH_2=CHCH_2CH_2CH_3)]^+$ . . . Re: Org.Comp.2-351
$C_{10}H_{10}O_6Re^-$ . . . . $[(CO)_4Re(C(=O)CH_3)C(=O)C_3H_7-i]^-$ . . . Re: Org.Comp.1-425
$C_{10}H_{10}O_6ReS_3$ . . . $(CO)_4Re[C_2H_5-OC(S)SC(S)O-C_2H_5]$, radical . . Re: Org.Comp.1-372
$C_{10}H_{10}O_6STi_2$ . . . . $[(C_5H_5)_2Ti_2(SO_4)O_2 \cdot 2\ H_2O]_n$ . . . Ti: Org.Comp.5-326
$C_{10}H_{10}O_6Sb_2$ . . . . 1,4-$[(HO)_2(O)Sb]_2C_{10}H_6$ . . . Sb: Org.Comp.5-316
$C_{10}H_{10}O_8Th$ . . . . . $Th(OH)_2[C_6H_4-1,3-(OCH_2COO)_2]$ . . . Th: SVol.C7-119
$C_{10}H_{10}S_4Ti_2$ . . . . . $[(C_5H_5)TiS_2]_2$ . . . Ti: Org.Comp.5-45
$C_{10}H_{10}Ti$ . . . . $[(C_{10}H_{10})Ti]_n$ . . . Ti: Org.Comp.5-322
$C_{10}H_{11}IN_2Si$ . . . . . $[SiH_3(NC_5H_4-C_5H_4N)]I = SiH_3I \cdot NC_5H_4-C_5H_4N$ Si: SVol.B4-330
$C_{10}H_{11}In$ . . . . $C_6H_5-C{\equiv}C-In(CH_3)_2$ . . . In: Org.Comp.1-101, 104

$C_{10}H_{11}InMoO_3$ ... $(CH_3)_2InMo(C_5H_5)(CO)_3$ .... In: Org.Comp.1-56, 329/31, 332, 334
$C_{10}H_{11}InO_3$ ...... $(CH_3)_2InOC(O)-C_6H_4-2-CHO$ .... In: Org.Comp.1-196, 198
$C_{10}H_{11}InO_3W$ .... $(CH_3)_2InW(C_5H_5)(CO)_3$ .... In: Org.Comp.1-56
$C_{10}H_{11}InO_4$ ...... $C_6H_5-In[OC(O)CH_3]_2$ .... In: Org.Comp.1-231, 233
$C_{10}H_{11}LiMoO_3$ ... $Li[(C_5H_5)Mo(CO)_2(-CH(C_2H_5)-O-)]$ .... Mo: Org.Comp.8-112, 126
$C_{10}H_{11}MnNO_5$ ... $Mn[H(OOC-CHCH_3-N{=}CH-C_6N_3(O)_2)(H_2O)] \cdot 1.5\ H_2O$
Mn: MVol.D6-63, 65
$C_{10}H_{11}MnN_2OPS_2$ $[Mn((CH_3)_2(C_6H_5)P{=}O)(NCS)_2]$ .... Mn: MVol.D8-101/2
$C_{10}H_{11}MnN_2PS_2$ .. $Mn[P(C_6H_5)(CH_3)_2](NCS)_2$ .... Mn: MVol.D8-38
$C_{10}H_{11}MnN_2PSe_2$ $Mn[P(C_6H_5)(CH_3)_2](NCSe)_2$ .... Mn: MVol.D8-38
$C_{10}H_{11}MnN_3O_3S_2$ $[Mn(-O-C_6H_4-C(CH_3){=}N-SN_2C_2-S-)(H_2O)_2]_n$ .. Mn: MVol.D6-87/8
$C_{10}H_{11}MnN_3O_4$ ... $[Mn(-O-2-C_6H_4-CH{=}N-N{=}C(C(O)NHCH_3)O-)(H_2O)]$
Mn: MVol.D6-287
$C_{10}H_{11}MnO_8S_4^-$ .. $[MnH((OC({=}O)CH_2S)_2CHCH(SCH_2C(O{=})O)_2)]^-$ Mn: MVol.D7-89/90
$C_{10}H_{11}MoNO_2$ ... $(C_5H_5)Mo(CO)(NO)(CH_2{=}CHCH{=}CH_2)$ .... Mo: Org.Comp.6-293
– .... $(C_5H_5)Mo(CO)_2[-C(CH_3){=}N(CH_3)-]$ .... Mo: Org.Comp.8-112, 122, 162
$C_{10}H_{11}MoNO_2S$ .. $(C_5H_5)Mo(CO)_2[-C(N(CH_3)_2){=}S-]$ .... Mo: Org.Comp.8-129
$C_{10}H_{11}MoNO_2SSe$ $(C_5H_5)Mo(CO)_2[-S-C(N(CH_3)_2)-Se-]$ .... Mo: Org.Comp.7-189, 198
$C_{10}H_{11}MoNO_2S_2$ .. $(C_5H_5)Mo(CO)_2[-S-C(N(CH_3)_2)-S-]$ .... Mo: Org.Comp.7-189, 195, 203/4
$C_{10}H_{11}MoNO_2Se_2$ $(C_5H_5)Mo(CO)_2[-Se-C(N(CH_3)_2)-Se-]$ .... Mo: Org.Comp.7-189, 199
$C_{10}H_{11}MoNO_3$ ... $(C_5H_5)Mo(CO)_2[-C({=}O)-CH_2CH_2-NH_2-]$ .... Mo: Org.Comp.8-111, 135, 167
– .... $(C_5H_5)Mo(CO)_2[N(-O){=}C(CH_3)_2]$ .... Mo: Org.Comp.7-207, 209/10, 241/2
$C_{10}H_{11}MoNO_3S$ .. $(C_5H_5)Mo(CO)_2[-O-C(N(CH_3)_2)-S-]$ .... Mo: Org.Comp.7-191
$C_{10}H_{11}MoNO_3Se$ $(C_5H_5)Mo(CO)_2[-O-C(N(CH_3)_2)-Se-]$ .... Mo: Org.Comp.7-198
$C_{10}H_{11}MoNO_4$ ... $(C_5H_5)Mo(CO)_2[-O-C({=}O)-CH(CH_3)-NH_2-]$ .... Mo: Org.Comp.7-208, 209, 230
– .... $(C_5H_5)Mo(CO)_2[-O-C({=}O)-CH_2-NH(CH_3)-]$ .... Mo: Org.Comp.7-208, 209, 232/3
$C_{10}H_{11}MoNO_5$ ... $(C_5H_5)Mo(CO)_2[-O-C({=}O)-CH(CH_2OH)-NH_2-]$ Mo: Org.Comp.7-208, 209, 231, 246, 247
$C_{10}H_{11}MoNO_5Pb$ $(CO)_5Mo[CN-CH_2-Pb(CH_3)_3]$ .... Mo: Org.Comp.5-6, 7
$C_{10}H_{11}MoNO_5Si$ .. $(CO)_5Mo[CN-CH_2-Si(CH_3)_3]$ .... Mo: Org.Comp.5-6, 7
$C_{10}H_{11}MoNO_5Sn$ $(CO)_5Mo[CN-CH_2-Sn(CH_3)_3]$ .... Mo: Org.Comp.5-6, 7
$C_{10}H_{11}MoN_2NaO$ $Na[C_5H_5Mo(CO)(CNCH_3)_2]$ .... Mo: Org.Comp.6-270, 271/2
$C_{10}H_{11}MoN_2O^-$ ... $[C_5H_5Mo(CO)(CNCH_3)_2]^-$ .... Mo: Org.Comp.6-270, 271/2
$C_{10}H_{11}MoN_3O_2S$ .. $(CH_2CHCH_2)Mo(CO)_2(NC-CH_3)_2(SCN)$ .... Mo: Org.Comp.5-235, 237, 243
$C_{10}H_{11}MoN_4O_4W^+$
$[C_5H_5Mo(NO)_2(H)W(NO)_2C_5H_5]^+$ .... Mo: Org.Comp.6-57/8
$C_{10}H_{11}MoO_3^-$ .... $Li[(C_5H_5)Mo(CO)_2(-CH(C_2H_5)-O-)]$ .... Mo: Org.Comp.8-112, 126
$C_{10}H_{11}NO_5S^-$ .... $[CH_3C(OC(O)CH_3)(C_6H_5)-N(O)-SO_2]^-$, radical anion
S: S-N Comp.8-326/8
$C_{10}H_{11}NO_7Re^+$ ... $[(CO)_5ReNH_2C(CH_3)(H)COOC_2H_5]^+$ .... Re: Org.Comp.2-151
$C_{10}H_{11}N_2Si^+$ ..... $[H_3SiC_5H_4N(2-C_5H_4N)-2]^+$ .... Si: SVol.B4-330
$C_{10}H_{11}N_5O_8S$ .... $C_2H_4N-1-S(O)-1-NC_2H_4 \cdot HO-C_6H_2-2,4,6-(NO_2)_3$
S: S-N Comp.8-346
$C_{10}H_{11}OTi$ .... $[(C_5H_5)_2Ti(OH)]_n$ .... Ti: Org.Comp.5-331
$C_{10}H_{11}O_3PSn$ .... $(CH_2{=}CH)_2SnOP(O)(C_6H_5)O$ .... Sn: Org.Comp.16-92
$C_{10}H_{11}O_6Re$ ..... cis-$(CO)_4Re[C({=}O)CH_3]{=}C(OH)C_3H_7$-i .... Re: Org.Comp.1-388/9, 404/5

$C_{10}H_{11}Ti$ .... $[(C_5H_5)_2TiH]_n$ .... Ti: Org.Comp.5-328/30
$C_{10}H_{12}IMoNO_2$ ... $(C_5H_5)Mo(CO)_2(I){=}C(CH_3)-NH-CH_3$ Mo:Org.Comp.8-16, 17, 18
$C_{10}H_{12}IMoNO_3$ ... $C_5H_5Mo(NO)(I)(-OC(CH_3){=}CHC(CH_3)O-)$ Mo:Org.Comp.6-47
$C_{10}H_{12}MnNO_3S^+$ $[Mn(SCH_2C({=}O)NHC_6H_3(OCH_3)_2\text{-}2,5)]^+$ Mn:MVol.D7-54
$C_{10}H_{12}MnN_2O_3S$ .. $[Mn(-O-2-C_6H_4-CH{=}N-N{=}C(OC_2H_5)S-)(H_2O)]$ Mn:MVol.D6-361
$C_{10}H_{12}MnO_8P_2^{2-}$ $[Mn(OOCCH_2)_2PCH_2CH_2P(CH_2COO)_2]^{2-}$ Mn:MVol.D8-81/2
$C_{10}H_{12}MnO_8S_4$ ... $[MnH_2((O_2C-CH_2S)_2CH{=}CH(SCH_2-CO_2)_2)]$ Mn:MVol.D7-89/90
– .... $[MnH_2((O_2C-CH_2S)_2CH{=}CH(SCH_2-CO_2)_2)] \cdot 6\ H_2O$ Mn:MVol.D7-89/90
$C_{10}H_{12}Mn_4N_2O_{12}P_4$ $Mn_4[1,2\text{-}((O_3PCH_2)_2N)_2-C_6H_4] \cdot 2\ H_2O$ Mn:MVol.D8-136
$C_{10}H_{12}MoNNaO_2$ $Na[(C_5H_5)Mo(CO)_2{=}C(CH_3)-NH-CH_3]$ Mo:Org.Comp.8-1, 2
$C_{10}H_{12}MoNO_2^+$ .. $[(C_5H_5)Mo(CO)(NO)(CH_2C(CH_3)CH_2)]^+$ Mo:Org.Comp.6-348, 352
– .... $[(C_5H_5)Mo(CO)(NO)(CH_2CHCH-CH_3)]^+$ Mo:Org.Comp.6-347, 352
$C_{10}H_{12}MoNO_2^-$ ... $Na[(C_5H_5)Mo(CO)_2{=}C(CH_3)-NH-CH_3]$ Mo:Org.Comp.8-1, 2
$C_{10}H_{12}MoNO_3P$ .. $(C_5H_5)Mo(CO)_2{=}[2\text{-}(1,3,2\text{-}ONPC_2H_4\text{-}3\text{-}CH_3)]$ .. Mo:Org.Comp.7-35
$C_{10}H_{12}MoN_2O_2$ ... $(C_5H_5)Mo(CO)_2[-N(CH_3){=}CH-N(CH_3)-]$ Mo:Org.Comp.7-157, 164
$C_{10}H_{12}MoN_2O_{11}^{4-}$ $[MoO_3(OOC-CH_2)_2N-CH_2CH_2-N(CH_2-COO)_2]^{4-}$ Mo:SVol.B3b-157/8, 168/70
$C_{10}H_{12}MoN_4O_2$ ... $(CO)_2Mo(CN-CH_3)_4$ Mo:Org.Comp.5-50
$C_{10}H_{12}MoN_5O_3^+$. $[Mo(C_{10}H_8N_2)(H_2NO)_2(NO)]^+$ Mo:SVol.B3b-199
$C_{10}H_{12}MoO_2$ ..... $(CH_2{=}CHCH{=}CH_2)_2Mo(CO)_2$ Mo:Org.Comp.5-349, 354
$C_{10}H_{12}MoO_3$ ..... $(C_5H_5)Mo(CO)_2[-CH(CH_3)-O(CH_3)-]$ Mo:Org.Comp.8-126
$C_{10}H_{12}MoO_4$ ..... $(CH_3-CH{=}CH_2)_2Mo(CO)_4$ Mo:Org.Comp.5-194
$C_{10}H_{12}MoO_4Si$ ... $[CH_2{=}CH-Si(CH_3)_2-CH{=}CH_2]Mo(CO)_4$ Mo:Org.Comp.5-312
$C_{10}H_{12}Mo_2N_2O_{13}$ $(C_5H_5NH)_2Mo_2O_{13} \cdot H_2O$ Mo:SVol.B3b-136
$C_{10}H_{12}Mo_2N_2O_{14}^{4-}$ $[MoO_3(OOC-CH_2)_2N-CH_2CH_2-N(CH_2-COO)_2MoO_3]^{4-}$ Mo:SVol.B3b-157/8, 168/70
$C_{10}H_{12}NO_4Re$ .... $(CO)_4Re[-CH_2-(1,1\text{-}NC_5H_{10})-]$ .... Re: Org.Comp.1-408
$C_{10}H_{12}NO_5Re$ .... cis-$(CO)_4Re[C(O)C_3H_7\text{-}i]C(CH_3){=}NH_2$ .... Re: Org.Comp.1-401/2
$C_{10}H_{12}N_2O_3S$ .... $O{=}S{=}NC_6(CH_3)_4\text{-}2,3,5,6\text{-}NO_2\text{-}4$ .... S: S-N Comp.6-158
$C_{10}H_{12}N_2O_8Po$ ... $Po[(OOCCH_2)_2NC_2H_4N(CH_2COO)_2]$ .... Po: SVol.1-332
$C_{10}H_{12}N_2O_8Th$ ... $Th[(OOC-CH_2)_2N-C_2H_4-N(CH_2-COO)_2]$ ...... Th: SVol.C7-123/4
Th: SVol.D1-93/5, 99/100
– .... $Th[(OOC-CH_2)_2N-C_2H_4-N(CH_2-COO)_2] \cdot 0.5\ H_2O$ Th: SVol.C7-123/4
– .... $Th[(OOC-CH_2)_2N-C_2H_4-N(CH_2-COO)_2] \cdot 1.45\ H_2O$ Th: SVol.C7-123/4
– .... $Th[(OOC-CH_2)_2N-C_2H_4-N(CH_2-COO)_2] \cdot 2\ H_2O$ Th: SVol.C7-123/4
– .... $Th[(OOC-CH_2)_2N-C_2H_4-N(CH_2-COO)_2] \cdot 3\ H_2O$ Th: SVol.C7-123/4
– .... $Th[(OOC-CH_2)_2N-C_2H_4-N(CH_2-COO)_2] \cdot 6\ H_2O$ Th: SVol.C7-123/4
– .... $Th[(OOC-CH_2)_2N-C_2H_4-N(CH_2-COO)_2] \cdot 9\ H_2O$ Th: SVol.C7-123, 124/5
$C_{10}H_{12}N_4O_2Re^+$ .. $[(CO)_2Re(CN-CH_3)_4]^+$ .... Re: Org.Comp.2-283
– .... $[(CO)_2Re(NC-CH_3)_4]^+$ .... Re: Org.Comp.1-107/8

$C_{10}H_{12}N_5O_4Th^{3+}$	$[Th(C_{10}H_{12}N_5O_4)]^{3+}$	Th: SVol.D1-121
$C_{10}H_{12}N_5O_5Th^{3+}$	$[Th(C_{10}H_{12}N_5O_5)]^{3+}$	Th: SVol.D1-121
$C_{10}H_{12}N_6Re$	$Re(CNCH_3)_4(CN)_2$	Re: Org.Comp.2-277, 279
$C_{10}H_{12}N_8O_{12}Th$	$Th(NO_3)_4 \cdot 2\ (2\text{-}NH_2\text{-}NC_5H_4)$	Th: SVol.D4-159
$C_{10}H_{12}O_8Th$	$Th(OOCC_3H_6COO)_2$	Th: SVol.D1-74/6, 78
$C_{10}H_{12}O_{14}Th$	$Th(OOC(CHOH)_3COO)_2 \cdot H_2O$	Th: SVol.C7-110/1
$C_{10}H_{13}IMoN_2O_2$	$[CH_2C(CH_3)CH_2]Mo(I)(CO)_2(NC\text{-}CH_3)_2$	Mo: Org.Comp.5-239
$C_{10}H_{13}IN_2Si$	$[SiH_3(NC_5H_5)_2]I = SiH_3I \cdot 2\ NC_5H_5$	Si: SVol.B4-327/9
–	$[SiD_3(NC_5H_5)_2]I = SiD_3I \cdot 2\ NC_5H_5$	Si: SVol.B4-327
$C_{10}H_{13}InO_2S_2$	$[\text{-}OC(CH_3)O\text{-}]In(CH_3)\text{-}S\text{-}C_6H_3\text{-}2\text{-}(SH)\text{-}4\text{-}(CH_3)$	In: Org.Comp.1-239
$C_{10}H_{13}InS_2$	$(C_6H_5)In[\text{-}S\text{-}CH_2CH_2\text{-}CH_2CH_2\text{-}S\text{-}]$	In: Org.Comp.1-246, 248
$C_{10}H_{13}MnNO_7$	$[Mn(\text{-}O\text{-}HC_5O(=O)(CH_3)\text{-}C(CH_3)=N\text{-}CH_2COO\text{-})(H_2O)_2]_n$	Mn: MVol.D6-96/7
$C_{10}H_{13}MnO_8P_2^-$	$[MnH(OOCCH_2)_2PCH_2CH_2P(CH_2COO)_2]^-$	Mn: MVol.D8-81/2
$C_{10}H_{13}MoNO$	$C_5H_5Mo(NO)(H_2C=C(CH_3)CH=CH_2)$	Mo: Org.Comp.6-185/6
$C_{10}H_{13}MoNO_2$	$(C_5H_5)Mo(CO)(NO)[CH_2=C(CH_3)_2]$	Mo: Org.Comp.6-293
–	$(C_5H_5)Mo(CO)_2[\text{-}CH_2\text{-}N(CH_3)_2\text{-}]$	Mo: Org.Comp.8-113
$C_{10}H_{13}MoNO_3$	$C_5H_5Mo(CO)(NO)CH_2=CHCH_2OCH_3$	Mo: Org.Comp.6-294
$C_{10}H_{13}MoNO_4$	$(CO)_4Mo\text{-}N(CH_3)_2\text{-}CH_2CH_2\text{-}CH=CH_2$	Mo: Org.Comp.5-164, 166
$C_{10}H_{13}MoN_2O^+$	$[C_5H_5Mo(NO)(NCCH_3)C_3H_5]^+$	Mo: Org.Comp.6-177, 182
$C_{10}H_{13}MoN_2O_{11}^{3-}$	$[MoO_3H((OOC\text{-}CH_2)_2N\text{-}CH_2CH_2\text{-}N(CH_2\text{-}COO)_2)]^{3-}$	Mo: SVol.B3b-157/8, 168/70
$C_{10}H_{13}MoO_2PS$	$(CH_3\text{-}C_5H_4)Mo(CO)_2[\text{-}S=P(CH_3)_2\text{-}]$	Mo: Org.Comp.7-207, 212, 242/3
$C_{10}H_{13}MoO_4P$	$(CO)_4Mo\text{-}P(CH_3)_2\text{-}CH_2CH_2\text{-}CH=CH_2$	Mo: Org.Comp.5-164, 165
$C_{10}H_{13}NO_3Re^+$	$[(C_5H_5)Re(CO)(NO)(OC_4H_8)][BF_4]$	Re: Org.Comp.3-155
–	$[(C_5H_5)Re(CO)(NO)(OC_4H_8)][PF_6]$	Re: Org.Comp.3-153, 154, 155
$C_{10}H_{13}NO_3SSn$	$(CH_3)_3SnN(COC_6H_4SO_2)$	Sn: Org.Comp.18-84, 94, 103
$C_{10}H_{13}N_2NaO_9Th$	$Na[Th(OH)((OOCCH_2)_2NC_2H_4N(CH_2COO)_2)] \cdot 4\ H_2O$	Th: SVol.C7-123, 125
$C_{10}H_{13}N_2O_3SSb$	$(4\text{-}CH_2=CHCH_2NHC(S)NHC_6H_4)Sb(O)(OH)_2$	Sb: Org.Comp.5-290
$C_{10}H_{13}N_2O_8Th^+$	$Th[HOOCCH_2N(CH_2COO)C_2H_4N(CH_2COO)_2]^+$	Th: SVol.D1-99/100
$C_{10}H_{13}N_2O_9Th^-$	$Th(OH)[C_2H_4(N(CH_2COO)_2)_2]^-$	Th: SVol.D1-97, 99/100
$C_{10}H_{13}N_3O_2Sn$	$(CH_3)_3Sn\text{-}N(CN)\text{-}C_6H_4\text{-}3\text{-}NO_2$	Sn: Org.Comp.18-58, 68
–	$(CH_3)_3Sn\text{-}N(CN)\text{-}C_6H_4\text{-}4\text{-}NO_2$	Sn: Org.Comp.18-58, 68
$C_{10}H_{13}N_5O_2Sn$	$(CH_3)_3SnN(N=NC(C_6H_4NO_2\text{-}4)=N)$	Sn: Org.Comp.18-84, 92/3
$C_{10}H_{13}NaO_{15}Th$	$NaTh(OH)(OOC(CHOH)_3COO)_2 \cdot H_2O$	Th: SVol.C7-110/1
$C_{10}H_{13}O_5SSb$	$(4\text{-}C_2H_5O_2CCH_2SC_6H_4)Sb(O)(OH)_2$	Sb: Org.Comp.5-288
$C_{10}H_{14}HgIMoO_5P$	$[(C_5H_5)Mo(CO)_2(P(OCH_3)_3)HgI]$	Mo: Org.Comp.7-122, 147
$C_{10}H_{14}IMoO_5P$	$[(C_5H_5)Mo(CO)_2(I)(P(OCH_3)_3)]$	Mo: Org.Comp.7-56/7, 86, 112, 113
$C_{10}H_{14}IORe$	$(CH_3C{\equiv}C(CH_2)_4C{\equiv}CCH_3)Re(O)I$	Re: Org.Comp.2-363
$C_{10}H_{14}I_3MoO_5PSn$	$(C_5H_5)Mo(CO)_2[P(OCH_3)_3]SnI_3$	Mo: Org.Comp.7-119, 129
$C_{10}H_{14}MnN_4O_4S_2$	$[Mn(SC(2\text{-}C_4H_3O)=NNH_2)_2(H_2O)_2]$	Mn: MVol.D7-218/20
$C_{10}H_{14}MnN_6O_4$	$[Mn(1,4\text{-}(O\text{-}C(NH_2)=N\text{-}N=CH)_2C_6H_4)(H_2O)_2]$	Mn: MVol.D6-333/4
$C_{10}H_{14}MnO_8P_2$	$[MnH_2(OOCCH_2)_2PCH_2CH_2P(CH_2COO)_2]$	Mn: MVol.D8-81/2
$C_{10}H_{14}MnS_4$	$[Mn(SC(CH_3)CHC(CH_3)S)_2]$	Mn: MVol.D7-55/7
$C_{10}H_{14}MoN_2O_2Si$	$C_5H_5Mo(CO)(NO)CNSi(CH_3)_3$	Mo: Org.Comp.6-263

$C_{10}H_{14}MoN_2O_2Sn$ $(CH_3)_3SnNC-Mo(CO)(NO)(C_5H_5-c)$ Mo:Org.Comp.6-263/4
Sn: Org.Comp.18-127/8

$C_{10}H_{14}MoNaO_5P$ $Na[(C_5H_5)Mo(CO)_2(P(OCH_3)_3)]$ Mo:Org.Comp.7-47, 50, 52

$C_{10}H_{14}MoO_2P^-$ ... $[(C_5H_5)Mo(CO)_2(P(CH_3)_3)]^-$ Mo:Org.Comp.7-47/9

$C_{10}H_{14}MoO_5P^-$ ... $[(C_5H_5)Mo(CO)_2(P(OCH_3)_3)]^-$ Mo:Org.Comp.7-47, 50, 52

$C_{10}H_{14}MoO_6$ ..... $MoO_2(CH(COCH_3)_2)_2$ Mo:SVol.B3b-177

$C_{10}H_{14}Mo_2N_2O_{13}{}^{2-}$
$[CH_3N(CH_2COO)_2O_2MoOMoO_2(OOCCH_2)_2NCH_3]^{2-}$
Mo:SVol.B3b-167

$C_{10}H_{14}Mo_2N_2O_{14}{}^{2-}$
$[H_2(MoO_3)_2((OOCCH_2)_2N-CH_2CH_2-N(CH_2COO)_2)]^{2-}$
Mo:SVol.B3b-169

$C_{10}H_{14}NO_2Re$ .... $(C_5H_5)Re(CO)_2(NH_2-C_3H_7-n)$ .... Re: Org.Comp.3-193/7, 198

$C_{10}H_{14}NO_3ReSn$ .. $C_3H_5Re(CO)_3CNSn(CH_3)_3$ .... Re: Org.Comp.2-375

$C_{10}H_{14}NO_5Sb$ .... 4-[n-C_3H_7-OC(=O)-NH]-C_6H_4-Sb(=O)$(OH)_2$ .. Sb: Org.Comp.5-290

– .... 4-[i-C_3H_7-OC(=O)-NH]-C_6H_4-Sb(=O)$(OH)_2$ ... Sb: Org.Comp.5-290

$C_{10}H_{14}NO_7Re$ .... (1,4-$O_2C_4H_8$)-1-$Re(CO)_3$[-NH_2-CH_2-CH_2-C(O)-O-]
Re: Org.Comp.1-135

$C_{10}H_{14}N_2OS$ ..... O=S=NN$(C_6H_5)(C_4H_9$-i) .... S: S-N Comp.6-63/8

$C_{10}H_{14}N_2O_2S_2$ ... C_6H_5-$S(O)_2$N=S=N-C_4H_9-t. .... S: S-N Comp.7-54

$C_{10}H_{14}N_2O_8Th$ ... $Th(OOCC_2H_4CH(NH_2)COO)_2$ .... Th: SVol.D1-82/3, 87/8

$C_{10}H_{14}N_2O_{10}Th$ .. $(NO_3)_2Th[CH_3-C(O)CHC(O)-CH_3]_2$ .... Th: SVol.D4-179

$C_{10}H_{14}N_2S$ ...... C_6H_5-N=S=N-C_4H_9-t .... S: S-N Comp.7-218

$C_{10}H_{14}N_2Sn$ ..... 1-$(CH_3)_3$Sn-1,3-$N_2C_7H_5$ .... Sn: Org.Comp.18-83, 89

– .... $(CH_3)_3$Sn-N=C=N-C_6H_5 .... Sn: Org.Comp.18-107, 110

– .... $(CH_3)_3$Sn-N(CN)-C_6H_5 .... Sn: Org.Comp.18-57/8, 66

– .... $[NC_4H_4-1-]_2Sn(CH_3)_2$ .... Sn: Org.Comp.19-70

$C_{10}H_{14}N_3O_7Th$ ... $Th[OC_2H_4N(CH_2COO)NC_2H_4N(CH_2COO)_2]$ · 2 H_2O
Th: SVol.C7-121/3

$C_{10}H_{14}N_3O_8Sb$ ... $(CH_3)_3Sb(OCH_3)OC_6H_2(NO_2)_3$-2,4,6 .... Sb: Org.Comp.5-45

$C_{10}H_{14}N_4OS$ ..... [3,5-$(CH_3)_2$-1,2-N_2C_3H-1$]_2$SO. .... S: S-N Comp.8-354, 361

$C_{10}H_{14}N_4PdS_2$ ... $Pd(SCN)_2(CNC_3H_7-i)_2$ .... Pd: SVol.B2-307

$C_{10}H_{14}N_4Sn$ ..... $(CH_3)_3$SnN(N=NC(C_6H_5)=N) .... Sn: Org.Comp.18-84, 92

$C_{10}H_{14}N_6O_{10}S_4Th$ $Th(NO_3)_2[CH_3-C(=O)-NH-N=C(-S)-S-C(=O)-CH_3]_2$
Th: SVol.D4-161

$C_{10}H_{14}Na_2O_{16}Th$ $Na_2Th(OH)_2(OOC(CHOH)_3COO)_2$. .... Th: SVol.C7-110/1

$C_{10}H_{14}O_2PRe$ .... $(C_5H_5)Re(CO)_2[P(CH_3)_3]$ .... Re: Org.Comp.3-210/1, 212, 215

$C_{10}H_{14}O_4Th^{2+}$ ... $Th[CH_3C(O)CHC(O)CH_3]_2{}^{2+}$. .... Th: SVol.D1-128

$C_{10}H_{14}O_5PRe$ .... $(CO)_4Re[-P(CH_3)_2-O-(CH_2)_4-]$ .... Re: Org.Comp.1-414

$C_{10}H_{14}O_5Th$ ..... $ThO[(CH_3)_3C_5H_5(COO)_2]$ · 3 H_2O .... Th: SVol.C7-112/4

$C_{10}H_{15}IMoN_2OS_2$ $C_5H_5Mo(NO)(I)S_2CN(C_2H_5)_2$. Mo:Org.Comp.6-51

$C_{10}H_{15}IMoN_6O$ ... $[Mo(CN-CH_3)_5(NO)]I$ Mo:Org.Comp.5-52/3

$C_{10}H_{15}I_2MnNOP$ .. $Mn(NO)[P(C_6H_5)(C_2H_5)_2]I_2$ Mn:MVol.D8-46, 61

$C_{10}H_{15}I_2MnP$ .... $Mn(I)_2(SO_2)_n-P(C_6H_5)(C_2H_5)_2$ Mn:MVol.D8-46, 57/8

– .... $Mn(I)_2-P(C_6H_5)(C_2H_5)_2$. Mn:MVol.D8-42/3

$C_{10}H_{15}I_2Sb$ ...... $(C_2H_5)_2(C_6H_5)SbI_2$ .... Sb: Org.Comp.5-68

$C_{10}H_{15}In$ .... $In[C_5(CH_3)_5]$ .... In: Org.Comp.1-379, 381, 383/4

$C_{10}H_{15}InOS_3$	CH_3-In[1,2-(S)$_2$-C_6H_3-4-CH_3] · $OS(CH_3)_2$	In: Org.Comp.1-246, 249, 250/1
$C_{10}H_{15}MnN_2O_3S^+$	[Mn((OOC(CH_2)$_4$)(O=)$C_5H_7N_2S$)]$^+$	Mn: MVol.D7-227
$C_{10}H_{15}MnN_2O_4S^+$	[Mn((OOC(CH_2)$_4$)(O=)$_2C_5H_7N_2S$)]$^+$	Mn: MVol.D7-227
$C_{10}H_{15}MnN_2O_5S^+$	[Mn((OOC(CH_2)$_4$)(O=)$_3C_5H_7N_2S$)]$^+$	Mn: MVol.D7-227
$C_{10}H_{15}MoN_2O_2{}^+$	[(C_5H_5)Mo(CO)$_2$(NH_2-CH(CH_3)CH_2-NH_2)]$^+$	Mo: Org.Comp.7-249, 251
$C_{10}H_{15}MoN_6O^+$	[Mo(CN-CH_3)$_5$(NO)]I	Mo: Org.Comp.5-52/3
–	[Mo(CN-CH_3)$_5$(NO)][PF_6]	Mo: Org.Comp.5-52/3
$C_{10}H_{15}MoO_2P$	[(C_5H_5)Mo(CO)$_2$(H)(P(CH_3)$_3$)]	Mo: Org.Comp.7-55, 57, 60/1, 103
$C_{10}H_{15}MoO_2PS$	(C_5H_5)Mo(CO)$_2$(SH)-P(CH_3)$_3$	Mo: Org.Comp.7-119, 140
$C_{10}H_{15}MoO_5P$	(C_5H_5)Mo(H)(CO)$_2$(P(OCH_3)$_3$)	Mo: Org.Comp.7-55, 57, 84, 111
$C_{10}H_{15}Mo_2N_2O_{14}{}^-$	[H_3(MoO_3)$_2$(($OOCCH_2$)$_2$N-CH_2CH_2-N(CH_2COO)$_2$)]$^-$	Mo: SVol.B3b-169
$C_{10}H_{15}NOS$	O=S=N$C_{10}H_{15}$	S: S-N Comp.6-105
$C_{10}H_{15}NOSn$	(CH_3)$_3$SnN(C_6H_5)CHO	Sn: Org.Comp.18-47, 49
$C_{10}H_{15}NO_2S$	(CH_3)$_2$N-S(O)O-CH(C_6H_5)CH_3	S: S-N Comp.8-306/7, 310
–	C_6H_5-N(C_2H_5)-S(O)O-C_2H_5	S: S-N Comp.8-320/1
$C_{10}H_{15}NO_6Sn$	n-C_4H_9-Sn[-O-C(=O)-CH_2-]$_3$N	Sn: Org.Comp.17-45
–	t-C_4H_9-Sn[-O-C(=O)-CH_2-]$_3$N	Sn: Org.Comp.17-52, 53
$C_{10}H_{15}N_2O_5Sb$	2-(n-C_4H_9-NH)-5-O_2N-C_6H_3-Sb(=O)(OH)$_2$	Sb: Org.Comp.5-297
–	2-(i-C_4H_9-NH)-5-O_2N-C_6H_3-Sb(=O)(OH)$_2$	Sb: Org.Comp.5-297
$C_{10}H_{15}N_2O_7Th^+$	Th[(HOC_2H_4)($OOCCH_2$)NC_2H_4N(CH_2COO)$_2$]$^+$	Th: SVol.D1-93/4, 97/9
$C_{10}H_{15}N_2O_{10}Po^-$	[Po(OH)$_2$(H($OOCCH_2$)$_2$NC_2H_4N(CH_2COO)$_2$)]$^-$	Po: SVol.1-354/5
$C_{10}H_{15}N_3Sn$	(C_2H_5)$_3$SnN=C=C(CN)$_2$	Sn: Org.Comp.18-151/2
$C_{10}H_{15}OPSn$	2,2-(CH_3)$_2$-3-C_6H_5-1,3,2-OPSnC_2H_4	Sn: Org.Comp.19-218, 224
$C_{10}H_{15}O_2ReSi$	(C_5H_5)Re(H)(CO)$_2$-Si(CH_3)$_3$	Re: Org.Comp.3-175/6, 178
$C_{10}H_{15}O_2ReSn$	(C_5H_5)ReH(CO)$_2$-Sn(CH_3)$_3$	Re: Org.Comp.3-175/6, 180
$C_{10}H_{15}O_3Re$	($CH_3C{\equiv}CCH_3$)$_2$Re(O)O_2CCH_3	Re: Org.Comp.2-362, 370
$C_{10}H_{15}O_5PSnW$	(CH_3)$_3$Sn-P(CH_3)$_2$W(CO)$_5$	Sn: Org.Comp.19-181/2
$C_{10}H_{15}O_5ReSi_2$	(CO)$_5$ReSi(CH_3)$_2$Si(CH_3)$_3$	Re: Org.Comp.2-4
$C_{10}H_{15}O_{10}Th^-$	[Th($OOCCH_3$)$_5$]$^-$	Th: SVol.D1-67/8
$C_{10}H_{15}O_{15}Th^-$	[Th($OOCCH_2OH$)$_5$]$^-$	Th: SVol.D1-71
$C_{10}H_{15}PSSn$	2,2-(CH_3)$_2$-3-C_6H_5-1,3,2-SPSnC_2H_4	Sn: Org.Comp.19-218, 224
$C_{10}H_{15}SSb$	(CH_3)(i-C_3H_7)(C_6H_5)SbS	Sb: Org.Comp.5-73, 74
$C_{10}H_{16}MnN_2O_2S_6$	[Mn($C_4H_8S_2$-1,4-(=O-1))$_2$(NCS)$_2$]	Mn: MVol.D7-105
$C_{10}H_{16}MnN_2O_4S_2$	Mn((OOC-4)(CH_3-2)C_3H_5NS-3,1)$_2$ · 2 H_2O	Mn: MVol.D7-230/2
$C_{10}H_{16}MnN_2S_4$	Mn(1-S_2C-NC_4H_8)$_2$	Mn: MVol.D7-166/7
–	[Mn(SC_4H_8)$_2$(NCS)$_2$]	Mn: MVol.D7-225/6
$C_{10}H_{16}MnN_3OS_4{}^+$	[Mn((SC(=S))NC_4H_8)$_2$(NO)]$^+$	Mn: MVol.D7-168
$C_{10}H_{16}MnN_4O_2$	Mn[(CH_3)$_2$C=N-N=C(O-)CH_2CH_2C(O-)=N-N=C(CH_3)$_2$]	Mn: MVol.D6-313/4
$C_{10}H_{16}MnN_4S_8Zn$	MnZn(SC(=S)NH$CH_2CH_2CH_2$NHC(=S)S)$_2$	Mn: MVol.D7-180/4
$C_{10}H_{16}MnN_6S_2$	Mn((2-S)(4-$NH_2CH_2CH_2$)$C_3H_2N_2$-1,3)$_2$	Mn: MVol.D7-58/9
$C_{10}H_{16}MnN_8O_2S_2{}^-$	Mn[-ON=C(CH_3)C(CH_3)=NN=C(NH_2)S-]$_2{}^-$	Mn: MVol.D6-344
$C_{10}H_{16}MoN_2O_3S_4$	[HC≡C-C(=O)O-CH_3]Mo(O)[S_2C-N(CH_3)$_2$]$_2$	Mo: Org.Comp.5-139/40, 142
$C_{10}H_{16}MoN_2S_4$	(HC≡CH)$_2$Mo[S_2C-N(CH_3)$_2$]$_2$	Mo: Org.Comp.5-185, 188

$C_{10}H_{16}MoO_3P^{+}$ .. $[(C_5H_5)Mo(CO)_2(P(CH_3)_3)(H_2O)]^{+}$ Mo:Org.Comp.7-283, 294
$C_{10}H_{16}Mo_2N_2O_{14}$ $H_4(MoO_3)_2((OOCCH_2)_2NC_2H_4N(CH_2COO)_2)$ ... Mo:SVol.B3b-169
$C_{10}H_{16}NO_3Sb$ .... $(4-(C_2H_5)_2NC_6H_4)Sb(O)(OH)_2$ .... Sb: Org.Comp.5-289
$C_{10}H_{16}NO_4Sb$ .... $(CH_3)_3Sb(OCH_3)OC_6H_4NO_2$-4 .... Sb: Org.Comp.5-45
$C_{10}H_{16}N_2O_2S_2Si$ .. $(CH_3)_3SiN=S=NS(O)_2C_6H_4CH_3$-4 .... S: S-N Comp.7-152
$C_{10}H_{16}N_2O_8Th$ ... $Th(OH)[(HOC_2H_4)(OOCCH_2)NC_2H_4N(CH_2COO)_2]$
Th: SVol.D1-99
$C_{10}H_{16}N_2SSi$ ..... $(CH_3)_3SiN=S=N-C_6H_4CH_3$-4. .... S: S-N Comp.7-141
$C_{10}H_{16}N_2S_2Si$ .... $(CH_3)_3SiN=S=NS-C_6H_4-CH_3$-4 .... S: S-N Comp.7-146
$C_{10}H_{16}N_4O_2PdS_4$ trans-$Pd(SCN)_2[SC(=NCH_3)OC_2H_5]_2$ .... Pd: SVol.B2-309/10
$C_{10}H_{16}N_4O_2S$ .... $((CH_3)_2N)_2S=N-C_6H_4$-4-NO_2 .... S: S-N Comp.8-207/9
$C_{10}H_{16}O_2Sn$ ..... $(C_2H_5)(C_6H_5)Sn(OCH_3)_2$ .... Sn: Org.Comp.16-211
$C_{10}H_{16}O_4PRe$ .... $(CO)_4Re(H)P(C_2H_5)_3$ .... Re: Org.Comp.1-464
$C_{10}H_{16}O_5Ti_2^{2+}$ ... $[((C_5H_5)Ti(H_2O)(OH))_2O]^{2+}$ .... Ti: Org.Comp.5-50
$C_{10}H_{17}I_2N_5ORe$ .. $[Re(CNCH_3)_5H_2O]I_2$ .... Re: Org.Comp.2-284, 287
$C_{10}H_{17}MnNO_2S_2$ .. $[Mn(SC(=S)N(C_2H_5)_2)(OC(CH_3)CHC(CH_3)O)]$ ... Mn:MVol.D7-143/4
$C_{10}H_{17}MnN_8O_2S_2^{-}$
$[Mn(O-N=C(CH_3)C(CH_3)=N-NHC(=S)NH_2)$
$(-ON=C(CH_3)C(CH_3)=NN=C(NH_2)S-)]^{-}$ Mn:MVol.D6-344
$C_{10}H_{17}MnO_7P_2^{-}$ .. $[Mn(O_7P_2CH_2CH=C(CH_3)CH_2CH_2CH=C(CH_3)_2)]^{-}$
Mn:MVol.D8-166/7
$C_{10}H_{17}Mn_2O_7P_2^{+}$ $[Mn_2(O_7P_2CH_2CH=C(CH_3)CH_2CH_2CH=C(CH_3)_2)]^{+}$
Mn:MVol.D8-166/7
$C_{10}H_{17}NO_2PRe$ ... $(C_5H_5)Re(NO)[P(CH_3)_3]-C(=O)CH_3$ .... Re: Org.Comp.3-45, 46
$C_{10}H_{17}NO_2SSn$ ... $(CH_3)_3SnNHSO_2C_6H_4CH_3$-4. .... Sn: Org.Comp.18-19, 21
$C_{10}H_{17}NO_2Sn$ .... $(C_2H_5)_3SnN(C(=O)CH)_2$. .... Sn: Org.Comp.18-142/3, 145
$C_{10}H_{17}NO_3PRe$ ... $(C_5H_5)Re(NO)[P(CH_3)_3]-C(=O)O-CH_3$ .... Re: Org.Comp.3-44/5, 46
$C_{10}H_{17}NSn$ ...... $(CH_3)_3Sn-N(CH_3)-C_6H_5$ .... Sn: Org.Comp.18-47
– .... $(CH_3)_3Sn-NH-CH_2-C_6H_5$ .... Sn: Org.Comp.18-16
– .... $(CH_3)_3Sn-NH-C_6H_4CH_3$-4 .... Sn: Org.Comp.18-15, 17
$C_{10}H_{17}N_2O_3ReS_2$ $(CH_3)_2NH-Re(CO)_3[-S-C(N(C_2H_5)_2)=S-]$. ..... Re: Org.Comp.1-128/9
$C_{10}H_{17}N_2O_9Th^{-}$ .. $Th(OH)_2[(HOC_2H_4)(OOCCH_2)NC_2H_4N(CH_2COO)_2]^{-}$
Th: SVol.D1-99
$C_{10}H_{17}N_3O_2S_2$ ... $((CH_3)_2N)_2S=N-S(O)_2-C_6H_5$ .... S: S-N Comp.8-207/9
$C_{10}H_{17}N_3O_9Th$ ... $[NH_4][Th(OH)((OOCCH_2)_2NC_2H_4N(CH_2COO)_2)] \cdot 4\ H_2O$
Th: SVol.C7-123, 125
$C_{10}H_{17}N_3O_{12}Th$ .. $[Th(C_5H_7NO_4)(C_5H_8NO_4)(H_2O)]NO_3 \cdot 3\ H_2O$ .. Th: SVol.C7-110/1
$C_{10}H_{17}N_3Sn$ ..... $(CH_3)_3SnN(C_6H_5)N=NCH_3$. .... Sn: Org.Comp.18-123, 125
$C_{10}H_{17}N_5ORe^{2+}$ .. $[Re(CNCH_3)_5H_2O]^{2+}$ .... Re: Org.Comp.2-284, 287
$C_{10}H_{17}N_7O_7Re$ ... $[Re(CNCH_3)_5H_2O][NO_3]_2$ .... Re: Org.Comp.2-284, 287
$C_{10}H_{18}In_2$ .... $[CH_3-C{\equiv}C-In(CH_3)_2]_2$ .... In: Org.Comp.1-112/4
$C_{10}H_{18}MnNO_3S_4^{+}$ $[Mn(SC(=S)O-C_4H_9-n)_2(NO)]^{+}$ Mn:MVol.D7-130/3
– .... $[Mn(SC(=S)O-C_4H_9-i)_2(NO)]^{+}$ Mn:MVol.D7-130/3
$C_{10}H_{18}MnN_2O_{12}P_4^{6-}$
$[Mn(1,4-((O_3PCH_2)_2N)_2-C_6H_{10}-c)]^{6-}$ Mn:MVol.D8-134/5
$C_{10}H_{18}MnN_3O_4S_2$ $[Mn((SC(=S))NC_5H_{10})(NH_2CH_2COO)_2]$ Mn:MVol.D7-177/8
$C_{10}H_{18}MnNa_2O_6S_8^{2-}$
$[Mn(SCH_2CH(S)CH_2SCH_2CH_2SO_3Na)_2]^{2-}$ Mn:MVol.D7-49/50
$C_{10}H_{18}MnNa_2O_8S_6^{2-}$
$[Mn(SCH_2CH(S)CH_2OCH_2CH_2SO_3Na)_2]^{2-}$ Mn:MVol.D7-49/50

$C_{10}H_{18}MnNa_2O_{10}S_8^{2-}$
$[Mn(SCH_2CH(S)CH_2SO_2CH_2CH_2SO_3Na)_2]^{2-}$ ... Mn: MVol.D7-49/50
$C_{10}H_{18}MnO_2S_4$ ... $Mn(SC(=S)OC_4H_9\text{-}i)_2$ ... Mn: MVol.D7-130/3
$C_{10}H_{18}N_2O_2S$ ... $((CH_3)_2N)_2SO \cdot C_6H_5\text{-}OH$ ... S: S-N Comp.8-341
$C_{10}H_{18}N_2O_2SSn$ ... $(CH_3)_3SnN(SO_2C_6H_4CH_3\text{-}4)NH_2$ ... Sn: Org.Comp.18-121/2, 124
$C_{10}H_{18}N_2O_6Th^{2+}$ $Th[C_2H_4(N(C_2H_4OH)CH_2COO)_2]^{2+}$ ... Th: SVol.D1-93/4, 97/8
$C_{10}H_{18}N_3OReS_6$ ... $(CO)Re[S_2CN(CH_3)_2]_3$ ... Re: Org.Comp.1-29
$C_{10}H_{18}N_4O_2PdS_2$ $Pd(SCN)_2(1,4\text{-}ONC_4H_9)_2$ ... Pd: SVol.B2-306
$C_{10}H_{18}N_4O_3Sn$ ... $CH_3\text{-}Sn[N(CH_3)\text{-}C(=O)\text{-}CH_2\text{-}]_3N$ ... Sn: Org.Comp.19-113
$C_{10}H_{18}N_4O_5S_3$ ... $O=S(N(CH_3)\text{-}C(O)O\text{-}N=C(CH_3)\text{-}SCH_3)_2$ ... S: S-N Comp.8-351
$C_{10}H_{18}N_4O_6Th^{2+}$ $[Th(H_2NC(O)C_2H_4CH(NH_2)COO)_2]^{2+}$ ... Th: SVol.D1-82/3, 88
$C_{10}H_{18}O_4ReSn_2^-$ $[(CO)_4Re(Sn(CH_3)_3)_2]^-$ ... Re: Org.Comp.1-344
$C_{10}H_{18}O_4Sn$ ... $(C_4H_9)_2SnOOCCOO$ ... Sn: Org.Comp.15-301
$C_{10}H_{18}O_5Si_2Sn$ ... $O(Si(CH_3)_2CH_2)_2Sn(OOCCH=CHCOO)$ ... Sn: Org.Comp.16-220
$C_{10}H_{18}O_5Ti_2^{4+}$ ... $[((C_5H_5)Ti(H_2O)_2)_2O]^{4+}$ ... Ti: Org.Comp.5-50
$C_{10}H_{18}O_6S_3Sn$ ... $C_4H_9Sn(OOCCH_2SH)_3$ ... Sn: Org.Comp.17-45
$C_{10}H_{18}O_6Sn$ ... $C_4H_9Sn(OOCCH_3)_3$ ... Sn: Org.Comp.17-44, 46/7
$C_{10}H_{19}IMoN_3O^+$ ... $[C_5H_5Mo(NO)(N(CH_3)(C_2H_5)N(CH_3)_2)I]^+$ ... Mo: Org.Comp.6-40
$C_{10}H_{19}InN_2$ ... $(CH_3)_3In \cdot NC_5H_4\text{-}4\text{-}N(CH_3)_2$ ... In: Org.Comp.1-27, 36
$C_{10}H_{19}NOS$ ... $O=S=NC_6H_{10}C_4H_9\text{-}t$ ... S: S-N Comp.6-106
$C_{10}H_{19}NO_2Sn$ ... $(C_2H_5)_3SnN(C(=O)CH_2)_2$ ... Sn: Org.Comp.18-142/3, 148/9
$C_{10}H_{19}NSn$ ... $(CH_2=CH)_3Sn\text{-}N(C_2H_5)_2$ ... Sn: Org.Comp.19-20
$C_{10}H_{19}N_2O_7Th^+$ ... $Th[C_2H_4(N(C_2H_4OH)CH_2COO)_2](OH)^+$ ... Th: SVol.D1-97/8
$C_{10}H_{19}N_3O_4S$ ... $(1,4\text{-}ONC_4H_8\text{-}4)_2S=N\text{-}C(O)OCH_3$ ... S: S-N Comp.8-207/8, 213
$C_{10}H_{19}O_6Sb$ ... $(C_2H_5)_2Sb(O_2CCH_3)_3$ ... Sb: Org.Comp.5-181
$C_{10}H_{20}I_4NO_2Re$ ... $[N(C_2H_5)_4][(CO)_2ReI_4]$ ... Re: Org.Comp.1-57
$C_{10}H_{20}In_2O_4$ ... $(C_2H_5)_2InOC(O)\text{-}C(O)OIn(C_2H_5)_2$ ... In: Org.Comp.1-207
$C_{10}H_{20}MnN_2O_4$ ... $[Mn(CH_3C(=O)CH=C(NH)CH_3)_2(H_2O)_2]$ ... Mn: MVol.D6-95
$C_{10}H_{20}MnN_2O_4S_4$ $Mn(SC(=S)N(C_2H_4OH)_2)_2$ ... Mn: MVol.D7-164/5
$C_{10}H_{20}MnN_2O_{12}P_4^{6-}$
$[Mn(O_3PCH_2)_2N(CH_2)_6N(CH_2PO_3)_2]^{6-}$ ... Mn: MVol.D8-134/5
$C_{10}H_{20}MnN_2S_4$ ... $Mn(SC(=S)N(C_2H_5)_2)_2$ ... Mn: MVol.D7-139/42
$C_{10}H_{20}MnN_2S_4^+$ ... $[Mn(SC(=S)N(C_2H_5)_2)_2]^+$ ... Mn: MVol.D7-146
$C_{10}H_{20}MnN_3OS_4^+$ $[Mn(SC(=S)N(C_2H_5)_2)_2(NO)]^+$ ... Mn: MVol.D7-144
$C_{10}H_{20}MnN_4O_4$ ... $[Mn((CH_3)_2C=N\text{-}N=C(O\text{-})CH_2CH_2C(O\text{-})=N$
$\text{-}N=C(CH_3)_2)(H_2O)_2]$ ... Mn: MVol.D6-313/4
$C_{10}H_{20}MnO_{10}P_2$ ... $[Mn(OP(=O)(OC_2H_5)C(=O)OC_2H_5)_2]$ ... Mn: MVol.D8-139/40
$C_{10}H_{20}MnO_{12}S_4$ ... $Mn[(O_3S)_2C_5H_{10}]_2$ ... Mn: MVol.D7-114/5
– ... $Mn[(O_3S)_2C_5H_{10}]_2 \cdot 5\ H_2O$ ... Mn: MVol.D7-114/5
$C_{10}H_{20}NO_2S$ ... $c\text{-}C_6H_{11}\text{-}N\text{-}S(O)O\text{-}C_4H_9\text{-}t$, radical ... S: S-N Comp.8-325/6
$C_{10}H_{20}N_2OS$ ... $C_5H_{10}N\text{-}1\text{-}S(O)\text{-}1\text{-}NC_5H_{10}$ ... S: S-N Comp.8-347, 359
$C_{10}H_{20}N_2O_4S_2Th^{2+}$
$[Th(CH_3SCH_2CH_2CH(NH_2)COO)_2]^{2+}$ ... Th: SVol.D1-82/3, 86/7
$C_{10}H_{20}N_2O_4Th^{2+}$ $[Th(i\text{-}C_3H_7CH(NH_2)COO)_2]^{2+}$ ... Th: SVol.D1-82/3, 85
$C_{10}H_{20}N_2Sn$ ... $1\text{-}(C_2H_5)_3Sn\text{-}2\text{-}CH_3\text{-}1,3\text{-}N_2C_3H_2$ ... Sn: Org.Comp.18-142, 148
– ... $(CH_3)_3Sn\text{-}N=C=N\text{-}C_6H_{11}\text{-}c$ ... Sn: Org.Comp.18-107, 110
$C_{10}H_{20}N_4O_{17}Th$ ... $Th(NO_3)_4 \cdot [\text{-}CH_2CH_2\text{-}O\text{-}]_5$ ... Th: SVol.D4-158
$C_{10}H_{20}O_3Sn$ ... $(C_2H_5)_2Sn(OC_2H_5)OOCCH=CHCH_3$ ... Sn: Org.Comp.16-181
$C_{10}H_{20}O_4Sn$ ... $(C_4H_9)_2Sn(OOCH)_2$ ... Sn: Org.Comp.15-103
$C_{10}H_{20}O_5Si_2Sn$ ... $O(Si(CH_3)_2CH_2)_2Sn[\text{-}OOCCH_2CH_2COO\text{-}]$ ... Sn: Org.Comp.16-220

$C_{10}H_{20}O_6Th$ ..... $Th(OH)_2(n\text{-}C_4H_9COO)_2 \cdot 4\ H_2O$ .... Th: SVol.C7-72/3
$C_{10}H_{20}Pb$ .... $Pb(\text{-}CH_2CH_2CH_2CH_2CH_2\text{-})_2$ .... Pb: Org.Comp.3-235/6, 238/9
– .... $Pb[\text{-}CH(CH_3)\text{-}CH_2CH_2CH_2\text{-}]_2$ .... Pb: Org.Comp.3-235/6
– .... $Pb[\text{-}CH_2\text{-}CH(CH_3)\text{-}CH_2CH_2\text{-}]_2$ .... Pb: Org.Comp.3-235/6
$C_{10}H_{21}IO_2Sn$..... $(C_4H_9)_2Sn(I)OOCCH_3$ .... Sn: Org.Comp.17-134
$C_{10}H_{21}InO_2$...... $(i\text{-}C_4H_9)_2InOC(O)\text{-}CH_3$.................. In: Org.Comp.1-199
$C_{10}H_{21}MnN_2O_{12}P_4{}^{5-}$
$[MnH(O_3PCH_2)_2N(CH_2)_6N(CH_2PO_3)_2]^{5-}$....... Mn: MVol.D8-134/5
$C_{10}H_{21}NOS$...... $O{=}S{=}N(CH_2)_9CH_3$ .... S: S-N Comp.6-105, 111
$C_{10}H_{21}NOSSn$.... $(C_4H_9)_2Sn(NCS)OCH_3$.................. Sn: Org.Comp.17-137
$C_{10}H_{21}NO_3Sn$.... $n\text{-}C_4H_9\text{-}Sn(\text{-}O\text{-}CH_2CH_2\text{-})_3N$ .... Sn: Org.Comp.17-34, 40
– .... $t\text{-}C_4H_9\text{-}Sn(\text{-}O\text{-}CH_2CH_2\text{-})_3N$.............. Sn: Org.Comp.17-51, 53
$C_{10}H_{21}NSn$ ...... $(C_2H_5)_3SnN{=}C{=}C(CH_3)_2$ .... Sn: Org.Comp.18-152/3
$C_{10}H_{21}N_3O_4S_2$ ... $CH_3S\text{-}C(CH_3)_2CH{=}N\text{-}OC(O)N(CH_3)\text{-}S(O)\text{-}N(CH_3)O\text{-}C_2H_5$
S: S-N Comp.8-350
$C_{10}H_{21}O_3P_2ReS$ .. $(CO)_3Re[P(CH_3)_3]_2SCH_3$................. Re: Org.Comp.1-256
$C_{10}H_{21}O_3P_2ReSe$ fac-$(CO)_3Re[P(CH_3)_3]_2SeCH_3$ .... Re: Org.Comp.1-259
$C_{10}H_{22}IN_3S$...... $[(1\text{-}C_4H_8N)_2S\text{-}N(CH_3)_2]I$.................. S: S-N Comp.8-230, 242
$C_{10}H_{22}InN$....... $(CH_3)_3In \cdot N(\text{-}CH_2CH_2\text{-})_3CH$............. In: Org.Comp.1-27, 33, 36
$C_{10}H_{22}In_2N_2$ ..... $2\ (CH_3)_3In \cdot 1,4\text{-}N_2C_4H_4$ .... In: Org.Comp.1-49, 51, 53
$C_{10}H_{22}In_2O_4$ ..... $[(CH_3)_2InOC(O)\text{-}C_2H_5]_2$ .... In: Org.Comp.1-197/8
$C_{10}H_{22}LiO_2Re$.... $Li[(t\text{-}C_4H_9CH_2)_2ReO_2]$ .... Re: Org.Comp.1-5
$C_{10}H_{22}MnN_2O_6P_2S^{2-}$
$[Mn((O_3PC(CH_3)_2NHCH_2CH_2)_2S)]^{2-}$.......... Mn: MVol.D8-132
$C_{10}H_{22}MnN_2O_7P_2{}^{2-}$
$[Mn((O_3PC(CH_3)_2NHCH_2CH_2)_2O)]^{2-}$.......... Mn: MVol.D8-132
$C_{10}H_{22}MnN_2O_{12}P_4{}^{4-}$
$[MnH_2(O_3PCH_2)_2N(CH_2)_6N(CH_2PO_3)_2]^{4-}$...... Mn: MVol.D8-134/5
$C_{10}H_{22}MnN_6O_4S_3$ $[Mn(C_2H_5C(CH_3){=}N\text{-}NHC({=}S)NH_2)_2(SO_4)]$..... Mn: MVol.D6-341
$C_{10}H_{22}MnN_6O_5S_4$ $[Mn(NH(C(NH_2){=}S)_2)_2(OC(CH_3){=}O)_2] \cdot C_2H_5OH$
Mn: MVol.D7-199/200
$C_{10}H_{22}MnN_6O_6S$.. $[Mn((CH_3)(C_2H_5)C{=}N\text{-}NHC(O)NH_2)_2SO_4]$...... Mn: MVol.D6-327
$C_{10}H_{22}MnO_6S_2$... $Mn(O_3S\text{-}C_5H_{11})_2$ .... Mn: MVol.D7-114/5
– .... $Mn(O_3S\text{-}C_5H_{11})_2 \cdot 2\ H_2O$ .... Mn: MVol.D7-114/5
$C_{10}H_{22}MoN_4O_2S_4$ $Mo(H_2NO)(NO)(S_2CN(C_2H_5)_2)_2$............. Mo: SVol.B3b-200
$C_{10}H_{22}N_2O_2S$.... $[C_5H_{10}NH_2][C_5H_{10}N\text{-}1\text{-}SO_2]$ .... S: S-N Comp.8-305
$C_{10}H_{22}N_2S$ ...... $t\text{-}C_4H_9\text{-}CH_2\text{-}N{=}S{=}N\text{-}CH_2\text{-}C_4H_9\text{-}t$ .... S: S-N Comp.7-210/1
$C_{10}H_{22}N_2Si$...... $SiH_2(\text{-}NC_5H_{10})_2$ .... Si: SVol.B4-188
$C_{10}H_{22}N_3O_5PS_3$ .. $CH_3S\text{-}C(CH_3){=}N\text{-}OC(O)\text{-}N(CH_3)\text{-}S(O)$
$\text{-}N(CH_3)\text{-}P(S)(OC_2H_5)_2$................. S: S-N Comp.8-356
– .... $CH_3S\text{-}C(CH_3){=}N\text{-}OC(O)\text{-}N(CH_3)\text{-}S(O)$
$\text{-}N(C_3H_7\text{-}n)\text{-}P(S)(OCH_3)_2$............... S: S-N Comp.8-356
– .... $CH_3S\text{-}C(CH_3){=}N\text{-}OC(O)\text{-}N(CH_3)\text{-}S(O)$
$\text{-}N(C_3H_7\text{-}i)\text{-}P(S)(OCH_3)_2$ .... S: S-N Comp.8-356
$C_{10}H_{22}N_3OsS_2Si_2{}^-$
$[N(C_4H_9\text{-}n)_4][((CH_3)_3SiCH_2)_2Os(N)(NCS)(SCN)]$ Os: Org.Comp.A1-21
– .... $[N(C_4H_9\text{-}n)_4][((CH_3)_3SiCH_2)_2Os(N)(NCS)_2]$.... Os: Org.Comp.A1-20/1
$C_{10}H_{22}N_3S^+$ ..... $[(1\text{-}C_4H_8N)_2S\text{-}N(CH_3)_2]^+$ .... S: S-N Comp.8-230, 242
$C_{10}H_{22}N_4OS$ ..... $4\text{-}CH_3\text{-}1,4\text{-}N_2C_4H_8\text{-}1\text{-}S(O)\text{-}1\text{-}1,4\text{-}N_2C_4H_8\text{-}4\text{-}CH_3$
S: S-N Comp.8-348

$C_{10}H_{22}N_4Si$	$(C_2H_5)_2NSi(-NC_2H_4)_3$	Si: SVol.B4-221
$C_{10}H_{22}NaO_2Re$	$Na[ReO_2(CH_2C_4H_9-t)_2]$	Re: Org.Comp.1-5/6
$C_{10}H_{22}O_2Re^-$	$[(t-C_4H_9CH_2)_2ReO_2]^-$	Re: Org.Comp.1-5/7
–	$[(t-C_4H_9CH_2)_2Re^{18}O_2]^-$	Re: Org.Comp.1-7
$C_{10}H_{22}O_2Sn$	$(n-C_4H_9)_2SnH-OC(=O)-CH_3$	Sn: Org.Comp.17-78/9
–	$(n-C_4H_9)_2Sn[-O-CH_2CH_2-O-]$	Sn: Org.Comp.15-48, 76/80
$C_{10}H_{22}O_2Ti$	$(CH_3)_2Ti(OCH_2CH(C_2H_5)CH(C_3H_7)O)$	Ti: Org.Comp.5-12
$C_{10}H_{22}O_3Sn$	$CH_3CH=CHCH_2-Sn(OC_2H_5)_3$	Sn: Org.Comp.17-60
–	$(C_2H_5)_2Sn(OC_4H_9-t)-OC(=O)-CH_3$	Sn: Org.Comp.16-181
–	$(n-C_4H_9)_2Sn(OH)-OC(=O)-CH_3$	Sn: Org.Comp.16-186, 197
–	$(t-C_4H_9)_2Sn(OH)-OC(=O)-CH_3$	Sn: Org.Comp.16-200
–	$(n-C_4H_9)_2Sn[-OO-CH(CH_3)-O-]$	Sn: Org.Comp.15-334, 336
$C_{10}H_{22}O_4SSn$	$(C_5H_{11})_2SnOSO_3$	Sn: Org.Comp.16-17
$C_{10}H_{22}O_5Si_2Sn$	$O(Si(CH_3)_2CH_2)_2Sn(OOCCH_3)_2$	Sn: Org.Comp.16-220
$C_{10}H_{23}I_2MnO_3P$	$Mn[P(C_2H_5)_3](OC_4H_8)(O_2)I_2$	Mn: MVol.D8-51
$C_{10}H_{23}InN_2$	$(CH_3)_2In-N(C_3H_7-i)-C(CH_3)=N-C_3H_7-i$	In: Org.Comp.1-269, 270/1
$C_{10}H_{23}InO$	$(C_2H_5)_3In \cdot OC_4H_8$	In: Org.Comp.1-70
–	$(t-C_4H_9)_2In-O-C_2H_5$	In: Org.Comp.1-181/2, 186/7
$C_{10}H_{23}InO_2$	$(C_2H_5)_3In \cdot 1,4-O_2C_4H_8$	In: Org.Comp.1-70
–	$C_2H_5-In(O-C_4H_9-t)_2$	In: Org.Comp.1-231, 233
$C_{10}H_{23}InS_2$	$n-C_4H_9-In(S-C_3H_7-i)_2$	In: Org.Comp.1-246, 247
–	$n-C_4H_9-In(S-C_3H_7-n)_2$	In: Org.Comp.1-246, 247
–	$i-C_4H_9-In(S-C_3H_7-n)_2$	In: Org.Comp.1-246, 248
$C_{10}H_{23}MnN_2O_{12}P_4^{3-}$	$[MnH_3(O_3PCH_2)_2N(CH_2)_6N(CH_2PO_3)_2]^{3-}$	Mn: MVol.D8-134/5
$C_{10}H_{23}MnN_3O_6P_2^{2-}$	$[Mn((O_3PC(CH_3)_2NHCH_2CH_2)_2NH)]^{2-}$	Mn: MVol.D8-132
$C_{10}H_{23}Mn_2N_3O_6P_2$	$Mn_2[(O_3PC(CH_3)_2NHCH_2CH_2)_2NH] \cdot 3\ H_2O$	Mn: MVol.D8-132
$C_{10}H_{23}NOSn$	$2,2-(n-C_4H_9)_2-1,3,2-ONSnC_2H_5$	Sn: Org.Comp.19-129, 134
$C_{10}H_{23}NO_2Sn$	$2-(n-C_4H_9)-2-(i-C_3H_7-O)-1,3,2-ONSnC_3H_7$	Sn: Org.Comp.19-158/9
–	$2-(n-C_4H_9)-2-(i-C_3H_7-O)-3-CH_3-1,3,2-ONSnC_2H_4$	Sn: Org.Comp.19-158/9
$C_{10}H_{23}NO_3SiSn$	$(CH_3)_3SiCH_2Sn(OCH_2CH_2)_3N$	Sn: Org.Comp.17-57
$C_{10}H_{23}NSn$	$1-(t-C_4H_9)-2,2-(CH_3)_2-1,2-NSnC_4H_8$	Sn: Org.Comp.19-53, 58
–	$(CH_3)_3Sn-N(C_3H_7-i)-CH=C(CH_3)_2$	Sn: Org.Comp.18-47, 49, 53
–	$(CH_3)_3Sn-N(C_4H_9-i)-CH=CHCH_3$	Sn: Org.Comp.18-47, 49, 53
$C_{10}H_{23}N_2S^+$	$[t-C_4H_9-N(C_2H_5)S=N-C_4H_9-t]^+$	S: S-N Comp.7-333/4
$C_{10}H_{23}O_3PTh^{2+}$	$Th(OH)[(C_2H_5CH(CH_3)CH_2)_2PO_2]^{2+}$	Th: SVol.D1-131
–	$Th(OH)[(n-C_5H_{11})_2PO_2]^{2+}$	Th: SVol.D1-131
$C_{10}H_{23}O_5PTh^{2+}$	$Th(OH)[O_2P(OC_5H_{11})_2]^{2+}$	Th: SVol.D1-131
$C_{10}H_{24}InN$	$(CH_3)_3In \cdot NC_5H_9-2,6-(CH_3)_2$	In: Org.Comp.1-27, 33, 35
–	$(i-C_3H_7)_2In-NH-C_4H_9-t$	In: Org.Comp.1-253/5, 257, 263
$C_{10}H_{24}InP$	$(CH_3)_2In-P(C_4H_9-t)_2$	In: Org.Comp.1-312, 313/4, 317
$C_{10}H_{24}InSb$	$(CH_3)_2In-Sb(C_4H_9-t)_2$	In: Org.Comp.1-316, 322/3
$C_{10}H_{24}In_2N_2$	$[(CH_3)_2In-N=C(CH_3)_2]_2$	In: Org.Comp.1-265/6
$C_{10}H_{24}In_2N_2O_2$	$[(CH_3)_2In-N(CH_3)-C(O)CH_3]_2$	In: Org.Comp.1-267
–	$[(CH_3)_2In-O-N=C(CH_3)_2]_2$	In: Org.Comp.1-211, 214

$C_{10}H_{24}In_2N_4$ ..... $(CH_3)_2In-N(CH_3)C(=N-CH_3)C(=N-CH_3)N(CH_3)-In(CH_3)_2$ In: Org.Comp.1-291/3

$C_{10}H_{24}MnN_2O_4S_4Se_2$ $[Mn(O=S(CH_3)_2)_4(NCSe)_2]$ Mn:MVol.D7-97/8

$C_{10}H_{24}MnN_2O_4S_6$ $[Mn(O=S(CH_3)_2)_4(NCS)_2]$ Mn:MVol.D7-97/8

$C_{10}H_{24}MnN_2O_6P_2S$ $[Mn(H_2(O_3PC(CH_3)_2NHCH_2CH_2)_2S)]$ Mn:MVol.D8-132

$C_{10}H_{24}MnN_2O_{12}P_4{}^{2-}$ $[MnH_4(O_3PCH_2)_2N(CH_2)_6N(CH_2PO_3)_2]^{2-}$ Mn:MVol.D8-134/5

$C_{10}H_{24}N_2OS$ ..... $(C_2H_5)_2N-S(O)-N(C_3H_7-i)_2$ .... S: S-N Comp.8-349

$C_{10}H_{24}N_2O_2Sn$ ... 2-(n-C_4H_9)-2-(CH_3-NH-CH_2CH_2-O)-3-CH_3-1,3,2-$ONSnC_2H_4$ .... Sn: Org.Comp.19-159

– .... 2-(n-C_4H_9)-2-($H_2N-CH_2CH_2CH_2$-O)-1,3,2-$ONSnC_3H_7$ Sn: Org.Comp.19-159

– .... $(C_2H_5)_3Sn-N(C_4H_9-t)-NO_2$ .... Sn: Org.Comp.18-141

– .... $(i-C_3H_7)_3Sn-N(CH_3)-NO_2$ .... Sn: Org.Comp.18-159

$C_{10}H_{24}N_2O_3SSi_2$ .. $CH_3-C(O)-N(Si(CH_3)_3)-S(O)-N(Si(CH_3)_3)-C(O)-CH_3$ S: S-N Comp.8-357

$C_{10}H_{24}N_4S_2Si$ .... $(CH_3)_2Si(N=S=N-C_4H_9-t)_2$ .... S: S-N Comp.7-140

$C_{10}H_{24}N_4S_2Sn$ ... $(CH_3)_2Sn(N=S=N-C_4H_9-t)_2$ .... S: S-N Comp.7-174; Sn: Org.Comp.19-74

$C_{10}H_{24}N_6O_{20}Th$ .. $[NH_4]_6[Th(C_2O_4)_5] \cdot H_2O$ .... Th: SVol.C7-89

– .... $[NH_4]_6[Th(C_2O_4)_5] \cdot 3\ H_2O$ .... Th: SVol.C7-89, 94

– .... $[NH_4]_6[Th(C_2O_4)_5] \cdot 7.5\ H_2O$ .... Th: SVol.C7-89, 94

$C_{10}H_{24}N_{10}NiPd$... $[Ni(NH_2CH_2CH_2NH_2)_3][Pd(CN)_4]$ .... Pd: SVol.B2-279

$C_{10}H_{24}N_{10}Pd_2$.... $[Pd(NH_2CH_2CH_2NH_2)_3][Pd(CN)_4]$ .... Pd: SVol.B2-279

$C_{10}H_{24}OSn$ ...... $(n-C_4H_9)_2Sn(H)OC_2H_5$ .... Sn: Org.Comp.17-78

$C_{10}H_{24}O_2Sn$ ..... $(C_2H_5)_2Sn(OC_2H_5)-OC_4H_9-t$ .... Sn: Org.Comp.16-181

– .... $(n-C_4H_9)_2Sn(OCH_3)_2$ .... Sn: Org.Comp.15-3, 8/14

– .... $(i-C_4H_9)_2Sn(OCH_3)_2$ .... Sn: Org.Comp.15-369, 371

– .... $(t-C_4H_9)_2Sn(OCH_3)_2$ .... Sn: Org.Comp.15-373

– .... $[C_2H_5-C(CH_3)_2]_2Sn(OH)_2$ .... Sn: Org.Comp.16-17

$C_{10}H_{24}O_2Ti$...... $(CH_3)_2Ti(OC_4H_9-i)_2$ .... Ti: Org.Comp.5-8, 12

– .... $(CH_3)_2Ti(OC_4H_9-t)_2$ .... Ti: Org.Comp.5-8

$C_{10}H_{24}O_3Sn$ ..... $CH_3Sn(OC_3H_7-i)_3$ .... Sn: Org.Comp.17-14

– .... $n-C_4H_9Sn(OC_2H_5)_3$ .... Sn: Org.Comp.17-33, 36/40

$C_{10}H_{24}O_4SSn$ .... $(C_4H_9)_2Sn(OCH_3)OSO_2CH_3$ .... Sn: Org.Comp.16-189

$C_{10}H_{24}O_4Se_2Sn$ .. $(C_4H_9)_2Sn(OSe(O)CH_3)_2$ .... Sn: Org.Comp.15-368

$C_{10}H_{24}O_5Sb_2$ .... $[(CH_3)_3SbO_2CCH_3]_2O$ .... Sb: Org.Comp.5-106

$C_{10}H_{24}O_6S_2Sn$ ... $(C_4H_9)_2Sn(OSO_2CH_3)_2$ .... Sn: Org.Comp.15-348

$C_{10}H_{24}O_6Sn$ ..... $C_4H_9Sn(OCH_2CH_2OH)_3$ .... Sn: Org.Comp.17-34

$C_{10}H_{25}I_3InN$ ..... $[N(C_2H_5)_4][C_2H_5-In(I)_3]$ .... In: Org.Comp.1-349/50, 355

$C_{10}H_{25}InN_2$ ...... $C_2H_5-In[N(C_2H_5)_2]_2$ .... In: Org.Comp.1-296

$C_{10}H_{25}InO$....... $(C_2H_5)_3In \cdot O(C_2H_5)_2$ .... In: Org.Comp.1-70/1

$C_{10}H_{25}InSeSi$ .... $(C_2H_5)_2In-Se-Si(C_2H_5)_3$ .... In: Org.Comp.1-252

$C_{10}H_{25}MnN_3O_6P_2$ $[Mn(H_2(O_3PC(CH_3)_2NHCH_2CH_2)_2NH)]$ Mn:MVol.D8-132

$C_{10}H_{25}NSn$ ...... $(C_2H_5)_3SnN(C_2H_5)_2$ .... Sn: Org.Comp.18-132/4

$C_{10}H_{25}O_3Sb$ ..... $(C_2H_5)_2Sb(OC_2H_5)_3$ .... Sb: Org.Comp.5-179

$C_{10}H_{25}PSn$ ...... $(C_2H_5)_3Sn-P(C_2H_5)_2$ .... Sn: Org.Comp.19-190

$C_{10}H_{26}MnN_2O_4Si_2$ $[Mn(N(Si(CH_3)_3)_2)(NO_3)(OC_4H_8)]$ Mn:MVol.D8-30

$C_{10}H_{26}MnN_6O_2S_2^{2+}$
$[Mn(C_2H_5C(CH_3)=N-NHC(=S)NH_2)_2(H_2O)_2]^{2+}$ .. Mn:MVol.D6-341
$C_{10}H_{26}Mn_2O_{11}P_4$ $[Mn_2(CH_3-PO_2-O-PO_2-CH_3)(CH_3-PO_2-O-C_3H_7-i)_2]$ Mn:MVol.D8-138
$C_{10}H_{26}NO_2OsSi_2^-$ $[N(C_4H_9-n)_4][((CH_3)_3SiCH_2)_2Os(N)(O-CH_2CH_2-O)]$ Os: Org.Comp.A1-21
$C_{10}H_{26}NOsS_2Si_2^-$ $[N(C_4H_9-n)_4][((CH_3)_3SiCH_2)_2Os(N)(S-CH_2CH_2-S)]$ Os: Org.Comp.A1-21
$C_{10}H_{26}N_2OSiSn$ .. $(CH_3)_3SnNHC(C_3H_7)=NOSi(CH_3)_3$ .... Sn: Org.Comp.18-15/6, 18
$C_{10}H_{26}N_2Sn$ ..... $(C_2H_5)_2N-Sn(CH_3)_2-N(C_2H_5)_2$ .... Sn: Org.Comp.19-64/5
– .... $(C_2H_5)_2N-Sn(CD_3)_2-N(C_2H_5)_2$ .... Sn: Org.Comp.19-64
– .... $(t-C_4H_9)_2Sn(NH-CH_3)_2$ .... Sn: Org.Comp.19-98/101
$C_{10}H_{26}N_6Si_2$ ..... $[(N(CH_3)C_2H_4(CH_3)N)SiN(CH_3)]_2$ .... Si: SVol.B4-228
$C_{10}H_{27}InN_2Si_2$ ... $(CH_3)_2In-N[Si(CH_3)_3]-C(CH_3)=N-Si(CH_3)_3$ .... In: Org.Comp.1-270/1
– .... $(CH_3)_3In \cdot (CH_3)_3Si-N=C=N-Si(CH_3)_3$ .... In: Org.Comp.1-27, 37
$C_{10}H_{27}In_2N_3$ ..... $(CH_3)_2In-N(CH_3)-N(CH_3)-C(CH_3)=N-CH_3 \cdot In(CH_3)_3$ In: Org.Comp.1-271
$C_{10}H_{27}MnNOSi_2$ .. $[Mn(N(Si(CH_3)_3)_2)(OC_4H_9-i)]$ Mn:MVol.D8-30
$C_{10}H_{27}MnN_3O_3PSi_2$
$Mn(NO)_3[P(C_4H_9-t)(Si(CH_3)_3)_2]$ Mn:MVol.D8-73/4
$C_{10}H_{27}MnN_3O_3PSn_2$
$Mn(NO)_3[P(C_4H_9-t)(Sn(CH_3)_3)_2]$ Mn:MVol.D8-73/4
$C_{10}H_{27}NO_2P_3Re$ .. $(CO)Re[P(CH_3)_3]_3(NO)$ .... Re: Org.Comp.1-47
$C_{10}H_{27}NPbSn$ .... $(CH_3)_3SnN(C_4H_9-t)Pb(CH_3)_3$ .... Sn: Org.Comp.18-57, 64
$C_{10}H_{27}NSiSn$ .... $(CH_3)_3SnN(C_4H_9-t)Si(CH_3)_3$ .... Sn: Org.Comp.18-57, 62/3, 73/4
$C_{10}H_{27}N_2PSn$ .... $(CH_3)_3SnN(CH_3)P(=NCH_3)(CH_3)C_4H_9-t$ .... Sn: Org.Comp.18-57, 60
$C_{10}H_{27}N_3Sn$ ..... $[(CH_3)_2N]_3Sn-C_4H_9-n$ .... Sn: Org.Comp.19-111, 113/4
$C_{10}H_{27}PSiSn$ ..... $(CH_3)_3Sn-P(C_4H_9-t)-Si(CH_3)_3$ .... Sn: Org.Comp.19-175
$C_{10}H_{28}InNO_4S_2Si_2$ $[(CH_3)_3Si-CH_2]_2In-N[S(=O)_2-CH_3]_2$ .... In: Org.Comp.1-257/8, 263/4
$C_{10}H_{28}In_2O$ ...... $2 (CH_3)_3In \cdot O(C_2H_5)_2$ .... In: Org.Comp.1-30
$C_{10}H_{28}N_2SSiSn$ ... $(CH_3)_3SnN(C_4H_9-t)SNHSi(CH_3)_3$ .... Sn: Org.Comp.18-62, 73
$C_{10}H_{28}N_2SiSn$ .... $(CH_3)_3SnN(Si(CH_3)_3)NHC_4H_9-t$ .... Sn: Org.Comp.18-121/2, 125
$C_{10}H_{28}N_2Si_2Sn$ ... 1,3-$[(CH_3)_3Si]_2$-2,2-$(CH_3)_2$-1,3,2-$N_2SnC_2H_4$ .. Sn: Org.Comp.19-75/6
$C_{10}H_{28}OOsSi_2$ ... $[(CH_3)_3Si-CH_2]_2Os(CH_3)_2=O$ .... Os: Org.Comp.A1-29/30
$C_{10}H_{28}O_3Sb_2$ .... $[(CH_3)_3SbOC_2H_5]_2O$ .... Sb: Org.Comp.5-106
$C_{10}H_{28}O_9S_2Sb_2$ .. $[(CH_3)_3SbOSO_2CH_2CH_2OH]_2O$ .... Sb: Org.Comp.5-106
$C_{10}H_{30}I_4N_4O_{10}Th_2$ $[ThI_2(H_2O)_3]_2(OH)_2 \cdot [CH_3-C(=N-O)-C(CH_3)=N-CH_2CH_2-N=C(CH_3)-C(=N-O)-CH_3]$ Th: SVol.D4-158
$C_{10}H_{30}In_2N_2$ ..... $(CH_3)_3In \cdot CH_3-NH-CH_2CH_2-NH-CH_3 \cdot In(CH_3)_3$ In: Org.Comp.1-49, 52/3
$C_{10}H_{30}In_2N_2O_2P_2$ $[(CH_3)_2InOP(=NCH_3)(CH_3)_2]_2$ .... In: Org.Comp.1-215
$C_{10}H_{30}In_2N_2P_2$ ... $[(CH_3)_2In-N=P(CH_3)_3]_2$ .... In: Org.Comp.1-285/6
$C_{10}H_{30}In_2O_2Si_2$ .. $[(CH_3)_2In-O-Si(CH_3)_3]_2$ .... In: Org.Comp.1-188
$C_{10}H_{30}In_2P_2$ ..... $(CH_3)_3In \cdot (CH_3)_2P-P(CH_3)_2 \cdot In(CH_3)_3$ ...... In: Org.Comp.1-49, 51/2
$C_{10}H_{30}MgMnSi_3$ .. $[(CH_3)_3Si]_3MnMgCH_3$ Mn:MVol.D8-22/3
$C_{10}H_{30}N_2SSi_2Sn$ .. $(CH_3)_3SnN(S(CH_3)=NSi(CH_3)_3)Si(CH_3)_3$ .... Sn: Org.Comp.18-76, 80
$C_{10}H_{30}N_2Si_2Sn$ ... $(CH_3)_2N-Sn(CH_3)_2-N[Si(CH_3)_3]_2$ .... Sn: Org.Comp.19-108
– .... $(CH_3)_3Si-N(CH_3)-Sn(CH_3)_2-N(CH_3)-Si(CH_3)_3$ .. Sn: Org.Comp.19-67/70
– .... $(CH_3)_3Sn-N(CH_3)-Si(CH_3)_2-N(CH_3)-Si(CH_3)_3$ .. Sn: Org.Comp.18-57, 61

$C_{10}H_{30}N_2Si_2Sn$ . . . $(CH_3)_3Sn-N[Si(CH_3)_3]-N(CH_3)-Si(CH_3)_3$ . . . . . . Sn: Org.Comp.18-121, 123
$C_{10}H_{30}N_6S_2Sn$ . . . $(CH_3)_2Sn[N(CH_3)-S(=NCH_3)_2-CH_3]_2$ . . . . Sn: Org.Comp.19-67/70
$C_{10}H_{30}N_8O_{22}Th_2$ . . $[Th(NO_3)_2(H_2O)_3]_2(OH)_2 \cdot [CH_3-C(=N-O)-CCH_3=N-CH_2CH_2-N=CCH_3-C(=N-O)-CH_3]$ Th: SVol.D4-158
$C_{10}H_{30}O_3S_2Sb_2^{2+}$ $[((CH_3)_3SbOS(CH_3)_2)_2O]^{2+}$ . . . . Sb: Org.Comp.5-101
$C_{10}H_{30}O_3Si_3Sn$ . . . $CH_3Sn(OSi(CH_3)_3)_3$ . . . . Sn: Org.Comp.17-16
$C_{10}H_{40}N_{12}Si_4$ . . . . $(CH_3NH)_2Si[-NSi(NHCH_3)_3-Si(NHCH_3)_2-NSi(NHCH_3)_3-]$ Si: SVol.B4-247/8
$C_{10}H_{45}N_5Si_5$ . . . . . $(SiH_3N(CH_3)_2)_5$ . . . . Si: SVol.B4-158/71
$C_{10}HgMnO_{10}Re$ . . $(CO)_5ReHgMn(CO)_5$ . . . . Re: Org.Comp.2-187, 203
$C_{10}K_6O_{24}Th_2$ . . . . $K_6Th_2(C_2O_4)_3(CO_3)_4 \cdot 6\ H_2O$ . . . . Th: SVol.C7-99, 101
$C_{10}MnO_9ReS_2$ . . . $(CO)_5ReCS_2Mn(CO)_4$ . . . . Re: Org.Comp.2-180, 183
$C_{10}MnO_9ReS_3$ . . . $(CO)_4Re(S)_2C-S-Mn(CO)_5$ . . . . Re: Org.Comp.1-488
$C_{10}MnO_{10}Re$. . . . . $(CO)_5ReMn(CO)_5$ . . . . Re: Org.Comp.2-199/200, 206/10
$C_{10}MoO_{10}Re^-$ . . . . $[(CO)_5ReMo(CO)_5]^-$ . . . . Re: Org.Comp.2-192, 204
$C_{10}Na_{12}O_{26}Th$ . . . $Na_{12}Th(C_2O_4)_2(CO_3)_6 \cdot 13\ H_2O$ . . . . Th: SVol.C7-98, 100
$C_{10}O_{10}ReTc$ . . . . . $(CO)_5ReTc(CO)_5$ . . . . Re: Org.Comp.2-200, 210
$C_{10}O_{10}ReW^-$ . . . . . $[(CO)_5ReW(CO)_5]^-$ . . . . Re: Org.Comp.2-197, 204
$C_{10}O_{20}Th^{6-}$ . . . . . . $[Th(C_2O_4)_5]^{6-}$ . . . . Th: SVol.C7-91
$C_{11}ClFFeH_9^+$ . . . . $[(3-F-C_6H_4-Cl)Fe(C_5H_5)][PF_6]$ . . . . Fe: Org.Comp.B19-1, 67
– . . . . $[(4-F-C_6H_4-Cl)Fe(C_5H_5)][PF_6]$ . . . . Fe: Org.Comp.B19-1, 67
$C_{11}ClFH_9MoN_3O$ . . $(C_5H_5)MoCl(NO)-N=N-C_6H_4-3-F$ . . . . Mo: Org.Comp.6-58
– . . . . $(C_5H_5)MoCl(NO)-N=N-C_6H_4-4-F$ . . . . Mo: Org.Comp.6-58
$C_{11}ClF_2H_{15}MoNO_2P$ $(C_5H_5)Mo(CO)_2(Cl)[F_2P-N(C_2H_5)_2]$ . . . . Mo: Org.Comp.7-56/7, 93
$C_{11}ClF_3H_{20}N_2S$ . . . $CF_3SCl-(1-NC_5H_{10})_2$ . . . . S: S-N Comp.8-401, 402
$C_{11}ClF_4FeH_5O_2$ . . $(C_5H_5)Fe(CO)_2C[=CCl-CF_2-CF_2-]$ . . . . Fe: Org.Comp.B13-196, 198, 238/9
$C_{11}ClF_5H_{18}O_2Sn$ . . $(C_4H_9)_2Sn(Cl)OOCC_2F_5$ . . . . Sn: Org.Comp.17-104
$C_{11}ClF_6FeH_9NO_2P$ $[(2-Cl-C_6H_4-NO_2)Fe(C_5H_5)][PF_6]$ . . . . Fe: Org.Comp.B19-4, 76
– . . . . $[(3-Cl-C_6H_4-NO_2)Fe(C_5H_5)][PF_6]$ . . . . Fe: Org.Comp.B19-4, 76
– . . . . $[(4-Cl-C_6H_4-NO_2)Fe(C_5H_5)][PF_6]$ . . . . Fe: Org.Comp.B19-4, 76/7
$C_{11}ClF_6FeH_{10}OP$ $[(2-Cl-C_6H_4-OH)Fe(C_5H_5)][PF_6]$ . . . . Fe: Org.Comp.B19-3, 68
$C_{11}ClF_6FeH_{10}P$ . . . $[(C_6H_6)Fe(C_5H_4-Cl)][PF_6]$ . . . . Fe: Org.Comp.B18-142/6, 151, 169, 188
– . . . . $[(Cl-C_6H_5)Fe(C_5H_5)][PF_6]$ . . . . Fe: Org.Comp.B18-142/6, 197, 200, 201, 245, 277/80
$C_{11}ClF_6FeH_{11}NP$. $[(2-Cl-C_6H_4-NH_2)Fe(C_5H_5)][PF_6]$ . . . . Fe: Org.Comp.B19-3, 70
– . . . . $[(3-Cl-C_6H_4-NH_2)Fe(C_5H_5)][PF_6]$ . . . . Fe: Org.Comp.B19-3, 70
– . . . . $[(4-Cl-C_6H_4-NH_2)Fe(C_5H_5)][PF_6]$ . . . . Fe: Org.Comp.B19-3, 71
$C_{11}ClF_6FeH_{12}HgO_2P$ $[(C_5H_5)Fe(CO)_2(CH_2=C(CH_3)CH_2HgCl)][PF_6]$. . . Fe: Org.Comp.B17-63
$C_{11}ClF_6FeH_{12}N_2P$ $[(2-Cl-C_6H_4-NH-NH_2)Fe(C_5H_5)][PF_6]$ . . . . Fe: Org.Comp.B19-3, 74
$C_{11}ClF_6H_7Mo$ . . . . $(C_5H_5)Mo(Cl)(HC{\equiv}C-CF_3)_2$ . . . . Mo: Org.Comp.6-133, 142
$C_{11}ClF_6H_8MoO_4^-$ $[As(C_6H_5)_4][(CH_2C(CH_3)CH_2)Mo(Cl)(CO)_2(CF_3-C(O)CHC(O)-CF_3)]$ . . . . Mo: Org.Comp.5-269/71
– . . . . $[As(C_6H_5)_4][(CH_3CHCHCH_2)Mo(Cl)(CO)_2(CF_3-C(O)CHC(O)-CF_3)]$ . . . . Mo: Org.Comp.5-269/71

$C_{11}ClF_7FeH_9P$	$[(3\text{-}F\text{-}C_6H_4\text{-}Cl)Fe(C_5H_5)][PF_6]$	Fe: Org.Comp.B19-1, 67
–	$[(4\text{-}F\text{-}C_6H_4\text{-}Cl)Fe(C_5H_5)][PF_6]$	Fe: Org.Comp.B19-1, 67
$C_{11}ClF_8N_3$	$NCFCClC(NNC_6F_5)CFCF$	F: PFHOrg.SVol.4-146, 156
$C_{11}ClFeH_7HgO_2S$	$(C_5H_5)Fe(CO)_2\text{-}5\text{-}(SC_4H_2\text{-}2\text{-}HgCl)$	Fe: Org.Comp.B14-2, 6, 13
$C_{11}ClFeH_8N$	$NC\text{-}C_5H_4FeC_5H_4\text{-}Cl$	Fe: Org.Comp.A9-113, 120
$C_{11}ClFeH_9HgO$	$(C_5H_5)Fe[C_5H_3(HgCl)\text{-}CH{=}O]$	Fe: Org.Comp.A10-146, 147
–	$(HgCl\text{-}C_5H_4)Fe(C_5H_4\text{-}CH{=}O)$	Fe: Org.Comp.A10-141, 143
$C_{11}ClFeH_9NO_2^+$	$[(2\text{-}Cl\text{-}C_6H_4\text{-}NO_2)Fe(C_5H_5)][PF_6]$	Fe: Org.Comp.B19-4, 76
–	$[(3\text{-}Cl\text{-}C_6H_4\text{-}NO_2)Fe(C_5H_5)][PF_6]$	Fe: Org.Comp.B19-4, 76
–	$[(4\text{-}Cl\text{-}C_6H_4\text{-}NO_2)Fe(C_5H_5)][PF_6]$	Fe: Org.Comp.B19-4, 76/7
$C_{11}ClFeH_9O_2$	$(C_5H_5)Fe(CO)_2\text{-}CH{=}CHCH{=}CHCl$	Fe: Org.Comp.B13-97, 126
–	$(C_5H_5)Fe(CO)_2\text{-}CH[\text{-}CHCl\text{-}CH{=}CH\text{-}]$	Fe: Org.Comp.B13-198, 237/8
$C_{11}ClFeH_9O_3$	$[(C_8H_9)Fe(CO)_3]Cl$	Fe: Org.Comp.B15-214, 227/9
$C_{11}ClFeH_9O_3S$	$ClO_2S\text{-}C_5H_4FeC_5H_4\text{-}CHO$	Fe: Org.Comp.A9-233, 241
$C_{11}ClFeH_9O_4S$	$ClO_2S\text{-}C_5H_4FeC_5H_4\text{-}C(O)OH$	Fe: Org.Comp.A9-233, 241
$C_{11}ClFeH_9O_7$	$[(C_8H_9)Fe(CO)_3][ClO_4]$	Fe: Org.Comp.B15-214, 227/9
$C_{11}ClFeH_{10}$	$(C_6H_6)Fe(C_5H_4\text{-}Cl)$	Fe: Org.Comp.B18-146/8, 170, 188
–	$(Cl\text{-}C_6H_5)Fe(C_5H_5)$	Fe: Org.Comp.B18-146/8, 200/1, 246
$C_{11}ClFeH_{10}^+$	$[(C_6H_6)Fe(C_5H_4\text{-}Cl)]^+$	Fe: Org.Comp.B18-142/6, 151, 169/70, 188
–	$[(Cl\text{-}C_6H_5)Fe(C_5H_5)]^+$	Fe: Org.Comp.B18-142/6, 197, 200, 201, 245/6, 277/80
$C_{11}ClFeH_{10}Li$	$(C_5H_5)Fe[C_5H_2(Li)(Cl)\text{-}CH_3]$	Fe: Org.Comp.A10-327
$C_{11}ClFeH_{10}NO_3S$	$ClO_2S\text{-}C_5H_4FeC_5H_4\text{-}CONH_2$	Fe: Org.Comp.A9-233, 241
$C_{11}ClFeH_{10}NO_5S$	$(C_5H_5)Fe(CO)_2CH[\text{-}CH_2\text{-}C({=}O)\text{-}N(SO_2Cl)\text{-}CH_2\text{-}]$	Fe: Org.Comp.B14-20, 27
$C_{11}ClFeH_{10}O^+$	$[(2\text{-}Cl\text{-}C_6H_4\text{-}OH)Fe(C_5H_5)][PF_6]$	Fe: Org.Comp.B19-3, 68
$C_{11}ClFeH_{11}$	$(C_5H_5)Fe(C_6H_6Cl\text{-}1)$	Fe: Org.Comp.B17-270
–	$(C_5H_5)Fe(C_6H_6Cl\text{-}2)$	Fe: Org.Comp.B17-270
–	$(C_5H_5)Fe(C_6H_6Cl\text{-}3)$	Fe: Org.Comp.B17-271
–	$[(C_6H_6)Fe(C_5H_5)]Cl$	Fe: Org.Comp.B18-142/6, 151/2, 154
$C_{11}ClFeH_{11}Hg$	$(C_5H_5)Fe[C_5H_3(HgCl)\text{-}CH_3]$	Fe: Org.Comp.A10-147, 151
$C_{11}ClFeH_{11}HgO$	$(HgCl\text{-}C_5H_4)Fe(C_5H_4\text{-}OCH_3)$	Fe: Org.Comp.A10-141, 142, 143
$C_{11}ClFeH_{11}N^+$	$[(2\text{-}Cl\text{-}C_6H_4\text{-}NH_2)Fe(C_5H_5)][PF_6]$	Fe: Org.Comp.B19-3, 70
–	$[(3\text{-}Cl\text{-}C_6H_4\text{-}NH_2)Fe(C_5H_5)][PF_6]$	Fe: Org.Comp.B19-3, 70
–	$[(4\text{-}Cl\text{-}C_6H_4\text{-}NH_2)Fe(C_5H_5)][PF_6]$	Fe: Org.Comp.B19-3, 71
$C_{11}ClFeH_{11}O_2$	$(C_5H_5)Fe(CO)_2\text{-}CH_2C(CH_2Cl){=}CH_2$	Fe: Org.Comp.B13-101
–	$(C_5H_5)Fe(CO)_2\text{-}CH_2CH{=}CCl\text{-}CH_3$	Fe: Org.Comp.B13-105, 133
$C_{11}ClFeH_{11}O_2S$	$(C_5H_5)Fe[C_5H_3(CH_3)\text{-}SO_2Cl]$	Fe: Org.Comp.A9-253
$C_{11}ClFeH_{11}O_3$	$(C_5H_5)Fe(CO)_2CO(CH_2)_3Cl$	Fe: Org.Comp.B13-10, 17, 36/7
$C_{11}ClFeH_{11}O_4$	$[(C_6H_6)Fe(C_5H_5)][ClO_4]$	Fe: Org.Comp.B18-142/6, 154, 155
$C_{11}ClFeH_{11}O_6$	$[(C_5H_5)Fe(CO)_2(CH_2{=}C{=}CHCH_3)][ClO_4]$	Fe: Org.Comp.B17-111
$C_{11}ClFeH_{12}HgO_2^+$	$[(C_5H_5)Fe(CO)_2(CH_2{=}C(CH_3)\text{-}CH_2\text{-}HgCl)]^+$	Fe: Org.Comp.B17-63, 78

$C_{11}ClFeH_{12}HgO_2^{+}$ $[(C_5H_5)Fe(CO)_2(CH_2{=}CH{-}CH(CH_3){-}HgCl)]^{+}$ ... Fe: Org.Comp.B17-28
$C_{11}ClFeH_{12}N_2^{+}$ .. $[(2{-}Cl{-}C_6H_4{-}NH{-}NH_2)Fe(C_5H_5)][PF_6]$ .... Fe: Org.Comp.B19-3, 74
$C_{11}ClFeH_{13}O_2$ ... $[(C_5H_5)Fe(CO)_2(CH_2{=}CH{-}C_2H_5)]Cl$ .... Fe: Org.Comp.B17-8/9
– .... $[(C_5H_5)Fe(CO)_2(CH_2{=}CH{-}CH_2CH_2D)]Cl$ .... Fe: Org.Comp.B17-11
$C_{11}ClFeH_{13}O_6$ ... $[(C_5H_5)Fe(CO)_2(CH_2{=}CH{-}C_2H_5)][ClO_4]$ .... Fe: Org.Comp.B17-9
– .... $[(C_5H_5)Fe(CO)_2(CH_2{=}CH{-}CH_2CH_2D)][ClO_4]$ ... Fe: Org.Comp.B17-11
$C_{11}ClFeH_{13}O_7$ ... $[(CH_2C(CH_3)C(CH_3)C(CH_3)CH_2)Fe(CO)_3][ClO_4]$ Fe: Org.Comp.B15-24
– .... $[(CH_3{-}CHC(CH_3)CHCHCH{-}CH_3)Fe(CO)_3][ClO_4]$ Fe: Org.Comp.B15-23, 30
– .... $[(CH_3{-}CHCHC(CH_3)CHCH{-}CH_3)Fe(CO)_3][ClO_4]$ Fe: Org.Comp.B15-14, 24
$C_{11}ClFeH_{14}N$ .... $[C_5H_5FeC_5H_3(NH_3)CH_3]Cl$ .... Fe: Org.Comp.A9-71
$C_{11}ClFeH_{14}N_3$.... $[C_5H_5Fe(CNCH_3)_3]Cl$ .... Fe: Org.Comp.B15-344
$C_{11}ClFeH_{15}O_2$ ... $[C_5H_5(CO)(C_4H_8O)Fe{=}CH_2]Cl$.............. Fe: Org.Comp.B16a-18, 24
$C_{11}ClGeH_{15}O$ .... $Ge(CH_3)_3CH_2COC_6H_4Cl{-}4$ .... Ge: Org.Comp.1-174, 176/7
$C_{11}ClGeH_{19}$ ..... $(C_2H_5)_2Ge[{-}CH{=}C(CH_3)CCl{=}C(CH_3)CH_2{-}]$..... Ge: Org.Comp.3-305/6
$C_{11}ClGeH_{21}$ ..... $Ge(C_2H_5)_3{-}C{\equiv}C{-}CCl(CH_3)_2$ .... Ge: Org.Comp.2-263
$C_{11}ClGeH_{23}O$ .... $Ge(C_2H_5)_3C_4H_8COCl$ .... Ge: Org.Comp.2-163
$C_{11}ClH_4O_4ReS_2$ .. $(CO)_4Re[{-}S{=}C(C_6H_4{-}4{-}Cl){-}S{-}]$ .... Re: Org.Comp.1-358
$C_{11}ClH_4O_5Re$ .... $(CO)_5Re{-}C_6H_4Cl{-}2$....................... Re: Org.Comp.2-129
– .... $(CO)_5Re{-}C_6H_4Cl{-}3$....................... Re: Org.Comp.2-129
– .... $(CO)_5Re{-}C_6H_4Cl{-}4$....................... Re: Org.Comp.2-129
$C_{11}ClH_5MoO_4$.... $C_6H_5{-}CMo(CO)_4(Cl)$ .... Mo: Org.Comp.5-97/100
$C_{11}ClH_6N_2O_3Re$ .. $(CO)_3ReCl[{-}1,8{-}(C_8H_6N_2{-}1,8){-}]$............. Re: Org.Comp.1-163
$C_{11}ClH_6N_4O_3Re$ .. $(CO)_3ReCl[{-}1,2{-}(C_4H_3N_2{-}1,3){-}2,1{-}(C_4H_3N_2{-}1,3){-}]$
Re: Org.Comp.1-163
$C_{11}ClH_6O_5Re$ .... cis-$(CO)_4Re(Cl)C(OH)C_6H_5$ .... Re: Org.Comp.1-381
$C_{11}ClH_7MoO_4$.... $[7{-}Cl{-}(2.2.1){-}C_7H_7{-}2,5]Mo(CO)_4$ .... Mo: Org.Comp.5-338, 339, 340/1
$C_{11}ClH_7NO_4ReS$.. $(CO)_4Re(Cl)[S{=}C(NH_2)C_6H_5]$ .... Re: Org.Comp.1-459
$C_{11}ClH_{10}MoN_3O$.. $C_5H_5Mo(NO)(N{=}NC_6H_5)Cl$ .... Mo: Org.Comp.6-58
$C_{11}ClH_{10}O_3ReSSe$ $(CO)_3ReCl(CH_3S{-}C_6H_4{-}2{-}SeCH_3)$........... Re: Org.Comp.1-200, 214/5
$C_{11}ClH_{10}O_3ReS_2$ $(CO)_3ReCl(CH_3S{-}C_6H_4{-}2{-}SCH_3)$............ Re: Org.Comp.1-199, 214/5
$C_{11}ClH_{10}O_5ReSi$.. $(CO)_5ReCH_2C{\equiv}CSi(CH_3)_2CH_2Cl$............. Re: Org.Comp.2-109
$C_{11}ClH_{10}STi$ ..... $[(C_5H_5)Ti(SC_6H_5)Cl]_n$ .... Ti: Org.Comp.5-37
$C_{11}ClH_{11}MoN_2O_2$ $[(C_5H_5)Mo(CO)_2(NC{-}CH_3)_2]Cl$ .... Mo: Org.Comp.7-283, 284
$C_{11}ClH_{11}MoN_2O_2S_2$
$[(C_5H_5)Mo(CO)_2({-}S{-}C(C(NCH_3)_2){-}S{-})]Cl$ ..... Mo: Org.Comp.7-189, 192
$C_{11}ClH_{12}NOSn$ ... $(CH_3)_2Sn(Cl)OC_9H_6N$ .... Sn: Org.Comp.17-82, 86/7
$C_{11}ClH_{12}NO_2Sn$ .. $(CH_3)_3SnN(CO)_2C_6H_3Cl$ .... Sn: Org.Comp.18-84, 88
$C_{11}ClH_{12}N_3O_4S$ .. $4{-}(4{-}NO_2{-}C_6H_4{-}C(O){-}N{=}S(Cl)){-}1,4{-}ONC_4H_8$ .. S: S-N Comp.8-185
$C_{11}ClH_{13}MoNO_4P$ $(C_5H_5)Mo(CO)_2(Cl)(1,7,4,8{-}O_2NPC_4H_8)$ .... Mo: Org.Comp.7-56/7, 94/5
$C_{11}ClH_{13}N_2OSn$ .. $(CH_3)_3SnN(COC_6H_4Cl{-}4)CN$................ Sn: Org.Comp.18-58, 69
$C_{11}ClH_{14}MoO_4^{-}$ .. $[N(C_2H_5)_4][(CH_3CHCHCH_2)Mo(Cl)(CO)_2$
$(CH_3{-}C(O)CHC(O){-}CH_3)]$ .... Mo: Org.Comp.5-269/71
– .... $[N(C_4H_9{-}t)_4][(CH_2C(CH_3)CH_2)Mo(Cl)(CO)_2$
$(CH_3{-}C(O)CHC(O){-}CH_3)]$ .... Mo: Org.Comp.5-269/71
– .... $[P(C_6H_5)_4][(CH_2CHCH_2)Mo(Cl)(CO)_2$
$(CH_3{-}C(O)C(CH_3)C(O){-}CH_3)]$............. Mo: Org.Comp.5-269/71
$C_{11}ClH_{14}MoO_5^{-}$ .. $[P(C_6H_5)_4][(CH_2CHCH_2)MoCl(CO)_3{-}OC(O){-}C_4H_9{-}t]$
Mo: Org.Comp.5-292/5
$C_{11}ClH_{14}NO_3S$ ... $3{-}(i{-}C_3H_7){-}C_6H_4{-}OC(O){-}N(CH_3){-}S(O)Cl$ ..... S: S-N Comp.8-268/70

$C_{11}ClH_{14}NO_5PS$	$(C_2H_5\text{-}O)P(O)\text{-}CH_2\text{-}N(CH_2C(O)OC_6H_5)\text{-}S(O)Cl$	S: S-N Comp.8-259, 268
–	$(C_6H_5\text{-}O)P(O)\text{-}CH_2\text{-}N(CH_2C(O)OC_2H_5)\text{-}S(O)Cl$	S: S-N Comp.8-259, 268
$C_{11}ClH_{15}MoN_2O_2$	$(CH_2CHCH_2)Mo(Cl)(CO)_2(CN\text{-}C_2H_5)_2$	Mo: Org.Comp.5-296/9
$C_{11}ClH_{15}N_2O_3S_2$	$4\text{-}(4\text{-}CH_3\text{-}C_6H_4\text{-}S(O)_2\text{-}N{=}S(Cl))\text{-}1,4\text{-}ONC_4H_8$	S: S-N Comp.8-186
$C_{11}ClH_{16}MoO_4P$	$(C_5H_5)Mo(CO)_2(H)[ClP(O\text{-}C_2H_5)_2]$	Mo: Org.Comp.7-57, 92
$C_{11}ClH_{16}N_2O_3Re$	$(CO)_3ReCl[\text{-}N(C_3H_7\text{-}i){=}CH\text{-}CH{=}N(C_3H_7\text{-}i)\text{-}]$	Re: Org.Comp.1-159, 177, 179
$C_{11}ClH_{17}MnN_7O_3^+$	$[MnCl(2,6\text{-}(H_2N\text{-}C(O)NH\text{-}N{=}C(CH_3))_2C_5H_3N)(H_2O)]^+$	Mn: MVol.D6-333, 334
$C_{11}ClH_{17}MoN_2O_2$	$(CH_2CHCH_2)Mo(Cl)(CO)_2(C_2H_5\text{-}N{=}CHCH{=}N\text{-}C_2H_5)$	Mo: Org.Comp.5-262/3
$C_{11}ClH_{17}O_3Sn$	$C_4H_9Sn(Cl)OC(CH_3){=}CHC({=}CHCOCH_3)O$	Sn: Org.Comp.17-151
$C_{11}ClH_{17}O_4Sn$	$CH_3SnCl(OC(CH_3){=}CHCOCH_3)_2$	Sn: Org.Comp.17-145, 147/8
$C_{11}ClH_{17}O_5PRe$	$(CO)_4Re(Cl)P(CH_3)_2OCH_2C_4H_9\text{-}t$	Re: Org.Comp.1-448
$C_{11}ClH_{18}MoNO$	$C_5(CH_3)_5Mo(NO)(Cl)CH_3$	Mo: Org.Comp.6-82
$C_{11}ClH_{18}N_2O_3Re$	cis-$(CO)_3ReCl(NC_4H_9)_2$	Re: Org.Comp.1-224
$C_{11}ClH_{19}MoO_3$	$(C_5H_5)MoCl(O)(O\text{-}C_3H_7\text{-}n)_2$	Mo: Org.Comp.6-10, 11/2
–	$(C_5H_5)MoCl(O)(O\text{-}C_3H_7\text{-}i)_2$	Mo: Org.Comp.6-10, 11/2
$C_{11}ClH_{19}O_2Sn$	$(C_2H_5)_2Sn(Cl)\text{-}OC(O)\text{-}C{\equiv}C\text{-}C_4H_9\text{-}n$	Sn: Org.Comp.17-95
$C_{11}ClH_{20}NO_2SSiSn$	$(CH_3)_2Sn(Cl)\text{-}N[S({=}O)_2\text{-}C_6H_5]\text{-}Si(CH_3)_3$	Sn: Org.Comp.19-117, 119
$C_{11}ClH_{20}O_3ReS_2$	cis-$(CO)_3ReCl[S(C_2H_5)_2]_2$	Re: Org.Comp.1-286
$C_{11}ClH_{20}O_3ReSe_2$	cis-$(CO)_3ReCl[Se(C_2H_5)_2]_2$	Re: Org.Comp.1-287
$C_{11}ClH_{20}O_3ReTe_2$	cis-$(CO)_3ReCl[Te(C_2H_5)_2]_2$	Re: Org.Comp.1-289
$C_{11}ClH_{21}O_2Sn$	$(C_4H_9)_2Sn(Cl)OOCCH{=}CH_2$	Sn: Org.Comp.17-105
$C_{11}ClH_{23}MoO_8P_2$	$(CH_2CHCH_2)Mo(Cl)(CO)_2[P(OCH_3)_3]_2$	Mo: Org.Comp.5-271, 272, 273, 279, 280
$C_{11}ClH_{23}O_2Sn$	$(C_4H_9)_2SnOCH(CH_2Cl)CH_2O$	Sn: Org.Comp.15-49
$C_{11}ClH_{24}NOSn$	$(C_4H_9)_2Sn(Cl)ON{=}C(CH_3)_2$	Sn: Org.Comp.17-106
$C_{11}ClH_{25}OSn$	$(n\text{-}C_4H_9)_2SnCl\text{-}OC_3H_7\text{-}i$	Sn: Org.Comp.17-99
–	$(i\text{-}C_4H_9)_2SnCl\text{-}OC_3H_7\text{-}n$	Sn: Org.Comp.17-113
$C_{11}ClH_{27}InNOSi$	$CH_3\text{-}InCl\text{-}N(C_4H_9\text{-}t)\text{-}Si(CH_3)_2\text{-}O\text{-}C_4H_9\text{-}t$	In: Org.Comp.1-297
$C_{11}ClH_{27}NPSn$	$(CH_3)_3SnN{=}P(C_4H_9\text{-}t)_2Cl$	Sn: Org.Comp.18-114, 118
$C_{11}ClH_{27}N_2SSn$	$(CH_3)_2Sn(Cl)\text{-}N(C_4H_9\text{-}t)\text{-}S(CH_3){=}N\text{-}C_4H_9\text{-}t$	Sn: Org.Comp.19-117/8
$C_{11}ClH_{27}OSiSn$	$(n\text{-}C_4H_9)_2SnCl\text{-}O\text{-}Si(CH_3)_3$	Sn: Org.Comp.17-107, 110
–	$n\text{-}C_4H_9\text{-}SnCl(C_4H_9\text{-}t)\text{-}O\text{-}Si(CH_3)_3$	Sn: Org.Comp.17-141, 142
$C_{11}ClH_{27}O_2P_3Re$	$(CO)_2ReCl[P(CH_3)_3]_3$	Re: Org.Comp.1-98, 102
$C_{11}ClH_{27}O_{11}P_3Re$	cis-$(CO)_2ReCl[P(OCH_3)_3]_3$	Re: Org.Comp.1-98
$C_{11}ClH_{28}NSiSn$	$(t\text{-}C_4H_9)_2Sn(Cl)\text{-}NH\text{-}Si(CH_3)_3$	Sn: Org.Comp.19-117, 121
$C_{11}ClH_{30}N_3O_9P_3Re$	mer-$Re(CNCH_3)(P(OCH_3)_3)_3(Cl)N_2$	Re: Org.Comp.2-229, 236
$C_{11}ClH_{31}InNSi_2$	$[(CH_3)_3Si\text{-}CH_2]_2InCl \cdot N(CH_3)_3$	In: Org.Comp.1-121
$C_{11}Cl_2F_6FeH_9P$	$[(1,2\text{-}Cl_2\text{-}C_6H_4)Fe(C_5H_5)][PF_6]$	Fe: Org.Comp.B19-1, 4/5, 17, 87/90
–	$[(1,3\text{-}Cl_2\text{-}C_6H_4)Fe(C_5H_5)][PF_6]$	Fe: Org.Comp.B19-1, 4/5, 18, 87/90
–	$[(1,4\text{-}Cl_2\text{-}C_6H_4)Fe(C_5H_5)][PF_6]$	Fe: Org.Comp.B19-1, 4/5, 18, 87/90
$C_{11}Cl_2F_6HN_3O_2$	$NCFCClC(NHC_6F_4NO_2\text{-}4)CClCF$	F: PFHOrg.SVol.4-150, 160
$C_{11}Cl_2F_7HN_2$	$NCFCClC(NHC_6F_5)CClCF$	F: PFHOrg.SVol.4-150, 160

$C_{11}Cl_2F_7N$	$C_{10}F_7$-2-N=CCl_2	F: PFHOrg.SVol.6-192, 206
$C_{11}Cl_2FeH_8O_3S$	ClO_2S-$C_5H_4FeC_5H_4$-COCl	Fe: Org.Comp.A9-233, 241
$C_{11}Cl_2FeH_9^+$	[(1,2-Cl_2-C_6H_4)Fe(C_5H_5)][1,4-((NC)$_2$C=)$_2$-C_6H_4]$_2$ · 4 [1,4-((NC)$_2$C=)$_2$-C_6H_4]	Fe: Org.Comp.B19-5, 18
–	[(1,2-Cl_2-C_6H_4)Fe(C_5H_5)][BF_4]	Fe: Org.Comp.B19-1, 4/5, 17/8, 87/90
–	[(1,2-Cl_2-C_6H_4)Fe(C_5H_5)][Cr(NH_3)$_2$(SCN)$_4$]	Fe: Org.Comp.B19-1, 5, 18
–	[(1,2-Cl_2-C_6H_4)Fe(C_5H_5)][PF_6]	Fe: Org.Comp.B19-1, 4/5, 17, 87/90
–	[(1,3-Cl_2-C_6H_4)Fe(C_5H_5)][1,4-((NC)$_2$C=)$_2$-C_6H_4]$_2$ · 4 [1,4-((NC)$_2$C=)$_2$-C_6H_4]	Fe: Org.Comp.B19-5, 18
–	[(1,3-Cl_2-C_6H_4)Fe(C_5H_5)][BF_4]	Fe: Org.Comp.B19-1, 4/5, 18, 87/90
–	[(1,3-Cl_2-C_6H_4)Fe(C_5H_5)][Cr(NH_3)$_2$(SCN)$_4$]	Fe: Org.Comp.B19-1, 5, 18
–	[(1,3-Cl_2-C_6H_4)Fe(C_5H_5)][PF_6]	Fe: Org.Comp.B19-1, 4/5, 18, 87/90
–	[(1,4-Cl_2-C_6H_4)Fe(C_5H_5)]$^+$	Fe: Org.Comp.B19-18
–	[(1,4-Cl_2-C_6H_4)Fe(C_5H_5)][1,4-((NC)$_2$C=)$_2$-C_6H_4]$_2$ · 4 [1,4-((NC)$_2$C=)$_2$-C_6H_4]	Fe: Org.Comp.B19-5, 19
–	[(1,4-Cl_2-C_6H_4)Fe(C_5H_5)][BF_4]	Fe: Org.Comp.B19-1, 4/5, 18, 87/90
–	[(1,4-Cl_2-C_6H_4)Fe(C_5H_5)][Cr(NH_3)$_2$(SCN)$_4$]	Fe: Org.Comp.B19-1, 5, 19
–	[(1,4-Cl_2-C_6H_4)Fe(C_5H_5)][PF_6]	Fe: Org.Comp.B19-1, 4/5, 18, 87/90
$C_{11}Cl_2FeH_{10}$	(C_5H_5)Fe($C_6H_5Cl_2$-1,2)	Fe: Org.Comp.B17-282
–	(C_5H_5)Fe($C_6H_5Cl_2$-1,3)	Fe: Org.Comp.B17-282
–	(C_5H_5)Fe($C_6H_5Cl_2$-1,4)	Fe: Org.Comp.B17-283
–	(C_5H_5)Fe($C_6H_5Cl_2$-1,5)	Fe: Org.Comp.B17-283
–	(C_5H_5)Fe($C_6H_5Cl_2$-2,3)	Fe: Org.Comp.B17-282
–	(C_5H_5)Fe($C_6H_5Cl_2$-2,4)	Fe: Org.Comp.B17-283
$C_{11}Cl_2FeH_{12}HgO_2$	[(C_5H_5)Fe(CO)$_2$(CH_2=C(CH_3)-CH_2-HgCl)]Cl	Fe: Org.Comp.B17-63, 78
–	[(C_5H_5)Fe(CO)$_2$(CH_2=CH-CH(CH_3)-HgCl)]Cl	Fe: Org.Comp.B17-28
$C_{11}Cl_2FeH_{13}N$	[(C_6H_6)Fe(C_5H_4-NH_3)]Cl_2	Fe: Org.Comp.B18-167
$C_{11}Cl_2Fe_3O_9$	(CO)$_9$$Fe_3$(C-Cl)$_2$	Fe: Org.Comp.C6a-237, 242/3
$C_{11}Cl_2GeH_{24}$	Ge(C_2H_5)$_3$$CH_2$C($CH_3$)$_2$$CHCl_2$	Ge: Org.Comp.2-148
$C_{11}Cl_2H_{10}MnN_4$	[$MnCl_2$(NC_5H_4-2-CH=NNH-2-C_5H_4N)]	Mn: MVol.D6-244, 246/7
$C_{11}Cl_2H_{10}MnN_4O$	$MnCl_2$[1-(O=)NC_5H_4-2-CH=N-NH-2-C_5H_4N]	Mn: MVol.D6-244, 247
$C_{11}Cl_2H_{10}N_3O_2Re$	(CO)$ReCl_2$(NC_5H_5)$_2$(NO)	Re: Org.Comp.1-41
$C_{11}Cl_2H_{11}InN_2$	CH_3-In(Cl)$_2$ · 2-(NC_5H_4-2)-NC_5H_4	In: Org.Comp.1-136, 139
$C_{11}Cl_2H_{12}MnN_4O$	[$MnCl_2$(NC_5H_4-2-CH=NNH-2-C_5H_4N)(H_2O)]	Mn: MVol.D6-244, 246
$C_{11}Cl_2H_{12}MnN_4O_3S$	Mn(4-$NH_2C_6H_4SO_2$NH-3-$C_4H_2N_2$-1,2-OCH_3-6)Cl_2	Mn: MVol.D7-116/23
–	Mn(4-$NH_2C_6H_4SO_2$NH-4-$C_4H_2N_2$-1,3-OCH_3-6)Cl_2	Mn: MVol.D7-116/23
$C_{11}Cl_2H_{12}NOSb$	(CH_3)$_2$Sb(Cl_2)OC_9H_6N	Sb: Org.Comp.5-191/2
$C_{11}Cl_2H_{12}N_2O_2S$	4-(4-Cl-C_6H_4-C(O)-N=S(Cl))-1,4-ONC_4H_8	S: S-N Comp.8-185
$C_{11}Cl_2H_{12}O_4Sn$	$CH_3OOCCH_2CH_2SnCl_2$(OC_6H_4CHO-2)	Sn: Org.Comp.17-179
$C_{11}Cl_2H_{13}InN_2$	CH_3-In(Cl)$_2$ · 2 NC_5H_5	In: Org.Comp.1-136, 139
$C_{11}Cl_2H_{13}MnN_3S$	$MnCl_2$[C_6H_5CH=CHC(CH_3)=N-NHC(=S)NH_2]	Mn: MVol.D6-342

$C_{11}Cl_2H_{15}MnN_7S_2$ $[MnCl(2,6\text{-}(NH_2C(=S)NH\text{-}N=C(CH_3))_2C_5H_3N)][Cl]$
Mn:MVol.D6-352/3

$C_{11}Cl_2H_{15}MnOP$ . . $Mn(CO)[P(C_6H_5)(C_2H_5)_2]Cl_2$ Mn:MVol.D8-47

$C_{11}Cl_2H_{15}MoO_4P$ $(C_5H_5)Mo(CO)_2(Cl)[ClP(O\text{-}C_2H_5)_2]$ Mo:Org.Comp.7-57, 92

$C_{11}Cl_2H_{15}N_3O_2S_2$ $(C_2H_5)_2N\text{-}CCl=N\text{-}S(Cl)=N\text{-}S(O)_2\text{-}C_6H_5$ . . . . S: S-N Comp.8-187/9

$C_{11}Cl_2H_{15}Sb$ . . . . . $(CH_2)_5(C_6H_5)SbCl_2$ . . . . Sb: Org.Comp.5-75

$C_{11}Cl_2H_{16}MoN_2O_2S_3$
$C_5H_5Mo(NO)(S_2CN(CH_3)_2)CH(\text{-}CH_2S(Cl)_2OCH_2\text{-})$
Mo:Org.Comp.6-86

$C_{11}Cl_2H_{17}MnN_3O_3S$
$Mn(4\text{-}NH_2C_6H_4SO_2NH\text{-}CONHC_4H_9)Cl_2$ Mn:MVol.D7-116/23

$C_{11}Cl_2H_{17}MnN_7O_3$ $[MnCl(2,6\text{-}(H_2N\text{-}C(O)NH\text{-}N=C(CH_3))_2C_5H_3N)$
$(H_2O)][Cl] \cdot 2\ H_2O$ Mn:MVol.D6-333, 334/5

$C_{11}Cl_2H_{19}MoO_2$ . . $(C_5H_5)Mo(O\text{-}C_3H_7\text{-}n)_2Cl_2$ Mo:Org.Comp.6-3, 6/7

– . . . . $(C_5H_5)Mo(O\text{-}C_3H_7\text{-}i)_2Cl_2$ Mo:Org.Comp.6-4, 6/7

$C_{11}Cl_2H_{19}NS_2Sn$ . . 1-$[n\text{-}C_4H_9\text{-}SnCl_2\text{-}NH]$-2-$[CH_3\text{-}S\text{-}C(=S)]\text{-}C_5H_6$
Sn: Org.Comp.19-157

$C_{11}Cl_2H_{21}N_2O_2Re$ $Re(CNC_4H_9\text{-}t)_2(OCH_3)(Cl_2)O$ . . . . Re: Org.Comp.2-259, 265

$C_{11}Cl_2H_{22}MoP_2$ . . $C_5H_5Mo(P(CH_3)_2CH_2CH_2P(CH_3)_2)Cl_2H$ Mo:Org.Comp.6-17

$C_{11}Cl_2H_{25}MnP$ . . . $Mn[P(C_3H_7)_3](CH_2=CH_2)Cl_2$ Mn:MVol.D8-46/7

$C_{11}Cl_2H_{27}InN_2$ . . . $t\text{-}C_4H_9\text{-}CH_2\text{-}In(Cl)_2 \cdot (CH_3)_2N\text{-}CH_2CH_2\text{-}N(CH_3)_2$
In: Org.Comp.1-136, 139

$C_{11}Cl_3FeH_5O_2S$ . . $(C_5H_5)Fe(CO)_2$-2-$(SC_4Cl_3$-3,4,5) . . . . Fe: Org.Comp.B14-6, 13

$C_{11}Cl_3FeH_9O_3$ . . . $(C_5H_5)Fe(CO)_2COCH_2CH_2CCl_3$ . . . . Fe: Org.Comp.B13-7

$C_{11}Cl_3FeH_{11}N_2O_2$ $C_5H_5Fe(CNCH_3)_2OC(O)CCl_3$ . . . . Fe: Org.Comp.B15-319

$C_{11}Cl_3GaH_9NO_2$ . . $GaCl_3[1\text{-}(4\text{-}CH_3\text{-}C_6H_4)\text{-}NC_4H_2(=O)_2\text{-}2,5]$ . . . . . Ga: SVol.D1-240/1

$C_{11}Cl_3GeH_{13}MoO_3$
$(C_5H_5)Mo(CO)_2(GeCl_3)=C(CH_3)\text{-}O\text{-}C_2H_5$ Mo:Org.Comp.8-16, 20

$C_{11}Cl_3GeH_{19}O_2$ . . $Ge(C_2H_5)_3C\equiv CCH_2OCH(OH)CCl_3$ . . . . Ge: Org.Comp.2-261

$C_{11}Cl_3H_6NO_8Th$ . . $Th(CH_3COO)(CCl_3COO)(NC_5H_3(COO)_2\text{-}2,6) \cdot 2\ H_2O$
Th: SVol.C7-154, 156

$C_{11}Cl_3H_8N_2ORe$ . . $(CO)ReCl_3[NC_5H_4\text{-}2\text{-}(2\text{-}C_5H_4N)]$ . . . . Re: Org.Comp.1-30

$C_{11}Cl_3H_{12}O_2Sb$. . . $(C_6H_5)Sb(Cl_3)OC(CH_3)=CHC(O)CH_3$. . . . . Sb: Org.Comp.5-268/9

$C_{11}Cl_3H_{13}N_2O_2S$. . 2-(2,4,6-$Cl_3\text{-}C_6H_2\text{-}N=$)-6-$(CH_3)$-1,3,2,6-$O_2SNC_4H_8$
S: S-N Comp.8-170/2

$C_{11}Cl_3H_{13}N_2O_3S_2$ 3,5-$(CH_3)_2\text{-}C_6H_3C(O)\text{-}NH\text{-}S(CCl_3)=N\text{-}S(O)_2CH_3$
S: S-N Comp.8-195

– . . . . $n\text{-}C_3H_7\text{-}C(O)\text{-}NH\text{-}S(CCl_3)=N\text{-}S(O)_2\text{-}C_6H_5$ . . . . S: S-N Comp.8-195

$C_{11}Cl_3H_{21}MoP_2$ . . $C_5H_5Mo(P(CH_3)_2CH_2CH_2P(CH_3)_2)Cl_3$ Mo:Org.Comp.6-17, 21

$C_{11}Cl_4GeH_{20}O$ . . . $Ge(C_2H_5)_3C(CH_2Cl)=CHCHOHCCl_3$ . . . . Ge: Org.Comp.2-213

$C_{11}Cl_4H_5IMoNOS_2^-$
$[C_5H_5Mo(NO)(I)(1,2\text{-}S_2C_6Cl_4)]^-$ Mo:Org.Comp.6-30/1

$C_{11}Cl_4H_{10}NOSb$ . . $C_6H_5SbCl_4 \cdot ONC_5H_5$ . . . . Sb: Org.Comp.5-244

$C_{11}Cl_4H_{11}O_2Sb$. . . $(C_6H_5)Sb(Cl_3)OC(CH_3)=CClC(O)CH_3$ . . . . Sb: Org.Comp.5-269

$C_{11}Cl_4H_{11}O_4Sb$. . . $(CH_3)Sb(\text{-}OC(CH_3)HCH(CH_3)O\text{-})(\text{-}1\text{-}OC_6Cl_4O\text{-}2\text{-})$
Sb: Org.Comp.5-313

$C_{11}Cl_4H_{12}N_2O_3S_2$ $n\text{-}C_3H_7\text{-}C(O)\text{-}NH\text{-}S(CCl_3)=N\text{-}S(O)_2\text{-}C_6H_4\text{-}4\text{-}Cl$
S: S-N Comp.8-195

$C_{11}Cl_4H_{13}O_4Sb$. . . $(CH_3)Sb(OC_2H_5)_2(1\text{-}OC_6Cl_4O\text{-}2)$ . . . . Sb: Org.Comp.5-312

$C_{11}Cl_5F_8NO$ . . . . . $NCF_2CF_2CF_2CF_2C(OC_6Cl_5)$ . . . . F: PFHOrg.SVol.4-143, 154

$C_{11}Cl_5H_9N_2O_2SSb^-$
$[(4-(2'-NC_5H_4NHSO_2)C_6H_4)SbCl_5]^-$ .... Sb: Org.Comp.5-253
$C_{11}Cl_5H_{10}N_2O_2Sb$ $[C_5H_5NH][3-O_2N-C_6H_4-SbCl_5]$ .... Sb: Org.Comp.5-254
– .... $[C_5H_5NH][4-O_2N-C_6H_4-SbCl_5]$ .... Sb: Org.Comp.5-254
$C_{11}Cl_5H_{11}NOSb$ .. $[C_5H_5NH][4-HO-C_6H_4-SbCl_5]$ .... Sb: Org.Comp.5-251
$C_{11}Cl_5H_{11}NSb$. ... $[C_5H_5NH][C_6H_5-SbCl_5]$. .... Sb: Org.Comp.5-245/6
$C_{11}Cl_5H_{12}N_2O_2SSb$
$[C_5H_5NH][(4-H_2NSO_2C_6H_4)SbCl_5]$ .... Sb: Org.Comp.5-253
$C_{11}Cl_5H_{13}N_2O_2Sb^-$
$[3-O_2N-4-(CH_2)_5NC_6H_3SbCl_5]^-$ .... Sb: Org.Comp.5-259
$C_{11}Cl_5H_{13}O_2SSb^-$ $[(4-i-C_3H_7O_2CCH_2SC_6H_4)SbCl_5]^-$ .... Sb: Org.Comp.5-253
$C_{11}Cl_5H_{13}O_2Sb^-$ $[(4-C_2H_5O_2CCH_2CH_2C_6H_4)SbCl_5]^-$ .... Sb: Org.Comp.5-257
$C_{11}Cl_5H_{14}N_2O_2Sb$ $H[(3-O_2N-4-(CH_2)_5NC_6H_3)SbCl_5]$ .... Sb: Org.Comp.5-259
$C_{11}Cl_5H_{16}NSb^-$ ... $[(4-(C_2H_5)_2NCH_2C_6H_4)SbCl_5]^-$ .... Sb: Org.Comp.5-256
$C_{11}Cl_5H_{17}NSb$. ... $H[(4-(C_2H_5)_2NCH_2C_6H_4)SbCl_5]$ .... Sb: Org.Comp.5-256
$C_{11}Cl_6FeH_9O_3Sb$ $[(C_8H_9)Fe(CO)_3][SbCl_6]$ .... Fe: Org.Comp.B15-214, 227/9
$C_{11}Cl_6Ga_2H_9NO_2$ $Ga_2Cl_6[1-(4-CH_3-C_6H_4)-NC_4H_2(=O)_2-2,5]$ .... Ga: SVol.D1-240/1
$C_{11}Cl_6H_4MnO_2S_4$ $Mn((SC(=S)OCH_2)_2C_7Cl_6)$. .... Mn: MVol.D7-133/4
$C_{11}Cl_6H_6MnO_2S_4$ $Mn[(SC(=S)OCH_2)_2C_7H_2Cl_6]$. .... Mn: MVol.D7-133/4
$C_{11}Cl_6H_{10}NSb$. ... $[C_5H_5NH][3-Cl-C_6H_4-SbCl_5]$ .... Sb: Org.Comp.5-250
– .... $[C_5H_5NH][4-Cl-C_6H_4-SbCl_5]$ .... Sb: Org.Comp.5-251
$C_{11}Cl_7F_2HN_2$. .... $4-(C_6Cl_5-NH)-3,5-Cl_2-2,6-F_2-NC_5$ .... F: PFHOrg.SVol.4-150, 160
$C_{11}Cl_7F_2NO$ ..... $4-(C_6Cl_5-O)-3,5-Cl_2-2,6-F_2-NC_5$ .... F: PFHOrg.SVol.4-143, 153
– .... $6-(C_6Cl_5-O)-3,5-Cl_2-2,4-F_2-NC_5$ .... F: PFHOrg.SVol.4-143, 153
$C_{11}CrH_{14}IO_3Sb$. .. $(CH_3)_3Sb(I)Cr(CO)_3C_5H_5$ = $[(CH_3)_3SbCr(CO)_3C_5H_5]I$
Sb: Org.Comp.5-32, 34
$C_{11}CrH_{14}N_2O_5S$ .. $[Cr(CO)_5(i-C_3H_7-N=S=N-C_3H_7-i)]$. .... S: S-N Comp.7-299/302
$C_{11}CrH_{14}O_3Sb^+$ .. $[(CH_3)_3SbCr(CO)_3C_5H_5]^+$ .... Sb: Org.Comp.5-34
$C_{11}CsFe_3HO_{11}$ ... $Cs[(H)Fe_3(CO)_{11}]$ .... Fe: Org.Comp.C6b-93, 96
$C_{11}FFeH_9O_4S$ .... $FO_2S-C_5H_4FeC_5H_4-C(O)OH$ .... Fe: Org.Comp.A9-232, 241
$C_{11}FFeH_{10}$ ...... $(F-C_6H_5)Fe(C_5H_5)$. .... Fe: Org.Comp.B18-146/8, 200/1, 244
$C_{11}FFeH_{10}^+$ ..... $[(F-C_6H_5)Fe(C_5H_5)]^+$ .... Fe: Org.Comp.B18-142/6, 197, 200, 201, 243/4, 276/7
$C_{11}FH_{13}N_2OSn$ ... $(CH_3)_3SnN(COC_6H_4F-4)CN$ .... Sn: Org.Comp.18-58, 69
$C_{11}FH_{20}N_3OS$ .... $(NC_5H_{10}-1)_2S=N-C(O)F$ .... S: S-N Comp.8-207/8, 210
– .... $NC_5H_{10}-1-S(F)=N-C(O)-1-NC_5H_{10}$ .... S: S-N Comp.8-179
$C_{11}FH_{28}NSiSn$ ... $(CH_3)_3SnNHSiF(C_4H_9-t)_2$ .... Sn: Org.Comp.18-16, 20
$C_{11}F_2FeH_9^+$ ..... $[(1,2-F_2-C_6H_4)Fe(C_5H_5)][BF_4]$ .... Fe: Org.Comp.B19-1, 17
– .... $[(1,2-F_2-C_6H_4)Fe(C_5H_5)][PF_6]$ .... Fe: Org.Comp.B19-1, 17
– .... $[(1,3-F_2-C_6H_4)Fe(C_5H_5)][BF_4]$ .... Fe: Org.Comp.B19-1, 17
– .... $[(1,3-F_2-C_6H_4)Fe(C_5H_5)][PF_6]$ .... Fe: Org.Comp.B19-1, 17
– .... $[(1,4-F_2-C_6H_4)Fe(C_5H_5)][BF_4]$ .... Fe: Org.Comp.B19-1, 17
– .... $[(1,4-F_2-C_6H_4)Fe(C_5H_5)][PF_6]$ .... Fe: Org.Comp.B19-1, 17
$C_{11}F_2Fe_3O_9$ ..... $(CO)_9Fe_3(C-F)_2$. .... Fe: Org.Comp.C6a-236/7, 241/2
$C_{11}F_2GeH_{20}$ ..... $(C_2H_5)_2Ge(C_5H_4F_2(CH_3)_2)$ .... Ge: Org.Comp.3-319
$C_{11}F_2H_{14}MoNO_3P$ $(C_5H_5)Mo(CO)_2[PF_2-N(CH_3)_2]-C(=O)CH_3$ ..... Mo: Org.Comp.8-45, 56
$C_{11}F_2H_{29}N_3SSi$. .. $[((CH_3)_2N)_2S-1-NC_4H_8][Si(CH_3)_3F_2]$ .... S: S-N Comp.8-230, 239

$C_{11}F_3FeH_5N_2O_2$ .. 4-[$(C_5H_5)Fe(CO)_2$]-1,2-$N_2C_4(F_3$-3,5,6) .... Fe: Org.Comp.B14-26, 30
$C_{11}F_3FeH_8NO_4$ ... $[(C_5H_5)Fe(CO)_2(CH_2{=}C{=}NH)][CF_3CO_2] \cdot CF_3CO_2H$
Fe: Org.Comp.B17-125
$C_{11}F_3FeH_9O_2$ .... $(C_5H_5)Fe(CO)_2C(CF_3){=}CHCH_3$ .... Fe: Org.Comp.B13-93, 120/1
$C_{11}F_3FeH_9O_2S$ ... $(C_5H_5)Fe(CO)_2C(CF_3){=}CHSCH_3$ .... Fe: Org.Comp.B13-92, 119/20
$C_{11}F_3FeH_{10}NO_6S$ $[C_5H_5(CO)_2Fe{=}C(-NHCH_2CH_2O-)][CF_3SO_3]$ ... Fe: Org.Comp.B16a-139, 151
$C_{11}F_3FeH_{11}N_2O_3S_2$
$[C_5H_5Fe(CS)(NCCH_3)_2][SO_3CF_3]$ .... Fe: Org.Comp.B15-264
$C_{11}F_3FeH_{11}O_5S$ .. $[C_5H_5(CO)_2Fe{=}C(CH_3)_2][CF_3SO_3]$ .... Fe: Org.Comp.B16a-91, 99
$C_{11}F_3FeH_{11}O_5S_2$. $[C_5H_5(CO)_2Fe{=}C(CH_3)SCH_3][CF_3SO_3]$ .... Fe: Org.Comp.B16a-114, 124
$C_{11}F_3FeH_{11}O_5S_3$. $[C_5H_5(CO)_2Fe{=}C(SCH_3)_2][CF_3SO_3]$ .... Fe: Org.Comp.B16a-140, 152/4
$C_{11}F_3FeH_{11}O_6S$ .. $[C_5H_5(CO)_2Fe{=}C(CH_3)OCH_3][CF_3SO_3]$ .... Fe: Org.Comp.B16a-108, 118/9
$C_{11}F_3FeH_{11}O_6S_2$. $[C_5H_5(CO)_2Fe{=}C(OCH_3)SCH_3][CF_3SO_3]$ ...... Fe: Org.Comp.B16a-137, 150
$C_{11}F_3FeH_{11}O_7S$ .. $[C_5H_5(CO)_2Fe{=}C(OCH_3)_2][CF_3SO_3]$ .... Fe: Org.Comp.B16a-136, 148
$C_{11}F_3FeH_{12}NO_6S$ $[C_5H_5(CO)_2Fe{=}C(OCH_3)NHCH_3][CF_3SO_3]$ ..... Fe: Org.Comp.B16a-137, 151
$C_{11}F_3FeH_{13}N_2O_5S$ $[C_5H_5(CO)(CH_3CN)Fe{=}C(OCH_3)NH_2][CF_3SO_3]$ Fe: Org.Comp.B16a-66
$C_{11}F_3FeH_{14}O_7PS_2$ $[(C_5H_5)Fe(CS)(CO)P(OCH_3)_3][SO_3CF_3]$ .... Fe: Org.Comp.B15-272
$C_{11}F_3Fe_3O_{11}P$.... $(CO)_{11}Fe_3(PF_3)$ .... Fe: Org.Comp.C6b-128, 129
$C_{11}F_3GeH_{10}MnO_5$ $Ge(CH_3)_3CH{=}C(Mn(CO)_5)CF_3$ .... Ge: Org.Comp.2-9
$C_{11}F_3GeH_{15}$ ..... $Ge(CH_3)_3CH_2C_6H_4CF_3$-3 .... Ge: Org.Comp.1-158
$C_{11}F_3H_{11}MoO$.... $(C_5H_5)Mo(CO)(CF_3)(H_3C{-}C{\equiv}C{-}CH_3)$ .... Mo: Org.Comp.6-312
$C_{11}F_3H_{11}MoOS$... $(C_5H_5)Mo(CO)(SCF_3)(H_3C{-}C{\equiv}C{-}CH_3)$ .... Mo: Org.Comp.6-277, 281
$C_{11}F_3H_{11}MoO_6$ ... $(CH_2CHCH_2)Mo(CO)_3[O{=}C(CH_3)_2]{-}OC(O)CF_3$ .. Mo: Org.Comp.5-292, 293
$C_{11}F_3H_{13}N_2S$ .... 3-CF_3-C_6H_4-N=S=N-C_4H_9-t .... S: S-N Comp.7-219
$C_{11}F_3H_{14}MoO_5PS$ $[(C_5H_5)Mo(CO)_2(P(CH_3)_3){-}OS({=}O)_2CF_3]$ ...... Mo: Org.Comp.7-56/7, 62
$C_{11}F_3H_{14}MoO_8PS$ $[(C_5H_5)Mo(CO)_2(P(OCH_3)_3){-}OS({=}O)_2CF_3]$ ..... Mo: Org.Comp.7-56/7, 84
$C_{11}F_3H_{14}NS$ ..... F_3S-1-NC_5H_9-4-C_6H_5 .... S: S-N Comp.8-386
$C_{11}F_3H_{15}MoO_6$ ... $(CH_2CHCH_2)Mo(CO)_2(CH_3O{-}CH_2CH_2{-}OCH_3){-}OC(O)CF_3$
Mo: Org.Comp.5-268/9
$C_{11}F_4GeH_{15}N$ .... 3-$(C_2H_5)_3Ge{-}NC_5F_4$ .... Ge: Org.Comp.2-293
– .... 4-$(C_2H_5)_3Ge{-}NC_5F_4$ .... Ge: Org.Comp.2-293
$C_{11}F_4GeH_{20}O_2$ ... $Ge(C_2H_5)_3CF(CF_3)COOC_2H_5$ .... Ge: Org.Comp.2-136
$C_{11}F_5FeH_5O_2$ .... $(C_5H_5)Fe(CO)_2{-}C(CF_3){=}C{=}CF_2$ .... Fe: Org.Comp.B13-111
– .... $(C_5H_5)Fe(CO)_2{-}CF{=}CFCF{=}CF_2$ .... Fe: Org.Comp.B13-97, 126
– .... $(C_5H_5)Fe(CO)_2{-}C[{=}CF{-}CF_2{-}CF_2{-}]$ .... Fe: Org.Comp.B13-194, 198, 238
$C_{11}F_5GeH_9$ ...... $Ge(CH_3)_3C{\equiv}CC_6F_5$ .... Ge: Org.Comp.2-55
$C_{11}F_5GeH_{15}$ ..... $Ge(C_2H_5)_3C_5F_5$ .... Ge: Org.Comp.2-275
$C_{11}F_5H_7IMoN_3O$ .. $(C_5H_5)Mo(NO)(I)NHN(C_6F_5)H$ .... Mo: Org.Comp.6-52
$C_{11}F_5H_8I_2MoN_3O$ $(C_5H_5)Mo(NO)(NH_2NHC_6F_5)I_2$ .... Mo: Org.Comp.6-37
$C_{11}F_5H_{10}MoNO_3$.. $(C_5H_5)Mo(NO)(O_2CC_2F_5)C_3H_5$ .... Mo: Org.Comp.6-173
$C_{11}F_5O_5Re$ ...... 2-$(CO)_5Re$-[2.2.0]-C_6F_5 .... Re: Org.Comp.2-142
– .... $(CO)_5Re{-}C_6F_5$ .... Re: Org.Comp.2-143
$C_{11}F_5O_5ReS$ ..... $(CO)_5ReSC_6F_5$ .... Re: Org.Comp.2-27, 30/1
$C_{11}F_6FeH_6O_2$ .... $(C_5H_5)Fe(CO)_2{-}C(CF_3){=}CHCF_3$ .... Fe: Org.Comp.B13-93, 94, 121
$C_{11}F_6FeH_9O_2P$ ... $[(C_5H_5)Fe(CO)_2(CH_2{=}C{=}C{=}CH_2)][PF_6]$ .... Fe: Org.Comp.B17-116, 119
– .... $[(C_5H_5)Fe(CO)_2(C_4H_4{-}c)][PF_6]$ .... Fe: Org.Comp.B17-86, 103

$C_{11}F_6FeH_9O_3P$ ... $[((5.1.0)\text{-}C_8H_9)Fe(CO)_3][PF_6]$ ... Fe: Org.Comp.B15-214, 227/9
– ... $[(6\text{-}(CH_2{=})C_7H_7)Fe(CO)_3][PF_6]$ ... Fe: Org.Comp.B15-209, 225
$C_{11}F_6FeH_9O_4P$ ... $[(6\text{-}(HOCH{=})C_7H_7)Fe(CO)_3][PF_6]$ ... Fe: Org.Comp.B15-209
$C_{11}F_6FeH_9O_5P$ ... $[(1\text{-}CH_3OC(O)\text{-}C_6H_6)Fe(CO)_3][PF_6]$ ... Fe: Org.Comp.B15-101, 136/8
– ... $[(6\text{-}CH_3OC(O)\text{-}C_6H_6)Fe(CO)_3][PF_6]$ ... Fe: Org.Comp.B15-109/10, 136/8

$C_{11}F_6FeH_{10}IP$ ... $[(I\text{-}C_6H_5)Fe(C_5H_5)][PF_6]$ ... Fe: Org.Comp.B18-142/6, 197, 248

$C_{11}F_6FeH_{10}NO_2P$ $[(O_2N\text{-}C_6H_5)Fe(C_5H_5)][PF_6]$ ... Fe: Org.Comp.B18-142/6, 201, 255, 281

$C_{11}F_6FeH_{10}N_3P$ ... $[(N_3\text{-}C_6H_5)Fe(C_5H_5)][PF_6]$ ... Fe: Org.Comp.B18-142/6, 198, 254/5

$C_{11}F_6FeH_{11}OP$ ... $[(HO\text{-}C_6H_5)Fe(C_5H_5)][PF_6]$ ... Fe: Org.Comp.B18-142/6, 198, 248

$C_{11}F_6FeH_{11}O_2P$ ... $[(C_5H_5)Fe(CO)_2(CH_2{=}C{=}CHCH_3)][PF_6]$ ... Fe: Org.Comp.B17-111
– ... $[(C_5H_5)Fe(CO)_2(CH_2{=}CHCH{=}CH_2)][PF_6]$ ... Fe: Org.Comp.B17-36
– ... $[(C_5H_5)Fe(CO)_2(H_3C\text{-}C{\equiv}C\text{-}CH_3)][PF_6]$ ... Fe: Org.Comp.B17-122
$C_{11}F_6FeH_{11}O_2PS_2$ $[C_5H_5(CO)_2Fe{=}C(SCH_2)_2CH_2][PF_6]$ ... Fe: Org.Comp.B16a-144, 156
$C_{11}F_6FeH_{11}O_3P$ ... $[(1\text{-}C_2H_5\text{-}C_6H_6)Fe(CO)_3][PF_6]$ ... Fe: Org.Comp.B15-99, 135/6
– ... $[(1\text{-}C_2H_5\text{-}C_6H_5D)Fe(CO)_3][PF_6]$ ... Fe: Org.Comp.B15-135/6
– ... $[(1,4\text{-}(CH_3)_2C_6H_5)Fe(CO)_3][PF_6]$ ... Fe: Org.Comp.B15-122, 175
– ... $[(2,3\text{-}(CH_3)_2C_6H_5)Fe(CO)_3][PF_6]$ ... Fe: Org.Comp.B15-123
– ... $[(2,4\text{-}(CH_3)_2C_6H_5)Fe(CO)_3][PF_6]$ ... Fe: Org.Comp.B15-124, 177
– ... $[(2,6\text{-}(CH_3)_2C_6H_5)Fe(CO)_3][PF_6]$ ... Fe: Org.Comp.B15-124, 178
– ... $[2\text{-}(C_5H_5)Fe(CO)_2\text{-}OC_4H_6][PF_6]$ ... Fe: Org.Comp.B16a-113, 122
– ... $[(6\text{-}CH_3\text{-}C_7H_8)Fe(CO)_3][PF_6]$ ... Fe: Org.Comp.B15-206, 222
– ... $[(C_8H_{11})Fe(CO)_3][PF_6]$ ... Fe: Org.Comp.B15-236/7
$C_{11}F_6FeH_{11}O_4P$ ... $[(1\text{-}CH_3\text{-}2\text{-}CH_3O\text{-}C_6H_5)Fe(CO)_3][PF_6]$ ... Fe: Org.Comp.B15-110
– ... $[(1\text{-}CH_3\text{-}4\text{-}CH_3O\text{-}C_6H_5)Fe(CO)_3][PF_6]$ ... Fe: Org.Comp.B15-113, 161/9
– ... $[(2\text{-}CH_3\text{-}3\text{-}CH_3O\text{-}C_6H_5)Fe(CO)_3][PF_6]$ ... Fe: Org.Comp.B15-123, 175/6
– ... $[(2\text{-}CH_3O\text{-}3\text{-}CH_3\text{-}C_6H_5)Fe(CO)_3][PF_6]$ ... Fe: Org.Comp.B15-123, 175/6
$C_{11}F_6FeH_{11}O_5P$ ... $[(2,4\text{-}(CH_3O)_2C_6H_5)Fe(CO)_3][PF_6]$ ... Fe: Org.Comp.B15-124
$C_{11}F_6FeH_{11}P$ ... $MnPS_3 \cdot 0.30\ [(C_6H_6)Fe(C_5H_5)][PF_6]$ ... Fe: Org.Comp.B18-180
– ... $TaS_2 \cdot 0.16\ [(C_6H_6)Fe(C_5H_5)][PF_6]$ ... Fe: Org.Comp.B18-180
– ... $ZnPS_3 \cdot 0.2\ [(C_6H_6)Fe(C_5H_5)][PF_6]$ ... Fe: Org.Comp.B18-180
– ... $[(C_6H_6)Fe(C_5H_5)][PF_6]$ ... Fe: Org.Comp.B18-142/6, 150/4, 156/7, 174/81

– ... $[(C_6H_6)Fe(C_5H_5)][PF_6] \cdot 2\ C_{36}H_{60}O_{30} \cdot 8\ H_2O$ Fe: Org.Comp.B18-180/1
– ... $[(C_6H_6)Fe(C_5H_5)][PF_6] \cdot 2\ C_{42}H_{70}O_{35} \cdot 7\ H_2O$ Fe: Org.Comp.B18-180
$C_{11}F_6FeH_{11}PS$ ... $[(HS\text{-}C_6H_5)Fe(C_5H_5)][PF_6]$ ... Fe: Org.Comp.B18-142/6, 198, 253

$C_{11}F_6FeH_{11}Sb$ ... $[(C_6H_6)Fe(C_5H_5)][SbF_6]$ ... Fe: Org.Comp.B18-142/6, 150/1, 154, 157, 174/81

$C_{11}F_6FeH_{12}NOPS_2$
$[C_5H_5(CO)(CH_3CN)Fe{=}C(SCH_2)_2][PF_6]$ ... Fe: Org.Comp.B16a-68, 72
$C_{11}F_6FeH_{12}NO_2P$ $[(C_5H_5)Fe(CO)_2\text{-}CN\text{-}C_3H_7\text{-}n][PF_6]$ ... Fe: Org.Comp.B15-313
– ... $[(C_5H_5)Fe(CO)_2\text{-}CN\text{-}C_3H_7\text{-}i][PF_6]$ ... Fe: Org.Comp.B15-313, 317
$C_{11}F_6FeH_{12}NO_3P$ $[(C_5H_5)Fe(CO)_2{=}C(\text{-}N(CH_3)\text{-}CH_2CH_2\text{-}O\text{-})][PF_6]$ Fe: Org.Comp.B16a-139, 151
– ... $[(C_5H_5)Fe(CO)_2{=}C(\text{-}NH\text{-}CH_2CH_2CH_2\text{-}O\text{-})][PF_6]$ Fe: Org.Comp.B16a-140, 151/2

$C_{11}F_6FeH_{12}NO_3P$ $[(C_5H_5)Fe(CO)_2-CN-CH_2CH_2CH_2-OH][PF_6]$ ... Fe: Org.Comp.B15-314
$C_{11}F_6FeH_{12}NP$ ... $[(C_6H_6)Fe(C_5H_4-NH_2)][PF_6]$ ... Fe: Org.Comp.B18-142/6, 167, 188
– ... $[(NH_2-C_6H_5)Fe(C_5H_5)][PF_6]$ ... Fe: Org.Comp.B18-142/6, 197, 255/6, 281/2
$C_{11}F_6FeH_{13}N_2O_2P$ $[(C_5H_5)Fe(CO)_2{=}C(-NH-CH_2CH(CH_3)-NH-)][PF_6]$ Fe: Org.Comp.B16a-147
– ... $[(C_5H_5)Fe(CO)_2{=}C(-NH-CH_2CH_2CH_2-NH-)][PF_6]$ Fe: Org.Comp.B16a-147
– ... $[(C_5H_5)Fe(CO)_2{=}C(-NH-CH_2CH_2-N(CH_3)-)][PF_6]$ Fe: Org.Comp.B16a-147
$C_{11}F_6FeH_{13}OP$ ... $[C_5H_5Fe(CH_2{=}CHCH{=}CHCH_3)CO][PF_6]$ ... Fe: Org.Comp.B17-187
$C_{11}F_6FeH_{13}O_2P$ .. $[(C_5H_5)Fe(CO)_2(CH_2{=}C(CH_3)_2)][PF_6]$ ... Fe: Org.Comp.B17-61
– ... $[(C_5H_5)Fe(CO)_2(CH_2{=}CH-C_2H_5)][PF_6]$ ... Fe: Org.Comp.B17-9
– ... $[(C_5H_5)Fe(CO)_2(CH_2{=}CH-CH_2CH_2D)][PF_6]$ ... Fe: Org.Comp.B17-11
– ... $[(C_5H_5)Fe(CO)_2(CH_3CH{=}CHCH_3)][PF_6]$ ... Fe: Org.Comp.B17-56, 72
$C_{11}F_6FeH_{13}O_2PS_2$ $[C_5H_5(CO)_2Fe{=}C(SCH_3)SC_2H_5][PF_6]$ ... Fe: Org.Comp.B16a-141
$C_{11}F_6FeH_{13}O_3P$ .. $[(CH_2CHCHCHCH-C_3H_7-n)Fe(CO)_3][PF_6]$ ... Fe: Org.Comp.B15-26
– ... $[(CH_3-CHCHCHCHC(CH_3)_2)Fe(CO)_3][PF_6]$ ... Fe: Org.Comp.B15-6, 8
– ... $[(C_5H_5)Fe(CO)_2{=}C(CH_3)-OC_2H_5][PF_6]$ ... Fe: Org.Comp.B16a-109, 119
– ... $[(C_5H_5)Fe(CO)_2(CH_2{=}C(CH_3)-OCH_3)][PF_6]$ ... Fe: Org.Comp.B17-67, 79
– ... $[(C_5H_5)Fe(CO)_2(CH_2{=}CH-OC_2H_5)][PF_6]$ ... Fe: Org.Comp.B17-21
$C_{11}F_6FeH_{13}O_4PS$ $[(C_5H_5)Fe(CO)_2(CH_2{=}CHCH_2SO_2CH_3)][PF_6]$ ... Fe: Org.Comp.B17-26
$C_{11}F_6FeH_{13}PS$ ... $[C_5H_5FeC_4H_2S((CH_3)_2-2,5)][PF_6]$ ... Fe: Org.Comp.B17-214
$C_{11}F_6FeH_{14}NOPS_2$
$[(C_5H_5)FeCO(CN-CH_3){=}C(SCH_3)_2][PF_6]$ ... Fe: Org.Comp.B16a-163, 167/8
– ... $[(C_5H_5)FeCO(NC-CH_3){=}C(SCH_3)_2][PF_6]$ ... Fe: Org.Comp.B16a-66, 71/2
$C_{11}F_6FeH_{14}NO_2P$ $[(C_5H_5)Fe(CO)_2{=}CH-NH-C_3H_7-i][PF_6]$ ... Fe: Org.Comp.B16a-114, 125
– ... $[(C_5H_5)Fe(CO)_2-CH_2-CH-N(CH_3)_2][PF_6]$ ... Fe: Org.Comp.B14-127/8
$C_{11}F_6FeH_{14}NO_2PS$
$[C_5H_5(CO)_2Fe{=}C(SCH_3)N(CH_3)_2][PF_6]$ ... Fe: Org.Comp.B16a-146, 158
$C_{11}F_6FeH_{14}NO_5PS_2$
$[(C_5H_5)Fe(CO)_2(CH_2{=}CHCH_2S(O)NH-SO_2-CH_3)][PF_6]$ Fe: Org.Comp.B14-38
$C_{11}F_6FeH_{14}N_3P$ .. $[(C_5H_5)Fe(CN-CH_3)_3][PF_6]$ ... Fe: Org.Comp.B15-344/5, 346/7
– ... $[(C_5H_5)Fe(CN-CH_3)_2-NC-CH_3][PF_6]$ ... Fe: Org.Comp.B15-327, 328
– ... $[(C_5H_5)Fe(NC-CH_3)_2-CN-CH_3][PF_6]$ ... Fe: Org.Comp.B15-281
$C_{11}F_6FeH_{15}N_2P$ .. $[(C_5H_5)Fe(CH_2{=}CH_2)(CN-CH_3)_2][PF_6]$ ... Fe: Org.Comp.B17-138
$C_{11}F_6FeH_{17}NO_3P_2S$
$[(C_5H_5)Fe(CS)(NC-CH_3)P(OCH_3)_3][PF_6]$ ... Fe: Org.Comp.B15-264
$C_{11}F_6FeH_{17}OPS_3$ $[C_5H_5(CO)(S(CH_3)_2)Fe{=}C(SCH_3)_2][PF_6]$ ... Fe: Org.Comp.B16a-66, 71
$C_{11}F_6FeH_{18}O_4P_2$. $[(C_5H_5)Fe(CH_2{=}CH_2)(CO)P(OCH_3)_3][PF_6]$ ... Fe: Org.Comp.B16b-81
$C_{11}F_6H_3MnO_8PReSe$
$(CO)_4ReP(CF_3)_2Se(CH_3)Mn(CO)_4$ ... Re: Org.Comp.1-488
$C_{11}F_6H_8MoOS$ ... $(C_5H_5)Mo(CO)(SCH_3)(F_3C-C{\equiv}C-CF_3)$ ... Mo: Org.Comp.6-277
$C_{11}F_6H_{10}MoNO_4P$ $[(C_5H_5)Mo(CO)(NO)(CH_2C_4H_3O-(=O)-c)][PF_6]$ Mo: Org.Comp.6-350, 353/4
$C_{11}F_6H_{11}MoN_2O_2P$
$[(C_5H_5)Mo(CO)_2(CN-CH_3)_2][PF_6]$ ... Mo: Org.Comp.8-183

$C_{11}F_6H_{11}MoN_2O_2P$
$[(C_5H_5)Mo(CO)_2(NC\text{-}CH_3)_2][PF_6]$ Mo:Org.Comp.7-283, 284
$C_{11}F_6H_{11}MoN_2O_2PS_2$
$[(C_5H_5)Mo(CO)_2(\text{-}S\text{-}C(C(NCH_3)_2)\text{-}S\text{-})][PF_6]$ . . . Mo:Org.Comp.7-189, 192
$C_{11}F_6H_{12}MoNO_3P$ $[H_3CC(O)C_5H_4Mo(CO)(NO)C_3H_5][PF_6]$ Mo:Org.Comp.6-347
$C_{11}F_6H_{14}MoNO_2P$ $[(C_5H_5)Mo(CO)(NO)C_3H_3(CH_3)_2\text{-}1,2][PF_6]$ Mo:Org.Comp.6-349
$C_{11}F_6H_{14}MoNO_2PS$
$[(C_5H_5)Mo(CO)_2(\text{-}C(N(CH_3)_2)=S(CH_3)\text{-})][PF_6]$ . . Mo:Org.Comp.8-130
$C_{11}F_6H_{14}MoN_3O_2P$
$[(CH_2CHCH_2)Mo(CO)_2(NC\text{-}CH_3)_3][PF_6]$ Mo:Org.Comp.5-200
$C_{11}F_6H_{15}N_5OPRe$ $[(CO)Re(CNCH_3)_5][PF_6]$ . . . . Re: Org.Comp.2-290
$C_{11}F_6H_{15}O_5P_2Re$ $[(CO)_5ReP(C_2H_5)_3][PF_6]$ . . . . Re: Org.Comp.2-155
$C_{11}F_6H_{16}O_4Sn$ . . . $(C_3H_7)(C_4H_9)Sn(OOCCF_3)_2$ . . . . Sn: Org.Comp.16-211, 214
$C_{11}F_6H_{18}MoN_3O_2P$
$[(C_5H_5)Mo(CO)(NO)H_2C=CHCH_2N(CH_3)_2NH_2][PF_6]$
Mo:Org.Comp.6-304
$C_{11}F_6H_{21}MoOP_3$ . . $[(C_5H_5)Mo(P(CH_3)_2CH_2CH_2P(CH_3)_2)O][PF_6]$ . . . Mo:Org.Comp.6-19
$C_{11}F_6H_{25}MoP_3$ . . . $[C_5H_5Mo(P(CH_3)_2CH_2CH_2P(CH_3)_2)H_4][PF_6]$. . . . Mo:Org.Comp.6-16, 20
$C_{11}F_7FeH_5O_2$ . . . . $(C_5H_5)Fe(CO)_2C(CF_3)=CFCF_3$. . . . . Fe: Org.Comp.B13-94, 122
$C_{11}F_7FeH_5O_3$ . . . . $(C_5H_5)Fe(CO)_2COC_3F_7\text{-}n$ . . . . Fe: Org.Comp.B13-16/7, 36
$C_{11}F_7FeH_{10}P$ . . . . $[(F\text{-}C_6H_5)Fe(C_5H_5)][PF_6]$ . . . . Fe: Org.Comp.B18-142/6, 197, 201, 243/4, 276/7
$C_{11}F_7GeH_{16}I$. . . . . $Ge(C_2H_5)_3C(I)=CHC_3F_7$ . . . . Ge: Org.Comp.2-222, 227
$C_{11}F_8FeH_9P$ . . . . . $[(1,2\text{-}F_2\text{-}C_6H_4)Fe(C_5H_5)][PF_6]$ . . . . Fe: Org.Comp.B19-1, 17
– . . . . $[(1,3\text{-}F_2\text{-}C_6H_4)Fe(C_5H_5)][PF_6]$ . . . . Fe: Org.Comp.B19-1, 17
– . . . . $[(1,4\text{-}F_2\text{-}C_6H_4)Fe(C_5H_5)][PF_6]$ . . . . Fe: Org.Comp.B19-1, 17
$C_{11}F_8HN_3O_2$ . . . . . $NCFCFC(NHC_6F_4NO_2\text{-}4)CFCF$ . . . . F: PFHOrg.SVol.4-149
$C_{11}F_8H_9NOSn$. . . . $(CH_3)_3SnN(C_6F_5)COCF_3$ . . . . Sn: Org.Comp.18-47, 51, 56
$C_{11}F_9HN_2$ . . . . $NCFCFC(NHC_6F_5)CFCF$ . . . . F: PFHOrg.SVol.4-149
$C_{11}F_9H_5MoOS$ . . . $(C_5H_5)Mo(CO)(SCF_3)(F_3C\text{-}C\equiv C\text{-}CF_3)$. Mo:Org.Comp.6-278, 281
$C_{11}F_9H_6MoNO_4$ . . $(C_5H_5)Mo(CO)_2NO \cdot t\text{-}C_4F_9OH$ Mo:Org.Comp.7-6, 9
$C_{11}F_9N$ . . . . $NCFCFC(C_6F_5)CFCF$. . . . . F: PFHOrg.SVol.4-119, 139
$C_{11}F_9NO$ . . . . $2\text{-}(C_6F_5\text{-}O)\text{-}NC_5F_4$ . . . . F: PFHOrg.SVol.4-143, 153
– . . . . $4\text{-}(C_6F_5\text{-}O)\text{-}NC_5F_4$ . . . . F: PFHOrg.SVol.4-143, 153
$C_{11}F_9N_2Na$ . . . . . . $Na[4\text{-}(C_6F_5\text{-}N)\text{-}NC_5F_4]$. F: PFHOrg.SVol.4-148
$C_{11}F_9O_5Re$ . . . . . . $(CO)_5ReC_6F_9\text{-}c$. . . . . Re: Org.Comp.2-141/2
$C_{11}F_{11}NO_2$ . . . . . . $N(C_6F_5)C(O)CF_2CF_2CF_2C(O)$ . . . . F: PFHOrg.SVol.4-142, 152, 162
$C_{11}F_{12}FeH_{10}N_2P_2$ $[(C_6H_6)Fe(C_5H_4\text{-}NN)][PF_6]_2$ . . . . Fe: Org.Comp.B18-142/6, 168
$C_{11}F_{12}H_{13}NO_2S$ . . $(C_2H_5)_2N\text{-}S(O)O\text{-}CH_2(CF_2)_6H$ . . . . S: S-N Comp.8-315/6
$C_{11}F_{13}O_8ReS$ . . . . $(CO)_5ReOS(O)_2C_6F_{13}$ . . . . Re: Org.Comp.2-25
$C_{11}F_{14}H_{11}NO_2S$ . . $HCF_2CF_2CF_2CF_2CF_2CF_2CH_2O\text{-}S(F_2)\text{-}4\text{-}1,4\text{-}ONC_4H_8$
S: S-N Comp.8-395/7
$C_{11}F_{14}H_{13}NOS$ . . . $HCF_2CF_2CF_2CF_2CF_2CF_2CH_2O\text{-}S(F_2)\text{-}N(C_2H_5)_2$ S: S-N Comp.8-395/7
$C_{11}F_{15}H_2N$ . . . . . . $(CF_3)_5C_6\text{-}NH_2$ . . . . F: PFHOrg.SVol.5-10, 41
$C_{11}F_{15}H_6NO_2$ . . . . $(CF_3)_3CNC(CF_3)_2C(O)OC(CH_3)_2$. . . . . F: PFHOrg.SVol.4-13
$C_{11}F_{15}N$ . . . . $N(C_6F_5)CF_2CF_2CF_2CF_2CF_2$. . . . . F: PFHOrg.SVol.4-121, 133
$C_{11}F_{16}N_2O_2$ . . . . . $NC\text{-}C(CF_3)_2\text{-}O\text{-}CF=CFCF_2\text{-}O\text{-}C(CF_3)_2\text{-}CN$. . . . F: PFHOrg.SVol.6-103/4
$C_{11}F_{17}N$ . . . . $2,4\text{-}(i\text{-}C_3F_7)_2\text{-}NC_5F_3$ . . . . F: PFHOrg.SVol.4-117/8, 125, 139

$C_{11}F_{17}N$	2,5-(i-C_3F_7)$_2$-NC_5F_3	F: PFHOrg.SVol.4-117/8, 126
$C_{11}F_{18}H_2N_4O_2$	$OC(CF_3)_2OC(CF_3)_2NC(NHC(NH)NC(CF_3)_2)$ · 0.5 H_2SO_4	F: PFHOrg.SVol.4-176, 181/2
$C_{11}F_{18}H_3N_3$	C_3F_7-C(=NH)-N=C(NH_2)-C_6F_{11}-c	F: PFHOrg.SVol.5-5
$C_{11}F_{18}H_4N_4O_3$	$OC(CF_3)_2OC(CF_3)_2NC(NHC(NH)NHC(OH)(CF_3)_2)$	F: PFHOrg.SVol.4-176, 181/2
$C_{11}F_{18}H_4N_4O_6S$	$OC(CF_3)_2OC(CF_3)_2NC(NHC(NH)NC(CF_3)_2)$ · 0,5 H_2SO_4	F: PFHOrg.SVol.4-176, 181/2
$C_{11}F_{18}N_2$	2-(i-C_3F_7)-5-(s-C_4F_9)-1,4-$N_2C_4F_2$	F: PFHOrg.SVol.4-199, 211
–	2-(i-C_3F_7)-6-(s-C_4F_9)-1,4-$N_2C_4F_2$	F: PFHOrg.SVol.4-199, 211
–	4-(i-C_3F_7)-5-(s-C_4F_9)-1,3-$N_2C_4F_2$	F: PFHOrg.SVol.4-192, 206
–	5-(i-C_3F_7)-4-(s-C_4F_9)-1,2-$N_2C_4F_2$	F: PFHOrg.SVol.4-185/6
–	5-(i-C_3F_7)-4-(s-C_4F_9)-1,3-$N_2C_4F_2$	F: PFHOrg.SVol.4-192, 205
$C_{11}F_{18}N_2O_2$	NC-CF(CF_3)-O-(CF_2)$_5$-O-CF(CF_3)-CN	F: PFHOrg.SVol.6-103, 123, 136
–	NC-CF(CF_3)-O-$CF_2CF(CF_3)$-O-(CF_2)$_4$-CN	F: PFHOrg.SVol.6-103, 123, 136
$C_{11}F_{19}N$	(i-C_3F_7)$_2$-NC_5F_5	F: PFHOrg.SVol.4-117/8, 127
$C_{11}F_{19}NO_3$	CF_2=CF[-O-$CF_2CF(CF_3)$]$_2$-O-CF_2CF_2-CN	F: PFHOrg.SVol.6-99, 143
$C_{11}F_{20}H_3N_3$	C_3F_7-C(=NH)-N=C(NH_2)-C_6F_{13}	F: PFHOrg.SVol.5-5
$C_{11}F_{21}N$	1-C_5F_9-2-C_2F_5-NC_4F_7	F: PFHOrg.SVol.4-28
–	1-C_5F_9-3-C_2F_5-NC_4F_7	F: PFHOrg.SVol.4-28
–	1-(C_5F_9-CF_2)-2-CF_3-NC_4F_7	F: PFHOrg.SVol.4-28/9
–	1-(C_5F_9-CF_2)-3-CF_3-NC_4F_7	F: PFHOrg.SVol.4-28/9
–	1-(C_5F_9-CF_2CF_2)-NC_4F_8	F: PFHOrg.SVol.4-28
–	1-C_6F_{11}-2-CF_3-NC_4F_7	F: PFHOrg.SVol.4-28/9
–	1-C_6F_{11}-3-CF_3-NC_4F_7	F: PFHOrg.SVol.4-28/9
–	1-(C_6F_{11}-CF_2)-NC_4F_8	F: PFHOrg.SVol.4-28
–	1-C_7F_{13}-NC_4F_8	F: PFHOrg.SVol.4-28
–	1-[(CF_3)$_2$-C_5F_7]-NC_4F_8	F: PFHOrg.SVol.4-28
–	1-[C_5F_9-CF(CF_3)]-NC_4F_8	F: PFHOrg.SVol.4-28
–	2,4-(i-C_3F_7)$_2$-NC_5F_7	F: PFHOrg.SVol.4-117/8
–	CF_3-C_5F_8-CF_2-NC_4F_8	F: PFHOrg.SVol.4-28
–	CF_3-C_5F_8-NC_4F_7-CF_3	F: PFHOrg.SVol.4-28/9
–	CF_3-C_6F_{10}-NC_4F_8	F: PFHOrg.SVol.4-28
–	C_2F_5-C_5F_8-NC_4F_8	F: PFHOrg.SVol.4-28
–	C_5F_9-$NC_4F_6(CF_3)_2$	F: PFHOrg.SVol.4-28
–	[2.2.1]-$C_7F_9(CF_3)_2$-$N(CF_3)_2$	F: PFHOrg.SVol.6-228
–	[2.2.1]-$C_7F_{10}(CF_3)$-N(CF_3)-C_2F_5	F: PFHOrg.SVol.6-230
–	[2.2.1]-$C_7F_{10}(C_2F_5)$-$N(CF_3)_2$	F: PFHOrg.SVol.6-228/9
–	[2.2.1]-C_7F_{11}-N(CF_3)-C_3F_7-n	F: PFHOrg.SVol.6-230
–	[2.2.1]-C_7F_{11}-N(CF_3)-C_3F_7-i	F: PFHOrg.SVol.6-230
–	[2.2.1]-C_7F_{11}-$N(C_2F_5)_2$	F: PFHOrg.SVol.6-228/9
$C_{11}F_{21}NO_3S$	1-(c-C_6F_{11})-2-(F-S(=O)$_2$-O)-NC_5F_9	F: PFHOrg.SVol.4-144, 155
$C_{11}F_{21}N_3O_3$	N_3-CF_2CF(O-C_3F_7-n)CF_2-O-CF(CF_3)CF_2-O-CF=CF_2	F: PFHOrg.SVol.5-204, 216, 224
$C_{11}F_{24}N_2O$	$OCF_2CF_2N(CF_2CF_2CF_2N(C_2F_5)_2)CF_2CF_2$	F: PFHOrg.SVol.4-173/4
$C_{11}F_{24}N_2O_4S$	$OCF_2CF(OSO_2F)N(CF_2CF_2CF_2N(C_2F_5)_2)CF_2CF_2$	F: PFHOrg.SVol.4-174, 182

$C_{11}F_{25}NO_2$ ...... n-C_3F_7-N(CF_2CF_2-O-C_2F_5)$_2$ .... F: PFHOrg.SVol.6-226/7
$C_{11}F_{25}NO_3$ ...... C_2F_5-O-CF_2CF_2-O-CF_2CF_2-N(CF_3)-CF_2CF_2-O-C_2F_5 F: PFHOrg.SVol.6-230
$C_{11}F_{27}HO_2P_3Re$ .. $(CO)_2ReH[P(CF_3)_3]_3$ .... Re: Org.Comp.1-100/1, 103
$C_{11}FeGeH_{14}O_3$ ... $(CH_3)_2Ge(-CH=C(CH_3)C(CH_3)=CH-)Fe(CO)_3$ ... Ge: Org.Comp.3-280, 285
$C_{11}FeGeH_{16}O_4S_2$ Ge$(CH_3)_3$CH(S$(CH_2)_3$S · Fe$(CO)_4$) .... Ge: Org.Comp.2-100, 106
$C_{11}FeH_7IO_2S$ .... $(C_5H_5)Fe(CO)_2$-2-(SC_4H_2-5-I) .... Fe: Org.Comp.B14-7, 14
$C_{11}FeH_7N_2O_2$ .... $(C_5H_5)Fe(CO)_2CH=CH(CN)_2$ .... Fe: Org.Comp.B13-85, 116
$C_{11}FeH_7NaO_2$ .... Na$[C_9H_7Fe(CO)_2]$ .... Fe: Org.Comp.B14-119/20
$C_{11}FeH_7O_2$ ...... $C_9H_7Fe(CO)_2$, radical .... Fe: Org.Comp.B14-125
$C_{11}FeH_7O_2^-$ ..... $[C_9H_7Fe(CO)_2]^-$ .... Fe: Org.Comp.B14-119/20
$C_{11}FeH_8IN$ ...... NC-$C_5H_4FeC_5H_4$I .... Fe: Org.Comp.A9-113
$C_{11}FeH_8KNO_3S$ .. K[O_3S-$C_5H_4FeC_5H_4$-CN] · 2 H_2O .... Fe: Org.Comp.A9-239
$C_{11}FeH_8NO_3S^-$ ... [O_3S-$C_5H_4FeC_5H_4$-CN]$^-$ .... Fe: Org.Comp.A9-239
$C_{11}FeH_8N_2O_2$ .... O_2N-$C_5H_4FeC_5H_4$-CN .... Fe: Org.Comp.A9-153
$C_{11}FeH_8O_2S$ ..... $(C_5H_5)Fe(CO)_2$-2-C_4H_3S .... Fe: Org.Comp.B14-2/3, 6, 12/3
$C_{11}FeH_8O_3$ ...... 2-$(C_5H_5)Fe(CO)_2$-OC_4H_3 .... Fe: Org.Comp.B14-2/3, 5, 12
– .... C_6H_6-CH=CH-C(=O)-$Fe(CO)_2$ .... Fe: Org.Comp.B18-32, 46/7
$C_{11}FeH_8O_3^+$ ..... $[(C_8H_8)Fe(CO)_3]^+$, radical cation .... Fe: Org.Comp.B15-201/2
$C_{11}FeH_9KO_4S$ .... K[O_3S-$C_5H_4FeC_5H_4$-CHO] · 2 H_2O .... Fe: Org.Comp.A9-238
$C_{11}FeH_9NO_3S$ .... HO_3S-$C_5H_4FeC_5H_4$-CN .... Fe: Org.Comp.A9-231, 239
$C_{11}FeH_9O_2^+$ ..... $[(C_5H_5)Fe(CO)_2(CH_2=C=C=CH_2)]^+$ .... Fe: Org.Comp.B17-116, 119
– .... $[(C_5H_5)Fe(CO)_2(C_4H_4\text{-c})]^+$ .... Fe: Org.Comp.B17-86, 103
$C_{11}FeH_9O_3^+$ ..... [((5.1.0)-C_8H_9)$Fe(CO)_3$]$^+$ .... Fe: Org.Comp.B15-214, 227/9, 238, 247/8
– .... [((5.1.0)-C_8H_8D)$Fe(CO)_3$]$^+$ .... Fe: Org.Comp.B15-228/9
– .... [((5.1.0)-C_8D_9)$Fe(CO)_3$]$^+$ .... Fe: Org.Comp.B15-229
– .... [(6-(CH_2=)C_7H_7)$Fe(CO)_3$]$^+$ .... Fe: Org.Comp.B15-209, 225
– .... [(6-(CHD=)C_7H_7)$Fe(CO)_3$]$^+$ .... Fe: Org.Comp.B15-225
– .... [($CH_3C_7H_6$)$Fe(CO)_3$]$^+$ .... Fe: Org.Comp.B15-219, 232
$C_{11}FeH_9O_4^+$ ..... [(1-CH_3-7-O=C_7H_6)$Fe(CO)_3$]$^+$ .... Fe: Org.Comp.B15-212, 226
– .... [(1-CH_3-7-O=$C_7H_4D_2$)$Fe(CO)_3$]$^+$ .... Fe: Org.Comp.B15-226
– .... [(7-CH_3-6-O=C_7H_6)$Fe(CO)_3$]$^+$ .... Fe: Org.Comp.B15-213, 227
– .... [(1-HO-7-CH_3-C_7H_5)$Fe(CO)_3$]$^+$ .... Fe: Org.Comp.B15-219
– .... [(2-OCH-C_7H_8)$Fe(CO)_3$]$^+$ .... Fe: Org.Comp.B15-205
– .... [(6-(HOCH=)C_7H_7)$Fe(CO)_3$]$^+$ .... Fe: Org.Comp.B15-209
– .... [(2-HO-C_8H_8)$Fe(CO)_3$]$^+$ .... Fe: Org.Comp.B15-220, 233, 248
– .... [(7-HO-C_8H_8)$Fe(CO)_3$]$^+$ .... Fe: Org.Comp.B15-248
– .... [(7-HO-C_8H_7D)$Fe(CO)_3$]$^+$ .... Fe: Org.Comp.B15-248
– .... [(8-O=C_8H_9)$Fe(CO)_3$]$^+$ .... Fe: Org.Comp.B15-244
– .... $[(C_5H_5)Fe(CO)_2(HC\equiv C\text{-COO-}CH_3)]^+$ .... Fe: Org.Comp.B17-121
$C_{11}FeH_9O_4S^-$ .... [O_3S-$C_5H_4FeC_5H_4$-CHO]$^-$ .... Fe: Org.Comp.A9-238
$C_{11}FeH_9O_5^+$ ..... [(1-CH_3OC(O)-C_6H_6)$Fe(CO)_3$]$^+$ .... Fe: Org.Comp.B15-97, 101, 136/8
– .... [(2-CH_3OC(O)-C_6H_6)$Fe(CO)_3$]$^+$ .... Fe: Org.Comp.B15-96/7, 105, 136/8
– .... [(3-CH_3OC(O)-C_6H_6)$Fe(CO)_3$]$^+$ .... Fe: Org.Comp.B15-96/7, 107, 136/8

$C_{11}FeH_9O_5^+$	$[(6\text{-}CH_3OC(O)\text{-}C_6H_6)Fe(CO)_3]^+$	Fe: Org.Comp.B15-96/7, 109/10, 136/8
–	$[(6\text{-}CH_3C(O)O\text{-}C_6H_6)Fe(CO)_3]^+$	Fe: Org.Comp.B15-107, 159
$C_{11}FeH_{10}I^+$	$[(I\text{-}C_6H_5)Fe(C_5H_5)]^+$	Fe: Org.Comp.B18-142/6, 197, 248
$C_{11}FeH_{10}INO$	$H_2NCO\text{-}C_5H_4FeC_5H_4I$	Fe: Org.Comp.A9-79
$C_{11}FeH_{10}K_2$	$Fe[C_{10}H_7(K)_2(CH_3)]$	Fe: Org.Comp.A10-328
$C_{11}FeH_{10}Li_2$	$(Li\text{-}C_5H_4)Fe[C_5H_3(Li)\text{-}CH_3]$	Fe: Org.Comp.A10-322/3, 325
$C_{11}FeH_{10}NO_2^+$	$[(C_5H_5)Fe(CO)_2(CH_2{=}C(CH_3)CN)]^+$	Fe: Org.Comp.B17-63
–	$[(C_5H_5)Fe(CO)_2(CH_3CH{=}CHCN)]^+$	Fe: Org.Comp.B17-59
–	$[(O_2N\text{-}C_6H_5)Fe(C_5H_5)]^+$	Fe: Org.Comp.B18-142/6, 201, 255, 281
$C_{11}FeH_{10}NO_3^+$	$[(C_5H_5)Fe(CO)_2(NCCH{=}CHOCH_3)]^+$	Fe: Org.Comp.B17-66, 79
$C_{11}FeH_{10}N_2$	$(C_5H_5)Fe[C_5H_3(CN)\text{-}NH_2]$	Fe: Org.Comp.A9-130, 133/4, 137, 139
$C_{11}FeH_{10}N_2^{2+}$	$[(C_6H_6)Fe(C_5H_4\text{-}NN)]^{2+}$	Fe: Org.Comp.B18-142/6, 168
$C_{11}FeH_{10}N_2O$	$(C_5H_5)Fe(CN)(CO)CN\text{-}CH_2CH{=}CH_2$	Fe: Org.Comp.B15-291
$C_{11}FeH_{10}N_3^+$	$[(N_3\text{-}C_6H_5)Fe(C_5H_5)]^+$	Fe: Org.Comp.B18-142/6, 198, 254/5
$C_{11}FeH_{10}Na_2$	$Fe[C_{10}H_7(Na)_2(CH_3)]$	Fe: Org.Comp.A10-328
$C_{11}FeH_{10}O$	$(C_5H_5)Fe[C_6H_5({=}O)]$	Fe: Org.Comp.B17-260, 307
–	$(C_5H_5)_2Fe(CO)$	Fe: Org.Comp.B17-160
$C_{11}FeH_{10}OS$	$C_5H_5Fe(CH_2C_4H_3S)CO$	Fe: Org.Comp.B17-157
$C_{11}FeH_{10}O_2$	$(C_5H_5)Fe(CO)\text{-}C({=}O)\text{-}C({=}CH_2)\text{-}CH{=}CH_2$	Fe: Org.Comp.B17-168
–	$(C_5H_5)Fe(CO)_2\text{-}C(CH{=}CH_2){=}CH_2$	Fe: Org.Comp.B13-78, 97, 124/6
–	$(C_5H_5)Fe(CO)_2\text{-}CH{=}C{=}CHCH_3$	Fe: Org.Comp.B13-111, 138
–	$(C_5H_5)Fe(CO)_2\text{-}CH{=}CHCH{=}CH_2$	Fe: Org.Comp.B13-97
–	$(C_5H_5)Fe(CO)_2\text{-}CH_2CH_2C{\equiv}CH$	Fe: Org.Comp.B13-155
–	$(C_5H_5)Fe(CO)_2\text{-}CH_2C{\equiv}CCH_3$	Fe: Org.Comp.B13-155, 159/60
–	$(C_5H_5)Fe(CO)_2\text{-}c\text{-}C_4H_5$	Fe: Org.Comp.B13-195, 198, 237
$C_{11}FeH_{10}O_2S_2$	$(C_5H_5)Fe(CO)_2C(S)SCH_2CH{=}CH_2$	Fe: Org.Comp.B13-56
$C_{11}FeH_{10}O_3$	$(C_5H_5)Fe(CO)_2\text{-}C({=}CH_2)\text{-}C({=}O)CH_3$	Fe: Org.Comp.B13-96, 123
–	$(C_5H_5)Fe(CO)_2\text{-}C({=}O)C(CH_3){=}CH_2$	Fe: Org.Comp.B13-10, 12, 18/9, 38
–	$(C_5H_5)Fe(CO)_2\text{-}C({=}O)CH{=}CHCH_3$	Fe: Org.Comp.B13-9/10, 12, 18, 38
–	$(C_5H_5)Fe(CO)_2\text{-}C({=}O)CH_2CH{=}CH_2$	Fe: Org.Comp.B13-19, 39
–	$(C_5H_5)Fe(CO)_2\text{-}C({=}O)C_3H_5\text{-}c$	Fe: Org.Comp.B13-10, 19, 39
–	$(C_5H_5)Fe(CO)_2\text{-}C(CH_2OH){=}C{=}CH_2$	Fe: Org.Comp.B13-111
–	$(C_5H_5)Fe(CO)_2\text{-}CH{=}CH\text{-}C({=}O)CH_3$	Fe: Org.Comp.B13-96, 123
–	$(C_5H_5)Fe(CO)_2\text{-}CH_2CH{=}CH\text{-}CHO$	Fe: Org.Comp.B13-105, 133
–	$C_5H_4\text{-}CH(CH_3)CH_2\text{-}C({=}O)\text{-}Fe(CO)_2$	Fe: Org.Comp.B18-18/20, 25
–	$C_5H_4\text{-}CH_2CH(CH_3)\text{-}C({=}O)\text{-}Fe(CO)_2$	Fe: Org.Comp.B18-18/20, 26
$C_{11}FeH_{10}O_4$	$(C_5H_5)Fe(CO)_2CH{=}CHCOOCH_3$	Fe: Org.Comp.B13-90
$C_{11}FeH_{10}O_4S$	$(C_5H_5)Fe(CO)_2\text{-}C[{=}C(CH_3)\text{-}S({=}O)\text{-}O\text{-}CH_2\text{-}]$	Fe: Org.Comp.B14-32, 36
–	$(C_5H_5)Fe(CO)_2\text{-}C[{=}CH\text{-}CH_2\text{-}SO_2\text{-}CH_2\text{-}]$	Fe: Org.Comp.B14-6
–	$(HO_3S\text{-}C_5H_4)Fe(C_5H_4\text{-}CHO)$	Fe: Org.Comp.A9-231, 238

$C_{11}FeH_{10}O_5S$	$(C_5H_5)Fe(CO)_2(CH_2{=}C{=}C(CH_3)SO_3)$	Fe: Org.Comp.B17-114
–	$(C_5H_5)Fe(CO)_2$-C[=C(CH_3)-S(=O)$_2$-O-CH_2-]	Fe: Org.Comp.B14-32, 36/7
–	$(HO_3S\text{-}C_5H_4)Fe(C_5H_4\text{-}CO_2H)$	Fe: Org.Comp.A9-231, 238
$C_{11}FeH_{10}O_6PbS_2$	$Pb[(O_3S\text{-}C_5H_4)Fe(C_5H_3(CH_3)\text{-}SO_3)] \cdot x\ H_2O$	Fe: Org.Comp.A10-297
$C_{11}FeH_{10}O_6S_2^{2-}$	$[(O_3S\text{-}C_5H_4)Fe(C_5H_3(CH_3)\text{-}SO_3)]^{2-}$	Fe: Org.Comp.A10-297
$C_{11}FeH_{10}S$	$C_5H_5FeC_6H_5(=S)$	Fe: Org.Comp.B17-260, 307
$C_{11}FeH_{11}$	$[(C_6H_6)Fe(C_5H_5)]$	Fe: Org.Comp.B18-146/8, 153, 154, 160, 181/5
$C_{11}FeH_{11}^+$	$[(C_6H_6)Fe(C_5H_5)]^+$	Fe: Org.Comp.B18-142/6, 150/60, 174/81
–	$[(C_6H_6)Fe(C_5D_5)]^+$	Fe: Org.Comp.B18-174
–	$[(C_6D_6)Fe(C_5H_5)]^+$	Fe: Org.Comp.B18-174
$C_{11}FeH_{11}^-$	$[(C_6H_6)Fe(C_5H_5)]^-$	Fe: Org.Comp.B18-148, 160, 185
$C_{11}FeH_{11}HgI_3$	$[(C_6H_6)Fe(C_5H_5)][HgI_3]$	Fe: Org.Comp.B18-142/6, 150/1, 154, 160
$C_{11}FeH_{11}I$	$[(C_6H_6)Fe(C_5H_5)]I$	Fe: Org.Comp.B18-142/6, 150/4, 155, 174/81
–	$[(C_6H_6)Fe(C_5H_5)][I_x]$	Fe: Org.Comp.B18-142/6, 150/1, 154, 156
$C_{11}FeH_{11}I_3$	$[(C_6H_6)Fe(C_5H_5)][I_3]$	Fe: Org.Comp.B18-142/6, 150/1, 154, 155
$C_{11}FeH_{11}I_x$	$[(C_6H_6)Fe(C_5H_5)][I_x]$	Fe: Org.Comp.B18-142/6, 150/1, 154, 156
$C_{11}FeH_{11}K$	$(C_5H_5)Fe[C_5H_3(K)\text{-}CH_3]$	Fe: Org.Comp.A10-131
–	$(K\text{-}C_5H_4)Fe(C_5H_4\text{-}CH_3)$	Fe: Org.Comp.A10-131
$C_{11}FeH_{11}Li$	$(Li\text{-}C_5H_4)Fe(C_5H_4\text{-}CH_3)$	Fe: Org.Comp.A10-113/5
$C_{11}FeH_{11}LiO$	$(C_5H_5)Fe[C_5H_3(Li)\text{-}OCH_3]$	Fe: Org.Comp.A10-118/23
$C_{11}FeH_{11}LiO_3$	$Li[(C_5H_4C(O)C_3H_7\text{-}i)Fe(CO)_2]$	Fe: Org.Comp.B14-118
$C_{11}FeH_{11}N$	$C_5H_5FeC_6H_5(=NH)$	Fe: Org.Comp.B17-259, 307
$C_{11}FeH_{11}NO_2$	$(C_5H_5)Fe(C_6H_6\text{-}NO_2\text{-}1)$	Fe: Org.Comp.B17-268
–	$(C_5H_5)Fe[C_5H_3(NH_2)\text{-}COOH]$	Fe: Org.Comp.A9-25, 71
–	$(O_2N\text{-}C_5H_4)Fe(C_5H_4\text{-}CH_3)$	Fe: Org.Comp.A9-153
$C_{11}FeH_{11}NO_3$	$(C_5H_5)Fe(CO)_2CH$[-CH_2-C(=O)-NH-CH_2-]	Fe: Org.Comp.B14-20, 27
$C_{11}FeH_{11}NO_4S$	$H_2NO_2S\text{-}C_5H_4FeC_5H_4\text{-}C(O)OH$	Fe: Org.Comp.A9-233/4, 242
$C_{11}FeH_{11}N_2^+$	$[(C_5H_5)Fe(C_5H_3(CN)\text{-}NH_3)]^+$	Fe: Org.Comp.A9-133, 139
$C_{11}FeH_{11}N_3O_2$	$(C_5H_5)Fe(CO)_2C(CH_2N_3){=}CHCH_3$	Fe: Org.Comp.B13-95
$C_{11}FeH_{11}Na$	$(C_5H_5)Fe[C_5H_3(Na)\text{-}CH_3]$	Fe: Org.Comp.A10-130/1
–	$(Na\text{-}C_5H_4)Fe(C_5H_4\text{-}CH_3)$	Fe: Org.Comp.A10-130/1
$C_{11}FeH_{11}NaO_2S$	$Na[C_5H_5FeC_5H_3(SO_2)CH_3] \cdot 2\ H_2O$	Fe: Org.Comp.A9-252
$C_{11}FeH_{11}O^+$	$[(HO\text{-}C_6H_5)Fe(C_5H_5)]^+$	Fe: Org.Comp.B18-142/6, 198, 248
$C_{11}FeH_{11}OP$	$(C_5H_5)Fe[PC_4H_3\text{-}2\text{-}C(O)CH_3]$	Fe: Org.Comp.B17-205
–	$(C_5H_5)Fe[PC_4H_3\text{-}3\text{-}C(O)CH_3]$	Fe: Org.Comp.B17-206
$C_{11}FeH_{11}O_2^+$	$[(C_5H_5)Fe(CO)_2{=}CH\text{-}C_3H_5\text{-}c]^+$	Fe: Org.Comp.B16a-82, 88, 96
–	$[(C_5H_5)Fe(CO)_2(CH_2{=}C{=}CHCH_3)]^+$	Fe: Org.Comp.B17-111/2, 117
–	$[(C_5H_5)Fe(CO)_2(CH_2{=}CHCH{=}CH_2)]^+$	Fe: Org.Comp.B17-36/7, 51/2
–	$[(C_5H_5)Fe(CO)_2(C_4H_6\text{-}c)]^+$	Fe: Org.Comp.B17-86
–	$[(C_5H_5)Fe(CO)_2(H_3C\text{-}C{\equiv}C\text{-}CH_3)]^+$	Fe: Org.Comp.B17-122

$C_{11}FeH_{11}O_2P$ .... $HO_2CC_5H_4FeC_4H_3P(CH_3\text{-}3)$ .... Fe: Org.Comp.B17-220
$C_{11}FeH_{11}O_2S^-$ ... $[C_5H_5FeC_5H_3(SO_2)CH_3]^-$ .... Fe: Org.Comp.A9-252
$C_{11}FeH_{11}O_2S_2^+$ .. $[C_5H_5(CO)_2Fe{=}C(SCH_2)_2CH_2]^+$ .... Fe: Org.Comp.B16a-144, 156
$C_{11}FeH_{11}O_3^+$ .... $[(1,3\text{-}(CH_3)_2C_6H_5)Fe(CO)_3]^+$ .... Fe: Org.Comp.B15-112, 161
– .... $[(1,4\text{-}(CH_3)_2C_6H_5)Fe(CO)_3]^+$ .... Fe: Org.Comp.B15-122, 175
– .... $[(2,3\text{-}(CH_3)_2C_6H_5)Fe(CO)_3]^+$ .... Fe: Org.Comp.B15-123
– .... $[(2,4\text{-}(CH_3)_2C_6H_5)Fe(CO)_3]^+$ .... Fe: Org.Comp.B15-124, 177
– .... $[(2,4\text{-}(CH_3)_2C_6H_4D)Fe(CO)_3]^+$ .... Fe: Org.Comp.B15-177
– .... $[(2,6\text{-}(CH_3)_2C_6H_5)Fe(CO)_3]^+$ .... Fe: Org.Comp.B15-124, 178
– .... $[(6,6\text{-}(CH_3)_2C_6H_5)Fe(CO)_3]^+$ .... Fe: Org.Comp.B15-126
– .... $[(1\text{-}C_2H_5\text{-}C_6H_6)Fe(CO)_3]^+$ .... Fe: Org.Comp.B15-99, 135/6
– .... $[(1\text{-}CH_3CHD\text{-}C_6H_6)Fe(CO)_3]^+$ .... Fe: Org.Comp.B15-135
– .... $[2\text{-}((C_5H_5)Fe(CO)_2{=})\text{-}OC_4H_6]^+$ .... Fe: Org.Comp.B16a-113, 122
– .... $[2\text{-}((C_5H_5)Fe(CO)_2{=})\text{-}OC_4H_5D\text{-}3]^+$ .... Fe: Org.Comp.B16a-113, 122
– .... $[(4\text{-}CH_3CH{=}C_6H_7)Fe(CO)_3]^+$ .... Fe: Org.Comp.B15-6/7, 9
– .... $[(6\text{-}CH_3\text{-}C_7H_8)Fe(CO)_3]^+$ .... Fe: Org.Comp.B15-206, 222
– .... $[(C_5H_5)Fe(CO)_2(CH_2{=}C{=}CH\text{-}CH_2\text{-}OH)]^+$ ...... Fe: Org.Comp.B17-112
– .... $[(C_5H_5)Fe(CO)_2(CH_2{=}C{=}CD\text{-}CH_2\text{-}OD)]^+$ ...... Fe: Org.Comp.B17-112
– .... $[(C_5H_5)Fe(CO)_2(CH_2{=}CH\text{-}COCH_3)]^+$ .... Fe: Org.Comp.B17-30, 50/1
– .... $[(C_5H_5)Fe(CO)_2(HC{\equiv}C\text{-}CH_2CH_2OH)]^+$ .... Fe: Org.Comp.B17-121
– .... $[(C_5H_5)Fe(CO)_2(H_3C\text{-}C{\equiv}C\text{-}CH_2OH)]^+$ .... Fe: Org.Comp.B17-122
– .... $[(C_5H_5)Fe(CO)_2(OC_4H_6)]^+$ .... Fe: Org.Comp.B17-93/4
– .... $[(C_5H_5)Fe(CO)_2(OC_4H_5\text{-}D\text{-}2)]^+$ .... Fe: Org.Comp.B17-93
– .... $[(C_5H_5)Fe(CO)_2(OC_4H_4\text{-}D_2\text{-}4,4)]^+$ .... Fe: Org.Comp.B17-93
– .... $[(C_8H_{11})Fe(CO)_3]^+$ .... Fe: Org.Comp.B15-236/7, 255/6
$C_{11}FeH_{11}O_3^-$ .... $[(C_5H_4C(O)C_3H_7\text{-}i)Fe(CO)_2]^-$ .... Fe: Org.Comp.B14-118
$C_{11}FeH_{11}O_3S^-$ ... $[C_5H_5FeC_5H_3(SO_3)CH_3]^-$ .... Fe: Org.Comp.A9-252
$C_{11}FeH_{11}O_4^+$ .... $[(1\text{-}CH_3\text{-}2\text{-}CH_3O\text{-}C_6H_5)Fe(CO)_3]^+$ .... Fe: Org.Comp.B15-110
– .... $[(1\text{-}CH_3\text{-}3\text{-}CH_3O\text{-}C_6H_5)Fe(CO)_3]^+$ .... Fe: Org.Comp.B15-112, 160/1
– .... $[(1\text{-}CH_3\text{-}4\text{-}CH_3O\text{-}C_6H_5)Fe(CO)_3]^+$ .... Fe: Org.Comp.B15-113, 161/9
– .... $[(2\text{-}CH_3\text{-}3\text{-}CH_3O\text{-}C_6H_5)Fe(CO)_3]^+$ .... Fe: Org.Comp.B15-123, 175/6
– .... $[(3\text{-}CH_3\text{-}1\text{-}CH_3O\text{-}C_6H_5)Fe(CO)_3]^+$ .... Fe: Org.Comp.B15-96/7, 111, 160
– .... $[(3\text{-}CH_3\text{-}2\text{-}CH_3O\text{-}C_6H_5)Fe(CO)_3]^+$ .... Fe: Org.Comp.B15-123, 175/6
– .... $[(4\text{-}CH_3\text{-}1\text{-}CH_3O\text{-}C_6H_5)Fe(CO)_3]^+$ .... Fe: Org.Comp.B15-96/7, 113
– .... $[(4\text{-}CH_3\text{-}2\text{-}CH_3O\text{-}C_6H_5)Fe(CO)_3]^+$ .... Fe: Org.Comp.B15-91
– .... $[(5\text{-}CH_3\text{-}1\text{-}CH_3O\text{-}C_6H_5)Fe(CO)_3]^+$ .... Fe: Org.Comp.B15-96/7, 123, 160
– .... $[(1\text{-}C_2H_5O\text{-}C_6H_6)Fe(CO)_3]^+$ .... Fe: Org.Comp.B15-98
– .... $[(1\text{-}HO\text{-}3,5\text{-}(CH_3)_2C_6H_4)Fe(CO)_3]^+$ .... Fe: Org.Comp.B15-91
– .... $[(6\text{-}HO\text{-}7\text{-}CH_3\text{-}C_7H_7)Fe(CO)_3]^+$ .... Fe: Org.Comp.B15-213
– .... $[(C_5H_5)Fe(CO)_2(1,4\text{-}O_2C_4H_6)]^+$ .... Fe: Org.Comp.B17-95, 105
$C_{11}FeH_{11}O_5^+$ .... $[(1,4\text{-}(CH_3O)_2C_6H_5)Fe(CO)_3]^+$ .... Fe: Org.Comp.B15-96/7, 113
– .... $[(2,4\text{-}(CH_3O)_2C_6H_5)Fe(CO)_3]^+$ .... Fe: Org.Comp.B15-124
$C_{11}FeH_{11}S^+$ ..... $[(HS\text{-}C_6H_5)Fe(C_5H_5)]^+$ .... Fe: Org.Comp.B18-142/6, 198, 253
$C_{11}FeH_{12}$ .... $(C_5H_5)Fe(C_6H_7)$ .... Fe: Org.Comp.B17-258/9, 306
– .... $(C_5H_5)Fe(C_6H_6D\text{-}6)$ .... Fe: Org.Comp.B17-259

$C_{11}FeH_{12}N^+$ ..... $[(C_6H_6)Fe(C_5H_4-NH_2)]^+$ .... Fe: Org.Comp.B18-142/6, 167, 188
– .... $[(NH_2-C_6H_5)Fe(C_5H_5)]^+$ .... Fe: Org.Comp.B18-142/6, 197, 255/7, 281/2
$C_{11}FeH_{12}NOS_2^+$ .. $[C_5H_5(CO)(CH_3CN)Fe=C(SCH_2)_2]^+$ .... Fe: Org.Comp.B16a-68, 72
$C_{11}FeH_{12}NO_2^+$ ... $[(C_5H_5)Fe(CO)_2-CN-C_3H_7-n]^+$ .... Fe: Org.Comp.B15-313
– .... $[(C_5H_5)Fe(CO)_2-CN-C_3H_7-i]^+$ .... Fe: Org.Comp.B15-313, 317
– .... $[(C_6H_7)Fe(CO)_2-CN-C_2H_5]^+$ .... Fe: Org.Comp.B15-350, 352
– .... $[(C_7H_9)Fe(CO)_2-CN-CH_3]^+$ .... Fe: Org.Comp.B15-351
$C_{11}FeH_{12}NO_3^+$ ... $[(C_5H_5)Fe(CNCH_2CH_2CH_2OH)(CO)_2]^+$ .... Fe: Org.Comp.B15-314
– .... $[(C_5H_5)Fe(CO)_2=C(-N(CH_3)-CH_2CH_2-O-)]^+$ ... Fe: Org.Comp.B16a-139, 151
– .... $[(C_5H_5)Fe(CO)_2=C(-NH-CH_2CH_2CH_2-O-)]^+$ ... Fe: Org.Comp.B16a-140, 151/2
$C_{11}FeH_{12}N_2O$ .... $(C_5H_5)Fe(CO)(CN)-CN-C_3H_7-n$ .... Fe: Org.Comp.B15-289, 295
– .... $(C_5H_5)Fe(CO)(CN)-CN-C_3H_7-i$ .... Fe: Org.Comp.B15-289, 295
$C_{11}FeH_{12}O$ ...... $(C_5H_5)Fe(CO)-CH=CHCH=CHCH_3$ .... Fe: Org.Comp.B17-173
– .... $(C_5H_5)Fe(CO)-CH_2CHCHCH=CH_2$ .... Fe: Org.Comp.B17-149/50
– .... $[(C_6H_6)Fe(C_5H_5)][OH]$ .... Fe: Org.Comp.B18-142/6, 151/2, 154, 156
$C_{11}FeH_{12}O_2$ ..... $(C_5H_5)Fe(CO)-C(=O)CH_2CH_2CH=CH_2$ .... Fe: Org.Comp.B17-168/9
– .... $(C_5H_5)Fe(CO)_2-C(CH_3)=CHCH_3$ .... Fe: Org.Comp.B13-92, 120
– .... $(C_5H_5)Fe(CO)_2-C(C_2H_5)=CH_2$ .... Fe: Org.Comp.B13-93, 120
– .... $(C_5H_5)Fe(CO)_2-CH_2C(CH_3)=CH_2$ .... Fe: Org.Comp.B13-100/1, 130
– .... $(C_5H_5)Fe(CO)_2-CH_2C(CH_3)=CD_2$ .... Fe: Org.Comp.B13-101, 131
– .... $(C_5H_5)Fe(CO)_2-CD_2C(CH_3)=CH_2$ .... Fe: Org.Comp.B13-101, 131
– .... $(C_5H_5)Fe(CO)_2-CH_2CH=CHCH_3$ .... Fe: Org.Comp.B13-103/4, 132/3
– .... $(C_5H_5)Fe(CO)_2-CHDCH=CHCH_3$ .... Fe: Org.Comp.B13-104
– .... $(C_5H_5)Fe(CO)_2-CH_2CH_2CH=CH_2$ .... Fe: Org.Comp.B13-108/9, 135/6
– .... $(C_5H_5)Fe(CO)_2-C_4H_7-c$ .... Fe: Org.Comp.B13-198, 237
– .... $C_5H_4-C(CH_3)_2CH_2-Fe(CO)_2$ .... Fe: Org.Comp.B18-18, 24
– .... $C_5H_4-CH(CH_3)CH(CH_3)-Fe(CO)_2$ .... Fe: Org.Comp.B18-18, 24
– .... $C_5H_4-CH_2CH_2-CH_2CH_2-Fe(CO)_2$ .... Fe: Org.Comp.B18-18/20, 30, 44
$C_{11}FeH_{12}O_2S$ .... $(C_5H_5)Fe[C_5H_3(CH_3)-SO_2H]$ .... Fe: Org.Comp.A9-252
$C_{11}FeH_{12}O_2S_2$ ... $(C_5H_5)Fe(CO)_2CH[-S-CH_2-CH_2-S-CH_2-]$ ..... Fe: Org.Comp.B14-9, 14, 15
$C_{11}FeH_{12}O_3$ ..... $(C_5H_5)Fe(CO)[=C(OCH_3)O-CH=C(CH_3)-]$ ...... Fe: Org.Comp.B16b-126, 137
– .... $(C_5H_5)Fe(CO)[=C(OCH_3)-O-CH_2-C(=CH_2)-]$ ... Fe: Org.Comp.B16b-126, 137
– .... $(C_5H_5)Fe(CO)_2-C(=O)-C_3H_7-n$ .... Fe: Org.Comp.B13-8, 16, 36
– .... $(C_5H_5)Fe(CO)_2-C(=O)-C_3H_7-i$ .... Fe: Org.Comp.B13-16, 36
– .... $(C_5H_5)Fe(CO)_2-C(=CH_2)CH(OH)CH_3$ .... Fe: Org.Comp.B13-95, 122
– .... $(C_5H_5)Fe(CO)_2-C(=CH_2)CH_2-OCH_3$ .... Fe: Org.Comp.B13-88
– .... $(C_5H_5)Fe(CO)_2-C(CH_2OH)=CHCH_3$ .... Fe: Org.Comp.B13-95, 122
– .... $(C_5H_5)Fe(CO)_2-CH_2C(=CH_2)OCH_3$ .... Fe: Org.Comp.B13-102, 131/2
– .... $(C_5H_5)Fe(CO)_2-CH_2CH=CHOCH_3$ .... Fe: Org.Comp.B13-102, 131
– .... $[C_2H_5C(O)C_5H_4]Fe(CO)_2-CH_3$ .... Fe: Org.Comp.B14-57, 64
$C_{11}FeH_{12}O_3S$ .... $(C_5H_5)Fe[C_5H_3(CH_3)-SO_3H]$ .... Fe: Org.Comp.A9-252
$C_{11}FeH_{12}O_4$ ..... $(C_5H_5)Fe(CO)_2-C(=O)-CH_2OC_2H_5$ .... Fe: Org.Comp.B13-10, 14
– .... $(C_5H_5)Fe(CO)_2-CH[-O-CH_2-CH_2-O-CH_2-]$ ... Fe: Org.Comp.B14-8

$C_{11}FeH_{12}O_4$	$(c-C_7H_9)Fe(CO)_2-COOCH_3$	Fe: Org.Comp.B14-80, 81/2
$C_{11}FeH_{12}O_4S$	$(C_5H_5)Fe(CO)_2(CH_2=C(CH_3)CH_2-SO_2)$	Fe: Org.Comp.B17-128
–	$(C_5H_5)Fe(CO)_2(CH_2=CHCH(CH_3)-SO_2)$	Fe: Org.Comp.B17-128
–	$(C_5H_5)Fe(CO)_2-C(=O)CH-S(=O)(CH_3)_2$	Fe: Org.Comp.B13-14, 33
–	$(C_5H_5)Fe(CO)_2-CH[-CH_2-CH_2-SO_2-CH_2-]$	Fe: Org.Comp.B14-6, 12
–	$(C_5H_5)Fe(CO)_2-CH_2CH=CHSO_2CH_3$	Fe: Org.Comp.B13-103
$C_{11}FeH_{12}O_6S_2$	$(HO_3S-C_5H_4)Fe[C_5H_3(CH_3)-SO_3H]$	Fe: Org.Comp.A10-297
$C_{11}FeH_{13}IN_2O$	$[(C_5H_5)Fe(CNCH_3)(CO)CN-C_2H_5]I$	Fe: Org.Comp.B15-333/4
$C_{11}FeH_{13}N$	$(CH_3C_5H_4)Fe(NC_4H_3-2-CH_3)$	Fe: Org.Comp.B17-216
–	$(C_5H_5)Fe[C_5H_3(NH_2)-CH_3]$	Fe: Org.Comp.A9-25, 71
–	$(C_5H_5)Fe[NC_4H_2-2,4-(CH_3)_2]$	Fe: Org.Comp.B17-202
–	$(C_5H_5)Fe[NC_4H_2-2,5-(CH_3)_2]$	Fe: Org.Comp.B17-202
$C_{11}FeH_{13}N^{2+}$	$[(C_6H_6)Fe(C_5H_4-NH_3)]^{2+}$	Fe: Org.Comp.B18-167
–	$[(NH_3-C_6H_5)Fe(C_5H_5)]^{2+}$	Fe: Org.Comp.B18-255
$C_{11}FeH_{13}NO_2$	$(C_5H_5)Fe(CNCH_3)(CO)C(O)C_2H_5$	Fe: Org.Comp.B15-305, 309
$C_{11}FeH_{13}NO_3$	$(C_5H_5)Fe(CO)_2CONHC_3H_7-i$	Fe: Org.Comp.B13-9, 25
$C_{11}FeH_{13}NO_5S_2$	$(C_5H_5)Fe(CO)_2CH[-CH_2-N(SO_2CH_3)-S(=O)-CH_2-]$	Fe: Org.Comp.B14-33, 37/8
$C_{11}FeH_{13}N_2^+$	$[(NH_2NH-C_6H_5)Fe(C_5H_5)]^+$	Fe: Org.Comp.B18-142/6, 198, 261/2
$C_{11}FeH_{13}N_2O^+$	$[(C_5H_5)Fe(CNCH_3)(CO)CN-C_2H_5]^+$	Fe: Org.Comp.B15-333/4
$C_{11}FeH_{13}N_2O_2^+$	$[(C_5H_5)Fe(CO)_2=C(-NH-CH_2CH_2CH_2-NH-)]^+$	Fe: Org.Comp.B16a-147
–	$[(C_5H_5)Fe(CO)_2=C(-NH-CH_2CH_2-N(CH_3)-)]^+$	Fe: Org.Comp.B16a-147
–	$[(C_5H_5)Fe(CO)_2=C(-NH-CH_2-CH(CH_3)-NH-)]^+$	Fe: Org.Comp.B16a-147
$C_{11}FeH_{13}O^+$	$[(C_5H_5)Fe(CO)(=CHCH_2CH_2CH=CH_2)]^+$	Fe: Org.Comp.B17-167
–	$[(C_5H_5)Fe(CO)(CH_2=CHCH=CHCH_3)]^+$	Fe: Org.Comp.B17-187
$C_{11}FeH_{13}O_2^+$	$[(C_5H_5)Fe(CO)_2=CH-C_3H_7-i]^+$	Fe: Org.Comp.B16a-87
–	$[(C_5H_5)Fe(CO)_2(CH_2=C(CH_3)_2)]^+$	Fe: Org.Comp.B17-61/2, 73/8
–	$[(C_5H_5)Fe(CO)_2(CD_2=C(CH_3)_2)]^+$	Fe: Org.Comp.B17-62
–	$[(C_5D_5)Fe(CO)_2(CH_2=C(CH_3)_2)]^+$	Fe: Org.Comp.B17-130
–	$[(C_5H_5)Fe(CO)_2(CH_2=CH-C_2H_5)]^+$	Fe: Org.Comp.B17-8/11
–	$[(C_5H_5)Fe(CO)_2(CH_2=CH-CH_2CH_2D)]^+$	Fe: Org.Comp.B17-11
–	$[(C_5H_5)Fe(CO)_2(CH_3CH=CHCH_3)]^+$	Fe: Org.Comp.B17-56, 72
$C_{11}FeH_{13}O_2S_2^+$	$[(C_5H_5)(CO)_2Fe=C(SCH_3)SC_2H_5]^+$	Fe: Org.Comp.B16a-141
$C_{11}FeH_{13}O_3^+$	$[(CH_2C(CH_3)C(CH_3)C(CH_3)CH_2)Fe(CO)_3]^+$	Fe: Org.Comp.B15-17, 24
–	$[(CH_2CHCHCHCH-C_3H_7-n)Fe(CO)_3]^+$	Fe: Org.Comp.B15-26
–	$[(CH_2CHCHCHCH-C_3H_7-i)Fe(CO)_3]^+$	Fe: Org.Comp.B15-2/3, 12/3, 20
–	$[((CH_3)_2CC(CH_2)C(CH_3)CH_2)Fe(CO)_3]^+$	Fe: Org.Comp.B15-33/4
–	$[(CH_3-CHC(CH_3)CHCHCH-CH_3)Fe(CO)_3]^+$	Fe: Org.Comp.B15-12, 14, 23, 30
–	$[(CH_3-CHCHC(CH_3)CHCH-CH_3)Fe(CO)_3]^+$	Fe: Org.Comp.B15-12, 14, 17, 24
–	$[(CH_3-CHCHCHCHC(CH_3)_2)Fe(CO)_3]^+$	Fe: Org.Comp.B15-6, 8, 23
–	$[(CH_3-CHCHCHCHCH-C_2H_5)Fe(CO)_3]^+$	Fe: Org.Comp.B15-4/5, 12/3, 22
–	$[(C_5H_5)(CO)_2Fe=C(CH_3)-OC_2H_5]^+$	Fe: Org.Comp.B16a-109, 119
–	$[(C_5H_5)(CO)_2Fe=C(C_2H_5)-OCH_3]^+$	Fe: Org.Comp.B16a-109, 120
–	$[(C_5H_5)Fe(CO)_2(CH_2=C(CH_3)-OCH_3)]^+$	Fe: Org.Comp.B17-67, 79
–	$[(C_5H_5)Fe(CO)_2(CH_2=CHCH_2-OCH_3)]^+$	Fe: Org.Comp.B17-25

$C_{11}FeH_{13}O_3^+$ .... $[(C_5H_5)Fe(CO)_2(CH_2{=}CH{-}OC_2H_5)]^+$ .... Fe: Org.Comp.B17-21/2, 49
– .... $[(C_5H_5)Fe(CO)_2(CH_3CH{=}CH{-}OCH_3)]^+$ .... Fe: Org.Comp.B17-64
$C_{11}FeH_{13}O_3^-$ .... $[(C_5H_5)Fe({=}C(O)OCH(CH_3)CH(CH_3){-})CO]^-$ .... Fe: Org.Comp.B16b-125, 136
$C_{11}FeH_{13}O_4^+$ .... $[(CH_3{-}CHC(CH_3)CHCHC(CH_3)OH)Fe(CO)_3]^+$ .. Fe: Org.Comp.B15-6, 9
– .... $[(CH_3{-}CHCHC(CH_3)CHC(CH_3)OH)Fe(CO)_3]^+$ .. Fe: Org.Comp.B15-6, 8
– .... $[(C_5H_5)Fe(CO)_2{=}C(OCH_3){-}OC_2H_5]^+$ .... Fe: Org.Comp.B16a-136, 149
– .... $[(C_5H_5)Fe(CO)_2(CH_2{=}C(OCH_3)_2)]^+$ .... Fe: Org.Comp.B17-69
– .... $[(C_5H_5)Fe(CO)_2(H_3C{-}O{-}CH{=}CH{-}O{-}CH_3)]^+$ .... Fe: Org.Comp.B17-69, 80/1
$C_{11}FeH_{13}O_4S^+$ ... $[(C_5H_5)Fe(CO)_2(CH_2{=}CHCH_2SO_2CH_3)]^+$ ...... Fe: Org.Comp.B17-26
$C_{11}FeH_{13}P$ ...... $C_5H_5FeC_4H_2P((CH_3)_2{-}3,4)$ .... Fe: Org.Comp.B17-206, 231/3
$C_{11}FeH_{13}S^+$ ..... $[C_5H_5FeC_4H_2S((CH_3)_2{-}2,5)]^+$ .... Fe: Org.Comp.B17-214
$C_{11}FeH_{14}$ .... $(CH_3{-}C_6H_5)Fe(CH_2{=}CHCH{=}CH_2)$ .... Fe: Org.Comp.B18-103, 110
– .... $(C_5H_5)Fe(CH_3{-}CHCHCHCHCH_2)$ .... Fe: Org.Comp.B17-253, 254
$C_{11}FeH_{14}INO$ .... $C_5H_5Fe(CO)(I)CN{-}C_4H_9{-}t$ .... Fe: Org.Comp.B15-290, 295/6
$C_{11}FeH_{14}IN_3$ ..... $[C_5H_5Fe(CNCH_3)_3]I$ .... Fe: Org.Comp.B15-344
$C_{11}FeH_{14}N^+$ ..... $[(C_5H_5)Fe(C_5H_3(NH_3){-}CH_3)]^+$ .... Fe: Org.Comp.A9-25, 71
$C_{11}FeH_{14}NOS_2^+$ .. $[(C_5H_5)Fe(CO)(CN{-}CH_3){=}C(SCH_3)_2]^+$ .... Fe: Org.Comp.B16a-163, 167/8
– .... $[(C_5H_5)Fe(CO)(NC{-}CH_3){=}C(SCH_3)_2]^+$ .... Fe: Org.Comp.B16a-66, 71/2
$C_{11}FeH_{14}NO_2^+$ ... $[(C_5H_5)Fe(CO)_2{=}CH{-}NH{-}C_3H_7{-}i]^+$ .... Fe: Org.Comp.B16a-114, 125
– .... $[(C_5H_5)Fe(CO)_2(CH_2{=}CH{-}N(CH_3)_2)]^+$ .... Fe: Org.Comp.B17-22/3, 49/50
– .... $[(C_5H_5)Fe(CO)_2{-}CH_2CH{=}N(CH_3)_2]^+$ .... Fe: Org.Comp.B14-127/8
$C_{11}FeH_{14}NO_2S^+$ .. $[(C_5H_5)Fe(CO)(NC{-}CH_3){=}C(OCH_3){-}SCH_3]^+$ ... Fe: Org.Comp.B16a-65
– .... $[(C_5H_5)Fe(CO)_2{=}C(SCH_3){-}N(CH_3)_2]^+$ .... Fe: Org.Comp.B16a-146, 158
$C_{11}FeH_{14}NO_3^+$ ... $[C_5H_5(CO)_2Fe{=}C(OCH_3)N(CH_3)_2]^+$ .... Fe: Org.Comp.B16a-138
$C_{11}FeH_{14}NO_4^+$ ... $[C_5H_5(CO)_2Fe{=}C(OCH_3)NHCH_2CH_2OH]^+$ ..... Fe: Org.Comp.B16a-138
$C_{11}FeH_{14}NO_5S_2^+$ $[(C_5H_5)Fe(CO)_2(CH_2{=}CH{-}CH_2{-}S({=}O){-}NH{-}S(O)_2{-}CH_3)]^+$ Fe: Org.Comp.B17-25/6
$C_{11}FeH_{14}N_3^+$ .... $[(C_5H_5)Fe(CN{-}CH_3)_3]^+$ .... Fe: Org.Comp.B15-344/5, 346/7
– .... $[(C_5H_5)Fe(CN{-}CH_3)_2{-}NC{-}CH_3]^+$ .... Fe: Org.Comp.B15-327, 328
– .... $[(C_5H_5)Fe(NC{-}CH_3)_2{-}CN{-}CH_3]^+$ .... Fe: Org.Comp.B15-281
$C_{11}FeH_{14}O_2$ ..... $(C_2H_5{-}C_5H_4)Fe(CO)_2C_2H_5$ .... Fe: Org.Comp.B14-56, 63
– .... $(C_5H_5)Fe(CO)(CH_2CHCH{-}OC_2H_5)$ .... Fe: Org.Comp.B17-153
– .... $(C_5H_5)Fe(CO)[{=}C(OCH_3){-}CH_2CH_2CH_2{-}]$ ...... Fe: Org.Comp.B16b-120, 134
– .... $(C_5H_5)Fe(CO)[{=}C(O{-}C_2H_5){-}CH_2CH_2{-}]$ .... Fe: Org.Comp.B16b-117,132
– .... $(C_5H_5)Fe(CO)[CH_2C(O{-}C_2H_5)CH_2]$ .... Fe: Org.Comp.B17-153
$C_{11}FeH_{14}O_2S_2Si$ .. $(C_5H_5)Fe(CO)_2C(S)SSi(CH_3)_3$ .... Fe: Org.Comp.B13-55, 57
$C_{11}FeH_{14}O_2S_2Sn$ $(C_5H_5)Fe(CO)_2C(S)SSn(CH_3)_3$ .... Fe: Org.Comp.B13-55, 57
$C_{11}FeH_{14}O_5S$ .... $[(C_5H_5)Fe(CO)_2(CH_2{=}CHCH_3)][CH_3SO_3]$ ...... Fe: Org.Comp.B17-6
$C_{11}FeH_{15}KO_3Si$ .. $K[(C_5H_4Si(CH_3)_2OC_2H_5)Fe(CO)_2]$ .... Fe: Org.Comp.B14-119
$C_{11}FeH_{15}LiO_2Si$ .. $Li[(CH_3{-}C_5H_3{-}2{-}Si(CH_3)_3)Fe(CO)_2]$ .... Fe: Org.Comp.B14-120
– .... $Li[(CH_3{-}C_5H_3{-}3{-}Si(CH_3)_3)Fe(CO)_2]$ .... Fe: Org.Comp.B14-120
$C_{11}FeH_{15}N_2^+$ .... $[(C_5H_5)Fe(CH_2{=}CH_2)(CNCH_3)_2]^+$ .... Fe: Org.Comp.B17-138
$C_{11}FeH_{15}N_2O_2P$ .. $[1,3{-}(CH_3)_2{-}1,3,2{-}N_2PC_2H_4][C_5H_5Fe(CO)_2]$ .... Fe: Org.Comp.B14-111
$C_{11}FeH_{15}N_3O$ .... $(C_5H_5)Fe(CO)(CN){=}C(NHCH_3){-}N(CH_3)_2$ .... Fe: Org.Comp.B16a-173, 174
– .... $(C_5H_5)Fe(CO)(CN){=}C(NHCH_3){-}NH{-}C_2H_5$ ...... Fe: Org.Comp.B16a-172, 173/4
$C_{11}FeH_{15}O_2^+$ .... $[C_5H_5(CO)(C_4H_8O)Fe{=}CH_2]^+$ .... Fe: Org.Comp.B16a-18, 24

$C_{11}FeH_{15}O_2Si^-$	$[(CH_3-C_5H_3-2-Si(CH_3)_3)Fe(CO)_2]^-$	Fe: Org.Comp.B14-120
–	$[(CH_3-C_5H_3-3-Si(CH_3)_3)Fe(CO)_2]^-$	Fe: Org.Comp.B14-120
$C_{11}FeH_{15}O_3Si^-$	$[(C_5H_4Si(CH_3)_2OC_2H_5)Fe(CO)_2]^-$	Fe: Org.Comp.B14-119
$C_{11}FeH_{16}$	$(CH_3C_6H_5)Fe(CH_2=CH_2)_2$	Fe: Org.Comp.B18-81/3, 90/1
$C_{11}FeH_{16}INO_2$	$[C_5H_5Fe(CO)_2CH_2N(CH_3)_3]I$	Fe: Org.Comp.B14-136, 143
$C_{11}FeH_{16}IN_3O$	$[C_5H_5(CH_3NC)(CO)Fe=C(NHCH_3)_2]I$	Fe: Org.Comp.B16a-163, 168
$C_{11}FeH_{16}LiNO_2Si$	$Li[(C_5H_4Si(CH_3)_2N(CH_3)_2)Fe(CO)_2]$	Fe: Org.Comp.B14-119
$C_{11}FeH_{16}NO_2^+$	$[(C_5H_5)Fe(CO)_2-CH_2CH_2-NH(CH_3)_2]^+$	Fe: Org.Comp.B14-149, 155
–	$[(C_5H_5)Fe(CO)_2-CH_2-N(CH_3)_3]^+$	Fe: Org.Comp.B14-136, 143
$C_{11}FeH_{16}NO_2Si^-$	$[(C_5H_4Si(CH_3)_2N(CH_3)_2)Fe(CO)_2]^-$	Fe: Org.Comp.B14-119
$C_{11}FeH_{16}N_3O^+$	$[C_5H_5(CH_3NC)(CO)Fe=C(NHCH_3)_2]^+$	Fe: Org.Comp.B16a-163, 168
$C_{11}FeH_{16}O_2Si$	$C_5H_4Si(CH_3)_3Fe(CO)_2CH_3$	Fe: Org.Comp.B14-59, 65
$C_{11}FeH_{16}O_4P^+$	$[C_5H_5(CO)(P(OCH_3)_3)Fe=C=CH_2]^+$	Fe: Org.Comp.B16a-207/8, 210
$C_{11}FeH_{16}O_5P^+$	$[C_5H_5Fe(CO)_2CH_2P(OCH_3)_3]^+$	Fe: Org.Comp.B14-138
$C_{11}FeH_{17}INOP$	$[C_5H_5Fe(CNCH_3)(CO)P(CH_3)_3]I$	Fe: Org.Comp.B15-300
$C_{11}FeH_{17}NOP^+$	$[C_5H_5Fe(CNCH_3)(CO)P(CH_3)_3]^+$	Fe: Org.Comp.B15-300
$C_{11}FeH_{17}NOSi$	$(C_5H_5)Fe(CN-C_4H_9-t)(CO)SiH_3$	Fe: Org.Comp.B15-290, 296
$C_{11}FeH_{17}NOSn$	$C_5H_5Fe(CNCH_3)(CO)Sn(CH_3)_3$	Fe: Org.Comp.B15-289, 295
$C_{11}FeH_{17}NO_2Sn$	$(CH_3)_3SnNC-Fe(CO)_2(CH_2=CHCH=CHCH_3)$	Sn: Org.Comp.18-128
$C_{11}FeH_{17}NO_3PS^+$	$[(C_5H_5)Fe(CS)(NCCH_3)P(OCH_3)_3]^+$	Fe: Org.Comp.B15-264
$C_{11}FeH_{17}OS_3^+$	$[C_5H_5(CO)(S(CH_3)_2)Fe=C(SCH_3)_2]^+$	Fe: Org.Comp.B16a-66, 71
$C_{11}FeH_{17}O_2P$	$[P(CH_3)_4][C_5H_5Fe(CO)_2]$	Fe: Org.Comp.B14-110/1
$C_{11}FeH_{18}OP^+$	$[C_5H_5(CO)(P(CH_3)_3)Fe=CH(CH_3)]^+$	Fe: Org.Comp.B16a-15, 18, 24/5
$C_{11}FeH_{18}OSn$	$(C_5H_5)Fe(CH_2=CH_2)(CO)Sn(CH_3)_3$	Fe: Org.Comp.B16b-73
$C_{11}FeH_{18}O_2P^+$	$[C_5H_5(CO)(P(CH_3)_3)Fe=C(CH_3)OH]^+$	Fe: Org.Comp.B16a-35, 51
$C_{11}FeH_{18}O_4P^+$	$[(C_5H_5)Fe(CH_2=CH_2)(CO)P(OCH_3)_3]^+$	Fe: Org.Comp.B16b-81
$C_{11}FeH_{18}O_5P^+$	$[(C_5H_5)Fe(CO)(P(OCH_3)_3)=C(CH_3)OH]^+$	Fe: Org.Comp.B16a-36
–	$[(C_5H_5)Fe(CO)(P(OCH_3)_3)(CH_2=CHOH)]^+$	Fe: Org.Comp.B16b-86
$C_{11}Fe_3Ge_2H_6O_9$	$(CO)_9Fe_3(Ge-CH_3)_2$	Fe: Org.Comp.C6a-282
$C_{11}Fe_3HKO_{11}$	$K[(H)Fe_3(CO)_{11}]$	Fe: Org.Comp.C6b-85, 92/3, 96, 107, 117
$C_{11}Fe_3HLiO_{11}$	$Li[(H)Fe_3(CO)_{11}]$	Fe: Org.Comp.C6b-85, 90, 93, 95
$C_{11}Fe_3HNaO_{11}$	$Na[(H)Fe_3(CO)_{11}]$	Fe: Org.Comp.C6b-85, 90/3, 95/6
$C_{11}Fe_3HO_{10}^-$	$[(CO)_{10}Fe_3(CH)]^-$	Fe: Org.Comp.C6b-60, 61, 63/4
$C_{11}Fe_3HO_{11}$	$(H)Fe_3(CO)_{11}$, radical	Fe: Org.Comp.C6b-125
$C_{11}Fe_3HO_{11}^-$	$[(H)Fe_3(CO)_{11}]^-$	
	Catalytical properties	Fe: Org.Comp.C6b-95, 116/8
	Chemical reaction	Fe: Org.Comp.C6b-106/16
	Formation	Fe: Org.Comp.C6b-85/90
	Physical properties	Fe: Org.Comp.C6b-90/4
	Preparation	Fe: Org.Comp.C6b-85/90
$C_{11}Fe_3HO_{11}Rb$	$Rb[(H)Fe_3(CO)_{11}]$	Fe: Org.Comp.C6b-93, 96
$C_{11}Fe_3H_2O_8S_4$	$(CO)_8Fe_3S_2(1,3-S_2C_3H_2)$	Fe: Org.Comp.C6a-54/5
$C_{11}Fe_3H_2O_9$	$(CO)_9Fe_3(C-H)_2$	Fe: Org.Comp.C6a-237, 243

$C_{11}Fe_3H_2O_{10}$ .... $(CO)_{10}(H)Fe_3(CH)$ .... Fe: Org.Comp.C6b-49, 50, 53/4

$C_{11}Fe_3H_2O_{11}$ .... $(CO)_{10}(H)Fe_3(C\text{-}OH)$ .... Fe: Org.Comp.C6b-49, 51, 55/6

– .... $(CO)_{10}(H)Fe_3(C\text{-}OD)$ .... Fe: Org.Comp.C6b-107

– .... $H_2Fe_3(CO)_{11}$ = $(CO)_{10}(H)Fe_3(C\text{-}OH)$ .... Fe: Org.Comp.C6b-107, 116

$C_{11}Fe_3H_3NO_9$ .... $(CO)_9Fe_3(NC\text{-}CH_3)$ .... Fe: Org.Comp.C6a-246

$C_{11}Fe_3H_3O_{10}^-$ .... $[(CO)_9Fe_3(OC\text{-}CH_3)]^-$ .... Fe: Org.Comp.C6a-290/4

– .... $[(CO)_9Fe_3(O\text{-}CH{=}CH_2)]^-$ .... Fe: Org.Comp.C6a-132

$C_{11}Fe_3H_3O_{10}P$ .... $(CO)_{10}Fe_3P\text{-}CH_3$ .... Fe: Org.Comp.C6b-27, 28, 31, 33

$C_{11}Fe_3H_4KNO_9$ .... $K[(CO)_9Fe_3(N{=}CH\text{-}CH_3)]$ .... Fe: Org.Comp.C6a-290

– .... $K[(CO)_9Fe_3(NH{=}C\text{-}CH_3)]$ .... Fe: Org.Comp.C6a-287, 288

$C_{11}Fe_3H_4NNaO_9$ .... $Na[(CO)_9Fe_3(N{=}CH\text{-}CH_3)]$ .... Fe: Org.Comp.C6a-289/90

– .... $Na[(CO)_9Fe_3(NH{=}C\text{-}CH_3)]$ .... Fe: Org.Comp.C6a-287, 288

$C_{11}Fe_3H_4NO_9^-$ .... $[(CO)_9Fe_3(N{=}CH\text{-}CH_3)]^-$ .... Fe: Org.Comp.C6a-289/90

– .... $[(CO)_9Fe_3(NH{=}C\text{-}CH_3)]^-$ .... Fe: Org.Comp.C6a-286/8

$C_{11}Fe_3H_4O_{10}$ .... $(CO)_9Fe_3[C(CH_3)\text{-}OH]$ .... Fe: Org.Comp.C6a-271

– .... $(CO)_9(H)Fe_3[C(CH_3)O]$ .... Fe: Org.Comp.C6a-263/4

$C_{11}Fe_3H_4O_{12}$ .... $[H_3O \cdot n\ H_2O][(H)Fe_3(CO)_{11}]$ .... Fe: Org.Comp.C6b-91/2, 95

$C_{11}Fe_3H_5NO_9$ .... $(CO)_9(H)Fe_3(N{=}CH\text{-}CH_3)$ .... Fe: Org.Comp.C6a-260/2

– .... $(CO)_9(D)Fe_3(N{=}CH\text{-}CH_3)$ .... Fe: Org.Comp.C6a-262

– .... $(CO)_9(H)Fe_3(N{=}CD\text{-}CH_3)$ .... Fe: Org.Comp.C6a-262

– .... $(CO)_9(H)Fe_3(N{=}CH\text{-}CD_3)$ .... Fe: Org.Comp.C6a-262

– .... $(CO)_9(H)Fe_3(NH{=}C\text{-}CH_3)$ .... Fe: Org.Comp.C6a-255/9

– .... $(CO)_9(D)Fe_3(NH{=}C\text{-}CH_3)$ .... Fe: Org.Comp.C6a-259

– .... $(CO)_9(H)Fe_3(ND{=}C\text{-}CH_3)$ .... Fe: Org.Comp.C6a-259

– .... $(CO)_9(D)Fe_3(ND{=}C\text{-}CH_3)$ .... Fe: Org.Comp.C6a-259

– .... $(CO)_9(H)Fe_3(NH{=}C\text{-}CD_3)$ .... Fe: Org.Comp.C6a-259

$C_{11}Fe_3H_5NO_{11}$ .... $[NH_4][(H)Fe_3(CO)_{11}]$ .... Fe: Org.Comp.C6b-96

$C_{11}Fe_3H_5NaO_9S$ .... $Na[(CO)_9Fe_3S\text{-}C_2H_5]$ .... Fe: Org.Comp.C6a-136

$C_{11}Fe_3H_5O_9S^-$ .... $[(CO)_9Fe_3S\text{-}C_2H_5]^-$ .... Fe: Org.Comp.C6a-136

$C_{11}Fe_3H_6NNaO_9$ .... $Na[(CO)_9(H)Fe_3N\text{-}C_2H_5]$ .... Fe: Org.Comp.C6a-223, 225

$C_{11}Fe_3H_6NO_9^-$ .... $[(CO)_9(H)Fe_3N\text{-}C_2H_5]^-$ .... Fe: Org.Comp.C6a-223, 225/6

$C_{11}Fe_3H_6N_2O_9$ .... $(CO)_9Fe_3(NH)(N\text{-}C_2H_5)$ .... Fe: Org.Comp.C6a-147, 150, 154

– .... $(CO)_9Fe_3(N\text{-}CH_3)_2$ .... Fe: Org.Comp.C6a-147, 148, 150/2

$C_{11}Fe_3H_6O_9$ .... $(CO)_9(H)_3Fe_3C\text{-}CH_3$ .... Fe: Org.Comp.C6a-249/54

$C_{11}Fe_3H_6O_9P_2$ .... $(CO)_9Fe_3(P\text{-}CH_3)_2$ .... Fe: Org.Comp.C6a-156/7, 158, 162

$C_{11}Fe_3H_6O_9S$ .... $(CO)_9(H)Fe_3S\text{-}C_2H_5$ .... Fe: Org.Comp.C6a-94, 96

$C_{11}Fe_3H_6O_9SSn$ .... $(CO)_9Fe_3S[Sn(CH_3)_2]$ .... Fe: Org.Comp.C6a-285

$C_{11}Fe_3H_6O_9Si_2$ .... $(CO)_9Fe_3(Si\text{-}CH_3)_2$ .... Fe: Org.Comp.C6a-282

$C_{11}Fe_3H_6O_{10}$ .... $(CO)_9(H)_3Fe_3C\text{-}O\text{-}CH_3$ .... Fe: Org.Comp.C6a-249, 251, 255

$C_{11}Fe_3H_7NO_9$ .... $(CO)_9(H)_2Fe_3N\text{-}C_2H_5$ .... Fe: Org.Comp.C6a-187, 188, 190/1

$C_{11}Fe_3H_8O_7S_4$ .... $(CO)_7Fe_3(S\text{-}CH_2CH_2\text{-}S)_2$ .... Fe: Org.Comp.C6a-26/7

$C_{11}Fe_3H_9O_{11}PS_2$ $(CO)_8Fe_3S_2[P(OCH_3)_3]$ . . . Fe: Org.Comp.C6a-42, 43, 46, 47
$C_{11}Fe_3H_9O_{11}PSe_2$ $(CO)_8Fe_3Se_2[P(OCH_3)_3]$ . . . Fe: Org.Comp.C6a-42, 44, 46, 48
$C_{11}Fe_3H_{18}N_6NiO_{11}$
$[Ni(NH_3)_6][Fe_3(CO)_{11}]$ . . . Fe: Org.Comp.C6b-86, 95, 104
$C_{11}Fe_3K_2O_{11}$ . . . $K_2[Fe_3(CO)_{11}]$ . . . Fe: Org.Comp.C6b-102, 107, 112, 118
$C_{11}Fe_3Na_2O_{11}$ . . . $Na_2[Fe_3(CO)_{11}]$ . . . Fe: Org.Comp.C6b-94/5, 102, 107, 115
$C_{11}Fe_3O_{11}$ . . . $Fe_3(CO)_{11}$ . . . Fe: Org.Comp.C6b-126/7
$C_{11}Fe_3O_{11}{}^+$ . . . $[Fe_3(CO)_{11}]^+$ . . . Fe: Org.Comp.C6a-4
$C_{11}Fe_3O_{11}{}^-$ . . . $[Fe_3(CO)_{11}]^-$, radical anion . . . Fe: Org.Comp.C6b-125/6
$C_{11}Fe_3O_{11}{}^{2-}$ . . . $[Fe_3(CO)_{11}]^{2-}$
Catalytical properties . . . Fe: Org.Comp.C6b-116/8
Chemical reaction . . . Fe: Org.Comp.C6b-106/16
Formation . . . Fe: Org.Comp.C6b-85/90
Physical properties . . . Fe: Org.Comp.C6b-94/5
Preparation . . . Fe: Org.Comp.C6b-85/90
$C_{11}Fe_4H_{18}N_6O_{11}$ . . . $[Fe(NH_3)_6][Fe_3(CO)_{11}]$ . . . Fe: Org.Comp.C6b-104/5
$C_{11}Fe_4O_{11}$ . . . $Fe[Fe_3(CO)_{11}]$ . . . Fe: Org.Comp.C6b-104
$C_{11}GaH_6N_3OS_2$ . . . $Ga(NCS)_2(1-NC_9H_6-8-O)$ . . . Ga: SVol.D1-284
$C_{11}GaH_6O_3{}^+$ . . . $[Ga(2-(OC(O))-3-(O)-C_{10}H_6)]^+$ . . . Ga: SVol.D1-183
$C_{11}GaH_7O_4{}^{2+}$ . . . $[Ga(7-O-8-(O{=}CH)-2-(O{=})-4-(CH_3)-1-OC_9H_3)]^{2+}$
Ga: SVol.D1-123
$C_{11}GaH_{10}NO_4$ . . . $Ga(OH)_2[1-NC_9H_5-5-(C(=O)CH_3)-8-O]$ . . . Ga: SVol.D1-286/90
$C_{11}GaH_{14}N_2O_5Re$ $(OC)_4Re[-N_2C_3H(CH_3)_2-Ga(CH_3)_2-O(H)-]$ . . . Re: Org.Comp.1-354
$C_{11}GaH_{16}N_3O_4S_3$ $Ga(1,4-O_2C_4H_8)_2(NCS)_3$ . . . Ga: SVol.D1-147
$C_{11}GaH_{21}O_4$ . . . $(i-C_3H_7-O)_2Ga[CH_3-C(O)CHC(O)-CH_3]$ . . . Ga: SVol.D1-68/9
$C_{11}GaH_{21}O_5$ . . . $Ga(O-C_3H_7-i)_2[CH_3-C(O)CHC(O)-O-CH_3]$ . . . Ga: SVol.D1-201, 202
$C_{11}GeH_{14}$ . . . $Ge(CH_3)_3C{\equiv}CC_6H_5$ . . . Ge: Org.Comp.2-54/5, 63/4
$C_{11}GeH_{14}O$ . . . $Ge(CH_3)_3C(=CHC_6H_4O)$ . . . Ge: Org.Comp.2-99, 105/6
$C_{11}GeH_{14}O_3$ . . . $(CH_3)_2Ge(C_8H_5O_3(CH_3))$ . . . Ge: Org.Comp.3-323
$C_{11}GeH_{15}N_3O_6$ . . . $Ge(CH_3)_3C_6((NO_2)_3-2,4,6)(CH_3)_2$ . . . Ge: Org.Comp.2-78
$C_{11}GeH_{16}$ . . . $Ge(CH_3)_3-C(C_6H_5){=}CH_2$ . . . Ge: Org.Comp.2-6, 22
– . . . $Ge(CH_3)_3-CH{=}CH-C_6H_5$ . . . Ge: Org.Comp.2-6, 22
– . . . $Ge(CH_3)_3-C_6H_4-CH{=}CH_2-4$ . . . Ge: Org.Comp.2-76, 86
– . . . $Ge(CH_3)_3-C_8H_7-c$ . . . Ge: Org.Comp.2-39, 48
$C_{11}GeH_{16}K_2$ . . . $K_2[Ge(CH_3)_3C_8H_7]$ . . . Ge: Org.Comp.2-48
$C_{11}GeH_{16}O$ . . . $Ge(CH_3)_3-C(=O)-CH_2-C_6H_5$ . . . Ge: Org.Comp.1-173
– . . . $Ge(CH_3)_3-CH_2-C(=O)-C_6H_5$ . . . Ge: Org.Comp.1-174, 176/7
– . . . $Ge(CH_3)_3-C_6H_4-4-C(=O)CH_3$ . . . Ge: Org.Comp.2-75, 86
$C_{11}GeH_{16}O_2$ . . . $Ge(CH_3)_3CH_2C_6H_4COOH-4$ . . . Ge: Org.Comp.1-158, 163
$C_{11}GeH_{17}N$ . . . $1,1,3-(CH_3)_3-3,1-NGeC_8H_8$ . . . Ge: Org.Comp.3-331
– . . . $1-C_6H_5-3,3-(CH_3)_2-1,3-NGeC_3H_6$ . . . Ge: Org.Comp.3-251
– . . . $C_6H_5-NH-CH_2-Ge(CH_3)_2CH{=}CH_2$ . . . Ge: Org.Comp.3-195
$C_{11}GeH_{17}N_3O_4$ . . . $Ge(CH_3)_3C(=CHCH{=}C(CH{=}NN(CH_2COOH)CONH_2)O)$
Ge: Org.Comp.2-97, 105
$C_{11}GeH_{18}$ . . . $(CH_2{=}CHCH_2)_2Ge[-CH_2-C(CH_3){=}CH-CH_2-]$ . . . Ge: Org.Comp.3-267, 273

$C_{11}GeH_{18}$ .... $CH_2{=}CHCH_2{-}Ge(CH_3)[-CH{=}CH{-}C(C_2H_5){=}CH{-}CH_2{-}]$ Ge: Org.Comp.3-306
– .... $n{-}C_3H_7{-}Ge(CH_3)_2{-}C_6H_5$ .... Ge: Org.Comp.3-202
– .... $Ge(CH_3)_3{-}CH_2CH_2{-}C_6H_5$ .... Ge: Org.Comp.1-173
– .... $Ge(CH_3)_3{-}CH_2{-}C_6H_4CH_3{-}4$ .... Ge: Org.Comp.1-157
– .... $Ge(CH_3)_3{-}C_6H_4{-}C_2H_5{-}4$ .... Ge: Org.Comp.2-74
– .... $Ge(C_2H_5)_3{-}C{\equiv}CC{\equiv}C{-}CH_3$ .... Ge: Org.Comp.2-240
– .... $[-CH_2{-}CH{=}C(CH_3){-}CH_2{-}]Ge[-CH_2{-}C(CH_3){=}C(CH_3){-}CH_2{-}]$ .... Ge: Org.Comp.3-344
$C_{11}GeH_{18}O$ .... $Ge(CH_3)_3CH(OCH_3)C_6H_5$ .... Ge: Org.Comp.1-159
$C_{11}GeH_{18}S$ .... $Ge(CH_3)_3CH(SCH_3)C_6H_5$ .... Ge: Org.Comp.1-159
$C_{11}GeH_{19}N$ .... $Ge(CH_3)_3C_6H_4N(CH_3)_2{-}4$ .... Ge: Org.Comp.2-72, 84
$C_{11}GeH_{19}N^+$ .... $[Ge(CH_3)_3C_6H_4(N(CH_3)_2{-}4)]^+$, radical cation .. Ge: Org.Comp.3-353
$C_{11}GeH_{20}$ .... $2{-}[(CH_3)_3Ge{-}CH_2]{-}[2.2.1]{-}C_7H_9$ .... Ge: Org.Comp.1-156
– .... $(CH_3)_2Ge[-CH{=}CH{-}C(C_4H_9{-}t){=}CH{-}CH_2{-}]$ .... Ge: Org.Comp.3-302
– .... $n{-}C_4H_9{-}Ge(CH_3)[-CH{=}C(CH_3){-}C({=}CH_2){-}CH_2{-}]$ Ge: Org.Comp.3-267
– .... $n{-}C_4H_9{-}Ge(CH_3)[-CH{=}C(CH_3){-}C(CH_3){=}CH{-}]$ .. Ge: Org.Comp.3-279
– .... $Ge(CH_2CH{=}CH_2)_3{-}C_2H_5$ .... Ge: Org.Comp.3-52
– .... $Ge(CH_3)_3{-}C[=CH{-}CH_2{-}C(CH_3){=}C(CH_3){-}CH_2{-}]$ Ge: Org.Comp.2-37
– .... $Ge(CH_3)_3{-}[4.1.0]{-}C_7H_8(CH_3)$ .... Ge: Org.Comp.2-41, 48
– .... $Ge(CH_3)_3{-}[4.1.0.0^{2,7}]{-}C_7H_8(CH_3)$ .... Ge: Org.Comp.1-198/9
– .... $Ge(C_2H_5)_3{-}CH(CH{=}CH)_2$ .... Ge: Org.Comp.2-231, 234
– .... $Ge(C_2H_5)_3{-}C{\equiv}C{-}C(CH_3){=}CH_2$ .... Ge: Org.Comp.2-239, 247
– .... $Ge(C_2H_5)_3{-}C{\equiv}C{-}CH{=}CHCH_3$ .... Ge: Org.Comp.2-240, 247/9
– .... $Ge(C_2H_5)_3{-}C{\equiv}C{-}CH_2{-}CH{=}CH_2$ .... Ge: Org.Comp.2-240
– .... $GeC_{11}H_{20}$ .... Ge: Org.Comp.2-48
$C_{11}GeH_{20}N_2O$ .... $3{-}(C_2H_5)_3Ge{-}4{-}[CH_3{-}C({=}O)]{-}1,2{-}N_2C_3H_2$ .... Ge: Org.Comp.2-290
– .... $3{-}[CH_3{-}C({=}O)]{-}4{-}(C_2H_5)_3Ge{-}1,2{-}N_2C_3H_2$ .... Ge: Org.Comp.2-290
$C_{11}GeH_{20}N_2O_2$ ... $Ge(C_2H_5)_3C({=}CHNHN{=}C(COOCH_3))$ .... Ge: Org.Comp.2-292
$C_{11}GeH_{20}O$ .... $1{-}[(CH_3)_2C(OH){-}C{\equiv}C]{-}1{-}C_2H_5{-}GeC_4H_8$ .... Ge: Org.Comp.3-249
$C_{11}GeH_{20}O_2$ ..... $(C_2H_5)_2Ge[-CH_2CH_2{-}C({=}CHCOOH){-}CH_2CH_2{-}]$ Ge: Org.Comp.3-291
– .... $Ge(C_2H_5)_3{-}C{\equiv}C{-}CH_2{-}OC({=}O)CH_3$ .... Ge: Org.Comp.2-261
$C_{11}GeH_{20}O_3$ ..... $(C_2H_5)_2Ge[-CH_2{-}CH(COO{-}CH_3){-}C({=}O){-}CH_2CH_2{-}]$ Ge: Org.Comp.3-291
– .... $Ge(CH_3)_3{-}CH{=}CHCH_2{-}O{-}CH_2{-}O{-}CH_2C{\equiv}C{-}CH_2OH$ Ge: Org.Comp.2-8
$C_{11}GeH_{21}$ .... $Ge(CH_3)_2(CH(-C(CH_3)_2CH_2{-}))CH{=}C(CH_3){=}CH_2$ Ge: Org.Comp.3-198
$C_{11}GeH_{21}Mo_{11}O_{41}P^{4-}$
$[PMo_{11}Ge((CH_2)_{10}COOH)O_{39}]^{4-}$ .... Mo: SVol.B3b-128/9
$C_{11}GeH_{21}Mo_{11}O_{41}Si^{5-}$
$[SiMo_{11}Ge((CH_2)_{10}COOH)O_{39}]^{5-}$ .... Mo: SVol.B3b-128/9
$C_{11}GeH_{21}NO_2$ .... $Ge(C_2H_5)_3CH(COOC_2H_5)CN$ .... Ge: Org.Comp.2-135
$C_{11}GeH_{21}NO_4$ .... $Ge(CH_3)_3C(COOCH_3){=}C(N(CH_3)_2)COOCH_3$ .... Ge: Org.Comp.2-11, 22
$C_{11}GeH_{21}N_3S$ .... $Ge(C_2H_5)_3{-}C{\equiv}C{-}C(CH_3){=}NNH{-}C(S)NH_2$ .... Ge: Org.Comp.2-265
$C_{11}GeH_{22}$ .... $3,3{-}(C_2H_5)_2{-}1,5{-}(CH_3)_2{-}[3.1.0]{-}3{-}GeC_5H_6$ .... Ge: Org.Comp.3-319
– .... $CH_3{-}GeC_{10}H_{19}$ .... Ge: Org.Comp.3-329
– .... $(C_2H_5)_2Ge[-CH_2{-}C(CH_3){=}C(CH_3){-}CH_2CH_2{-}]$ .. Ge: Org.Comp.3-300
– .... $Ge(CH_3)_3{-}C(C_2H_5){=}C{=}CH{-}C_3H_7{-}n$ .... Ge: Org.Comp.2-20
– .... $Ge(CH_3)_3{-}CH[-CH{=}C(CH_3){-}CH_2{-}CH(CH_3){-}CH_2{-}]$ Ge: Org.Comp.2-36, 47

$C_{11}GeH_{22}$	$Ge(CH_3)_3-CH[-CH=CH-C(CH_3)_2-CH_2CH_2-]$	Ge: Org.Comp.2-36, 47
–	$Ge(CH_3)_3-CH_2CH_2-C_6H_9$-c	Ge: Org.Comp.1-173
–	$Ge(CH_3)_3-C[=CH-CH_2-C(CH_3)_2-CH_2CH_2-]$	Ge: Org.Comp.2-34
–	$Ge(CH_3)_3-C\equiv C-C_6H_{13}$-n	Ge: Org.Comp.2-60
–	$Ge(CH_3)_3-C_8H_{13}$-c	Ge: Org.Comp.2-38
–	$Ge(C_2H_5)_3-C(=CH_2)C(CH_3)=CH_2$	Ge: Org.Comp.2-172
–	$Ge(C_2H_5)_3-CH=CHC(CH_3)=CH_2$	Ge: Org.Comp.2-175
–	$Ge(C_2H_5)_3-CH=CHCH=CHCH_3$	Ge: Org.Comp.2-177
–	$Ge(C_2H_5)_3-CH_2CH_2CH_2-C\equiv CH$	Ge: Org.Comp.2-240, 249
–	$Ge(C_2H_5)_3-C\equiv C-C_3H_7$-i	Ge: Org.Comp.2-238
–	$Ge(C_2H_5)_3-C\equiv C-C_3H_7$-n	Ge: Org.Comp.2-239
–	$Ge(C_2H_5)_3-C_5H_7$-c	Ge: Org.Comp.2-230, 231, 234
$C_{11}GeH_{22}NSi$	$Ge(CH_3)_3-C[-CH=CH-N(Si(CH_3)_3)-CHCH=]$, radical	Ge: Org.Comp.3-352
–	$Ge(CH_3)_3-C[-CHCH=CH-N(Si(CH_3)_3)-CH=]$, radical	Ge: Org.Comp.3-352
$C_{11}GeH_{22}O$	1-$(CH_3)_3Ge$-4,4-$(CH_3)_2$-[4.1.0]-7-OC_6H_7	Ge: Org.Comp.2-98, 105
–	$CH_3-GeC_{10}H_{18}-OH$	Ge: Org.Comp.3-329
–	$(C_2H_5)_2Ge[-CH(CH_3)-CH_2-C(=O)-CH_2-CH(CH_3)-]$	Ge: Org.Comp.3-293/4
–	$(n-C_3H_7)_2Ge[-CH_2CH_2-C(=O)-CH_2CH_2-]$	Ge: Org.Comp.3-296
–	$Ge(CH_3)_3-CH[-CH_2-C(=O)-C(CH_3)_2-CH_2CH_2-]$	Ge: Org.Comp.1-197
–	$Ge(C_2H_5)_3-C(C_3H_7-i)=C=O$	Ge: Org.Comp.2-198/9
–	$Ge(C_2H_5)_3-C\equiv C-C(OH)(CH_3)_2$	Ge: Org.Comp.2-263, 270
$C_{11}GeH_{22}O_2$	1-$(CH_3)_3Ge$-1-$[C_2H_5-OC(=O)]$-2,2-$(CH_3)_2-C_3H_2$	Ge: Org.Comp.1-194
–	$Ge(CH_3)_3-CH(COO-C_2H_5)-CH_2C(CH_3)=CH_2$	Ge: Org.Comp.2-16
–	$Ge(CH_3)_3-CH(COO-C_2H_5)-CH_2CH=CHCH_3$	Ge: Org.Comp.2-18
–	$Ge(C_2H_5)_3-CH=CHCH_2-OC(=O)CH_3$	Ge: Org.Comp.2-206
–	$Ge(C_2H_5)_3-C\equiv C-CH(OCH_3)_2$	Ge: Org.Comp.2-260/1
–	$Ge(C_2H_5)_3-C\equiv C-CH(OH)-OC_2H_5$	Ge: Org.Comp.2-260
$C_{11}GeH_{22}Si$	$Ge(CH_3)_3C(Si(CH_3)_3)(CH=CH)_2$	Ge: Org.Comp.2-33, 47
$C_{11}GeH_{22}Sn$	$Ge(CH_3)_3C(Sn(CH_3)_3)(CH=CH)_2$	Ge: Org.Comp.2-33, 47
$C_{11}GeH_{23}N$	$Ge(C_2H_5)_3C_4H_8CN$	Ge: Org.Comp.2-163
$C_{11}GeH_{24}$	$(n-C_4H_9)_2Ge[-CH_2CH_2CH_2-]$	Ge: Org.Comp.3-235, 238/40
–	$Ge(C_2H_5)_3-CH_2C(CH_3)=CHCH_3$	Ge: Org.Comp.2-174
–	$Ge(C_2H_5)_3-CH_2CH=C(CH_3)_2$	Ge: Org.Comp.2-173/4, 184
–	$Ge(C_2H_5)_3-CH_2CH=CHC_2H_5$	Ge: Org.Comp.2-176
–	$Ge(C_2H_5)_3-CH_2CH_2CH_2-CH=CH_2$	Ge: Org.Comp.2-177
–	$Ge(C_2H_5)_3-CH_2CH_2-CH=CHCH_3$	Ge: Org.Comp.2-176
–	$Ge(C_2H_5)_3-C_5H_9$-c	Ge: Org.Comp.2-168
$C_{11}GeH_{24}NO$	$Ge(CH_3)_3-CH[-C(CH_3)_2-CH_2-C(CH_3)_2-N(O)-]$, radical	Ge: Org.Comp.2-104
$C_{11}GeH_{24}NP$	$Ge(CH_3)_3C(CH_3)(CN)CH_2P(C_2H_5)_2$	Ge: Org.Comp.1-170
$C_{11}GeH_{24}N_2$	$Ge(C_2H_5)_3CH(CH_2N(CH_3)_2)CN$	Ge: Org.Comp.2-134
$C_{11}GeH_{24}O$	$Ge(CH_3)_3-CH[-CH_2CH_2-C(CH_3)_2-CH(OH)-CH_2-]$	Ge: Org.Comp.1-198, 200
–	$Ge(CH_3)_3-CH[-CH_2CH_2-C(CH_3)_2-CH_2-CH(OH)-]$	Ge: Org.Comp.1-197

$C_{11}GeH_{24}O$ ...... $Ge(CH_3)_3$-C(OH)[-CH_2CH_2-$C(CH_3)_2$-CH_2CH_2-] Ge: Org.Comp.1-197
– ...... $Ge(C_2H_5)_3$-C(=O)C_4H_9-t ...... Ge: Org.Comp.2-152, 155
– ...... $Ge(C_2H_5)_3$-$C(CH_3)$=CH-O-C_2H_5 ...... Ge: Org.Comp.2-193, 200
– ...... $Ge(C_2H_5)_3$-$C(C_2H_5)$=CH-OCH_3 ...... Ge: Org.Comp.2-193, 200
– ...... $Ge(C_2H_5)_3$-$CH_2CH_2CH_2$-C(=O)CH_3 ...... Ge: Org.Comp.2-163
– ...... $Ge(C_2H_5)_3$-$C(OCH_3)$=CHC_2H_5 ...... Ge: Org.Comp.2-213, 218
– ...... $Ge(C_2H_5)_3$-C(O-C_2H_5)=$CHCH_3$ ...... Ge: Org.Comp.2-207
– ...... $Ge(C_3H_7\text{-}n)_3$-C(=O)CH_3 ...... Ge: Org.Comp.3-2
$C_{11}GeH_{24}OSi$ .... $Ge(C_2H_5)_3C(Si(CH_3)_3)$=C=O ...... Ge: Org.Comp.2-198
$C_{11}GeH_{24}OSn$.... $Ge(CH_3)_3C(Sn(C_2H_5)_3)$=C=O ...... Ge: Org.Comp.2-7
$C_{11}GeH_{24}O_2$ ..... $Ge(CH_3)_3$-C(OH)[-CH_2CH_2-$C(CH_3)_2$-CH_2-CH(OH)-] Ge: Org.Comp.1-197, 200
– ...... $Ge(C_2H_5)_3$-$CH(CH_3)$-COO-C_2H_5 ...... Ge: Org.Comp.2-133/4
– ...... $Ge(C_2H_5)_3$-$CH(C_2H_5)$-COO-CH_3 ...... Ge: Org.Comp.2-149
– ...... $Ge(C_2H_5)_3$-$CH_2CH(CH_3)$-$COOCH_3$ ...... Ge: Org.Comp.2-150
– ...... $Ge(C_2H_5)_3$-$CH_2CH_2CH_2CH_2$-COOH ...... Ge: Org.Comp.2-163
– ...... $Ge(C_2H_5)_3$-$CH_2CH_2CH_2$-OC(=O)-CH_3 ...... Ge: Org.Comp.2-145
– ...... $Ge(C_2H_5)_3$-CH_2CH_2-COO-C_2H_5 ...... Ge: Org.Comp.2-150
– ...... $Ge(C_2H_5)_3$-CH_2CH_2-O-CH_2-CH[-CH_2-O-] ...... Ge: Org.Comp.2-125
$C_{11}GeH_{24}O_2S$ .... $Ge(C_2H_5)_3CH_2CH_2CH_2SCH_2COOH$ ...... Ge: Org.Comp.2-146
$C_{11}GeH_{24}O_3$ ..... 2-[$(CH_3)_3Ge$-$CH_2CH_2CH_2$-O-CH_2]-1,4-$O_2C_4H_7$ Ge: Org.Comp.1-181
– ...... $Ge(C_2H_5)_3$-$CH_2CH_2CH_2$-OC(=O)O-CH_3 ...... Ge: Org.Comp.2-145
$C_{11}GeH_{24}SSi$ .... $Ge(C_2H_5)_3C(Si(CH_3)_3)$=C=S ...... Ge: Org.Comp.2-199
$C_{11}GeH_{24}S_2$ ..... $Ge(C_2H_5)_3C(CH_3)((SCH_2)_2CH_2)$ ...... Ge: Org.Comp.2-289
$C_{11}GeH_{24}Si$ ..... $Ge(CH_3)_3$-C≡C-$Si(C_2H_5)_3$ ...... Ge: Org.Comp.2-54
– ...... $Ge(C_2H_5)_3$-C≡C-$Si(CH_3)_3$ ...... Ge: Org.Comp.2-253, 258
$C_{11}GeH_{24}Sn$ ..... $Ge(CH_3)_3$-C≡C-$Sn(C_2H_5)_3$ ...... Ge: Org.Comp.2-54
– ...... $Ge(C_2H_5)_3$-C≡C-$Sn(CH_3)_3$ ...... Ge: Org.Comp.2-254, 258
$C_{11}GeH_{25}$ .... $GeCH_3(CH_2C_4H_9\text{-}t)_2$, radical ...... Ge: Org.Comp.3-360, 361, 363
$C_{11}GeH_{25}NO$..... $Ge(C_2H_5)_3CON(C_2H_5)_2$ ...... Ge: Org.Comp.2-119
$C_{11}GeH_{25}NO_2S$... $Ge(C_2H_5)_3CH(CH_2CH_2CH_2N(CH_3)S(O)_2)$ ...... Ge: Org.Comp.2-290
$C_{11}GeH_{25}O_2P$ .... $Ge(CH_3)_3CH(COOCH_3)CH_2P(C_2H_5)_2$ ...... Ge: Org.Comp.1-169
$C_{11}GeH_{25}O_2PS$... $Ge(CH_3)_3CH(COOCH_3)CH_2P(S)(C_2H_5)_2$ ...... Ge: Org.Comp.1-170
$C_{11}GeH_{25}O_3P$ .... $Ge(CH_3)_3CH(COOCH_3)CH_2P(O)(C_2H_5)_2$ ...... Ge: Org.Comp.1-170
$C_{11}GeH_{25}O_4P$ .... $Ge(CH_3)_3CH[P(OC_2H_5)_2]COOC_2H_5$ ...... Ge: Org.Comp.1-172
$C_{11}GeH_{26}$ .... CH_3-$Ge(C_4H_9\text{-}n)_2$-C_2H_5 ...... Ge: Org.Comp.3-213
– ...... n-C_4H_9-$Ge(C_2H_5)_2$-C_3H_7-n ...... Ge: Org.Comp.3-209, 212
– ...... n-C_4H_9-$Ge(C_3H_7\text{-}n)_2$-CH_3 ...... Ge: Org.Comp.3-213
– ...... $Ge(C_2H_5)_3$-C_5H_{11}-n ...... Ge: Org.Comp.2-111, 113
– ...... $Ge(C_2H_5)_3$-C_5H_{11}-neo ...... Ge: Org.Comp.2-111
– ...... $Ge(C_3H_7\text{-}n)_3$-C_2H_5 ...... Ge: Org.Comp.3-1, 6/7
– ...... $Ge(C_3H_7\text{-}i)_3$-C_2H_5 ...... Ge: Org.Comp.3-14
$C_{11}GeH_{26}O$...... $Ge(C_2H_5)_3$-CH_2CH_2-CH_2CH_2-CH_2OH ...... Ge: Org.Comp.2-161
– ...... $Ge(C_2H_5)_3$-C(OH)$(C_2H_5)_2$ ...... Ge: Org.Comp.2-149
– ...... $Ge(C_3H_7\text{-}n)_3$-CH_2CH_2OH ...... Ge: Org.Comp.3-2
$C_{11}GeH_{26}OS$..... $Ge(C_2H_5)_3CH_2CH_2CH_2SCH_2CH_2OH$ ...... Ge: Org.Comp.2-146
$C_{11}GeH_{26}O_3S$.... $Ge(C_2H_5)_3CH(SO_2OCH_3)C_3H_7$ ...... Ge: Org.Comp.2-159
$C_{11}GeH_{26}S$ ...... $Ge(C_2H_5)_3$-$CH_2CH_2CH_2$-S-C_2H_5 ...... Ge: Org.Comp.2-145

$C_{11}GeH_{26}S$	$Ge(C_2H_5)_3$-CH_2CHDCH_2-S-C_2H_5	Ge: Org.Comp.2-146
$C_{11}GeH_{27}MnN_3O_3P$		
	$Mn(NO)_3[P(C_4H_9\text{-}t)_2Ge(CH_3)_3]$	Mn: MVol.D8-73/4
$C_{11}GeH_{27}N$	$Ge(C_2H_5)_3C_5H_{10}NH_2$	Ge: Org.Comp.2-163
$C_{11}GeH_{27}NO_2S$	$Ge(C_2H_5)_3$-CH_2-$S(=O)_2$-$N(C_2H_5)_2$	Ge: Org.Comp.2-117
–	$Ge(C_2H_5)_3$-CH_2-$S(=O)_2$-NH-C_4H_9-t	Ge: Org.Comp.2-117
$C_{11}GeH_{28}Si$	$Ge(C_2H_5)_3CH_2CH_2Si(CH_3)_3$	Ge: Org.Comp.2-126, 138
$C_{11}GeH_{29}O_3PSi_2$	$Ge(CH_3)_3CH{=}C(OSi(CH_3)_3)P(H)(O)OSi(CH_3)_3$	Ge: Org.Comp.2-4
$C_{11}GeH_{30}NPSn$	$(CH_3)_3SnN(P(CH_3)C_4H_9\text{-}t)Ge(CH_3)_3$	Sn: Org.Comp.18-76, 80
$C_{11}H_3MnNO_9Re$	$(CO)_4Re(NCCD_3)Mn(CO)_5$	Re: Org.Comp.1-492/3
–	$(CO)_5ReMn(CO)_4NCCD_3$	Re: Org.Comp.2-198
$C_{11}H_3N_3O_{10}Re_2S_2^{2+}$		
	$[(CO)_5Re\text{-}NS\text{-}N(CH_3)\text{-}SN\text{-}Re(CO)_5]^{2+}$	S: S-N Comp.5-256
$C_{11}H_3O_9ReW$	$(CO)_5ReW(CO)_4{\equiv}CCH_3$	Re: Org.Comp.2-193
$C_{11}H_4IO_5Re$	$(CO)_5ReC_6H_4I$-4	Re: Org.Comp.2-129
$C_{11}H_4NO_7ReS$	$(CO)_5ReSC_6H_4NO_2$-4	Re: Org.Comp.2-27
$C_{11}H_4O_5ReS_4^+$	$[(CO)_5ReC_3H_2S_2{=}C_3H_2S_2]^+$	Re: Org.Comp.2-158
$C_{11}H_4O_9ReW^-$	$[(CO)_4Re(\text{-}CH(CH_3)\text{-}W(CO)_5\text{-})]^-$	Re: Org.Comp.1-504
$C_{11}H_5IMoO_4$	C_6H_5-$CMo(CO)_4(I)$	Mo: Org.Comp.5-97/100
$C_{11}H_5NO_6Re$	$(CO)_5ReN(O)C_6H_5$, radical	Re: Org.Comp.2-170
$C_{11}H_5O_4ReS_2$	$(CO)_4Re[\text{-}S{=}C(C_6H_5)\text{-}S\text{-}]$	Re: Org.Comp.1-358, 366, 367
$C_{11}H_5O_5Re$	$(CO)_5Re$-C_6H_5	Re: Org.Comp.2-130, 131
–	$(CO)_4(^{13}CO)Re$-C_6H_5	Re: Org.Comp.2-130
$C_{11}H_5O_7ReS$	$(CO)_5ReOS(O)C_6H_5$	Re: Org.Comp.2-25
$C_{11}H_5O_8ReS$	$(CO)_5ReOS(O)_2C_6H_5$	Re: Org.Comp.2-25/6
$C_{11}H_5O_9OsRe$	$(CO)_5ReOs(CO)_4C_2H_5$	Re: Org.Comp.2-202
$C_{11}H_6NO_4Re$	$(CO)_4Re[\text{-}(1,2\text{-}NC_5H_4)\text{-}CH{=}CH\text{-}]$	Re: Org.Comp.1-408
$C_{11}H_6NO_4ReS$	$(CO)_4Re[\text{-}S{=}C(C_6H_5)\text{-}NH\text{-}]$	Re: Org.Comp.1-356
$C_{11}H_6N_3O_6Re$	O_2NO-$Re(CO)_3[\text{-}1,8\text{-}(1,8\text{-}N_2C_8H_6)\text{-}]$	Re: Org.Comp.1-155
$C_{11}H_6O_4Th$	$ThO(2\text{-}OC_{10}H_6\text{-}3\text{-}COO) \cdot 3\ H_2O$	
	$= Th(OH)_2(2\text{-}OC_{10}H_6\text{-}3\text{-}COO) \cdot 2\ H_2O$	Th: SVol.C7-149/50
$C_{11}H_7IMoO_2$	$(C_9H_7)Mo(CO)_2I$	Mo: Org.Comp.7-2
$C_{11}H_7K_2MoN_5$	$K_2[CH_3C_5H_4Mo(CN)_5]$	Mo: Org.Comp.6-5
$C_{11}H_7MoN_3O_2$	$(CH_3\text{-}C_5H_4)Mo(CO)_2(CN)_3$	Mo: Org.Comp.7-3
$C_{11}H_7MoN_5^-$	$[CH_3C_5H_4Mo(CN)_5]^-$	Mo: Org.Comp.6-5
$C_{11}H_8MnNO_2S^+$	$[Mn(2\text{-}(C_6H_5\text{-}N(O)C(O))\text{-}SC_4H_3)]^+$	Mn: MVol.D7-222/5
$C_{11}H_8MnNO_3S_2^+$	$Mn[2\text{-}(O_3S\text{-}C_6H_4\text{-}2\text{-}N{=}CH)C_4H_3S]^+$	Mn: MVol.D6-75/6
$C_{11}H_8MnNO_4S^+$	$Mn[2\text{-}(O_3S\text{-}C_6H_4\text{-}N{=}CH)C_4H_3O]^+$	Mn: MVol.D6-75/6
$C_{11}H_8MnN_3O_3S^+$	$[Mn(1,3\text{-}N_2C_4H\text{-}2\text{-}(=S)\text{-}3\text{-}(C_6H_4\text{-}2\text{-}CH_3)\text{-}5\text{-}$	
	$(=N\text{-}O)\text{-}4,6\text{-}(=O)_2)]^+$	Mn: MVol.D7-75/8
–	$[Mn(1,3\text{-}N_2C_4H\text{-}2\text{-}(=S)\text{-}3\text{-}(C_6H_4\text{-}3\text{-}CH_3)\text{-}5\text{-}$	
	$(=N\text{-}O)\text{-}4,6\text{-}(=O)_2)]^+$	Mn: MVol.D7-75/8
–	$[Mn(1,3\text{-}N_2C_4H\text{-}2\text{-}(=S)\text{-}3\text{-}(C_6H_4\text{-}4\text{-}CH_3)\text{-}5\text{-}$	
	$(=N\text{-}O)\text{-}4,6\text{-}(=O)_2)]^+$	Mn: MVol.D7-75/8
$C_{11}H_8MnN_3O_5S^+$	$Mn[O_3S\text{-}C_{10}H_5(=O\text{-}4){=}N\text{-}NHC(O)NH_2\text{-}3]^+$	Mn: MVol.D6-331/2
$C_{11}H_8MoN_2O_3$	$(C_5H_5)Mo(CO)_2[\text{-}CH{=}C(CN)\text{-}C(OH){=}NH\text{-}]$	Mo: Org.Comp.8-112, 132
–	$(C_5H_5)Mo(CO)_2[\text{-}CH{=}C(CN)\text{-}C(OD){=}ND\text{-}]$	Mo: Org.Comp.8-132/3
$C_{11}H_8MoO_4$	$[(2.2.1)\text{-}C_7H_8\text{-}2,5]Mo(CO)_4$	Mo: Org.Comp.5-317/38
–	$[(2.2.1)\text{-}C_7H_8\text{-}2,5]Mo(CO)_{4-n}(^{13}CO)_n$	Mo: Org.Comp.5-317/8
$C_{11}H_8NO_6Th^+$	$Th[2\text{-}(OOC)C_6H_4(N(CH_2COO)_2)]^+$	Th: SVol.D1-93/4, 98

$C_{11}H_8N_3O_2Th^{3+}$ .. $Th[(HO)(O)C_6H_3NNC_5H_4N]^{3+}$ ... Th: SVol.D1-122
$C_{11}H_8N_4O_2S$ ..... 1-(4-NN-C_6H_4)-2-(OS(O)-N=)-NC_5H_4 ... S: S-N Comp.8-305/6
$C_{11}H_8O_5Re^+$ ..... $[(CO)_5Re(C_6H_8$-1,4-c$)]^+$ ... Re: Org.Comp.2-351
$C_{11}H_8O_5Th$ ...... $Th(OH)_2$(2-$OC_{10}H_6$-3-COO) · 2 H_2O
= ThO(2-$OC_{10}H_6$-3-COO) · 3 H_2O ... Th: SVol.C7-149/50
$C_{11}H_9IMoNOS_2^-$ .. $[C_5H_5Mo(NO)(I)(1,2\text{-}S_2C_6H_4)]^-$ ... Mo: Org.Comp.6-31
$C_{11}H_9LiMoO_6$ .... $Li[(C_5H_5)Mo(CO)_2$(CH=O)-C(=O)-C(=O)O-CH_3] Mo: Org.Comp.8-176, 177
$C_{11}H_9MnN_2O_2S^+$ [Mn(1,3-$N_2C_4H_2$-2-(=S)-3-(C_6H_4-2-CH_3)-4,6-(=O$)_2$)$]^+$
Mn: MVol.D7-75/8
– ... [Mn(1,3-$N_2C_4H_2$-2-(=S)-3-(C_6H_4-3-CH_3)-4,6-(=O$)_2$)$]^+$
Mn: MVol.D7-75/8
– ... [Mn(1,3-$N_2C_4H_2$-2-(=S)-3-(C_6H_4-4-CH_3)-4,6-(=O$)_2$)$]^+$
Mn: MVol.D7-75/8
$C_{11}H_9MnN_3O_3^{2+}$. Mn[HO-2-$C_{10}H_5$(=O-1)=N-NHC(O)NH_2-4$]^{2+}$ .. Mn: MVol.D6-331/2
$C_{11}H_9MnN_4O_2S^+$ [Mn(NCCH_2C(=O)N_2HC(=S)NHC(=O)C_6H_5)$]^+$ ... Mn: MVol.D7-202
$C_{11}H_9MnN_5O_2S_4$.. Mn(4-$NH_2C_6H_4SO_2$NH-2-C_3H_2NS-3,1)$(NCS)_2$ Mn: MVol.D7-116/23
$C_{11}H_9MoO_6^-$ ..... $Li[(C_5H_5)Mo(CO)_2$(CH=O)-C(=O)-C(=O)O-CH_3] Mo: Org.Comp.8-176, 177
$C_{11}H_9NO_8Th$ ..... $Th(OH)_3$(1-$HOC_{10}H_5$-4-NO_2-2-COO) · 2 H_2O Th: SVol.C7-148/50
$C_{11}H_9N_3O_2S_2$ .... 2-[C_6H_5-S(O)$_2$N=S=N]-NC_5H_4 ... S: S-N Comp.7-60
$C_{11}H_{10}IO_3ReSSe$ $(CO)_3$ReI(CH_3S-C_6H_4-2-SeCH_3) ... Re: Org.Comp.1-210, 214/5
$C_{11}H_{10}IO_3ReS_2$... $(CO)_3$ReI(CH_3S-C_6H_4-2-SCH_3) ... Re: Org.Comp.1-210, 214/5
$C_{11}H_{10}I_2MnN_4$ ... [MnI_2(NC_5H_4-2-CH=NNH-2-C_5H_4N)] ... Mn: MVol.D6-244, 246/7
$C_{11}H_{10}KMoNO_3$ .. $K[(C_5H_5)Mo(CO)_2$(CN)-C(=O)-C_2H_5] ... Mo: Org.Comp.8-1, 4
$C_{11}H_{10}MnN_2O_5S$.. Mn[OC(CH_3)C(C(O)CH_3)=N-N-2-C_6H_4-SO_3] .. Mn: MVol.D6-256/7
– ... Mn[OC(CH_3)C(C(O)CH_3)=N-N-3-C_6H_4-SO_3] .. Mn: MVol.D6-256/7
– ... Mn[OC(CH_3)C(C(O)CH_3)=N-N-4-C_6H_4-SO_3] .. Mn: MVol.D6-256/7
$C_{11}H_{10}MnN_3O_2S^+$ [Mn(2-(4-NH_2-C_6H_4-SO_2=N)-NC_5H_4)$]^+$ ... Mn: MVol.D7-116/23
$C_{11}H_{10}MnN_4^{2+}$ ... [Mn(NC_5H_4-2-CH=NNH-2-C_5H_4N)$]^{2+}$ ... Mn: MVol.D6-244
$C_{11}H_{10}MoNO_3^-$ ... $K[(C_5H_5)Mo(CO)_2$(CN)-C(=O)-C_2H_5] ... Mo: Org.Comp.8-1, 4
– ... $[As(C_6H_5)_4][(C_5H_5)Mo(CO)_2$(CN)-C(=O)-$C_2H_5$] Mo: Org.Comp.8-1, 4
$C_{11}H_{10}MoNO_4^+$ .. [C_5H_5Mo(CO)(NO)$CH_2C_4H_3$O-c-(=O)$]^+$ ... Mo: Org.Comp.6-350, 353/4
$C_{11}H_{10}MoN_2O_2$... $(C_5H_5)Mo(CO)_2$(CN-CH_3)-CH_2-CN ... Mo: Org.Comp.8-179
– ... $(C_5H_5)Mo(NO)_2$-C_6H_5 ... Mo: Org.Comp.6-96
$C_{11}H_{10}MoN_2O_2S$.. $(C_5H_5)Mo(NO)_2SC_6H_5$ ... Mo: Org.Comp.6-57
$C_{11}H_{10}MoO_3$ ..... $(C_5H_5)Mo(CO)_2$[-CH=CH-C(CH_3)=O-] ... Mo: Org.Comp.8-112, 146
$C_{11}H_{10}MoO_3S_2$... $(C_5H_5)Mo(CO)_2(CS_2)$-C(=O)-C_2H_5 ... Mo: Org.Comp.8-176, 177/8
$C_{11}H_{10}MoO_4$ ..... $(C_5H_5)Mo(CO)_2$-C(=O)O-CH_2-CH=CH_2 ... Mo: Org.Comp.8-202, 204/5
$C_{11}H_{10}MoO_5$ ..... (C_2H_5-C≡C-C_2H_5)Mo$(CO)_5$ ... Mo: Org.Comp.5-168/9
$C_{11}H_{10}NO_6Re$.... $(CO)_5$ReC(O)N$(CH_2)_5$ ... Re: Org.Comp.2-117
$C_{11}H_{10}OSTi$ ..... $[(C_5H_5)Ti(SC_6H_5)O]_n$ ... Ti: Org.Comp.5-326
$C_{11}H_{10}O_3Ti^-$ ..... $[(C_5H_5)_2TiCO_3]^-$ ... Ti: Org.Comp.5-153
$C_{11}H_{10}O_5Re^+$ .... $[(CO)_5Re(CH_2$=CH-CH_2CH_2-CH=$CH_2)]^+$ ... Re: Org.Comp.2-351, 353
– ... $[(CO)_5Re(C_6H_{10}$-c$)]^+$ ... Re: Org.Comp.2-351
$C_{11}H_{10}O_6Th$ ..... $Th(OH)_3$(1-$HOC_{10}H_6$-2-COO) · 2 H_2O ... Th: SVol.C7-148/50
$C_{11}H_{10}O_8OsPRe$.. $(CO)_5$ReOs(P$(CH_3)_3$)(H)$(CO)_3$ ... Re: Org.Comp.2-202
$C_{11}H_{10}O_9Th$ ..... Th(C_6H_5CHO)$(HCOO)_4$
Photoemission spectra ... Th: SVol.A4-130
$C_{11}H_{10}S_2Ti$ ...... $[(C_5H_5)_2Ti(CS_2)]_n$ ... Ti: Org.Comp.5-332
$C_{11}H_{11}IMoN_2O_2$ .. $[(C_5H_5)Mo(CO)_2(CN\text{-}CH_3)_2]I$ ... Mo: Org.Comp.8-183
$C_{11}H_{11}IMoN_4O_3$ .. (C_5H_5)Mo(NO)(I)NHN($C_6H_4NO_2$-4)H ... Mo: Org.Comp.6-52

$C_{11}H_{11}IMoO_3$ .... $(C_5H_5)Mo(CO)_2(I)$=C(-O-$CH_2CH_2CH_2$-)....... Mo:Org.Comp.8-17, 26, 30, 31

$C_{11}H_{11}I_2InN_2$..... CH_3-In$(I)_2$ · 2-$(NC_5H_4$-2)-NC_5H_4 .... In: Org.Comp.1-161, 163

$C_{11}H_{11}MnN_2O_3{}^+$. Mn[H(OC(CH_3)C(C(O)CH_3)=N-N-C_6H_4-2-O)]$^+$ Mn:MVol.D6-256/7

$C_{11}H_{11}MnN_2O_5{}^-$ .. Mn[-O-2-C_6H_4-CH=NCH_2C(O)O-]$(H_2NCH_2COO)^-$ Mn:MVol.D6-34/5

$C_{11}H_{11}MnN_4O_2S^+$ [Mn(4-NH_2-C_6H_4-SO_2=N-2-$C_4H_2N_2$-1,3-CH_3-4)]$^+$ Mn:MVol.D7-116/23

$C_{11}H_{11}MnN_4O_3S^+$ [Mn(4-NH_2-C_6H_4-SO_2=N-3-$C_4H_2N_2$-1,2-OCH_3-6)]$^+$ Mn:MVol.D7-116/23

$C_{11}H_{11}MnN_5S_4$... [Mn(NC_5H_4-2-CH=N-N(CH_3)C(=S)SCH_3)$(NCS)_2]_n$ Mn:MVol.D6-361, 362

$C_{11}H_{11}MoNO_3$ ... $(C_5H_5)Mo(CO)_2$(CN-CH_3)-C(=O)-CH_3 .... Mo:Org.Comp.8-179/80

– .... $(C_5H_5)Mo(CO)_2$N(-O)=C(CH_3)CH=CH_2........ Mo:Org.Comp.7-207, 209, 210/1

$C_{11}H_{11}MoNO_4$ ... (C_5H_5)Mo(CO)(NO)[-O-CH_2-CH=C(CH_3)-C(=O)-] Mo:Org.Comp.6-299

– .... $(C_5H_5)Mo(CO)_2$[-O-C(=O)-C(C_2H_5)=NH-]..... Mo:Org.Comp.7-208/9, 238

$C_{11}H_{11}MoNO_5$ ... $(C_5H_5)Mo(CO)_2$[-O-C(=O)-C(C_2H_5)=N(OH)-] .. Mo:Org.Comp.7-208, 240

$C_{11}H_{11}MoN_2O_2$... [$(C_5H_5)Mo(CO)_2$(NC-$CH_3)_2$]............... Mo:Org.Comp.7-283, 284

$C_{11}H_{11}MoN_2O_2{}^+$. [$(C_5H_5)Mo(CO)_2$(CN-$CH_3)_2$]I Mo:Org.Comp.8-183

– .... [$(C_5H_5)Mo(CO)_2$(CN-$CH_3)_2$][PF_6]............ Mo:Org.Comp.8-183

$C_{11}H_{11}MoN_2O_2S_2{}^+$ [$(C_5H_5)Mo(CO)_2$(-S-C(C$(NCH_3)_2$)-S-)]$^+$ Mo:Org.Comp.7-189, 192

$C_{11}H_{11}MoNaO_3$ .. Na[$(C_5H_5)Mo(CO)_2$=C(-$CH_2CH_2CH_2$-O-)]..... Mo:Org.Comp.8-1, 2

$C_{11}H_{11}MoO_2{}^+$ ... [$(C_5H_5)Mo(CO)_2$($(CH_2)_3$C)][BF_4]............ Mo:Org.Comp.8-305/6, 310, 316

– .... [$(C_5H_5)Mo(CO)_2$(CH_2=CHCH=CH_2)][BF_4]...... Mo:Org.Comp.8-305/6, 316

– .... [$(C_5H_5)Mo(CO)_2$(CH_2=C_3H_4-c)][BF_4] Mo:Org.Comp.8-186

– .... [$(C_5H_5)Mo(CO)_2$(CH_3-C≡C-CH_3)][BF_4] Mo:Org.Comp.8-186

$C_{11}H_{11}MoO_3{}^-$ .... Na[$(C_5H_5)Mo(CO)_2$=C(-$CH_2CH_2CH_2$-O-)]..... Mo:Org.Comp.8-1, 2

$C_{11}H_{11}MoO_4$..... $(C_5H_5)Mo(CO)_2$[CH_3C(O)C(O)CH_3], radical.... Mo:Org.Comp.7-53

$C_{11}H_{11}NO_3S_2$ .... 1,3-$(O)_2$-C_8H_4N-2-S(O)-S-C_3H_7-i .... S: S-N Comp.8-333

$C_{11}H_{11}N_2O_5SSb$ .. (4-(2'-C_5H_4NNHSO_2)C_6H_4)Sb(O)$(OH)_2$ .... Sb: Org.Comp.5-288

$C_{11}H_{12}IMoN_3O$ ... C_5H_5Mo(NO)(I)NHN(C_6H_5)H Mo:Org.Comp.6-52, 54

$C_{11}H_{12}INOSn$ .... $(CH_3)_2$Sn(I)OC_9H_6N .... Sn: Org.Comp.17-133

$C_{11}H_{12}I_2MoN_2O$ .. (C_5H_5)Mo(NO)($NH_2C_6H_5$)I_2................ Mo:Org.Comp.6-33

$C_{11}H_{12}I_2MoN_4O_3$ (C_5H_5)Mo(NO)($NH_2NHC_6H_4NO_2$-4)I_2 .... Mo:Org.Comp.6-37

$C_{11}H_{12}InNO$ ..... $(CH_3)_2$In(1-NC_9H_6-8-O).................. In: Org.Comp.1-190, 192

$C_{11}H_{12}MnN_4O_3S^{2+}$ [Mn(4-NH_2-C_6H_4-SO_2-NH-3-(1,2-$N_2C_4H_2$-6-OCH_3))]$^{2+}$ Mn:MVol.D7-116/23

– .... [Mn(4-NH_2-C_6H_4-SO_2-NH-4-(1,3-$N_2C_4H_2$-6-OCH_3))]$^{2+}$ Mn:MVol.D7-116/23

$C_{11}H_{12}MnN_5PS_2$.. Mn[P$(CH_2CH_2CN)_3$]$(NCS)_2$................ Mn:MVol.D8-69/70

$C_{11}H_{12}MoNO_3{}^+$ .. [H_3CC(O)C_5H_4Mo(CO)(NO)C_3H_5]$^+$ Mo:Org.Comp.6-347

$C_{11}H_{12}MoN_2O_3$... $(C_5H_5)Mo(CO)_2$[-C(=O)-N(CH_3)-CH=N(CH_3)-] Mo:Org.Comp.8-112, 141

$C_{11}H_{12}MoN_2O_4$... $(CO)_4$Mo(CN-CH_3)(CN-C_4H_9-t) Mo:Org.Comp.5-25, 29

– .... $(C_5H_5)Mo(CO)_2$[-O-C(=O)-C(C_2H_5)=N(NH_2)-].. Mo:Org.Comp.7-208, 239

$C_{11}H_{12}MoN_2O_5$... [-N(CH_3)-$CH_2CH_2CH_2$-N(CH_3)-]C=Mo$(CO)_5$... Mo:Org.Comp.5-100, 107

$C_{11}H_{12}MoO_2$	$(C_5H_5)Mo(CO)_2(CH_2CHCH\text{-}CH_3)$	Mo:Org.Comp.8-205, 207, 211/2, 242/3
–	$(C_5H_5)Mo(CO)_2(CH_2CHCH\text{-}CHD_2)$	Mo:Org.Comp.8-205, 207, 212
–	$(C_5H_5)Mo(CO)_2[CH_2C(CH_3)CH_2]$	Mo:Org.Comp.8-205, 207, 218/9, 245/6
$C_{11}H_{12}MoO_3$	$(C_5H_5)Mo(CO)_2[1\text{-}OC_4H_6(D\text{-}2)\text{-}2\text{-}]$	Mo:Org.Comp.8-112, 127
–	$(C_5H_5)Mo(CO)_2[1\text{-}OC_4H_7\text{-}2\text{-}]$	Mo:Org.Comp.8-112, 126/7
$C_{11}H_{12}MoO_4$	$(C_5H_5)Mo(CO)_2[CH_2C(OC_2H_5)O]$	Mo:Org.Comp.8-186, 195
$C_{11}H_{12}NO_5Re$	$(CO)_4Re(NC\text{-}CD_3)\text{-}C(=O)\text{-}C_4H_9\text{-}i$	Re: Org.Comp.1-472
–	$(CO)_4Re[\text{-}C(=N(CH_3)_2)\text{-}C(CH_3)_2\text{-}C(=O)\text{-}]$	Re: Org.Comp.2-330, 335
–	$(CO)_5Re\text{-}C(N(CH_3)_2)=C(CH_3)_2$	Re: Org.Comp.2-127
$C_{11}H_{12}NO_6Re$	$(CO)_4Re(CNC_4H_9\text{-}t)COOCH_3$	Re: Org.Comp.2-247
$C_{11}H_{12}NO_7Re$	$(CO)_4Re[C(O)CH_3]C(CH_3)=NH\text{-}CH(CH_3)COOH$	Re: Org.Comp.1-401
$C_{11}H_{12}N_3O_5SSb$	$(4\text{-}(2'\text{-}C_5H_5N_2NHSO_2)C_6H_4)Sb(O)(OH)_2$	Sb: Org.Comp.5-288
$C_{11}H_{13}IMoN_3O^+$	$[(C_5H_5)Mo(NO)(NH_2NHC_6H_5)I]^+$	Mo:Org.Comp.6-39, 43
$C_{11}H_{13}I_2MoN_3O$	$(C_5H_5)Mo(NO)(NH_2NHC_6H_5)I_2$	Mo:Org.Comp.6-36
$C_{11}H_{13}MnN_5O_6S$	$[Mn(C_6H_5CH=CHC(CH_3)=N\text{-}NHC(=S)NH_2)(NO_3)_2]$	Mn:MVol.D6-342
$C_{11}H_{13}MnN_5O_7$	$[Mn(C_6H_5CH=CHC(CH_3)=N\text{-}NHC(O)NH_2)(NO_3)_2]$	Mn:MVol.D6-328
$C_{11}H_{13}MoNO_2$	$(CH_3\text{-}C_5H_4)Mo(CO)_2[\text{-}C(CH_3)=N(CH_3)\text{-}]$	Mo:Org.Comp.8-112, 122, 162
$C_{11}H_{13}MoNO_2S_2$	$(C_5H_5)Mo(CO)_2[\text{-}S\text{-}C(NH\text{-}C_3H_7\text{-}i)\text{-}S\text{-}]$	Mo:Org.Comp.7-188, 194
$C_{11}H_{13}MoNO_3$	$(CH_3\text{-}C(=O)\text{-}C_5H_4)Mo(CO)(NO)(CH_2=CHCH_3)$	Mo:Org.Comp.6-293
–	$(C_5H_5)Mo(CO)_2[\text{-}C(=O)\text{-}CH_2\text{-}CH(CH_3)\text{-}NH_2\text{-}]$	Mo:Org.Comp.8-111, 135
$C_{11}H_{13}MoNO_3S$	$(C_5H_5)Mo(CO)(NO)CH_2=CHCH_2SC(O)CH_3$	Mo:Org.Comp.6-295
$C_{11}H_{13}MoNO_4$	$(C_5H_5)Mo(CO)(NO)[CH_2=CH\text{-}CH_2\text{-}OC(=O)CH_3]$	Mo:Org.Comp.6-294
–	$(C_5H_5)Mo(CO)_2[\text{-}O\text{-}C(=O)\text{-}C(CH_3)_2\text{-}NH_2\text{-}]$	Mo:Org.Comp.7-208, 209, 232
–	$(C_5H_5)Mo(CO)_2[\text{-}O\text{-}C(=O)\text{-}CH_2\text{-}N(CH_3)_2\text{-}]$	Mo:Org.Comp.7-208, 209, 233
$C_{11}H_{13}MoNO_4S$	$(C_5H_5)Mo(CO)_2[\text{-}O\text{-}C(O)\text{-}CH(CH_2\text{-}SCH_3)\text{-}NH_2\text{-}]$	Mo:Org.Comp.7-208, 209, 233
–	$(C_5H_5)Mo(CO)_2[\text{-}S\text{-}CH_2\text{-}CH(COO\text{-}CH_3)\text{-}NH_2\text{-}]$	Mo:Org.Comp.7-234/5
$C_{11}H_{13}MoNO_5$	$(C_5H_5)Mo(CO)_2[\text{-}O\text{-}C(=O)\text{-}CH(CH(CH_3)OH)\text{-}NH_2\text{-}]$	Mo:Org.Comp.7-208, 209, 231
$C_{11}H_{13}MoN_3O_4$	$(CH_2CHCH_2)Mo(CO)_2[\text{-}N_2C_3H_3\text{-}CH_2CH(NH_2)\text{-}C(=O)O\text{-}]$	Mo:Org.Comp.5-286, 292
$C_{11}H_{13}NO_2Sn$	$(CH_3)_3SnN(CO)_2C_6H_4$	Sn: Org.Comp.18-84, 88, 100/1
$C_{11}H_{13}N_3O_3PRe$	$(CO)_3Re(N_2)[P(CH_3)_2C_6H_5]NH_2$	Re: Org.Comp.1-290/1
$C_{11}H_{13}N_3O_3Sn$	$(CH_3)_3SnN(COC_6H_4NO_2\text{-}4)CN$	Sn: Org.Comp.18-58, 70
$C_{11}H_{13}O_2ReS$	$(C_5H_5)Re(CO)(CS)(OC_4H_8)$	Re: Org.Comp.3-144
$C_{11}H_{13}O_3Re$	$(C_5H_5)Re(CO)_2(OC_4H_8)$	Re: Org.Comp.3-187
$C_{11}H_{13}Ti^+$	$[(C_5H_5)_2Ti(CH_3)]^+$	Ti: Org.Comp.5-353/4
$C_{11}H_{14}IMoO_3Sb$	$(CH_3)_3Sb(I)\text{-}Mo(CO)_3(C_5H_5)$	Sb: Org.Comp.5-32, 34
–	$[(CH_3)_3Sb\text{-}Mo(CO)_3(C_5H_5)]I$	Sb: Org.Comp.5-34
$C_{11}H_{14}LiN_2O_5Re$	$Li[(CO)_4Re(\text{-}N(C_3H_7\text{-}i)\text{-}C(=O)\text{-}N(C_3H_7\text{-}i)\text{-})]$	Re: Org.Comp.1-369
$C_{11}H_{14}MnNO_2Sn$	$(CH_3)_3SnNC\text{-}Mn(CO)_2(C_5H_5\text{-}c)$	Sn: Org.Comp.18-127
$C_{11}H_{14}MnN_2O_2S_2$	$[Mn(SC(=S)N(C_2H_5)_2)(NC_5H_4COO\text{-}4)]$	Mn:MVol.D7-143/4
$C_{11}H_{14}MnN_2O_4S$	$[Mn(S=C(NH_2)NHC_6H_5)(OC(CH_3)=O)_2] \cdot H_2O$	Mn:MVol.D7-195/6
$C_{11}H_{14}MnN_2O_6S$	$[Mn(NC_4H_3\text{-}2\text{-}CH=N\text{-}2\text{-}C_6H_4\text{-}SO_3)(H_2O)_3]$	Mn:MVol.D6-83
$C_{11}H_{14}MoNO_2^+$	$[(C_5H_5)Mo(CO)(NO)C_3H_3(CH_3)_2\text{-}1,2]^+$	Mo:Org.Comp.6-349

$C_{11}H_{14}MoNO_2S^+$ [$(C_5H_5)Mo(CO)_2(-C(N(CH_3)_2)=S(CH_3)-)$][$BF_4$] . . Mo:Org.Comp.8-129
– . . . . [$(C_5H_5)Mo(CO)_2(-C(N(CH_3)_2)=S(CH_3)-)$][$PF_6$] . . Mo:Org.Comp.8-130
$C_{11}H_{14}MoNO_3P$ . . $(C_5H_5)Mo(CO)_2=[2-(1,3,2-ONPC_2H_4-3-C_2H_5)]$ Mo:Org.Comp.7-35
$C_{11}H_{14}MoN_2O_2$ . . . $(C_5H_5)Mo(CO)(NO)CNC_4H_9-t$ Mo:Org.Comp.6-262
$C_{11}H_{14}MoN_2O_2S$ . . $(C_5H_5)Mo(CO)_2[-S-C(N(CH_3)_2)=N(CH_3)-]$ Mo:Org.Comp.7-207/10, 226
$C_{11}H_{14}MoN_2O_3$ . . . $[(CH_3)_2NCH(CH_3)C_5H_4]Mo(CO)_2(NO)$ Mo:Org.Comp.7-6, 10, 22
$C_{11}H_{14}MoN_2O_5S$ . . $[Mo(CO)_5(i-C_3H_7-N=S=N-C_3H_7-i)]$ . . . . S: S-N Comp.7-302/3
$C_{11}H_{14}MoN_3O_2^+$. $[(CH_2CHCH_2)Mo(CO)_2(NC-CH_3)_3][BF_4]$ Mo:Org.Comp.5-200
– . . . . $[(CH_2CHCH_2)Mo(CO)_2(NC-CH_3)_3][PF_6]$ Mo:Org.Comp.5-200
$C_{11}H_{14}MoO$ . . . . . . $(C_5H_5)Mo(CO)(CH_3)(H_3C-C{\equiv}C-CH_3)$ Mo:Org.Comp.6-312
$C_{11}H_{14}MoOS$ . . . . $(C_5H_5)Mo(CO)(SCH_3)(H_3C-C{\equiv}C-CH_3)$ Mo:Org.Comp.6-277
$C_{11}H_{14}MoO_2$ . . . . . $(C_5H_5)Mo(CO)(OCH_3)(H_3C-C{\equiv}C-CH_3)$ Mo:Org.Comp.6-277
– . . . . $(C_5H_5)Mo(CO)(OCD_3)(H_3C-C{\equiv}C-CH_3)$ Mo:Org.Comp.6-277
$C_{11}H_{14}MoO_3$ . . . . . $(C_5H_5)Mo(CO)_2(H)(C_4H_8O)$ Mo:Org.Comp.7-102, 115
$C_{11}H_{14}MoO_3Sb^+$ $[(CH_3)_3SbMo(CO)_3C_5H_5]^+$ . . . . Sb: Org.Comp.5-30, 34
$C_{11}H_{14}NO_3Re$ . . . . $(C_5H_5)Re(CO)(NO)-C(=O)C_4H_9$ . . . . Re: Org.Comp.3-158/60, 161
$C_{11}H_{14}NO_5Re$ . . . . $(CO)_4Re[C(O)CH_3]C(CH_3)=NHC_3H_7-i$ . . . . Re: Org.Comp.1-391
– . . . . $(CO)_4Re[C(O)C_3H_7-i]-C(CH_3)NHCH_3$ . . . . Re: Org.Comp.1-402
$C_{11}H_{14}N_2OSn$ . . . . $(CH_3)_3SnN(COC_6H_5)CN$ . . . . Sn: Org.Comp.18-58, 69
$C_{11}H_{14}N_2O_2S_2$ . . . $C_6H_5-NH-S(O)-SCH=C(CH_3)-NH-C(O)CH_3$ . . . S: S-N Comp.8-331
$C_{11}H_{14}N_2O_5Re^-$ . . $[(CO)_4Re(-N(C_3H_7-i)-C(=O)-N(C_3H_7-i)-)]^-$ . . . . Re: Org.Comp.1-369
$C_{11}H_{14}N_2O_5SW$ . . . $[W(CO)_5(i-C_3H_7-N=S=N-C_3H_7-i)]$ . . . . S: S-N Comp.7-300, 302/5
$C_{11}H_{14}N_2O_8Th$ . . . $Th[(OOCCH_2)_2NCH(CH_3)CH_2N(CH_2COO)_2] \cdot x\ H_2O$
Th: SVol.C7-124, 125/6
$C_{11}H_{14}O_3SbW^+$ . . $[(CH_3)_3SbW(CO)_3C_5H_5]^+$ . . . . Sb: Org.Comp.5-30
$C_{11}H_{14}O_6Sn$ . . . . . $c-C_5H_5Sn(OOCCH_3)_3$ . . . . Sn: Org.Comp.17-60, 61
$C_{11}H_{15}I_2InO_2$ . . . . $C_6H_5-CH_2-In(I)_2 \cdot 1,4-O_2C_4H_8$ . . . . In: Org.Comp.1-161, 163
$C_{11}H_{15}InMoO_3$ . . . $(CH_3)_3In \cdot (C_5H_5)Mo(CO)_3H$ . . . . In: Org.Comp.1-56
$C_{11}H_{15}InN_3^+$ . . . . . $[(CH_3)_2In(2-(NC_5H_4-2)-1,3-N_2C_3H_2-1-CH_3)]^+$ In: Org.Comp.1-223, 226
$C_{11}H_{15}InN_4O_3$ . . . . $[(CH_3)_2In(2-(NC_5H_4-2)-1,3-N_2C_3H_2-1-CH_3)][NO_3]$
In: Org.Comp.1-223, 226
$C_{11}H_{15}InOS$ . . . . . $(C_2H_5)_2InOC(S)-C_6H_5$ . . . . In: Org.Comp.1-205
$C_{11}H_{15}InO_3W$ . . . . $(CH_3)_3In \cdot (C_5H_5)W(CO)_3H$ . . . . In: Org.Comp.1-56
$C_{11}H_{15}Mo$ . . . . $C_5H_5Mo(C_3H_5)_2$ Mo:Org.Comp.6-183
$C_{11}H_{15}MoNO$ . . . . $C_5H_5Mo(NO)(H_2C=C(CH_3)C(CH_3)=CH_2)$ Mo:Org.Comp.6-186/7
$C_{11}H_{15}MoNO_2S$ . . $C_5H_5Mo(CO)(NO)CH_2=CHCH_2SC_2H_5$ Mo:Org.Comp.6-294
$C_{11}H_{15}MoNO_3$ . . . $C_5H_5Mo(CO)(NO)CH_2=C(CH_3)CH_2OCH_3$ Mo:Org.Comp.6-294
$C_{11}H_{15}MoNO_3S_4$ . . $C_5H_5Mo(NO)(SC(S)OC_2H_5)S_2COC_2H_5$ Mo:Org.Comp.6-49
$C_{11}H_{15}MoN_2O_2P$ . . $(C_5H_5)Mo(CO)_2=[2-(1,3,2-N_2PC_2H_4(CH_3)_2-1,3)]$
Mo:Org.Comp.7-32/4
$C_{11}H_{15}MoN_3O_2$ . . . $C_5H_5Mo(CO)(NO)=C(-N(CH_3)CH_2CH_2N(CH_3)-)$ Mo:Org.Comp.6-265
$C_{11}H_{15}MoO_2PS$ . . $(C_5H_5)Mo(CO)_2[-S=P(C_2H_5)_2-]$ Mo:Org.Comp.7-207, 212
$C_{11}H_{15}MoO_2PS_2$ . . $(C_5H_5)Mo(CO)_2[-S-P(C_2H_5)_2-S-]$ Mo:Org.Comp.7-199
$C_{11}H_{15}MoO_4PS_2$ . . $(C_5H_5)Mo(CO)_2[-S-P(O-C_2H_5)_2-S-]$ Mo:Org.Comp.7-189, 200
$C_{11}H_{15}MoO_4S_4^-$ . . $[(CH_2CHCH_2)Mo(CO)_2(SC(=S)-O-C_2H_5)_2]^-$ Mo:Org.Comp.5-283, 287, 292
$C_{11}H_{15}NOS$ . . . . . . $O=S=NC_6H_4(C_5H_{11}-s)-4$ . . . . S: S-N Comp.6-205
$C_{11}H_{15}NO_3S$ . . . . . $C_5H_5N-1-S(O)O-C_6H_{10}-2-O$ . . . . S: S-N Comp.8-322/3
$C_{11}H_{15}N_2O_5Re$ . . . $cis-(CO)_4Re[C(O)CH_3]C(CH_3)=NH-CH_2CH_2NHCH_3$
Re: Org.Comp.1-391/2
$C_{11}H_{15}N_2O_5Sb$ . . . $4-(CH_2)_5N-3-O_2N-C_6H_3Sb(O)(OH)_2$ . . . . Sb: Org.Comp.5-298

$C_{11}H_{15}N_2O_8Re$ . . . $(1,4\text{-}O_2C_4H_8)\text{-}1\text{-}Re(CO)_3[\text{-}NH_2\text{-}CH_2\text{-}C(O)\text{-}NH\text{-}CH_2\text{-}C(O)\text{-}O\text{-}]$ . . . Re: Org.Comp.1-135

$C_{11}H_{15}N_3Sn$ . . . $(CH_3)_3SnN(N{=}C(C_6H_5)CH{=}N)$ . . . Sn: Org.Comp.18-84, 91

$C_{11}H_{15}N_5ORe^+$ . . . $[(CO)Re(CNCH_3)_5]^+$ . . . Re: Org.Comp.2-290

$C_{11}H_{15}N_5Sn$ . . . $(CH_3)_3SnN{=}C(C(CN){=}C(CN)_2)N(CH_3)_2$ . . . Sn: Org.Comp.18-107/8

$C_{11}H_{15}N_6Re^+$ . . . $[Re(CNCH_3)_5CN]^+$ . . . Re: Org.Comp.2-284, 287

$C_{11}H_{15}OSb$ . . . $(CH_2)_5(C_6H_5)SbO$ . . . Sb: Org.Comp.5-83

$C_{11}H_{15}O_3Re$ . . . $(C_5H_5)Re(CO)_2[O(C_2H_5)_2]$ . . . Re: Org.Comp.3-186

$C_{11}H_{15}O_5PRe^+$ . . . $[(CO)_5ReP(C_2H_5)_3]^+$ . . . Re: Org.Comp.2-155

$C_{11}H_{16}IMoO_5P$ . . . $[(CH_3C_5H_4)Mo(CO)_2(I)(P(OCH_3)_3)]$ . . . Mo: Org.Comp.7-57, 86, 112, 113

$C_{11}H_{16}IO_3ReS_2$ . . . $cis\text{-}(CO)_3ReI(C_4H_8S)_2$ . . . Re: Org.Comp.1-287

$C_{11}H_{16}I_2O_4Sn$ . . . $ICH_2SnI(OC(CH_3){=}CHCOCH_3)_2$ . . . Sn: Org.Comp.17-171, 172

$C_{11}H_{16}I_3O_5Re$ . . . $(CO)_3Re(OC_4H_8)_2I_3$ . . . Re: Org.Comp.1-111

$C_{11}H_{16}MoN_2OS_2$ . . . $(C_5H_5)Mo(NO)(S_2CN(CH_3)_2)CH_2CH{=}CH_2$ . . . Mo: Org.Comp.6-84, 91/2

$C_{11}H_{16}MoN_2O_2Si$ $(C_5H_5)Mo(CO)_2\text{-}N{=}N\text{-}CH_2Si(CH_3)_3$ . . . Mo: Org.Comp.7-6, 15

$C_{11}H_{16}MoN_2O_3S_3$ $(C_5H_5)Mo(NO)(S_2CN(CH_3)_2)CH[\text{-}CH_2S(O)OCH_2\text{-}]$ Mo: Org.Comp.6-86

$C_{11}H_{16}NO_4Re$ . . . $(CO)_4Re[\text{-}CH_2\text{-}N(C_3H_7\text{-}n)_2\text{-}]$ . . . Re: Org.Comp.1-408

$C_{11}H_{16}NO_5Sb$ . . . $4\text{-}[n\text{-}C_4H_9\text{-}OC({=}O)\text{-}NH]\text{-}C_6H_4\text{-}Sb({=}O)(OH)_2$ . . Sb: Org.Comp.5-290

– . . . $4\text{-}[i\text{-}C_4H_9\text{-}OC({=}O)\text{-}NH]\text{-}C_6H_4\text{-}Sb({=}O)(OH)_2$ . . . Sb: Org.Comp.5-290

$C_{11}H_{16}N_2O_2SSn$ . . $(CH_3)_3SnN(CN)SO_2C_6H_4CH_3\text{-}4$ . . . Sn: Org.Comp.18-77/9

$C_{11}H_{16}N_2O_2S_2$ . . . $4\text{-}CH_3\text{-}C_6H_4\text{-}S(O)_2N{=}S{=}N\text{-}C_4H_9\text{-}t$ . . . S: S-N Comp.7-61

$C_{11}H_{16}N_4OSn$ . . . $(CH_3)_3SnN(N{=}NC(C_6H_4OCH_3\text{-}4){=}N)$ . . . Sn: Org.Comp.18-84, 93

$C_{11}H_{17}INO_2Re$ . . . $[N(CH_3)_4][(C_5H_5)Re(CO)_2I]$ . . . Re: Org.Comp.3-171, 173

$C_{11}H_{17}IO_4Sn$ . . . $CH_3SnI(OC(CH_3){=}CHCOCH_3)_2$ . . . Sn: Org.Comp.17-170, 172

$C_{11}H_{17}I_2MoN$ . . . $(C_5H_5)Mo(I)_2C({=}N(CH_3)_2){=}C(CH_3)_2$ . . . Mo: Org.Comp.6-64

$C_{11}H_{17}MnNO_2S_2$ . . $[Mn((SC({=}S))NC_5H_{10})OC(CH_3)CHC(CH_3)O]$ . . . Mn: MVol.D7-168

$C_{11}H_{17}MnN_2O_3S^+$ $[Mn((OOC(CH_2)_5)(O{=})C_5H_7N_2S)]^+$ . . . Mn: MVol.D7-227

$C_{11}H_{17}MnN_3O_3S^{2+}$ $[Mn(4\text{-}NH_2\text{-}C_6H_4\text{-}SO_2\text{-}NH\text{-}C(O)NH\text{-}C_4H_9\text{-}n)]^{2+}$ Mn: MVol.D7-116/23

$C_{11}H_{17}MoN_2O_2P$ . . $(C_5H_5)Mo(CO)_2{=}P(N(CH_3)_2)_2$ . . . Mo: Org.Comp.7-31/2

$C_{11}H_{17}MoN_2O_2PS$ $(C_5H_5)Mo(CO)_2[\text{-}S{=}P(N(CH_3)_2)_2\text{-}]$ . . . Mo: Org.Comp.7-208, 212/3

$C_{11}H_{17}MoN_2O_2S_4^-$ $[(CH_2CHCH_2)Mo(CO)_2(SC({=}S)\text{-}NH\text{-}C_2H_5)_2]^-$ . . . Mo: Org.Comp.5-283, 287/8

$C_{11}H_{17}MoO_2P$ . . . $(C_5H_5)Mo(CO)_2[P(CH_3)_3]CH_3$ . . . Mo: Org.Comp.8-77, 79, 99

$C_{11}H_{17}MoO_2S_2^+$ . . $[(C_5H_5)Mo(CO)_2(CH_3\text{-}S\text{-}CH_3)_2]^+$ . . . Mo: Org.Comp.7-283, 296

$C_{11}H_{17}MoO_2Se_2^+$ $[(C_5H_5)Mo(CO)_2(CH_3\text{-}Se\text{-}CH_3)_2]^+$ . . . Mo: Org.Comp.7-283, 296

$C_{11}H_{17}MoO_2Te_2^+$ $[(C_5H_5)Mo(CO)_2(CH_3\text{-}Te\text{-}CH_3)_2]^+$ . . . Mo: Org.Comp.7-283, 296

$C_{11}H_{17}MoO_5P$ . . . $(C_5H_5)Mo(CO)_2[P(OCH_3)_3]CH_3$ . . . Mo: Org.Comp.8-77/8, 81

$C_{11}H_{17}NOSn$ . . . $(CH_3)_3SnN(CH_3)COC_6H_5$ . . . Sn: Org.Comp.18-48, 52

$C_{11}H_{17}NO_2S$ . . . $C_6H_5\text{-}N(C_2H_5)\text{-}S(O)O\text{-}C_3H_7\text{-}n$ . . . S: S-N Comp.8-320/1

$C_{11}H_{17}NO_3Sn$ . . . $c\text{-}C_5H_5Sn(OCH_2CH_2)_3N$ . . . Sn: Org.Comp.17-60, 61

$C_{11}H_{17}NO_5PRe$ . . . $(CO)_4Re[P(C_2H_5)_3]C(O)NH_2$ . . . Re: Org.Comp.1-469

$C_{11}H_{17}N_2O_3Sb$ . . . $3\text{-}H_2N\text{-}4\text{-}(CH_2)_5NC_6H_3Sb(O)(OH)_2$ . . . Sb: Org.Comp.5-297

$C_{11}H_{17}N_2O_5Re$ . . . $(CO)_4Re(NH_2C_3H_7\text{-}i)C(O)NHC_3H_7\text{-}i$ . . . Re: Org.Comp.1-470

$C_{11}H_{17}N_2O_5Sb$ . . . $2\text{-}(n\text{-}C_5H_{11}\text{-}NH)\text{-}5\text{-}O_2N\text{-}C_6H_3\text{-}Sb({=}O)(OH)_2$ . . Sb: Org.Comp.5-298

– . . . $2\text{-}(i\text{-}C_5H_{11}\text{-}NH)\text{-}5\text{-}O_2N\text{-}C_6H_3\text{-}Sb({=}O)(OH)_2$ . . . Sb: Org.Comp.5-298

$C_{11}H_{17}N_3OS_2$ . . . $((CH_3)_2N)_2SO \cdot C_6H_5\text{-}NCS$ . . . S: S-N Comp.8-342

$C_{11}H_{17}N_3O_2S$	$((CH_3)_2N)_2SO \cdot C_6H_5$-NCO	S: S-N Comp.8-342
$C_{11}H_{17}O_2ReSi$	$(C_5H_5)Re(H)(CO)_2$-$SiH(C_2H_5)_2$	Re: Org.Comp.3-175/6, 178
$C_{11}H_{17}O_3Sb$	$(CH_3)_3Sb(OCH_3)$-$OC(=O)C_6H_5$	Sb: Org.Comp.5-45
–	$(CH_3)_3Sb(OCH_3)$-O-C_6H_4CHO-2	Sb: Org.Comp.5-46
$C_{11}H_{17}O_4Sb$	$(CH_3)Sb(OC_2H_5)_2$(1-OC_6H_4O-2)	Sb: Org.Comp.5-312
$C_{11}H_{17}O_5ReSi_2$	$(CO)_5ReCH_2Si(CH_3)_2Si(CH_3)_3$	Re: Org.Comp.2-106
$C_{11}H_{17}PSn$	1-C_6H_5-2,2-$(CH_3)_2$-1,2-$PSnC_3H_6$	Sn: Org.Comp.19-202, 206
$C_{11}H_{18}IMoNO$	$C_5(CH_3)_5Mo(NO)(I)CH_3$	Mo: Org.Comp.6-82
$C_{11}H_{18}InN$	$(CH_3)_2$In-CH_2-C_6H_4-2-$N(CH_3)_2$	In: Org.Comp.1-101, 107
–	$(CH_3)_2$In-C_6H_4-2-[CH_2-$N(CH_3)_2$]	In: Org.Comp.1-101, 107
$C_{11}H_{18}InNS_2$	CH_3-In[1,2-$(S)_2$-C_6H_3-4-CH_3] $\cdot$ $N(CH_3)_3$	In: Org.Comp.1-246, 249, 250/1
$C_{11}H_{18}InN_4O^+$	$[(CH_3)_2In((1$-CH_3-1,3-$N_2C_3H_2$-2$)_2CHOH)]^+$	In: Org.Comp.1-223, 226
$C_{11}H_{18}InN_5O_4$	$[(CH_3)_2In((1$-CH_3-1,3-$N_2C_3H_2$-2$)_2CHOH)][NO_3]$	In: Org.Comp.1-223, 226
$C_{11}H_{18}MoN_2O_3S_3$	$(C_5H_5)Mo(NO)(OS(O)C_3H_7)S_2CN(CH_3)_2$	Mo: Org.Comp.6-50
$C_{11}H_{18}MoN_3O_2{}^+$	$[(C_5H_5)Mo(CO)(NO)H_2C{=}CHCH_2N(CH_3)_2NH_2]^+$	Mo: Org.Comp.6-304
$C_{11}H_{18}MoN_4O_3S_2$	$(CH_2CHCH_2)Mo(CO)_2[NH_2$-N=$C(CH_3)C(CH_3)$=N-$NH_2]$-SC(S)-$OCH_3$	Mo: Org.Comp.5-250, 257
$C_{11}H_{18}N_2O_3SSn$	$(CH_3)_3SnN(COCH_3)SO_2C_6H_4NH_2$-2	Sn: Org.Comp.18-58, 68
$C_{11}H_{18}N_2SSi$	$(CH_3)_3$SiN=S=N-$C_6H_3(CH_3)_2$-2,6	S: S-N Comp.7-141/2
$C_{11}H_{18}N_2SSn$	$(CH_3)_3$SnN=S=N-$C_6H_3(CH_3)_2$-2,6	S: S-N Comp.7-174
$C_{11}H_{19}IMoNO_3PS_2$	$(C_5H_5)Mo(NO)(I)S_2P(OC_3H_7$-i$)_2$	Mo: Org.Comp.6-48
$C_{11}H_{19}Mn_2N_2S_8{}^{3-}$	$[Mn_2(SCH_2CH_2S)_4(C_3H_3N_2$-1,3$)]^{3-}$	Mn: MVol.D7-46/7
$C_{11}H_{19}MoNOS_2$	$(C_5H_5)Mo(NO)(SC_3H_7$-n$)_2$	Mo: Org.Comp.6-31
–	$(C_5H_5)Mo(NO)(SC_3H_7$-i$)_2$	Mo: Org.Comp.6-31
$C_{11}H_{19}NOSSn$	$(CH_3)_3SnN(CH_3)S(=O)C_6H_4CH_3$-4	Sn: Org.Comp.18-58
$C_{11}H_{19}NO_2Sn$	$(CH_3)_3SnNC_8H_{10}(O)_2$	Sn: Org.Comp.18-84, 88
$C_{11}H_{19}N_3O_4Sn$	$(CH_3)_3SnN(N{=}C(COOC_2H_5)C(COOC_2H_5){=}N)$	Sn: Org.Comp.18-84, 91
$C_{11}H_{20}IMoN_7O$	$[(CH_3$-$NH)_2C{=}Mo(CN$-$CH_3)_4(NO)]I$	Mo: Org.Comp.5-118/9
$C_{11}H_{20}IO_3ReS_2$	$(CO)_3ReI[S(C_2H_5)_2]_2$	Re: Org.Comp.1-287
$C_{11}H_{20}IO_3ReSe_2$	$(CO)_3ReI[Se(C_2H_5)_2]_2$	Re: Org.Comp.1-288
$C_{11}H_{20}MoN_7O^+$	$[(CH_3$-$NH)_2C{=}Mo(CN$-$CH_3)_4(NO)]I$	Mo: Org.Comp.5-118/9
$C_{11}H_{20}NPSSn$	$(CH_3)_3SnN(C_6H_5)P(=S)(CH_3)_2$	Sn: Org.Comp.18-58, 67
$C_{11}H_{20}NPSn$	$(CH_3)_3SnN(C_6H_5)P(CH_3)_2$	Sn: Org.Comp.18-58, 66
$C_{11}H_{20}N_2OSSn$	$(CH_3)_2$N-S(=O)-$N(C_6H_5)$-$Sn(CH_3)_3$	S: S-N Comp.8-358 Sn: Org.Comp.18-58, 66
$C_{11}H_{20}N_2O_2S_3$	S=S=NS-$NC_7H_8O_2(CH_3)_4$	S: S-N Comp.6-314/7
$C_{11}H_{20}N_2Sn$	$(C_2H_5)_3$SnN=C=C(CN)C_2H_5	Sn: Org.Comp.18-152/3
$C_{11}H_{20}O_4Sn$	$(C_4H_9)_2SnOOCCH_2COO$	Sn: Org.Comp.15-322
$C_{11}H_{20}O_7P_2Re^+$	$[(C_5H_5)Re(CO)(P(OCH_3)_3)$-$P(=O)(OCH_3)_2]Br$	Re: Org.Comp.3-143, 146
$C_{11}H_{21}InSi_2$	$In[C_5H_3(Si(CH_3)_3)_2]$	In: Org.Comp.1-379, 380
$C_{11}H_{21}MnN_2OPS_2$	$[Mn((C_3H_7)_3P{=}O)(NCS)_2]$	Mn: MVol.D8-87
$C_{11}H_{21}MnN_2O_2PS_2$	$Mn[P(C_3H_7)_3](O_2)(NCS)_2$	Mn: MVol.D8-56
$C_{11}H_{21}MnN_2O_5Si$	$[SiH_3(N(CH_3)_3)_2][Mn(CO)_5]$ = $SiH_3Mn(CO)_5 \cdot 2\ N(CH_3)_3$	Si: SVol.B4-324/5
$C_{11}H_{21}MnN_2PS_2$	$Mn[P(C_3H_7)_3](NCS)_2$	Mn: MVol.D8-38

$C_{11}H_{21}MoOP_2^+$	$[C_5H_5Mo(P(CH_3)_2CH_2CH_2P(CH_3)_2)O]^+$	Mo: Org.Comp.6-19
$C_{11}H_{21}NOPRe$	$(C_5H_5)ReH(NO)[P(C_2H_5)_3]$	Re: Org.Comp.3-15, 18, 24
$C_{11}H_{21}NOSSn$	$(CH_3)_3SnC_8H_{12}NOS$	Sn: Org.Comp.18-84, 93, 103
$C_{11}H_{21}NO_3Sn$	$(C_2H_5)_3SnN(C(=O)CH(OCH_3)CH_2C(=O))$	Sn: Org.Comp.18-142/4
$C_{11}H_{21}N_3O_4S$	$(1,4\text{-}ONC_4H_8\text{-}4)_2S=N\text{-}C(O)OC_2H_5$	S: S-N Comp.8-207/8, 213
$C_{11}H_{21}N_3O_5S_2$	$CH_3S\text{-}C(CH_3)=N\text{-}OC(O)\text{-}N(CH_3)\text{-}S(O)$ $\text{-}N(C_2H_5)\text{-}C(O)O\text{-}C_3H_7\text{-}n$	S: S-N Comp.8-351
$C_{11}H_{21}O_4P_2ReS$	fac-$(CO)_3Re[P(CH_3)_3]_2SC(O)CH_3$	Re: Org.Comp.1-257
$C_{11}H_{21}PSn$	1-$(CH_3)_3Sn$-2,3,4,5-$(CH_3)_4$-PC_4	Sn: Org.Comp.19-177/8
$C_{11}H_{22}N_2O_2Sn$	$(CH_3)_3SnN(CH_3)COON=C_6H_{10}\text{-}c$	Sn: Org.Comp.18-47/8
$C_{11}H_{22}N_3O_5PS_3$	$CH_3S\text{-}C(CH_3)=N\text{-}OC(O)\text{-}N(CH_3)\text{-}S(O)$ $\text{-}N(CH_3)\text{-}P(S)(\text{-}OCH_2C(CH_3)_2CH_2O\text{-})$	S: S-N Comp.8-356
$C_{11}H_{22}N_{10}O_{16}Th$	$[C(NH_2)_3]_3[NH_4][Th(C_2O_4)_4] \cdot 3\,H_2O$	Th: SVol.C7-88, 94
$C_{11}H_{22}O_3Sn$	$(C_4H_9)_2SnOOCCH(CH_3)O$	Sn: Org.Comp.15-331
$C_{11}H_{22}O_5Sn$	$C_4H_9Sn(OOCCH_3)_2OC_3H_7\text{-}i$	Sn: Org.Comp.17-73
$C_{11}H_{23}IMoNOP_2^+$	$[C_5H_5Mo(NO)(P(CH_3)_3)_2I]^+$	Mo: Org.Comp.6-38
$C_{11}H_{23}IMoNO_7P_2^+$	$[C_5H_5Mo(NO)(P(OCH_3)_3)_2I]^+$	Mo: Org.Comp.6-39
$C_{11}H_{23}IMoO_8P_2$	$(CH_2CHCH_2)Mo(I)(CO)_2[P(OCH_3)_3]_2$	Mo: Org.Comp.5-271, 272, 273
$C_{11}H_{23}I_2MoNOP_2$	$[C_5H_5Mo(NO)(P(CH_3)_3)_2I]I$	Mo: Org.Comp.6-38
$C_{11}H_{23}I_2MoNO_7P_2$	$[C_5H_5Mo(NO)(P(OCH_3)_3)_2I]I$	Mo: Org.Comp.6-39
$C_{11}H_{23}I_2MoN_4OP$	$C_5H_5Mo(NO)(P(N(CH_3)_2)_3)I_2$	Mo: Org.Comp.6-35
$C_{11}H_{23}InO_2$	$(n\text{-}C_4H_9)_2InOC(O)\text{-}C_2H_5$	In: Org.Comp.1-196, 199, 203
$C_{11}H_{23}MnN_4O_2S^+$	$[Mn(4\text{-}CH_3C_6H_4SO_2)(NH_2CH_2CH_2NH_2)_2]^+$	Mn: MVol.D7-109/11
$C_{11}H_{23}MoNOP_2$	$C_5H_5Mo(NO)(P(CH_3)_3)_2$	Mo: Org.Comp.6-40
$C_{11}H_{23}MoNO_7P_2$	$C_5H_5Mo(NO)(P(OCH_3)_3)_2$	Mo: Org.Comp.6-41, 44
$C_{11}H_{23}MoOP_2^+$	$[C_5H_5Mo(P(CH_3)_3)_2O]^+$	Mo: Org.Comp.6-14
$C_{11}H_{23}N_3O_2Sn$	$(C_2H_5)_2Sn[\text{-}N(CH_3)\text{-}C(=O)\text{-}CH_2\text{-}N(CH_3)$ $\text{-}CH_2\text{-}C(=O)\text{-}N(CH_3)\text{-}]$	Sn: Org.Comp.19-83
$C_{11}H_{24}I_2MoP_2$	$C_5H_5Mo(P(CH_3)_3)_2I_2H$	Mo: Org.Comp.6-13/4
$C_{11}H_{24}LiP_2Re$	$(C_5H_5)Re[P(CH_3)_3]_2(Li)H$	Re: Org.Comp.3-9
$C_{11}H_{24}MoP_2$	$C_5H_5Mo(P(CH_3)_2CH_2CH_2P(CH_3)_2)H_3$	Mo: Org.Comp.6-16, 20/1
$C_{11}H_{24}N_3O_4PS_3$	$CH_3S\text{-}C(CH_3)=N\text{-}OC(O)\text{-}N(CH_3)\text{-}S(O)$ $\text{-}N(CH_3)\text{-}P(O)(CH_3)\text{-}SC_4H_9\text{-}t$	S: S-N Comp.8-356
$C_{11}H_{24}N_3O_5PS_3$	$CH_3S\text{-}C(CH_3)=N\text{-}OC(O)\text{-}N(CH_3)\text{-}S(O)$ $\text{-}N(C_2H_5)\text{-}P(S)(OC_2H_5)_2$	S: S-N Comp.8-356
–	$CH_3S\text{-}C(CH_3)=N\text{-}OC(O)\text{-}N(CH_3)\text{-}S(O)$ $\text{-}N(C_4H_9\text{-}s)\text{-}P(S)(OCH_3)_2$	S: S-N Comp.8-356
$C_{11}H_{24}OSn$	$(C_4H_9)_2Sn(H)OCH_2CH=CH_2$	Sn: Org.Comp.17-78
$C_{11}H_{24}O_2Sn$	$(C_4H_9)_2Sn[\text{-}O\text{-}CH(CH_3)\text{-}CH_2\text{-}O\text{-}]$	Sn: Org.Comp.15-49, 80/2
–	$(C_4H_9)_2Sn[\text{-}O\text{-}CH_2CH_2CH_2\text{-}O\text{-}]$	Sn: Org.Comp.15-53/4, 85
$C_{11}H_{24}O_3P_3ReS$	$(CO)_3Re[P(CH_3)_3]_2SP(CH_3)_2$	Re: Org.Comp.1-257
$C_{11}H_{24}O_3P_3ReSSe$	$(CO)_3Re[P(CH_3)_3]_2SP(Se)(CH_3)_2$	Re: Org.Comp.1-258
$C_{11}H_{24}O_3P_3ReS_2$	$(CO)_3Re[P(CH_3)_3]_2SP(S)(CH_3)_2$	Re: Org.Comp.1-257/8
$C_{11}H_{24}O_3Sn$	$(C_4H_9)_2Sn(OCH_3)\text{-}OC(=O)CH_3$	Sn: Org.Comp.16-188
–	$(C_4H_9)_2Sn[\text{-}OO\text{-}CH(C_2H_5)\text{-}O\text{-}]$	Sn: Org.Comp.15-334, 336
$C_{11}H_{25}In$	$CH_3\text{-}In(CH_2\text{-}C_4H_9\text{-}t)_2$	In: Org.Comp.1-101, 102/3
$C_{11}H_{25}InN_2$	$CH_3\text{-}In[\text{-}CH_2CH_2CH_2\text{-}N(CH_3)\text{-}CH_2$ $CH_2\text{-}N(CH_3)\text{-}CH_2CH_2CH_2\text{-}]$	In: Org.Comp.1-101, 110

$C_{11}H_{25}InO$	$(t\text{-}C_4H_9\text{-}CH_2)_2In\text{-}OCH_3$	In: Org.Comp.1-183
$C_{11}H_{25}InS$	$(n\text{-}C_4H_9)_2In\text{-}S\text{-}C_3H_7\text{-}n$	In: Org.Comp.1-240, 241, 244
–	$(n\text{-}C_4H_9)_2In\text{-}S\text{-}C_3H_7\text{-}i$	In: Org.Comp.1-240, 242, 244
–	$(i\text{-}C_4H_9)_2In\text{-}S\text{-}C_3H_7\text{-}n$	In: Org.Comp.1-240, 242, 244
–	$(i\text{-}C_4H_9)_2In\text{-}S\text{-}C_3H_7\text{-}i$	In: Org.Comp.1-240, 242, 244
$C_{11}H_{25}MoP_2^+$	$[C_5H_5Mo(P(CH_3)_2CH_2CH_2P(CH_3)_2)H_4]^+$	Mo: Org.Comp.6-16, 20
$C_{11}H_{25}NOSn$	$2,2\text{-}(n\text{-}C_4H_9)_2\text{-}3\text{-}CH_3\text{-}1,3,2\text{-}ONSnC_2H_4$	Sn: Org.Comp.19-129, 135
–	$(C_2H_5)_3Sn\text{-}N(C_4H_9\text{-}n)\text{-}CHO$	Sn: Org.Comp.18-137/8
–	$(C_2H_5)_3Sn\text{-}N(C_4H_9\text{-}t)\text{-}CHO$	Sn: Org.Comp.18-137/8
$C_{11}H_{25}NO_2Os$	$(CH_3)_3Os(N)(OC_4H_8)_2$	Os: Org.Comp.A1-26
$C_{11}H_{25}NO_3Sn$	$C_4H_9Sn((OCH_2CH_2\text{-})_2NH)OC_3H_7\text{-}i$	Sn: Org.Comp.17-74
$C_{11}H_{25}NSn$	$1\text{-}(C_2H_5)_3Sn\text{-}NC_5H_{10}$	Sn: Org.Comp.18-142/3
–	$(CH_3)_3Sn\text{-}N(C_4H_9\text{-}i)\text{-}CH{=}C(CH_3)_2$	Sn: Org.Comp.18-47, 49, 53
$C_{11}H_{25}P_2Re$	$(C_5H_5)ReH_2[P(CH_3)_3]_2$	Re: Org.Comp.3-6, 7, 10
$C_{11}H_{26}InN$	$(C_2H_5)_2In\text{-}CH_2CH_2CH_2\text{-}N(C_2H_5)_2$	In: Org.Comp.1-101, 108
–	$(C_2H_5)_3In \cdot NC_5H_{11}$	In: Org.Comp.1-71
–	$(n\text{-}C_3H_7)_2In\text{-}CH_2CH_2CH_2\text{-}N(CH_3)_2$	In: Org.Comp.1-101, 109
–	$(i\text{-}C_3H_7)_2In\text{-}CH_2CH_2CH_2\text{-}N(CH_3)_2$	In: Org.Comp.1-101, 109
$C_{11}H_{26}O_2Sn$	$(C_4H_9)_2Sn(OCH_3)OC_2H_5$	Sn: Org.Comp.16-187
$C_{11}H_{26}O_3Sn$	$C_2H_5Sn(OC_3H_7\text{-}i)_3$	Sn: Org.Comp.17-25, 28
$C_{11}H_{27}IN_2SiSn$	$[\text{-}N(C_4H_9\text{-}t)\text{-}Si(CH_3)_2\text{-}N(C_4H_9\text{-}t)\text{-}]Sn(I)\text{-}CH_3$	Sn: Org.Comp.19-149, 152
$C_{11}H_{27}InN_2$	$CH_3\text{-}In[CH_2CH_2CH_2\text{-}N(CH_3)_2]_2$	In: Org.Comp.1-110
$C_{11}H_{27}InN_2Si$	$CH_3\text{-}In[\text{-}N(C_4H_9\text{-}t)\text{-}Si(CH_3)_2\text{-}N(C_4H_9\text{-}t)\text{-}]$	In: Org.Comp.1-310/1
$C_{11}H_{27}MnN_3O_3PSi$	$Mn(NO)_3[P(C_4H_9\text{-}t)_2Si(CH_3)_3]$	Mn: MVol.D8-73/4
$C_{11}H_{27}MnN_3O_3PSn$	$(CH_3)_3Sn\text{-}P(C_4H_9\text{-}t)_2Mn(NO)_3$	Mn: MVol.D8-73/4
		Sn: Org.Comp.19-181/3
$C_{11}H_{27}MnN_7O_2P_2S_2$	$[Mn((O{=}P(N(CH_3)_2)_2)_2NCH_3)(NCS)_2]$	Mn: MVol.D8-180
$C_{11}H_{27}NO_2Sn$	$(CH_3)_2Sn(O\text{-}C_3H_7\text{-}i)\text{-}O\text{-}CH_2CH_2\text{-}N(C_2H_5)_2$	Sn: Org.Comp.16-173
–	$(C_2H_5)_2Sn(O\text{-}C_3H_7\text{-}i)\text{-}O\text{-}CH_2CH_2\text{-}N(CH_3)_2$	Sn: Org.Comp.16-181
–	$(C_2H_5)_2Sn(O\text{-}C_3H_7\text{-}i)\text{-}O\text{-}N(C_2H_5)_2$	Sn: Org.Comp.16-181
$C_{11}H_{27}NSn$	$(CH_3)_3Sn\text{-}N(C_4H_9\text{-}n)_2$	Sn: Org.Comp.18-23/4
–	$(n\text{-}C_3H_7)_3Sn\text{-}N(CH_3)_2$	Sn: Org.Comp.18-156/7
$C_{11}H_{27}N_2PSn$	$(CH_3)_3Sn[\text{-}N(C_4H_9\text{-}t)\text{-}P\text{-}N(C_4H_9\text{-}t)\text{-}]$	Sn: Org.Comp.18-129
		Sn: Org.Comp.19-54, 58
$C_{11}H_{27}N_2Re$	$(CH_3)_3Re({=}N\text{-}C_4H_9\text{-}t)_2$	Re: Org.Comp.1-21
$C_{11}H_{27}N_2SSi$	$(CH_3)_3SiN(C_4H_9\text{-}t)\text{-}SN\text{-}C_4H_9\text{-}t$, radical	S: S-N Comp.7-331/2
$C_{11}H_{27}O_3Sb$	$(C_4H_9)_2Sb(OCH_3)_3$	Sb: Org.Comp.5-179
$C_{11}H_{27}PSn$	$(CH_3)_3Sn\text{-}P(C_4H_9\text{-}t)_2$	Sn: Org.Comp.19-165/7
$C_{11}H_{27}P_3Sn$	$3\text{-}(CH_3)_3Sn\text{-}P_3\text{-}1,2\text{-}(C_4H_9\text{-}t)_2$	Sn: Org.Comp.19-178/9
$C_{11}H_{27}SbSn$	$(CH_3)_3Sn\text{-}Sb(C_4H_9\text{-}n)_2$	Sn: Org.Comp.19-243/6
–	$(CH_3)_3Sn\text{-}Sb(C_4H_9\text{-}t)_2$	Sn: Org.Comp.19-243/6
$C_{11}H_{28}NPSSn$	$(CH_3)_3SnNHP({=}S)(C_4H_9\text{-}t)_2$	Sn: Org.Comp.18-20/1
$C_{11}H_{28}NPSn$	$(CH_3)_3SnNHP(C_4H_9\text{-}t)_2$	Sn: Org.Comp.18-16, 19
$C_{11}H_{28}N_2SSn$	$(CH_3)_3SnN(C_4H_9\text{-}t)SNHC_4H_9\text{-}t$	Sn: Org.Comp.18-62, 73
$C_{11}H_{28}O_2P_3ReS$	$(CO)_2Re[P(CH_3)_3]_3SH$	Re: Org.Comp.1-97
$C_{11}H_{29}NOsS_2Si_2$	$[(CH_3)_3SiCH_2]_2Os(N)(S\text{-}CH_2CH_2\text{-}SCH_3)$	Os: Org.Comp.A1-21
$C_{11}H_{30}NPSiSn$	$(CH_3)_3Sn\text{-}P(CH_3)(C_4H_9\text{-}t){=}N\text{-}Si(CH_3)_3$	Sn: Org.Comp.19-180

$C_{11}H_{30}N_2SSiSn$	$(CH_3)_3SnN(C_4H_9\text{-}t)S(CH_3){=}NSi(CH_3)_3$	Sn: Org.Comp.18-57, 62
$C_{11}H_{30}N_2Si_2Sn$	1,3-$[(CH_3)_3Si]_2$-2,2-$(CH_3)_2$-1,3,2-$N_2SnC_3H_6$	Sn: Org.Comp.19-76
$C_{11}H_{32}In_2N_2$	$(CH_3)_3In \cdot N(CH_3)_2\text{-}CH_2\text{-}N(CH_3)_2 \cdot In(CH_3)_3$	In: Org.Comp.1-49
$C_{11}H_{33}InNOPSi_2$	$(CH_3)_3In \cdot (CH_3)_3Si\text{-}O\text{-}Si(CH_3)_2\text{-}N{=}P(CH_3)_3$	In: Org.Comp.1-27, 38
$C_{11}H_{33}InN_2P_2Si$	$(CH_3)_3In \cdot (CH_3)_3P{=}N\text{-}Si(CH_3)_2\text{-}N{=}P(CH_3)_3$	In: Org.Comp.1-27, 38, 44
$C_{11}H_{33}NO_3Si_4Sn$	$(CH_3)_3SnN(Si(CH_3)_2OSi(CH_3)_2)_2O$	Sn: Org.Comp.18-84, 93, 103
$C_{11}H_{33}N_3Si_2Sn$	$[(CH_3)_2N]_2Sn(CH_3)\text{-}N[Si(CH_3)_3]_2$	Sn: Org.Comp.19-116
$C_{11}H_{36}N_{18}O_{15}Th$	$[C(NH_2)_3]_6[Th(CO_3)_5]$	Th: SVol.C7-6
–	$[C(NH_2)_3]_6[Th(CO_3)_5] \cdot H_2O$	Th: SVol.C7-6, 10
–	$[C(NH_2)_3]_6[Th(CO_3)_5] \cdot 3\ H_2O$	Th: SVol.C7-6, 10/1
–	$[C(NH_2)_3]_6[Th(CO_3)_5] \cdot 4\ H_2O$	Th: SVol.C7-6, 8, 10/1
$C_{11}H_{55}N_{11}Si_{10}$	$CH_3HN(SiH_2N(CH_3))_{10}H$	Si: SVol.B4-260
$C_{11}K_8O_{27}Th_2$	$K_8Th_2(C_2O_4)_3(CO_3)_5 \cdot 13\ H_2O$	Th: SVol.C7-99, 101
–	$K_8Th_2(C_2O_4)_3(CO_3)_5 \cdot 16\ H_2O$	Th: SVol.C7-99, 100
$C_{11}K_{10}O_{29}Th_2$	$K_{10}Th_2(C_2O_4)_2(CO_3)_7 \cdot 8\ H_2O$	Th: SVol.C7-99, 100/1
–	$K_{10}Th_2(C_2O_4)_2(CO_3)_7 \cdot 12\ H_2O$	Th: SVol.C7-99, 100/1
–	$K_{10}Th_2(C_2O_4)_2(CO_3)_7 \cdot 14\ H_2O$	Th: SVol.C7-99, 100
$C_{11}O_{11}Re_2$	$[Re(CO)_6][Re(CO)_5]$	Re: Org.Comp.2-219
$C_{12}ClCuH_{16}S_2Ti$	$[(C_5H_5)_2Ti((SCH_3)_2CuCl)]_n$	Ti: Org.Comp.5-346/7
$C_{12}ClF_5H_4N_2S$	4-Cl-C_6H_4-N=S=N-C_6F_5	S: S-N Comp.7-260
$C_{12}ClF_6FeH_5O_2$	$(C_5H_5)Fe(CO)_2C({=}CClCF_2CF_2CF_2\text{-})$	Fe: Org.Comp.B13-194, 213, 246
$C_{12}ClF_6FeH_9NP$	[(2-Cl-C_6H_4-CN)$Fe(C_5H_5)][PF_6]$	Fe: Org.Comp.B19-3, 43
–	[(3-Cl-C_6H_4-CN)$Fe(C_5H_5)][PF_6]$	Fe: Org.Comp.B19-3, 43
–	[(4-Cl-C_6H_4-CN)$Fe(C_5H_5)][PF_6]$	Fe: Org.Comp.B19-3, 43
–	$[(Cl\text{-}C_6H_5)Fe(C_5H_4\text{-}CN)][PF_6]$	Fe: Org.Comp.B18-142/6, 247
$C_{12}ClF_6FeH_9NaO_2P$	[(4-Cl-C_6H_4-COO)$Fe(C_5H_5)] \cdot Na[PF_6]$	Fe: Org.Comp.B19-43/4
$C_{12}ClF_6FeH_{10}O_2P$	[(3-Cl-C_6H_4-COOH)$Fe(C_5H_5)][PF_6]$	Fe: Org.Comp.B19-5, 44
–	[(4-Cl-C_6H_4-COOH)$Fe(C_5H_5)][PF_6]$	Fe: Org.Comp.B19-5, 44
–	$[(Cl\text{-}C_6H_5)Fe(C_5H_4\text{-}COOH)][PF_6]$	Fe: Org.Comp.B18-142/6, 247
$C_{12}ClF_6FeH_{10}PS$	[(1-Cl-C_7H_5=S-7)$Fe(C_5H_5)][PF_6]$	Fe: Org.Comp.B19-340, 342
$C_{12}ClF_6FeH_{11}NOP$	[(3-Cl-C_6H_4-C(=O)-NH_2)$Fe(C_5H_5)][PF_6]$	Fe: Org.Comp.B19-45
–	[(4-Cl-C_6H_4-C(=O)-NH_2)$Fe(C_5H_5)][PF_6]$	Fe: Org.Comp.B19-45
–	$[(Cl\text{-}C_6H_5)Fe(C_5H_4\text{-}C({=}O)\text{-}NH_2)][PF_6]$	Fe: Org.Comp.B18-142/6, 247
$C_{12}ClF_6FeH_{11}NO_2P$	[(2-Cl-C_6H_4-CH_2-NO_2)$Fe(C_5H_5)][PF_6]$	Fe: Org.Comp.B19-3, 37
$C_{12}ClF_6FeH_{12}OP$	[(2-Cl-C_6H_4-OCH_3)$Fe(C_5H_5)][PF_6]$	Fe: Org.Comp.B19-68
–	[(3-Cl-C_6H_4-OCH_3)$Fe(C_5H_5)][PF_6]$	Fe: Org.Comp.B19-1, 3, 5, 68
–	[(4-Cl-C_6H_4-OCH_3)$Fe(C_5H_5)][PF_6]$	Fe: Org.Comp.B19-1, 3, 5, 68
$C_{12}ClF_6FeH_{12}P$	[(2-Cl-C_6H_4-CH_3)$Fe(C_5H_5)][PF_6]$	Fe: Org.Comp.B19-1, 4, 5, 33, 90/2
–	[(3-Cl-C_6H_4-CH_3)$Fe(C_5H_5)][PF_6]$	Fe: Org.Comp.B19-1, 4/6, 33, 90/2
–	[(4-Cl-C_6H_4-CH_3)$Fe(C_5H_5)][PF_6]$	Fe: Org.Comp.B19-1, 4/6, 33/4, 90/2
$C_{12}ClF_6FeH_{13}NP$	[(2-Cl-C_6H_4-NH-CH_3)$Fe(C_5H_5)][PF_6]$	Fe: Org.Comp.B19-3, 71
$C_{12}ClF_6H_{14}MoP$	$(C_5H_5)Mo[P(CH_3)_3](Cl)(F_3C\text{-}C{\equiv}C\text{-}CF_3)$	Mo: Org.Comp.6-139
$C_{12}ClF_7H_{18}O_2Sn$	$(C_4H_9)_2Sn(Cl)OOCC_3F_7$	Sn: Org.Comp.17-104
$C_{12}ClF_9HN$	4-Cl-C_6F_4-NH-C_6F_5	F: PFHOrg.SVol.5-16, 47, 84

$C_{12}ClF_9HNO$ . . . . . 2,4-C_6F_4(=O)-1-(NH-C_6F_5)-3-Cl-6 . . . . F: PFHOrg.SVol.5-15/6, 51
$C_{12}ClF_9N^-$ . . . . [N(C_6F_5)-C_6F_4-Cl-4]$^-$ . . . . F: PFHOrg.SVol.5-16, 47
$C_{12}ClF_9NNa$ . . . . . Na[N(C_6F_5)-C_6F_4-Cl-4] . . . . F: PFHOrg.SVol.5-16, 47
$C_{12}ClF_9N_2$ . . . . NCFCFC(NCClC_6F_5)CFCF . . . . F: PFHOrg.SVol.4-150, 161, 165
$C_{12}ClF_{20}NO_4$ . . . . . CF_3-CF(CN)-O-CF_2CF(CF_3)-O-CF_2CF_2
CF_2-O-CF(CF_3)-CCl=O . . . . F: PFHOrg.SVol.6-101, 121, 140
$C_{12}ClF_{21}N_4O_2$ . . . . $(CF_3)_2$C(CN)-N(CF_3)-O-CF_2CFCl-O-N(CF_3)-C$(CF_3)_2$-CN
F: PFHOrg.SVol.5-116, 127
$C_{12}ClFeH_8NO$ . . . . (C_5H_5)Fe[C_5H_3(CN)-CCl=O] . . . . Fe: Org.Comp.A9-130, 133, 137, 139
$C_{12}ClFeH_9N^+$ . . . . [(2-Cl-C_6H_4-CN)Fe(C_5H_5)][PF_6] . . . . Fe: Org.Comp.B19-3, 43
– . . . . [(3-Cl-C_6H_4-CN)Fe(C_5H_5)][PF_6] . . . . Fe: Org.Comp.B19-3, 43
– . . . . [(4-Cl-C_6H_4-CN)Fe(C_5H_5)][PF_6] . . . . Fe: Org.Comp.B19-3, 43
– . . . . [(Cl-C_6H_5)Fe(C_5H_4-CN)]$^+$ . . . . Fe: Org.Comp.B18-142/6, 247
$C_{12}ClFeH_9O_2$ . . . . [(4-Cl-C_6H_4-COO)Fe(C_5H_5)] · Na[PF_6] . . . . . . Fe: Org.Comp.B19-43/4
$C_{12}ClFeH_{10}N$ . . . . $C_5H_5FeC_6H_5$Cl-1(CN-6) . . . . Fe: Org.Comp.B17-258
$C_{12}ClFeH_{10}NO_5S$ (C_5H_5)Fe$(CO)_2$C[=C(CH_3)-C(=O)-N(SO_2Cl)-CH_2-]
Fe: Org.Comp.B14-18, 19, 24, 29
$C_{12}ClFeH_{10}O^+$ . . . [(ClC(=O)C_6H_5)Fe(C_5H_5)]$^+$ . . . . Fe: Org.Comp.B18-142/6, 202
$C_{12}ClFeH_{10}O_2{}^+$ . . [(3-Cl-C_6H_4-COOH)Fe(C_5H_5)][PF_6] . . . . Fe: Org.Comp.B19-5, 44
– . . . . [(4-Cl-C_6H_4-COOH)Fe(C_5H_5)][PF_6] . . . . Fe: Org.Comp.B19-5, 44
– . . . . [(Cl-C_6H_5)Fe(C_5H_4-COOH)]$^+$ . . . . Fe: Org.Comp.B18-142/6, 247
$C_{12}ClFeH_{10}S^+$ . . . [(1-Cl-C_7H_5=S-7)Fe(C_5H_5)][PF_6] . . . . Fe: Org.Comp.B19-340, 342
$C_{12}ClFeH_{11}HgO$ . . (C_5H_5)Fe[C_5H_3(HgCl)-C(=O)CH_3] . . . . Fe: Org.Comp.A10-146, 147, 149
– . . . . (HgCl-C_5H_4)Fe[C_5H_4-C(=O)CH_3] . . . . Fe: Org.Comp.A10-141, 143, 144, 145/6
$C_{12}ClFeH_{11}HgO_2$ (C_5H_5)Fe[C_5H_3(HgCl)-C(=O)OCH_3] . . . . Fe: Org.Comp.A10-146/9
– . . . . (C_5H_5)Fe[C_5H_3(HgCl)-OC(=O)CH_3] . . . . Fe: Org.Comp.A10-147
– . . . . (HgCl-C_5H_4)Fe[C_5H_4-C(=O)OCH_3] . . . . Fe: Org.Comp.A10-141/6
– . . . . (HgCl-C_5H_4)Fe[C_5H_4-OC(=O)CH_3] . . . . Fe: Org.Comp.A10-141, 142, 143, 145
$C_{12}ClFeH_{11}NO^+$ . . [(3-Cl-C_6H_4-C(O)-NH_2)Fe(C_5H_5)][PF_6] . . . . Fe: Org.Comp.B19-45
– . . . . [(4-Cl-C_6H_4-C(O)-NH_2)Fe(C_5H_5)][PF_6] . . . . Fe: Org.Comp.B19-45
– . . . . [(Cl-C_6H_5)Fe(C_5H_4-C(=O)NH_2)]$^+$ . . . . Fe: Org.Comp.B18-142/6, 247
$C_{12}ClFeH_{11}NO_2{}^+$ [(2-Cl-C_6H_4-CH_2-NO_2)Fe(C_5H_5)][PF_6] . . . . Fe: Org.Comp.B19-3, 37
$C_{12}ClFeH_{11}O_2$ . . . (C_5H_5)Fe[C_5H_2(Cl)(CH_3)-C(=O)OH] . . . . Fe: Org.Comp.A10-275/6
– . . . . [(C_5H_5)Fe$(CO)_2$(C_5H_6-c)]Cl . . . . Fe: Org.Comp.B17-86
$C_{12}ClFeH_{11}O_4S$ . . ClO_2S-C_5H_4FeC_5H_4-C(O)OCH_3 . . . . Fe: Org.Comp.A9-233, 241
$C_{12}ClFeH_{11}O_6$ . . . [(C_5H_5)Fe$(CO)_2$(C_5H_6-c)][ClO_4] . . . . Fe: Org.Comp.B17-87
$C_{12}ClFeH_{12}$ . . . . . . (4-Cl-C_6H_4-CH_3)Fe(C_5H_5) . . . . Fe: Org.Comp.B19-34
$C_{12}ClFeH_{12}{}^+$ . . . . [(2-Cl-C_6H_4-CH_3)Fe(C_5H_5)][BF_4] . . . . Fe: Org.Comp.B19-1, 4, 5, 33, 90/2
– . . . . [(2-Cl-C_6H_4-CH_3)Fe(C_5H_5)][Cr$(NH_3)_2(SCN)_4$] . . Fe: Org.Comp.B19-1, 5, 33
– . . . . [(2-Cl-C_6H_4-CH_3)Fe(C_5H_5)][PF_6] . . . . Fe: Org.Comp.B19-1, 4, 5, 33, 90/2

$C_{12}ClFeH_{12}^+$ .... $[(3\text{-}Cl\text{-}C_6H_4\text{-}CH_3)Fe(C_5H_5)][BF_4]$ .... Fe: Org.Comp.B19-1, 4/6, 33, 90/2
– .... $[(3\text{-}Cl\text{-}C_6H_4\text{-}CH_3)Fe(C_5H_5)][Cr(NH_3)_2(SCN)_4]$.. Fe: Org.Comp.B19-1, 5/6, 33
– .... $[(3\text{-}Cl\text{-}C_6H_4\text{-}CH_3)Fe(C_5H_5)][PF_6]$ .... Fe: Org.Comp.B19-1, 4/6, 33, 90/2
– .... $[(4\text{-}Cl\text{-}C_6H_4\text{-}CH_3)Fe(C_5H_5)][BF_4]$ .... Fe: Org.Comp.B19-1, 4/6, 34, 90/2
– .... $[(4\text{-}Cl\text{-}C_6H_4\text{-}CH_3)Fe(C_5H_5)][Cr(NH_3)_2(SCN)_4]$.. Fe: Org.Comp.B19-1, 5/6, 34
– .... $[(4\text{-}Cl\text{-}C_6H_4\text{-}CH_3)Fe(C_5H_5)][O\text{-}C_6H_2\text{-}2,4,6\text{-}(NO_2)_3]$ Fe: Org.Comp.B19-1, 5/6, 34
– .... $[(4\text{-}Cl\text{-}C_6H_4\text{-}CH_3)Fe(C_5H_5)][PF_6]$ .... Fe: Org.Comp.B19-1, 4/6, 33/4, 90/2
$C_{12}ClFeH_{12}NO$ ... $CH_3CONH\text{-}C_5H_4FeC_5H_4\text{-}Cl$ .... Fe: Org.Comp.A9-83
$C_{12}ClFeH_{12}NO_5S$ $(C_5H_5)Fe(CO)_2(CH_2{=}C(CH_3)CH_2C({=}O)NSO_2Cl)$ Fe: Org.Comp.B17-128
$C_{12}ClFeH_{12}O^+$ ... $[(2\text{-}Cl\text{-}C_6H_4\text{-}OCH_3)Fe(C_5H_5)][PF_6]$ .... Fe: Org.Comp.B19-68
– .... $[(3\text{-}Cl\text{-}C_6H_4\text{-}OCH_3)Fe(C_5H_5)][BF_4]$ .... Fe: Org.Comp.B19-1, 5, 68
– .... $[(3\text{-}Cl\text{-}C_6H_4\text{-}OCH_3)Fe(C_5H_5)][Cr(NH_3)_2(SCN)_4]$ Fe: Org.Comp.B19-1, 5, 68
– .... $[(3\text{-}Cl\text{-}C_6H_4\text{-}OCH_3)Fe(C_5H_5)][PF_6]$ .... Fe: Org.Comp.B19-1, 3, 5, 68
– .... $[(4\text{-}Cl\text{-}C_6H_4\text{-}OCH_3)Fe(C_5H_5)][PF_6]$ .... Fe: Org.Comp.B19-1, 3, 5, 68
$C_{12}ClFeH_{13}$...... $(C_5H_5)Fe(1\text{-}Cl\text{-}2\text{-}CH_3\text{-}C_6H_5)$ .... Fe: Org.Comp.B17-280
– .... $(C_5H_5)Fe(1\text{-}Cl\text{-}3\text{-}CH_3\text{-}C_6H_5)$ .... Fe: Org.Comp.B17-281
– .... $(C_5H_5)Fe(1\text{-}Cl\text{-}4\text{-}CH_3\text{-}C_6H_5)$ .... Fe: Org.Comp.B17-281
– .... $(C_5H_5)Fe(1\text{-}Cl\text{-}4\text{-}CH_3\text{-}6\text{-}D\text{-}C_6H_4)$ .... Fe: Org.Comp.B17-284
– .... $(C_5H_5)Fe(1\text{-}Cl\text{-}5\text{-}CH_3\text{-}C_6H_5)$ .... Fe: Org.Comp.B17-281
– .... $(C_5H_5)Fe(1\text{-}Cl\text{-}6\text{-}CH_3\text{-}C_6H_5)$ .... Fe: Org.Comp.B17-272/3
– .... $(C_5H_5)Fe(2\text{-}Cl\text{-}1\text{-}CH_3\text{-}C_6H_5)$ .... Fe: Org.Comp.B17-280
– .... $(C_5H_5)Fe(2\text{-}Cl\text{-}3\text{-}CH_3\text{-}C_6H_5)$ .... Fe: Org.Comp.B17-280
– .... $(C_5H_5)Fe(2\text{-}Cl\text{-}6\text{-}CH_3\text{-}C_6H_5)$ .... Fe: Org.Comp.B17-273
– .... $(C_5H_5)Fe(3\text{-}Cl\text{-}1\text{-}CH_3\text{-}C_6H_5)$ .... Fe: Org.Comp.B17-280
– .... $(C_5H_5)Fe(4\text{-}Cl\text{-}1\text{-}CH_3\text{-}C_6H_5)$ .... Fe: Org.Comp.B17-281
– .... $(C_5H_5)Fe(4\text{-}Cl\text{-}1\text{-}CH_3\text{-}6\text{-}D\text{-}C_6H_4)$ .... Fe: Org.Comp.B17-284
– .... $(C_5H_5)Fe[C_5H_2(Cl)(CH_3)_2]$ .... Fe: Org.Comp.A10-216
$C_{12}ClFeH_{13}HgO$ .. $(C_5H_5)Fe[C_5H_3(HgCl)\text{-}CH_2\text{-}OCH_3]$........... Fe: Org.Comp.A10-147
– .... $(HgCl\text{-}C_5H_4)Fe(C_5H_4\text{-}CH_2\text{-}OCH_3)$ .... Fe: Org.Comp.A10-143, 145
$C_{12}ClFeH_{13}N^+$ ... $[(2\text{-}Cl\text{-}C_6H_4\text{-}NH\text{-}CH_3)Fe(C_5H_5)][PF_6]$ .... Fe: Org.Comp.B19-3, 71
$C_{12}ClFeH_{13}O$ .... $(C_5H_5)Fe(1\text{-}Cl\text{-}3\text{-}CH_3O\text{-}C_6H_5)$. .... Fe: Org.Comp.B17-282
– .... $(C_5H_5)Fe(5\text{-}Cl\text{-}1\text{-}CH_3O\text{-}C_6H_5)$. .... Fe: Org.Comp.B17-282
– .... $(C_5H_5)Fe(6\text{-}Cl\text{-}6\text{-}CH_3O\text{-}C_6H_5)$. .... Fe: Org.Comp.B17-263
– .... $[(CH_3O\text{-}C_6H_5)Fe(C_5H_5)]Cl$ .... Fe: Org.Comp.B18-142/6, 201, 248
$C_{12}ClFeH_{13}O_2S$ .. $(C_5H_5)Fe[C_5H_3(SO_2Cl)C_2H_5]$ .... Fe: Org.Comp.A9-253
$C_{12}ClFeH_{14}LiSi$... $(Li\text{-}C_5H_4)Fe[C_5H_4\text{-}Si(CH_3)_2Cl]$ .... Fe: Org.Comp.A10-103/4
$C_{12}ClFeH_{16}O_3P$ .. $[C_5H_5Fe(CO)_2C(O)CH_2P(CH_3)_3]Cl$ .... Fe: Org.Comp.B14-154
$C_{12}ClFeH_{20}PS$ ... $C_5H_5Fe(CS)(P(C_2H_5)_3)Cl$ .... Fe: Org.Comp.B15-261
$C_{12}ClGaH_{30}O_3$ ... $GaCl \cdot 3\ O(C_2H_5)_2$ .... Ga: SVol.D1-99/100
$C_{12}ClGeH_{19}$ ..... $CH_2Cl\text{-}CH_2CH_2CH_2\text{-}Ge(CH_3)_2\text{-}C_6H_5$.......... Ge: Org.Comp.3-202/3
– .... $Ge(C_2H_5)_3\text{-}C_6H_4Cl\text{-}3$ .... Ge: Org.Comp.2-275
– .... $Ge(C_2H_5)_3\text{-}C_6H_4Cl\text{-}4$ .... Ge: Org.Comp.2-275
$C_{12}ClGeH_{19}O_2S$ .. $Ge(C_2H_5)_3C_6H_4SO_2Cl\text{-}4$.................. Ge: Org.Comp.2-276
$C_{12}ClGeH_{21}O_2$ ... $Ge(C_2H_5)_3C({=}C(OC_2H_5)CHClC({=}O))$. .... Ge: Org.Comp.2-230, 233

$C_{12}ClGeH_{21}Si$	$Ge(CH_3)_2(CH_2Cl)CH_2Si(CH_3)_2C_6H_5$	Ge: Org.Comp.3-193
$C_{12}ClGeH_{23}$	$Ge(C_2H_5)_3$-$C(CH_2Cl)$=CH-$C(CH_3)$=CH_2	Ge: Org.Comp.2-215
–	$Ge(C_2H_5)_3$-C≡C-$CCl(CH_3)$-C_2H_5	Ge: Org.Comp.2-266
$C_{12}ClGeH_{25}O$	$Ge(C_3H_7)_3CH_2CH_2COCl$	Ge: Org.Comp.3-4
$C_{12}ClGeH_{25}O_2$	cis- and trans-$Ge(C_2H_5)_3CH$=$CHCH_2OCH_2OCH_2CH_2Cl$	Ge: Org.Comp.2-204
$C_{12}ClGeH_{27}$	$Ge(C_3H_7)_3CH_2CH_2CH_2Cl$	Ge: Org.Comp.3-3
$C_{12}ClH_4O_6Re$	$(CO)_5Re$-C(=O)-C_6H_4Cl	Re: Org.Comp.2-121
–	$(CO)_5Re$-C(=O)-C_6H_4Cl-2	Re: Org.Comp.2-121
–	$(CO)_5Re$-C(=O)-C_6H_4Cl-3	Re: Org.Comp.2-121
–	$(CO)_5Re$-C(=O)-C_6H_4Cl-4	Re: Org.Comp.2-122, 124
$C_{12}ClH_6O_5Re$	$(CO)_5ReC(Cl)(H)C_6H_5$	Re: Org.Comp.2-111
$C_{12}ClH_7MoN_2O_8S$	MoO_2(1-(O)-2-(SO_3H)-4-(Cl)C_6H_2-6-N=N -C_6H_3-4-(OH)-2-(O))	Mo: SVol.B3b-166
$C_{12}ClH_7MoO_4$	4-CH_3-C_6H_4-$CMo(CO)_4(Cl)$	Mo: Org.Comp.5-97/100
$C_{12}ClH_8I_2OSb$	$(4\text{-}IC_6H_4)_2Sb(O)Cl$	Sb: Org.Comp.5-204
$C_{12}ClH_8N_2O_3Re$	$(CO)_3ReCl$[-1,8-(1,8-$N_2C_8H_5$-CH_3-2)-]	Re: Org.Comp.1-163
$C_{12}ClH_8N_3O_2S$	3-Cl-C_6H_4-N=S=N-C_6H_4-NO_2-4	S: S-N Comp.7-259
–	4-Cl-C_6H_4-N=S=N-C_6H_4-NO_2-4	S: S-N Comp.7-259, 264
$C_{12}ClH_8N_3O_2S_3$	4-$NO_2C_6H_4$-SN=S=NS-C_6H_4Cl-4	S: S-N Comp.7-40
$C_{12}ClH_9NO_4Sb$	(4-ClC_6H_4)(4-$O_2NC_6H_4$)Sb(O)OH	Sb: Org.Comp.5-223
$C_{12}ClH_9N_2O_2S_2$	C_6H_5-$S(O)_2$-N=S=N-C_6H_4Cl-3	S: S-N Comp.7-55/60
–	C_6H_5-$S(O)_2$-N=S=N-C_6H_4Cl-4	S: S-N Comp.7-55/60
$C_{12}ClH_9N_2O_2S_3$	C_6H_5-$S(O)_2$N=S=NS-C_6H_4-Cl-4	S: S-N Comp.7-66
$C_{12}ClH_9N_2O_4S_3$	C_6H_5-$S(O)_2$N=S=$NS(O)_2$-C_6H_4-Cl-4	S: S-N Comp.7-100/1
$C_{12}ClH_9N_6O_{12}Th$	$Th(NO_3)_4$ · NC_5H_4-2-(CH=N-C_6H_4-3-Cl)	Th: SVol.D4-141
$C_{12}ClH_{10}In$	$(C_6H_5)_2InCl$	In: Org.Comp.1-130/1
$C_{12}ClH_{10}MnN_3OS$	MnCl[O-3-$C_{10}H_6$-2-CH=N-NHC(=S)NH_2]	Mn: MVol.D6-339, 340
$C_{12}ClH_{10}MoNO_2$	[$(C_5H_5)Mo(CO)_2(NC_5H_5)(Cl)$]	Mo: Org.Comp.7-56/7, 59
$C_{12}ClH_{10}N_2O_4Sb$	4-(3'-Cl-4'-HOC_6H_3N=N)$C_6H_4Sb(O)(OH)_2$	Sb: Org.Comp.5-291
$C_{12}ClH_{10}OSb$	$(C_6H_5)_2Sb(O)Cl$	Sb: Org.Comp.5-203/4
$C_{12}ClH_{10}O_2Sb$	3-Cl-C_6H_4-Sb(=O)(OH)-C_6H_5	Sb: Org.Comp.5-222
–	4-Cl-C_6H_4-Sb(=O)(OH)-C_6H_5	Sb: Org.Comp.5-222
$C_{12}ClH_{11}OSn$	$(C_6H_5)_2Sn(Cl)OH$	Sn: Org.Comp.17-118, 121
$C_{12}ClH_{11}O_4PRe$	cis-$(CO)_4Re(Cl)P(CH_3)_2C_6H_5$	Re: Org.Comp.1-440
$C_{12}ClH_{12}MnN_3O_3$	[MnCl(O-2-$C_{10}H_6$-1-CH=N-NHC(O)NH_2)(H_2O)]	Mn: MVol.D6-325, 326
$C_{12}ClH_{12}N_2O_3Re$	$(CO)_3ReCl$[-N(C_3H_7-i)=CH-(2,1-C_5H_4N)-]	Re: Org.Comp.1-160
$C_{12}ClH_{12}O_6ReSi$	$(CO)_5ReCH_2OCH_2C$≡$CSi(CH_3)_2CH_2Cl$	Re: Org.Comp.2-106
$C_{12}ClH_{13}Mn_2N_2Na_2O_8P_2$	Na_2[Mn_2(O_3P-CH(CH_3)N=CH-C_6H_2(Cl)(O) -CH=NCH(CH_3)PO_3)(OH)] · 2 CH_3OH	Mn: MVol.D6-211, 214
–	Na_2[Mn_2(O_3P-CH_2CH_2N=CH-C_6H_2(Cl)(O) -CH=NCH_2CH_2-PO_3)(OH)] · CH_3OH · H_2O	Mn: MVol.D6-211, 214
$C_{12}ClH_{13}Mn_2N_2O_8S_2$	[Mn_2(-O_3S-CH_2CH_2-N=CH-C_6H_2(Cl)(O) -CH=N-CH_2CH_2-SO_3-)(OH)] · CH_3OH · H_2O	Mn: MVol.D6-211, 213/4
$C_{12}ClH_{13}O_2Sn$	$(c\text{-}C_5H_5)_2Sn(Cl)OOCCH_3$	Sn: Org.Comp.17-117
$C_{12}ClH_{14}InN_2$	$(CH_3)_2InCl$ · 2-(NC_5H_4-2)-NC_5H_4	In: Org.Comp.1-118, 121

$C_{12}ClH_{14}MoNO_2$ . . $(C_5H_5)Mo(CO)_2(Cl)(CN-C_4H_9-t)$ Mo:Org.Comp.8-9, 12
$C_{12}ClH_{14}MoO_5P$ . . $(C_5H_5)Mo(CO)_2(Cl)[P(-OCH_2-)_3CCH_3]$ Mo:Org.Comp.7-56/7, 89, 114
$C_{12}ClH_{14}NO_4S$ . . . 2,2-$(CH_3)_2$-OC_8H_5-7-OC(O)-N(CH_3)-S(O)Cl . . S: S-N Comp.8-268/70
$C_{12}ClH_{15}NOSb$ . . . $(CH_3)_3Sb(Cl)OC_9H_6N$ . . . . Sb: Org.Comp.5-10, 14
$C_{12}ClH_{15}NTi$ . . . . . $[(C_5H_5)_2TiCl]NC_2H_5$ . . . . Ti: Org.Comp.5-203
$C_{12}ClH_{15}N_3O_7Re$ $[(CO)_3Re(NC-C_2H_5)_3][ClO_4]$ . . . . Re: Org.Comp.1-309
$C_{12}ClH_{16}MoO_4^-$ . . $[P(C_6H_5)_4][(CH_2C(CH_3)CH_2)Mo(Cl)(CO)_2$
$(CH_3-C(O)C(CH_3)C(O)-CH_3)]$ Mo:Org.Comp.5-269/71
$C_{12}ClH_{16}MoO_5^-$ . . $[P(C_6H_5)_4][(CH_2C(CH_3)CH_2)MoCl(CO)_3$-OC(O)-$C_4H_9$-t]
Mo:Org.Comp.5-292, 294, 295
$C_{12}ClH_{16}NO_5PS$ . . $(C_2H_5O)P(O)-CH_2-N(CH_2C(O)OCH_2C_6H_5)$-S(O)Cl
S: S-N Comp.8-259, 268
$C_{12}ClH_{17}MnNO_4$ . . $[Mn(CH_3COCHCOCH_3)_2Cl] \cdot CH_3CN$ Mn:MVol.D7-6
$C_{12}ClH_{17}MnN_4O_4PS$
$[Mn(3-(2-CH_3-4-NH_2-1,3-N_2C_4H-5-CH_2)-4-$
$CH_3-1,3-SNC_3H-5-CH_2CH_2-O-PO_3H)]Cl$ Mn:MVol.D8-154
$C_{12}ClH_{17}MnN_4O_4PS^+$
$[MnH(CH_3-NH_2-C_4HN_2-CH_2-NC_3HS(CH_3)$
$-CH_2CH_2OPO_3)Cl]^+$ Mn:MVol.D8-154
$C_{12}ClH_{17}N_2O_2S_2$ . . $(1-C_5H_{10}N)S(Cl)=N-S(O)_2-C_6H_4-4-CH_3$ . . . . . . S: S-N Comp.8-186
$C_{12}ClH_{18}MnN_2O_2$ $MnCl[-O-C(CH_3)=CH-C(CH_3)=N$
$-C_2H_4-N=C(CH_3)CH=C(CH_3)O-]$ Mn:MVol.D6-203, 204/6
– . . . . $MnCl[-O-C(CH_3)=CH-C(CH_3)=N-C_2H_4-N=$
$C(CH_3)CH=C(CH_3)O-] \cdot 0.5\ H_2O$ Mn:MVol.D6-203, 206/7
$C_{12}ClH_{18}NO_2Sn$ . . $(C_2H_5)_2Sn(Cl)-N(C_6H_5)-C(=O)OCH_3$ . . . . Sn: Org.Comp.19-117, 120
$C_{12}ClH_{18}N_6O_7ReS_3$
$[(CO)_3Re(1,3-N_2C_3H_6(=S)-2)_3][ClO_4]$ . . . . Re: Org.Comp.1-310/1
$C_{12}ClH_{19}N_2O_2S$ . . 4-$(CH_3)_2N-C_6H_4-N(CH_2CH_2Cl)-S(O)O-C_2H_5$ . . S: S-N Comp.8-321
$C_{12}ClH_{19}O_4Sn$ . . . $C_2H_5SnCl(OC(CH_3)=CHCOCH_3)_2$ . . . . Sn: Org.Comp.17-149
$C_{12}ClH_{20}I_3MoN_4O$ $[Mo(CN-C_2H_5)_4(O)(Cl)][I_3]$ Mo:Org.Comp.5-39, 41
$C_{12}ClH_{20}MnN_2S_4$ $[Mn((SC(=S))NC_5H_{10})_2Cl]$ Mn:MVol.D7-177
$C_{12}ClH_{20}MnN_3OS_4$
$[Mn((SC(=S))NC_5H_{10})_2(NO)Cl]$ Mn:MVol.D7-168
$C_{12}ClH_{20}MoN_4O^+$ $[Mo(CN-C_2H_5)_4(O)(Cl)][I_3]$ Mo:Org.Comp.5-39, 41
$C_{12}ClH_{20}NO_2SSn$ $(C_2H_5)_3SnNHSO_2C_6H_4Cl$-4. . . . . Sn: Org.Comp.18-129/30
$C_{12}ClH_{21}MoOP_2$ . . $C_5H_5Mo(CO)(P(CH_3)_2CH_2CH_2P(CH_3)_2)Cl$ Mo:Org.Comp.6-230
$C_{12}ClH_{21}OSn$ . . . . $(CH_2=CHCH_2)_2Sn(Cl)OC(CH_3)_2CH_2CH=CH_2$ . . . Sn: Org.Comp.17-117
$C_{12}ClH_{22}NO_2S_2Si_2$
$((CH_3)_3Si)_2S=N-S(O)_2-C_6H_4-4-Cl$ . . . . S: S-N Comp.8-251
$C_{12}ClH_{23}MoOP_2$ . . $C_5H_5Mo(CO)(P(CH_3)_3)_2Cl$. Mo:Org.Comp.6-221
$C_{12}ClH_{23}MoO_7P_2$ $C_5H_5Mo(CO)(P(OCH_3)_3)_2Cl$ Mo:Org.Comp.6-225, 233
$C_{12}ClH_{24}N_3O_3S$ . . $[(1,4-ONC_4H_8-4-)_3S]Cl$ . . . . S: S-N Comp.8-230, 243
$C_{12}ClH_{24}O_2Sn$ . . . $(C_4H_9)_2Sn(Cl)OC(CH_3)=C(CH_3)O$, radical . . . . . Sn: Org.Comp.17-102, 108
$C_{12}ClH_{25}MoO_8P_2$ $[CH_2C(CH_3)CH_2]Mo(Cl)(CO)_2[P(OCH_3)_3]_2$ Mo:Org.Comp.5-271, 278/9
$C_{12}ClH_{25}O_2Sn$ . . . $(C_4H_9)_2Sn(Cl)OCH(CH_3)COCH_3$ . . . . Sn: Org.Comp.17-99/100
$C_{12}ClH_{26}NO_2Sn$ . . $(n-C_4H_9)_2Sn(Cl)-N(C_2H_5)-C(=O)O-CH_3$ . . . . . . Sn: Org.Comp.19-117, 120
$C_{12}ClH_{27}NO_2Sn$ . . $(C_4H_9)_2Sn(Cl)ON(O)C_4H_9$-t, radical. . . . . Sn: Org.Comp.17-106, 110
$C_{12}ClH_{27}OSn$ . . . . $(n-C_4H_9)_2SnCl-O-C_4H_9$-n . . . . Sn: Org.Comp.17-99
– . . . . $(n-C_4H_9)_2SnCl-O-C_4H_9$-t . . . . Sn: Org.Comp.17-99, 108
– . . . . $(i-C_4H_9)_2SnCl-O-C_4H_9$-n . . . . Sn: Org.Comp.17-113

$C_{12}ClH_{27}OSn$	$(i\text{-}C_4H_9)_2SnCl\text{-}O\text{-}C_4H_9\text{-}i$	Sn: Org.Comp.17-113
–	$t\text{-}C_4H_9\text{-}SnCl(C_4H_9\text{-}n)\text{-}O\text{-}C_4H_9\text{-}t$	Sn: Org.Comp.17-141, 142
$C_{12}ClH_{28}O_3PSn$	$(C_4H_9)_2Sn(Cl)OP(OC_2H_5)_2$	Sn: Org.Comp.17-106
$C_{12}ClH_{29}N_2Sn$	$[(C_2H_5)_2N]_2Sn(Cl)\text{-}C_4H_9\text{-}n$	Sn: Org.Comp.19-149/50
$C_{12}ClH_{30}IInN$	$[N(CH_3)_4][(C_4H_9)_2In(Cl)(I)]$	In: Org.Comp.1-349/50, 356
$C_{12}ClH_{30}NOP_2PtS$	$PtCl(N{=}S{=}O)(P(C_2H_5)_3)_2$	S: S-N Comp.6-259
$C_{12}ClH_{30}N_3S$	$[((C_2H_5)_2N)_3S]Cl$	S: S-N Comp.8-230, 240
$C_{12}ClH_{30}O_{11}PPo$	$[Po(OH)_3(ClO_4)((n\text{-}C_4H_9O)_3PO)]$	Po: SVol.1-347
$C_{12}ClH_{32}N_3O_9P_3Re$	$Re(CNC_2H_5)(P(OCH_3)_3)_3(Cl)N_2$	Re: Org.Comp.2-230
$C_{12}ClH_{33}NOsSi_3^-$	$[N(C_4H_9\text{-}n)_4][((CH_3)_3Si\text{-}CH_2)_3Os(N)Cl]$	Os: Org.Comp.A1-26
$C_{12}Cl_2Cu_2GeH_{24}$	$Ge(C_4H_9\text{-}n)_2(CH{=}CH_2 \cdot CuCl)_2$	Ge: Org.Comp.3-181
$C_{12}Cl_2FH_{10}Sb$	$(C_6H_5)_2Sb(Cl_2)F$	Sb: Org.Comp.5-183
$C_{12}Cl_2FH_{14}N_3OS_2$	$3\text{-}FC_6H_4\text{-}NHC(O)CCl_2SN{=}S{=}N\text{-}C_4H_9\text{-}t$	S: S-N Comp.7-24
$C_{12}Cl_2F_3H_8Sb$	$(C_6H_4Cl\text{-}4)_2SbF_3 \cdot H_2O$	Sb: Org.Comp.5-137
$C_{12}Cl_2F_4H_8Sb^-$	$[(C_6H_4Cl\text{-}4)_2SbF_4]^-$	Sb: Org.Comp.5-137
$C_{12}Cl_2F_5H_{16}InN_2$	$C_6F_5\text{-}In(Cl)_2 \cdot (CH_3)_2N\text{-}CH_2CH_2\text{-}N(CH_3)_2$	In: Org.Comp.1-136/7, 140
$C_{12}Cl_2F_6FeH_{11}P$	$[(2,6\text{-}Cl_2\text{-}C_6H_3\text{-}CH_3)Fe(C_5H_5)][PF_6]$	Fe: Org.Comp.B19-99, 129, 138
–	$[(3,4\text{-}Cl_2\text{-}C_6H_3\text{-}CH_3)Fe(C_5H_5)][PF_6]$	Fe: Org.Comp.B19-99, 130
$C_{12}Cl_2F_6HN_3$	$NCFCClC(NHC_6F_4CN\text{-}4)CClCF$	F: PFHOrg.SVol.4-150, 160
$C_{12}Cl_2F_9HN_2$	$CF_3C_6F_4NHC_5F_2Cl_2N$	F: PFHOrg.SVol.4-100
$C_{12}Cl_2F_{12}GeH_6$	$Ge(CH_3)_2[C(\text{-}(CF_2)_3CCl{=})]_2$	Ge: Org.Comp.3-153/4
$C_{12}Cl_2FeH_9N$	$(C_5H_5)Fe(C_6H_4\text{-}1,2\text{-}Cl_2\text{-}6\text{-}CN)$	Fe: Org.Comp.B17-258
–	$(C_5H_5)Fe(C_6H_4\text{-}1,3\text{-}Cl_2\text{-}6\text{-}CN)$	Fe: Org.Comp.B17-258
–	$(C_5H_5)Fe(C_6H_4\text{-}1,4\text{-}Cl_2\text{-}6\text{-}CN)$	Fe: Org.Comp.B17-258
–	$(C_5H_5)Fe(C_6H_4\text{-}1,5\text{-}Cl_2\text{-}6\text{-}CN)$	Fe: Org.Comp.B17-258
–	$(C_5H_5)Fe(C_6H_4\text{-}1,6\text{-}Cl_2\text{-}6\text{-}CN)$	Fe: Org.Comp.B17-258
$C_{12}Cl_2FeH_{10}^{2+}$	$[(Cl\text{-}C_6H_5)_2Fe][O\text{-}C_6H_2\text{-}2,4,6\text{-}(NO_2)_3]_2$	Fe: Org.Comp.B19-358
$C_{12}Cl_2FeH_{11}^+$	$[(2,6\text{-}Cl_2\text{-}C_6H_3\text{-}CH_3)Fe(C_5H_5)][PF_6]$	Fe: Org.Comp.B19-99, 129, 138
–	$[(3,4\text{-}Cl_2\text{-}C_6H_3\text{-}CH_3)Fe(C_5H_5)][PF_6]$	Fe: Org.Comp.B19-99, 130
$C_{12}Cl_2FeH_{12}O_4S_2$	$(ClO_2S\text{-}C_5H_4)Fe[C_5H_3(C_2H_5)\text{-}SO_2Cl]$	Fe: Org.Comp.A10-298
$C_{12}Cl_2FeH_{14}PdS_2$	$Fe(C_5H_4SCH_3)_2 \cdot PdCl_2$	Fe: Org.Comp.A9-210
$C_{12}Cl_2FeH_{14}PtS_2$	$Fe(C_5H_4SCH_3)_2 \cdot PtCl_2$	Fe: Org.Comp.A9-210
$C_{12}Cl_2GaH_{24}O_6^+$	$[(C_{12}H_{24}O_6)GaCl_2][GaCl_4]$	Ga: SVol.D1-150/1
$C_{12}Cl_2GaH_{28}O_6^+$	$[GaCl_2(CH_3\text{-}O\text{-}CH_2CH_2\text{-}O\text{-}CH_2CH_2\text{-}O\text{-}CH_3)_2][GaCl_4]$	Ga: SVol.D1-110
$C_{12}Cl_2Ga_2H_{18}I_6MnN_6$	$[Mn(NCCH_3)_6][GaClI_3]_2$	Mn: MVol.D7-7/9
$C_{12}Cl_2GeH_{14}Ti$	$(CH_3)_2Ge(C_5H_4)_2TiCl_2$	Ge: Org.Comp.3-330
$C_{12}Cl_2GeH_{20}O_2$	$Ge(C_2H_5)_3\text{-}C[{=}C(OC_2H_5)\text{-}CCl_2\text{-}C({=}O)\text{-}]$	Ge: Org.Comp.2-230, 233
$C_{12}Cl_2H_6MnO_2S$	$Mn(S(2\text{-}C_6H_3(O\text{-}1)(Cl\text{-}4))_2)$	Mn: MVol.D7-81/2
$C_{12}Cl_2H_7IN_2S$	$4\text{-}I\text{-}C_6H_4\text{-}N{=}S{=}N\text{-}C_6H_3\text{-}Cl_2\text{-}2,4$	S: S-N Comp.7-260
$C_{12}Cl_2H_8MnN_2S$	$[Mn(2\text{-}(2\text{-}C_5H_4N)C_7H_4NS\text{-}3,1)Cl_2]$	Mn: MVol.D7-233/6
$C_{12}Cl_2H_8MnN_6$	$Mn[NC_5H_3(Cl\text{-}4)\text{-}2\text{-}N\text{-}N{=}CH\text{-}CH{=}N\text{-}N\text{-}2\text{-}C_5H_3N(Cl\text{-}4)]$	Mn: MVol.D6-264/5
$C_{12}Cl_2H_8MnO_4Se_2$	$Mn(O_2Se\text{-}C_6H_4Cl\text{-}3)_2$	Mn: MVol.D7-244/6
–	$Mn(O_2Se\text{-}C_6H_4Cl\text{-}4)_2$	Mn: MVol.D7-244/6
–	$[Mn(O_2Se\text{-}C_6H_4Cl\text{-}3)_2(H_2O)_2]$	Mn: MVol.D7-244/6

$C_{12}Cl_2H_8MnO_4Se_2$ $[Mn(O_2Se-C_6H_4Cl-4)_2(H_2O)_2] \cdot H_2O$ Mn:MVol.D7-244/6

$C_{12}Cl_2H_8MnO_6S_2$ $Mn(O_3S-C_6H_4Cl-4)_2$ Mn:MVol.D7-114/5

– .... $Mn(O_3S-C_6H_4Cl-4)_2 \cdot 2\ H_2O$ Mn:MVol.D7-114/5

$C_{12}Cl_2H_8NO_2Re$ .. $(C_5H_5)Re(CO)_2(NC_5H_3-3,5-Cl_2)$ .... Re: Org.Comp.3-193/7, 198

$C_{12}Cl_2H_8N_2O_2S_3$ .. $4-Cl-C_6H_4-S(O)_2N=S=NS-C_6H_4-Cl-4$ .... S: S-N Comp.7-66

$C_{12}Cl_2H_8N_2O_4S_3$ .. $4-Cl-C_6H_4-S(O)_2N=S=NS(O)_2-C_6H_4-Cl-4$ .... S: S-N Comp.7-95/8

$C_{12}Cl_2H_8N_2O_4Th$. $ThCl_2(NC_5H_4-2-COO)_2$ .... Th: SVol.C7-151/3
Th: SVol.D4-148

$C_{12}Cl_2H_8N_2O_6Th$. $ThCl_2(OC_6H_4-2-NO_2)_2$ .... Th: SVol.C7-29/32

– .... $ThCl_2(OC_6H_4-4-NO_2)_2$ .... Th: SVol.C7-29/32

$C_{12}Cl_2H_8N_2S$ ..... $2-Cl-C_6H_4-N=S=N-C_6H_4-2-Cl$ .... S: S-N Comp.7-236, 252, 254

– .... $3-Cl-C_6H_4-N=S=N-C_6H_4-3-Cl$ .... S: S-N Comp.7-236, 254

– .... $4-Cl-C_6H_4-N=S=N-C_6H_4-4-Cl$ .... S: S-N Comp.7-236/7, 252/5

$C_{12}Cl_2H_8N_2S_3$ .... $4-ClC_6H_4-S-N=S=N-S-C_6H_4Cl-4$ .... S: S-N Comp.7-30/3, 36

$C_{12}Cl_2H_8N_3S_4^+$ .. $[4-Cl-C_6H_4-S-N=SNS=N-S-C_6H_4-4-Cl]^+$ .... S: S-N Comp.7-42/4

$C_{12}Cl_2H_8Po$ ...... $(4-ClC_6H_4)_2Po$ .... Po: SVol.1-334/40

$C_{12}Cl_2H_9I_2OSb$ ... $(4-IC_6H_4)_2Sb(Cl_2)OH$ .... Sb: Org.Comp.5-186

$C_{12}Cl_2H_9NO_2Sn$ .. $C_6H_5SnCl_2(OOCC_5H_4N-2)$ .... Sn: Org.Comp.17-181

$C_{12}Cl_2H_9O_2Sb$ ... $(2-ClC_6H_4)_2Sb(=O)OH$ .... Sb: Org.Comp.5-208

– .... $(3-ClC_6H_4)_2Sb(=O)OH$ .... Sb: Org.Comp.5-208

– .... $(4-ClC_6H_4)_2Sb(=O)OH$ .... Sb: Org.Comp.5-208

$C_{12}Cl_2H_{10}MnN_2O_2S_2$
$MnCl_2(O=S=N-C_6H_5)_2$ Mn:MVol.D7-107
S: S-N Comp.6-271

$C_{12}Cl_2H_{10}MnN_3S_2$ $[MnCl_2(1-(S-C(SCH_3)=NN=CH)-2-NC_9H_6)]_n$ ... Mn:MVol.D6-361, 363

– .... $[MnCl_2(2-(S-C(SCH_3)=NN=CH)-1-NC_9H_6)]_n$ ... Mn:MVol.D6-361, 363

$C_{12}Cl_2H_{10}NO_3PS$ $Cl_2S=N-P(O)(O-C_6H_5)_2$ .... S: S-N Comp.8-94/5

$C_{12}Cl_2H_{10}Po$ ..... $(C_6H_5)_2PoCl_2$ .... Po: SVol.1-334/40

$C_{12}Cl_2H_{11}MnN_3$ .. $[Mn(NC_5H_4-2-CH=NCH_2-2-C_5H_4N)Cl_2]$ Mn:MVol.D6-80/2

$C_{12}Cl_2H_{11}MnN_3O_2$ $MnCl_2[NC_5H_4-2-C(CH_3)=N-NHC(O)-2-C_4H_3O]$ Mn:MVol.D6-307, 308

$C_{12}Cl_2H_{12}MnN_4$ .. $MnCl_2[NC_5H_3(CH_3-6)-2-CH=NNH-2-NC_5H_4]$ .. Mn:MVol.D6-244, 246/7

– .... $MnCl_2[NC_5H_4-2-C(CH_3)=N-NH-2-NC_5H_4]$ Mn:MVol.D6-250, 251/2

– .... $MnCl_2[NC_5H_4-2-CH=NNH-2-NC_5H_3-6-CH_3]$ .. Mn:MVol.D6-244, 246/7

$C_{12}Cl_2H_{12}MnN_4S_2$ $[Mn(S=C(4-C_5H_4N)NH_2)_2Cl_2]$ Mn:MVol.D7-212/3

$C_{12}Cl_2H_{12}MnN_6$ .. $[MnCl_2(NC_5H_4-2-NH-N=CH-CH=N-NH-2-C_5H_4N)]$
Mn:MVol.D6-264, 266/7

$C_{12}Cl_2H_{12}MnO_6Se_2$
$[Mn(O_2Se-C_6H_4Cl-3)_2(H_2O)_2]$ Mn:MVol.D7-244/6

– .... $[Mn(O_2Se-C_6H_4Cl-4)_2(H_2O)_2] \cdot H_2O$ Mn:MVol.D7-244/6

$C_{12}Cl_2H_{12}MoO$ ... $C_5H_5Mo(O)(Cl)_2(C_7H_7-c)$ Mo:Org.Comp.6-64

$C_{12}Cl_2H_{12}N_2O_2Th$ $ThCl_2(O-C_6H_4-2-NH_2)_2$ .... Th: SVol.C7-29/32

$C_{12}Cl_2H_{12}O_{15}Th_2$ $Th_2Cl_2[OOCCH(OH)CH_2COO]_3$ .... Th: SVol.C7-105/6

$C_{12}Cl_2H_{13}MnN_2P$ $Mn[(C_6H_5)P(CH_2CH_2CN)_2]Cl_2$ Mn:MVol.D8-69/70

$C_{12}Cl_2H_{14}MnN_2O_{12}S_2$
$[Mn(O=S(CH_3)(C_5H_4NO))_2(H_2O)_2][ClO_4]_2$ Mn:MVol.D7-106

$C_{12}Cl_2H_{14}MnN_4O_2S$
$Mn(4-NH_2C_6H_4SO_2NH-2-C_4HN_2-1,3-(CH_3)_2-4,6)Cl_2$
Mn:MVol.D7-116/23

$C_{12}Cl_2H_{14}MnN_4O_4S$
$Mn(4-NH_2C_6H_4SO_2NH-4-C_4HN_2-1,3-(OCH_3)_2-2,6)Cl_2$ ... Mn:MVol.D7-116/23
$C_{12}Cl_2H_{15}N_3OS_2$ .. $C_6H_5-NHC(O)CCl_2SN=S=N-C_4H_9-t$ ... S: S-N Comp.7-24
$C_{12}Cl_2H_{15}Sb$ ..... $(CH_2=CHCH_2)_2(C_6H_5)SbCl_2$ ... Sb: Org.Comp.5-56
$C_{12}Cl_2H_{16}MnN_4O_4S_2$
$Mn(4-NH_2C_6H_4SO_2NH_2)_2Cl_2$ ... Mn:MVol.D7-116/23
$C_{12}Cl_2H_{16}MnN_4S_2$ $[Mn(1,3-N_2C_4H_2(=S-2)((CH_3)_2-1,6))_2Cl_2]$ ... Mn:MVol.D7-78/9
$C_{12}Cl_2H_{16}N_2OS$ .. $ClS(O)-N(C_6H_5)-CHCl-1-NC_5H_{10}$ ... S: S-N Comp.8-259, 267
$C_{12}Cl_2H_{16}N_2Ti_2$ .. $[(C_5H_5)Ti(NCH_3)Cl]_2$ ... Ti: Org.Comp.5-54
$C_{12}Cl_2H_{17}NO_2Sn$ $C_2H_5-SnCl_2-N(C_6H_5)-C(=O)O-C_3H_7-i$ ... Sn: Org.Comp.19-157
$C_{12}Cl_2H_{17}N_3O_2S_2$ $(C_2H_5)_2N-CCl=N-S(Cl)=N-S(O)_2-C_6H_4-4-CH_3$ S: S-N Comp.8-187/9
$C_{12}Cl_2H_{18}I_6In_2MnN_6$
$[Mn(NCCH_3)_6][InClI_3]_2$ ... Mn:MVol.D7-7/9
$C_{12}Cl_2H_{18}MnN_2O_3S$
$Mn(4-CH_3C_6H_4SO_2NH-CONHC_4H_9)Cl_2$ ... Mn:MVol.D7-124/5
$C_{12}Cl_2H_{18}MnN_2O_{14}S_2$
$[Mn(O=S(CH_3)(C_5H_4NO))_2(H_2O)_2][ClO_4]_2$ ... Mn:MVol.D7-106
$C_{12}Cl_2H_{18}MnN_6O_8$ $[Mn(NCCH_3)_6][ClO_4]_2$ ... Mn:MVol.D7-6/7
$C_{12}Cl_2H_{18}O_4Sn$ ... $(n-C_4H_9)_2Sn[-OC(=O)-CCl=CCl-C(=O)O-]$ ... Sn: Org.Comp.15-325
$C_{12}Cl_2H_{19}InN_2$ ... $2,6-[(CH_3)_2N-CH_2]_2-C_6H_3-In(Cl)_2$ ... In: Org.Comp.1-136, 140
$C_{12}Cl_2H_{19}MnN_4OS^+$
$[Mn(S=C(NHC_6H_5)NHNHCOCH_2N(CH_3)_3)Cl_2]^+$ Mn:MVol.D7-206
$C_{12}Cl_2H_{19}MnP$ ... $Mn[P(C_6H_5)(C_2H_5)_2](CH_2=CH_2)Cl_2$ ... Mn:MVol.D8-46/7
– ... $Mn[P(C_6H_5)(C_3H_7)_2]Cl_2$ ... Mn:MVol.D8-42/3
$C_{12}Cl_2H_{20}InN$ .... $2,4,6-(CH_3)_3-C_6H_2-In(Cl)_2 \cdot N(CH_3)_3$ ... In: Org.Comp.1-140
$C_{12}Cl_2H_{20}MnN_4O_8S_8$
$[Mn(1,3-SNC_3H_5(=S)-2)_4][ClO_4]_2 \cdot 2 H_2O$ ... Mn:MVol.D7-61/3
$C_{12}Cl_2H_{20}MnN_4S_8$ $[Mn(1,3-SNC_3H_5(=S)-2)_4Cl_2] \cdot 4 H_2O$ ... Mn:MVol.D7-61/3
$C_{12}Cl_2H_{20}O_4Sn$ ... $C_4H_9OOCCH_2CH_2SnCl_2(OC(CH_3)=CHCOCH_3)$. Sn: Org.Comp.17-180
$C_{12}Cl_2H_{22}MnN_6O_2$ $[Mn(c-C_5H_8=N-NHC(O)NH_2)_2Cl_2]$ ... Mn:MVol.D6-328/9
$C_{12}Cl_2H_{22}MnN_6S_2$ $MnCl_2[c-C_5H_8=N-NHC(=S)NH_2]_2$ ... Mn:MVol.D6-342/4
$C_{12}Cl_2H_{22}MnN_8O_2P_2S_2$
$[Mn(S=P(OC_6H_5)(NHNH_2)_2)_2Cl_2]$ ... Mn:MVol.D8-200
$C_{12}Cl_2H_{22}N_2O_2Th$ $ThCl_2[CH_3-CH(-O)-CH_2-C(CH_3)=N-CH_2$
$CH_2-N=C(CH_3)-CH_2-CH(-O)-CH_3]$ ... Th: SVol.D4-163
$C_{12}Cl_2H_{22}O_4Sn$ ... $(C_4H_9)_2Sn(OOCCH_2Cl)_2$ ... Sn: Org.Comp.15-255
$C_{12}Cl_2H_{26}OSn$ ... $C_{12}H_{25}SnCl_2(OH)$ ... Sn: Org.Comp.17-179
$C_{12}Cl_2H_{27}MnNOP$ $Mn(NO)[P(C_4H_9)_3]Cl_2$ ... Mn:MVol.D8-46, 61
$C_{12}Cl_2H_{27}MnP$ ... $Mn[P(C_4H_9)_3]Cl_2$ ... Mn:MVol.D8-38
$C_{12}Cl_2H_{28}In_2$ ..... $[(i-C_3H_7)_2InCl]_2$ ... In: Org.Comp.1-119
$C_{12}Cl_2H_{28}MnN_4O_8S_4$
$[Mn(S=CHN(CH_3)_2)_4][ClO_4]_2$ ... Mn:MVol.D7-212
$C_{12}Cl_2H_{28}N_2PtSSe$
$[((C_2H_5)_2Se)PtCl_2(t-C_4H_9-N=S=N-C_4H_9-t)]$ ... S: S-N Comp.7-316/8
$C_{12}Cl_2H_{28}N_2PtSTe$ $[((C_2H_5)_2Te)PtCl_2(t-C_4H_9-N=S=N-C_4H_9-t)]$ ... S: S-N Comp.7-316/8
$C_{12}Cl_2H_{29}N_2PPtS$ $[((C_2H_5)_3P)PtCl_2(i-C_3H_7-N=S=N-C_3H_7-i)]$ ... S: S-N Comp.7-316/7, 319, 321/2
$C_{12}Cl_2H_{29}N_3PtS$ .. $[PtCl_2(t-C_4H_9-N=S=N-C_4H_9-t)(NH_2-C_4H_9-t)]$ .. S: S-N Comp.7-322/3
$C_{12}Cl_2H_{29}O_{14}PPo$ $[Po(OH)_2(ClO_4)_2((n-C_4H_9O)_3PO)]$ ... Po: SVol.1-347

$C_{12}Cl_2H_{30}MnO_8P_2$ $[Mn(O=P(OC_2H_5)_3)_2Cl_2]$ Mn:MVol.D8-160/1
$C_{12}Cl_2H_{30}MnO_{14}S_6$
$[Mn(O=S(CH_3)-CH_2CH_2-S(CH_3)=O)_3][ClO_4]_2$ .. Mn:MVol.D7-108/9
$C_{12}Cl_2H_{30}OSb_2$... $[(C_2H_5)_3SbCl]_2O$ .... Sb: Org.Comp.5-89
$C_{12}Cl_2H_{31}NSb_2$... $[(C_2H_5)_3SbCl]_2NH$ .... Sb: Org.Comp.5-120
$C_{12}Cl_2H_{32}MnN_8O_8S_4$
$[Mn(S=C(NHCH_3)_2)_4][ClO_4]_2$ Mn:MVol.D7-194/5
$C_{12}Cl_2H_{36}In_3Sb$ .. $[Sb(CH_3)_4][(CH_3)_2In(Cl)_2(In(CH_3)_3)_2]$ .... In: Org.Comp.1-363
$C_{12}Cl_2H_{36}MnN_6O_2P_2$
$[Mn(O=P(N(CH_3)_2)_3)_2Cl_2]$ Mn:MVol.D8-174/5
$C_{12}Cl_2H_{36}MnN_{12}O_8S_6$
$[Mn(S=C(NH_2)NHCH_3)_6][ClO_4]_2$ Mn:MVol.D7-194/5
$C_{12}Cl_2H_{36}MnO_{14}S_6$
$[Mn(O=S(CH_3)_2)_6][ClO_4]_2$ Mn:MVol.D7-95/6
$C_{12}Cl_2H_{36}MnO_{14}Se_6$
$[Mn(OSe(CH_3)_2)_6][ClO_4]_2$ Mn:MVol.D7-243
$C_{12}Cl_2H_{36}MnO_{16}P_4$
$[Mn(O=P(CH_2OH)_3)_4Cl_2]$ Mn:MVol.D8-85/6
$C_{12}Cl_3FH_9Sb$..... $(C_6H_5)(4-FC_6H_4)SbCl_3 \cdot H_2O$ .... Sb: Org.Comp.5-222, 229
$C_{12}Cl_3F_2H_8Sb$.... $(4-FC_6H_4)_2SbCl_3$ .... Sb: Org.Comp.5-163
$C_{12}Cl_3F_7HN$ ..... 2,4,6-Cl_3-C_6F_2-NH-C_6F_5 .... F: PFHOrg.SVol.5-16, 48, 84
$C_{12}Cl_3F_7N^-$ ...... $[N(C_6F_5)-C_6F_2-Cl_3-2,4,6]^-$ .... F: PFHOrg.SVol.5-16, 48
$C_{12}Cl_3F_7NNa$ .... $Na[N(C_6F_5)-C_6F_2-Cl_3-2,4,6]$ .... F: PFHOrg.SVol.5-16, 48
$C_{12}Cl_3FeH_{11}$ ..... $C_5H_5FeC_6H_6(CCl_3-6)$ .... Fe: Org.Comp.B17-261/2
$C_{12}Cl_3FeH_{11}O_3$... $(C_5H_5)Fe(CO)_2COCHCH_3CH_2CCl_3$ .... Fe: Org.Comp.B13-7
$C_{12}Cl_3GaH_{10}O$ ... $GaCl_3(C_6H_5-O-C_6H_5)$ .... Ga: SVol.D1-105, 106/8
– .... $GaCl_3[2-(CH_3-C(=O))-C_{10}H_7]$ .... Ga: SVol.D1-39/42
$C_{12}Cl_3GaH_{11}N$ ... $GaCl_3[NH(C_6H_5)_2]$ .... Ga: SVol.D1-222
$C_{12}Cl_3GeH_{21}O_2$ .. $Ge(C_2H_5)_3C(COOC_2H_5)=CHCCl_3$ .... Ge: Org.Comp.2-208
$C_{12}Cl_3H_7N_2S$..... 4-Cl-C_6H_4-N=S=N-C_6H_3-Cl_2-2,4 .... S: S-N Comp.7-259
– .... 4-Cl-C_6H_4-N=S=N-C_6H_3-Cl_2-3,4 .... S: S-N Comp.7-259
$C_{12}Cl_3H_7N_2S_2$.... C_6H_5-SN=S=N-$C_6H_2Cl_3$-2,4,6 .... S: S-N Comp.7-25/7
$C_{12}Cl_3H_8N_2O_2Re$ $(CO)_2ReCl_3[NC_5H_4-2-(2-C_5H_4N)]$ .... Re: Org.Comp.1-60
$C_{12}Cl_3H_8N_2O_4Sb$ $(3-O_2N-C_6H_4)_2SbCl_3$ .... Sb: Org.Comp.5-164
– .... $(4-O_2N-C_6H_4)_2SbCl_3$ .... Sb: Org.Comp.5-164
$C_{12}Cl_3H_8N_3S_4$.... [4-Cl-C_6H_4-S-N=SNS=N-S-C_6H_4-4-Cl]Cl .... S: S-N Comp.7-42/4
$C_{12}Cl_3H_8OSb$ .... $(2-C_6H_4OC_6H_4-2')SbCl_3$ .... Sb: Org.Comp.5-170
$C_{12}Cl_3H_8Sb$...... $(2-C_6H_4C_6H_4-2')SbCl_3$ .... Sb: Org.Comp.5-169, 171
$C_{12}Cl_3H_9ISb$ ..... $(C_6H_5)(4-IC_6H_4)SbCl_3$ .... Sb: Org.Comp.5-222
$C_{12}Cl_3H_9NO_2Sb$ .. $(C_6H_5)(4-O_2NC_6H_4)SbCl_3$ .... Sb: Org.Comp.5-223
$C_{12}Cl_3H_{10}N_2O_3S_2V$
$V(O)Cl_3(O=S=NC_6H_5)_2$ .... S: S-N Comp.6-270
$C_{12}Cl_3H_{10}N_3Sb^-$ .. $[(C_6H_5)_2Sb(Cl_3)N_3]^-$ .... Sb: Org.Comp.5-158
$C_{12}Cl_3H_{10}O_2Sb$... $(2-ClC_6H_4)_2Sb(OH)_2Cl$ .... Sb: Org.Comp.5-214
$C_{12}Cl_3H_{10}Sb$..... $(C_6H_5)_2SbCl_3$ .... Sb: Org.Comp.5-146/52
– .... $(C_6H_5)_2SbCl_3 \cdot H_2O$ .... Sb: Org.Comp.5-148/9
$C_{12}Cl_3H_{11}NSb$.... $(C_6H_5)(3-H_2NC_6H_4)SbCl_3$ .... Sb: Org.Comp.5-222
$C_{12}Cl_3H_{14}N_3OS_2$.. 4-ClC_6H_4-NHC(O)CCl_2SN=S=N-C_4H_9-t ...... S: S-N Comp.7-24
$C_{12}Cl_3H_{14}O_2Sb$... $(3-CH_3-C_6H_4)SbCl_3[-OC(CH_3)CHC(CH_3)O-]$ .. Sb: Org.Comp.5-269
– .... $(4-CH_3-C_6H_4)SbCl_3[-OC(CH_3)CHC(CH_3)O-]$ .. Sb: Org.Comp.5-269

$C_{12}Cl_3H_{14}O_2Sb$ . . . $(C_6H_5)SbCl_3[-OC(CH_3)C(CH_3)C(CH_3)O-]$ . . . . . Sb: Org.Comp.5-269
$C_{12}Cl_3H_{15}N_2O_3S_2$ $n-C_3H_7-C(O)-NH-S(CCl_3)=N-S(O)_2-C_6H_4-4-CH_3$ S: S-N Comp.8-196
– . . . . $n-C_4H_9-C(O)-NH-S(CCl_3)=N-S(O)_2-C_6H_5$ . . . . S: S-N Comp.8-195
– . . . . $i-C_4H_9-C(O)-NH-S(CCl_3)=N-S(O)_2-C_6H_5$ . . . . . S: S-N Comp.8-195
$C_{12}Cl_3H_{19}MnN_4OS$
$[Mn(S=C(NHC_6H_5)NHNHCOCH_2N(CH_3)_3)Cl_2]Cl$ Mn: MVol.D7-206
$C_{12}Cl_3H_{20}MoOP$ . . $C_5H_5Mo(CO)(P(C_2H_5)_3)Cl_3$ . . . . Mo: Org.Comp.6-219
$C_{12}Cl_3H_{20}N_4S^+$ . . $[(1-C_5H_{10}N)-CCl=N-S(Cl)-N=CCl-(NC_5H_{10}-1)]^+$ S: S-N Comp.8-193
$C_{12}Cl_3H_{22}NO_2SSn$ $(C_4H_9)_2Sn(NCS)OCH(CCl_3)OCH_3$ . . . . Sn: Org.Comp.17-137
$C_{12}Cl_3H_{24}MoO_3$ . . $MoCl_3 \cdot 3\ C_4H_8O$ . . . . Mo: SVol.B5-324
$C_{12}Cl_3H_{24}NO_2Sn$ $CH_3-O-Sn(C_4H_9-n)_2-N=C(CCl_3)-O-CH_3$ . . . . . . Sn: Org.Comp.19-129, 134
$C_{12}Cl_3H_{25}O_3Sn$ . . . $(C_4H_9)_2Sn(OCH_3)OCH(CCl_3)OCH_3$ . . . . Sn: Org.Comp.16-187
$C_{12}Cl_3H_{27}MnO_4P^-$
$[Mn(O=P(OC_4H_9)_3)Cl_3]^-$ . . . . Mn: MVol.D8-162/4
$C_{12}Cl_3H_{28}O_{17}PPo$ $[Po(OH)(ClO_4)_3((n-C_4H_9O)_3PO)]$ . . . . Po: SVol.1-334, 347
$C_{12}Cl_3H_{31}MoN_3$ . . $MoCl_3(NH(CH_2)_3CH_3)_2 \cdot NH_2(CH_2)_3CH_3$ . . . . . . Mo: SVol.B5-373
$C_{12}Cl_3H_{36}MnO_{18}S_6$
$[Mn(O=S(CH_3)_2)_6][ClO_4]_3$ . . . . Mn: MVol.D7-99/100
$C_{12}Cl_4F_2H_{14}N_2Pd$ $(4-F-C_6H_4-NH_3)_2[PdCl_4]$ . . . . Pd: SVol.B2-123
$C_{12}Cl_4F_3H_4N_3O_2$ . . $(CF_3)(NO_2)C_6H_3NHC_5Cl_4N$ . . . . F: PFHOrg.SVol.4-101
$C_{12}Cl_4F_4HN_3$ . . . . . $4-(2-Cl-4-CN-C_6F_3NH)-2-F-C_5Cl_3N$ . . . . F: PFHOrg.SVol.4-150, 160
$C_{12}Cl_4F_4H_6Sb^-$ . . . $[(C_6H_3Cl_2-3,4)_2SbF_4]^-$ . . . . Sb: Org.Comp.5-137
$C_{12}Cl_4F_8N_2$ . . . . . . $4-NCl_2-C_6F_4-C_6F_4-NCl_2-4$ . . . . F: PFHOrg.SVol.6-5/6, 25
$C_{12}Cl_4FeH_{14}Si_2$ . . $Fe(C_5H_4SiCl_2CH_3)_2$ . . . . Fe: Org.Comp.A9-302
$C_{12}Cl_4GaH_{14}N_2^-$ $[C_6H_5-NH_3 \cdot C_6H_5-NH_2][GaCl_4(NH_2-C_6H_5)_2]$. Ga: SVol.D1-217/8
$C_{12}Cl_4GaH_{21}O_2$ . . $H[GaCl_4(O=C(CH_3)-CH=C(CH_3)_2)_2]$ . . . . Ga: SVol.D1-36/7
– . . . . $H[GaCl_4(O=C_6H_{10})_2]$ . . . . Ga: SVol.D1-42
$C_{12}Cl_4GaH_{25}O_2$ . . $H[GaCl_4(O=C(CH_3)-C_4H_9-i)_2]$ . . . . Ga: SVol.D1-36/7
$C_{12}Cl_4GaH_{25}O_6$ . . $HGaCl_4 \cdot 3\ (1,4-O_2C_4H_8)$ . . . . Ga: SVol.D1-147
$C_{12}Cl_4Ga_2H_{14}N_2$ . . $Ga_2Cl_4(NC_5H_4-3-CH_3)_2$ . . . . Ga: SVol.D1-255/6
– . . . . $Ga_2Cl_4(NC_5H_4-4-CH_3)_2$ . . . . Ga: SVol.D1-255/6
$C_{12}Cl_4Ga_2H_{16}N_4$ . . $[Ga(1,2-(NH_2)_2-C_6H_4)_2][GaCl_4] \cdot C_6H_6$ . . . . . . Ga: SVol.D1-238/9
$C_{12}Cl_4H_4MnO_2S$ . . $Mn(S(2-C_6H_2(O-1)(Cl_2-3,4))_2)$ . . . . Mn: MVol.D7-81/2
$C_{12}Cl_4H_6N_2S$ . . . . . $3,5-Cl_2-C_6H_3-N=S=N-C_6H_3-3,5-Cl_2$ . . . . S: S-N Comp.7-244, 252/3
– . . . . $4-Cl-C_6H_4-N=S=N-C_6H_2-2,4,6-Cl_3$ . . . . S: S-N Comp.7-259/60
$C_{12}Cl_4H_6N_2S_3$ . . . . $(2,5-Cl_2C_6H_3-SN=)_2S$ . . . . S: S-N Comp.7-32/7
$C_{12}Cl_4H_7N_2O_3Sb$ $3-(4',5'-Cl_2C_6H_3)N=N-4,5-Cl_2-C_6H_2Sb(O)-(OH)_2$ Sb: Org.Comp.5-299
$C_{12}Cl_4H_8NO_2Sb$ . . $(4-ClC_6H_4)(4-O_2NC_6H_4)SbCl_3$ . . . . Sb: Org.Comp.5-223
$C_{12}Cl_4H_8N_2O_4Sb^-$ $[(4-O_2NC_6H_4)_2SbCl_4]^-$ . . . . Sb: Org.Comp.5-165
$C_{12}Cl_4H_8OSb^-$ . . . $[(2-C_6H_4OC_6H_4-2')SbCl_4]^-$ . . . . Sb: Org.Comp.5-171
$C_{12}Cl_4H_8Sb^-$ . . . . . $[(2-C_6H_4C_6H_4-2')SbCl_4]^-$ . . . . Sb: Org.Comp.5-170
$C_{12}Cl_4H_9NO_2Sb^-$ $[(C_6H_5)(4-O_2NC_6H_4)SbCl_4]^-$ . . . . Sb: Org.Comp.5-226
$C_{12}Cl_4H_9Sb$ . . . . . . $3-ClC_6H_4-SbCl_3-C_6H_5$ . . . . Sb: Org.Comp.5-222
– . . . . $4-ClC_6H_4-SbCl_3-C_6H_5$ . . . . Sb: Org.Comp.5-222
– . . . . $4-ClC_6H_4-SbCl_3-C_6H_5 \cdot H_2O$ . . . . Sb: Org.Comp.5-222
$C_{12}Cl_4H_{10}Mn_2N_4$. $[Mn_2(NC_5H_4-2-CH=NN=CH-2-C_5H_4N)Cl_4]$ . . . . Mn: MVol.D6-261, 262
$C_{12}Cl_4H_{10}N_2O_2S_2Ti$
$TiCl_4(O=S=NC_6H_5)_2$ . . . . S: S-N Comp.6-270

$C_{12}Cl_4H_{10}N_2O_4Th$	$ThCl_4 \cdot 2\ NC_5H_4\text{-}2\text{-}COOH$	Th: SVol.D4-162
$C_{12}Cl_4H_{10}N_2Pd$	$(1,10\text{-}N_2C_{12}H_{10})[PdCl_4]$	Pd: SVol.B2-124
$C_{12}Cl_4H_{10}Sb^-$	$[(C_6H_5)_2SbCl_4]^-$	Sb: Org.Comp.5-154/7
$C_{12}Cl_4H_{12}NOSb$	$4\text{-}CH_3\text{-}C_6H_4\text{-}SbCl_4\text{-}O\text{-}1\text{-}NC_5H_5$	Sb: Org.Comp.5-255, 260
$C_{12}Cl_4H_{13}N_2Sb$	$(3\text{-}H_2N\text{-}C_6H_4)_2SbCl_3 \cdot HCl \cdot 2\ H_2O$	Sb: Org.Comp.5-164, 166
$C_{12}Cl_4H_{13}O_2Sb$	$4\text{-}CH_3\text{-}C_6H_4\text{-}SbCl_3[\text{-}OC(CH_3)CClC(CH_3)O\text{-}]$	Sb: Org.Comp.5-269
–	$4\text{-}ClC_6H_4\text{-}SbCl_3[\text{-}OC(CH_3)C(CH_3)C(CH_3)O\text{-}]$	Sb: Org.Comp.5-270
$C_{12}Cl_4H_{14}N_2O_2Th$	$ThCl_4 \cdot 2\ (2\text{-}CH_3\text{-}1\text{-}O\text{-}NC_5H_4)$	Th: SVol.D4-162
–	$ThCl_4 \cdot 2\ (3\text{-}CH_3\text{-}1\text{-}O\text{-}NC_5H_4)$	Th: SVol.D4-162
–	$ThCl_4 \cdot 2\ (4\text{-}CH_3\text{-}1\text{-}O\text{-}NC_5H_4)$	Th: SVol.D4-162
$C_{12}Cl_4H_{14}N_2O_3S_2$	$n\text{-}C_4H_9\text{-}C(O)\text{-}NH\text{-}S(CCl_3)\text{=}N\text{-}S(O)_2\text{-}C_6H_4\text{-}4\text{-}Cl$	S: S-N Comp.8-195
–	$i\text{-}C_4H_9\text{-}C(O)\text{-}NH\text{-}S(CCl_3)\text{=}N\text{-}S(O)_2\text{-}C_6H_4\text{-}4\text{-}Cl$	S: S-N Comp.8-195
$C_{12}Cl_4H_{14}N_2Pd$	$[1\text{-}CH_3\text{-}NC_5H_4\text{-}4\text{-}(4\text{-}NC_5H_4(CH_3\text{-}1))][PdCl_4]$	Pd: SVol.B2-126
$C_{12}Cl_4H_{14}N_4O_4Pd$	$(3\text{-}NO_2\text{-}C_6H_4\text{-}NH_3)_2[PdCl_4]$	Pd: SVol.B2-123
$C_{12}Cl_4H_{16}N_2Pd$	$(2\text{-}CH_3\text{-}NC_5H_5)_2[PdCl_4]$	Pd: SVol.B2-123
–	$(3\text{-}CH_3\text{-}NC_5H_5)_2[PdCl_4]$	Pd: SVol.B2-123
–	$(4\text{-}CH_3\text{-}NC_5H_5)_2[PdCl_4]$	Pd: SVol.B2-123
–	$(C_6H_5\text{-}NH_3)_2[PdCl_4]$	Pd: SVol.B2-123
$C_{12}Cl_4H_{16}N_2Ti_2$	$[(C_5H_5)TiCl_2]_2(N(CH_3)N(CH_3))$	Ti: Org.Comp.5-56
$C_{12}Cl_4H_{18}InMnN_6$	$[Mn(NCCH_3)_6][InCl_4]$	Mn: MVol.D7-7/9
$C_{12}Cl_4H_{18}MnN_4O_4S_2$	$(4\text{-}NH_3C_6H_4SO_2NH_2)_2[MnCl_4]$	Mn: MVol.D7-116/23
$C_{12}Cl_4H_{20}N_4S$	$[(1\text{-}C_5H_{10}N)\text{-}CCl\text{=}N\text{-}S(Cl)\text{-}N\text{=}CCl\text{-}(NC_5H_{10}\text{-}1)]Cl$	S: S-N Comp.8-193
$C_{12}Cl_4H_{20}O_4Sn$	$(C_4H_9)_2Sn(OOCCHCl_2)_2$	Sn: Org.Comp.15-255
$C_{12}Cl_4H_{23}N_3OPSb$	$C_6H_5SbCl_4 \cdot OP[N(CH_3)_2]_3$	Sb: Org.Comp.5-244
$C_{12}Cl_4H_{27}O_{20}PPo$	$Po(ClO_4)_4[(n\text{-}C_4H_9O)_3PO]$	Po: SVol.1-333/4, 347
$C_{12}Cl_4H_{28}N_2Pd$	$(C_6H_{11}NH_3)_2[PdCl_4]$	Pd: SVol.B2-123
$C_{12}Cl_4H_{28}N_4O_4Th$	$ThCl_4 \cdot 4\ CH_3\text{-}C(\text{=}O)\text{-}NH\text{-}CH_3$	Th: SVol.D4-161
$C_{12}Cl_4H_{32}O_4Th$	$ThCl_4 \cdot 4\ i\text{-}C_3H_7\text{-}OH$	Th: SVol.D4-130, 132, 145, 171, 187, 189
$C_{12}Cl_4H_{36}Mn_2O_6S_6$	$[Mn(O\text{=}S(CH_3)_2)_6][MnCl_4]$	Mn: MVol.D7-94/5
$C_{12}Cl_4H_{36}N_6O_2P_2Th$	$ThCl_4 \cdot 2\ [(CH_3)_2N]_3P\text{=}O$	Th: SVol.D4-148, 207
$C_{12}Cl_4H_{36}O_6S_6Th$	$ThCl_4 \cdot 6\ (CH_3)_2SO$	Th: SVol.D4-162
$C_{12}Cl_5H_8NO_2Sb^-$	$[(4\text{-}ClC_6H_4)(4\text{-}O_2NC_6H_4)SbCl_4]^-$	Sb: Org.Comp.5-227
$C_{12}Cl_5H_8Sb$	$(3\text{-}ClC_6H_4)_2SbCl_3$	Sb: Org.Comp.5-163
–	$(4\text{-}ClC_6H_4)_2SbCl_3$	Sb: Org.Comp.5-164, 166
$C_{12}Cl_5H_9IN_2Sb$	$[4\text{-}IC_6H_4N_2][C_6H_5SbCl_5]$	Sb: Org.Comp.5-246/7
$C_{12}Cl_5H_9N_2OSb^-$	$[(4\text{-}(4'\text{-}HOC_6H_4N_2)C_6H_4)SbCl_5]^-$	Sb: Org.Comp.5-255
$C_{12}Cl_5H_9N_3O_2Sb$	$[2\text{-}O_2N\text{-}C_6H_4\text{-}N_2][C_6H_5\text{-}SbCl_5]$	Sb: Org.Comp.5-246/7
–	$[3\text{-}O_2N\text{-}C_6H_4\text{-}N_2][C_6H_5\text{-}SbCl_5]$	Sb: Org.Comp.5-246/7
–	$[4\text{-}O_2N\text{-}C_6H_4\text{-}N_2][C_6H_5\text{-}SbCl_5]$	Sb: Org.Comp.5-246/7
–	$[C_6H_5\text{-}N_2][4\text{-}O_2N\text{-}C_6H_4\text{-}SbCl_5]$	Sb: Org.Comp.5-254
$C_{12}Cl_5H_9N_3O_3Sb$	$[4\text{-}O_2NC_6H_4N_2][(4\text{-}HOC_6H_4)SbCl_5]$	Sb: Org.Comp.5-251
$C_{12}Cl_5H_9Sb^-$	$[4\text{-}ClC_6H_4\text{-}SbCl_4\text{-}C_6H_5]^-$	Sb: Org.Comp.5-226
–	$[C_6H_5\text{-}C_6H_4\text{-}2\text{-}SbCl_5]^-$	Sb: Org.Comp.5-257
$C_{12}Cl_5H_{10}N_2O_4Sb$	$[C_5H_5NH][(3\text{-}O_2N\text{-}4\text{-}HO_2CC_6H_3)SbCl_5]$	Sb: Org.Comp.5-259

$C_{12}Cl_5H_{10}N_2Sb$	$[C_6H_5-N_2][C_6H_5-SbCl_5]$	Sb: Org.Comp.5-246/7
–	$[NC_5H_6][4-NC-C_6H_4-SbCl_5]$	Sb: Org.Comp.5-251
$C_{12}Cl_5H_{11}NO_2Sb$	$[NC_5H_6][3-HO_2C-C_6H_4-SbCl_5]$	Sb: Org.Comp.5-256
–	$[NC_5H_6][4-HO_2C-C_6H_4-SbCl_5]$	Sb: Org.Comp.5-256
$C_{12}Cl_5H_{11}NO_3Sb$	$[C_5H_5NH][(3-HO-4-HO_2CC_6H_3)SbCl_5]$	Sb: Org.Comp.5-259
$C_{12}Cl_5H_{12}N_2O_2Sb$	$[C_5H_5NH][(3-O_2N-4-CH_3C_6H_3)SbCl_5]$	Sb: Org.Comp.5-259
$C_{12}Cl_5H_{13}NOSb$	$[C_5H_5NH][(4-CH_3OC_6H_4)SbCl_5]$	Sb: Org.Comp.5-252
$C_{12}Cl_5H_{13}NSb$	$[NC_5H_6][2-CH_3-C_6H_4-SbCl_5]$	Sb: Org.Comp.5-255
–	$[NC_5H_6][3-CH_3-C_6H_4-SbCl_5]$	Sb: Org.Comp.5-255
–	$[NC_5H_6][4-CH_3-C_6H_4-SbCl_5]$	Sb: Org.Comp.5-256
–	$[NH_4][C_6H_5-C_6H_4-2-SbCl_5]$	Sb: Org.Comp.5-257
$C_{12}Cl_5H_{13}N_3OSb$	$[C_5H_5NH][(4-H_2NNHC(O)C_6H_4)SbCl_5]$	Sb: Org.Comp.5-256
$C_{12}Cl_5H_{18}NOSb^-$	$[(4-(C_2H_5)_2NCH_2CH_2OC_6H_4)SbCl_5]^-$	Sb: Org.Comp.5-252
$C_{12}Cl_5H_{18}NSSb^-$	$[(4-(C_2H_5)_2NCH_2CH_2SC_6H_4)SbCl_5]^-$	Sb: Org.Comp.5-253
$C_{12}Cl_5H_{18}NSb^-$	$[(4-(C_2H_5)_2NCH_2CH_2C_6H_4)SbCl_5]^-$	Sb: Org.Comp.5-256
$C_{12}Cl_5H_{19}NOSb$	$H[(4-(C_2H_5)_2NCH_2CH_2OC_6H_4)SbCl_5]$	Sb: Org.Comp.5-252
$C_{12}Cl_5H_{19}NSSb$	$H[(4-(C_2H_5)_2NCH_2CH_2SC_6H_4)SbCl_5]$	Sb: Org.Comp.5-253
$C_{12}Cl_5H_{19}NSb$	$H[(4-(C_2H_5)_2NCH_2CH_2C_6H_4)SbCl_5]$	Sb: Org.Comp.5-256
$C_{12}Cl_5H_{22}N_3OPSb$	$(4-ClC_6H_4)SbCl_4 \cdot OP[N(CH_3)_2]_3$	Sb: Org.Comp.5-251, 259
$C_{12}Cl_5H_{27}N_3O_3Th^-$	$[((CH_3)_2N-C(=O)-CH_3)_3ThCl_5]^-$	Th: SVol.D4-205
$C_{12}Cl_6F_{10}N_2P_2$	$1,3-(C_6F_5)_2-1,3,2,4-N_2P_2-2,2,2,4,4,4-Cl_6$	F: PFHOrg.SVol.6-81, 88
$C_{12}Cl_6FeH_{11}O_2Sb$	$[(C_5H_5)Fe(CO)_2(C_5H_6-c)][SbCl_6]$	Fe: Org.Comp.B17-87
$C_{12}Cl_6FeH_{12}Sb$	$[(CH_2C_6H_5)Fe(C_5H_5)][SbCl_6]$	Fe: Org.Comp.B18-270
$C_{12}Cl_6Ga_2H_{24}O_3$	$Ga_2Cl_6 \cdot 3\ OC_4H_8$	Ga: SVol.D1-113
$C_{12}Cl_6Ga_2H_{24}O_6$	$[(C_{12}H_{24}O_6)GaCl_2][GaCl_4]$	Ga: SVol.D1-150/1
$C_{12}Cl_6Ga_2H_{28}O_6$	$[GaCl_2(CH_3-O-CH_2CH_2-O-CH_2CH_2-O-CH_3)_2][GaCl_4]$	Ga: SVol.D1-110
$C_{12}Cl_6H_4N_2S_3$	$(2,4,5-Cl_3C_6H_2-SN=)_2S$	S: S-N Comp.7-32/7
$C_{12}Cl_6H_8Sb^-$	$[(3-ClC_6H_4)_2SbCl_4]^-$	Sb: Org.Comp.5-165
–	$[(4-ClC_6H_4)_2SbCl_4]^-$	Sb: Org.Comp.5-165
$C_{12}Cl_6H_9N_2Sb$	$[4-ClC_6H_4-N_2][C_6H_5-SbCl_5]$	Sb: Org.Comp.5-246/7
–	$[C_6H_5-N_2][4-ClC_6H_4-SbCl_5]$	Sb: Org.Comp.5-251
$C_{12}Cl_6H_{14}N_2Pd$	$(4-Cl-C_6H_4-NH_3)_2[PdCl_4]$	Pd: SVol.B2-123
$C_{12}Cl_6H_{18}MnN_6Sn$	$[Mn(NCCH_3)_6][SnCl_6]$	Mn: MVol.D7-7/9
$C_{12}Cl_6H_{18}N_6PtRe$	$[Re(CNCH_3)_6][PtCl_6]$	Re: Org.Comp.2-291
$C_{12}Cl_6H_{18}O_4Sn$	$(C_4H_9)_2Sn(OOCCCl_3)_2$	Sn: Org.Comp.15-256
$C_{12}Cl_8Fe_2H_{18}MnN_6$	$[Mn(NCCH_3)_6][FeCl_4]_2$	Mn: MVol.D7-7/9
$C_{12}Cl_8Ga_2H_{18}MnN_6$	$[Mn(NCCH_3)_6][GaCl_4]_2$	Mn: MVol.D7-7/9
$C_{12}Cl_8H_{12}Pb$	$Pb(c-C_3H_3-Cl_2-2,2)_4$	Pb: Org.Comp.3-65
$C_{12}Cl_8H_{18}MnN_6Tl_2$	$[Mn(NCCH_3)_6][TlCl_4]_2$	Mn: MVol.D7-7/9
$C_{12}Cl_8MnNOS_4{}^{2-}$	$[Mn(1,2-S_2C_6Cl_4)_2NO]^{2-}$	Mn: MVol.D7-45
$C_{12}Cl_9H_{36}Mo_2N_3$	$[(CH_3)_4N]_3Mo_2Cl_9$	Mo: SVol.B5-304/5
$C_{12}Cl_{12}H_{12}O_4Th$	$Th[OC(CH_3)CCl_3]_4$	Th: SVol.C7-28
$C_{12}Cl_{12}H_{14}Mo_6N_2$	$Mo_6Cl_{12} \cdot 2\ CH_3C_5H_4N$	Mo: SVol.B5-267
$C_{12}Cl_{12}H_{18}MnN_6Sb_2$	$[Mn(NCCH_3)_6][SbCl_6]_2$	Mn: MVol.D7-7/9

$C_{12}Cl_{12}H_{30}Mo_6N_2$ $Mo_6Cl_{12} \cdot 2\,(C_2H_5)_3N$ Mo:SVol.B5-266
$C_{12}Cl_{12}H_{30}Mo_6P_2$ $Mo_6Cl_{12}((C_2H_5)_3P)_2$ Mo:SVol.B5-267
$C_{12}Cl_{12}H_{32}Mo_6N_4$ $[(Mo_6Cl_8)Cl_2((CH_3)_2NC_2H_4N(CH_3)_2)_2]Cl_2$ Mo:SVol.B5-267
$C_{12}Cl_{18}K_2O_{12}Th$ .. $K_2[Th(CCl_3COO)_6]$ .... Th: SVol.C7-59, 63
$C_{12}Cl_{18}Na_2O_{12}Th$ $Na_2[Th(CCl_3COO)_6]$ .... Th: SVol.C7-59, 63
$C_{12}CoF_{36}N_{12}O_6S_6^{2+}$
$[Co(NS-ON(CF_3)_2)_6]^{2+}$ .... S: S-N Comp.5-253/4
$C_{12}CoFeH_5O_7$ .... $[(C_5H_5)Fe(CO)_3][Co(CO)_4]$ .... Fe: Org.Comp.B15-38, 47
$C_{12}CoFeH_9O_4S_2$ .. $(C_5H_5)(CO)Fe(CO)(C(SCH_2)_2)Co(CO)_2$ .... Fe: Org.Comp.B16b-158
$C_{12}CoFeH_{11}O_4S_2$ $(C_5H_5)(CO)Fe(CO)(C(SCH_3)_2)Co(CO)_2$ .... Fe: Org.Comp.B16b-156/7
$C_{12}CoFeH_{12}^{+}$ .... $[(C_5H_4-CH_2)_2CoFe]^{+}$ .... Fe: Org.Comp.B18-2
– .... $[(C_6H_6)_2FeCo]^{+}$ .... Fe: Org.Comp.B19-354
$C_{12}Co_2H_{28}N_{20}O_2Pd_3$
$[(NC)_3Pd-CN-Co(NH_3)_4(H_2O)]_2[Pd(CN)_4] \cdot 2\,H_2O$
Pd: SVol.B2-287
$C_{12}Co_2H_{30}N_{22}Pd_3$ $[(NC)_3Pd-CN-Co(NH_3)_5]_2[Pd(CN)_4]$ .... Pd: SVol.B2-287
$C_{12}Co_2H_{34}N_{22}O_2Pd_3$
$[Co(NH_3)_5(H_2O)]_2[Pd(CN)_4]_3$ .... Pd: SVol.B2-279
$C_{12}CrGeH_{14}O_3$ ... $Ge(CH_3)_3C_6H_5 \cdot Cr(CO)_3$ .... Ge: Org.Comp.2-73
$C_{12}CrH_6O_{10}PRe$ .. $(CO)_5ReP(CH_3)_2-Cr(CO)_5$ .... Re: Org.Comp.2-174
$C_{12}CrH_{15}NO_2S$ ... $(C_5(CH_3)_5)Cr(NS)(CO)_2$ .... S: S-N Comp.5-55
$C_{12}CrH_{42}Mo_6N_3O_{24}$
$(t-C_4H_9NH_3)_3[Cr(OH)_6Mo_6O_{18}]$ Mo:SVol.B3b-254
$C_{12}Cr_2H_{48}Mo_3N_{12}S_{12}$
$[Cr(NH_2-CH_2CH_2-NH_2)_3]_2[MoS_4]_3$ Mo:SVol.B7-289
$C_{12}Cr_2H_{48}Mo_3N_{24}O_{12}S_{12}$
$[Cr((NH_2)_2C=O)_6]_2[MoS_4]_3$ Mo:SVol.B7-289
$C_{12}Cr_4H_{54}Mo_3N_{12}O_6S_{12}$
$[Cr_4(OH)_6(NH_2-CH_2CH_2-NH_2)_6][MoS_4]_3$ Mo:SVol.B7-285
$C_{12}CsFH_{10}N_2O_4S_3$ $Cs[C_6H_5-S(O)_2-NS(F)N-S(O)_2-C_6H_5]$ .... S: S-N Comp.8-182
$C_{12}CsF_4N_4$ ...... $Cs[2,5-C_6F_4-1,4-(=C(CN)_2)_2]$, radical .... F: PFHOrg.SVol.6-106, 112/3
$C_{12}CuF_4N_4$ ...... $Cu[2,5-C_6F_4-1,4-(=C(CN)_2)_2]$, radical .... F: PFHOrg.SVol.6-106, 112/3
$C_{12}CuFe_3H_{10}O_9PS$
$(CO)_9(H)Fe_3S[Cu-P(CH_3)_3]$ .... Fe: Org.Comp.C6a-314, 316
$C_{12}FFeH_{10}O_2^{+}$ ... $[(F-C_6H_5)Fe(C_5H_4-COOH)]^{+}$ .... Fe: Org.Comp.B18-142/6, 197/8, 245
$C_{12}FFeH_{11}O_4S$ ... $FO_2S-C_5H_4FeC_5H_4-C(O)OCH_3$ .... Fe: Org.Comp.A9-232
$C_{12}FFeH_{12}$ ...... $(4-F-C_6H_4-CH_3)Fe(C_5H_5)$ .... Fe: Org.Comp.B19-32
$C_{12}FFeH_{12}^{+}$ ..... $[(3-F-C_6H_4-CH_3)Fe(C_5H_5)][PF_6]$ .... Fe: Org.Comp.B19-1, 32
– .... $[(4-F-C_6H_4-CH_3)Fe(C_5H_5)][PF_6]$ .... Fe: Org.Comp.B19-1, 32
$C_{12}FFeH_{13}O$ ..... $C_5H_5FeC_6H_5F-6(OCH_3-6)$ .... Fe: Org.Comp.B17-263
$C_{12}FFeH_{16}NO_5S$ .. $[C_5H_5(CO)_2Fe=CH(N(CH_3)C_3H_7-i)][FSO_3]$ ..... Fe: Org.Comp.B16a-115, 126
$C_{12}FFeH_{20}O_5PS$ .. $[C_5H_5(CO)(P(CH_3)_3)Fe=C(CH_3)OCH_3][FSO_3]$ ... Fe: Org.Comp.B16a-36, 52
$C_{12}FGeH_{19}$ ...... $Ge(C_2H_5)_3C_6H_4F-4$ .... Ge: Org.Comp.2-275
$C_{12}FGeH_{33}Si_3$ .... $Ge(CH_3)_3C[Si(CH_3)_3]_2Si(CH_3)_2F$ .... Ge: Org.Comp.1-144
$C_{12}FH_8N_3O_2S$ .... $4-F-C_6H_4-N=S=N-C_6H_4-NO_2-4$ .... S: S-N Comp.7-258/9
$C_{12}FH_{10}N_2O_4S_3^{-}$ $[C_6H_5-S(O)_2-NS(F)N-S(O)_2-C_6H_5]^{-}$ .... S: S-N Comp.8-182
$C_{12}FH_{10}O_2Sb$ .... $(C_6H_5)(4-FC_6H_4)Sb(O)OH$ .... Sb: Org.Comp.5-222
$C_{12}FH_{20}NSn$ ..... $(C_2H_5)_2N-Sn(CH_3)_2-C_6H_4-4-F$ .... Sn: Org.Comp.19-47/8, 53
$C_{12}F_2FeH_{10}^{2+}$ ... $[(F-C_6H_5)_2Fe][O-C_6H_2-2,4,6-(NO_2)_3]_2$ .... Fe: Org.Comp.B19-358

$C_{12}F_2FeH_{10}O$	$C_5H_5FeCOCF_2C_5H_5$	Fe: Org.Comp.B17-254, 255/6
$C_{12}F_2GeH_{17}P$	$(CH_3)_2Ge(-CF_2P(C_6H_5)CH_2CH_2CH_2-)$	Ge: Org.Comp.3-294
$C_{12}F_2H_8N_2O_4S_3$	$4-F-C_6H_4-S(O)_2N=S=NS(O)_2-C_6H_4-F-4$	S: S-N Comp.7-95/8
$C_{12}F_2H_8N_2S$	$4-F-C_6H_4-N=S=N-C_6H_4-4-F$	S: S-N Comp.7-236
$C_{12}F_2H_9O_2Sb$	$(4-FC_6H_4)_2Sb(O)OH$	Sb: Org.Comp.5-208
$C_{12}F_3FeH_7N_2O_2$	$2-[(C_5H_5)Fe(CO)_2]-4-NH_2-NC_5F_3-3,5,6$	Fe: Org.Comp.B14-26
$C_{12}F_3FeH_{10}NO_6S$	$[(C_5H_5)Fe(CO)_2(NCCH=CHOCH_3)][CF_3SO_3]$	Fe: Org.Comp.B17-66, 79
$C_{12}F_3FeH_{11}O$	$C_5H_5Fe(C(CF_3)=CHCH_2CH=CH_2)CO$	Fe: Org.Comp.B17-172
$C_{12}F_3FeH_{11}O_4$	$[(C_5H_5)Fe(CO)_2(CH_2=CHCH_3)][CF_3COO]$	Fe: Org.Comp.B17-6
$C_{12}F_3FeH_{11}O_5$	$[(C_5H_5)Fe(CO)_2(CH_2=C(CH_3)OH)][CF_3CO_2]$	Fe: Org.Comp.B17-67
$C_{12}F_3FeH_{11}O_5S$	$[C_5H_5(CO)_2Fe=CH(C_3H_5-c)][CF_3SO_3]$	Fe: Org.Comp.B16a-88, 96
$C_{12}F_3FeH_{13}O_4S$	$[C_5H_5Fe(=CHCH_2CH_2CH=CH_2)CO][CF_3SO_3]$	Fe: Org.Comp.B17-167
$C_{12}F_3FeH_{13}O_5S$	$[(C_5H_5)(CO)_2Fe=CH-C_3H_7-i][CF_3SO_3]$	Fe: Org.Comp.B16a-87
$C_{12}F_3FeH_{13}O_6S$	$[(C_5H_5)Fe(CO)_2=C(C_2H_5)OCH_3][CF_3SO_3]$	Fe: Org.Comp.B16a-109, 120
–	$[(C_5H_5)Fe(CO)_2(CH_3CH=CH-O-CH_3)][CF_3SO_3]$	Fe: Org.Comp.B17-64
$C_{12}F_3FeH_{13}O_7S$	$[C_5H_5(CO)_2Fe=C(OCH_3)OC_2H_5][CF_3SO_3]$	Fe: Org.Comp.B16a-136, 149
$C_{12}F_3FeH_{14}NO_4S_3$	$[C_5H_5(CO)(CH_3CN)Fe=C(SCH_3)_2][CF_3SO_3]$	Fe: Org.Comp.B16a-66, 71/2
$C_{12}F_3FeH_{14}NO_5S_2$	$[C_5H_5(CO)(CH_3CN)Fe=C(OCH_3)SCH_3][CF_3SO_3]$	Fe: Org.Comp.B16a-65
$C_{12}F_3FeH_{14}NO_6S$	$[C_5H_5(CO)_2Fe=C(OCH_3)N(CH_3)_2][CF_3SO_3]$	Fe: Org.Comp.B16a-138
$C_{12}F_3FeH_{14}NO_7S$	$[C_5H_5(CO)_2Fe=C(OCH_3)NHCH_2CH_2OH][CF_3SO_3]$	Fe: Org.Comp.B16a-138
$C_{12}F_3FeH_{16}O_7PS$	$[C_5H_5(CO)(P(OCH_3)_3)Fe=C=CH_2][CF_3SO_3]$	Fe: Org.Comp.B16a-207, 210
$C_{12}F_3FeH_{18}O_4PS$	$[C_5H_5(CO)(P(CH_3)_3)Fe=CH(CH_3)][CF_3SO_3]$	Fe: Org.Comp.B16a-18, 24/5
$C_{12}F_3FeH_{18}O_5PS$	$[C_5H_5(CO)(P(CH_3)_3)Fe=C(CH_3)OH][CF_3SO_3]$	Fe: Org.Comp.B16a-35, 51
$C_{12}F_3FeH_{18}O_8PS$	$[(C_5H_5)Fe(CO)(CH_2=CHOH)-P(OCH_3)_3][CF_3SO_3]$	Fe: Org.Comp.B16b-86
–	$[(C_5H_5)Fe(CO)(P(OCH_3)_3)=C(CH_3)OH][CF_3SO_3]$	Fe: Org.Comp.B16a-36
$C_{12}F_3GaH_{27}N$	$GaF_3[N(C_4H_9-n)_3]$	Ga: SVol.D1-233
$C_{12}F_3H_4O_6ReS$	$(CO)_4Re[-O=C(2-H_3C_4S)-CH=C(CF_3)-O-]$	Re: Org.Comp.1-356
$C_{12}F_3H_9MoO_4$	$(C_5H_5)Mo(CO)_2[-O-C(CH_3)CHC(CF_3)-O-]$	Mo: Org.Comp.7-189, 199
$C_{12}F_3H_{10}Sb$	$(C_6H_5)_2SbF_3$	Sb: Org.Comp.5-134
–	$(C_6H_5)_2SbF_3 \cdot H_2O$	Sb: Org.Comp.5-135/6
$C_{12}F_3H_{11}NO_9Re$	$[(CO)_5ReNH_2C(CH_3)(H)COOC_2H_5][O_2CCF_3]$	Re: Org.Comp.2-151
$C_{12}F_3H_{12}InO_2$	$(CH_3)_2In[-OC(CF_3)CHC(C_6H_5)O-]$	In: Org.Comp.1-190, 191
$C_{12}F_3H_{13}MoO_6$	$(CH_2CHCH_2)Mo(CO)_3(OC_4H_8)-OC(=O)CF_3$	Mo: Org.Comp.5-292, 293
–	$(CH_2CHCH_2)Mo(CO)_3[O=C(CH_3)-C_2H_5]-OC(=O)CF_3$	Mo: Org.Comp.5-293
$C_{12}F_3H_{13}N_2O_2S$	$1,4-ONC_4H_8-4-S(O)-N=C(C_6H_5)CF_3$	S: S-N Comp.8-353, 360
$C_{12}F_4FeH_5NO_2$	$4-(C_5H_5)Fe(CO)_2-NC_5F_4$	Fe: Org.Comp.B14-26, 29
$C_{12}F_4FeH_8$	$(1,4-F_2-C_6H_4)_2Fe$	Fe: Org.Comp.B19-348, 361
$C_{12}F_4H_3O_5ReS$	$(CO)_5ReC_6F_4SCH_3-4$	Re: Org.Comp.2-142
$C_{12}F_4H_{10}KSb$	$K[(C_6H_5)_2SbF_4]$	Sb: Org.Comp.5-136
$C_{12}F_4H_{10}NaSb$	$Na[(C_6H_5)_2SbF_4]$	Sb: Org.Comp.5-136
$C_{12}F_4H_{10}Sb^-$	$[(C_6H_5)_2SbF_4]^-$	Sb: Org.Comp.5-136, 138
$C_{12}F_4H_{11}Sb$	$H[(C_6H_5)_2SbF_4]$	Sb: Org.Comp.5-136
$C_{12}F_4H_{18}O_4Sn$	$(C_4H_9)_2SnOOCCF_2CF_2COO$	Sn: Org.Comp.15-322
$C_{12}F_4H_{27}P_3Sn$	$(n-C_4H_9)_3Sn-P(PF_2)_2$	Sn: Org.Comp.19-192, 194
$C_{12}F_4KN_4$	$K[2,5-C_6F_4-1,4-(=C(CN)_2)_2]$, radical	F: PFHOrg.SVol.6-106, 112/3
$C_{12}F_4LiN_4$	$Li[2,5-C_6F_4-1,4-(=C(CN)_2)_2]$, radical	F: PFHOrg.SVol.6-106, 112/3
$C_{12}F_4NO_5Re$	$(CO)_5ReC_6F_4CN-4$	Re: Org.Comp.2-143

$C_{12}F_4N_4$ .... 2,5-C_6F_4-1,4-[=C(CN)$_2$]$_2$ .... F: PFHOrg.SVol.6-106, 112/3, 146/8

$C_{12}F_4N_4^-$........ [2,5-C_6F_4-1,4-(=C(CN)$_2$)$_2$]$^-$, radical anion .... F: PFHOrg.SVol.6-106, 112/3, 146

$C_{12}F_4N_4Na$ ...... Na[2,5-C_6F_4-1,4-(=C(CN)$_2$)$_2$], radical........ F: PFHOrg.SVol.6-106, 112/3

$C_{12}F_4N_4Rb$ ...... Rb[2,5-C_6F_4-1,4-(=C(CN)$_2$)$_2$], radical........ F: PFHOrg.SVol.6-106, 112/3

$C_{12}F_5H_4IN_2S$..... 4-I-C_6H_4-N=S=N-C_6F_5 .... S: S-N Comp.7-260

$C_{12}F_5H_4N_3O_2S$ ... 4-NO_2-C_6H_4-N=S=N-C_6F_5................. S: S-N Comp.7-262

$C_{12}F_5H_5N_2S$ ..... C_6H_5-N=S=N-C_6F_5...................... S: S-N Comp.7-258, 263/4

– .... C_6H_5-^{15}N=S=N-C_6F_5 .... S: S-N Comp.7-258

$C_{12}F_5H_5N_2S_2$ .... C_6F_5-S-N=S=N-C_6H_5.................... S: S-N Comp.7-25/7

– .... C_6H_5-S-N=S=N-C_6F_5.................... S: S-N Comp.7-25/7

$C_{12}F_5H_5N_2S_3$ .... C_6F_5-SN=S=NS-C_6H_5.................... S: S-N Comp.7-39

$C_{12}F_5H_8Sb$ ...... $(C_6H_4F\text{-}4)_2SbF_3 \cdot H_2O$ .... Sb: Org.Comp.5-137

$C_{12}F_5H_{18}NOSSi_2$. $((CH_3)_3Si)_2$N-S(O)-C_6F_5................. S: S-N Comp.8-298

$C_{12}F_5O_6Re$ ...... $(CO)_5ReC(O)C_6F_5$......................... Re: Org.Comp.2-123

$C_{12}F_6FeH_7O_3P$ ... [$(C_9H_7)Fe(CO)_3$][PF_6] .... Fe: Org.Comp.B15-53

$C_{12}F_6FeH_9O_3P$ ... [(8-CH_2=C_8H_7)Fe$(CO)_3$][PF_6]................ Fe: Org.Comp.B15-241, 258

$C_{12}F_6FeH_{10}KO_2P$ K[(OOC-C_6H_5)Fe(C_5H_5)][PF_6] .... Fe: Org.Comp.B18-142/6, 202

$C_{12}F_6FeH_{10}NP$ ... [(NC-C_6H_5)Fe(C_5H_5)][PF_6]................. Fe: Org.Comp.B18-142/6, 198, 201, 202, 268

$C_{12}F_6FeH_{11}N_2P$ .. [(1,3-$N_2C_7H_6$)Fe(C_5H_5)][PF_6]............... Fe: Org.Comp.B19-216, 289

$C_{12}F_6FeH_{11}OP$ ... [(c-C_7H_6=O-7)Fe(C_5H_5)][PF_6] .... Fe: Org.Comp.B19-340/1

$C_{12}F_6FeH_{11}O_2P$ .. [$(C_5H_5)Fe(CO)_2(C_5H_6$-c)][PF_6].............. Fe: Org.Comp.B17-87

– .... [(C_6H_6)Fe(C_5H_4-COOH)][PF_6] .... Fe: Org.Comp.B18-142/6, 151, 154, 160, 185

– .... [(HOOC-C_6H_5)Fe(C_5H_5)][PF_6] .... Fe: Org.Comp.B18-142/6, 203, 268/9

$C_{12}F_6FeH_{11}O_3P$ .. [(2-CH_3CH-(2.2.1)-C_7H_7)Fe$(CO)_3$][PF_6] ...... Fe: Org.Comp.B15-257/8

– .... [(6-(CH_3CH=)C_7H_7)Fe$(CO)_3$][PF_6] .... Fe: Org.Comp.B15-209, 225

$C_{12}F_6FeH_{11}O_4P$ .. [(2-CH_3C(O)-C_7H_8)Fe$(CO)_3$][PF_6] .... Fe: Org.Comp.B15-206

– .... [(6-CH_3C(O)-C_7H_8)Fe$(CO)_3$][PF_6] .... Fe: Org.Comp.B15-208, 223

– .... [(6-CH_3C(OH)=C_7H_7)Fe$(CO)_3$][PF_6].......... Fe: Org.Comp.B15-210

$C_{12}F_6FeH_{11}O_5P$ .. [(1-CH_3OC(O)-2-CH_3-C_6H_5)Fe$(CO)_3$][PF_6].... Fe: Org.Comp.B15-110, 159/60

– .... [(1-CH_3OC(O)-3-CH_3-C_6H_5)Fe$(CO)_3$][PF_6].... Fe: Org.Comp.B15-112, 161

$C_{12}F_6FeH_{11}O_6P$ .. [(1-CH_3OC(O)-2-CH_3O-C_6H_5)Fe$(CO)_3$][PF_6] .. Fe: Org.Comp.B15-110, 159

$C_{12}F_6FeH_{11}PS$ ... [(c-C_7H_6=S-7)Fe(C_5H_5)][PF_6] .... Fe: Org.Comp.B19-340, 341/2

$C_{12}F_6FeH_{12}NOP$ .. [(NH_2-C(=O)C_6H_5)Fe(C_5H_5)][PF_6] .... Fe: Org.Comp.B18-142/6, 203

$C_{12}F_6FeH_{12}NO_2P$ [(CH_3-C_6H_4-2-NO_2)Fe(C_5H_5)][PF_6].......... Fe: Org.Comp.B19-4/6, 63

– .... [(CH_3-C_6H_4-3-NO_2)Fe(C_5H_5)][PF_6].......... Fe: Org.Comp.B19-4/6, 63

– .... [(CH_3-C_6H_4-4-NO_2)Fe(C_5H_5)][PF_6].......... Fe: Org.Comp.B19-4/6, 63/4

– .... [(O_2N-C_6H_5)Fe(C_5H_4-CH_3)][PF_6]............ Fe: Org.Comp.B18-142/6, 197/8, 255

$C_{12}F_6FeH_{12}NO_4P$ [(C_5H_5)Fe(CNCH(CH_3)COOCH_3)$(CO)_2$][PF_6] ... Fe: Org.Comp.B15-314

$C_{12}F_6FeH_{12}N_3P$ .. [(2-CH_3-C_6H_4-N_3)Fe(C_5H_5)][PF_6] .... Fe: Org.Comp.B19-2, 62

– .... [(3-CH_3-C_6H_4-N_3)Fe(C_5H_5)][PF_6] .... Fe: Org.Comp.B19-2, 62

– .... [(4-CH_3-C_6H_4-N_3)Fe(C_5H_5)][PF_6] .... Fe: Org.Comp.B19-2, 62/3

$C_{12}F_6FeH_{12}P$ .... [$(C_6H_6)_2$Fe][PF_6]........................ Fe: Org.Comp.B19-348, 355

$C_{12}F_6FeH_{13}N_2O_4P$ $[C_5H_5(CO)_2Fe{=}C(-N(C(=O)NHCH_3)CH_2CH_2O-)][PF_6]$ Fe: Org.Comp.B16a-139/40, 151

$C_{12}F_6FeH_{13}OP$ ... $[(CH_3O-C_6H_5)Fe(C_5H_5)][PF_6]$ ... Fe: Org.Comp.B18-142/6, 197/9, 201, 248/9

– ... $[(C_6H_6)Fe(C_5H_4-CH_2OH)][PF_6]$ ... Fe: Org.Comp.B18-142/6, 162

– ... $[(C_6H_6)Fe(C_5H_4-OCH_3)][PF_6]$ ... Fe: Org.Comp.B18-142/6, 168/9

– ... $[(HOCH_2-C_6H_5)Fe(C_5H_5)][PF_6]$ ... Fe: Org.Comp.B18-142/6, 209

$C_{12}F_6FeH_{13}O_2P$ .. $[(C_5H_5)Fe(CO)_2(CH_2{=}CHC(CH_3){=}CH_2)][PF_6]$ ... Fe: Org.Comp.B17-37

– ... $[(C_5H_5)Fe(CO)_2(CH_2{=}CHCH{=}CHCH_3)][PF_6]$ ... Fe: Org.Comp.B17-37

– ... $[(C_5H_5)Fe(CO)_2(CH_2{=}CHCH_2CH{=}CH_2)][PF_6]$ ... Fe: Org.Comp.B17-38

– ... $[(C_5H_5)Fe(CO)_2(C_5H_8-c)][PF_6]$ ... Fe: Org.Comp.B17-87

$C_{12}F_6FeH_{13}O_2PS$ $[(CH_3-SO_2-C_6H_5)Fe(C_5H_5)][PF_6]$ ... Fe: Org.Comp.B18-142/6, 254

– ... $[(C_5H_5)Fe(CO)_2-C(=CH-CH_2-S(CH_3)-CH_2-)][PF_6]$ Fe: Org.Comp.B14-149, 155

$C_{12}F_6FeH_{13}O_3P$ .. $[(1,3,5-(CH_3)_3-C_6H_4)Fe(CO)_3][PF_6]$ ... Fe: Org.Comp.B15-129, 181/2

– ... $[(1-(i-C_3H_7)-C_6H_6)Fe(CO)_3][PF_6]$ ... Fe: Org.Comp.B15-99, 135/6

$C_{12}F_6FeH_{13}O_3Sb$ $[(C_5H_5)Fe(CO)_2(CH_2{=}CHCH_2COCH_3)][SbF_6]$ ... Fe: Org.Comp.B17-33

$C_{12}F_6FeH_{13}O_4P$ .. $[(1,2-(CH_3)_2-4-CH_3O-C_6H_4)Fe(CO)_3][PF_6]$ ... Fe: Org.Comp.B15-126, 179

– ... $[(1,3-(CH_3)_2-4-CH_3O-C_6H_4)Fe(CO)_3][PF_6]$ ... Fe: Org.Comp.B15-129, 181

– ... $[(1,5-(CH_3)_2-3-CH_3O-C_6H_4)Fe(CO)_3][PF_6]$ ... Fe: Org.Comp.B15-129, 181

– ... $[(1-C_2H_5-4-CH_3O-C_6H_5)Fe(CO)_3][PF_6]$ ... Fe: Org.Comp.B15-114, 169/70

– ... $[(C_5H_5)Fe(CO)_2(CH_3CH{=}CH-COO-CH_3)][PF_6]$ .. Fe: Org.Comp.B17-59

$C_{12}F_6FeH_{13}O_5P$ .. $[C_5H_5(CO)Fe(=C(OCH_2)_2)_2][PF_6]$ ... Fe: Org.Comp.B16a-193/4

$C_{12}F_6FeH_{13}P$ .... $[(CH_3C_6H_5)Fe(C_5H_5)][PF_6]$ ... Fe: Org.Comp.B18-142/6, 197, 200, 201, 204/5, 269/72

– ... $[(C_6H_6)Fe(C_5H_4-CH_3)][PF_6]$ ... Fe: Org.Comp.B18-142/6, 151, 154, 161, 185/6

– ... $[(C_6H_6)Fe(C_6H_7)][PF_6]$ ... Fe: Org.Comp.B18-142/6, 153, 172, 189/90

– ... $[(C_6H_6)Fe(C_6H_6D)][PF_6]$ ... Fe: Org.Comp.B18-189

– ... $[(C_6D_6)Fe(C_6D_6H)][PF_6]$ ... Fe: Org.Comp.B18-189

– ... $[(c-C_7H_8)Fe(C_5H_5)][PF_6]$ ... Fe: Org.Comp.B19-340/1, 343

$C_{12}F_6FeH_{13}Sb$ ... $[(CH_3-C_6H_5)Fe(C_5H_5)][SbF_6]$ ... Fe: Org.Comp.B18-142/6, 197, 201, 205, 269/72

$C_{12}F_6FeH_{14}NOPS_2$
$[C_5H_5(CO)(CH_3CN)Fe{=}C(SCH_2)_2CH_2][PF_6]$ ... Fe: Org.Comp.B16a-68

$C_{12}F_6FeH_{14}NO_2P$ $[2-(C_5H_5)Fe(CO)-1,3-ONC_3H_4-3-CH_2-CH{=}CH_2][PF_6]$ Fe: Org.Comp.B17-167

– ... $[(C_5H_5)Fe(CO)_2-CN-C_4H_9-t][PF_6]$ ... Fe: Org.Comp.B15-313, 317

$C_{12}F_6FeH_{14}NP$ ... $[(2-CH_3-C_6H_4-NH_2)Fe(C_5H_5)][PF_6]$ ... Fe: Org.Comp.B19-1, 5/6, 53, 92/3

– ... $[(3-CH_3-C_6H_4-NH_2)Fe(C_5H_5)][PF_6]$ ... Fe: Org.Comp.B19-1, 5/6, 53, 92/3

– ... $[(4-CH_3-C_6H_4-NH_2)Fe(C_5H_5)][PF_6]$ ... Fe: Org.Comp.B19-1, 5/6, 54, 92/3

$C_{12}F_6FeH_{14}NP$ . . . $[(CH_3\text{-}NH\text{-}C_6H_5)Fe(C_5H_5)][PF_6]$. . . . . Fe: Org.Comp.B18-142/6, 198, 257/8

– . . . . $[(NH_2\text{-}C_6H_5)Fe(C_5H_4\text{-}CH_3)][PF_6]$. . . . . Fe: Org.Comp.B18-142/6, 197/8, 257

$C_{12}F_6FeH_{14}P_2$. . . . $[(C_6H_6)Fe(PC_4H_2(CH_3)_2\text{-}3,4)][PF_6]$. . . . . Fe: Org.Comp.B18-130

$C_{12}F_6FeH_{15}N_2OP$ $[(C_5H_5)Fe(CN\text{-}C_2H_5)_2CO][PF_6]$ . . . . Fe: Org.Comp.B15-335

$C_{12}F_6FeH_{15}N_2O_2P$ $[(C_5H_5)Fe(CN\text{-}CH_2CH_2N(CH_3)_2)(CO)_2][PF_6]$ . . . Fe: Org.Comp.B15-314

– . . . . $[(C_5H_5)Fe(CO)_2{=}C(\text{-}NH\text{-}CH_2\text{-}C(CH_3)_2\text{-}NH\text{-})][PF_6]$ Fe: Org.Comp.B16a-147

$C_{12}F_6FeH_{15}OP$ . . . $[(C_5H_5)Fe(C(CH_3)H{=}CHCH{=}CHCH_3)CO][PF_6]$ . . Fe: Org.Comp.B17-187, 191

$C_{12}F_6FeH_{15}O_2P$ . . $[(C_5H_5)Fe(CO)_2(CH_2{=}CHC_3H_7\text{-}i)][PF_6]$ . . . . Fe: Org.Comp.B17-11

$C_{12}F_6FeH_{15}O_3P$ . . $[(CH_2CHCHCHCH\text{-}C_4H_9\text{-}n)Fe(CO)_3][PF_6]$ . . . . . Fe: Org.Comp.B15-26

– . . . . $[(C_5H_5)Fe(CO)_2(CH_3CH{=}CH\text{-}O\text{-}C_2H_5)][PF_6]$ . . . Fe: Org.Comp.B17-65, 78

$C_{12}F_6FeH_{15}O_3PSi$ $[(3\text{-}(CH_3)_3Si\text{-}C_6H_6)Fe(CO)_3][PF_6]$ . . . . Fe: Org.Comp.B15-107, 158

$C_{12}F_6FeH_{15}O_4P$ . . $[(C_5H_5)Fe(CO)_2(CH_2{=}C(OCH_3)OC_2H_5)][PF_6]$ . . . Fe: Org.Comp.B17-69, 81

$C_{12}F_6FeH_{17}N_2O_6PS_3$
$[(C_5H_5)Fe(CO)_2(CH_2{=}CHCH_2\text{-}N(SO_2CH_3)S$
$\text{-}NH\text{-}SO_2CH_3)][PF_6]$ . . . . Fe: Org.Comp.B17-27, 50

$C_{12}F_6FeH_{17}N_2PS_2$ $[(C_5H_5)(CH_3NC)_2Fe{=}C(SCH_3)_2][PF_6]$ . . . . Fe: Org.Comp.B16a-169/70

$C_{12}F_6FeH_{18}NO_2P$ $[(C_5H_5)Fe(CO)_2\text{-}CH_2CH_2\text{-}N(CH_3)_3][PF_6]$ . . . . . . Fe: Org.Comp.B14-150, 155

$C_{12}F_6FeH_{18}NO_3P$ $[(C_5H_5)Fe(CO)_2\text{-}CH_2\text{-}O\text{-}CH_2CH_2\text{-}NH(CH_3)_2][PF_6]$ Fe: Org.Comp.B14-152, 157

$C_{12}F_6FeH_{18}N_3OP$ $[(C_5H_5)(CH_3NC)(CO)Fe{=}C(NHCH_3)N(CH_3)_2][PF_6]$ Fe: Org.Comp.B16a-165, 168

$C_{12}F_6FeH_{20}O_4P_2S_2$
$[(C_5H_5)(CO)(P(OCH_3)_3)Fe{=}C(SCH_3)_2][PF_6]$ . . . . Fe: Org.Comp.B16a-67, 72

$C_{12}F_6FeH_{20}O_5P_2$. $[(C_5H_5)(CO)(P(OCH_3)_3)Fe{=}C(CH_3)OCH_3][PF_6]$. . Fe: Org.Comp.B16a-37, 52

$C_{12}F_6FeH_{23}P_3$. . . . $[(C_6H_6)Fe(H)((CH_3)_2P\text{-}CH_2CH_2\text{-}P(CH_3)_2)][PF_6]$ Fe: Org.Comp.B18-5

$C_{12}F_6FeH_{25}P_3$. . . . $[(C_6H_6)Fe(P(CH_3)_3)_2H][PF_6]$. . . . . Fe: Org.Comp.B18-4

$C_{12}F_6GaH_{24}K_3O_{12}$ $K_3[GaF_6]\cdot 6\ HOC(O)\text{-}CH_3$ . . . . Ga: SVol.D1-154

$C_{12}F_6GeH_9MnO_5$ $Ge(CH_3)_3[\text{-}CF\text{-}CF_2CF_2\text{-}CF(Mn(CO)_5)\text{-}]$. . . . . . . Ge: Org.Comp.1-195

– . . . . $Ge(CH_3)_3\text{-}C(CF_3){=}C(CF_3)\text{-}Mn(CO)_5$. . . . . Ge: Org.Comp.2-11

– . . . . $Ge(CH_3)_3\text{-}CF_2CF_2\text{-}CF{=}CF\text{-}Mn(CO)_5$ . . . . Ge: Org.Comp.2-15/6

$C_{12}F_6GeH_{26}Sn_2$ . . $(CH_3)_2Ge[CF(CHF_2)\text{-}Sn(CH_3)_3]_2$ . . . . Ge: Org.Comp.3-149

– . . . . $(CH_3)_2Ge[CHF\text{-}CF_2\text{-}Sn(CH_3)_3]_2$. . . . . Ge: Org.Comp.3-149

– . . . . $(CH_3)_3Sn\text{-}CF_2\text{-}CHF\text{-}Ge(CH_3)_2\text{-}CF(CHF_2)\text{-}Sn(CH_3)_3$ Ge: Org.Comp.3-196

$C_{12}F_6H_4N_2S$ . . . . . $4\text{-}F\text{-}C_6H_4\text{-}N{=}S{=}N\text{-}C_6F_5$ . . . . S: S-N Comp.7-259

$C_{12}F_6H_5NO_5PRe$. . $[(CO)_5Re(CNC_6H_5)][PF_6]$ . . . . Re: Org.Comp.2-253

$C_{12}F_6H_6MoO_3$. . . . $(C_5H_5)Mo(CO)_2[{=}C(CF_3)\text{-}CH(CF_3)\text{-}C({=}O)\text{-}]$ . . . . Mo: Org.Comp.8-196, 200/1

$C_{12}F_6H_8Sb^-$. . . . . . $[(C_6H_4F\text{-}4)_2SbF_4]^-$ . . . . Sb: Org.Comp.5-137

$C_{12}F_6H_{10}MoOS$. . . $(C_5H_5)Mo(CO)(SC_2H_5)(F_3C\text{-}C{\equiv}C\text{-}CF_3)$. . . . . Mo: Org.Comp.6-277

$C_{12}F_6H_{13}MoN_2O_5P$
$[(C_5H_5)Mo(CO)_2(\text{-}C(OH){=}C(COO\text{-}C_2H_5)\text{-}NH\text{-}NH\text{-})][PF_6]$ Mo: Org.Comp.8-139/40

$C_{12}F_6H_{14}MoNO_2P$ $[(C_5H_5)Mo(CO)(NO)C_6H_9\text{-}c][PF_6]$. . . . . Mo: Org.Comp.6-350

$C_{12}F_6H_{14}MoNO_3P$ $[(H_3CC(O)\text{-}C_5H_4)Mo(CO)(NO)(CH_2C(CH_3)CH_2)][PF_6]$ Mo: Org.Comp.6-348

– . . . . $[(H_3CC(O)\text{-}C_5H_4)Mo(CO)(NO)(CH_2CHCH\text{-}CH_3)][PF_6]$ Mo: Org.Comp.6-348

$C_{12}F_6H_{14}MoN_3OP$	$[(C_5H_5)Mo(CO)(CNCH_3)_3][PF_6]$	Mo: Org.Comp.6-274
$C_{12}F_6H_{14}N_2Sn$	$(CH_3)_3SnN{=}C{=}C(CN)C(CF_3)_2CH_2CH{=}CH_2$	Sn: Org.Comp.18-107, 109
$C_{12}F_6H_{18}NPS$	$[SN][PF_6] \cdot (CH_3)_6C_6$	S: S-N Comp.5-47/9
$C_{12}F_6H_{18}NSSb$	$[SN][SbF_6] \cdot (CH_3)_6C_6$	S: S-N Comp.5-47/9
$C_{12}F_6H_{18}N_6PRe$	$[Re(CNCH_3)_6][PF_6]$	Re: Org.Comp.2-291, 297/8
$C_{12}F_6H_{18}O_4Sn$	$(C_4H_9)_2Sn(OOCCF_3)_2$	Sn: Org.Comp.15-255
$C_{12}F_6H_{18}O_6PRe$	$[(CO)_3Re(OC(CH_3)_2)_3][PF_6]$	Re: Org.Comp.1-310
$C_{12}F_6H_{18}O_6S_2Sn$	$(C_4H_9)_2Sn(OSO_2CFCF_2)_2$	Sn: Org.Comp.15-348
$C_{12}F_6H_{22}O_2Sn$	$(C_4H_9)_2Sn(OCH_2CF_3)_2$	Sn: Org.Comp.15-5, 22
$C_{12}F_6H_{36}N_6S_2Si$	$[((CH_3)_2N)_3S]_2[SiF_6]$	S: S-N Comp.8-230, 238
$C_{12}F_7FeH_5O_2$	$(C_5H_5)Fe(CO)_2C({=}CFCF_2CF_2CF_2{-})$	Fe: Org.Comp.B13-213, 246
$C_{12}F_7FeH_{10}O_2P$	$[(F{-}C_6H_5)Fe(C_5H_4{-}COOH)][PF_6]$	Fe: Org.Comp.B18-142/6, 197/8, 245
$C_{12}F_7FeH_{12}P$	$[(3{-}F{-}C_6H_4{-}CH_3)Fe(C_5H_5)][PF_6]$	Fe: Org.Comp.B19-1, 32
–	$[(4{-}F{-}C_6H_4{-}CH_3)Fe(C_5H_5)][PF_6]$	Fe: Org.Comp.B19-1, 32
$C_{12}F_7O_5Re$	$(CO)_5ReC_7F_7$	Re: Org.Comp.2-142
$C_{12}F_8HNO$	$(CF)_4CNHC(CF)_4COC$	F: PFHOrg.SVol.4-293/4, 307
$C_{12}F_8HN_3O_4$	$(4{-}NO_2{-}C_6F_4)_2NH$	F: PFHOrg.SVol.5-16
$C_{12}F_8H_2N_2S$	$2,3,5,6{-}F_4{-}C_6H{-}N{=}S{=}N{-}C_6H{-}F_4{-}2,3,5,6$	S: S-N Comp.7-246
$C_{12}F_8H_4N_2$	$4{-}H_2N{-}C_6F_4{-}C_6F_4{-}NH_2{-}4$	F: PFHOrg.SVol.5-12, 42, 83
$C_{12}F_8H_6N_4$	$H_2N{-}NH{-}4{-}C_6F_4{-}C_6F_4{-}NH{-}NH_2{-}4$	F: PFHOrg.SVol.5-205, 226
$C_{12}F_8H_6Sb^-$	$[(C_6H_3F_2{-}3,4)_2SbF_4]^-$	Sb: Org.Comp.5-137
$C_{12}F_8N_2O_4S$	$(4{-}NO_2{-}C_6F_4)_2S$	F: PFHOrg.SVol.5-183, 192
$C_{12}F_8N_4O_4$	$4{-}NO_2{-}C_6F_4{-}N{=}N{-}C_6F_4{-}NO_2{-}4$	F: PFHOrg.SVol.5-210, 220
$C_{12}F_9FeH_5O_2S$	$(C_5H_5)Fe(CO)_2C(CF_3){=}C(CF_3)SCF_3$	Fe: Org.Comp.B13-94, 122
$C_{12}F_9HN_2O_2$	$4{-}NO_2{-}C_6F_4{-}NH{-}C_6F_5$	F: PFHOrg.SVol.5-48
$C_{12}F_9H_2NO$	$C_6F_5{-}NH{-}3{-}C_6F_4{-}OH$	F: PFHOrg.SVol.5-16, 47
–	$C_6F_5{-}O{-}C_6F_4{-}NH_2{-}2$	F: PFHOrg.SVol.5-10, 41, 82
–	$C_6F_5{-}O{-}C_6F_4{-}NH_2{-}4$	F: PFHOrg.SVol.5-10, 41
$C_{12}F_9H_5O_6Sn$	$C_6H_5Sn(OOCCF_3)_3$	Sn: Org.Comp.17-65
$C_{12}F_9H_6N_3O_3$	$C_6{-}(C({=}O)NH_2)_3{-}1,3,5{-}(CF_3)_3{-}2,4,6$	F: PFHOrg.SVol.5-22, 55
$C_{12}F_9NO_2$	$2{-}NO_2{-}C_6F_4{-}C_6F_5$	F: PFHOrg.SVol.5-182, 186
$C_{12}F_9NO_3$	$2{-}NO_2{-}C_6F_4{-}O{-}C_6F_5$	F: PFHOrg.SVol.5-183, 197
$C_{12}F_9N_3$	$1,3,5{-}(NC)_3{-}C_6(CF_3)_3$	F: PFHOrg.SVol.6-107/8, 127
$C_{12}F_{10}HN$	$(C_6F_5)_2NH$	F: PFHOrg.SVol.5-16, 46, 84
$C_{12}F_{10}HN_2O^+$	$[C_6F_5{-}N(OH){=}N{-}C_6F_5]^+$	F: PFHOrg.SVol.5-202
$C_{12}F_{10}H_2N_2$	$C_6F_5{-}NH{-}NH{-}C_6F_5$	F: PFHOrg.SVol.5-206, 217
$C_{12}F_{10}H_2N_2O^{2+}$	$[C_6F_5{-}N(OH){=}NH{-}C_6F_5]^{2+}$	F: PFHOrg.SVol.5-202
$C_{12}F_{10}IIn$	$(C_6F_5)_2InI$	In: Org.Comp.1-156, 157, 159
$C_{12}F_{10}LiN$	$Li[N(C_6F_5)_2]$	F: PFHOrg.SVol.5-16
$C_{12}F_{10}N^-$	$[N(C_6F_5)_2]^-$	F: PFHOrg.SVol.5-16, 46
$C_{12}F_{10}NNa$	$Na[N(C_6F_5)_2]$	F: PFHOrg.SVol.5-16, 46
$C_{12}F_{10}N_2$	$4{-}(C_6F_5{-}CF{=}N){-}NC_5F_4$	F: PFHOrg.SVol.4-150, 161
–	$C_6F_5{-}N{=}N{-}C_6F_5$	F: PFHOrg.SVol.5-210, 220, 232
–	$(NC_6F_5){=}(NC_6F_5)$	F: PFHOrg.SVol.4-274
$C_{12}F_{10}N_2^{2+}$	$[C_6F_5{-}N{=}N{-}C_6F_5]^{2+}$	F: PFHOrg.SVol.5-210
$C_{12}F_{10}N_2O$	$C_6F_5{-}N(O){=}N{-}C_6F_5$	F: PFHOrg.SVol.5-202, 213, 223

$C_{12}F_{10}N_2S$ ...... C_6F_5-N=S=N-C_6F_5 ...... F: PFHOrg.SVol.6-53, 67, 76/7
S: S-N Comp.7-246/7, 249, 252
– ...... C_6F_5-^{15}N=S=^{15}N-C_6F_5 ...... S: S-N Comp.7-247
$C_{12}F_{10}N_2S_2$ ...... C_6F_5-SN=S=N-C_6F_5 ...... S: S-N Comp.7-25/7
$C_{12}F_{10}N_2S_3$ ...... C_6F_5-SN=S=NS-C_6F_5 ...... S: S-N Comp.7-37/8
$C_{12}F_{10}N_4S_3$ ...... C_6F_5-N=S=NSN=S=N-C_6F_5 ...... S: S-N Comp.7-52
$C_{12}F_{10}N_4S_5$ ...... $(C_6F_5$-SN=S=N$)_2$S ...... S: S-N Comp.7-41/2
$C_{12}F_{11}H_{18}N_3S$ .... $[((CH_3)_2N)_3S][C_4F_5$-1,3-$(CF_3)_2]$ ...... S: S-N Comp.8-230, 236, 248/9
$C_{12}F_{11}O_5Re$ ..... $(CO)_5ReC(=C(CF_3)_2)C(CF_3)=CF_2$ ...... Re: Org.Comp.2-140, 145/6
$C_{12}F_{12}Fe$ .... $(C_6F_6)_2Fe$ ...... Fe: Org.Comp.B19-348, 374
$C_{12}F_{12}FeH_{12}P_2$ ... $[(C_6H_6)_2Fe][PF_6]_2$ ...... Fe: Org.Comp.B19-347, 354/5, 375/6
– ...... $[(C_6D_6)_2Fe][PF_6]_2$ ...... Fe: Org.Comp.B19-375
$C_{12}F_{12}FeH_{23}N_7P_2Ru$
$[(NH_3)_5RuNC$-$C_5H_4FeC_5H_4$-$CN][PF_6]_2$ ...... Fe: Org.Comp.A9-127/8
$C_{12}F_{12}Ge$ .... $Ge(C{\equiv}C$-$CF_3)_4$ ...... Ge: Org.Comp.1-104
$C_{12}F_{12}GeH_{18}N_2Sn$ 1,3-$[CH_3$-$C(CF_3)_2]_2$-2,2,4,4-$(CH_3)_4$-1,3,2,4-N_2GeSn
Sn: Org.Comp.19-75, 78
$C_{12}F_{12}H_{18}N_2SiSn$ 1,3-$[CH_3$-$C(CF_3)_2]_2$-2,2,4,4-$(CH_3)_4$-1,3,2,4-N_2SiSn
Sn: Org.Comp.19-75, 77
$C_{12}F_{12}MnO_8P_2Re$ $(CO)_4Re[$-$P(CF_3)_2$-$Mn(CO)_4$-$P(CF_3)_2$-$]$ ...... Re: Org.Comp.1-487
$C_{12}F_{12}Pb$ .... $Pb(C{\equiv}C$-$CF_3)_4$ ...... Pb: Org.Comp.3-92, 93, 96
$C_{12}F_{14}GeH_6$ ..... $Ge(CH_3)_2[C(-CF_2C(C_2F_5)=)]_2$ ...... Ge: Org.Comp.3-153
$C_{12}F_{16}N_2O_3$ ..... $[NC$-CF_2CF_2-CF_2CF_2-$C(=O)]_2O$ ...... F: PFHOrg.SVol.6-100/1
$C_{12}F_{17}NO_2$ ...... 6-NO_2-C_6F_3-$(C_3F_7$-i$)_2$-3,5 ...... F: PFHOrg.SVol.5-181, 191
$C_{12}F_{18}H_8N_2O_{12}Th$ $[NH_4]_2[Th(CF_3COO)_6]$ ...... Th: SVol.C7-56
$C_{12}F_{18}H_{18}Mo_3N_6Tl$
$Tl[MoF_6]_3 \cdot 6\ CH_3CN$ ...... Mo: SVol.B5-176, 185
$C_{12}F_{18}N_2$ .... 1,5-(i-$C_3F_7)_2$-[4.2.0]-3,7-$N_2C_6F_4$ ...... F: PFHOrg.SVol.4-282, 297
– ...... 1,5-(i-$C_3F_7)_2$-$[4.2.0.0^{2,5}]$-3,7-$N_2C_6F_4$ ...... F: PFHOrg.SVol.4-293, 307
– ...... 3,7-(i-$C_3F_7)_2$-1,5-$N_2C_6F_4$ ...... F: PFHOrg.SVol.4-277/8
$C_{12}F_{18}N_3$ .... $[NC(CF_2)_3NC(CF_2)_3NC(CF_2)_3]_n$ ...... F: PFHOrg.SVol.4-268
$C_{12}F_{19}N$ .... $NCFC(CF_3)C(i$-$C_3F_7)CFC(i$-$C_3F_7)$ ...... F: PFHOrg.SVol.4-118/9, 128
$C_{12}F_{20}N_2$ .... 2,5-(s-$C_4F_9)_2$-1,4-$N_2C_4F_2$ ...... F: PFHOrg.SVol.4-199
– ...... 4,5-(s-$C_4F_9)_2$-1,2-$N_2C_4F_2$ ...... F: PFHOrg.SVol.4-185/6, 201
– ...... 4,5-(s-$C_4F_9)_2$-1,3-$N_2C_4F_2$ ...... F: PFHOrg.SVol.4-192, 205
$C_{12}F_{20}N_2O_2$ ..... NC-$CF(CF_3)$-O-$(CF_2)_6$-O-$CF(CF_3)$-CN ...... F: PFHOrg.SVol.6-103/4, 133
$C_{12}F_{21}GaO_6$ ..... $Ga[OC(O)$-C_3F_7-n$]_3$ ...... Ga: SVol.D1-156, 158
$C_{12}F_{21}NO_2$ ...... $N(n$-$C_7F_{15})C(O)CF_2CF_2CF_2C(O)$ ...... F: PFHOrg.SVol.4-142, 152
$C_{12}F_{21}N_3$ .... 2,4,6-(n-$C_3F_7)_3$-1,3,5-N_3C_3 ...... F: PFHOrg.SVol.4-228/9, 241, 253
– ...... 2,4,6-(i-$C_3F_7)_3$-1,3,5-N_3C_3 ...... F: PFHOrg.SVol.4-228/9
– ...... 3,5,6-(i-$C_3F_7)_3$-1,2,4-N_3C_3 ...... F: PFHOrg.SVol.4-232/3, 245
$C_{12}F_{21}N_3O$ ...... $NC(CF_3)NC(n$-$C_3F_7)NC(CF(CF_3)OC_3F_7)$ ...... F: PFHOrg.SVol.4-228/9, 242
$C_{12}F_{21}N_3O_3$ ..... $N_3C_3(CF_2CF_2OCF_3)_3$ ...... F: PFHOrg.SVol.4-228/9, 241
$C_{12}F_{22}H_3N_3O_2$ ... C_3F_7-O-$CF(CF_3)C(=NH)$-N=$C(NH_2)CF(CF_3)$-O-C_3F_7
F: PFHOrg.SVol.5-5, 37, 75/6

$C_{12}F_{22}N_4$ .... $(CF_2CF_2N(CF_3)CF_2N)C_4F_4(NCF_2N(CF_3)CF_2CF_2)$ F: PFHOrg.SVol.4-36/7, 62

$C_{12}F_{23}H_3N_2O_3$ ... C_3F_7-O-[CF(CF_3)CF_2-O]$_2$-CF(CF_3)C(=NH)-NH_2 F: PFHOrg.SVol.5-3, 72

$C_{12}F_{23}H_3N_2O_4$ ... C_3F_7-O-[CF(CF_3)CF_2-O]$_2$-CF(CF_3)-C(N=OH)-NH_2 F: PFHOrg.SVol.5-4

$C_{12}F_{23}N$ .... N(4-$CF_3C_6F_{10}$)$CF_2CF_2CF_2CF_2CF_2$ .... F: PFHOrg.SVol.4-121, 123

$C_{12}F_{23}NO_3$ ...... C_3F_7-O[-CF(CF_3)CF_2-O]$_2$-CF(CF_3)-CN ...... F: PFHOrg.SVol.6-98/9, 117/8, 135

$C_{12}F_{23}NO_4$ ...... CF_3[-O-CF_2CF(CF_3)]$_3$-O-CF_2-CN. .... F: PFHOrg.SVol.6-98/9, 117

$C_{12}F_{23}N_3O_4$ ..... N_3-CF_2CF(O-C_3F_7-n)-[CF_2-O-CF(CF_3)]$_2$-CF=O F: PFHOrg.SVol.5-204, 216

$C_{12}F_{24}FeH_{38}N_{12}P_4Ru_2$
[Fe(C_5H_4-CNRu(NH_3)$_5$)$_2$][PF_6]$_4$ .... Fe: Org.Comp.A9-126/7

$C_{12}F_{24}N_2$ .... C_5F_{11}-CF=NN=CF-C_5F_{11} .... F: PFHOrg.SVol.5-208, 227

$C_{12}F_{24}N_4S$ ...... $(CF_3)_2C$=NC$(CF_3)_2$N=S=N-C$(CF_3)_2$N=C$(CF_3)_2$ .. S: S-N Comp.7-267

$C_{12}F_{24}N_4Si$ ...... Si(NC$(CF_3)_2$)$_4$ .... Si: SVol.B4-218

$C_{12}F_{24}N_6$ .... 1,4-N_2C_4-2,3,5,6-[N$(CF_3)_2$]$_4$ .... F: PFHOrg.SVol.6-227, 231, 239

$C_{12}F_{27}N$ .... (n-C_4F_9)$_3$N .... F: PFHOrg.SVol.6-223, 232/3, 240/2

$C_{12}F_{27}NO_3$ ...... C_2F_5-O-CF_2CF_2-O-CF_2CF_2-N(C_2F_5)-CF_2CF_2-O-C_2F_5 F: PFHOrg.SVol.6-230

$C_{12}F_{28}N_2$ .... (n-C_3F_7)$_2$N-N(C_3F_7-n)$_2$ .... F: PFHOrg.SVol.5-206

$C_{12}F_{36}FeN_{12}O_6S_6^{2+}$
$[Fe(NS\text{-}ON(CF_3)_2)_6]^{2+}$ .... S: S-N Comp.5-253/4

$C_{12}F_{36}N_{12}NiO_6S_6^{2+}$
$[Ni(NS\text{-}ON(CF_3)_2)_6]^{2+}$ .... S: S-N Comp.5-253/4

$C_{12}FeGeH_{12}O_4$ ... Ge(CH=CH_2)$_3$(CH=CH_2)Fe(CO)$_4$. .... Ge: Org.Comp.3-52

$C_{12}FeGeH_{18}O_4$ ... Ge(C_2H_5)$_3$CH=CH_2 · Fe(CO)$_4$ .... Ge: Org.Comp.2-199

$C_{12}FeH_5MnO_6S_2$ .. C_5H_5(CO)$_2$Fe=CS_2Mn(CO)$_4$ .... Fe: Org.Comp.B16a-183, 188

$C_{12}FeH_5O_6ReS_2$ .. (C_5H_5)Fe(CO)$_2$=C[-S-Re(CO)$_4$-S-] .... Fe: Org.Comp.B16a-183, 188; Re: Org.Comp.1-488

$C_{12}FeH_5O_7Re$ .... (CO)$_5$ReFe(CO)$_2C_5H_5$. .... Re: Org.Comp.2-200

$C_{12}FeH_7MnO_6$ ... (C_5H_5)(CO)Fe(CO)(CH_2)Mn(CO)$_4$. .... Fe: Org.Comp.B16b-151/2

$C_{12}FeH_7O_3$ ...... (C_9H_7)Fe(CO)$_3$, radical. .... Fe: Org.Comp.B14-125

$C_{12}FeH_7O_3^+$ ..... $[(C_9H_7)Fe(CO)_3]^+$ .... Fe: Org.Comp.B15-53

$C_{12}FeH_8^+$ .... $[(C_{12}H_8)Fe]^+$. .... Fe: Org.Comp.B18-2

$C_{12}FeH_8Hg_2N_2S_2$ Fe(C_5H_4-HgSCN)$_2$ .... Fe: Org.Comp.A10-135/6

$C_{12}FeH_8NO_2^-$ .... $[(C_5H_5)Fe(C_5H_3(CN)\text{-}COO)]^-$ .... Fe: Org.Comp.A9-133, 138

$C_{12}FeH_8N_2$ ...... (C_5H_5)Fe[C_5H_3(CN)$_2$] .... Fe: Org.Comp.A9-129, 131

– .... Fe(C_5H_4-CN)$_2$. .... Fe: Org.Comp.A9-108/9, 117/8

$C_{12}FeH_8N_2^+$ ..... $[Fe(C_5H_4\text{-}CN)_2]^+$ .... Fe: Org.Comp.A9-118

$C_{12}FeH_8N_2O_2$ .... Fe(C_5H_4-NCO)$_2$ .... Fe: Org.Comp.A9-99/100

$C_{12}FeH_8N_2S_2$ .... Fe(C_5H_4SCN)$_2$ .... Fe: Org.Comp.A9-226

$C_{12}FeH_8N_4O$ ..... (C_5H_5)Fe[C_5H_3(CO-N_3)CN] .... Fe: Org.Comp.A9-148/9

$C_{12}FeH_8N_6O_2$ .... Fe(C_5H_4-CO-N_3)$_2$. .... Fe: Org.Comp.A9-145, 147

$C_{12}FeH_8O_3S$ ..... (C_5H_5)Fe(CO)$_2$C(=O)-2-C_4H_3S. .... Fe: Org.Comp.B13-8, 10/1, 23, 43

$C_{12}FeH_8O_4$ ...... $(C_5H_5)Fe(CO)_2C(=O)$-2-C_4H_3O .... Fe: Org.Comp.B13-8, 10, 23, 43

$C_{12}FeH_9NO_2$ ..... $(C_5H_5)Fe[C_5H_3(CN)$-COOH] .... Fe: Org.Comp.A9-130, 133, 136, 138

– .... $(NC$-$C_5H_4)Fe(C_5H_4$-COOH) .... Fe: Org.Comp.A9-114, 122

$C_{12}FeH_9O_3^+$ ..... $[((3.2.2)$-$C_9H_9)Fe(CO)_3]^+$ .... Fe: Org.Comp.B15-253/4

– .... $[((3.2.2)$-$C_9H_8D)Fe(CO)_3]^+$ .... Fe: Org.Comp.B15-253

– .... $[(8$-$CH_2{=}C_8H_7)Fe(CO)_3]^+$ .... Fe: Org.Comp.B15-241, 258

$C_{12}FeH_9O_4^+$ ..... $[(HO$-$C_9H_8)Fe(CO)_3]^+$ .... Fe: Org.Comp.B15-254

– .... $[(O{=}C_9H_9)Fe(CO)_3]^+$ .... Fe: Org.Comp.B15-253

$C_{12}FeH_{10}KO_2^+$ ... $K[(OOC$-$C_6H_5)Fe(C_5H_5)]^+$ .... Fe: Org.Comp.B18-142/6, 202

$C_{12}FeH_{10}N$ ...... $(NC$-$C_6H_5)Fe(C_5H_5)$ .... Fe: Org.Comp.B18-146/8, 203, 268

$C_{12}FeH_{10}N^+$ ..... $[(C_6H_6)Fe(C_5H_4$-$CN)]^+$ .... Fe: Org.Comp.B18-142/6, 151, 154, 161

– .... $[(NC$-$C_6H_5)Fe(C_5H_5)]^+$ .... Fe: Org.Comp.B18-142/6, 198, 201, 202/3, 268

$C_{12}FeH_{10}NOS^+$ .. $[C_5H_5Fe(CS)(CO)NC_5H_5]^+$ .... Fe: Org.Comp.B15-271

$C_{12}FeH_{10}N_2O_2$ ... $C_5H_5FeC_6H_5NO_2$-1(CN-6) .... Fe: Org.Comp.B17-276, 308

$C_{12}FeH_{10}N_2O_3$ ... $(C_9H_7)Fe(CO)_2C(O)NHNH_2$ .... Fe: Org.Comp.B14-67, 74

$C_{12}FeH_{10}O_2$ ..... $(C_5H_5)Fe(CO)_2$-C_5H_5 .... Fe: Org.Comp.B13-195, 213/4, 246/51

– .... $(C_9H_7)Fe(CO)_2$-CH_3 .... Fe: Org.Comp.B14-67, 74

$C_{12}FeH_{10}O_2S$ .... 2-$[(C_5H_5)Fe(CO)_2$-$CH_2]$-SC_4H_3 .... Fe: Org.Comp.B13-2, 5

– .... 3-$[(C_5H_5)Fe(CO)_2$-$CH_2]$-SC_4H_3 .... Fe: Org.Comp.B13-2, 5

$C_{12}FeH_{10}O_3$ ..... 2-$(C_5H_5)Fe(CO)_2$-5-CH_3-OC_4H_2 .... Fe: Org.Comp.B14-2/3, 6, 12

– .... $(C_5H_5)Fe(CO)_2$-C[=CH-C(=O)-CH_2CH_2-] ..... Fe: Org.Comp.B13-195, 210

– .... C_5H_4-CH_2-C(=$CHCH_3$)-C(=O)-$Fe(CO)_2$ ...... Fe: Org.Comp.B18-18/20, 26

– .... $[(4.4.1.0^{2,5})$-11-$OC_{10}H_{10}]Fe(CO)_2$ .... Fe: Org.Comp.B18-18, 32

$C_{12}FeH_{11}KO_4S$ ... $K[O_3S$-$C_5H_4FeC_5H_4$-$COCH_3]$ · 2 H_2O .... Fe: Org.Comp.A9-238

$C_{12}FeH_{11}N$ ...... $(C_5H_5)Fe(C_6H_6CN$-1) .... Fe: Org.Comp.B17-266

– .... $(C_5H_5)Fe(C_6H_6CN$-2) .... Fe: Org.Comp.B17-266

– .... $(C_5H_5)Fe(C_6H_6CN$-3) .... Fe: Org.Comp.B17-266

– .... $(C_5H_5)Fe[C_5H_3(CN)$-$CH_3]$ .... Fe: Org.Comp.A9-129/30, 132, 136, 137/8

– .... $FeC_{10}H_8(CN)(CH_3)$ .... Fe: Org.Comp.A9-140

– .... $(NC$-$C_5H_4)Fe(C_5H_4$-$CH_3)$ .... Fe: Org.Comp.A9-113, 120

$C_{12}FeH_{11}NO$ ..... $(C_5H_5)Fe[C_5H_3(CN)$-$CH_2OH]$ .... Fe: Org.Comp.A9-130

– .... $(C_5H_5)Fe[C_5H_3(NCO)$-$CH_3]$ .... Fe: Org.Comp.A9-104

$C_{12}FeH_{11}NO_3$ .... $(C_5H_5)Fe(CO)_2$-C[=C(CH_3)-C(=O)-NH-CH_2-] .. Fe: Org.Comp.B14-18/9, 24, 28/9

– .... $(H_2NCO$-$C_5H_4)Fe(C_5H_4$-COOH) .... Fe: Org.Comp.A9-79, 82

$C_{12}FeH_{11}NO_4$ .... $(C_5H_5)Fe(CO)_2$CH[-O-CH_2-CH_2-O-CH(CN)-] .. Fe: Org.Comp.B14-2, 8

$C_{12}FeH_{11}NO_6S$ ... $(C_5H_5)Fe(CO)_2$C[=CH-O-C(OCH_3)=NS(=O)$_2CH_2$-] Fe: Org.Comp.B14-35, 39

$C_{12}FeH_{11}NS$ ..... $(C_5H_5)Fe[C_5H_3(SCN)$-$CH_3]$ .... Fe: Org.Comp.A9-229

– .... $(NCS$-$C_5H_4)Fe(C_5H_4$-$CH_3)$ .... Fe: Org.Comp.A9-227

$C_{12}FeH_{11}N_2^+$ .... $[(1,3$-$N_2C_7H_6)Fe(C_5H_5)][PF_6]$ .... Fe: Org.Comp.B19-216, 289

$C_{12}FeH_{11}N_3O$ .... $(C_5H_5)Fe[C_5H_3(CH_3)$-C(O)-$N_3]$ .... Fe: Org.Comp.A9-148, 149

$C_{12}FeH_{11}O^+$ ..... $[(c$-$C_7H_6{=}O$-7$)Fe(C_5H_5)][PF_6]$ .... Fe: Org.Comp.B19-340/1

$C_{12}FeH_{11}O_2$ — $(C_6H_6)Fe(C_5H_4\text{-}COOH)$ — Fe: Org.Comp.B18-146/8, 154, 161, 185
— $(HOOC\text{-}C_6H_5)Fe(C_5H_5)$ — Fe: Org.Comp.B18-146/8, 203, 268/9
$C_{12}FeH_{11}O_2^+$ — $[(C_5H_5)Fe(CO)_2(C_5H_6\text{-}c)]^+$ — Fe: Org.Comp.B17-86/7
— $[(C_6H_6)Fe(C_5H_4\text{-}COOH)]^+$ — Fe: Org.Comp.B18-142/6, 151, 154, 160, 185
— $[(HOOC\text{-}C_6H_5)Fe(C_5H_5)]^+$ — Fe: Org.Comp.B18-142/6, 203, 268/9
$C_{12}FeH_{11}O_3^+$ — $[(2\text{-}CH_3CH\text{-}(2.2.1)\text{-}C_7H_7)Fe(CO)_3]^+$ — Fe: Org.Comp.B15-257/8
— $[(3\text{-}CH_3\text{-}(5.1.0)\text{-}C_8H_8)Fe(CO)_3]^+$ — Fe: Org.Comp.B15-216, 239
— $[(4\text{-}CH_3\text{-}(5.1.0)\text{-}C_8H_8)Fe(CO)_3]^+$ — Fe: Org.Comp.B15-216, 239
— $[(6\text{-}(CH_3CH{=})C_7H_7)Fe(CO)_3]^+$ — Fe: Org.Comp.B15-209, 225
— $[(C_9H_{11})Fe(CO)_3]^+$ — Fe: Org.Comp.B15-111, 123, 246
$C_{12}FeH_{11}O_4^+$ — $[(1\text{-}CH_3O\text{-}C_8H_7D)Fe(CO)_3]^+$ — Fe: Org.Comp.B15-239
— $[(2\text{-}CH_3C(O)\text{-}C_7H_8)Fe(CO)_3]^+$ — Fe: Org.Comp.B15-206
— $[(6\text{-}CH_3C(O)\text{-}C_7H_8)Fe(CO)_3]^+$ — Fe: Org.Comp.B15-208, 223
— $[(2\text{-}CH_3\text{-}8\text{-}(O{=})C_8H_8)Fe(CO)_3]^+$ — Fe: Org.Comp.B15-245
— $[(2\text{-}CH_3\text{-}8\text{-}(O{=})C_8H_7D)Fe(CO)_3]^+$ — Fe: Org.Comp.B15-245
— $[(2\text{-}CH_3\text{-}8\text{-}(O{=})C_8H_4D_4)Fe(CO)_3]^+$ — Fe: Org.Comp.B15-245
— $[(2\text{-}HO\text{-}C_8H_7\text{-}8\text{-}CH_3)Fe(CO)_3]^+$ — Fe: Org.Comp.B15-220, 233
— $[(6\text{-}CH_3\text{-}8\text{-}(O{=})C_8H_8)Fe(CO)_3]^+$ — Fe: Org.Comp.B15-244
— $[(6\text{-}CH_3C(OH){=}C_7H_7)Fe(CO)_3]^+$ — Fe: Org.Comp.B15-210
$C_{12}FeH_{11}O_4S^-$ — $[O_3SC_5H_4FeC_5H_4\text{-}COCH_3]^-$ — Fe: Org.Comp.A9-238
$C_{12}FeH_{11}O_5^+$ — $[(1\text{-}CH_3OC(O)\text{-}2\text{-}CH_3\text{-}C_6H_5)Fe(CO)_3]^+$ — Fe: Org.Comp.B15-110, 159/60
— $[(1\text{-}CH_3OC(O)\text{-}3\text{-}CH_3\text{-}C_6H_5)Fe(CO)_3]^+$ — Fe: Org.Comp.B15-112, 161
— $[(1\text{-}CH_3OC(O)\text{-}C_7H_8)Fe(CO)_3]^+$ — Fe: Org.Comp.B15-201
— $[(1\text{-}CH_3OC(O)CH_2\text{-}C_6H_6)Fe(CO)_3]^+$ — Fe: Org.Comp.B15-96/7, 100, 136
— $[(6\text{-}CH_3OC(O)CH_2\text{-}C_6H_6)Fe(CO)_3]^+$ — Fe: Org.Comp.B15-108, 159
— $[(6\text{-}CH_3OC(O)CH_2\text{-}C_6H_5D)Fe(CO)_3]^+$ — Fe: Org.Comp.B15-159
— $[(6\text{-}CH_3OC(O)CHD\text{-}C_6H_6)Fe(CO)_3]^+$ — Fe: Org.Comp.B15-159
$C_{12}FeH_{11}O_6^+$ — $[(1\text{-}CH_3O\text{-}3\text{-}CH_3OC(O)\text{-}C_6H_5)Fe(CO)_3]^+$ — Fe: Org.Comp.B15-111
— $[(2\text{-}CH_3O\text{-}1\text{-}CH_3OC(O)\text{-}C_6H_5)Fe(CO)_3]^+$ — Fe: Org.Comp.B15-110, 159
— $[(2\text{-}CH_3O\text{-}3\text{-}CH_3OC(O)\text{-}C_6H_5)Fe(CO)_3]^+$ — Fe: Org.Comp.B15-123
— $[(3\text{-}CH_3O\text{-}1\text{-}CH_3OC(O)\text{-}C_6H_5)Fe(CO)_3]^+$ — Fe: Org.Comp.B15-112
$C_{12}FeH_{11}S^+$ — $[(c\text{-}C_7H_6{=}S\text{-}7)Fe(C_5H_5)][PF_6]$ — Fe: Org.Comp.B19-340, 341/2
$C_{12}FeH_{12}$ — $(C_5H_5)Fe(C_6H_5{=}CH_2)$ — Fe: Org.Comp.B17-259, 306
— $(C_6H_6)_2Fe$ — Fe: Org.Comp.B18-106
Fe: Org.Comp.B19-348, 355, 376/7
— $(C_6D_6)_2Fe$ — Fe: Org.Comp.B19-376
$C_{12}FeH_{12}^+$ — $[(CH_2C_6H_5)Fe(C_5H_5)]^+$ — Fe: Org.Comp.B18-270
— $[(C_6H_6)_2Fe]^+$ — Fe: Org.Comp.B19-355
— $[(C_6H_6)_2Fe][PF_6]$ — Fe: Org.Comp.B19-348, 355
$C_{12}FeH_{12}^{2+}$ — $[(C_6H_6)_2Fe]^{2+}$ — Fe: Org.Comp.B19-354
— $[(C_6H_6)_2Fe]I_2$ — Fe: Org.Comp.B19-347, 354
— $[(C_6H_6)_2Fe][BF_4]_2$ — Fe: Org.Comp.B19-355

$C_{12}FeH_{12}^{2+}$ ..... $[(C_6H_6)_2Fe][B_{12}H_{12}]$ ..... Fe: Org.Comp.B19-347, 355
– ..... $[(C_6H_6)_2Fe][Cr(SCN)_4(NH_3)_2]_2$ ..... Fe: Org.Comp.B19-348, 355
– ..... $[(C_6H_6)_2Fe][PF_6]_2$ ..... Fe: Org.Comp.B19-347, 354/5, 375/6
– ..... $[(C_6D_6)_2Fe][PF_6]_2$ ..... Fe: Org.Comp.B19-375
$C_{12}FeH_{12}I_2$ ..... $[(C_6H_6)_2Fe]I_2$ ..... Fe: Org.Comp.B19-347, 354
$C_{12}FeH_{12}K_2$ ..... $Fe[C_{10}H_7(K)_2(C_2H_5)]$ ..... Fe: Org.Comp.A10-328
$C_{12}FeH_{12}K_2O_6S_2$ $K_2[(O_3S-C_5H_4)Fe(C_5H_3(C_2H_5)-SO_3)] \cdot 4\ H_2O$ Fe: Org.Comp.A10-297
$C_{12}FeH_{12}LiNO$ ... $(C_5H_5)Fe[C_5H_3(Li)-C(=O)NH-CH_3]$ ..... Fe: Org.Comp.A10-119, 122, 126
$C_{12}FeH_{12}Li_2$ ..... $(Li-C_5H_4)Fe[C_5H_3(Li)-C_2H_5]$ ..... Fe: Org.Comp.A10-322/3, 325
$C_{12}FeH_{12}Li_2O$ .... $(LiCH_2-C_5H_4)Fe(C_5H_4-CH_2-OLi)$ ..... Fe: Org.Comp.A10-113/5
$C_{12}FeH_{12}NO^+$ .... $[(NH_2-C(=O)C_6H_5)Fe(C_5H_5)]^+$ ..... Fe: Org.Comp.B18-142/6, 203
$C_{12}FeH_{12}NO_2^+$ ... $[(CH_3-C_6H_4-2-NO_2)Fe(C_5H_5)][PF_6]$ ..... Fe: Org.Comp.B19-4/6, 63
– ..... $[(CH_3-C_6H_4-3-NO_2)Fe(C_5H_5)][PF_6]$ ..... Fe: Org.Comp.B19-4/6, 63
– ..... $[(CH_3-C_6H_4-4-NO_2)Fe(C_5H_5)][PF_6]$ ..... Fe: Org.Comp.B19-4/6, 63/4
– ..... $[(C_8H_9)Fe(CNCH_3)(CO)_2]^+$ ..... Fe: Org.Comp.B15-351
– ..... $[(O_2N-C_6H_5)Fe(C_5H_4-CH_3)]^+$ ..... Fe: Org.Comp.B18-142/6, 197/8, 255
$C_{12}FeH_{12}NO_4^+$ ... $[(C_5H_5)Fe(CN-CH(CH_3)COOCH_3)(CO)_2]^+$ ..... Fe: Org.Comp.B15-314
$C_{12}FeH_{12}N_2O_2$ ... $(C_5H_5)Fe(CO)(CN)(2-CN-OC_4H_7)$ ..... Fe: Org.Comp.B15-292, 297
– ..... $(C_5H_5)Fe[C_5H_3(CH=N-OH)_2]$ ..... Fe: Org.Comp.A9-165/6
– ..... $Fe(C_5H_4-CH=N-OH)_2$ ..... Fe: Org.Comp.A9-160
– ..... $Fe[C_5H_4-C(=O)NH_2]_2$ ..... Fe: Org.Comp.A9-77, 81
$C_{12}FeH_{12}N_2O_4$ ... $(C_5H_5)Fe(CO)C_5H_4N_2O(CH_3)(=O)_2$ ..... Fe: Org.Comp.B16b-126, 138
– ..... $Fe[C_5H_4-C(=O)-NHOH]_2$ ..... Fe: Org.Comp.A9-159
$C_{12}FeH_{12}N_3^+$ .... $[(2-CH_3-C_6H_4-N_3)Fe(C_5H_5)][PF_6]$ ..... Fe: Org.Comp.B19-2, 62
– ..... $[(3-CH_3-C_6H_4-N_3)Fe(C_5H_5)][PF_6]$ ..... Fe: Org.Comp.B19-2, 62
– ..... $[(4-CH_3-C_6H_4-N_3)Fe(C_5H_5)][PF_6]$ ..... Fe: Org.Comp.B19-2, 62/3
$C_{12}FeH_{12}N_6$ ..... $Fe(C_5H_4-CH_2-N_3)_2$ ..... Fe: Org.Comp.A9-145
$C_{12}FeH_{12}Na_2$ .... $Fe[C_{10}H_7(Na)_2(C_2H_5)]$ ..... Fe: Org.Comp.A10-328
$C_{12}FeH_{12}O_2$ ..... $(C_5H_5)Fe(CO)(CH_2C_5H_5=O)$ ..... Fe: Org.Comp.B17-157
– ..... $(C_5H_5)Fe(CO)-C(O)CH_2CH_2CH=C=CH_2$ ..... Fe: Org.Comp.B17-169
– ..... $(C_5H_5)Fe(CO)_2-CH(-CH=CH-CH_2CH_2-)$ ..... Fe: Org.Comp.B13-195, 209, 244/5
– ..... $(C_5H_5)Fe(CO)_2-CH=CHC(CH_3)=CH_2$ ..... Fe: Org.Comp.B13-97, 124
– ..... $(C_5H_5)Fe(CO)_2-CH=CHCH=CHCH_3$ ..... Fe: Org.Comp.B13-98, 126
– ..... $(C_5H_5)Fe(CO)_2-CH_2CH_2CH=C=CH_2$ ..... Fe: Org.Comp.B13-112
– ..... $(C_5H_5)Fe(CO)_2-CH_2CH_2C{\equiv}CCH_3$ ..... Fe: Org.Comp.B13-156
$C_{12}FeH_{12}O_3$ ..... $(C_5H_5)Fe(CO)[CH_2C(COO-CH_3)C=CH_2]$ ..... Fe: Org.Comp.B17-162
– ..... $(C_5H_5)Fe(CO)_2-C(=C=CH_2)CH_2-OCH_3$ ..... Fe: Org.Comp.B13-112
– ..... $(C_5H_5)Fe(CO)_2-C(=O)-CH=C(CH_3)_2$ ..... Fe: Org.Comp.B13-12, 25, 46
– ..... $(C_5H_5)Fe(CO)_2-C(=O)-CH_2CH_2CH=CH_2$ ..... Fe: Org.Comp.B13-19, 39
– ..... $(C_5H_5)Fe(CO)_2-C(=O)-C_4H_7-c$ ..... Fe: Org.Comp.B13-20, 40
– ..... $(C_5H_5)Fe(CO)_2-CH=CHCH=CH-OCH_3$ ..... Fe: Org.Comp.B13-98, 126
– ..... $(C_5H_5)Fe(CO)_2-CH_2CH=CH-C(=O)-CH_3$ ..... Fe: Org.Comp.B13-106, 134
– ..... $C_5H_4-C(CH_3)_2CH_2C(=O)-Fe(CO)_2$ ..... Fe: Org.Comp.B18-18, 26
– ..... $C_5H_4-CH(CH_3)CH(CH_3)-C(=O)-Fe(CO)_2$ ..... Fe: Org.Comp.B18-18, 27
– ..... $C_5H_4-CH_2C(CH_3)_2C(=O)-Fe(CO)_2$ ..... Fe: Org.Comp.B18-18, 26
– ..... $(c-C_4H_3-CH_2-O-CH_2CH_2-C{\equiv}C-CH_3)Fe(CO)_2$ .. Fe: Org.Comp.B18-19

$C_{12}FeH_{12}O_4$	$(C_5H_5)Fe(CO)_2$-C(=$CHCH_3$)-OC(=O)CH_3	Fe: Org.Comp.B13-90, 119
–	$(C_5H_5)Fe(CO)_2$-C(=CH_2)-CH_2-OC(=O)CH_3	Fe: Org.Comp.B13-88, 118
–	$(C_5H_5)Fe(CO)_2$-CH=CH-COO-C_2H_5	Fe: Org.Comp.B13-90/1
–	$(C_5H_5)Fe(CO)_2$-CH_2CH=CH-COO-CH_3	Fe: Org.Comp.B13-105, 133/4
$C_{12}FeH_{12}O_4S$	$(C_5H_5)Fe(CO)_2$-CH[-CH=C(CH_3)-CH_2-SO_2-]	Fe: Org.Comp.B14-6, 12
–	(HO_2S-C_5H_4)Fe[C_5H_4-C(O)OCH_3]	Fe: Org.Comp.A9-230
–	(HO_3S-C_5H_4)Fe[C_5H_4-C(O)CH_3]	Fe: Org.Comp.A9-231, 238
$C_{12}FeH_{12}O_4S^+$	[HO_3S-$C_5H_4FeC_5H_4$-$COCH_3$]$^+$	Fe: Org.Comp.A9-238
$C_{12}FeH_{12}O_5S$	$(C_5H_5)Fe(CO)_2$-CH[-CH=CHCH_2CH(SO_3H)-]	Fe: Org.Comp.B13-210
–	(HO_3S-C_5H_4)Fe(C_5H_4-C(O)OCH_3)	Fe: Org.Comp.A9-231, 238/9
$C_{12}FeH_{12}O_6PbS_2$	Pb[(O_3S-C_5H_4)Fe(C_5H_3(C_2H_5)-SO_3)] · x H_2O	Fe: Org.Comp.A10-297
$C_{12}FeH_{12}O_6S_2{}^{2-}$	[(O_3S-C_5H_4)Fe(C_5H_3(C_2H_5)-SO_3)]$^{2-}$	Fe: Org.Comp.A10-297
$C_{12}FeH_{13}$	(CH_3-C_6H_5)Fe(C_5H_5)	Fe: Org.Comp.B18-146/8, 200/1, 207, 272/3
–	(C_6H_6)Fe(C_5H_4-CH_3)	Fe: Org.Comp.B18-146/8, 153, 154, 162, 186
$C_{12}FeH_{13}{}^+$	[(CH_3-C_6H_5)Fe(C_5H_5)]$^+$	Fe: Org.Comp.B18-142/6, 197, 200, 201, 204/7, 269/72
–	[(C_6H_6)Fe(C_5H_4-CH_3)]$^+$	Fe: Org.Comp.B18-142/6, 151, 154, 161, 185/6
–	[(C_6H_6)Fe(C_6H_7)]$^+$	Fe: Org.Comp.B18-142/6, 153, 172, 189/90
–	[(C_6H_6)Fe(C_6H_6D)]$^+$	Fe: Org.Comp.B18-189
–	[(C_6D_6)Fe(C_6D_6H)]$^+$	Fe: Org.Comp.B18-189
–	[(C_7H_8)Fe(C_5H_5)][BF_4]	Fe: Org.Comp.B19-340/1, 343
–	[(C_7H_8)Fe(C_5H_5)][PF_6]	Fe: Org.Comp.B19-340/1, 343
$C_{12}FeH_{13}I$	[($CH_3C_6H_5$)Fe(C_5H_5)]I	Fe: Org.Comp.B18-142/6, 201, 204
–	[($CH_3C_6H_5$)Fe(C_5H_5)][I_x]	Fe: Org.Comp.B18-142/6, 201, 204
$C_{12}FeH_{13}IO$	[(CH_3O-C_6H_5)Fe(C_5H_5)]I	Fe: Org.Comp.B18-142/6, 199, 201, 248
$C_{12}FeH_{13}IS$	[(CH_3S-C_6H_5)Fe(C_5H_5)]I	Fe: Org.Comp.B18-142/6, 199, 253
$C_{12}FeH_{13}I_3$	[($CH_3C_6H_5$)Fe(C_5H_5)][I_3]	Fe: Org.Comp.B18-142/6, 197, 201, 204
$C_{12}FeH_{13}I_x$	[($CH_3C_6H_5$)Fe(C_5H_5)][I_x]	Fe: Org.Comp.B18-142/6, 201, 204
$C_{12}FeH_{13}KO_3S$	K[$C_5H_5FeC_5H_3$(SO_3)C_2H_5] · 2 H_2O	Fe: Org.Comp.A9-253
$C_{12}FeH_{13}Li$	(CH_3-C_5H_4)Fe[C_5H_3(Li)-CH_3]	Fe: Org.Comp.A10-322/3, 325, 326
–	(Li-C_5H_4)Fe(C_5H_4-C_2H_5)	Fe: Org.Comp.A10-113/5
$C_{12}FeH_{13}LiO$	(C_5H_5)Fe[C_5H_3(Li)-CH_2-OCH_3]	Fe: Org.Comp.A10-118, 120/2
–	(Li-C_5H_4)Fe(C_5H_4-CH_2-OCH_3)	Fe: Org.Comp.A10-113/5
$C_{12}FeH_{13}LiO_2$	(CH_3O-C_5H_4)Fe[C_5H_3(Li)-OCH_3]	Fe: Org.Comp.A10-322/4
$C_{12}FeH_{13}NO$	(C_5H_5)Fe[C_5H_3(CH_3)-CH=N-OH]	Fe: Org.Comp.A9-165, 168
–	(C_5H_5)Fe[C_5H_3(CH_3)-C(O)NH_2]	Fe: Org.Comp.A9-90, 93, 96/7
–	(HO-N=CH-C_5H_4)Fe(C_5H_4-CH_3)	Fe: Org.Comp.A9-160, 162
–	(H_2N-C_5H_4)Fe[C_5H_4-C(O)CH_3]	Fe: Org.Comp.A9-3, 4

$C_{12}FeH_{13}NO$ . . . . . $[H_2NC(O)-C_5H_4]Fe(C_5H_4-CH_3)$ . . . . Fe: Org.Comp.A9-79, 82
$C_{12}FeH_{13}NO^+$ . . . . $[C_5H_5FeC_5H_3(CH_3)-CONH_2]^+$ . . . . Fe: Org.Comp.A9-97
$C_{12}FeH_{13}NO_2$ . . . . $(C_5H_5)Fe(1-NO_2-2-CH_3-C_6H_5)$ . . . . Fe: Org.Comp.B17-277
– . . . . $(C_5H_5)Fe(1-NO_2-6-CH_3-C_6H_5)$ . . . . Fe: Org.Comp.B17-272
– . . . . $(C_5H_5)Fe(2-NO_2-1-CH_3-C_6H_5)$ . . . . Fe: Org.Comp.B17-277
– . . . . $(C_5H_5)Fe(2-NO_2-3-CH_3-C_6H_5)$ . . . . Fe: Org.Comp.B17-278
– . . . . $(C_5H_5)Fe(3-NO_2-2-CH_3-C_6H_5)$ . . . . Fe: Org.Comp.B17-278
– . . . . $(C_5H_5)Fe[C_5H_3(CH=N-OH)-CH_2OH]$ . . . . Fe: Org.Comp.A9-165
– . . . . $(O_2N-C_5H_4)Fe(C_5H_4-C_2H_5)$ . . . . Fe: Org.Comp.A9-153, 157
$C_{12}FeH_{13}NO_3$ . . . . 1-$[(C_5H_5)Fe(CO)_2-C(=O)]-NC_4H_8$ . . . . Fe: Org.Comp.B13-9, 25
– . . . . $(C_5H_5)Fe(CO)_2-CH[-CH(CH_3)-C(=O)-NH-CH_2-]$
Fe: Org.Comp.B14-18, 21
– . . . . $(O_2N-C_5H_4)Fe(C_5H_4-CH_2OCH_3)$ . . . . Fe: Org.Comp.A9-153
$C_{12}FeH_{13}NO_4S$ . . . $H_2NO_2S-C_5H_4FeC_5H_4-C(O)OCH_3$ . . . . Fe: Org.Comp.A9-234, 242
$C_{12}FeH_{13}NO_6S$ . . . $(C_5H_5)Fe(CO)_2-CH[-CH_2-C(=O)-N(SO_2-OCH_3)-CH_2-]$
Fe: Org.Comp.B14-20, 27
– . . . . $(C_5H_5)Fe(CO)_2-CH[-CH_2-N(COOCH_3)-S(=O)_2-CH_2-]$
Fe: Org.Comp.B14-33, 38
$C_{12}FeH_{13}N_2O_4^+$ . . $[C_5H_5(CO)_2Fe=C(-N(C(=O)NHCH_3)CH_2CH_2O-)]^+$
Fe: Org.Comp.B16a-139/40, 151
$C_{12}FeH_{13}N_3O$ . . . . $N_3-CH_2-C_5H_4FeC_5H_4-CH_2OH$ . . . . Fe: Org.Comp.A9-145
$C_{12}FeH_{13}Na$ . . . . . $(C_5H_5)Fe[C_5H_3(Na)-C_2H_5]$ . . . . Fe: Org.Comp.A10-130/1
– . . . . $(Na-C_5H_4)Fe(C_5H_4-C_2H_5)$ . . . . Fe: Org.Comp.A10-130/1
$C_{12}FeH_{13}O$ . . . . . . $(CH_3O-C_6H_5)Fe(C_5H_5)$ . . . . Fe: Org.Comp.B18-146/8, 249, 250
– . . . . $(C_6H_6)Fe(C_5H_4-CH_2OH)$ . . . . Fe: Org.Comp.B18-146/8, 162
– . . . . $(C_6H_6)Fe(C_5H_4-OCH_3)$ . . . . Fe: Org.Comp.B18-146/8, 169
– . . . . $(HOCH_2-C_6H_5)Fe(C_5H_5)$ . . . . Fe: Org.Comp.B18-146/8, 209
$C_{12}FeH_{13}O^+$ . . . . . $[(CH_3O-C_6H_5)Fe(C_5H_5)]^+$ . . . . Fe: Org.Comp.B18-142/6, 197/9, 201, 248/50
– . . . . $[(C_6H_6)Fe(C_5H_4-CH_2OH)]^+$ . . . . Fe: Org.Comp.B18-142/6, 162
– . . . . $[(C_6H_6)Fe(C_5H_4-OCH_3)]^+$ . . . . Fe: Org.Comp.B18-142/6, 168/9
– . . . . $[(HOCH_2-C_6H_5)Fe(C_5H_5)]^+$ . . . . Fe: Org.Comp.B18-142/6, 209
$C_{12}FeH_{13}O^{2+}$ . . . . $[(CH_3O-C_6H_5)Fe(C_5H_5)]^{2+}$ . . . . Fe: Org.Comp.B18-249/50
$C_{12}FeH_{13}OP$ . . . . . $(C_5H_5)Fe[PC_4H-2-CHO-3,4-(CH_3)_2]$ . . . . Fe: Org.Comp.B17-209
– . . . . $(C_5H_5)Fe[PC_4H_2-2-C(=O)CH_3-3-CH_3]$ . . . . Fe: Org.Comp.B17-207
– . . . . $(C_5H_5)Fe[PC_4H_2-2-C(=O)CH_3-4-CH_3]$ . . . . Fe: Org.Comp.B17-207
– . . . . $(C_5H_5)Fe[PC_4H_2-5-C(=O)CH_3-2-CH_3]$ . . . . Fe: Org.Comp.B17-207
$C_{12}FeH_{13}O_2^+$ . . . . $[(C_5H_5)Fe(CO)_2=CH(CH=C(CH_3)_2)]^+$ . . . . Fe: Org.Comp.B16a-88, 95
– . . . . $[(C_5H_5)Fe(CO)_2=CH(CH=CHC_2H_5)]^+$ . . . . Fe: Org.Comp.B16a-88, 95
– . . . . $[(C_5H_5)Fe(CO)_2=CHCH_2CH=CHCH_3]^+$ . . . . Fe: Org.Comp.B16a-95
– . . . . $[(C_5H_5)Fe(CO)_2=CH-C_4H_7-c]^+$ . . . . Fe: Org.Comp.B16a-89
– . . . . $[(C_5H_5)Fe(CO)_2(CH_2=C=C(CH_3)_2)]^+$ . . . . Fe: Org.Comp.B17-113
– . . . . $[(C_5H_5)Fe(CO)_2(CH_2=CHC(CH_3)=CH_2)]^+$ . . . . . . Fe: Org.Comp.B17-37
– . . . . $[(C_5H_5)Fe(CO)_2(CH_2=CHCH=CHCH_3)]^+$ . . . . Fe: Org.Comp.B17-37/8
– . . . . $[(C_5H_5)Fe(CO)_2(CH_2=CHCH_2CH=CH_2)]^+$ . . . . . . Fe: Org.Comp.B17-38
– . . . . $[(C_5H_5)Fe(CO)_2(CH_2=CH-C_3H_5-c)]^+$ . . . . Fe: Org.Comp.B17-15
– . . . . $[(C_5H_5)Fe(CO)_2(CH_2=CD-C_3H_5-c)]^+$ . . . . Fe: Org.Comp.B17-15

$C_{12}FeH_{13}O_2^+$ .... $[(C_5H_5)Fe(CO)_2(CD_2{=}CH{-}C_3H_5{-}c)]^+$ .... Fe: Org.Comp.B17-15
– .... $[(C_5H_5)Fe(CO)_2(C_5H_8{-}c)]^+$ .... Fe: Org.Comp.B17-87/8, 103
– .... $[(C_5H_5)Fe(CO)_2(HC{\equiv}C{-}C_3H_7{-}n)]^+$ .... Fe: Org.Comp.B17-121
– .... $[(C_5H_5)Fe(CO)_2(H_3CCH{=}C{=}CHCH_3)]^+$ .... Fe: Org.Comp.B17-113/4
$C_{12}FeH_{13}O_2P$ .... $HO_2CC_5H_4FeC_4H_2P((CH_3)_2{-}3,4)$ .... Fe: Org.Comp.B17-220
$C_{12}FeH_{13}O_2S^+$ ... $[(CH_3{-}SO_2{-}C_6H_5)Fe(C_5H_5)]^+$ .... Fe: Org.Comp.B18-142/6, 254
– .... $[(C_5H_5)Fe(CO)_2{-}C({=}CH{-}CH_2{-}S(CH_3){-}CH_2{-})]^+$ Fe: Org.Comp.B14-149, 155
$C_{12}FeH_{13}O_3^+$ .... $[(1,3,5{-}(CH_3)_3C_6H_4)Fe(CO)_3]^+$ .... Fe: Org.Comp.B15-129, 181/2
– .... $[(1,5{-}(CD_3)_2{-}3{-}CH_3{-}C_6H_4)Fe(CO)_3]^+$ .... Fe: Org.Comp.B15-182
– .... $[(1{-}CH_2D{-}3,5{-}(CH_3)_2{-}C_6H_4)Fe(CO)_3]^+$ .... Fe: Org.Comp.B15-182
– .... $[(1{-}(i{-}C_3H_7){-}C_6H_6)Fe(CO)_3]^+$ .... Fe: Org.Comp.B15-99, 135/6
– .... $[(1{-}(CH_3)_2CD{-}C_6H_6)Fe(CO)_3]^+$ .... Fe: Org.Comp.B15-135/6
– .... $[((CH_3)_2C_7H_7)Fe(CO)_3]^+$ .... Fe: Org.Comp.B15-7, 9, 212, 226
– .... $[(C_5H_5)Fe(CO)_2(2{-}(CH_2{=}){-}OC_4H_6)]^+$ .... Fe: Org.Comp.B17-70
– .... $[(C_5H_5)Fe(CO)_2(2{-}CH_3{-}OC_4H_5)]^+$ .... Fe: Org.Comp.B17-94
– .... $[(C_5H_5)Fe(CO)_2{=}C(C_3H_5{-}c)OCH_3]^+$ .... Fe: Org.Comp.B16a-111
– .... $[(C_5H_5)Fe(CO)_2(CH_2{=}C{=}CHCH_2CH_2OH)]^+$ ..... Fe: Org.Comp.B17-112
– .... $[(C_5H_5)Fe(CO)_2(CH_2{=}CHCH_2COCH_3)]^+$ .... Fe: Org.Comp.B17-33
– .... $[(c{-}C_9H_{13})Fe(CO)_3]^+$ .... Fe: Org.Comp.B15-246
$C_{12}FeH_{13}O_3S^-$ ... $[C_5H_5FeC_5H_3(SO_3)C_2H_5]^-$ .... Fe: Org.Comp.A9-252/3
$C_{12}FeH_{13}O_4$ ..... $(c{-}C_8H_{10})Fe(CO)_2C(O)OCH_3$ .... Fe: Org.Comp.B14-83
$C_{12}FeH_{13}O_4^+$ .... $[(1,2{-}(CH_3)_2{-}4{-}CH_3O{-}C_6H_4)Fe(CO)_3]^+$ .... Fe: Org.Comp.B15-126, 179
– .... $[(1,3{-}(CH_3)_2{-}4{-}CH_3O{-}C_6H_4)Fe(CO)_3]^+$ .... Fe: Org.Comp.B15-129, 181
– .... $[(1,5{-}(CH_3)_2{-}3{-}CH_3O{-}C_6H_4)Fe(CO)_3]^+$ .... Fe: Org.Comp.B15-97, 129, 181
– .... $[(3,5{-}(CH_3)_2{-}1{-}CH_3O{-}C_6H_4)Fe(CO)_3]^+$ .... Fe: Org.Comp.B15-96/7, 129
– .... $[(1{-}C_2H_5{-}4{-}CH_3O{-}C_6H_5)Fe(CO)_3]^+$ .... Fe: Org.Comp.B15-114, 169/70
– .... $[(2{-}(i{-}C_3H_7O){-}C_6H_6)Fe(CO)_3]^+$ .... Fe: Org.Comp.B15-103
– .... $[(CH_3{-}CHCHC_6H_7{-}OH)Fe(CO)_3]^+$ .... Fe: Org.Comp.B15-7, 9
– .... $[(C_5H_5)Fe(CO)_2(1,4{-}O_2C_4H_5{-}CH_3)]^+$ .... Fe: Org.Comp.B17-96
– .... $[(C_5H_5)Fe(CO)_2(CH_2{=}CHCH_2{-}C({=}O)O{-}CH_3)]^+$. Fe: Org.Comp.B17-33
– .... $[(C_5H_5)Fe(CO)_2(CH_2{=}CHCH_2{-}OC({=}O){-}CH_3)]^+$. Fe: Org.Comp.B17-25
– .... $[(C_5H_5)Fe(CO)_2(CH_3CH{=}CH{-}C({=}O)O{-}CH_3)]^+$ .. Fe: Org.Comp.B17-59
$C_{12}FeH_{13}O_5^+$ .... $[(C_5H_5)(CO)Fe({=}C(OCH_2)_2)_2]^+$ .... Fe: Org.Comp.B16a-193/4
$C_{12}FeH_{13}O_5S^+$ ... $[(C_5H_5)Fe(CO)_2(C_5H_7(SO_3H{-}3))]^+$ .... Fe: Org.Comp.B17-88
$C_{12}FeH_{13}S^+$ ..... $[(CH_3S{-}C_6H_5)Fe(C_5H_5)]^+$ .... Fe: Org.Comp.B18-142/6, 199, 253
$C_{12}FeH_{14}$ .... $(C_5H_5)Fe(C_6H_6{-}1{-}CH_3)$ .... Fe: Org.Comp.B17-263/4, 307
– .... $(C_5H_5)Fe(C_6H_6{-}2{-}CH_3)$ .... Fe: Org.Comp.B17-263/4, 307
– .... $(C_5H_5)Fe(C_6H_6{-}3{-}CH_3)$ .... Fe: Org.Comp.B17-263/4, 307
– .... $(C_5H_5)Fe(C_6H_6{-}6{-}CH_3)$ .... Fe: Org.Comp.B17-260/1, 307
– .... $(C_5H_5)Fe(C_7H_9)$ .... Fe: Org.Comp.B17-304
– .... $(C_6H_6)Fe(C_6H_8)$ .... Fe: Org.Comp.B18-103, 105, 106/7, 131
– .... $Fe(C_6H_7)_2$ .... Fe: Org.Comp.B17-315
$C_{12}FeH_{14}IN$ ...... $[(C_6H_6)Fe(C_5H_4{-}NHCH_3)]I$ .... Fe: Org.Comp.B18-142/6, 167
$C_{12}FeH_{14}N^+$ ..... $[(2{-}CH_3{-}C_6H_4{-}NH_2)Fe(C_5H_5)][PF_6]$ .... Fe: Org.Comp.B19-1, 5/6, 53, 92/3

$C_{12}FeH_{14}N^+$	$[(3\text{-}CH_3\text{-}C_6H_4\text{-}NH_2)Fe(C_5H_5)][PF_6]$	Fe: Org.Comp.B19-1, 5/6, 53, 92/3
–	$[(4\text{-}CH_3\text{-}C_6H_4\text{-}NH_2)Fe(C_5H_5)][PF_6]$	Fe: Org.Comp.B19-1, 5/6, 54, 92/3
–	$[(CH_3\text{-}NH\text{-}C_6H_5)Fe(C_5H_5)]^+$	Fe: Org.Comp.B18-142/6, 198, 257/8
–	$[(C_6H_6)Fe(C_5H_4\text{-}NHCH_3)]^+$	Fe: Org.Comp.B18-142/6, 167
–	$[(NH_2\text{-}C_6H_5)Fe(C_5H_4\text{-}CH_3)]^+$	Fe: Org.Comp.B18-142/6, 197/8, 257
$C_{12}FeH_{14}NOS_2{}^+$	$[C_5H_5(CO)(CH_3CN)Fe{=}C(SCH_2)_2CH_2]^+$	Fe: Org.Comp.B16a-68
$C_{12}FeH_{14}NO_2{}^+$	$[2\text{-}(C_5H_5)Fe(CO)\text{-}1,3\text{-}ONC_3H_4\text{-}3\text{-}CH_2\text{-}CH{=}CH_2]^+$	Fe: Org.Comp.B17-167
–	$[(C_5H_5)Fe(CO)_2(CN\text{-}C_4H_9\text{-}t)]^+$	Fe: Org.Comp.B15-313, 317/8
$C_{12}FeH_{14}N_2{}^{2+}$	$[(NH_2\text{-}C_6H_5)_2Fe][O\text{-}C_6H_2\text{-}2,4,6\text{-}(NO_2)_3]_2$	Fe: Org.Comp.B19-358
$C_{12}FeH_{14}N_4$	$Fe(C_5H_4\text{-}N{=}N\text{-}CH_3)_2$	Fe: Org.Comp.A9-144, 146
–	$[(C_5H_5)Fe(CNCH_3)_3][CN]$	Fe: Org.Comp.B15-344
$C_{12}FeH_{14}N_4O_2$	$Fe(C_5H_4\text{-}CONHNH_2)_2$	Fe: Org.Comp.A9-143, 146
$C_{12}FeH_{14}O$	$(C_5H_5)Fe(CO)(CH_2CHCHCH{=}CHCH_3)$	Fe: Org.Comp.B17-150, 154
–	$(C_5H_5)Fe(CO)(CH_2C_5H_7)$	Fe: Org.Comp.B17-156/7
–	$(C_5H_5)Fe(C_6H_6\text{-}1\text{-}CH_2OH)$	Fe: Org.Comp.B17-265
–	$(C_5H_5)Fe(C_6H_6\text{-}1\text{-}OCH_3)$	Fe: Org.Comp.B17-268/9
–	$(C_5H_5)Fe(C_6H_6\text{-}2\text{-}OCH_3)$	Fe: Org.Comp.B17-268/9
–	$(C_5H_5)Fe(C_6H_6\text{-}3\text{-}OCH_3)$	Fe: Org.Comp.B17-268/9
–	$[(CH_3\text{-}C_6H_5)Fe(C_5H_5)][OH]$	Fe: Org.Comp.B18-142/6, 201, 204
–	$[(C_6H_6)Fe(C_5H_5)][O\text{-}CH_3]$	Fe: Org.Comp.B18-142/6, 151/2, 154, 156
$C_{12}FeH_{14}O_2$	$(CH_3\text{-}C_5H_4)Fe(CO)_2\text{-}CH_2CH{=}CHCH_3$	Fe: Org.Comp.B14-54, 62
–	$(C_5H_5)Fe(CO)_2\text{-}CH_2CH{=}C(CH_3)_2$	Fe: Org.Comp.B13-104, 133
–	$(C_5H_5)Fe(CO)_2\text{-}CH_2CH{=}CHC_2H_5$	Fe: Org.Comp.B13-106, 134
–	$(C_5H_5)Fe(CO)_2\text{-}CH_2CH_2C(CH_3){=}CH_2$	Fe: Org.Comp.B13-109, 136
–	$(C_5H_5)Fe(CO)_2\text{-}CH_2CH_2CH{=}CHCH_3$	Fe: Org.Comp.B13-109
–	$(C_5H_5)Fe(CO)_2\text{-}CH_2CH_2CH_2\text{-}CH{=}CH_2$	Fe: Org.Comp.B13-110, 138
–	$(C_5H_5)Fe(CO)_2\text{-}C_5H_9\text{-}c$	Fe: Org.Comp.B13-199
$C_{12}FeH_{14}O_2S$	$(C_5H_5)Fe(CO)_2CH_2CH{=}C(CH_3)SCH_3$	Fe: Org.Comp.B13-106
$C_{12}FeH_{14}O_3$	$(C_5H_5)Fe(CO)[{=}C(OC_2H_5)\text{-}CH_2CH_2\text{-}C({=}O)\text{-}]$	Fe: Org.Comp.B16b-121, 135
–	$(C_5H_5)Fe(CO)[CH_2CHC(CH_3)COO\text{-}CH_3]$	Fe: Org.Comp.B17-150
–	$(C_5H_5)Fe(CO)_2\text{-}(C{=}O)\text{-}C_4H_9\text{-}n$	Fe: Org.Comp.B13-17
–	$(C_5H_5)Fe(CO)_2\text{-}(C{=}O)\text{-}C_4H_9\text{-}i$	Fe: Org.Comp.B13-17
–	$(C_5H_5)Fe(CO)_2\text{-}(C{=}O)\text{-}C_4H_9\text{-}t$	Fe: Org.Comp.B13-17, 37
–	$(C_5H_5)Fe(CO)_2\text{-}C(CH_2OCH_3){=}CHCH_3$	Fe: Org.Comp.B13-95, 122/3
–	$(C_5H_5)Fe(CO)_2\text{-}CH{=}CH\text{-}C(CH_3)_2\text{-}OH$	Fe: Org.Comp.B13-94/5, 122
–	$(C_5H_5)Fe(CO)_2\text{-}CH_2C(OC_2H_5){=}CH_2$	Fe: Org.Comp.B13-102, 132
–	$[i\text{-}C_3H_7C(O)\text{-}C_5H_4]Fe(CO)_2CH_3$	Fe: Org.Comp.B14-57, 64
$C_{12}FeH_{14}O_3S$	$(C_5H_5)Fe[C_5H_3(SO_3H)\text{-}C_2H_5]$	Fe: Org.Comp.A9-245, 252
–	$(HO_3S\text{-}C_5H_4)Fe(C_5H_4\text{-}C_2H_5)$	Fe: Org.Comp.A9-231, 237/8
$C_{12}FeH_{14}O_4$	$(C_5H_5)Fe(CO)_2\text{-}C({=}O)\text{-}CH_2CH(OH)\text{-}C_2H_5$	Fe: Org.Comp.B13-17
–	$(C_5H_5)Fe(CO)_2\text{-}CH[\text{-}O\text{-}CH_2\text{-}CH_2\text{-}O\text{-}CH(CH_3)\text{-}]$	Fe: Org.Comp.B14-2, 8
–	$(c\text{-}C_7H_9)Fe(CO)_2\text{-}COO\text{-}C_2H_5$	Fe: Org.Comp.B14-80, 81/2

$C_{12}FeH_{14}O_4S$	$(C_5H_5)Fe(CO)_2(CH_2{=}CHC(CH_3)_2S({=}O)O)$	Fe: Org.Comp.B17-128
$C_{12}FeH_{14}O_6S_2$	$Fe[C_5H_4\text{-}S({=}O)_2\text{-}OCH_3]_2$	Fe: Org.Comp.A9-232, 240
–	$(HO_3S\text{-}C_5H_4)Fe[C_5H_3(C_2H_5)\text{-}SO_3H]$	Fe: Org.Comp.A10-297
$C_{12}FeH_{14}P^+$	$[(C_6H_6)Fe(PC_4H_2(CH_3)_2\text{-}3,4)]^+$	Fe: Org.Comp.B18-130
$C_{12}FeH_{14}S$	$(CH_3S\text{-}C_5H_4)Fe(C_5H_4\text{-}CH_3)$	Fe: Org.Comp.A9-202, 206
–	$(C_5H_5)Fe[C_5H_3(SCH_3)\text{-}CH_3]$	Fe: Org.Comp.A9-216, 224
$C_{12}FeH_{14}S_2$	$(C_5H_5)Fe[C_5H_3(CH_2SH)_2]$	Fe: Org.Comp.A9-224/5
–	$Fe(C_5H_4\text{-}SCH_3)_2$	Fe: Org.Comp.A9-199, 204
$C_{12}FeH_{14}Se_2$	$Fe(C_5H_4SeCH_3)_2$	Fe: Org.Comp.A9-271
$C_{12}FeH_{15}IN_2O$	$[(C_5H_5)Fe(CNCH_3)(CN\text{-}C_3H_7\text{-}n)CO]I$	Fe: Org.Comp.B15-334
–	$[(C_5H_5)Fe(CNCH_3)(CN\text{-}C_3H_7\text{-}i)CO]I$	Fe: Org.Comp.B15-334
–	$[(C_5H_5)Fe(CN\text{-}C_2H_5)_2CO]I$	Fe: Org.Comp.B15-335
$C_{12}FeH_{15}KO_2$	$K[(C_5(CH_3)_5)Fe(CO)_2]$	Fe: Org.Comp.B14-120/2
$C_{12}FeH_{15}N$	$(C_5H_5)Fe[C_5H_3(CH_3)\text{-}CH_2NH_2]$	Fe: Org.Comp.A9-25/6
–	$(H_2N\text{-}C_5H_4)Fe(C_5H_4\text{-}C_2H_5)$	Fe: Org.Comp.A9-3, 4
$C_{12}FeH_{15}NO_2$	$(C_5H_5)Fe(CO)_2CH_2\text{-}2\text{-}NC_4H_8$	Fe: Org.Comp.B13-2
$C_{12}FeH_{15}NO_3$	$4\text{-}[(C_5H_5)Fe(CO)_2\text{-}CH_2]\text{-}1,4\text{-}ONC_4H_8$	Fe: Org.Comp.B13-3
–	$(C_5H_5)Fe(CO)_2\text{-}C({=}O)\text{-}N(C_2H_5)_2$	Fe: Org.Comp.B13-24, 45/6
–	$(C_5H_5)Fe(CO)_2\text{-}C({=}O)\text{-}NH\text{-}C_4H_9\text{-}n$	Fe: Org.Comp.B13-25
–	$(C_5H_5)Fe(CO)_2\text{-}C({=}O)\text{-}NH\text{-}C_4H_9\text{-}t$	Fe: Org.Comp.B13-25, 46
$C_{12}FeH_{15}NO_5S_2$	$(C_5H_5)Fe(CO)_2\text{-}CH[\text{-}CH_2\text{-}N(SO_2CH_3)\text{-}S({=}O)\text{-}CH_2CH_2\text{-}]$	Fe: Org.Comp.B14-34, 38
–	$(C_5H_5)Fe(CO)_2\text{-}N[S(O)_2CH_3]\text{-}S(O)CH_2\text{-}C_3H_5\text{-}c$	Fe: Org.Comp.B14-38
$C_{12}FeH_{15}N_2O^+$	$[(C_5H_5)Fe(CNCH_3)(CN\text{-}C_3H_7\text{-}n)CO]^+$	Fe: Org.Comp.B15-334
–	$[(C_5H_5)Fe(CNCH_3)(CN\text{-}C_3H_7\text{-}i)CO]^+$	Fe: Org.Comp.B15-334
–	$[(C_5H_5)Fe(CN\text{-}C_2H_5)_2CO]^+$	Fe: Org.Comp.B15-335
$C_{12}FeH_{15}N_2O_2{}^+$	$[(C_5H_5)Fe(CO)_2{=}C(\text{-}NH\text{-}CH_2\text{-}C(CH_3)_2\text{-}NH\text{-})]^+$	Fe: Org.Comp.B16a-147
–	$[(C_5H_5)Fe(CO)_2CN\text{-}CH_2CH_2\text{-}N(CH_3)_2]^+$	Fe: Org.Comp.B15-314
$C_{12}FeH_{15}NaO_2$	$Na[(C_5(CH_3)_5)Fe(CO)_2]$	Fe: Org.Comp.B14-120/2
$C_{12}FeH_{15}O^+$	$[C_5H_5Fe(C(CH_3)H{=}CHCH{=}CHCH_3)CO]^+$	Fe: Org.Comp.B17-187, 191
$C_{12}FeH_{15}OP$	$C_5H_5FeC_4HP(CH_2OH\text{-}2)((CH_3)_2\text{-}3,4)$	Fe: Org.Comp.B17-208/9
$C_{12}FeH_{15}O_2$	$[C_5(CH_3)_5]Fe(CO)_2$, radical	Fe: Org.Comp.B14-125
$C_{12}FeH_{15}O_2{}^+$	$[(C_5(CH_3)_5)Fe(CO)_2]^+$	Fe: Org.Comp.B14-124
–	$[(C_5H_5)Fe(CO)_2{=}CH\text{-}C_4H_9\text{-}t]^+$	Fe: Org.Comp.B16a-88, 96
–	$[(C_5H_5)Fe(CO)_2(CH_2{=}CH\text{-}C_3H_7\text{-}n)]^+$	Fe: Org.Comp.B17-12, 47
–	$[(C_5H_5)Fe(CO)_2(CH_2{=}CH\text{-}C_3H_7\text{-}i)]^+$	Fe: Org.Comp.B17-11/2
–	$[(C_5H_5)Fe(CO)_2(CH_3CH{=}C(CH_3)_2)]^+$	Fe: Org.Comp.B17-85
–	$[(C_5H_5)Fe(CO)_2(CH_3CH{=}CH\text{-}C_2H_5)]^+$	Fe: Org.Comp.B17-57/8
$C_{12}FeH_{15}O_2{}^-$	$[(C_5(CH_3)_5)Fe(CO)_2]^-$	Fe: Org.Comp.B14-120/2
$C_{12}FeH_{15}O_3{}^+$	$[(CH_2(CH)_4\text{-}C_4H_9\text{-}n)Fe(CO)_3]^+$	Fe: Org.Comp.B15-26
–	$[(CH_2(CH)_4\text{-}C_4H_9\text{-}t)Fe(CO)_3]^+$	Fe: Org.Comp.B15-2/3, 12/3, 20
–	$[(CH_3(CH)_5\text{-}C_3H_7\text{-}i)Fe(CO)_3]^+$	Fe: Org.Comp.B15-4/5, 12/3, 22
–	$[(C_5H_5)Fe(CO)_2{=}C(C_3H_7\text{-}i)\text{-}OCH_3]^+$	Fe: Org.Comp.B16a-109, 120
–	$[(C_5H_5)Fe(CO)_2(CH_2{=}C(CH_3)\text{-}OC_2H_5)]^+$	Fe: Org.Comp.B17-67, 79/80
–	$[(C_5H_5)Fe(CO)_2(CH_3CH{=}CH\text{-}OC_2H_5)]^+$	Fe: Org.Comp.B17-65, 78
$C_{12}FeH_{15}O_3P$	$(C_5H_5)Fe(CO)_2COCH{=}P(CH_3)_3$	Fe: Org.Comp.B13-14, 33/4
$C_{12}FeH_{15}O_3Si^+$	$[(1\text{-}(CH_3)_3Si\text{-}C_6H_6)Fe(CO)_3]^+$	Fe: Org.Comp.B15-102, 138
–	$[(2\text{-}(CH_3)_3Si\text{-}C_6H_6)Fe(CO)_3]^+$	Fe: Org.Comp.B15-105, 138

$C_{12}FeH_{15}O_3Si^+$	$[(3\text{-}(CH_3)_3Si\text{-}C_6H_6)Fe(CO)_3]^+$	Fe: Org.Comp.B15-107, 158
$C_{12}FeH_{15}O_4^+$	$[(C_5H_5)Fe(CO)_2{=}C(CH_2OCH_3)\text{-}OC_2H_5]^+$	Fe: Org.Comp.B16a-110, 120/1
–	$[(C_5H_5)Fe(CO)_2(CH_2{=}C(OCH_3)\text{-}OC_2H_5)]^+$	Fe: Org.Comp.B17-69, 81
$C_{12}FeH_{15}O_4Si^+$	$[(1\text{-}(CH_3)_3SiO\text{-}C_6H_6)Fe(CO)_3]^+$	Fe: Org.Comp.B15-91/2
–	$[(2\text{-}(CH_3)_3SiO\text{-}C_6H_6)Fe(CO)_3]^+$	Fe: Org.Comp.B15-91/2
$C_{12}FeH_{16}$	$C_5H_5Fe(CH_2C(CH_3)CHC(CH_3)CH_2)$	Fe: Org.Comp.B17-253, 254
$C_{12}FeH_{16}INO$	$(CH_3C_5H_4)Fe(CO)(I)CN\text{-}C_4H_9\text{-}t$	Fe: Org.Comp.B15-349
$C_{12}FeH_{16}N^+$	$[C_5H_5FeC_5H_3(CH_3)\text{-}CH_2NH_3]^+$	Fe: Org.Comp.A9-26
$C_{12}FeH_{16}NO_2^+$	$[(C_5H_5)Fe(CO)_2{=}CH\text{-}N(CH_3)\text{-}C_3H_7\text{-}i]^+$	Fe: Org.Comp.B16a-115, 126
–	$[(C_5H_5)Fe(CO)_2{=}CH\text{-}N(C_2H_5)_2]^+$	Fe: Org.Comp.B16a-115/6, 126
–	$[(C_5H_5)Fe(CO)_2{=}CH\text{-}NH\text{-}C_4H_9\text{-}t]^+$	Fe: Org.Comp.B16a-115, 125
–	$[(C_5H_5)Fe(CO)_2(CH_2{=}C(CH_3)\text{-}N(CH_3)_2)]^+$	Fe: Org.Comp.B17-68
–	$[(C_5H_5)Fe(CO)_2\text{-}CH(\text{-}CH_2\text{-}CH_2\text{-}CH_2\text{-}NH_2\text{-}CH_2\text{-})]^+$	Fe: Org.Comp.B14-151, 155
–	$[(C_5H_5)Fe(CO)_2\text{-}CH_2CH(\text{-}CH_2\text{-}CH_2\text{-}CH_2\text{-}NH_2\text{-})]^+$	Fe: Org.Comp.B14-151, 155/6
$C_{12}FeH_{16}N_2O_6S_3$	$(C_5H_5)Fe(CO)_2CH[\text{-}CH_2N(SO_2CH_3)\text{-}S\text{-}N(SO_2CH_3)CH_2\text{-}]$	Fe: Org.Comp.B14-35, 38/9
$C_{12}FeH_{16}O$	$C_5H_5Fe(CH_2C(CH_3)_2CH{=}CH_2)CO$	Fe: Org.Comp.B17-169
$C_{12}FeH_{16}O_2$	$(C_5H_5)Fe(CO)[{=}C(OCH_3)\text{-}CH(CH_3)\text{-}CH(CH_3)\text{-}]$	Fe: Org.Comp.B16b-117, 132
–	$(C_5H_5)Fe(CO)[CH_2OC(CH_3)_2CH{=}CH_2]$	Fe: Org.Comp.B17-169
–	$[C_5(CH_3)_5]Fe(CO)_2H$	Fe: Org.Comp.B14-75
$C_{12}FeH_{16}O_2S_2Sn$	$C_5H_4CH_3Fe(CO)_2C(S)SSn(CH_3)_3$	Fe: Org.Comp.B14-55
$C_{12}FeH_{16}O_3P^+$	$[C_5H_5Fe(CO)_2C(O)CH_2P(CH_3)_3]^+$	Fe: Org.Comp.B14-154
$C_{12}FeH_{17}LiO_2Si$	$Li[(C_5H_4Si(CH_3)_2C_3H_7\text{-}n)Fe(CO)_2]$	Fe: Org.Comp.B14-119
$C_{12}FeH_{17}NO$	$(C_5H_5)Fe(CO)(CN\text{-}C_4H_9\text{-}t)\text{-}CH_3$	Fe: Org.Comp.B15-306, 309
–	$(C_5H_5)Fe(CO)[{=}C(N(CH_3)_2)\text{-}CH_2CH_2CH_2\text{-}]$	Fe: Org.Comp.B16b-120/1, 134
$C_{12}FeH_{17}NO_2^{2+}$	$[(C_5H_5)Fe(CO)_2(CH_2{=}CH(CH_2)_3NH_3)]^{2+}$	Fe: Org.Comp.B17-35/6
$C_{12}FeH_{17}NO_2Sn$	$(CH_3)_3SnNC\text{-}Fe(CO)_2(C_6H_8\text{-}c)$	Sn: Org.Comp.18-128
$C_{12}FeH_{17}NSi$	$(C_5H_5)Fe[2\text{-}(CH_3)_3Si\text{-}NC_4H_3]$	Fe: Org.Comp.B17-202
–	$[Si(CH_3)_3\text{-}C_5H_4]Fe(NC_4H_4)$	Fe: Org.Comp.B17-217
$C_{12}FeH_{17}N_2O_6S_3^+$	$[(C_5H_5)Fe(CO)_2(CH_2{=}CHCH_2N(SO_2CH_3)SNHSO_2CH_3)]^+$	Fe: Org.Comp.B17-27, 50
$C_{12}FeH_{17}N_2S_2^+$	$[C_5H_5(CH_3NC)_2Fe{=}C(SCH_3)_2]^+$	Fe: Org.Comp.B16a-169/70
$C_{12}FeH_{17}N_3O$	$(C_5H_5)Fe(CO)(CN){=}C(NHCH_3)\text{-}NH\text{-}C_3H_7\text{-}n$	Fe: Org.Comp.B16a-172/4
–	$(C_5H_5)Fe(CO)(CN){=}C(NHCH_3)\text{-}NH\text{-}C_3H_7\text{-}i$	Fe: Org.Comp.B16a-173/4
$C_{12}FeH_{17}NaO_2Si$	$Na[(C_5H_4Si(CH_3)_2C_3H_7\text{-}n)Fe(CO)_2]$	Fe: Org.Comp.B14-119
$C_{12}FeH_{17}O_2Si^+$	$[((CH_3)_3Si\text{-}C_5H_4)Fe(CO)_2(CH_2{=}CH_2)]^+$	Fe: Org.Comp.B17-132
–	$[(C_5H_5)Fe(CO)_2(CH_2{=}CH\text{-}Si(CH_3)_3)]^+$	Fe: Org.Comp.B17-23
$C_{12}FeH_{17}O_2Si^-$	$[(C_5H_4Si(CH_3)_2C_3H_7\text{-}n)Fe(CO)_2]^-$	Fe: Org.Comp.B14-119
$C_{12}FeH_{18}$	$[1,4\text{-}(CH_3)_2\text{-}C_6H_4]Fe(CH_2{=}CH_2)_2$	Fe: Org.Comp.B18-81, 83, 90/1
–	$[CH_2C(CH_3)CHCHCH_2]_2Fe$	Fe: Org.Comp.B17-247, 249
–	$[CH_2CHC(CH_3)CHCH_2]_2Fe$	Fe: Org.Comp.B17-247, 250
$C_{12}FeH_{18}IN_3O$	$[(C_5H_5)Fe(CO)(CN\text{-}CH_3){=}C(NHCH_3)\text{-}NHC_2H_5]I$	Fe: Org.Comp.B16a-164, 168
–	$[(C_5H_5)Fe(CO)(CN\text{-}C_2H_5){=}C(NHCH_3)_2]I$	Fe: Org.Comp.B16a-163
$C_{12}FeH_{18}NO_2^+$	$[(C_5H_5)Fe(CO)_2\text{-}CH_2CH(CH_3)\text{-}NH(CH_3)_2]^+$	Fe: Org.Comp.B14-150

$C_{12}FeH_{18}NO_2^+$ ... $[(C_5H_5)Fe(CO)_2-CH_2CH_2-N(CH_3)_3]^+$ ... Fe: Org.Comp.B14-149, 155
$C_{12}FeH_{18}NO_3^+$ ... $[(C_5H_5)Fe(CO)_2CH_2OCH_2CH_2NH(CH_3)_2]^+$ ... Fe: Org.Comp.B14-152, 157
$C_{12}FeH_{18}N_3O^+$ ... $[(C_5H_5)Fe(CO)(CN-CH_3)=C(NHCH_3)-N(CH_3)_2]^+$ Fe: Org.Comp.B16a-165, 168
– ... $[(C_5H_5)Fe(CO)(CN-CH_3)=C(NHCH_3)-NHC_2H_5]^+$ Fe: Org.Comp.B16a-164, 168
– ... $[(C_5H_5)Fe(CO)(CN-C_2H_5)=C(NHCH_3)_2]^+$ ... Fe: Org.Comp.B16a-163
$C_{12}FeH_{18}O_2Si$ ... $(C_5H_4-CH_3)Fe(CO)_2-CH_2Si(CH_3)_3$ ... Fe: Org.Comp.B14-55
– ... $(C_5H_4-Si(CH_3)_3)Fe(CO)_2-C_2H_5$ ... Fe: Org.Comp.B14-59, 65
$C_{12}FeH_{18}O_4P^+$ ... $[(C_5H_5)(CO)(P(OCH_3)_3)Fe=C=CHCH_3]^+$ ... Fe: Org.Comp.B16a-208, 210
$C_{12}FeH_{19}INOP$ ... $[(C_5H_5)Fe(CNC_2H_5)(CO)P(CH_3)_3]I$ ... Fe: Org.Comp.B15-302
$C_{12}FeH_{19}NOP^+$ ... $[(C_5H_5)Fe(CNC_2H_5)(CO)P(CH_3)_3]^+$ ... Fe: Org.Comp.B15-302
$C_{12}FeH_{19}NO_2Si$ ... $[C_5H_4Si(CH_3)_2N(CH_3)_2]Fe(CO)_2CH_3$ ... Fe: Org.Comp.B14-60, 65
$C_{12}FeH_{19}NO_2Sn$ ... $(CH_3)_3SnNC-Fe(CO)_2(CH_2=C(CH_3)C(CH_3)=CH_2)$
Sn: Org.Comp.18-128
– ... $(CH_3)_3SnNC-Fe(CO)_2(CH_2=CHCH=CHC_2H_5)$ ... Sn: Org.Comp.18-128
$C_{12}FeH_{19}O_2P$ ... $[P(CH_3)_3C_2H_5][C_5H_5Fe(CO)_2]$ ... Fe: Org.Comp.B14-110/1
$C_{12}FeH_{20}IO_2P$ ... $[C_5H_5(CO)(P(CH_3)_3)Fe=C(CH_3)OCH_3]I$ ... Fe: Org.Comp.B16a-36, 52
$C_{12}FeH_{20}IPS$ ... $(C_5H_5)Fe(CS)(P(C_2H_5)_3)I$ ... Fe: Org.Comp.B15-262
$C_{12}FeH_{20}OSn$ ... $(C_5H_5)Fe(CH_2=CHCH_3)(CO)Sn(CH_3)_3$ ... Fe: Org.Comp.B16b-74, 78
$C_{12}FeH_{20}O_2P^+$ ... $[(C_5H_5)(CO)(P(CH_3)_3)Fe=C(CH_3)OCH_3]^+$ ... Fe: Org.Comp.B16a-36, 52
$C_{12}FeH_{20}O_4PS_2^+$ $[(C_5H_5)(CO)(P(OCH_3)_3)Fe=C(SCH_3)_2]^+$ ... Fe: Org.Comp.B16a-67, 72
$C_{12}FeH_{20}O_5P^+$ ... $[(C_5H_5)(CO)(P(OCH_3)_3)Fe=C(CH_3)OCH_3]^+$ ... Fe: Org.Comp.B16a-37, 52
$C_{12}FeH_{20}O_6P^+$ ... $[(C_5H_5)(CO)(P(OCH_3)_3)Fe=C(CH_2OCH_3)OH]^+$ ... Fe: Org.Comp.B16a-42, 56
$C_{12}FeH_{21}O_3P$ ... $(C_5H_5)Fe(CH_2C(CH_3)CH_2)-P(OCH_3)_3$ ... Fe: Org.Comp.B17-142
– ... $(C_5H_5)Fe(CH_2CHCH-CH_3)-P(OCH_3)_3$ ... Fe: Org.Comp.B17-141
$C_{12}FeH_{21}P$ ... $(C_5H_5)FeCH_2CHCHCH_3P(CH_3)_3$ ... Fe: Org.Comp.B17-141
$C_{12}FeH_{22}IN_2OP$ ... $[(C_5H_5)(CO)(P(CH_3)_3)Fe=C(NHCH_3)_2]I$ ... Fe: Org.Comp.B16a-68, 72/3
$C_{12}FeH_{22}N_2OP^+$ ... $[(C_5H_5)(CO)(P(CH_3)_3)Fe=C(NHCH_3)_2]^+$ ... Fe: Org.Comp.B16a-68, 72/3
$C_{12}FeH_{22}P_2$ ... $(C_6H_6)Fe[(CH_3)_2P-CH_2CH_2-P(CH_3)_2]$ ... Fe: Org.Comp.B18-5, 12
$C_{12}FeH_{23}N_7Ru^{2+}$ $[(NH_3)_5RuNCC_5H_4FeC_5H_4-CN]^{2+}$ ... Fe: Org.Comp.A9-127/8
$C_{12}FeH_{23}N_7Ru^{3+}$ $[(NH_3)_5RuNC-C_5H_4FeC_5H_4-CN]^{3+}$ ... Fe: Org.Comp.A9-127/8
$C_{12}FeH_{23}O_6P_2S^+$ $[(C_5H_5)Fe(CS)(P(OCH_3)_3)_2]^+$ ... Fe: Org.Comp.B15-264/5
$C_{12}FeH_{23}P_2^+$ ... $[(C_6H_6)Fe(H)((CH_3)_2P-CH_2CH_2-P(CH_3)_2)]^+$ ... Fe: Org.Comp.B18-5
$C_{12}FeH_{24}O_3P_2$ ... $(C_6H_6)Fe[P(CH_3)_3][P(OCH_3)_3]$ ... Fe: Org.Comp.B18-4
$C_{12}FeH_{24}O_6P_2$ ... $(C_6H_6)Fe[P(OCH_3)_3]_2$ ... Fe: Org.Comp.B18-2/3, 4, 12
$C_{12}FeH_{24}P_2$ ... $(C_6H_6)Fe[P(CH_3)_3]_2$ ... Fe: Org.Comp.B18-2, 3/4, 11/2
$C_{12}FeH_{25}P_2^+$ ... $[(C_6H_6)Fe(P(CH_3)_3)_2H]^+$ ... Fe: Org.Comp.B18-4
$C_{12}FeH_{27}N_2O_3PSn$
$(CH_3)_3Sn-P(C_4H_9-t)_2Fe(CO)(NO)_2$ ... Sn: Org.Comp.19-181, 183
$C_{12}FeH_{38}N_{12}Ru_2^{4+}$
$[Fe(C_5H_4-CNRu(NH_3)_5)_2]^{4+}$ ... Fe: Org.Comp.A9-126/7
$C_{12}FeH_{38}N_{12}Ru_2^{5+}$
$[(NH_3)_5RuNC-C_5H_4FeC_5H_4-CNRu(NH_3)_5]^{5+}$ ... Fe: Org.Comp.A9-127
$C_{12}FeH_{38}N_{12}Ru_2^{6+}$
$[Fe(C_5H_4-CNRu(NH_3)_5)_2]^{6+}$ ... Fe: Org.Comp.A9-126/7
$C_{12}Fe_3HNiO_{12}^-$ ... $[Fe_3Ni(CO)_{12}(H)]^-$ ... Fe: Org.Comp.C6b-150/4
$C_{12}Fe_3HO_{12}Rh^{2-}$ $[Fe_3Rh(CO)_{12}(H)]^{2-}$ ... Fe: Org.Comp.C6b-76/7
$C_{12}Fe_3H_3NO_{11}$ ... $Fe_3(CO)_{10}(NO)(C-CH_3)$ ... Fe: Org.Comp.C6b-82
$C_{12}Fe_3H_3O_{10}^-$ ... $[(CO)_{10}Fe_3(C-CH_3)]^-$ ... Fe: Org.Comp.C6b-60, 61, 64/5

$C_{12}Fe_3H_3O_{11}^-$ .... $[(CO)_{10}Fe_3(C-O-CH_3)]^-$ .... Fe: Org.Comp.C6b-60/1, 62, 66
$C_{12}Fe_3H_4NO_{10}^-$ .. $[(CO)_{10}(H)Fe_3(CN-CH_3)]^-$ .... Fe: Org.Comp.C6b-21
$C_{12}Fe_3H_4O_9S_3$ ... $(CO)_9Fe_3S(1,3-S_2C_3H_4)$ .... Fe: Org.Comp.C6a-271/3
$C_{12}Fe_3H_4O_{10}$ .... $(CO)_9Fe_3(C-CH_3)(C-OH)$ .... Fe: Org.Comp.C6a-234, 235
– .... $(CO)_{10}(H)Fe_3(C-CH_3)$ .... Fe: Org.Comp.C6b-49, 51, 54/5
$C_{12}Fe_3H_4O_{11}$ .... $(CO)_{10}(H)Fe_3(C-O-CH_3)$ .... Fe: Org.Comp.C6b-49, 51/2, 56/8
$C_{12}Fe_3H_4O_{11}P^-$ ... $[(CH_3)Fe_2P(H)(CO)_7-Fe(CO)_4]^-$ .... Fe: Org.Comp.C6b-157/8
$C_{12}Fe_3H_5NO_{10}$ ... $(CO)_{10}Fe_3N-C_2H_5$ .... Fe: Org.Comp.C6b-28, 31, 32
– .... $(CO)_{10}(H)Fe_3(C-NH-CH_3)$ .... Fe: Org.Comp.C6b-50, 52
$C_{12}Fe_3H_5O_{10}^-$ .... $[(CO)_9Fe_3C(C_2H_5)O]^-$ .... Fe: Org.Comp.C6a-290/2, 294/5
– .... $[(CO)_9Fe_3(O-CH_2-CH=CH_2)]^-$ .... Fe: Org.Comp.C6a-132
$C_{12}Fe_3H_5O_{10}P$ ... $(CO)_{10}Fe_3P-C_2H_5$ .... Fe: Org.Comp.C6b-27, 28, 31
$C_{12}Fe_3H_6NO_9P$ ... $(CO)_9(H)_2Fe_3P-CH_2CH_2-CN$ .... Fe: Org.Comp.C6a-187, 189, 190
$C_{12}Fe_3H_6N_2O_9$ ... $(CO)_9Fe_3[1,2-N_2C-3,3-(CH_3)_2]$ .... Fe: Org.Comp.C6a-181, 183
$C_{12}Fe_3H_6O_{10}$ .... $(CO)_9Fe_3(C-CH_3)(O-CH_3)$ .... Fe: Org.Comp.C6a-271, 272
– .... $(CO)_9(H)Fe_3C(C_2H_5)O$ .... Fe: Org.Comp.C6a-264
$C_{12}Fe_3H_7NO_{11}$ ... $[NH_3-CH_3][(H)Fe_3(CO)_{11}]$ .... Fe: Org.Comp.C6b-96
$C_{12}Fe_3H_7O_9PSe$ .. $(CO)_9Fe_3SeP-C_3H_7-i$ .... Fe: Org.Comp.C6a-203, 206
$C_{12}Fe_3H_7O_9PTe$ .. $(CO)_9Fe_3TeP-C_3H_7-i$ .... Fe: Org.Comp.C6a-203, 206, 210
$C_{12}Fe_3H_7O_9S^-$ ... $[(CO)_9Fe_3S-C_3H_7-i]^-$ .... Fe: Org.Comp.C6a-136
$C_{12}Fe_3H_8O_9$ ..... $(CO)_9(H)_3Fe_3C-C_2H_5$ .... Fe: Org.Comp.C6a-249, 250, 251
$C_{12}Fe_3H_8O_9S$ .... $(CO)_9(H)Fe_3S-C_3H_7-i$ .... Fe: Org.Comp.C6a-94, 95, 96/7
$C_{12}Fe_3H_9O_9P$ .... $(CO)_9(H)_2Fe_3P-C_3H_7-i$ .... Fe: Org.Comp.C6a-187/8, 190
$C_{12}Fe_3H_{10}O_9P_2$ .. $(CO)_9(H)Fe_3(CH_3-P-CH_2-PH-CH_3)$ .... Fe: Org.Comp.C6a-198/9
– .... $(CO)_9(H)Fe_3[P(CH_3)_2](P-CH_3)$ .... Fe: Org.Comp.C6a-193/5
$C_{12}Fe_3H_{11}NO_9Si$ .. $(CO)_9(H)_2Fe_3N-Si(CH_3)_3$ .... Fe: Org.Comp.C6a-187, 188, 191
$C_{12}Fe_3H_{14}O_8P_2$ .. $(CO)_8(H)_2Fe_3[P(CH_3)_2]_2$ .... Fe: Org.Comp.C6a-56
$C_{12}Fe_3MnO_{13}^-$ ... $[(CO)_9Fe_3(O)(Mn(CO)_3)]^-$ .... Fe: Org.Comp.C6a-302/3
$C_{12}Fe_3NiO_{12}^{2-}$ ... $[Fe_3Ni(CO)_{12}]^{2-}$ .... Fe: Org.Comp.C6b-150/4
$C_{12}Fe_3O_{12}^+$ ..... $[Fe_3(CO)_{12}]^+$ .... Fe: Org.Comp.C6a-4
$C_{12}Fe_3O_{12}^-$ ..... $[Fe_3(CO)_{12}]^-$, radical anion .... Fe: Org.Comp.C6b-126
$C_{12}GaH_4O_{16}S_4^{5-}$ $[Ga(4,5-(O)_2C_6H_2(SO_3)_2-1,3)_2]^{5-}$ .... Ga: SVol.D1-25/8
$C_{12}GaH_8N_2O_4^+$ .. $[Ga(NC_5H_4-2-C(O)O)_2]^+$ .... Ga: SVol.D1-255, 256
$C_{12}GaH_8O_4^-$ ..... $[Ga(1,2-(O)_2C_6H_4)_2]^-$ .... Ga: SVol.D1-25/8
$C_{12}GaH_8O_6^-$ ..... $[Ga((O)_2C_6H_3-OH)_2]^-$ .... Ga: SVol.D1-32/3
$C_{12}GaH_9N_2O^{2+}$ .. $[(HO)Ga(1,10-N_2C_{12}H_8)]^{2+}$ .... Ga: SVol.D1-308, 309
$C_{12}GaH_9O_5^{2-}$ .... $[(HO)Ga(1,2-(O)_2C_6H_4)_2]^{2-}$ .... Ga: SVol.D1-25/6
$C_{12}GaH_{10}NO_3$ .... $(NC_5H_5)Ga(O)[OC(O)-C_6H_5]$ .... Ga: SVol.D1-253
$C_{12}GaH_{10}N_2O_2^+$ .. $[(HO)_2Ga(1,10-N_2C_{12}H_8)]^+$ .... Ga: SVol.D1-308, 309
$C_{12}GaH_{10}O_6^+$ .... $[Ga(OC_4H_2-2-(CO-CH_3)-3-O)_2]^+$ .... Ga: SVol.D1-114
$C_{12}GaH_{10}O_{14}^{3-}$ ... $[Ga(H(OOC-CH_2-C(O)(COO)-CH_2-COO))_2]^{3-}$ Ga: SVol.D1-187/91

$C_{12}GaH_{11}K_2O_6$	$K_2[Ga(1,2\text{-}(O)_2C_6H_4)_2(H_2O)(OH)] \cdot 3\ H_2O$	Ga: SVol.D1-25/6, 32
$C_{12}GaH_{11}N_2O_3$	$Ga(OH)_3(1,10\text{-}N_2C_{12}H_8)$	Ga: SVol.D1-308, 309
$C_{12}GaH_{11}O_6^{2-}$	$K_2[Ga(1,2\text{-}(O)_2C_6H_4)_2(H_2O)(OH)] \cdot 3\ H_2O$	Ga: SVol.D1-25/6, 32
$C_{12}GaH_{12}N_2O_4^{+}$	$[Ga(NC_5H_3\text{-}2\text{-}CH_3\text{-}3,4\text{-}(O)_2)_2]^+$	Ga: SVol.D1-258
$C_{12}GaH_{12}N_2O_4^{-}$	$[(HO)_4Ga(1,10\text{-}N_2C_{12}H_8)]^-$	Ga: SVol.D1-308, 309
$C_{12}GaH_{12}N_4O_4Re$	$(OC)_4Re(1,2\text{-}N_2C_3H_3)_2Ga(CH_3)_2$	Re: Org.Comp.1-353
$C_{12}GaH_{12}O_{12}^{3-}$	$[Ga(OC(O)\text{-}CH(CH_3)\text{-}C(O)O)_3]^{3-}$	Ga: SVol.D1-166/7
$C_{12}GaH_{12}O_{15}^{3-}$	$[Ga(OC(O)\text{-}CH(OH)CH_2\text{-}C(O)O)_3]^{3-}$	Ga: SVol.D1-183/4
$C_{12}GaH_{12}O_{18}^{3-}$	$[(n\text{-}C_8H_{17})_3NH]_3[Ga(OC(O)\text{-}CH(OH)CH(OH)\text{-}C(O)O)_3]$	Ga: SVol.D1-184/5
$C_{12}GaH_{14}I_4N_2^{-}$	$[C_6H_5\text{-}NH_3 \cdot C_6H_5\text{-}NH_2][GaI_4(NH_2\text{-}C_6H_5)_2]$	Ga: SVol.D1-217/8
$C_{12}GaH_{14}N_4O_3ReS$	$(CO)_3Re[\text{-}(N_2C_3H_3)\text{-}Ga(CH_3)_2\text{-}S(N_2C_3H_2\text{-}1\text{-}CH_3)\text{-}]$	Re: Org.Comp.1-118
$C_{12}GaH_{16}N_4^{+}$	$[Ga(1,2\text{-}(NH_2)_2\text{-}C_6H_4)_2][GaCl_4] \cdot C_6H_6$	Ga: SVol.D1-238/9
$C_{12}GaH_{16}N_4^{3+}$	$[Ga(NC_5H_4\text{-}2\text{-}CH_2NH_2)_2]^{3+}$	Ga: SVol.D1-255, 256
$C_{12}GaH_{16}N_4O_2Se_4^{-}$	$Ba[Ga(OC_4H_8)_2(NCSe)_4]_2 \cdot 2\ OC_4H_8$	Ga: SVol.D1-111
$C_{12}GaH_{19}O_5$	$(C_2H_5\text{-}O)Ga[CH_3\text{-}C(O)CHC(O)\text{-}CH_3]_2$	Ga: SVol.D1-68/9
$C_{12}GaH_{19}O_{18}^{2-}$	$[(H_2O)(HO)Ga(OC(O)\text{-}CH(OH)CH(OH)\text{-}CH(OH)CH(OH)\text{-}C(O)O)_2]^{2-}$	Ga: SVol.D1-186/7
$C_{12}GaH_{20}MoN_3O_3$	$(CH_2CHCH_2)Mo(CO)_2[NH_2\text{-}CH_2CH_2\text{-}O\text{-}Ga(CH_3)_2\text{-}1,2\text{-}N_2C_3H_3]$	Mo: Org.Comp.5-205/6
$C_{12}GaH_{21}N_5O^{3+}$	$[Ga(HO\text{-}C_2H_5)(NC\text{-}CH_3)_5]^{3+}$	Ga: SVol.D1-19
$C_{12}GaH_{22}O_4^{+}$	$[Ga(OC(O)\text{-}C_5H_{11}\text{-}n)_2]^+$	Ga: SVol.D1-155
$C_{12}GaH_{23}O_5$	$Ga(O\text{-}C_3H_7\text{-}i)_2[CH_3\text{-}C(O)CHC(O)\text{-}O\text{-}C_2H_5]$	Ga: SVol.D1-201, 202
$C_{12}GaH_{24}N_4O_2^{3+}$	$[Ga(HO\text{-}C_2H_5)_2(NC\text{-}CH_3)_4]^{3+}$	Ga: SVol.D1-19
$C_{12}GaH_{27}N_3O_3^{3+}$	$[Ga(HO\text{-}C_2H_5)_3(NC\text{-}CH_3)_3]^{3+}$	Ga: SVol.D1-19
$C_{12}GaH_{30}N$	$GaH_3[N(C_4H_9\text{-}n)_3]$	Ga: SVol.D1-233
$C_{12}GaH_{30}N_2O_4^{3+}$	$[Ga(HO\text{-}C_2H_5)_4(NC\text{-}CH_3)_2]^{3+}$	Ga: SVol.D1-19
$C_{12}GaH_{32}O_8^{3+}$	$[Ga(HO\text{-}CH_2\text{-}CH(OH)\text{-}CH_3)_4]^{3+}$	Ga: SVol.D1-23
$C_{12}GaH_{34}O_9^{3+}$	$[(H_2O)Ga(HO\text{-}CH_2\text{-}CH(OH)\text{-}CH_3)_4]^{3+}$	Ga: SVol.D1-23
$C_{12}GaH_{36}O_6^{3+}$	$[Ga(HO\text{-}C_2H_5)_6]^{3+}$	Ga: SVol.D1-16, 18
–	$[Ga(HO\text{-}C_3H_7\text{-}n)_4(H_2O)_2]^{3+}$	Ga: SVol.D1-16, 18
$C_{12}Ga_2H_6O_{12}$	$Ga_2[OC(O)\text{-}CH\text{=}CH\text{-}C(O)O]_3 \cdot 2\ H_2O$	Ga: SVol.D1-169
$C_{12}Ga_2H_{12}O_{15}$	$Ga_2[OC(O)\text{-}CH(OH)CH_2\text{-}C(O)O]_3 \cdot 3\ H_2O$	Ga: SVol.D1-183, 184
$C_{12}Ga_2H_{12}O_{18}$	$Ga_2[OC(O)\text{-}CH(OH)CH(OH)\text{-}C(O)O]_3 \cdot 4\ H_2O$	Ga: SVol.D1-184, 185
$C_{12}Ga_2H_{24}N_6O_{24}S_3$	$Ga_2[(OC(O))_6C_6] \cdot 3\ [NH_4]_2[SO_4]$	Ga: SVol.D1-171
–	$Ga_2[(OC(O))_6C_6] \cdot 3\ [NH_4]_2[SO_4] \cdot 40\ H_2O$	Ga: SVol.D1-171
$C_{12}Ga_2O_{12}$	$Ga_2[(OC(O))_6C_6] \cdot 6\ H_2O$	Ga: SVol.D1-170/1
–	$Ga_2[(OC(O))_6C_6] \cdot 12\ H_2O$	Ga: SVol.D1-170/1
$C_{12}GeH_8$	$Ge(C{\equiv}CH)_3\text{-}C_6H_5$	Ge: Org.Comp.3-54
$C_{12}GeH_{10}$	$Ge(C_6H_5)_2$	Ge: Org.Comp.3-368
$C_{12}GeH_{12}$	$C_2H_5\text{-}Ge(C{\equiv}CH)_2\text{-}C_6H_5$	Ge: Org.Comp.3-217
–	$Ge(CH\text{=}C\text{=}CH_2)_4$	Ge: Org.Comp.1-97
–	$Ge(CH_2\text{-}C{\equiv}CH)_4$	Ge: Org.Comp.1-104
–	$Ge(C{\equiv}C\text{-}CH_3)_4$	Ge: Org.Comp.1-103, 105
$C_{12}GeH_{13}N$	$(CH_3)_2Ge(\text{-}C_4H_3NC_6H_4\text{-})$	Ge: Org.Comp.3-322
$C_{12}GeH_{14}$	$Ge(C_5H_4CH_3)_2$	Ge: Org.Comp.3-373, 377/8

$C_{12}GeH_{14}S_5Ti$ .... $(CH_3)_2Ge(C_5H_4)_2TiS_5$-c .... Ge: Org.Comp.3-330
$C_{12}GeH_{15}Mo_{11}O_{39}P^{4-}$
$[PMo_{11}Ge(C_{12}H_{15}\text{-n})O_{39}]^{4-}$ .... Mo: SVol.B3b-128/9
$C_{12}GeH_{15}Mo_{11}O_{39}Si^{5-}$
$[SiMo_{11}Ge(C_{12}H_{15}\text{-n})O_{39}]^{5-}$ .... Mo: SVol.B3b-128/9
$C_{12}GeH_{16}$ .... 1-$(CH_3)_3Ge$-C_9H_7 .... Ge: Org.Comp.2-42
– .... 2-$(CH_3)_3Ge$-C_9H_7 .... Ge: Org.Comp.2-41
– .... 3-$(CH_3)_3Ge$-C_9H_7 .... Ge: Org.Comp.2-42, 48
– .... $Ge(CH_3)_2(C_5H_5)_2$ .... Ge: Org.Comp.3-153, 161
$C_{12}GeH_{16}N_2O_3$ .... $Ge(CH_3)_3C(=CHCH=C(CH=NCH(C(=O)NHC(=O)CH_2))O)$
Ge: Org.Comp.2-97, 105
$C_{12}GeH_{16}O_2^-$ .... $Ge(CH_3)_3[C_6H_4COCOCH_3\text{-}4]^-$, radical anion .. Ge: Org.Comp.2-75, 86
$C_{12}GeH_{16}O_3$ .... $(CH_3)_2Ge(C_8H_4O_3(CH_3)_2)$ .... Ge: Org.Comp.3-323
$C_{12}GeH_{17}IO_2Ru$ .. $Ge(CH_3)_3C_7H_8(Ru(CO)_2I)$ .... Ge: Org.Comp.2-38
$C_{12}GeH_{17}MoNO_2$ $(C_5H_5)Mo(CO)_2(CN\text{-}CH_3)\text{-}Ge(CH_3)_3$ .... Mo: Org.Comp.8-9, 10
$C_{12}GeH_{17}NO_3S_2$ .. $(CH_3)_2Ge(\text{-}CH=CHCH(S(O)NHSO_2C_6H_5)CH_2\text{-})$ Ge: Org.Comp.3-263/4
$C_{12}GeH_{18}$ .... 1,2-$(CH_3)_2$-1-C_6H_5-GeC_4H_7 .... Ge: Org.Comp.3-250, 254
– .... 1-C_6H_5-1-C_2H_5-GeC_4H_8 .... Ge: Org.Comp.3-248
– .... 9,9-$(C_2H_5)_2$-[4.2.1]-9-GeC_8H_8 .... Ge: Org.Comp.3-329, 337
– .... $(CH_2=CH)_3Ge$-C_6H_9 .... Ge: Org.Comp.3-51/2
– .... $(CH_3)_3Ge$-CH_2-C_6H_4-4-$CH=CH_2$ .... Ge: Org.Comp.1-161
– .... $(CH_3)_3Ge$-C_6H_4-4-$C(CH_3)=CH_2$ .... Ge: Org.Comp.2-76, 86
– .... $(CH_3)_3Ge$-C_9H_9 .... Ge: Org.Comp.2-39, 42
– .... $Ge(CH=CH_2)_3$-$CH=CHCH_2CH_2CH=CH_2$ .... Ge: Org.Comp.3-51, 53
– .... $Ge(CH=CH_2)_3$-$CH_2CH=CHCH_2CH=CH_2$ .... Ge: Org.Comp.3-51
– .... $Ge(CH=CH_2)_3$-$CH_2CH_2CH=CHCH=CH_2$ .... Ge: Org.Comp.3-51
– .... $Ge(CH_3)_3$-$C(CH_3)=CHC_6H_5$ .... Ge: Org.Comp.2-6
$C_{12}GeH_{18}KN_3O_6$ .. $KGe(C_2H_5)_3C_6H_3(NO_2)_3$ .... Ge: Org.Comp.2-233/4
$C_{12}GeH_{18}LiN_3O_6$ $LiGe(C_2H_5)_3C_6H_3(NO_2)_3$ .... Ge: Org.Comp.2-233/4
$C_{12}GeH_{18}O$ .... $Ge(CH_3)_3$-CH_2-$C(=O)$-C_6H_4-4-CH_3 .... Ge: Org.Comp.1-174, 176/7
– .... $Ge(CH_3)_3$-$C(OCH_3)=CHC_6H_5$ .... Ge: Org.Comp.2-6, 7
– .... $Ge(CH_3)_3$-C_6H_4-4-$C(=O)C_2H_5$ .... Ge: Org.Comp.2-75
$C_{12}GeH_{18}O_2$ .... $Ge(CH_3)_3CH_2COC_6H_4OCH_3$-4 .... Ge: Org.Comp.1-174, 176/7
$C_{12}GeH_{19}I$ .... $Ge(C_2H_5)_3C_6H_4I$-4 .... Ge: Org.Comp.2-275
$C_{12}GeH_{19}Li$ .... $Ge(C_2H_5)_3C_6H_4Li$-4 .... Ge: Org.Comp.2-280, 282
$C_{12}GeH_{19}N$ .... 1-$(C_6H_5$-$CH_2)$-3,3-$(CH_3)_2$-1,3-$NGeC_3H_6$ .... Ge: Org.Comp.3-251
– .... $CH_2=CH$-$Ge(CH_3)_2$-CH_2-NH-CH_2-C_6H_5 .... Ge: Org.Comp.3-195
$C_{12}GeH_{19}NO_2$ .... $Ge(C_2H_5)_3$-C_6H_4-NO_2-3 .... Ge: Org.Comp.2-277
– .... $Ge(C_2H_5)_3$-C_6H_4-NO_2-4 .... Ge: Org.Comp.2-277, 282
$C_{12}GeH_{19}NO_2^-$ .... $[Ge(C_2H_5)_3\text{-}C_6H_4\text{-}NO_2\text{-}4]^-$, radical anion .... Ge: Org.Comp.2-277, 282
$C_{12}GeH_{19}NaO_3S$ .. $Na[Ge(C_2H_5)_3C_6H_4SO_3\text{-}4]$ .... Ge: Org.Comp.2-282
$C_{12}GeH_{20}$ .... 1-$(CH_2=CH$-$CH_2)$-1-CH_3-4-(i-$C_3H_7)$-GeC_5H_5 Ge: Org.Comp.3-306
– .... 9,9-$(C_2H_5)_2$-[3.3.1]-9-GeC_8H_{10} .... Ge: Org.Comp.3-325
– .... n-C_4H_9-$Ge(CH_3)_2$-C_6H_5 .... Ge: Org.Comp.3-202
– .... t-C_4H_9-$Ge(CH_3)_2$-C_6H_5 .... Ge: Org.Comp.3-202, 207
– .... $Ge(CH=CH\text{-}CH_3)_4$ .... Ge: Org.Comp.1-96
– .... $Ge(CH_3)_3$-CH_2-C_6H_3-3,5-$(CH_3)_2$ .... Ge: Org.Comp.1-158
– .... $Ge(CH_3)_3$-CH_2-C_6H_4-2-C_2H_5 .... Ge: Org.Comp.1-158, 163
– .... $Ge(CH_3)_3$-CH_2-C_6H_4-4-C_2H_5 .... Ge: Org.Comp.1-158, 163
– .... $Ge(CH_3)_3$-C_6H_2-2,4,6-$(CH_3)_3$ .... Ge: Org.Comp.2-77

$C_{12}GeH_{20}$ .... $Ge(CH_2{-}CH{=}CH_2)_4$ .... Ge: Org.Comp.1-97, 99
– .... $Ge(C_2H_5)_3{-}CH{=}C{=}CH{-}C{\equiv}C{-}CH_3$ .... Ge: Org.Comp.2-242
– .... $Ge(C_2H_5)_3{-}C{\equiv}C{-}CH{=}C{=}CH{-}CH_3$ .... Ge: Org.Comp.2-242
– .... $Ge(C_2H_5)_3{-}C{\equiv}C{-}CH_2{-}C{\equiv}C{-}CH_3$ .... Ge: Org.Comp.2-242
– .... $Ge(C_2H_5)_3{-}C{\equiv}C{-}C{\equiv}C{-}C_2H_5$ .... Ge: Org.Comp.2-241,249
– .... $Ge(C_2H_5)_3{-}C_6H_5$ .... Ge: Org.Comp.2-274, 281/2
– .... $Ge(C_2H_5)_3{-}C_6H_3D_2{-}2,5$ .... Ge: Org.Comp.2-274
– .... $Ge(C_3H_5{-}c)_4$ .... Ge: Org.Comp.1-86/7
– .... $Ge(C_4H_9{-}n)_2(C{\equiv}CH)_2$ .... Ge: Org.Comp.3-179, 181
– .... $Ge[C(CH_3){=}CH_2]_4$ .... Ge: Org.Comp.1-96
– .... $[{-}CH_2{-}C(CH_3){=}C(CH_3){-}CH_2{-}]Ge[{-}CH_2{-}C(CH_3){=}C(CH_3){-}CH_2{-}]$ .... Ge: Org.Comp.3-344, 346
$C_{12}GeH_{20}N_2O_3$ ... $5{-}(C_2H_5)_3Ge{-}3{-}[CH_3{-}OC({=}O)]{-}4{-}C({=}O)H{-}1,2{-}N_2C_3H$ Ge: Org.Comp.2-291/2, 294
– .... $(C_2H_5)_3Ge{-}C{\equiv}C{-}CH(OH){-}C({=}N_2){-}COO{-}CH_3$ ... Ge: Org.Comp.2-267/8, 271
$C_{12}GeH_{20}N_2O_5$ ... $(CH_3)_3Ge{-}C_4H_2N_2({=}O)_2[C_4H_5O(CH_2OH)(OH)]$ .. Ge: Org.Comp.2-103, 107
$C_{12}GeH_{20}O$ ...... $1{-}CH_3{-}1{-}(CH_2{=}CH{-}CH_2){-}4{-}CH_3O{-}4{-}C_2H_5{-}GeC_5H_4$ Ge: Org.Comp.3-306
– .... $Ge(C_2H_5)_3{-}C_6H_4{-}4{-}OH$ .... Ge: Org.Comp.2-275
$C_{12}GeH_{20}O_2$ ..... $Ge(CH_3)_2[C{\equiv}C{-}C(CH_3)_2{-}OH]_2$ .... Ge: Org.Comp.3-155
$C_{12}GeH_{20}O_3$ ..... $(C_2H_5)_2Ge[{-}CH_2CH_2C({-}C({=}O)OC({=}O)CH_2{-})CH_2CH_2{-}]$ Ge: Org.Comp.3-292
$C_{12}GeH_{20}O_3S$ .... $Ge(C_2H_5)_3C_6H_4SO_3H{-}4$ .... Ge: Org.Comp.2-276, 282
$C_{12}GeH_{20}O_8$ ..... $Ge(CH_2COOCH_3)_4$ .... Ge: Org.Comp.1-89, 92
$C_{12}GeH_{20}S$ ...... $SC_4H_3{-}2{-}Ge(CH_3)_2{-}C(C_4H_9{-}t){=}CH_2$ .... Ge: Org.Comp.3-204, 207
– .... $SC_4H_3{-}2{-}Ge(CH_3)_2{-}CH{=}CH{-}C_4H_9{-}t$ .... Ge: Org.Comp.3-205
$C_{12}GeH_{20}S_4$ ..... $1,1{-}(1,3{-}S_2C_4H_7{-}2)_2{-}GeC_4H_6$ .... Ge: Org.Comp.3-258, 268
$C_{12}GeH_{21}N$ ...... $Ge(C_2H_5)_3C_6H_4NH_2{-}4$ .... Ge: Org.Comp.2-276
$C_{12}GeH_{21}NO_2$ .... $Ge(CH_3)_3C({=}CHCH{=}C(CON(C_2H_5)_2)O)$ .... Ge: Org.Comp.2-97
$C_{12}GeH_{21}NO_2S$ ... $Ge(C_2H_5)_3C_6H_4SO_2NH_2{-}4$ .... Ge: Org.Comp.2-276
$C_{12}GeH_{21}P$ ...... $Ge(CH_3)_3C_3H_6P(H)C_6H_5$ .... Ge: Org.Comp.1-181
$C_{12}GeH_{22}$ .... $2,2{-}(CH_3)_2{-}C_3H_3{-}1{-}Ge(CH_3)_2{-}1{-}C_3H{-}2,2{-}(CH_3)_2$ Ge: Org.Comp.3-198/9
– .... $2,2{-}(CH_3)_2{-}C_3H_3{-}1{-}Ge(CH_3)_2{-}C({=}CH_2){-}C(CH_3){=}CH_2$ Ge: Org.Comp.3-197
– .... $(CH_3)_2Ge[{-}(CH_2)_4{-}C{\equiv}C{-}(CH_2)_4{-}]$ .... Ge: Org.Comp.3-313
– .... $Ge(C_2H_5)_3{-}CH{=}C{=}C{=}C(CH_3)_2$ .... Ge: Org.Comp.2-177
– .... $Ge(C_2H_5)_3{-}CH_2{-}C{\equiv}C{-}CH_2{-}CH{=}CH_2$ .... Ge: Org.Comp.2-241, 249
– .... $Ge(C_2H_5)_3{-}C{\equiv}C{-}CH{=}CH{-}C_2H_5$ .... Ge: Org.Comp.2-241
– .... $Ge(C_2H_5)_3{-}C_6H_7$ .... Ge: Org.Comp.2-232
$C_{12}GeH_{22}N_2$ ..... $Ge(CH_3)_2(CH_2CH_2CH_2CH_2{-}CN)_2$ .... Ge: Org.Comp.3-150
– .... $Ge(C_3H_7{-}n)_2(CH_2CH_2{-}CN)_2$ .... Ge: Org.Comp.3-176
$C_{12}GeH_{22}N_2O_4$ ... $1,1{-}(CH_3)_2{-}3{-}[C_2H_5{-}OOC{-}N(NH{-}COO{-}C_2H_5)]{-}GeC_4H_5$ Ge: Org.Comp.3-264
$C_{12}GeH_{22}O$ ...... $1,1{-}(CH_3)_2{-}4{-}CH_3O{-}4{-}(t{-}C_4H_9){-}GeC_5H_4$ ..... Ge: Org.Comp.3-303
– .... $1,1{-}(C_2H_5)_2{-}4{-}[CH_3{-}C({=}O){-}CH{=}]{-}GeC_5H_8$ .... Ge: Org.Comp.3-291
$C_{12}GeH_{22}OSi$ .... $Ge(CH_3)_3{-}C_6H_4{-}4{-}[O{-}Si(CH_3)_3]$ .... Ge: Org.Comp.2-79
$C_{12}GeH_{22}O_2$ ..... $3,3{-}(C_2H_5)_2{-}6{-}[C_2H_5{-}OC({=}O)]{-}[3.1.0]{-}3{-}GeC_5H_7$ Ge: Org.Comp.3-319
– .... $(CH_3)_2Ge[{-}(CH_2)_4{-}C({=}O){-}C({=}O){-}(CH_2)_4{-}]$ ..... Ge: Org.Comp.3-312, 314

$C_{12}GeH_{22}O_3$ ..... $2\text{-}(CH_3)_3Ge\text{-}5\text{-}(C_2H_5O)_2CH\text{-}OC_4H_2$ ..... Ge: Org.Comp.2-96, 105
$C_{12}GeH_{22}O_4$ ..... $1,1\text{-}(C_2H_5)_2\text{-}4\text{-}[HO\text{-}C(=O)]\text{-}4\text{-}[HO\text{-}C(=O)\text{-}CH_2]\text{-}GeC_5H_8$ Ge: Org.Comp.3-292
– ..... $Ge(C_2H_5)_3\text{-}C(COOCH_3)=CH\text{-}COOCH_3$ ..... Ge: Org.Comp.2-208
$C_{12}GeH_{22}S$ ..... $2,2,6,6,8,8\text{-}(CH_3)_6\text{-}[5.1.0]\text{-}4,8\text{-}SGeC_6H_4$ ..... Ge: Org.Comp.3-317, 331/2
$C_{12}GeH_{22}Si$ ..... $(CH_3)_4C_8GeH_{10}Si$ ..... Ge: Org.Comp.3-324/5
– ..... $C_6H_5\text{-}Ge(CH_3)_2\text{-}CH_2\text{-}Si(CH_3)_3$ ..... Ge: Org.Comp.3-201
– ..... $Ge(CH_3)_3\text{-}C_6H_4\text{-}Si(CH_3)_3\text{-}2$ ..... Ge: Org.Comp.2-73, 84
– ..... $Ge(CH_3)_3\text{-}C_6H_4\text{-}Si(CH_3)_3\text{-}3$ ..... Ge: Org.Comp.2-73, 84
– ..... $Ge(CH_3)_3\text{-}C_6H_4\text{-}Si(CH_3)_3\text{-}4$ ..... Ge: Org.Comp.2-73, 84
$C_{12}GeH_{23}NSn$ ..... $(CH_3)_3Sn\text{-}N(C_6H_5)\text{-}Ge(CH_3)_3$ ..... Sn: Org.Comp.18-57, 67, 74
$C_{12}GeH_{23}PSn$ ..... $(CH_3)_3Sn\text{-}P(C_6H_5)\text{-}Ge(CH_3)_3$ ..... Sn: Org.Comp.19-176
$C_{12}GeH_{24}$ ..... $1,1\text{-}(C_2H_5)_2\text{-}6\text{-}CH_3\text{-}GeC_7H_{11}$ ..... Ge: Org.Comp.3-310/1, 314
– ..... $(CH_3)_2Ge[\text{-}(CH_2)_4\text{-}CH=CH\text{-}(CH_2)_4\text{-}]$ ..... Ge: Org.Comp.3-312
– ..... $(t\text{-}C_4H_9)_2Ge[\text{-}CH_2\text{-}CH=CH\text{-}CH_2\text{-}]$ ..... Ge: Org.Comp.3-257
– ..... $Ge(CH_3)_3\text{-}CH=C=C(CH_3)\text{-}CH_2\text{-}C_4H_9\text{-}n$ ..... Ge: Org.Comp.2-20
– ..... $Ge(CH_3)_3\text{-}CH=C=C(CH_3)\text{-}CH_2\text{-}C_4H_9\text{-}t$ ..... Ge: Org.Comp.2-20
– ..... $Ge(C_2H_5)_2(CH_2CH=CHCH_3)_2$ ..... Ge: Org.Comp.3-170
– ..... $Ge(C_2H_5)_3\text{-}CH=C=CH\text{-}C_3H_7\text{-}n$ ..... Ge: Org.Comp.2-178
– ..... $Ge(C_2H_5)_3\text{-}CH_2CH=CH\text{-}CH_2\text{-}CH=CH_2$ ..... Ge: Org.Comp.2-178
– ..... $Ge(C_2H_5)_3\text{-}CH_2CH_2CH_2CH_2\text{-}C\equiv CH$ ..... Ge: Org.Comp.2-241, 249
– ..... $Ge(C_2H_5)_3\text{-}CH_2CH_2\text{-}CH=CH\text{-}CH=CH_2$ ..... Ge: Org.Comp.2-178
– ..... $Ge(C_2H_5)_3\text{-}C\equiv C\text{-}C_4H_9\text{-}n$ ..... Ge: Org.Comp.2-241
– ..... $Ge(C_2H_5)_3\text{-}C\equiv C\text{-}C_4H_9\text{-}t$ ..... Ge: Org.Comp.2-238, 247
– ..... $Ge(C_2H_5)_3\text{-}C_6H_9\text{-}c$ ..... Ge: Org.Comp.2-231
– ..... $Ge(C_3H_7\text{-}n)_2(CH=CHCH_3)_2$ ..... Ge: Org.Comp.3-176
– ..... $Ge(C_3H_7\text{-}i)_2(CH=CHCH_3)_2$ ..... Ge: Org.Comp.3-177
– ..... $Ge(C_3H_7\text{-}n)_2[C(CH_3)=CH_2]_2$ ..... Ge: Org.Comp.3-176
– ..... $Ge(C_3H_7\text{-}i)_2[C(CH_3)=CH_2]_2$ ..... Ge: Org.Comp.3-177
– ..... $Ge(C_4H_9\text{-}n)_2(CH=CH_2)_2$ ..... Ge: Org.Comp.3-179, 181
– ..... $[2,2\text{-}(CH_3)_2\text{-}C_3H_3\text{-}1]_2Ge(CH_3)_2$ ..... Ge: Org.Comp.3-150, 160
$C_{12}GeH_{24}N_2O_4$ ..... $Ge(CH_3)_3\text{-}CH=CH\text{-}CH_2\text{-}N(COOC_2H_5)\text{-}NH\text{-}COOC_2H_5$ Ge: Org.Comp.2-9
$C_{12}GeH_{24}O$ ..... $(CH_3)_2Ge[\text{-}(CH_2)_4\text{-}C(=O)\text{-}(CH_2)_5\text{-}]$ ..... Ge: Org.Comp.3-312, 314/5
– ..... $Ge(C_2H_5)_3\text{-}C(CH_2CH_2OH)=CHCH=CH_2$ ..... Ge: Org.Comp.2-215/6
– ..... $Ge(C_2H_5)_3\text{-}C(C_4H_9\text{-}s)=C=O$ ..... Ge: Org.Comp.2-199, 201
– ..... $Ge(C_2H_5)_3\text{-}C(C_4H_9\text{-}t)=C=O$ ..... Ge: Org.Comp.2-199, 201
– ..... $Ge(C_2H_5)_3\text{-}C[CH(OH)\text{-}CH_3]=CHCH=CH_2$ ..... Ge: Org.Comp.2-216
– ..... $Ge(C_2H_5)_3\text{-}C\equiv C\text{-}CH(OH)\text{-}C_3H_7\text{-}n$ ..... Ge: Org.Comp.2-268, 271
– ..... $Ge(C_2H_5)_3\text{-}C\equiv C\text{-}C(OH)(CH_3)\text{-}C_2H_5$ ..... Ge: Org.Comp.2-266/7
– ..... $Ge(C_2H_5)_3\text{-}C\equiv C\text{-}O\text{-}C_4H_9\text{-}n$ ..... Ge: Org.Comp.2-252
$C_{12}GeH_{24}O_2$ ..... $(CH_3)_2Ge[\text{-}(CH_2)_4\text{-}CH(OH)\text{-}C(=O)\text{-}(CH_2)_4\text{-}]$ ..... Ge: Org.Comp.3-312
– ..... $Ge(C_2H_5)_3\text{-}C\equiv C\text{-}CH(OH)\text{-}O\text{-}C_3H_7\text{-}n$ ..... Ge: Org.Comp.2-260
– ..... $Ge(C_2H_5)_3\text{-}C\equiv C\text{-}CH(OH)\text{-}O\text{-}C_3H_7\text{-}i$ ..... Ge: Org.Comp.2-260
$C_{12}GeH_{24}O_2S$ ..... $1\text{-}(n\text{-}C_3H_7)\text{-}1\text{-}[HO\text{-}C(=O)\text{-}CH_2\text{-}S\text{-}CH_2CH_2CH_2]\text{-}GeC_4H_8$ Ge: Org.Comp.3-249
$C_{12}GeH_{24}O_2S_2$ ..... $Ge(C_2H_5)_3\text{-}CH[S\text{-}C(=O)CH_3]\text{-}CH_2\text{-}S\text{-}C(=O)CH_3$ Ge: Org.Comp.2-128
$C_{12}GeH_{24}O_2Si$ ..... $2\text{-}(CH_3)_3Ge\text{-}2\text{-}[(CH_3)_3Si\text{-}C\equiv C]\text{-}1,3\text{-}O_2C_4H_6$ ..... Ge: Org.Comp.2-99, 106
– ..... $2\text{-}[(CH_3)_3Ge\text{-}C(Si(CH_3)_3)=C=]\text{-}1,3\text{-}O_2C_4H_6$ ..... Ge: Org.Comp.2-13

$C_{12}GeH_{24}O_4$	$Ge(CH_3)_2(CH_2CH_2CH_2CH_2-COOH)_2$	Ge: Org.Comp.3-150
–	$Ge(C_2H_5)_2(CH_2CH_2-COOCH_3)_2$	Ge: Org.Comp.3-167, 172
–	$Ge(C_2H_5)_3-CH_2CH_2CH_2CH(COOH)_2$	Ge: Org.Comp.2-163
–	$Ge(C_3H_7-n)_2(CH_2-COOCH_3)_2$	Ge: Org.Comp.3-176
$C_{12}GeH_{24}S$	$Ge(C_2H_5)_3-CH=C=CHCH_2-S-C_2H_5$	Ge: Org.Comp.2-218
–	$Ge(C_2H_5)_3-C(S-C_2H_5)=CHCH=CH_2$	Ge: Org.Comp.2-215, 219
–	$Ge(C_2H_5)_3-C\equiv C-CH_2CH_2-S-C_2H_5$	Ge: Org.Comp.2-264
$C_{12}GeH_{24}Sn_2$	$Ge(CH_3)_2[C\equiv C-Sn(CH_3)_3]_2$	Ge: Org.Comp.3-158
$C_{12}GeH_{25}IO$	$Ge(C_2H_5)_3-C(C_2H_5)=C(I)-OC_2H_5$	Ge: Org.Comp.2-197
$C_{12}GeH_{25}N$	$Ge(C_2H_5)_3-CH_2CH_2CH_2-CH_2CH_2-CN$	Ge: Org.Comp.2-165
–	$Ge(C_2H_5)_3-C\equiv C-N(C_2H_5)_2$	Ge: Org.Comp.2-252, 258
–	$Ge(C_3H_7-n)_3-CH_2CH_2CN$	Ge: Org.Comp.3-5
$C_{12}GeH_{25}NOSi$	$Ge(C_2H_5)_3-CH=C(CN)-O-Si(CH_3)_3$	Ge: Org.Comp.2-209
$C_{12}GeH_{26}$	$(CH_3)_2Ge[-(CH_2)_{10}-]$	Ge: Org.Comp.3-311
–	$(n-C_4H_9)_2Ge[-CH_2CH(CH_3)CH_2-]$	Ge: Org.Comp.3-236
–	$(n-C_4H_9)_2Ge[-CH_2CH_2CH_2CH_2-]$	Ge: Org.Comp.3-242
–	$n-C_4H_9-Ge(C_2H_5)_2-CH_2CH_2CH=CH_2$	Ge: Org.Comp.3-210
–	$n-C_6H_{13}-Ge(CH_3)[-CH(CH_3)CH_2CH_2CH_2-]$	Ge: Org.Comp.3-250
–	$Ge(C_2H_5)_3-C(CH_3)_2-C(CH_3)=CH_2$	Ge: Org.Comp.2-172, 184
–	$Ge(C_2H_5)_3-C(C_2H_5)=C(CH_3)_2$	Ge: Org.Comp.2-171, 183
–	$Ge(C_2H_5)_3-CH=CH-C_4H_9-n$	Ge: Org.Comp.2-178
–	$Ge(C_2H_5)_3-CH(C_2H_5)-C(CH_3)=CH_2$	Ge: Org.Comp.2-172
–	$Ge(C_2H_5)_3-CH_2CH=CH-C_3H_7-n$	Ge: Org.Comp.2-178
–	$Ge(C_2H_5)_3-CH_2-C(CH_3)=C(CH_3)_2$	Ge: Org.Comp.2-174
–	$Ge(C_2H_5)_3-C_6H_{11}-c$	Ge: Org.Comp.2-168
–	$Ge(C_3H_7-n)_3-C(CH_3)=CH_2$	Ge: Org.Comp.3-9, 11
–	$Ge(C_3H_7-i)_3-C(CH_3)=CH_2$	Ge: Org.Comp.3-14, 15
–	$Ge(C_3H_7-n)_3-CH=CHCH_3$	Ge: Org.Comp.3-10, 11
–	$Ge(C_3H_7-i)_3-CH=CHCH_3$	Ge: Org.Comp.3-14, 15
–	$Ge(C_3H_7-n)_3-CH_2CH=CH_2$	Ge: Org.Comp.3-10
–	$Ge(C_3H_7-i)_3-CH_2CH=CH_2$	Ge: Org.Comp.3-15
$C_{12}GeH_{26}O$	$(CH_3)_2Ge[-(CH_2)_4-CH(OH)-(CH_2)_5-]$	Ge: Org.Comp.3-311
–	$Ge(C_2H_5)_3-C(C_2H_5)=CH-O-C_2H_5$	Ge: Org.Comp.2-193, 200
–	$Ge(C_2H_5)_3-CH_2CH_2CH_2CH_2-C(=O)CH_3$	Ge: Org.Comp.2-164
–	$Ge(C_2H_5)_3-CH_2CH_2CH_2-C(=O)C_2H_5$	Ge: Org.Comp.2-165/6
–	$Ge(C_2H_5)_3-CH_2-C(=O)-C_4H_9-t$	Ge: Org.Comp.2-161
–	$Ge(C_2H_5)_3-C(O-C_2H_5)=CHC_2H_5$	Ge: Org.Comp.2-213, 218
–	$Ge(C_3H_7-n)_3-CH_2-C(=O)CH_3$	Ge: Org.Comp.3-4
$C_{12}GeH_{26}O_2$	$2-[(C_2H_5)_3Ge-CH_2CH_2CH_2-O-CH_2]-OC_2H_3$	Ge: Org.Comp.2-144
–	$(CH_3)_2Ge[-(CH_2)_4-CH(OH)-CH(OH)-(CH_2)_4-]$	Ge: Org.Comp.3-311, 314
–	$(CH_3)_3Ge-C(OCH_3)[-CH_2CH_2-C(CH_3)_2-CH_2-CH(OH)-]$	Ge: Org.Comp.1-197
–	$Ge(C_2H_5)_3-CH(CH_3)-COO-C_3H_7-i$	Ge: Org.Comp.2-134
–	$Ge(C_2H_5)_3-CH_2CH_2CH_2-CH_2CH_2-COOH$	Ge: Org.Comp.2-165
–	$Ge(C_2H_5)_3-CH_2CH_2CH_2-COO-C_2H_5$	Ge: Org.Comp.2-160
–	$Ge(C_2H_5)_3-CH_2-COO-C_4H_9-n$	Ge: Org.Comp.2-131
–	$Ge(C_3H_7-n)_3-CH_2CH_2-COOH$	Ge: Org.Comp.3-4
–	$Ge(C_3H_7-n)_3-CH_2-COO-CH_3$	Ge: Org.Comp.3-2
$C_{12}GeH_{26}O_3Si$	$2-[(CH_3)_3Ge-C(Si(CH_3)_3)=CH]-2-HO-1,3-O_2C_4H_6$	Ge: Org.Comp.2-9

$C_{12}GeH_{26}O_3Si$ . . . $(CH_3)_3Ge-C[Si(CH_3)_3]=CH-COO-CH_2CH_2CH_2OH$ Ge: Org.Comp.2-10

$C_{12}GeH_{27}$ . . . . $Ge(C_4H_9)_3$, radical . . . . . . Ge: Org.Comp.3-359

$C_{12}GeH_{27}NO$ . . . . . $Ge(C_2H_5)_3CH_2CON(C_2H_5)_2$ . . . . . . Ge: Org.Comp.2-132

$C_{12}GeH_{27}NO_2$ . . . . $Ge(C_2H_5)_3C(OH)(CH_3)CH_2CON(CH_3)_2$ . . . . . . Ge: Org.Comp.2-151, 154

$C_{12}GeH_{27}O_2P$ . . . . $Ge(CH_3)_3C(CH_3)(COOCH_3)CH_2P(C_2H_5)_2$ . . . . . . Ge: Org.Comp.1-170

$C_{12}GeH_{27}O_4P$ . . . . $Ge(C_2H_5)_3CH_2COP(O)(OC_2H_5)_2$ . . . . . . Ge: Org.Comp.2-132

$C_{12}GeH_{27}O_5P$ . . . . $Ge(CH_3)_3-C(CH_3)(COOCH_3)-CH_2-P(=O)(OC_2H_5)_2$ Ge: Org.Comp.1-170

– . . . . $Ge(CH_3)_3-CH(COOC_2H_5)-CH_2-P(=O)(OC_2H_5)_2$ Ge: Org.Comp.1-169

$C_{12}GeH_{28}$ . . . . $n-C_3H_7-Ge(C_4H_9-n)_2-CH_3$ . . . . . . Ge: Org.Comp.3-213

– . . . . $n-C_4H_9-Ge(C_3H_7-n)_2-C_2H_5$ . . . . . . Ge: Org.Comp.3-213

– . . . . $Ge(C_2H_5)_2(C_4H_9-n)_2$ . . . . . . Ge: Org.Comp.3-167, 172

– . . . . $Ge(C_2H_5)_3-C(CH_3)_2-C_3H_7-i$ . . . . . . Ge: Org.Comp.2-111

– . . . . $Ge(C_2H_5)_3-CH(C_2H_5)-C_3H_7-i$ . . . . . . Ge: Org.Comp.2-111

– . . . . $Ge(C_2H_5)_3-C_6H_{13}-n$ . . . . . . Ge: Org.Comp.2-111

– . . . . $Ge(C_3H_7-n)_2(C_3H_7-i)_2$ . . . . . . Ge: Org.Comp.3-176

– . . . . $Ge(C_3H_7-n)_4$ . . . . . . Ge: Org.Comp.1-65/71

– . . . . $Ge(C_3H_7-i)_4$ . . . . . . Ge: Org.Comp.1-72

– . . . . $Ge(C_3H_7-n)_3-C_3H_7-i$ . . . . . . Ge: Org.Comp.3-2

– . . . . $Ge(C_3H_7-i)_3-C_3H_7-n$ . . . . . . Ge: Org.Comp.3-14, 16

$C_{12}GeH_{28}IO_2P$ . . . $[Ge(CH_3)_3-CH(COOCH_3)-CH_2-P(C_2H_5)_2-CH_3]I$ Ge: Org.Comp.1-170

$C_{12}GeH_{28}O$ . . . . . . $Ge(C_2H_5)_3-CH(CH_3)-CH_2-CH(OH)-C_2H_5$ . . . . . Ge: Org.Comp.2-162

– . . . . $Ge(C_2H_5)_3-CH_2CH_2CH_2-CH_2CH_2CH_2-OH$ . . . . . Ge: Org.Comp.2-164

– . . . . $Ge(C_2H_5)_3-CH_2CH_2CH_2-CH_2CH_2-OCH_3$ . . . . . . Ge: Org.Comp.2-161

– . . . . $Ge(C_2H_5)_3-CH_2CH_2-C(OH)(CH_3)-C_2H_5$ . . . . . . Ge: Org.Comp.2-161

– . . . . $Ge(C_2H_5)_3-CH_2CH_2-O-C_4H_9-n$ . . . . . . Ge: Org.Comp.2-124

– . . . . $Ge(C_2H_5)_3-C(OH)(CH_3)-C_4H_9-t$ . . . . . . Ge: Org.Comp.2-148, 154

– . . . . $Ge(C_3H_7-n)_3-CH_2CH(OH)-CH_3$ . . . . . . Ge: Org.Comp.3-3

– . . . . $Ge(C_3H_7-n)_3-CH_2CH_2CH_2-OH$ . . . . . . Ge: Org.Comp.3-3

$C_{12}GeH_{28}OSi$ . . . . $Ge(C_2H_5)_3-CH=C(CH_3)-O-Si(CH_3)_3$ . . . . . . Ge: Org.Comp.2-206/7

– . . . . $Ge(C_2H_5)_3-CH[Si(CH_3)_3]-C(=O)CH_3$ . . . . . . Ge: Org.Comp.2-152

$C_{12}GeH_{28}O_2$ . . . . . $Ge(C_2H_5)_2(CH_2CH_2CH_2CH_2OH)_2$ . . . . . . Ge: Org.Comp.3-167/8

– . . . . $Ge(C_4H_9-n)_2(CH_2CH_2OH)_2$ . . . . . . Ge: Org.Comp.3-178

$C_{12}GeH_{28}O_2Si$ . . . $Ge(CH_3)_3-CH[Si(C_2H_5)_3]-COO-CH_3$ . . . . . . Ge: Org.Comp.1-172

– . . . . $Ge(C_2H_5)_3-CH_2CH_2-COO-Si(CH_3)_3$ . . . . . . Ge: Org.Comp.2-150

$C_{12}GeH_{28}O_3S$ . . . . $Ge(C_2H_5)_3-C(CH_3)(C_3H_7-n)-S(=O)_2-OCH_3$ . . . . Ge: Org.Comp.2-159

$C_{12}GeH_{29}N$ . . . . . . $Ge(C_2H_5)_3-CH_2CH_2CH_2-CH_2CH_2CH_2-NH_2$ . . . . Ge: Org.Comp.2-164

– . . . . $Ge(C_2H_5)_3-CH_2CH_2-N(C_2H_5)_2$ . . . . . . Ge: Org.Comp.2-126

$C_{12}GeH_{29}NOSn$ . . $(C_2H_5)_3Sn-N(CH_3)-C(=O)-CH_2-Ge(CH_3)_3$ . . . . . Ge: Org.Comp.1-171
Sn: Org.Comp.18-137

$C_{12}GeH_{29}O_3P$ . . . . $Ge(C_2H_5)_3CH_2CH_2P(O)(OC_2H_5)_2$ . . . . . . Ge: Org.Comp.2-126

$C_{12}GeH_{29}O_4PSi$ . . $Ge(CH_3)_3CH=C(OSi(CH_3)_3)P(O)(OC_2H_5)_2$ . . . . . Ge: Org.Comp.2-4

$C_{12}GeH_{30}OSi$ . . . . $Ge(C_2H_5)_3CH_2CH_2CH_2OSi(CH_3)_3$ . . . . . . Ge: Org.Comp.2-144

$C_{12}GeH_{30}OSi_2$ . . . $(CH_3)_2Ge(-C(Si(CH_3)_3)_2CH(OCH_3)CH_2-)$ . . . . . . Ge: Org.Comp.3-236, 240

$C_{12}GeH_{30}Si_2$ . . . . . $Ge(CH_3)_2(CH(Si(CH_3)_3)_2)CH_2CH=CH_2$ . . . . . . Ge: Org.Comp.3-195

$C_{12}GeH_{31}NSi$ . . . . $Ge(C_2H_5)_3CH_2CH_2CH_2NHSi(CH_3)_3$ . . . . . . Ge: Org.Comp.2-146

$C_{12}GeH_{31}NSi_2$ . . . $(CH_3)_2Ge(-CH_2Si(CH_3)_2N(C_4H_9)Si(CH_3)_2CH_2-)$ Ge: Org.Comp.3-295

$C_{12}GeH_{33}ISi_3$ . . . . $Ge(CH_3)_3C[Si(CH_3)_3]_2Si(CH_3)_2I$ . . . . . . Ge: Org.Comp.1-145

$C_{12}GeH_{33}InNP$ . . . $(CH_3)_3In \cdot (CH_3)_3Ge-N=P(C_2H_5)_3$ . . . . . . In: Org.Comp.1-27, 38

$C_{12}GeH_{33}NSiSn$	$(CH_3)_3SnN(Si(C_2H_5)_3)Ge(CH_3)_3$	Sn: Org.Comp.18-76, 81
$C_{12}GeH_{33}Si_3$	$Ge(CH_2Si(CH_3)_3)_3$, radical	Ge: Org.Comp.3-358
$C_{12}GeH_{34}Si_3$	$Ge(CH_3)_3C[Si(CH_3)_3]_2Si(CH_3)_2H$	Ge: Org.Comp.1-144
$C_{12}GeH_{36}N_2Si_2Sn$	$(CH_3)_3SnN(Si(CH_3)_2N(CH_3)Si(CH_3)_3)Ge(CH_3)_3$	Sn: Org.Comp.18-76, 81
$C_{12}Ge_2H_{18}MoN_2O_4$	$(CO)_4Mo[CN-Ge(CH_3)_3]_2$	Mo: Org.Comp.5-25, 29
$C_{12}Ge_3H_{36}N_2Sn$	$(CH_3)_3SnN(Ge(CH_3)_3)N(Ge(CH_3)_3)_2$	Sn: Org.Comp.18-121, 123, 125
$C_{12}HO_{12}Os_2Re$	$(CO)_4Re[-(OC)_4Os(H)Os(CO)_4-]$	Re: Org.Comp.1-501
$C_{12}H_3O_{11}ReW$	$(CO)_5ReOC(CH_3)=W(CO)_5$	Re: Org.Comp.2-176
$C_{12}H_4IO_6Re$	$(CO)_5ReC(O)C_6H_4I-4$	Re: Org.Comp.2-122
$C_{12}H_4MnO_{10}Re$	$(CO)_5ReCH_2CH_2-Mn(CO)_5$	Re: Org.Comp.2-180, 183
$C_{12}H_4NO_5Th^+$	$Th[1,3,7,9-(O)_4-5,10-NOC_{12}H_4]^+$	Th: SVol.D1-122
$C_{12}H_4N_6O_{14}Th^{2+}$	$Th[2,4,6-(NO_2)_3C_6H_2O]_2^{2+}$	Th: SVol.D1-103/4, 109
$C_{12}H_4N_8S_7$	$[=(2,5-SC_4(CN)_2-3,4)=NSN=S=NS-(1,4-C_6H_4)-SN=S=NSN=]_n$	S: S-N Comp.7-283, 286/7
$C_{12}H_5MoNO_5$	$(CO)_5Mo(CN-C_6H_5)$	Mo: Org.Comp.5-6, 8, 10
$C_{12}H_5NO_5Re^+$	$[(CO)_5Re(CNC_6H_5)]^+$	Re: Org.Comp.2-253, 254
$C_{12}H_5N_2O_5Re$	$(CO)_4Re(CNC_6H_5)NCO$	Re: Org.Comp.2-248
$C_{12}H_5O_6Re$	$(CO)_5ReC(O)C_6H_5$	Re: Org.Comp.2-121
$C_{12}H_5O_7Re$	$(CO)_5ReOC(O)C_6H_5$	Re: Org.Comp.2-23
$C_{12}H_5O_7ReRu$	$(CO)_5ReRu(CO)_2C_5H_5$	Re: Org.Comp.2-201
$C_{12}H_5O_{10}OsRe$	$(CO)_5ReOs(CO)_4C(O)C_2H_5$	Re: Org.Comp.2-202
$C_{12}H_5O_{12}Os_3Re$	$(CO)_3ReH_5Os_3(CO)_9$	Re: Org.Comp.1-336, 338, 340
$C_{12}H_6K_2O_6Th$	$K_2[Th(1,2,3-C_6H_3(O)_3)_2] \cdot 7\ H_2O$	Th: SVol.C7-35/6
$C_{12}H_6MnO_{10}Re$	$CH_3OC(CH_3)=Re(CO)_4-Mn(CO)_5$	Re: Org.Comp.1-496, 497
$C_{12}H_6MnO_{12}S_4$	$Mn[(O_3S)_2C_6H_3-1,3]_2 \cdot 3.5\ H_2O$	Mn: MVol.D7-114/5
$C_{12}H_6MoO_6$	$C_6H_5-C(OH)=Mo(CO)_5$	Mo: Org.Comp.5-100, 102
$C_{12}H_6N_4O_{10}Th^{2+}$	$Th[2,4-(NO_2)_2C_6H_3O]_2^{2+}$	Th: SVol.D1-109
$C_{12}H_6N_6O_8S_3$	$(2,4-(NO_2)_2C_6H_3-SN=)_2S$	S: S-N Comp.7-32/7
$C_{12}H_6N_8O_8S_5$	$(2,4-(NO_2)_2C_6H_3-SN=S=N)_2S$	S: S-N Comp.7-41/2
$C_{12}H_6Na_2O_6Th$	$Na_2[Th(1,2,3-C_6H_3(O)_3)_2] \cdot 7\ H_2O$	Th: SVol.C7-35/6
$C_{12}H_6O_6Re^+$	$[(CO)_5ReOC(C_6H_5)H]^+$	Re: Org.Comp.2-157
$C_{12}H_6O_{10}PReW$	$(CO)_5ReP(CH_3)_2-W(CO)_5$	Re: Org.Comp.2-174
$C_{12}H_7MoNO_5$	$H_2N-C(C_6H_5)=Mo(CO)_5$	Mo: Org.Comp.5-101, 109
$C_{12}H_7MoN_3O_2$	$(C_5H_5)Mo(CO)_2[=C(NH_2)-C(CN)=C(CN)-]$	Mo: Org.Comp.8-185, 196
–	$(C_5H_5)Mo(CO)_2[=C(ND_2)-C(CN)=C(CN)-]$	Mo: Org.Comp.8-196
$C_{12}H_7O_4ReS_2$	$(CO)_4Re[-S=C(CH_2C_6H_5)-S-]$	Re: Org.Comp.1-358
–	$(CO)_4Re[-S=C(C_6H_4-4-CH_3)-S-]$	Re: Org.Comp.1-358/9
$C_{12}H_7O_5Re$	$(CO)_4Re[-O=C(CH_3)-(1,2-C_6H_4)-]$	Re: Org.Comp.1-418
–	$(CO)_5Re-CH_2-C_6H_5$	Re: Org.Comp.2-109
–	$(CO)_5Re-C_6H_4CH_3-3$	Re: Org.Comp.2-130
–	$(CO)_5Re-C_6H_4CH_3-4$	Re: Org.Comp.2-130
–	$(CO)_5Re-C_7H_7-c$	Re: Org.Comp.2-110, 113
$C_{12}H_7O_7ReS$	$(CO)_5ReS(O)_2CH_2C_6H_5$	Re: Org.Comp.2-27/8
$C_{12}H_7O_8ReS$	$(CO)_5ReOS(O)_2C_6H_4CH_3-4$	Re: Org.Comp.2-26
$C_{12}H_8I_2N_2S$	$4-I-C_6H_4-N=S=N-C_6H_4-4-I$	S: S-N Comp.7-238, 253
$C_{12}H_8I_3Sb$	$(2-C_6H_4C_6H_4-2')SbI_3$	Sb: Org.Comp.5-176
$C_{12}H_8KN_4O_4S$	$K[(4-NO_2-C_6H_4-N)_2S]$, radical	S: S-N Comp.7-334/6

$C_{12}H_8K_2N_2O_6PdS_4$		
	$K_2[Pd(S_2O_3)_2(1,10\text{-}N_2C_{12}H_8)]$	Pd: SVol.B2-239
$C_{12}H_8K_4O_{20}Th$	$K_4[Th((CHOH)_2(COO)_2)_2(C_2O_4)_2] \cdot 3\ H_2O$	Th: SVol.C7-106/7
$C_{12}H_8LiO_5Re$	$Li[(CO)_4Re(CH_2C_6H_5)C(O)H]$	Re: Org.Comp.1-424
$C_{12}H_8MnNO_2S^+$	$Mn[2\text{-}(OOC\text{-}C_6H_4\text{-}2\text{-}N{=}CH)C_4H_3S]^+$	Mn: MVol.D6-75/6
$C_{12}H_8MnN_2O_2S$	$Mn(SC(2\text{-}C_4H_3O){=}NN{=}CHC_6H_4O\text{-}2) \cdot H_2O$	Mn: MVol.D7-218/20
$C_{12}H_8MnN_2O_{12}P_2{}^{2-}$		
	$[Mn(O_3PO\text{-}C_6H_4\text{-}NO_2\text{-}4)_2]^{2-}$	Mn: MVol.D8-146/7
$C_{12}H_8MnO_4S_2{}^{2-}$	$Mn[HOC_6H_3(O)S]_2{}^{2-}$	Mn: MVol.D7-29/30
$C_{12}H_8MoN_6O_2S_2$	$Mo(NO)_2(NCS)_2(C_{10}H_8N_2)$	Mo: SVol.B3b-200
$C_{12}H_8MoO_4$	$(c\text{-}C_8H_8)Mo(CO)_4$	Mo: Org.Comp.5-312/6
$C_{12}H_8MoO_4{}^{2+}$	$[MoO_4(C_6H_4)_2]^{2+}$	Mo: SVol.B3b-162/3
$C_{12}H_8MoO_6{}^{2-}$	$[MoO_2(1,2\text{-}(O)_2C_6H_4)_2]^{2-}$	Mo: SVol.B3b-148/50, 162/3
$C_{12}H_8N_2O_2S_2$	$4\text{-}O{=}S{=}NC_6H_4\text{-}4\text{-}(C_6H_4N{=}S{=}O\text{-}4)$	S: S-N Comp.6-210
$C_{12}H_8N_2O_4PdS$	$Pd(SO_4)(1,10\text{-}N_2C_{12}H_8)$	Pd: SVol.B2-238
$C_{12}H_8N_2O_6STh^{2+}$	$Th[(HO)_2(O)C_6H_2NNC_6H_4SO_3]^{2+}$	Th: SVol.D1-122
$C_{12}H_8N_2O_6Th$	$Th(OO)(NC_5H_4\text{-}2\text{-}COO)_2$	Th: SVol.C7-151/3
$C_{12}H_8N_2O_7Th$	$(O_2)Th[2,6\text{-}(OOC)_2\text{-}NC_5H_3] \cdot 1\text{-}O\text{-}NC_5H_5$	Th: SVol.D4-158
$C_{12}H_8N_4O_4S$	$O_2N\text{-}2\text{-}C_6H_4\text{-}N{=}S{=}N\text{-}C_6H_4\text{-}2\text{-}NO_2$	S: S-N Comp.7-240, 253
–	$O_2N\text{-}3\text{-}C_6H_4\text{-}N{=}S{=}N\text{-}C_6H_4\text{-}3\text{-}NO_2$	S: S-N Comp.7-240, 253/4
–	$O_2N\text{-}3\text{-}C_6H_4\text{-}N{=}S{=}N\text{-}C_6H_4\text{-}4\text{-}NO_2$	S: S-N Comp.7-262
–	$O_2N\text{-}4\text{-}C_6H_4\text{-}N{=}S{=}N\text{-}C_6H_4\text{-}4\text{-}NO_2$	S: S-N Comp.7-240/1, 248, 252/5
$C_{12}H_8N_4O_4S^-$	$[(4\text{-}NO_2\text{-}C_6H_4\text{-}N)_2S]^-$, radical anion	S: S-N Comp.7-334/6
$C_{12}H_8N_4O_4S_3$	$(2\text{-}NO_2\text{-}C_6H_4\text{-}SN{=})_2S$	S: S-N Comp.7-32/7
–	$(4\text{-}NO_2\text{-}C_6H_4\text{-}SN{=})_2S$	S: S-N Comp.7-32/7
$C_{12}H_8N_4O_6S_2$	$4\text{-}NO_2\text{-}C_6H_4\text{-}S(O)_2N{=}S{=}N\text{-}C_6H_4\text{-}NO_2\text{-}4$	S: S-N Comp.7-60/1
$C_{12}H_8N_4O_6S_5$	$4\text{-}O{=}S{=}N\text{-}C_6H_4\text{-}S(O)_2N{=}S{=}NS(O)_2\text{-}C_6H_4\text{-}N{=}S{=}O\text{-}4$	S: S-N Comp.7-95/8
$C_{12}H_8N_4O_8S_3$	$2\text{-}NO_2\text{-}C_6H_4\text{-}S(O)_2N{=}S{=}NS(O)_2\text{-}C_6H_4\text{-}NO_2\text{-}2$	S: S-N Comp.7-95/8
–	$3\text{-}NO_2\text{-}C_6H_4\text{-}S(O)_2N{=}S{=}NS(O)_2\text{-}C_6H_4\text{-}NO_2\text{-}3$	S: S-N Comp.7-95/8
–	$4\text{-}NO_2\text{-}C_6H_4\text{-}S(O)_2N{=}S{=}NS(O)_2\text{-}C_6H_4\text{-}NO_2\text{-}4$	S: S-N Comp.7-95/8
$C_{12}H_8N_4O_8Th$	$Th(OOC\text{-}CH_2\text{-}CN)_4$	Th: SVol.D4-166
$C_{12}H_8N_4Pd$	$Pd(CN)_2(NC_5H_4\text{-}C_5H_4N)$	Pd: SVol.B2-286
$C_{12}H_8N_4PdS_2$	$Pd(SCN)_2[2\text{-}NC_5H_4\text{-}(2\text{-}C_5H_4N)]$	Pd: SVol.B2-304
$C_{12}H_8N_4S_4$	$[\text{-}(1,4\text{-}C_6H_4)\text{-}SN{=}S{=}N\text{-}(1,4\text{-}C_6H_4)\text{-}N{=}S{=}NS\text{-}]_n$	S: S-N Comp.7-283, 285
$C_{12}H_8N_6OS$	$1,2,3\text{-}N_3C_6H_4\text{-}1\text{-}S(O)\text{-}1\text{-}(1,2,3\text{-}N_3C_6H_4)$	S: S-N Comp.8-354
$C_{12}H_8N_6O_4PdS_2$	$Pd(SCN)_2(O_2N\text{-}4\text{-}C_5H_4N)_2$	Pd: SVol.B2-302/3
$C_{12}H_8N_6O_4S_5$	$(2\text{-}NO_2\text{-}C_6H_4\text{-}SN{=}S{=}N)_2S$	S: S-N Comp.7-41/2
–	$(4\text{-}NO_2\text{-}C_6H_4\text{-}SN{=}S{=}N)_2S$	S: S-N Comp.7-41/2
$C_{12}H_8N_{10}PdS_2$	$Pd(SCN)_2[N_3\text{-}4\text{-}C_5H_4N]_2$	Pd: SVol.B2-302
$C_{12}H_8O_5Re^-$	$[(CO)_4Re(CH_2C_6H_5)C(O)H]^-$	Re: Org.Comp.1-424
–	$[(CO)_4Re(CH_3)C(O)C_6H_5]^-$	Re: Org.Comp.1-424, 427/8
–	$[(CO)_4Re(C_6H_5)C(O)CH_3]^-$	Re: Org.Comp.1-424, 427/8
$C_{12}H_8O_6Th$	$Th[H(1,2,3\text{-}(O)_3C_6H_3)]_2$	Th: SVol.D1-109/10
$C_{12}H_8O_8Th$	$Th(OH)_2(1,2,3\text{-}C_6H_3(O)_3)_2$	Th: SVol.C7-35/6
$C_{12}H_9I_2O_2Sb$	$(4\text{-}IC_6H_4)_2Sb(O)OH$	Sb: Org.Comp.5-209
$C_{12}H_9InS_3$	$(SC_4H_3\text{-}2)_3In$	In: Org.Comp.1-89, 91
$C_{12}H_9K_3MoO_3$	$K_3[(CH_3\text{-}C{\equiv}C)_3Mo(CO)_3]$	Mo: Org.Comp.5-134
$C_{12}H_9MoN_3O_3$	$(CH_2{=}CH\text{-}CN)_3Mo(CO)_3$	Mo: Org.Comp.5-198

$C_{12}H_9MoO_3^{3-}$	$K_3[(CH_3\text{-}C{\equiv}C)_3Mo(CO)_3]$	Mo:Org.Comp.5-134
$C_{12}H_9N_2O_6Sb$	$(3\text{-}O_2N\text{-}C_6H_4)_2Sb(=O)OH$	Sb: Org.Comp.5-209
–	$(4\text{-}O_2N\text{-}C_6H_4)_2Sb(=O)OH$	Sb: Org.Comp.5-209
–	$(4\text{-}O_2N\text{-}C_6H_4)_2Sb(=O)OH \cdot H_2O$	Sb: Org.Comp.5-209
$C_{12}H_9N_3OS_2$	$2\text{-}(O{=}S{=}N\text{-}N{=})\text{-}3\text{-}(CH_3)C_{11}H_6NS\text{-}3,1$	S: S-N Comp.6-69/71
$C_{12}H_9N_3O_2S$	$C_6H_5\text{-}N{=}S{=}N\text{-}C_6H_4\text{-}NO_2\text{-}4$	S: S-N Comp.7-257/8
$C_{12}H_9N_3O_2S_2$	$C_6H_5\text{-}SN{=}S{=}N\text{-}C_6H_4NO_2\text{-}4$	S: S-N Comp.7-25/7
$C_{12}H_9N_3O_4S_2$	$C_6H_5\text{-}S(O)_2N{=}S{=}N\text{-}C_6H_4\text{-}3\text{-}NO_2$	S: S-N Comp.7-55/60
–	$C_6H_5\text{-}S(O)_2N{=}S{=}N\text{-}C_6H_4\text{-}4\text{-}NO_2$	S: S-N Comp.7-55/60
$C_{12}H_9N_3O_4S_3$	$C_6H_5\text{-}S(O)_2N{=}S{=}NS\text{-}C_6H_4\text{-}NO_2\text{-}2$	S: S-N Comp.7-66
$C_{12}H_9N_3O_6S_3$	$C_6H_5\text{-}S(O)_2N{=}S{=}NS(O)_2\text{-}C_6H_4\text{-}NO_2\text{-}2$	S: S-N Comp.7-100/1
–	$C_6H_5\text{-}S(O)_2N{=}S{=}NS(O)_2\text{-}C_6H_4\text{-}NO_2\text{-}3$	S: S-N Comp.7-100/1
–	$C_6H_5\text{-}S(O)_2N{=}S{=}NS(O)_2\text{-}C_6H_4\text{-}NO_2\text{-}4$	S: S-N Comp.7-100/1
$C_{12}H_9N_6O_3PRe^+$	$[(CO)_3Re(\text{-}N_2C_3H_3\text{-})_3P]^+$	Re: Org.Comp.1-217
$C_{12}H_9O_9ReRuSi$	$(CO)_5ReRu(CO)_4Si(CH_3)_3$	Re: Org.Comp.2-201
$C_{12}H_{10}IIn$	$(C_6H_5)_2InI$	In: Org.Comp.1-156, 157, 159
$C_{12}H_{10}IMoNO_3$	$(C_5H_5)Mo(NO)(I)O_2CC_6H_5$	Mo:Org.Comp.6-46
$C_{12}H_{10}IO_2Sb$	$(C_6H_5)(4\text{-}IC_6H_4)Sb(O)OH$	Sb: Org.Comp.5-222
$C_{12}H_{10}KN_2S$	$K[(C_6H_5\text{-}N)_2S]$, radical	S: S-N Comp.7-334/6
$C_{12}H_{10}MnNOS^+$	$[Mn(SCH_2C(=O)NHC_{10}H_7\text{-}2)]^+$	Mn:MVol.D7-54
$C_{12}H_{10}MnNO_2S^+$	$[Mn(2\text{-}(4\text{-}CH_3\text{-}C_6H_4\text{-}N(O)C(O))\text{-}SC_4H_3)]^+$	Mn:MVol.D7-222/5
$C_{12}H_{10}MnN_2^{2+}$	$[Mn(NC_5H_4\text{-}2\text{-}CH{=}N\text{-}C_6H_5)]^{2+}$	Mn:MVol.D6-78
$C_{12}H_{10}MnN_2O_3^{2+}$	$Mn[CH_3C(O)NH\text{-}N{=}CH\text{-}3\text{-}(1\text{-}OC_9H_5(=O\text{-}4))]^{2+}$	Mn:MVol.D6-288/9
$C_{12}H_{10}MnN_2O_4$	$Mn[OC(CH_3)C(C(O)CH_3){=}N\text{-}N\text{-}2\text{-}C_6H_4COO]$	Mn:MVol.D6-256/7
$C_{12}H_{10}MnN_3O_3Se$	$[Mn(NO)_3((C_6H_5)_2Se)]$	Mn:MVol.D7-242
$C_{12}H_{10}MnN_3O_3Te$	$[Mn(NO)_3((C_6H_5)_2Te)]$	Mn:MVol.D7-242
$C_{12}H_{10}MnN_4O_2S$	$Mn(SC(NHNH_2)NNHC(=O)C_{10}H_6(O))$	Mn:MVol.D7-207/9
$C_{12}H_{10}MnN_4O_2S_2Se_2$	$[Mn(SC_4H_3(CH{=}NOH\text{-}2))_2(NCSe)_2]$	Mn:MVol.D7-222/5
$C_{12}H_{10}MnN_4O_2S_4$	$[Mn(SC_4H_3(CH{=}NOH\text{-}2))_2(NCS)_2]$	Mn:MVol.D7-222/5
$C_{12}H_{10}MnN_5O_6S_2$	$[Mn(S\text{-}C(SCH_3){=}NN{=}CH\text{-}1\text{-}(2\text{-}NC_9H_6))(NO_3)_2]_n$	Mn:MVol.D6-361, 363
–	$[Mn(S\text{-}C(SCH_3){=}NN{=}CH\text{-}2\text{-}(1\text{-}NC_9H_6))(NO_3)_2]_n$	Mn:MVol.D6-361, 363
$C_{12}H_{10}MnN_6$	$[Mn(NC_5H_4\text{-}2\text{-}N\text{-}N{=}CH\text{-}CH{=}N\text{-}N\text{-}2\text{-}C_5H_4N)]$	Mn:MVol.D6-264/5
$C_{12}H_{10}MnO_4S_2$	$[Mn(C_6H_5SO_2)_2(H_2O)_2]$	Mn:MVol.D7-109/11
$C_{12}H_{10}MnO_4Se_2$	$Mn(O_2Se\text{-}C_6H_5)_2$	Mn:MVol.D7-244/6
–	$[Mn(O_2Se\text{-}C_6H_5)_2(H_2O)_2]$	Mn:MVol.D7-244/6
$C_{12}H_{10}MnO_8P_2^{2-}$	$[Mn(O_3POC_6H_5)_2]^{2-}$	Mn:MVol.D8-146/7
$C_{12}H_{10}Mn_2N_2O_2S_8$	$Mn_2(C_4H_5N(=O\text{-}2)(C(=S)S)_2\text{-}1,3)_2$	Mn:MVol.D7-167
$C_{12}H_{10}MoN_6O_2S_2$	$Mo(H_2NO)(NO)(NCS)_2(C_{10}H_8N_2)$	Mo:SVol.B3b-200
$C_{12}H_{10}MoN_{10}O_2$	$Mo(H_2NO)(NO)(N_3)_2(C_{12}H_8N_2)$	Mo:SVol.B3b-200
$C_{12}H_{10}MoO_2S$	$(C_5H_5)Mo(CO)_2[(CH_2)SC_4H_3]$	Mo:Org.Comp.8-205, 236
$C_{12}H_{10}MoO_4$	$(C_5H_5)Mo(CO)_2[(CH_2)OC_4H_3(=O)]$	Mo:Org.Comp.8-206, 235
–	$[(2.2.2)\text{-}C_8H_{10}]Mo(CO)_4$	Mo:Org.Comp.5-338, 340
–	$(c\text{-}C_8H_{10})Mo(CO)_4$	Mo:Org.Comp.5-312, 313, 315
$C_{12}H_{10}MoO_5Sb^-$	$[(C_6H_5)_2Sb(O)MoO_4]_2^{2-}$	Sb: Org.Comp.5-234
$C_{12}H_{10}Mo_5O_{21}P_2^{4-}$	$[(C_6H_5PO_3)_2Mo_5O_{15}]^{4-}$	Mo:SVol.B3b-124

$C_{12}H_{10}NO_2PS$ .... $O{=}S{=}NP(O)(C_6H_5)_2$ .... S: S-N Comp.6-79
$C_{12}H_{10}NO_2Re$ .... $(C_5H_5)Re(CO)(NO){-}C_6H_5$ .... Re: Org.Comp.3-160, 164, 169
– .... $(C_5H_5)Re(CO)_2(NC_5H_5)$ .... Re: Org.Comp.3-193/7, 198
$C_{12}H_{10}NO_4PS$ .... $O{=}S{=}NP(O)(OC_6H_5)_2$ .... S: S-N Comp.6-76/9
$C_{12}H_{10}NO_4Sb$ .... $(C_6H_5)(4{-}O_2NC_6H_4)Sb(O)OH$ .... Sb: Org.Comp.5-223
$C_{12}H_{10}NO_5Re$ .... cis-$(CO)_4Re(NH_2{-}C_6H_5)C(O)CH_3$ .... Re: Org.Comp.1-472/3
$C_{12}H_{10}NO_5SSb$ .... $4{-}(4'{-}O_2NC_6H_4S)C_6H_4)Sb(O)(OH)_2$ .... Sb: Org.Comp.5-288
$C_{12}H_{10}NO_7SSb$ .... $(4{-}(4'{-}O_2NC_6H_4SO_2)C_6H_4)Sb(O)(OH)_2$ .... Sb: Org.Comp.5-288
$C_{12}H_{10}NSb$ .... $(C_6H_5)_2SbN$ .... Sb: Org.Comp.5-182
$C_{12}H_{10}N_2OS$ .... $O{=}S{=}N{-}N(C_6H_5)_2$ .... S: S-N Comp.6-63/8
– .... $O{=}S{=}N{-}NH{-}C_6H_4{-}C_6H_5{-}4$ .... S: S-N Comp.6-57/60
$C_{12}H_{10}N_2O_2S$ .... $O{=}S{=}NNHC(O)CH_2C_{10}H_7$ .... S: S-N Comp.6-61, 62
$C_{12}H_{10}N_2O_2S_2$ .... $C_6H_5{-}S(O)_2N{=}S{=}N{-}C_6H_5$ .... S: S-N Comp.7-55/60
$C_{12}H_{10}N_2O_2S_3$ .... $C_6H_5{-}S(O)_2N{=}S{=}NS{-}C_6H_5$ .... S: S-N Comp.7-66
$C_{12}H_{10}N_2O_3S_2$ .... $4{-}NO_2{-}C_6H_4{-}NH{-}S(O){-}S{-}C_6H_5$ .... S: S-N Comp.8-331
– .... $O{=}S{=}N{-}NH{-}C_6H_4{-}SO_2{-}C_6H_5{-}4$ .... S: S-N Comp.6-57/60
$C_{12}H_{10}N_2O_4S_3$ .... $C_6H_5{-}S(O)_2N{=}S{=}NS(O)_2{-}C_6H_5$ .... S: S-N Comp.7-75/9
$C_{12}H_{10}N_2O_6Sn$ .... $(C_6H_5)_2Sn(ONO_2)_2$ .... Sn: Org.Comp.16-138, 141
$C_{12}H_{10}N_2O_6Th$ .... $Th(OH)_2(NC_5H_4{-}4{-}COO)_2 \cdot 8\ H_2O$ .... Th: SVol.C7-152/3
$C_{12}H_{10}N_2S$ .... $C_6H_5{-}N{=}S{=}N{-}C_6H_5$ .... S: S-N Comp.7-220/33
– .... $C_6H_5{-}^{15}N{=}S{=}^{15}N{-}C_6H_5$ .... S: S-N Comp.7-220/1, 225/6
$C_{12}H_{10}N_2S^-$ .... $[(C_6H_5{-}N)_2S]^-$, radical anion .... S: S-N Comp.7-334/6
$C_{12}H_{10}N_2SSe_2$ .... $C_6H_5{-}SeN{=}S{=}NSe{-}C_6H_5$ .... S: S-N Comp.7-102
$C_{12}H_{10}N_2S_2$ .... $C_6H_5{-}SN{=}S{=}N{-}C_6H_5$ .... S: S-N Comp.7-25/7
$C_{12}H_{10}N_2S_3$ .... $C_6H_5{-}SN{=}S{=}NS{-}C_6H_5$ .... S: S-N Comp.7-29/32
$C_{12}H_{10}N_2Si$ .... $Si({=}NC_6H_5)_2$ .... Si: SVol.B4-154/5
$C_{12}H_{10}N_3O_6Sb$ .... $4{-}(3'{-}O_2N{-}4'{-}HOC_6H_3N{=}N)C_6H_4Sb(O)(OH)_2$ .... Sb: Org.Comp.5-291
$C_{12}H_{10}N_4O_2Pd$ .... $Pd(NCO)_2(C_5H_5N)_2$ .... Pd: SVol.B2-288
$C_{12}H_{10}N_4O_4S_5$ .... $C_6H_5{-}S(O)_2N{=}S{=}NSN{=}S{=}NS(O)_2{-}C_6H_5$ .... S: S-N Comp.7-68/9
$C_{12}H_{10}N_4O_4Th^{2+}$ .... $Th[C_6H_5N(NO)O]_2^{2+}$ .... Th: SVol.D1-112
$C_{12}H_{10}N_4Pd$ .... $Pd(CN)_2(C_5H_5N)_2$ .... Pd: SVol.B2-285
$C_{12}H_{10}N_4PdS_2$ .... $Pd(NCS)_2(NC_5H_5)_2$ .... Pd: SVol.B2-302
– .... $Pd(SCN)_2(NC_5H_5)_2$ .... Pd: SVol.B2-302
$C_{12}H_{10}N_4S_3$ .... $C_6H_5{-}N{=}S{=}NSN{=}S{=}N{-}C_6H_5$ .... S: S-N Comp.7-52
$C_{12}H_{10}N_6O_{12}Th$ .... $Th(NO_3)_4 \cdot NC_5H_4{-}2{-}(CH{=}N{-}C_6H_5)$ .... Th: SVol.D4-140
$C_{12}H_{10}N_8O_2PdS_2$ $Pd(SCN)_2[2{-}NH_2C(O){-}(1,4{-}N_2C_4H_3)]_2$ .... Pd: SVol.B2-306
$C_{12}H_{10}N_{12}Sb^-$ .... $[(C_6H_5)_2Sb(N_3)_4]^-$ .... Sb: Org.Comp.5-177
$C_{12}H_{10}O_3SSn$ .... $(C_6H_5)_2Sn{-}O{-}SO_2$ .... Sn: Org.Comp.16-143/4
– .... $[(C_6H_5)_2SnOSO_2]_n$ .... Sn: Org.Comp.16-144
$C_{12}H_{10}O_4Sn$ .... $(CH_2{=}CH)_2Sn[{-}OOC{-}(1,2{-}C_6H_4){-}COO{-}]$ .... Sn: Org.Comp.16-90/1
$C_{12}H_{10}O_{14}Th^{2-}$ .... $[Th(OOCCH_2C(OH)(COO)CH_2COO)_2]^{2-}$ .... Th: SVol.D1-74/6, 80
$C_{12}H_{10}Po$ .... $(C_6H_5)_2Po$ .... Po: SVol.1-334/40
$C_{12}H_{10}Ti$ .... $[(C_6H_5)_2Ti]_n$ .... Ti: Org.Comp.5-323/4
$C_{12}H_{11}IO_4PRe$ .... $(CO)_4Re(I)P(CH_3)_2C_6H_5$ .... Re: Org.Comp.1-443/4
$C_{12}H_{11}I_2MnN_3$ .... $[Mn(NC_5H_4{-}2{-}CH{=}NCH_2{-}2{-}C_5H_4N)I_2]$ .... Mn: MVol.D6-80/2
$C_{12}H_{11}KMoN_2O_2$. $K[(C_5H_5)Mo(CO)_2(CN){-}CH_2CH_2CH_2{-}CN]$ .... Mo: Org.Comp.8-5
$C_{12}H_{11}MnN_2NaO_8S_2$
$Na[Mn((5{-}S{=})(2{-}(4{-}C_6H_4SO_3))C_2HN_2O{-}4,3,1)$
$(H_2O)_3](OC({=}O)CH_3)_2$ .... Mn: MVol.D7-63/4
$C_{12}H_{11}MoNO_2$ .... $[(C_5H_5)Mo(CO)_2(NC_5H_5)(H)]$ .... Mo: Org.Comp.7-56/7, 59

$C_{12}H_{11}MoNO_3$ . . . $(C_5H_5)Mo(CO)_2(CN)=C(-O-CH_2CH_2CH_2-)$ . . . Mo:Org.Comp.8-17, 24
$C_{12}H_{11}MoNO_5$ . . . $(CO)_5Mo(CN-C_6H_{11}-c)$ . . . Mo:Org.Comp.5-6, 8
$C_{12}H_{11}MoN_2NaO_5$ $Na[(C_5H_5)Mo(CO)_2(-C(=O)-C(COO-C_2H_5)=N-NH-)]$ Mo:Org.Comp.8-138
– . . . $Na[(C_5H_5)Mo(CO)_2(-C(OH)=C(COO-C_2H_5)-N=N-)]$ Mo:Org.Comp.8-138
$C_{12}H_{11}MoN_2O_2^-$ . . $K[(C_5H_5)Mo(CO)_2(CN)-CH_2CH_2CH_2-CN]$ . . . Mo:Org.Comp.8-5
$C_{12}H_{11}MoN_2O_5^-$ . . $Na[(C_5H_5)Mo(CO)_2(-C(=O)-C(COO-C_2H_5)=N-NH-)]$ Mo:Org.Comp.8-138
– . . . $Na[(C_5H_5)Mo(CO)_2(-C(OH)=C(COO-C_2H_5)-N=N-)]$ Mo:Org.Comp.8-138
– . . . $[N(C_4H_9-n)_4][(C_5H_5)Mo(CO)_2(-C(OH)=C(COO-C_2H_5)-N=N-)]$ . . . Mo:Org.Comp.8-138
$C_{12}H_{11}NO_5Re^+$ . . . $[(CO)_5Re(CNC_6H_{11})]^+$ . . . Re: Org.Comp.2-253/4
$C_{12}H_{11}NO_6Sn$ . . . $C_6H_5Sn(OOCCH_2)_3N$ . . . Sn: Org.Comp.17-65
$C_{12}H_{11}N_2O_3Sb$ . . . $4-C_6H_5N=NC_6H_4Sb(O)(OH)_2$ . . . Sb: Org.Comp.5-300
$C_{12}H_{11}N_2O_4Sb$ . . . $(4-(4'-HOC_6H_4N=N)C_6H_4)Sb(O)(OH)_2 \cdot H_2O$ . . . Sb: Org.Comp.5-291
$C_{12}H_{11}N_2O_5Sb$ . . . $2-(C_6H_5-NH)-5-O_2N-C_6H_3-Sb(=O)(OH)_2$ . . . Sb: Org.Comp.5-298
– . . . $4-[2,4-(HO)_2-C_6H_3-N=N]-C_6H_4-Sb(=O)(OH)_2$ . . . Sb: Org.Comp.5-291
$C_{12}H_{11}N_5S_4Si$ . . . $Si(NCS)_4 \cdot (CH_3)_2NC_6H_5$ . . . Si: SVol.B4-300/1
$C_{12}H_{11}N_7O_{13}Th$ . . $Th(NO_3)_4 \cdot 2-(6-NH_2-NC_5H_3-2-N=CH)-C_6H_4-OH$ Th: SVol.D4-139
$C_{12}H_{11}O_2Sb$ . . . $(C_6H_5)_2Sb(O)OH$ . . . Sb: Org.Comp.5-206/8
$C_{12}H_{11}O_3Sb$ . . . $C_6H_5-C_6H_4-2-Sb(=O)(OH)_2$ . . . Sb: Org.Comp.5-295, 303
– . . . $C_6H_5-C_6H_4-4-Sb(=O)(OH)_2$ . . . Sb: Org.Comp.5-295
$C_{12}H_{11}O_3Ti$ . . . $(C_5H_5)Ti(C_7H_6O_3)$ . . . Ti: Org.Comp.5-40/1
$C_{12}H_{11}O_4Sb$ . . . $4-(4'-HOC_6H_4)C_6H_4Sb(O)(OH)_2$ . . . Sb: Org.Comp.5-300
$C_{12}H_{11}O_5PTh^{2+}$ . . $Th(OH)[O_2P(OC_6H_5)_2]^{2+}$ . . . Th: SVol.D1-131
$C_{12}H_{11}O_5Re$ . . . $(CO)_5Re-CH[-C(C_4H_9-t)=CH]$ . . . Re: Org.Comp.2-110
– . . . $(CO)_5Re-CD[-C(C_4H_9-t)=CH]$ . . . Re: Org.Comp.2-110
$C_{12}H_{11}O_5Sb_3$ . . . $(C_6H_5)_2Sb(O)OH \cdot Sb_2O_3$ . . . Sb: Org.Comp.5-208
$C_{12}H_{11}Ti$ . . . $[(C_5H_5)_2Ti(CCH)]_n$ . . . Ti: Org.Comp.5-352
$C_{12}H_{12}I_2MnN_4S_2$ . . $[Mn(S=C(2-C_5H_4N)NH_2)_2I_2] \cdot H_2O$ . . . Mn:MVol.D7-212/3
$C_{12}H_{12}I_3MoN_2^-$ . . . $[C_5H_5Mo(N=NC_6H_4CH_3-4)I_3]^-$ . . . Mo:Org.Comp.6-27
$C_{12}H_{12}K_2MoO_8$ . . . $K_2MoO_4 \cdot 2\ C_6H_4(OH)_2$ . . . Mo:SVol.B3b-162
$C_{12}H_{12}MnN_2OS_2$ . . $[Mn(SC(=S)N(CH_3)_2)(NC_9H_6O-8)]$ . . . Mn:MVol.D7-143/4
$C_{12}H_{12}MnN_2O_2S_2$ $Mn((2-ON=CH)(5-CH_3)C_4H_2S)_2$ . . . Mn:MVol.D7-222/5
$C_{12}H_{12}MnN_2O_5S$ . . $Mn[OC(CH_3)C(C(O)CH_3)=N-N-2-C_6H_3(CH_3-5)SO_3]$ Mn:MVol.D6-256/7
$C_{12}H_{12}MnN_2S_2$ . . . $[Mn(SC_6H_4NH_2-2)_2]$ . . . Mn:MVol.D7-28/9
$C_{12}H_{12}MoN_2O_4$ . . . $1-(CH_2=CHCH_2)-2,4-(CH_3)_2-1,2-N_2C_3H-2-Mo(CO)_4$ Mo:Org.Comp.5-164/7
$C_{12}H_{12}MoN_2O_5$ . . . $(C_5H_5)Mo(CO)_2[-C(OH)=C(COO-C_2H_5)-NHN-]$ Mo:Org.Comp.8-112, 139, 168/9
$C_{12}H_{12}MoN_8$ . . . $Mo(CN-CH_3)_4(CN)_4$ . . . Mo:Org.Comp.5-38, 39, 44, 45
$C_{12}H_{12}MoO_2$ . . . $(C_5H_5)Mo(CO)_2(CH_2CHCH-CH=CH_2)$ . . . Mo:Org.Comp.8-205, 207, 214, 243, 244
– . . . $(C_5H_5)Mo(CO)_2(C_5H_7)$ . . . Mo:Org.Comp.8-254, 297
– . . . $(C_5H_5)Mo(CO)_2(C_5H_5D_2)$ . . . Mo:Org.Comp.8-255
$C_{12}H_{12}MoO_3$ . . . $(CH_3-C_5H_4)Mo(CO)_2[-CH=CH-C(CH_3)=O-]$ . . . Mo:Org.Comp.8-112, 146/7

$C_{12}H_{12}MoO_3$	$(C_5H_5)Mo(CO)_2[-C(CH_3)=CH-C(CH_3)=O-]$	Mo:Org.Comp.8-112, 148
–	$(C_5H_5)Mo(CO)_2[-CH=C(CH_3)-C(CH_3)=O-]$	Mo:Org.Comp.8-112, 148
–	$[CH_3C(=O)-C_5H_4]Mo(CO)_2(CH_2CHCH_2)$	Mo:Org.Comp.8-205, 208, 241, 242
–	$[CD_3C(=O)-C_5H_4]Mo(CO)_2(CH_2CHCH_2)$	Mo:Org.Comp.8-205, 209
$C_{12}H_{12}MoO_3S$	$(C_5H_5)Mo(CO)_2[CH_2C(CH_2)-C(=O)-S-CH_3]$	Mo:Org.Comp.8-205, 221
$C_{12}H_{12}MoO_4$	$(CH_2=CHCH=CH_2)_2Mo(CO)_4$	Mo:Org.Comp.5-194/5
–	$(C_5H_5)Mo(CO)_2[-C(CH_3)=CH-C(OCH_3)=O-]$	Mo:Org.Comp.8-113, 149/50, 170
–	$(C_5H_5)Mo(CO)_2[CH_2C(CH_2)-COO-CH_3]$	Mo:Org.Comp.8-205, 220
–	$(C_5H_5)Mo(CO)_2[CH_2C(CH-CH_3)-COOH]$	Mo:Org.Comp.8-205, 223/4
–	$[(CH_3)_4-c-C_4]Mo(CO)_4$	Mo:Org.Comp.5-312, 313
–	$(c-C_8H_{12})Mo(CO)_4$	Mo:Org.Comp.5-312/5
$C_{12}H_{12}MoO_8$	$[CH_3-OC(=O)-CH=CH_2]_2Mo(CO)_4$	Mo:Org.Comp.5-195
$C_{12}H_{12}Mo_7O_{48}{}^{6-}$	$[(Mo_7O_{24} \cdot 6\ H_2C_2O_4)]^{6-}$	Mo:SVol.B3b-174
$C_{12}H_{12}NOPS$	$C_6H_5-NH-S(O)-PH-C_6H_5$	S: S-N Comp.8-371
$C_{12}H_{12}NO_2Sb$	$(C_6H_5)(3-H_2NC_6H_4)Sb(O)OH \cdot H_2O$	Sb: Org.Comp.5-222
$C_{12}H_{12}NO_3SSb$	$(4-(4'-H_2NC_6H_4S)C_6H_4)Sb(O)(OH)_2$	Sb: Org.Comp.5-288
$C_{12}H_{12}NO_3Sb$	$4-(4'-H_2NC_6H_4)C_6H_4Sb(O)(OH)_2$	Sb: Org.Comp.5-295
$C_{12}H_{12}NO_5SSb$	$4-[4-H_2N-C_6H_4-S(O)_2-]C_6H_4-Sb(=O)(OH)_2$	Sb: Org.Comp.5-288
–	$4-[C_6H_5-S(O)_2-NH]-C_6H_4-Sb(=O)(OH)_2$	Sb: Org.Comp.5-289
$C_{12}H_{12}NO_6Re$	$(CO)_4Re(CN-C_6H_{11})COOH$	Re: Org.Comp.2-248
–	$(CO)_5ReC(O)NH-C_6H_{11}-c$	Re: Org.Comp.2-117
$C_{12}H_{12}N_2Na_4O_{21}Th$	$Na_2[Th(O(CH_2COO)_2)_3] \cdot 2\ NaNO_3$	Th: SVol.C7-115, 118/9
$C_{12}H_{12}N_2OS$	$O=S(NH-C_6H_5)_2$	S: S-N Comp.8-350
$C_{12}H_{12}N_2OSSn$	$(CH_3)_2Sn(NCS)OC_9H_6N$	Sn: Org.Comp.17-137
$C_{12}H_{12}N_2O_4PoS_2$	$(4-NH_2SO_2C_6H_4)_2Po$	Po: SVol.1-334/40
$C_{12}H_{12}N_2O_4Th$	$Th(OO)(OC_6H_4-2-NH_2)_2$	Th: SVol.C7-29/32
$C_{12}H_{12}N_2O_{12}Th^{2-}$	$[Th(N(CH_2COO)_3)_2]^{2-}$	Th: SVol.D1-98
$C_{12}H_{12}N_3O_4Sb$	$4-(3'-H_2N-4'-HOC_6H_3N=N)C_6H_4Sb(O)(OH)_2$	Sb: Org.Comp.5-291
$C_{12}H_{12}N_4O_2S_2$	$[-NH-C_6H_4-NH-S(O)-NH-C_6H_4-NH-S(O)-]_n$	S: S-N Comp.8-368/9, 371
$C_{12}H_{12}N_4O_4S_2$	$4-NO_2-C_6H_4-S(O)_2N=S=NC-(CH_2)_4-CN$	S: S-N Comp.7-60/1
$C_{12}H_{12}N_4S_2Si$	$SiH_2(NCS)_2 \cdot 2\ C_5H_5N$	Si: SVol.B4-297
$C_{12}H_{12}N_6PdS_2$	$Pd(SCN)_2(NH_2-2-C_5H_4N)_2$	Pd: SVol.B2-302
–	$Pd(SCN)_2(NH_2-4-C_5H_4N)_2$	Pd: SVol.B2-302/3
$C_{12}H_{12}N_{10}O_{12}Th$	$Th(NO_3)_4 \cdot 3\ (1,4-N_2C_4H_4)$	Th: SVol.D4-135
$C_{12}H_{12}Na_2O_{15}Th$	$Na_2[Th(O(CH_2COO)_2)_3] \cdot 2\ H_2O$	Th: SVol.C7-115, 118
–	$Na_2[Th(O(CH_2COO)_2)_3] \cdot 3\ H_2O$	Th: SVol.C7-115, 118
$C_{12}H_{12}OTi$	$[(C_5H_5)_2Ti(OC=CH_2)]_n$	Ti: Org.Comp.5-352/3
$C_{12}H_{12}O_2Sn$	$(C_6H_5)_2Sn(OH)_2$	Sn: Org.Comp.16-97
$C_{12}H_{12}O_4Re^+$	$[C_8H_{12}Re(CO)_4]^+$	Re: Org.Comp.2-420
$C_{12}H_{12}O_5Th$	$Th(OH)_3(1-C_{10}H_7CH_2COO)$	Th: SVol.C7-65, 67
$C_{12}H_{12}O_6SSb_2$	$(HO)_2(O)Sb(4-C_6H_4SC_6H_4-4')Sb(O)(OH)_2$	Sb: Org.Comp.5-315
$C_{12}H_{12}O_6S_2Sb_2$	$(HO)_2(O)Sb(4-C_6H_4SSC_6H_4-4')Sb(O)(OH)_2$	Sb: Org.Comp.5-315
$C_{12}H_{12}O_6Sb_2$	$(HO)_2(O)Sb(4,4-C_6H_4C_6H_4)Sb(O)(OH)_2$	Sb: Org.Comp.5-314
$C_{12}H_{12}O_8SSb_2$	$(HO)_2(O)Sb(4-C_6H_4S(O)_2C_6H_4-4')Sb(O)(OH)_2$	Sb: Org.Comp.5-315
$C_{12}H_{12}O_8Th$	$Th(OH)_2[C_6H_3(O)(OH)_2]_2$	Th: SVol.C7-35/6
$C_{12}H_{12}O_{15}Th^{2-}$	$Th[O(CH_2COO)_2]_3{}^{2-}$	Th: SVol.C7-118/9
$C_{12}H_{12}O_{24}Th_3$	$Th_3(HCOO)_{12} \cdot 9\ H_2O$	Th: SVol.C7-39

$C_{12}H_{12}Pb$ .... $Pb(C{\equiv}C{-}CH_3)_4$ .... Pb: Org.Comp.3-92/6
$C_{12}H_{13}IMoO_3$ .... $(CH_3{-}C_5H_4)Mo(CO)_2(I){=}C({-}O{-}CH_2CH_2CH_2{-})$ .. Mo:Org.Comp.8-17, 26/7
$C_{12}H_{13}I_2MnN_2P$ .. $Mn[(C_6H_5)P(CH_2CH_2CN)_2]I_2$ Mn:MVol.D8-69/70
$C_{12}H_{13}MnN_4O_2S^+$ $[Mn(4,6{-}(CH_3)_2{-}2{-}(4{-}NH_2{-}C_6H_4{-}SO_2{=}N){-}1,3{-}N_2C_4H)]^+$
Mn:MVol.D7-116/23
$C_{12}H_{13}MnN_4O_4S^+$ $[Mn(2,6{-}(CH_3O)_2{-}4{-}(4{-}NH_2{-}C_6H_4{-}SO_2{=}N){-}1,3{-}N_2C_4H)]^+$
Mn:MVol.D7-116/23
$C_{12}H_{13}MoNO_3$ ... $(C_5H_5)Mo(CO)_2{-}C({=}O){-}C({=}CH_2){-}N(CH_3)_2$ Mo:Org.Comp.8-202/3
$C_{12}H_{13}MoNO_3S_2$ .. $(CH_2CHCH_2)Mo(CO)_2(NC_5H_5){-}SC(S){-}OCH_3$ ... Mo:Org.Comp.5-282, 283, 284, 289
– .... $(C_5H_5)Mo(CO)_2[{-}S{-}C(4{-}(1,4{-}ONC_4H_8)){-}S{-}]$ ... Mo:Org.Comp.7-189, 197
$C_{12}H_{13}MoNO_4$ ... $(C_5H_5)Mo(CO)_2[{-}1,2{-}NC_4H_8{-}C({=}O){-}O{-}]$ Mo:Org.Comp.7-208, 209, 233
– .... $(C_5H_5)Mo(CO)_2[{-}O{-}C({=}O){-}C(C_2H_5){=}N(CH_3){-}]$ .. Mo:Org.Comp.7-208, 239
$C_{12}H_{13}MoNO_5$ ... $(C_5H_5)Mo(CO)(NO)[{-}OCH_2CH{=}C(CH_2OCH_3)C({=}O){-}]$
Mo:Org.Comp.6-299
$C_{12}H_{13}MoN_2O_3^+$. $[(C_5H_5)Mo(NO)(NCCD_3)CH{=}C({-}C_2H_4OC(O){-})]^+$ Mo:Org.Comp.6-178
$C_{12}H_{13}MoN_2O_5^+$. $[(C_5H_5)Mo(CO)_2({-}C(OH){=}C(COO{-}C_2H_5){-}NH{-}NH{-})][BF_4]$
Mo:Org.Comp.8-139
– .... $[(C_5H_5)Mo(CO)_2({-}C(OH){=}C(COO{-}C_2H_5){-}NH{-}NH{-})][PF_6]$
Mo:Org.Comp.8-139/40
$C_{12}H_{13}MoO^+$ .... $[(C_5H_5)Mo(CO)(HC{\equiv}C{-}CH_3)_2]^+$ Mo:Org.Comp.6-316, 322
$C_{12}H_{13}MoO_2^+$ ... $[(C_5H_5)Mo(CO)_2(CH_2{=}C(CH_3){-}CH{=}CH_2)][BF_4]$ .. Mo:Org.Comp.8-305/6, 307, 308, 316
$C_{12}H_{13}NO_3Sn$ .... $(CH_3)_2Sn(OH)[O{-}C_{10}H_6(NO)]$ .... Sn: Org.Comp.16-172
$C_{12}H_{13}NSi$ .... $SiH_3N(C_6H_5)_2$ .... Si: SVol.B4-176
$C_{12}H_{13}N_2O_2Sb$ ... $(3{-}H_2NC_6H_4)_2Sb(O)OH \cdot 0.5\ H_2O$ .... Sb: Org.Comp.5-209
$C_{12}H_{13}N_2O_5SSb$ .. $(4{-}(4'{-}H_2NC_6H_4SO_2NH)C_6H_4)Sb(O)(OH)_2$ ..... Sb: Org.Comp.5-289
$C_{12}H_{13}N_2O_6Re$ ... $[{-}O{-}C(O){-}CH(CH(OH)CH_3){-}NH_2{-}]Re(CO)_3{-}1{-}NC_5H_5$
Re: Org.Comp.1-126
$C_{12}H_{13}N_2O_{13}Th^{3-}$ $Th(OH)[N(CH_2COO)_3]_2^{3-}$ .... Th: SVol.D1-97
$C_{12}H_{13}N_3O_2S_2$ ... $C_6H_5{-}S(O)_2N{=}S{=}NC_5H_8(CN{-}1)$ .... S: S-N Comp.7-54
$C_{12}H_{13}N_3O_9Th$ ... $Th(NO_3)_2[OOC{-}CH(C_3H_7{-}i){-}N{=}CH{-}C_6H_4{-}2{-}O] \cdot 3\ H_2O$
Th: SVol.D4-143
$C_{12}H_{13}N_3Si$ ...... $SiH({-}NC_4H_4)_3$ .... Si: SVol.B4-196
$C_{12}H_{14}IInN_2$ ..... $(CH_3)_2InI \cdot 2{-}(NC_5H_4{-}2){-}NC_5H_4$ .... In: Org.Comp.1-156, 158
$C_{12}H_{14}IMoNO_2$ ... $(C_5H_5)Mo(CO)_2(I){=}C[{-}N(CH_3){-}CH_2CH_2CH_2{-}]$ .. Mo:Org.Comp.8-17, 23, 29/30
– .... $(C_5H_5)Mo(CO)_2(I){-}CN{-}C_4H_9{-}t$ Mo:Org.Comp.8-9, 13
$C_{12}H_{14}IMoN_3O$ ... $(C_5H_5)Mo(NO)(I)NHNH{-}C_6H_4CH_3{-}4$ Mo:Org.Comp.6-52
$C_{12}H_{14}IMoO_5P$ ... $(C_5H_5)Mo(CO)_2(I)[P({-}OCH_2{-})_3CCH_3]$ Mo:Org.Comp.7-56/7, 90, 114
$C_{12}H_{14}InN_2^+$ ..... $[(CH_3)_2In(2{-}(NC_5H_4{-}2){-}NC_5H_4)]^+$ .... In: Org.Comp.1-223, 225/6
$C_{12}H_{14}InN_3O_3$ .... $[(CH_3)_2In(2{-}(NC_5H_4{-}2){-}NC_5H_4)][NO_3]$ .... In: Org.Comp.1-223, 225/6
$C_{12}H_{14}KMoNO_4S$ $K[(C_5H_5)Mo(CO)_2({-}S{-}C(CH_3)_2{-}CH(COO){-}NH_2{-})]$
Mo:Org.Comp.7-207/9, 234
$C_{12}H_{14}MnN_2NaO_6S$
$Na[Mn(OC(CH_3)C(C(O)CH_3){=}N$
${-}N{-}2{-}C_6H_3(CH_3{-}5)SO_3)(H_2O)]$ Mn:MVol.D6-256, 258
$C_{12}H_{14}MnN_2O_2S_2$ $[Mn((SC({=}S))NC_5H_{10})NC_5H_4C({=}O)O]$ Mn:MVol.D7-168
$C_{12}H_{14}MnN_2O_4S_2^{2+}$
$[Mn(O{=}S(CH_3)(C_5H_4NO))_2(H_2O)_2]^{2+}$ Mn:MVol.D7-106

$C_{12}H_{14}MnN_4O_2S^{2+}$
$[Mn(4,6\text{-}(CH_3)_2\text{-}2\text{-}(4\text{-}NH_2\text{-}C_6H_4\text{-}SO_2\text{-}NH)\text{-}1,3\text{-}N_2C_4H)]^{2+}$ ... Mn:MVol.D7-116/23
$C_{12}H_{14}MnN_4O_4S^{2+}$
$[Mn(2,6\text{-}(CH_3O)_2\text{-}4\text{-}(4\text{-}NH_2\text{-}C_6H_4\text{-}SO_2\text{-}NH)\text{-}1,3\text{-}N_2C_4H)]^{2+}$ ... Mn:MVol.D7-116/23
$C_{12}H_{14}MnN_4O_4S_2$ $Mn(4\text{-}NH_2C_6H_4SO_2NH)_2$ ... Mn:MVol.D7-116/23
$C_{12}H_{14}MnN_6O_4S_2{}^{2-}$
$[Mn(2\text{-}(S{=})\text{-}1,3\text{-}N_2H_2C_3\text{-}4\text{-}CH_2CH(NH_2)\text{-}COO)_2]^{2-}$ Mn:MVol.D7-58/9
$C_{12}H_{14}MnN_8O_4S_2$ $Mn(1,3\text{-}N_2C_4(SCH_3\text{-}2)(CH_3\text{-}3)({=}O\text{-}4)(NO\text{-}5)(NH\text{-}6))_2$ Mn:MVol.D7-80/1
$C_{12}H_{14}MnO_6S_2$ ... $[Mn(C_6H_5SO_2)_2(H_2O)_2]$ ... Mn:MVol.D7-109/11
$C_{12}H_{14}MnO_6Se_2$ ... $[Mn(O_2SeC_6H_5)_2(H_2O)_2]$ ... Mn:MVol.D7-244/6
$C_{12}H_{14}MnO_6Si_2$ ... $Mn[OSi(C_6H_5)(OH)_2]_2$ ... Mn:MVol.D8-23/4
$C_{12}H_{14}MoNNaO_2$ $Na[(C_5H_5)Mo(CO)_2{=}C(\text{-}CH_2CH_2CH_2\text{-}N(CH_3)\text{-})]$ Mo:Org.Comp.8-1, 2, 5/6
– ... $Na[(C_5H_5)Mo(CO)_2\text{-}CN\text{-}C_4H_9\text{-}t]$ ... Mo:Org.Comp.8-1, 3
$C_{12}H_{14}MoNO_2{}^+$ ... $[(C_5H_5)Mo(CO)(NO)C_6H_9\text{-}c]^+$ ... Mo:Org.Comp.6-350
$C_{12}H_{14}MoNO_2{}^-$ ... $Na[(C_5H_5)Mo(CO)_2{=}C(\text{-}CH_2CH_2CH_2\text{-}N(CH_3)\text{-})]$ Mo:Org.Comp.8-1, 2, 5/6
– ... $Na[(C_5H_5)Mo(CO)_2\text{-}CN\text{-}C_4H_9\text{-}t]$ ... Mo:Org.Comp.8-1, 3
$C_{12}H_{14}MoNO_3{}^+$ ... $[(H_3CC(O)\text{-}C_5H_4)Mo(CO)(NO)(CH_2C(CH_3)CH_2)]^+$ Mo:Org.Comp.6-348
– ... $[(H_3CC(O)\text{-}C_5H_4)Mo(CO)(NO)(CH_2CHCH\text{-}CH_3)]^+$ Mo:Org.Comp.6-348
$C_{12}H_{14}MoNO_4S^-$ $[(C_5H_5)Mo(CO)_2(\text{-}S\text{-}C(CH_3)_2\text{-}CH(COO)\text{-}NH_2\text{-})]^-$ Mo:Org.Comp.7-20/97, 234
$C_{12}H_{14}MoN_2O_2S_2$ $(CH_2CHCH_2)Mo(CO)_2(NC_5H_5)\text{-}SC(S)\text{-}NHCH_3$ ... Mo:Org.Comp.5-282, 283, 284/5
$C_{12}H_{14}MoN_2O_4$ ... $(CO)_4Mo(CN\text{-}C_3H_7\text{-}i)_2$ ... Mo:Org.Comp.5-25, 26
$C_{12}H_{14}MoN_2O_5$ ... $[\text{-}N(C_2H_5)\text{-}CH_2CH_2\text{-}N(C_2H_5)\text{-}]C{=}Mo(CO)_5$ ... Mo:Org.Comp.5-100, 106, 113
$C_{12}H_{14}MoN_3O^+$ ... $[(C_5H_5)Mo(CO)(CNCH_3)_3]^+$ ... Mo:Org.Comp.6-274
$C_{12}H_{14}MoN_6O_4$ ... $[(\text{-}N(CH_3)\text{-}N{=}CH\text{-}N(CH_3)\text{-})C{=}]_2Mo(CO)_4$ ... Mo:Org.Comp.5-124/5, 128/9, 132
$C_{12}H_{14}MoNaO_5P$ $Na[(C_5H_5)Mo(CO)_2P(\text{-}OCH_2\text{-})_3CCH_3]$ ... Mo:Org.Comp.7-47, 50
$C_{12}H_{14}MoO$ ... $(C_5H_5)Mo(CO)(C_5H_6\text{-}c)CH_3$ ... Mo:Org.Comp.6-364
$C_{12}H_{14}MoO_2$ ... $(CH_3\text{-}C_5H_4)Mo(CO)_2(CH_2CHCH\text{-}CH_3)$ ... Mo:Org.Comp.8-207, 212
– ... $(C_5H_5)Mo(CO)[CH_2CHCHCH_2C(O)CH_3]$ ... Mo:Org.Comp.6-341/2
– ... $(C_5H_5)Mo(CO)_2(CH_2CHCH\text{-}C_2H_5)$ ... Mo:Org.Comp.8-206, 207, 213
– ... $(C_5H_5)Mo(CO)_2(CH_3\text{-}CHCHCH\text{-}CH_3)$ ... Mo:Org.Comp.8-205, 227/8
– ... $(C_5H_5)Mo(CO)_2[CH_2CHC(CH_3)_2]$ ... Mo:Org.Comp.8-205, 222
$C_{12}H_{14}MoO_3$ ... $(CH_3\text{-}C_5H_4)Mo(CO)_2[\text{-}1\text{-}OC_4H_7\text{-}2\text{-}]$ ... Mo:Org.Comp.8-112, 127, 165
$C_{12}H_{14}MoO_4$ ... $(C_5H_5)Mo(CO)_2[\text{-}O\text{-}C(C_4H_9\text{-}t)\text{-}O\text{-}]$ ... Mo:Org.Comp.7-188, 190
$C_{12}H_{14}MoO_4S_3W$ $(C_5H_5)Mo(CO)(SCH_3)_3W(CO)_3$ ... Mo:Org.Comp.6-374
$C_{12}H_{14}MoO_5P^-$ ... $[(C_5H_5)Mo(CO)_2P(\text{-}OCH_2\text{-})_3CCH_3]^-$ ... Mo:Org.Comp.7-47, 50
$C_{12}H_{14}NO_6Re$ ... $(CO)_4Re(CNC_4H_9\text{-}t)COOC_2H_5$ ... Re: Org.Comp.2-247
$C_{12}H_{14}NO_6Sb$ ... $4\text{-}(HO_2C)(CH_3COCNH)CHCH_2C_6H_4Sb(O)(OH)_2$ Sb: Org.Comp.5-295
$C_{12}H_{14}NO_7Re$ ... $(CO)_4Re[C(O)CH_3]C(CH_3){=}NH\text{-}CH_2COOC_2H_5$ ... Re: Org.Comp.1-396
$C_{12}H_{14}N_2O_2S_2Ti$ ... $(C_5H_4CH_3)_2Ti(N{=}S{=}O)_2$ ... S: S-N Comp.6-255
$C_{12}H_{14}N_2O_8P_2Th^{2+}$
$Th[NC_5H_4CH(OH)PO_3H]_2{}^{2+}$ ... Th: SVol.D1-132

$C_{12}H_{14}N_2Si$	$SiH_2(NHC_6H_5)_2$	Si: SVol.B4-184
$C_{12}H_{14}N_6O_4S_6Th$	$Th(SCN)_2[CH_3-C(=O)-NH-N=C(-S)-S-C(=O)-CH_3]_2$	Th: SVol.D4-166
$C_{12}H_{14}N_6O_{12}Th$	$Th(NO_3)_4 \cdot 2\ (2-CH_3-NC_5H_4)$	Th: SVol.D4-159
$C_{12}H_{14}O_2Ti$	$[(C_5H_5)_2Ti(OCH_2CH_2O)]_n$	Ti: Org.Comp.5-335
$C_{12}H_{14}O_6Sn$	$C_6H_5Sn(OOCCH_3)_3$	Sn: Org.Comp.17-64
$C_{12}H_{14}O_{14}Th_2$	$(ThOH)_2[C_2H_4(COO)_2]_3 \cdot 8\ H_2O$	Th: SVol.C7-104
$C_{12}H_{14}O_{16}Po^{2-}$	$[Po(OH)_2(H(OOC-CH_2C(OH)(COO)CH_2-COO))_2]^{2-}$	Po: SVol.1-354/5
$C_{12}H_{14}O_{17}Th_2$	$(ThOH)_2[OOCCH(OH)CH_2COO]_3 \cdot 4\ H_2O$	Th: SVol.C7-105/6
$C_{12}H_{14}S_2Ti$	$[(C_5H_5)_2Ti(S(CH_2)_2S)]_n$	Ti: Org.Comp.5-347
$C_{12}H_{15}I_2MoN_3O$	$C_5H_5Mo(NO)(NH_2NHC_6H_4CH_3-4)I_2$	Mo: Org.Comp.6-37
$C_{12}H_{15}InMoO_3$	$(C_2H_5)_2InMo(C_5H_5)(CO)_3$	In: Org.Comp.1-329/31, 332, 334
$C_{12}H_{15}InO_4$	$C_6H_5-In[OC(O)C_2H_5]_2$	In: Org.Comp.1-231, 233
$C_{12}H_{15}MnN_2PS_2$	$Mn[P(C_6H_5)(C_2H_5)_2](NCS)_2$	Mn: MVol.D8-38
$C_{12}H_{15}MoNO_2$	$(C_5H_5)Mo(CO)_2(H)=C[-N(CH_3)-CH_2CH_2CH_2-]$	Mo: Org.Comp.8-16, 22
–	$(C_5H_5)Mo(CO)_2[-1-NC_4H_7-1-CH_3-2-]$	Mo: Org.Comp.8-112, 121
$C_{12}H_{15}MoNO_2S_2$	$(C_5H_5)Mo(CO)_2[-S-C(N(C_2H_5)_2)-S-]$	Mo: Org.Comp.7-189, 195/6, 204
$C_{12}H_{15}MoNO_3$	$(C_5H_5)Mo(CO)_2[-C(=O)-CH_2-C(CH_3)_2-NH_2-]$	Mo: Org.Comp.8-111, 136
–	$(C_5H_5)Mo(CO)_2[-C(=O)-CH_2-CH(C_2H_5)-NH_2-]$	Mo: Org.Comp.8-111, 136
–	$(H_3CC(O)-C_5H_4)Mo(CO)(NO)[CH_2=C(CH_3)_2]$	Mo: Org.Comp.6-293
–	$[(CH_3)_5C_5]Mo(CO)_2(NO)$	Mo: Org.Comp.7-5/6, 10/1, 22/3
$C_{12}H_{15}MoNO_4$	$(C_5H_5)Mo(CO)_2[-O-C(=O)-CH(C_3H_7-i)-NH_2-]$	Mo: Org.Comp.7-208, 209, 231
$C_{12}H_{15}MoNO_5Si$	$(CO)_5Mo[CN-Si(CH_3)_2-C_4H_9-t]$	Mo: Org.Comp.5-6, 9
$C_{12}H_{15}MoNO_6$	$(C_2H_5)_2N-C(O-C_2H_5)=Mo(CO)_5$	Mo: Org.Comp.5-104
$C_{12}H_{15}MoN_3O_3$	$(CO)_3Mo(CN-C_2H_5)_3$	Mo: Org.Comp.5-34, 35
$C_{12}H_{15}MoO^+$	$[(C_5H_5)Mo(CO)(H_3C-C{\equiv}C-CH_3)(H_2C=CH_2)]^+$	Mo: Org.Comp.6-315, 322
$C_{12}H_{15}MoO_2P_3$	$[(CH_3)_5C_5]Mo(CO)_2(P_3)$	Mo: Org.Comp.7-39
$C_{12}H_{15}NO_3PReS_2$	$C_5H_5N-1-Re(CO)_3[-S-P(C_2H_5)_2=S-]$	Re: Org.Comp.1-129
$C_{12}H_{15}N_2O_4Sb$	$(CH_3)_3Sb(NO_3)OC_9H_6N$	Sb: Org.Comp.5-36/7
$C_{12}H_{15}N_2O_{11}Th^-$	$Th[((OOCCH_2)_2N)_2C_2H_4](HOCH_2COO)^-$	Th: SVol.D1-136
$C_{12}H_{15}N_3O_3Re^+$	$[(CO)_3Re(NCC_2H_5)_3]^+$	Re: Org.Comp.1-309
$C_{12}H_{15}N_3O_4PRe$	$(CO)_3Re(NH_2NH_2)[P(CH_3)_2C_6H_5]NCO$	Re: Org.Comp.1-291
–	$(CO)_3Re(ND_2ND_2)[P(CH_3)_2C_6H_5]NCO$	Re: Org.Comp.1-291
$C_{12}H_{15}N_3O_{12}Th^{2-}$	$[Th(OOCCH_2CH(NH_2)COO)_3]^{2-}$	Th: SVol.D1-82/3, 87
$C_{12}H_{15}N_5Sn$	$(CH_3)_3SnN(C(NHC_6H_5)=NCN)CN$	Sn: Org.Comp.18-71, 75
$C_{12}H_{16}I_4NO_2Re$	$[(CH_3)_3NCH_2C_6H_5][(CO)_2ReI_4]$	Re: Org.Comp.1-58
$C_{12}H_{16}MnN_4O_4S_2{}^{2+}$	$[Mn(4-NH_2-C_6H_4-SO_2-NH_2)_2]^{2+}$	Mn: MVol.D7-116/23
$C_{12}H_{16}MnN_4O_7P_2S$	$[Mn(CH_3-NH_2-C_4HN_2-CH_2-NC_3HS-CH_3-CH_2CH_2P_2O_7)]$	Mn: MVol.D8-167/8
$C_{12}H_{16}MnO_8S_4{}^{2-}$	$[Mn(OC(=O)-CH_2CH_2-SS-CH_2CH_2-C(=O)O)_2]^{2-}$	Mn: MVol.D7-90/1
–	$[Mn(OC(=O)-CH_2-S-CH_2CH_2-S-CH_2-C(=O)O)_2]^{2-}$	Mn: MVol.D7-86/8
$C_{12}H_{16}MoN_2O_2S_2$	$(C_5H_5)Mo(CO)(NO)CH_2=CHCH_2SC(S)N(CH_3)_2$	Mo: Org.Comp.6-295, 306

$C_{12}H_{16}MoN_2O_5$ . . . $[-N(CH_3)-CH_2CH_2-N(CH_3)-]C{=}Mo(CO)_4{=}C(CH_3)-OCH_3$ Mo:Org.Comp.5-132/3
$C_{12}H_{16}MoO$ . . . $(C_5H_5)Mo(CO)(C_2H_5)(H_3C-C{\equiv}C-CH_3)$ . . . Mo:Org.Comp.6-312
$C_{12}H_{16}MoOS$ . . . $(C_5H_5)Mo(CO)(SC_2H_5)(H_3C-C{\equiv}C-CH_3)$ . . . Mo:Org.Comp.6-277
$C_{12}H_{16}MoO_2$ . . . $[CH_2{=}C(CH_3)-CH{=}CH_2]_2Mo(CO)_2$ . . . Mo:Org.Comp.5-349
– . . . $[(CH_3)_5C_5]Mo(CO)_2H$ . . . Mo:Org.Comp.7-1
– . . . $[(CH_3)_5C_5]Mo(CO)_2D$ . . . Mo:Org.Comp.7-1/2
$C_{12}H_{16}MoO_3$ . . . $(C_5H_5)Mo(CO)_2(CH_3)-1-OC_4H_8$ . . . Mo:Org.Comp.8-77, 84
$C_{12}H_{16}MoO_6Si$ . . . $(CH_3)_3SiCH_2-C(O-C_2H_5){=}Mo(CO)_5$ . . . Mo:Org.Comp.5-100, 103
$C_{12}H_{16}MoO_{18}^{2-}$ . . . $[MoO_4 \cdot 2\ HOOCCH_2C(OH)(COOH)CH_2COOH]^{2-}$ Mo:SVol.B3b-175
$C_{12}H_{16}NO_2Re$ . . . $(C_5H_5)Re(CO)_2(NC_5H_{11})$ . . . Re: Org.Comp.3-193/7, 198
$C_{12}H_{16}NO_2Sb$ . . . $(CH_3)_3Sb(OH)OC_9H_6N$ . . . Sb: Org.Comp.5-39
$C_{12}H_{16}NO_5Re$ . . . $(CO)_4Re[C(O)CH_3]-C(CH_3){=}NH-C_4H_9\text{-n}$ . . . Re: Org.Comp.1-391
– . . . $(CO)_4Re[C(O)CH_3]-C(CH_3){=}NH-C_4H_9\text{-i}$ . . . Re: Org.Comp.1-391
– . . . $(CO)_4Re[C(O)CH_3]-C(C_4H_9\text{-i}){=}NH-CH_3$ . . . Re: Org.Comp.1-402
$C_{12}H_{16}N_2OSn$ . . . $(CH_3)_3Sn-N(CN)-C({=}O)CH_2-C_6H_5$ . . . Sn: Org.Comp.18-58, 68
– . . . $(CH_3)_3Sn-N(CN)-C({=}O)-C_6H_4-4-CH_3$ . . . Sn: Org.Comp.18-58, 70
$C_{12}H_{16}N_2O_2Os$ . . . $(CH_3)_2Os(O)_2(NC_5H_5)_2$ . . . Os: Org.Comp.A1-2, 14/5, 39/40
$C_{12}H_{16}N_2O_2Sn$ . . . $(CH_3)_3SnN(COC_6H_4OCH_3-4)CN$ . . . Sn: Org.Comp.18-58, 70
$C_{12}H_{16}N_2O_3S$ . . . $1,4-ONC_4H_8-4-S(O)-N{=}C(C_6H_5)-OCH_3$ . . . S: S-N Comp.8-353, 360
$C_{12}H_{16}N_2O_8STh$ . . . $Th[S(C_2H_4N(CH_2COO)_2)_2]$ . . . Th: SVol.D1-93/4, 97/100
$C_{12}H_{16}N_2O_9Th$ . . . $Th[O(C_2H_4N(CH_2COO)_2)_2]$ . . . Th: SVol.D1-93/5, 100
$C_{12}H_{16}N_2O_{14}Po^{2-}$ $[Po(OH)_2(H(N(CH_2COO)_3))_2]^{2-}$ . . . Po: SVol.1-354/5
$C_{12}H_{16}N_2Ti$ . . . $[(C_5H_5)_2Ti(NH(CH_2)_2NH)]_n$ . . . Ti: Org.Comp.5-349
$C_{12}H_{16}N_3O_{10}Th^-$ $Th[((OOCCH_2)_2N)_2C_2H_4](H_2NCH_2COO)^-$ . . . Th: SVol.D1-136
$C_{12}H_{16}N_4O_2S_6Th$ $Th(SCN)_4 \cdot 2\ [SC_4H_8({=}O)-1]$ . . . Th: SVol.D4-155, 176
$C_{12}H_{16}N_4Pb$ . . . $Pb(CH_2CH_2-CN)_4$ . . . Pb: Org.Comp.3-67
$C_{12}H_{16}N_4S$ . . . $1-NC-c-C_5H_8-N{=}S{=}N-c-C_5H_8-1-CN$ . . . S: S-N Comp.7-211
$C_{12}H_{16}N_4Sn$ . . . $(C_2H_5)_3SnN{=}C{=}C(CN)CH(CN)_2$ . . . Sn: Org.Comp.18-152/3
$C_{12}H_{16}N_8O_{12}Th$ . . . $Th(NO_3)_4 \cdot 2\ C_6H_4-1,2-(NH_2)_2$ . . . Th: SVol.D4-134
$C_{12}H_{16}O_2SiTi$ . . . $[(C_5H_5)_2Ti(OSi(CH_3)_2O)]_n$ . . . Ti: Org.Comp.5-333
$C_{12}H_{16}O_8Th$ . . . $Th[OOC(CH_2)_4COO]_2$ . . . Th: SVol.C7-112/4
$C_{12}H_{17}IMoN_2O$ . . . $(C_5H_5)Mo(CO)(I)C(N(CH_3)_2)C(CH_3)NCH_3$ . . . Mo:Org.Comp.6-279/80, 284
$C_{12}H_{17}MnN_2NaO_{11}S_2$
$Na[Mn((5-S{=})(2-(4-C_6H_4SO_3))C_2HN_2O-4,3,1)(H_2O)_3](OC({=}O)CH_3)_2$ . . . Mn:MVol.D7-63/4
$C_{12}H_{17}MnN_4O_7P_2S^+$
$[Mn(3-(2-CH_3-4-NH_2-1,3-N_2C_4H-5-CH_2)-4-CH_3-1,3-SNC_3H-5-CH_2CH_2-O-PO_2-O-PO_3H)]^+$ Mn:MVol.D8-167/8
$C_{12}H_{17}MoNO_2P^+$ $[(C_5H_5)Mo(CO)_2(NC-CH_3)(P(CH_3)_3)]^+$ . . . Mo:Org.Comp.7-283, 293
$C_{12}H_{17}MoNO_2Sn$ $(C_5H_5)Mo(CO)_2(CN-CH_3)-Sn(CH_3)_3$ . . . Mo:Org.Comp.8-9, 10
$C_{12}H_{17}MoNS_2$ . . . $(C_5H_5)Mo(S_2CN(CH_3)_2)(H_3C-C{\equiv}C-CH_3)$ . . . Mo:Org.Comp.6-109
$C_{12}H_{17}MoN_2O_2PS_2$
$(C_5H_5)Mo(CO)_2[-P(N(CH_3)_2)_2-C(S)-S-]$ . . . Mo:Org.Comp.7-230
$C_{12}H_{17}MoO_2P$ . . . $(C_5H_5)Mo(CO)[P(CH_3)_3][C(O)CCH_3]$ . . . Mo:Org.Comp.6-280, 284
$C_{12}H_{17}MoO_3P$ . . . $(C_5H_5)Mo(CO)_2[P(CH_3)_3]-C({=}O)CH_3$ . . . Mo:Org.Comp.8-45, 49/50
$C_{12}H_{17}MoO_4P$ . . . $[(C_5H_5)Mo(CO)_2(P(CH_3)_3)-OC(O)CH_3]$ . . . Mo:Org.Comp.7-56/7, 62

$C_{12}H_{17}MoO_6P$	$(C_5H_5)Mo(CO)_2[P(OCH_3)_3]$-$C(=O)CH_3$	Mo:Org.Comp.8-45, 52, 73
$C_{12}H_{17}NO_3Sn$	$C_6H_5Sn(OCH_2CH_2)_3N$	Sn: Org.Comp.17-63/4, 66
$C_{12}H_{17}N_2O_8STh^+$	$Th[H(S(C_2H_4N(CH_2COO)_2)_2)]^+$	Th: SVol.D1-100
$C_{12}H_{17}N_2O_9STh^-$	$Th[S(C_2H_4N(CH_2COO)_2)_2](OH)^-$	Th: SVol.D1-100
$C_{12}H_{17}N_2O_9Th^+$	$Th[H(O(C_2H_4N(CH_2COO)_2)_2)]^+$	Th: SVol.D1-100
$C_{12}H_{17}N_2O_{10}Th^-$	$Th[O(C_2H_4N(CH_2COO)_2)_2](OH)^-$	Th: SVol.D1-97
$C_{12}H_{17}N_3O_6Sn$	$C_6H_5Sn(ONHCOCH_3)_3$	Sn: Org.Comp.17-65
$C_{12}H_{17}NaO_{17}Th$	$NaTh(OH)[OOC(CHOH)_4COO]_2 \cdot 6\ H_2O$	Th: SVol.C7-112/4
$C_{12}H_{18}IMnN_2O_2$	$MnI[$-O-$C(CH_3)=CH$-$C(CH_3)=N$-C_2H_4 -$N=C(CH_3)CH=C(CH_3)O$-$]$	Mn:MVol.D6-203, 204/6
–	$MnI[$-O-$C(CH_3)=CH$-$C(CH_3)=N$-C_2H_4 -$N=C(CH_3)CH=C(CH_3)O$-$] \cdot 2\ H_2O$	Mn:MVol.D6-203, 204, 206/7
$C_{12}H_{18}IMoN_6^+$	$[(I)Mo(CN$-$CH_3)_6][BF_4]$	Mo:Org.Comp.5-71, 73
$C_{12}H_{18}I_2N_6Re$	$[Re(CNCH_3)_6]I_2 \cdot 8\ AgI$	Re: Org.Comp.2-291, 297
$C_{12}H_{18}I_2OSb_2$	$[(CH_2=CH)_3SbI]_2O$	Sb: Org.Comp.5-91, 95
$C_{12}H_{18}I_4Mn_2N_6$	$[Mn(NCCH_3)_6][MnI_4]$	Mn:MVol.D7-7/9
$C_{12}H_{18}MnN_2O_3S^{2+}$	$[Mn(4$-CH_3-C_6H_4-SO_2-NH-$C(O)NH$-C_4H_9-$n)]^{2+}$	Mn:MVol.D7-124/5
$C_{12}H_{18}MnN_2O_6S_2^{2+}$	$[Mn(O=S(CH_3)(C_5H_4NO))_2(H_2O)_2]^{2+}$	Mn:MVol.D7-106
$C_{12}H_{18}MnN_3O_4$	$[Mn($-O-$C(CH_3)=CH$-$C(CH_3)=N$-C_2H_4 -$N=C(CH_3)CH=C(CH_3)O$-$)(NO_2)]$	Mn:MVol.D6-203, 204/6
$C_{12}H_{18}MnN_5O_2$	$[Mn($-O-$C(CH_3)=CH$-$C(CH_3)=N$-C_2H_4 -$N=C(CH_3)CH=C(CH_3)O$-$)(N_3)]$	Mn:MVol.D6-203, 204/6
$C_{12}H_{18}MnN_6^{2+}$	$[Mn(NCCH_3)_6]^{2+}$	Mn:MVol.D7-4/9
$C_{12}H_{18}Mn_2N_2O_6P_2$	$Mn_2[(O_3PC(CH_3)_2NH)_2C_6H_4] \cdot 2\ H_2O$	Mn:MVol.D8-130/1
$C_{12}H_{18}Mo$	$(CH_2=CHCH=CH_2)_3Mo$	Mo:Org.Comp.5-357/9
$C_{12}H_{18}MoN_2O_5S_4$	$[CH_3$-$OC(=O)$-CC-$C(=O)O$-$CH_3]Mo(O)[S_2C$-$N(CH_3)_2]_2$	Mo:Org.Comp.5-139/41, 143
$C_{12}H_{18}MoO_2$	$[(CH_3)_5C_5]Mo(CO)_2(H)_3$	Mo:Org.Comp.7-4/5
–	$[(CH_3)_5C_5]Mo(CO)_2(H)_2(D)$	Mo:Org.Comp.7-4/5
–	$[(CH_3)_5C_5]Mo(CO)_2(H)(D)_2$	Mo:Org.Comp.7-4/5
–	$[(CH_3)_5C_5]Mo(CO)_2(D)_3$	Mo:Org.Comp.7-4/5
$C_{12}H_{18}MoO_3$	$[CH_3$-$C(O)CHCH_2]_3Mo$	Mo:Org.Comp.5-299
$C_{12}H_{18}Mo_7O_{42}^{6-}$	$[Mo_7O_{24} \cdot 3\ HOOC(CHOH)_2COOH]^{6-}$	Mo:SVol.B3b-174/5
$C_{12}H_{18}NO_4Re$	$(CO)_3Re[$-O-$1,1$-$(NC_5H_6(CH_3)_4$-$2,2,6,6)$-$]$, radical	Re: Org.Comp.1-124
$C_{12}H_{18}NO_5Re$	$(C_2H_5)_2NH$-$Re(CO)_3[$-O-$C(CH_3)=CH$-$C(CH_3)=O$-$]$	Re: Org.Comp.1-127
$C_{12}H_{18}NO_6Re$	$(CO)_4Re[C(=O)CH_3]=C(OH)CH_3 \cdot NH_2C_4H_9$-i	Re: Org.Comp.1-390
$C_{12}H_{18}N_2OSn$	$(CH_3)_3Sn$-$N(C_6H_5)$-$C(=O)$-$N[$-CH_2CH_2-$]$	Sn: Org.Comp.18-47, 50, 55
–	$(CH_3)_3Sn$-$N[$-$C(=O)$-$N(C_6H_5)$-CH_2CH_2-$]$	Sn: Org.Comp.18-84, 88, 101
$C_{12}H_{18}N_6O_3ReS_3^+$	$[(CO)_3Re(1,3$-$N_2C_3H_6(=S)$-$2)_3]^+$	Re: Org.Comp.1-310/1
$C_{12}H_{18}N_6Re^+$	$[Re(CNCH_3)_6]^+$	Re: Org.Comp.2-291, 297/8
$C_{12}H_{18}N_6Re^{2+}$	$[Re(CNCH_3)_6]^{2+}$	Re: Org.Comp.2-291, 297
$C_{12}H_{18}N_8O_6Re$	$[Re(CNCH_3)_6][NO_3]_2$	Re: Org.Comp.2-291, 297
$C_{12}H_{18}Na_2O_{18}Th$	$Na_2Th(OH)_2[OOC(CHOH)_4COO]_2 \cdot 10\ H_2O$	Th: SVol.C7-112/4

$C_{12}H_{18}O_4Sn$ ..... $(n-C_4H_9)_2Sn[-OC(O)-C{\equiv}C-C(O)O-]$ .... Sn: Org.Comp.15-326
$C_{12}H_{18}O_6Re^+$ .... $[(CO)_3Re(OC(CH_3)_2)_3]^+$ .... Re: Org.Comp.1-310
$C_{12}H_{18}O_{12}Th^{2-}$ ... $[Th(OOCCH_3)_6]^{2-}$ .... Th: SVol.D1-67/8
$C_{12}H_{19}IO_4Sn$ ..... $C_2H_5SnI(OC(CH_3){=}CHCOCH_3)_2$ .... Sn: Org.Comp.17-171, 172
$C_{12}H_{19}I_2MnO_3P$ .. $Mn(I)_2(O_2)(OC_4H_8)-P(C_6H_5)(CH_3)_2$ .... Mn: MVol.D8-51
$C_{12}H_{19}I_2MnP$ .... $Mn(I)_2(SO_2)-P(C_6H_5)(C_3H_7-n)_2$ .... Mn: MVol.D8-46, 57/8
– .... $Mn(I)_2-P(C_6H_5)(C_3H_7-n)_2$ .... Mn: MVol.D8-42/3
$C_{12}H_{19}InO_3S_3$ .... $[-OC(CH_3)O-]In(CH_3)-S-C_6H_3-2-SH-5-CH_3 \cdot OS(CH_3)_2$
In: Org.Comp.1-239
$C_{12}H_{19}MoN_2O_2P$ .. $(C_5H_5)Mo(CO)_2[-CH_2-P(N(CH_3)_2)_2-]$ .... Mo: Org.Comp.8-112, 125
$C_{12}H_{19}MoO_2P$ .... $(C_5H_5)Mo(CO)_2[P(CH_3)_3]C_2H_5$ .... Mo: Org.Comp.8-77, 84
$C_{12}H_{19}N_3Sn$ ..... $(C_2H_5)_3SnN(N{=}NC_6H_4-1,2)$ .... Sn: Org.Comp.18-142, 148
$C_{12}H_{19}O_3Sb$ ..... $(CH_3)_3Sb(OCH_3)OC_6H_4C(O)CH_3-2$ .... Sb: Org.Comp.5-46
$C_{12}H_{20}IMoN_5O$ ... $(I)Mo(CN-C_2H_5)_4(NO)$ .... Mo: Org.Comp.5-39, 41
$C_{12}H_{20}IORe$ ..... $(CH{\equiv}C-C_4H_9-t)_2Re(O)I$ .... Re: Org.Comp.2-357, 365
– .... $(C_2H_5C{\equiv}CC_2H_5)_2Re(O)I$ .... Re: Org.Comp.2-362, 371
$C_{12}H_{20}I_2MnN_4S_8$ .. $[Mn((2-S{=})C_3H_5NS-3,1)_4I_2]$ .... Mn: MVol.D7-61/3
$C_{12}H_{20}I_2MoN_4O_2$ $[(-N(CH_3)-CH_2CH_2-N(CH_3)-)C{=}]_2Mo(CO)_2(I)_2$ Mo: Org.Comp.5-120/1
$C_{12}H_{20}I_2NO_4Re$ .. $[N(C_2H_5)_4][(CO)_4ReI_2]$ .... Re: Org.Comp.1-343
$C_{12}H_{20}I_3MoOP$ ... $(C_5H_5)Mo(CO)(P(C_2H_5)_3)I_3$ .... Mo: Org.Comp.6-219
$C_{12}H_{20}InN$ .... $(CH_3)_2In-CH_2-C_6H_4-2-[CH_2-N(CH_3)_2]$ .... In: Org.Comp.1-101, 107
– .... $(CH_3)_2In-C_6H_4-2-[CH(CH_3)-N(CH_3)_2]$ .... In: Org.Comp.1-101, 108, 115
$C_{12}H_{20}In_2O_4$ ..... $(C_2H_5)_2In[1,2-(O)_2-C_4-3,4-(O)_2]In(C_2H_5)_2$ .... In: Org.Comp.1-209
$C_{12}H_{20}MnN_2O_6S_2$ $[Mn((3-O{=})C_4H_7NS-4,1)_2(OC(CH_3){=}O)_2] \cdot 2\ H_2O$
Mn: MVol.D7-240/1
$C_{12}H_{20}MnN_3O_4S_4$ $[Mn((SC({=}S))NC_4H_8O)_2(NH_2CH_2COO)]$ .... Mn: MVol.D7-177/8
$C_{12}H_{20}MnN_4S_8^{2+}$ $[Mn((2-S{=})C_3H_5NS-3,1)_4]^{2+}$ .... Mn: MVol.D7-61/3
$C_{12}H_{20}MnN_4S_8Zn$ $MnZn(SC({=}S)NHCH_2CH_2CH_2CH_2NHC({=}S)S)_2$ .. Mn: MVol.D7-180/4
$C_{12}H_{20}MnO_{24}P_4^{6-}$
$[Mn(OC_4H_3-(OH)_3-(CH_2OPO_3)_2)_2]^{6-}$ .... Mn: MVol.D8-148
$C_{12}H_{20}Mn_2S_{12}^{2-}$ $[Mn_2(SCH_2CH(S)CH_2SSCH_2CH(S)CH_2S)_2]^{2-}$ ... Mn: MVol.D7-42/4
$C_{12}H_{20}Mo$ .... $(CH_2CHCH_2)_4Mo$ .... Mo: Org.Comp.5-299/302
$C_{12}H_{20}MoNO_2P$ .. $(C_5H_5)Mo(P(C_2H_5)_3)(NO)CO$ .... Mo: Org.Comp.6-237
$C_{12}H_{20}MoO_8P_2$ ... $(C_5H_5)Mo(CO)_2[P(OCH_3)_3]-P({=}O)(OCH_3)_2$ ..... Mo: Org.Comp.7-120, 131
$C_{12}H_{20}N_2OSn$ .... $(CH_3)_3SnN(C_6H_5)CON(CH_3)_2$ .... Sn: Org.Comp.18-47, 49, 53/5
$C_{12}H_{20}N_2O_2Sn$ ... $(C_2H_5)_3SnN(C_6H_5)NO_2$ .... Sn: Org.Comp.18-141
$C_{12}H_{20}N_2SSn$ .... $(CH_3)_3SnN(C_6H_5)CSN(CH_3)_2$ .... Sn: Org.Comp.18-47, 50, 53/5
$C_{12}H_{20}N_2Sn$ ..... $(CH_3)_3Sn-N{=}C(C_6H_5)-N(CH_3)_2$ .... Sn: Org.Comp.18-107/8, 115/6
– .... $(CH_3)_3Sn-N(CH_3)-C(C_6H_5){=}NCH_3$ .... Sn: Org.Comp.18-47/8, 52
$C_{12}H_{20}N_4O_8Th$ ... $Th[C_2H_4(N(CH_2COO)_2)_2](H_2NC_2H_4NH_2)$ .... Th: SVol.D1-138
$C_{12}H_{20}N_4O_{12}Th$ .. $[NH_4]_2[Th(N(CH_2COO)_3)_2] \cdot 4\ H_2O$ .... Th: SVol.C7-121/3
$C_{12}H_{20}N_4O_{16}Th$ .. $[NH_3-CH_2CH_2-NH_3]_2[Th(C_2O_4)_4] \cdot 2.5\ H_2O$ ... Th: SVol.C7-94
$C_{12}H_{20}N_4PdS_2$ ... $Pd(SCN)_2(NC_5H_{10})_2$ .... Pd: SVol.B2-306
$C_{12}H_{20}N_8O_4Si$ .... $Si[NHCON(-CH_2-CH_2-)]_4$ .... Si: SVol.B4-314
$C_{12}H_{20}O_3Sn$ ..... $C_6H_5Sn(OC_2H_5)_3$ .... Sn: Org.Comp.17-63
$C_{12}H_{20}O_4PRe^{2+}$ .. $[(C_5H_5)Re(CO)(P(O-C_2H_5)_3)]Br_2$ .... Re: Org.Comp.3-143, 145
$C_{12}H_{20}O_4S_2Sn$ ... $(CH_2{=}CHCH_2)_2Sn(OS(O)CH_2CH{=}CH_2)_2$ .... Sn: Org.Comp.16-90, 93
$C_{12}H_{20}O_4Sn$ ..... $(n-C_4H_9)_2Sn[-OC({=}O)-CH{=}CH-C({=}O)O-]$ ..... Sn: Org.Comp.15-301/21
– .... $(i-C_4H_9)_2Sn[-OC({=}O)-CH{=}CH-C({=}O)O-]$ ...... Sn: Org.Comp.15-371
$C_{12}H_{20}O_8Sn$ ..... $(CH_3OOCCH_2CH_2)_2Sn(OOCCH_3)_2$ .... Sn: Org.Comp.16-86

$C_{12}H_{20}O_8Th$	$Th(C_2H_5COO)_4$	Th: SVol.C7-68/9
$C_{12}H_{20}O_{12}Th$	$Th(OOCCH(OH)CH_3)_4$	Th: SVol.D1-71
$C_{12}H_{20}Pb$	$Pb(CH=CH-CH_3)_4$	Pb: Org.Comp.3-89/91
–	$Pb(CH=CH-CH_3)_3-C(CH_3)=CH_2$	Pb: Org.Comp.3-89/91
–	$Pb[C(CH_3)=CH_2]_2(CH=CH-CH_3)_2$	Pb: Org.Comp.3-89/91
–	$Pb[C(CH_3)=CH_2]_3-CH=CH-CH_3$	Pb: Org.Comp.3-89/91
–	$Pb[C(CH_3)=CH_2]_4$	Pb: Org.Comp.3-89/91
–	$Pb(C_3H_5-c)_4$	Pb: Org.Comp.3-62/3
$C_{12}H_{21}I_3In_2O_6$	$[C_2H_5-O-C(=O)-CH_2]_2In(-I-)_2InI-CH_2-C(=O)-O-C_2H_5$	In: Org.Comp.1-167, 170
$C_{12}H_{21}MoNO$	$[C_5(CH_3)_5]Mo(NO)(CH_3)_2$	Mo: Org.Comp.6-98
$C_{12}H_{21}MoN_2O_3PS_4$	$(C_5H_5)Mo(NO)(SP(S)(OC_2H_5)_2)S_2CN(CH_3)_2$	Mo: Org.Comp.6-50
$C_{12}H_{21}MoO_2PSi$	$(C_5H_5)Mo(CO)_2[P(CH_3)_3]SiH(CH_3)_2$	Mo: Org.Comp.7-123, 147
$C_{12}H_{21}NO_2SSn$	$(C_2H_5)_3SnNHSO_2C_6H_5$	Sn: Org.Comp.18-129/30
$C_{12}H_{21}NSn$	$(C_2H_5)_2N-Sn(CH_3)_2-C_6H_5$	Sn: Org.Comp.19-47/8, 52
–	$(C_2H_5)_3Sn-NH-C_6H_5$	Sn: Org.Comp.18-130
$C_{12}H_{21}N_2O_5Re$	$(C_2H_5)_2NH-Re(CO)_3[-O=C(N(C_2H_5)_2)-O-]$	Re: Org.Comp.1-127
$C_{12}H_{21}N_6O_9Th^+$	$[Th(H_2NC(O)CH_2CH(NH_2)COO)_3]^+$	Th: SVol.D1-82/3, 87
$C_{12}H_{21}O_5ReSi_3$	$(CO)_5Re-Si(CH_3)[Si(CH_3)_3]_2$	Re: Org.Comp.2-4
–	$(CO)_5Re-Si(CH_3)_2-Si(CH_3)_2-Si(CH_3)_3$	Re: Org.Comp.2-4
$C_{12}H_{21}O_6Th^+$	$[Th(OOCC_3H_7-n)_3]^+$	Th: SVol.D1-68
$C_{12}H_{21}O_9Th^+$	$[Th(OOC-CH_2CH(OH)CH_3)_3]^+$	Th: SVol.D1-72
–	$[Th(OOC-C(OH)(CH_3)_2)_3]^+$	Th: SVol.D1-72
$C_{12}H_{21}PSn$	$(CH_3)_3Sn-P(C_6H_5)-C_3H_7-i$	Sn: Org.Comp.19-173/4
$C_{12}H_{22}IMoN_7O$	$[CH_3-NH-C(NH-C_2H_5)=Mo(CN-CH_3)_4(NO)]I$	Mo: Org.Comp.5-118/9
$C_{12}H_{22}I_2MnN_6O_2$	$[Mn(c-C_5H_8=N-NHC(O)NH_2)_2I_2]$	Mn: MVol.D6-328/9
$C_{12}H_{22}I_2MnN_6S_2$	$MnI_2[c-C_5H_8=N-NHC(=S)NH_2]_2$	Mn: MVol.D6-342/4
$C_{12}H_{22}In_2S_4$	$[n-C_4H_9-In(-S-CH=CH-S-)]_2$	In: Org.Comp.1-248
$C_{12}H_{22}MnO_{18}P_2^{2-}$	$[Mn(OC_5H_5-(OH)_4-CH_2OPO_3)_2]^{2-}$	Mn: MVol.D8-148
$C_{12}H_{22}MoN_2OS_4$	$(HC{\equiv}CH)Mo(O)[S_2C-N(C_2H_5)_2]_2$	Mo: Org.Comp.5-139/40, 142
$C_{12}H_{22}MoN_7O^+$	$[CH_3-NH-C(NH-C_2H_5)=Mo(CN-CH_3)_4(NO)]I$	Mo: Org.Comp.5-118/9
$C_{12}H_{22}NO_4Re$	$[N(C_2H_5)_4][(CO)_4ReH_2]$	Re: Org.Comp.1-344, 345/6
$C_{12}H_{22}N_2OSn$	$(C_2H_5)_3SnN=C=C(CN)CH_2OC_2H_5$	Sn: Org.Comp.18-152/3
$C_{12}H_{22}N_2O_4S_2Si_2$	$((CH_3)_3Si)_2S=N-S(O)_2-C_6H_4-4-NO_2$	S: S-N Comp.8-251
$C_{12}H_{22}N_2S$	$c-C_6H_{11}-N=S=N-c-C_6H_{11}$	S: S-N Comp.7-212
$C_{12}H_{22}N_2Sn$	$(C_2H_5)_3SnN=C=C(CN)C_3H_7-i$	Sn: Org.Comp.18-152/3
$C_{12}H_{22}N_4O_2S$	$(1,4-ONC_4H_8-4)_2S=N-C(CH_3)_2CN$	S: S-N Comp.8-207/8, 211
$C_{12}H_{22}N_4S_2Si_2$	$1,4-[(CH_3)_3SiN=S=N]_2-C_6H_4$	S: S-N Comp.7-143
$C_{12}H_{22}N_4S_4Si_2$	$1,4-[(CH_3)_3SiN=S=NS]_2C_6H_4$	S: S-N Comp.7-146/7
$C_{12}H_{22}N_6Sn$	$[1,2,4-N_3C_2H_2-1-]_2Sn(C_4H_9-n)_2$	Sn: Org.Comp.19-93/6
$C_{12}H_{22}O_2Po$	$Po[CH_2C(O)C_4H_9-i]_2$	Po: SVol.1-330
$C_{12}H_{22}O_3Po$	$Po[CH_2C(O)C_4H_9-i]_2O$	Po: SVol.1-330
$C_{12}H_{22}O_4Sn$	$(C_4H_9)_2SnOOCCH_2CH_2COO$	Sn: Org.Comp.15-322
$C_{12}H_{22}O_5Sn$	$(C_4H_9)_2Sn[-OCH_2C(O)OCOCH_2O-]$	Sn: Org.Comp.15-70
$C_{12}H_{22}O_7Th$	$Th(OH)(n-C_3H_7COO)_3$	Th: SVol.C7-71/2
$C_{12}H_{23}IMoN_6O$	$[Mo(CN-CH_3)_4(NO)(NH_2-C_4H_9-t)]I$	Mo: Org.Comp.5-39, 40
$C_{12}H_{23}MoN_6O^+$	$[Mo(CN-CH_3)_4(NO)(NH_2-C_4H_9-t)]I$	Mo: Org.Comp.5-39, 40

$C_{12}H_{23}NOS_2Sn$ . . . 1-[i-C_3H_7-O-Sn$(CH_3)_2$-NH]-2-[CH_3-S-C(=S)]-C_5H_6
Sn: Org.Comp.19-129, 130

$C_{12}H_{23}NO_2S_2Si_2$ . . $((CH_3)_3Si)_2S$=N-S(O)$_2$-C_6H_5 . . . . S: S-N Comp.8-251

$C_{12}H_{23}NO_3Sn$ . . . . $(C_2H_5)_3SnN(C(=O)CH(OC_2H_5)CH_2C(=O))$ . . . . . . Sn: Org.Comp.18-142/4

$C_{12}H_{23}NPbSn$ . . . . $(CH_3)_3SnN(C_6H_5)Pb(CH_3)_3$ . . . . Sn: Org.Comp.18-57, 67

$C_{12}H_{23}NSiSn$ . . . . $(CH_3)_3SnN(C_6H_5)Si(CH_3)_3$ . . . . Sn: Org.Comp.18-57, 67, 74

$C_{12}H_{23}N_3O_2S$ . . . . $(1-C_5H_{10}N)_2S$=N-C(O)OCH_3 . . . . S: S-N Comp.8-207/8, 210

$C_{12}H_{23}N_3Sn$ . . . . . $[(CH_3)_2N]_3Sn-C_6H_5$ . . . . Sn: Org.Comp.19-111, 114

$C_{12}H_{23}O_6Sb$ . . . . . $(C_3H_7)_2Sb(O_2CCH_3)_3$ . . . . Sb: Org.Comp.5-181

$C_{12}H_{23}PSiSn$ . . . . . $(CH_3)_3Sn-P(C_6H_5)-Si(CH_3)_3$ . . . . Sn: Org.Comp.19-176

$C_{12}H_{24}IN_3S$ . . . . . . $[(1-C_4H_8N)_3S]I$ . . . . S: S-N Comp.8-230, 242

$C_{12}H_{24}I_4N_4S_4Si$ . . . $[Si(N(-CH_2-CH_2-S-CH_2-))_4]I_4$
= $[I_2Si(N(-CH_2-CH_2-S-CH_2-))_4]I_2$ . . . . Si: SVol.B4-343/4

$C_{12}H_{24}I_6N_8Pd_2S_8$ $[Pd(NH_2C(=S)CH_2C(=S)NH_2)_2][Pd(NH_2C(=S)$
$CH_2C(=S)NH_2)_2I_2]I_4$ . . . . Pd: SVol.B2-212

$C_{12}H_{24}In_2N_2O_2$ . . . $[1-(CH_3)_2In-NC_4H_6(=O)-2]_2$ . . . . In: Org.Comp.1-267

$C_{12}H_{24}In_2N_4NiO_4$ $Ni[(CH_3)_2In(-O-NC(CH_3)C(CH_3)N-O-)_2In(CH_3)_2]$
In: Org.Comp.1-228/30

$C_{12}H_{24}In_4N_4$ . . . . . $[(CH_3)_2In(CN)]_4$ . . . . In: Org.Comp.1-177

$C_{12}H_{24}MnN_2O_4$ . . . $[Mn(N(CH_3)-C(CH_3)=CHC(O)CH_3)_2(H_2O)_2]$ . . . . Mn: MVol.D6-95

$C_{12}H_{24}MnN_2O_9S_6$ $[Mn(C_4H_8S_2-1,4-(=O-1))_3(NO_3)_2]$ . . . . Mn: MVol.D7-105

$C_{12}H_{24}MnN_4O_{12}P_4^{6-}$
$[Mn(1,4,7,10-N_4C_8H_{16}-(CH_2PO_3)_4-1,4,7,10)]^{6-}$ Mn: MVol.D8-136/7

$C_{12}H_{24}MoN_2O_6$ . . . $MoO_2(C_5H_{11}CONHO)_2$ . . . . Mo: SVol.B3b-176/7

$C_{12}H_{24}MoOP_2$ . . . . $C_5H_5Mo(CO)(P(CH_3)_3)_2H$ . . . . Mo: Org.Comp.6-221

$C_{12}H_{24}MoO_8P_2$ . . . $(CH_2=CHCH=CH_2)Mo(CO)_2[P(OCH_3)_3]_2$ . . . . Mo: Org.Comp.5-306/10

$C_{12}H_{24}NO_2P_2Re$ . . $(CO)ReH[P(CH_3)_3]_2(NO)-C_5H_5$ . . . . Re: Org.Comp.1-50/1

$C_{12}H_{24}N_2O_2Sn$ . . . $(C_2H_5)_3SnN(C(=O)CH(N(CH_3)_2)CH_2C(=O))$ . . . . . Sn: Org.Comp.18-143/4

$C_{12}H_{24}N_2O_3Sn$ . . . 2,2-(n-C_4H_9)$_2$-3-[NH_2-CH_2-C(=O)]
-1,3,2-$ONSnC_2H_2$(=O)-5 . . . . Sn: Org.Comp.19-129, 138

$C_{12}H_{24}N_2O_4Sn$ . . . 1,2-[C_2H_5-O-C(=O)]$_2$-3,3-$(CH_3)_2$-1,2,3-$N_2SnC_4H_8$
Sn: Org.Comp.19-48, 54

$C_{12}H_{24}N_2O_4Th^{2+}$ $[Th(i-C_4H_9CH(NH_2)COO)_2]^{2+}$ . . . . Th: SVol.D1-82/3, 85

– . . . . $[Th(s-C_4H_9CH(NH_2)COO)_2]^{2+}$ . . . . Th: SVol.D1-82/3, 85

$C_{12}H_{24}N_2O_8Th^{2+}$ $[Th(OOCCH_2N(C_2H_4OH)_2)_2]^{2+}$ . . . . Th: SVol.D1-82/3, 86

$C_{12}H_{24}N_2SiSn$ . . . . $(CH_3)_3Sn-N(C_6H_5)-NH-Si(CH_3)_3$ . . . . Sn: Org.Comp.18-121/2

– . . . . $(CH_3)_3Sn-N[Si(CH_3)_3]-NH-C_6H_5$ . . . . Sn: Org.Comp.18-121/2

$C_{12}H_{24}N_2Sn$ . . . . . $(n-C_3H_7)_3Sn$-N[-CH=N-CH=CH-] . . . . Sn: Org.Comp.18-156/7

– . . . . $(i-C_3H_7)_3Sn$-N[-N=CH-CH=CH-] . . . . Sn: Org.Comp.18-159

$C_{12}H_{24}N_3O_3S^+$ . . . $[(1,4-ONC_4H_8-4)_3S]^+$ . . . . S: S-N Comp.8-222/4, 228/9, 243

$C_{12}H_{24}N_3O_5PS$ . . . 1,4-ONC_4H_8-4-S(O)O-P(4-1,4-ONC_4H_8)$_2$ . . . . S: S-N Comp.8-319

$C_{12}H_{24}N_3O_5PS_3$ . . $CH_3S-C(CH_3)$=N-OC(O)-N(CH_3)-S(O)
-N(C_2H_5)-P(S)[-$OCH_2C(CH_3)_2CH_2O$-] . . . . . . S: S-N Comp.8-356

– . . . . $CH_3S-C(CH_3)$=N-OC(O)-N(CH_3)-S(O)
-N(C_5H_9-c)-P(S)$(OCH_3)_2$ . . . . S: S-N Comp.8-356

$C_{12}H_{24}N_3O_6Th^+$ . . $[Th(C_2H_5CH(NH_2)COO)_3]^+$ . . . . Th: SVol.D1-82/3, 85

$C_{12}H_{24}N_3S^+$ . . . . . $[(1-C_4H_8N)_3S]^+$ . . . . S: S-N Comp.8-230, 241/2

$C_{12}H_{24}N_4O_4Si$ . . . . $Si(N(CH_3)COCH_3)_4$ . . . . Si: SVol.B4-310

$C_{12}H_{24}N_4O_{18}Th$ . . $Th(NO_3)_4 \cdot 1.5\ [-CH_2CH_2-O-]_4$ . . . . Th: SVol.D4-158

$C_{12}H_{24}N_4O_{18}Th$ .. $Th(NO_3)_4 \cdot [-CH_2CH_2-O-]_6$ .... Th: SVol.D4-158
$C_{12}H_{24}N_6O_{14}Th$ .. $Th(NO_3)_4 \cdot [CH_3-CH(OH)-CH_2-C(CH_3)=N$
$-CH_2CH_2-N=C(CH_3)-CH_2-CH(OH)-CH_3]$ .... Th: SVol.D4-160
$C_{12}H_{24}N_6Pd$ ..... $[(CH_3)_4N]_2[Pd(CN)_4]$ .... Pd: SVol.B2-295
– .... $[(C_2H_5)_2NH_2]_2[Pd(CN)_4]$ .... Pd: SVol.B2-270, 271
$C_{12}H_{24}N_{12}O_{16}Th$.. $[C(NH_2)_3]_4[Th(C_2O_4)_4] \cdot 2\ H_2O$ .... Th: SVol.C7-88, 94
– .... $[C(NH_2)_3]_4[Th(C_2O_4)_4] \cdot 2.06\ H_2O$ .... Th: SVol.C7-88
– .... $[C(NH_2)_3]_4[Th(C_2O_4)_4] \cdot 2.14\ H_2O$ .... Th: SVol.C7-88
– .... $[C(NH_2)_3]_4[Th(C_2O_4)_4] \cdot 2.2\ H_2O$ .... Th: SVol.C7-88
$C_{12}H_{24}O_2Sn$ ..... 2,2-$(n-C_4H_9)_2$-1,3,2-$O_2SnC_4H_6$ .... Sn: Org.Comp.15-58
– .... 2,2-$(n-C_4H_9)_2$-4,5-$(CH_3)_2$-1,3,2-O_2SnC_2 ..... Sn: Org.Comp.15-57
– .... 3,3-$(n-C_4H_9)_2$-[3.2.0]-2,4,3-$O_2SnC_4H_6$ .... Sn: Org.Comp.15-52
– .... $(c-C_6H_{11})_2Sn(OH)_2$ .... Sn: Org.Comp.16-66
$C_{12}H_{24}O_4Sn$ ..... $(n-C_4H_9)_2Sn(OOCCH_3)_2$ .... Sn: Org.Comp.15-104/26
– .... $(i-C_4H_9)_2Sn(OOCCH_3)_2$ .... Sn: Org.Comp.15-370
– .... $(t-C_4H_9)_2Sn(OOCCH_3)_2$ .... Sn: Org.Comp.15-376
$C_{12}H_{24}O_6S_2Sn$ ... $(C_4H_9)_2Sn(OSO_2CHCH_2)_2$ .... Sn: Org.Comp.15-348
$C_{12}H_{24}O_6Sn$ ..... $(n-C_4H_9)_2Sn(OOCOCH_3)_2$ .... Sn: Org.Comp.15-332
– .... $(n-C_4H_9)_2Sn(OOCCH_2OH)_2$ .... Sn: Org.Comp.15-256
$C_{12}H_{24}O_6Th$ ..... $Th(OH)_2(n-C_5H_{11}COO)_2 \cdot 4\ H_2O$ .... Th: SVol.C7-73/5
$C_{12}H_{25}MnN_4O_{12}P_4{}^{5-}$
$[MnH(1,4,7,10-N_4C_8H_{16}-(CH_2PO_3)_4-1,4,7,10)]^{5-}$
Mn: MVol.D8-136/7
$C_{12}H_{25}NOS$ ...... $O=S=N-C(CH_3)_2-C_9H_{19}-n$ .... S: S-N Comp.6-105
– .... $O=S=N-C_{12}H_{25}-n$ .... S: S-N Comp.6-105
$C_{12}H_{25}NOSSn$.... $(C_4H_9)_2Sn(NCS)OC_3H_7-i$ .... Sn: Org.Comp.17-137
$C_{12}H_{25}NOSn$ ..... $(C_4H_9)_2Sn(CN)OC_3H_7-i$ .... Sn: Org.Comp.17-136
$C_{12}H_{25}NO_2S$ ..... $(CH_3)_2N-S(O)O-C_6H_9-2-(C_3H_7-i)-5-CH_3$ ..... S: S-N Comp.8-306/7, 309
$C_{12}H_{25}NO_4Sn$ .... $(C_4H_9)_2SnOCH_2C(CH_3)(NO_2)CH_2O$ .... Sn: Org.Comp.15-56
$C_{12}H_{25}NSn$ ...... $(CH_3)_3SnN(C_6H_{11}-c)CH=CHCH_3$ .... Sn: Org.Comp.18-47, 49, 53
$C_{12}H_{25}N_3Si$ ...... $SiH(-NC_4H_8)_3$ .... Si: SVol.B4-197/8
$C_{12}H_{25}N_4O_7PS_2$ .. $(CH_3)_2N-C(O)-C(SCH_3)=N-OC(O)-N(CH_3)$
$-S(O)-N(CH_3)-P(O)(OC_2H_5)_2$ .... S: S-N Comp.8-356
$C_{12}H_{25}O_3Sb$ ..... $(C_4H_9)_2(C_2H_5O_2CCH_2)SbO$ .... Sb: Org.Comp.5-71
$C_{12}H_{26}INO_2Sn$ ... $(n-C_4H_9)_2Sn(I)-N(C_2H_5)-C(=O)O-CH_3$ .... Sn: Org.Comp.19-127
$C_{12}H_{26}In_2N_2O_2$ ... $(C_2H_5)_2In-O-C(=N-CH_3)-C(=O)-N(CH_3)-In(C_2H_5)_2$
In: Org.Comp.1-290/1
$C_{12}H_{26}MnN_4O_{12}P_4{}^{4-}$
$[MnH_2(1,4,7,10-N_4C_8H_{16}-(CH_2PO_3)_4-1,4,7,10)]^{4-}$
Mn: MVol.D8-136/7
$C_{12}H_{26}MnN_6O_2S_2{}^{2+}$
$[Mn(c-C_5H_8=N-NHC(=S)NH_2)_2(H_2O)_2]^{2+}$ ...... Mn: MVol.D6-342, 343
$C_{12}H_{26}MnN_6O_4{}^{2+}$ $[Mn(c-C_5H_8=N-NHC(O)NH_2)_2(H_2O)_2]^{2+}$ ...... Mn: MVol.D6-328/9
$C_{12}H_{26}MoN_8S_4$ ... $[N_4C_6H_{13}]_2[MoS_4]$ .... Mo: SVol.B7-280/1
$C_{12}H_{26}N_2O_2Sn$ ... $(C_4H_9)_2Sn(ONCHCH_3)_2$ .... Sn: Org.Comp.15-342
$C_{12}H_{26}N_2O_4Sn$ ... $(C_4H_9)_2Sn(ONHCOCH_3)_2$ .... Sn: Org.Comp.15-339
$C_{12}H_{26}N_2O_{12}Th$ .. $[NH_4]_2[Th(CH_3COO)_6]$ .... Th: SVol.C7-51/2
$C_{12}H_{26}N_3O_2S^+$ ... $[(1,4-ONC_4H_8-4-)_2S-N(C_2H_5)_2]^+$ .... S: S-N Comp.8-230, 243
$C_{12}H_{26}N_3O_4PS_3$ .. $CH_3S-C(CH_3)=N-OC(O)-N(CH_3)-S(O)$
$-N(CH_3)-P(O)(C_2H_5)-SC_4H_9-t$ .... S: S-N Comp.8-356

$C_{12}H_{26}N_3O_5PS$ . . . $(1,4\text{-}ONC_4H_8\text{-}4)_2S{=}N\text{-}P(O)(OC_2H_5)_2$ . . . . S: S-N Comp.8-207/8, 213
$C_{12}H_{26}N_3O_5PS_3$ . . $CH_3S\text{-}C(CH_3){=}N\text{-}OC(O)\text{-}N(CH_3)\text{-}S(O)$
$\text{-}N(C_3H_7\text{-}n)\text{-}P(S)(OC_2H_5)_2$ . . . . S: S-N Comp.8-356
– . . . . $CH_3S\text{-}C(CH_3){=}N\text{-}OC(O)\text{-}N(CH_3)\text{-}S(O)$
$\text{-}N(C_3H_7\text{-}i)\text{-}P(S)(OC_2H_5)_2$ . . . . S: S-N Comp.8-356
$C_{12}H_{26}N_6Os^{2+}$ . . . $[2,6\text{-}(CH_3)_2\text{-}NC_5H_3\text{-}4\text{-}Os(NH_3)_4(NC_5H_5)]^{2+}$ . . . Os: Org.Comp.A1-12
$C_{12}H_{26}OSn$ . . . . . . $(C_4H_9)_2Sn(H)OCH_2C(CH_3){=}CH_2$ . . . . Sn: Org.Comp.17-78
$C_{12}H_{26}O_2SSn$ . . . . $(n\text{-}C_4H_9)_2Sn[\text{-}O\text{-}CH_2CH_2\text{-}S\text{-}CH_2CH_2\text{-}O\text{-}]$ . . . . Sn: Org.Comp.15-73
– . . . . $(t\text{-}C_4H_9)_2Sn[\text{-}O\text{-}CH_2CH_2\text{-}S\text{-}CH_2CH_2\text{-}O\text{-}]$ . . . . Sn: Org.Comp.15-375, 376/8
$C_{12}H_{26}O_2Sn$ . . . . . $(n\text{-}C_4H_9)_2Sn[\text{-}O\text{-}C(CH_3)_2\text{-}CH_2\text{-}O\text{-}]$ . . . . Sn: Org.Comp.15-49, 82
– . . . . $(n\text{-}C_4H_9)_2Sn[\text{-}O\text{-}CH(CH_3)\text{-}CH(CH_3)\text{-}O\text{-}]$ . . . . . Sn: Org.Comp.15-50, 83
– . . . . $(n\text{-}C_4H_9)_2Sn[\text{-}O\text{-}CH(CH_3)\text{-}CH_2CH_2\text{-}O\text{-}]$ . . . . . . Sn: Org.Comp.15-54
– . . . . $(n\text{-}C_4H_9)_2Sn[\text{-}O\text{-}CH_2CH_2CH_2CH_2\text{-}O\text{-}]$ . . . . Sn: Org.Comp.15-57
$C_{12}H_{26}O_3Sn$ . . . . . $(n\text{-}C_4H_9)_2Sn[\text{-}O\text{-}CH_2CH_2\text{-}O\text{-}CH_2CH_2\text{-}O\text{-}]$ . . . . Sn: Org.Comp.15-71
– . . . . $(n\text{-}C_4H_9)_2Sn[\text{-}O\text{-}CH_2\text{-}C(CH_3)(OH)\text{-}CH_2\text{-}O\text{-}]$ . . Sn: Org.Comp.15-55
– . . . . $(n\text{-}C_4H_9)_2Sn[\text{-}OO\text{-}C(CH_3)(C_2H_5)\text{-}O\text{-}]$ . . . . Sn: Org.Comp.15-335, 336
$C_{12}H_{26}O_4Sn$ . . . . . $C_4H_9Sn(OC_3H_7\text{-}i)_2OOCCH_3$ . . . . Sn: Org.Comp.17-72
$C_{12}H_{27}I_2MnNOP$ . . $Mn(I)_2(NO)\text{-}P(C_4H_9\text{-}n)_3$ . . . . Mn: MVol.D8-46, 61
$C_{12}H_{27}I_2MnOP$ . . . $Mn(I)_2\text{-}O\text{-}P(C_4H_9\text{-}n)_3$ . . . . Mn: MVol.D8-55, 87
$C_{12}H_{27}I_2MnO_2P$ . . $Mn(I)_2(O_2)\text{-}P(C_4H_9\text{-}n)_3$ . . . . Mn: MVol.D8-55/7
$C_{12}H_{27}I_2MnP$ . . . . $Mn(I)_2(SO_2)_n\text{-}P(C_4H_9\text{-}n)_3$ . . . . Mn: MVol.D8-46, 57/8
– . . . . $Mn(I)_2\text{-}P(C_4H_9\text{-}n)_3$ . . . . Mn: MVol.D8-38
– . . . . $Mn[P(C_4H_9\text{-}t)_3]I_2$ . . . . Mn: MVol.D8-38
$C_{12}H_{27}I_3In_2$ . . . . . . $(n\text{-}C_4H_9)_2In(\text{-}I\text{-})_2InI\text{-}C_4H_9\text{-}n$ . . . . In: Org.Comp.1-167/8, 170
$C_{12}H_{27}In$ . . . . $In(C_4H_9\text{-}n)_3$ . . . . In: Org.Comp.1-73/5, 78/9
– . . . . $In(C_4H_9\text{-}i)_3$ . . . . In: Org.Comp.1-73/5, 79
– . . . . $In(C_4H_9\text{-}s)_3$ . . . . In: Org.Comp.1-73/5
– . . . . $In(C_4H_9\text{-}t)_3$ . . . . In: Org.Comp.1-73/5, 79
$C_{12}H_{27}InO$ . . . . $(n\text{-}C_4H_9)_2In\text{-}O\text{-}C_4H_9\text{-}t$ . . . . In: Org.Comp.1-181, 183, 186
$C_{12}H_{27}InO_2$ . . . . . . $(t\text{-}C_4H_9)_2In\text{-}O\text{-}O\text{-}C_4H_9\text{-}t$ . . . . In: Org.Comp.1-210
$C_{12}H_{27}In_3$ . . . . $[CH_2{=}CH\text{-}In(CH_3)_2]_3$ . . . . In: Org.Comp.1-103
$C_{12}H_{27}MnN_3O_3P$ . . $Mn(NO)_3[P(C_4H_9\text{-}n)_3]$ . . . . Mn: MVol.D8-64/7
– . . . . $Mn(NO)_3[P(C_4H_9\text{-}t)_3]$ . . . . Mn: MVol.D8-64/7
$C_{12}H_{27}MnN_4O_{12}P_4^{3-}$
$[MnH_3(1,4,7,10\text{-}N_4C_8H_{16}\text{-}(CH_2PO_3)_4\text{-}1,4,7,10)]^{3-}$
Mn: MVol.D8-136/7
$C_{12}H_{27}NOSSn$ . . . . $O{=}S{=}NSn(C_4H_9\text{-}n)_3$ . . . . S: S-N Comp.6-86/7
$C_{12}H_{27}NOSn$ . . . . . $2,2\text{-}(n\text{-}C_4H_9)_2\text{-}3\text{-}C_2H_5\text{-}1,3,2\text{-}ONSnC_2H_4$ . . . . . Sn: Org.Comp.19-135
$C_{12}H_{27}NO_2Sn$ . . . . $(n\text{-}C_4H_9)_2Sn(O\text{-}C_2H_5)\text{-}O\text{-}N{=}CHCH_3$ . . . . Sn: Org.Comp.16-190
– . . . . $(n\text{-}C_4H_9)_2Sn[\text{-}O\text{-}CH_2CH_2\text{-}NH\text{-}CH_2CH_2\text{-}O\text{-}]$ . . . Sn: Org.Comp.15-71
– . . . . $(i\text{-}C_4H_9)_2Sn[\text{-}O\text{-}CH_2CH_2\text{-}NH\text{-}CH_2CH_2\text{-}O\text{-}]$ . . . Sn: Org.Comp.15-370
– . . . . $(t\text{-}C_4H_9)_2Sn[\text{-}O\text{-}CH_2CH_2\text{-}NH\text{-}CH_2CH_2\text{-}O\text{-}]$ . . . Sn: Org.Comp.15-374, 376/7
– . . . . $(n\text{-}C_4H_9)_2Sn[\text{-}O\text{-}CH_2\text{-}C(CH_3)(NH_2)\text{-}CH_2\text{-}O\text{-}]$ . . Sn: Org.Comp.15-55
$C_{12}H_{27}NSn$ . . . . . . $(CH_3)_3Sn\text{-}N{=}C(C_4H_9\text{-}t)_2$ . . . . Sn: Org.Comp.18-107
– . . . . $(CH_3)_3Sn\text{-}N(C_2H_5)CH{=}CH\text{-}C_5H_{11}\text{-}n$ . . . . Sn: Org.Comp.18-47, 49, 53
$C_{12}H_{27}N_2PS$ . . . . . $(t\text{-}C_4H_9)_2PN{=}S{=}N\text{-}C_4H_9\text{-}t$ . . . . S: S-N Comp.7-103
$C_{12}H_{27}N_7O_{15}Th$ . . $Th(NO_3)_4 \cdot 3\ CH_3\text{-}C({=}O)\text{-}N(CH_3)_2 \cdot 3\ H_2O$ . . . Th: SVol.D4-134
– . . . . $Th(NO_3)_4 \cdot 3\ CH_3\text{-}C({=}O)\text{-}NH\text{-}C_2H_5$ . . . . Th: SVol.D4-159
$C_{12}H_{27}O_3P_2ReSSn$
$(CO)_3Re[P(CH_3)_3]_2SSn(CH_3)_3$ . . . . Re: Org.Comp.1-257

$C_{12}H_{27}O_5PTh^{2+}$	$Th(OH)[O_2P(OC_6H_{13})_2]^{2+}$	Th: SVol.D1-131
$C_{12}H_{27}O_6SSb_2$	$[((H_2O)(CH_3)_3Sb)_2O][O_3S-C_6H_5]$	Sb: Org.Comp.5-107, 111, 112
$C_{12}H_{27}P_2Re$	$(C_5H_5)ReH[P(CH_3)_3]_2-CH_3$	Re: Org.Comp.3-40/2
$C_{12}H_{28}InN$	$(CH_3)_3In \cdot NC_5H_7-2,2,6,6-(CH_3)_4$	In: Org.Comp.1-27, 33, 35, 42/3
$C_{12}H_{28}InP$	$(C_2H_5)_2In-P(C_4H_9-t)_2$	In: Org.Comp.1-312, 314, 318
$C_{12}H_{28}LiPSn$	$(t-C_4H_9)_3Sn-PH-Li$	Sn: Org.Comp.19-193/4
$C_{12}H_{28}MnN_4O_{12}P_4^{2-}$	$[MnH_4(1,4,7,10-N_4C_8H_{16}-(CH_2PO_3)_4-1,4,7,10)]^{2-}$	Mn: MVol.D8-136/7
$C_{12}H_{28}MnO_4P_2S_4$	$Mn[S_2P(OC_3H_7-i)_2]_2$	Mn: MVol.D8-193/5
$C_{12}H_{28}MnO_8P_2$	$[Mn(O_2P(O-C_3H_7-n)_2)_2]$	Mn: MVol.D8-155/6
–	$[Mn(O_2P(O-C_3H_7-n)_2)_2]_n$	Mn: MVol.D8-155
$C_{12}H_{28}Mo_4O_{25}^{2-}$	$[Mo_4O_{13}(C_6H_{14}O_6)_2]^{2-}$	Mo: SVol.B3b-170/1
$C_{12}H_{28}N_2OS$	$(n-C_3H_7)_2N-S(O)-N(C_3H_7-n)_2$	S: S-N Comp.8-345
–	$(i-C_3H_7)_2N-S(O)-N(C_3H_7-i)_2$	S: S-N Comp.8-345, 359
$C_{12}H_{28}N_2O_3PS$	$OP(O-C_2H_5)_2N(C_4H_9-t)-SN-C_4H_9-t$, radical	S: S-N Comp.7-331/2
$C_{12}H_{28}N_2O_3Si_2Sn$	$O(Si(CH_3)_2CH_2)_2Sn(ON{=}C(CH_3)_2)_2$	Sn: Org.Comp.16-220
$C_{12}H_{28}N_2P_2S$	$(t-C_4H_9)_2PN{=}S{=}NP(C_2H_5)_2$	S: S-N Comp.7-110/2
$C_{12}H_{28}N_2Sn$	$1,3-(CH_3)_2-2,2-(n-C_4H_9)_2-1,3,2-N_2SnC_2H_4$	Sn: Org.Comp.19-96/8
$C_{12}H_{28}N_3OS^+$	$[((C_2H_5)_2N)_2S-4-1,4-ONC_4H_8]^+$	S: S-N Comp.8-230, 241
$C_{12}H_{28}N_4Si$	$Si(-NC_2H_5-CH_2-CH_2-NC_2H_5-)_2$	Si: SVol.B4-225/6
$C_{12}H_{28}N_6O_3Si$	$SiH[N(CH_3)C(O)N(CH_3)_2]_3$	Si: SVol.B4-312
$C_{12}H_{28}O_2Sn$	$(C_4H_9)_2Sn(OC_2H_5)_2$	Sn: Org.Comp.15-3, 14, 15/7
$C_{12}H_{28}O_2Ti$	$(C_2H_5)_2Ti(O-C_4H_9-n)_2$	Ti: Org.Comp.5-9
–	$(n-C_4H_9)_2Ti(O-C_2H_5)_2$	Ti: Org.Comp.5-9
$C_{12}H_{28}O_3Sn$	$(C_4H_9)_2Sn(OCH_3)OCH(CH_3)OCH_3$	Sn: Org.Comp.16-187
$C_{12}H_{28}O_4Th$	$Th(O-C_3H_7-i)_4$	Th: SVol.C7-25/9
$C_{12}H_{28}O_4Ti_2$	$[(CH_3)_2Ti(OCH(CH_3)CH_2CH_2O)]_2$	Ti: Org.Comp.5-11
$C_{12}H_{28}Pb$	$Pb(C_3H_7-n)_4$	Pb: Org.Comp.3-5/20
–	$Pb(C_3H_7-i)_4$	Pb: Org.Comp.3-20/6
$C_{12}H_{28}S_2Sb_2$	$[(C_2H_5)_2SbS]_2(CH_2)_4$	Sb: Org.Comp.5-127
$C_{12}H_{28}S_4Sb_2$	$[(CH_3)_2SbS]_2[-(CH_2)_3S(CH_2)_2S(CH_2)_3-]$	Sb: Org.Comp.5-127
$C_{12}H_{29}InSn$	$(CH_3)_3Sn-CH_2-In(C_4H_9-t)_2$	In: Org.Comp.1-101, 104
$C_{12}H_{29}NOSiSn$	$(C_2H_5)_3SnN(CH_3)COCH_2Si(CH_3)_3$	Sn: Org.Comp.18-137
$C_{12}H_{29}NSn$	$(n-C_4H_9)_3Sn-NH_2$	Sn: Org.Comp.18-160
–	$(t-C_4H_9)_3Sn-NH_2$	Sn: Org.Comp.18-233/4
$C_{12}H_{29}O_3Sb$	$(C_3H_7)_2Sb(OC_2H_5)_3$	Sb: Org.Comp.5-179
$C_{12}H_{29}PSn$	$(t-C_4H_9)_3Sn-PH_2$	Sn: Org.Comp.19-190, 192
$C_{12}H_{30}IInN_4$	$[N(CH_3)_4][(C_4H_9)_2In(N_3)(I)]$	In: Org.Comp.1-363
$C_{12}H_{30}IMnN_2O_2P_2$	$Mn(NO)_2[P(C_2H_5)_3]_2I$	Mn: MVol.D8-61/4
$C_{12}H_{30}IMnN_2O_8P_2$	$[Mn(NO)_2(P(OC_2H_5)_3)_2I]$	Mn: MVol.D8-145
$C_{12}H_{30}IN_3S$	$[((C_2H_5)_2N)_3S]I$	S: S-N Comp.8-230, 240
$C_{12}H_{30}I_2MnP_2$	$Mn[P(C_2H_5)_3]_2I_2$	Mn: MVol.D8-49/50
$C_{12}H_{30}InNOSi$	$(CH_3)_2In[-N(C_4H_9-t)-Si(CH_3)_2-O(C_4H_9-t)-]$	In: Org.Comp.1-283, 286/7
$C_{12}H_{30}InN_3$	$(CH_3)_3In \cdot 1,3,5-N_3C_3H_6-1,3,5-(C_2H_5)_3$	In: Org.Comp.1-27, 33, 36, 44
$C_{12}H_{30}In_2N_4$	$2\ (CH_3)_3In \cdot N_4C_6H_{12}$	In: Org.Comp.1-49, 50
–	$[(CH_3)_2In-N(CH_3)-C(CH_3){=}NCH_3]_2$	In: Org.Comp.1-268, 270
$C_{12}H_{30}In_2O_2$	$[-In(CH_3)_2-O(C_4H_9-t)-]_2$	In: Org.Comp.1-181/2
$C_{12}H_{30}MnN_3O_4P_2$	$Mn(NO)_2[P(C_2H_5)_3]_2NO_2$	Mn: MVol.D8-61/4

$C_{12}H_{30}MnO_6S_6^{2+}$	$[Mn(O{=}S(CH_3)CH_2CH_2S(CH_3){=}O)_3]^{2+}$	Mn:MVol.D7-108/9
$C_{12}H_{30}MnO_{15}P_4$	$[Mn(O_5P_2(OC_2H_5)_2)(O_2P(OC_2H_5)_2)_2]$	Mn:MVol.D8-168/9
$C_{12}H_{30}Mn_2S_6^{2-}$	$[Mn_2(SC_2H_5)_6]^{2-}$	Mn:MVol.D7-24/7
$C_{12}H_{30}N_2O_2P_2PtS_2$	$Pt(N{=}S{=}O)_2(P(C_2H_5)_3)_2$	S: S-N Comp.6-259/64
$C_{12}H_{30}N_2O_7Sb_2$	$[(C_2H_5)_3SbNO_3]_2O$	Sb: Org.Comp.5-100
$C_{12}H_{30}N_2SSn$	$(CH_3)_3SnN(C_4H_9\text{-}t)S(CH_3){=}NC_4H_9\text{-}t$	Sn: Org.Comp.18-57, 62, 73
$C_{12}H_{30}N_2SiSn$	1,3-(t-C_4H_9)$_2$-2,2,4,4-(CH_3)$_4$-1,3,2,4-N_2SiSn	Sn: Org.Comp.19-75, 77, 80
$C_{12}H_{30}N_2Sn$	$(CH_3)_2N\text{-}Sn(C_4H_9\text{-}n)_2\text{-}N(CH_3)_2$	Sn: Org.Comp.19-86/91
–	$(CH_3)_2N\text{-}Sn(C_4H_9\text{-}t)_2\text{-}N(CH_3)_2$	Sn: Org.Comp.19-98/101
–	$(C_2H_5)_2N\text{-}Sn(C_2H_5)_2\text{-}N(C_2H_5)_2$	Sn: Org.Comp.19-81/2
$C_{12}H_{30}N_3S^+$	$[((C_2H_5)_2N)_3S]^+$	S: S-N Comp.8-230, 240/1
$C_{12}H_{30}N_4O_{15}S_3Th$	$Th(NO_3)_4 \cdot 3\,(C_2H_5)_2SO$	Th: SVol.D4-141
$C_{12}H_{30}N_6O_{12}Th$	$Th(NO_3)_4 \cdot 2\,(C_2H_5)_3N \cdot 2\,H_2O$ Photoemission spectra	Th: SVol.A4-131
$C_{12}H_{30}O_6P_2S_2Sn$	$(C_4H_9)_2Sn(OP(S)(OCH_3)_2)_2$	Sn: Org.Comp.15-356, 357
$C_{12}H_{31}InN_2Si$	$(CH_3)_2In[\text{-}N(C_4H_9\text{-}t)\text{-}Si(CH_3)_2\text{-}NH(C_4H_9\text{-}t)\text{-}]$	In: Org.Comp.1-284, 287/8
–	$(CH_3)_2In[\text{-}N(C_4H_9\text{-}t)\text{-}Si(CH_3)_2\text{-}ND(C_4H_9\text{-}t)\text{-}]$	In: Org.Comp.1-284, 287/8
$C_{12}H_{31}InOSi_2$	$[(CH_3)_3Si\text{-}CH_2]_2In\text{-}O\text{-}C_4H_9\text{-}t$	In: Org.Comp.1-183
$C_{12}H_{31}N_3Si$	$SiH(NH\text{-}C_4H_9\text{-}t)_3$	Si: SVol.B4-190
–	$SiH[N(C_2H_5)_2]_3$	Si: SVol.B4-194/6
$C_{12}H_{32}I_2MnP_4$	$Mn((CH_3)_2PCH_2CH_2P(CH_3)_2)_2I_2$	Mn:MVol.D8-77
$C_{12}H_{32}InPSi_2$	$[(CH_3)_3Si\text{-}CH_2]_2In\text{-}PH\text{-}C_4H_9\text{-}t$	In: Org.Comp.1-312, 315, 319
$C_{12}H_{32}In_2N_2$	$[(CH_3)_2In\text{-}N(C_2H_5)_2]_2$	In: Org.Comp.1-260
$C_{12}H_{32}In_2P_2$	$[\text{-}In(CH_3)_2\text{-}CH_2\text{-}P(CH_3)_2\text{-}CH_2\text{-}In(CH_3)_2\text{-}CH_2\text{-}P(CH_3)_2\text{-}CH_2\text{-}]$	In: Org.Comp.1-367
$C_{12}H_{32}MnN_2S_4$	$[N(CH_3)_4]_2[Mn(SCH_2CH_2S)_2] \cdot 0.5\,NCCH_3$	Mn:MVol.D7-38/9
$C_{12}H_{32}MnN_8S_4^{2+}$	$[Mn(S{=}C(NHCH_3)_2)_4]^{2+}$	Mn:MVol.D7-194/5
$C_{12}H_{32}MoN_2S_4$	$[(C_2H_5)_3NH]_2[MoS_4]$	Mo:SVol.B7-289
$C_{12}H_{32}NPSiSn$	$(CH_3)_3SnN(P(C_3H_7\text{-}i)_2)Si(CH_3)_3$	Sn: Org.Comp.18-76, 80
$C_{12}H_{32}N_2Si_2Sn$	1,3-[$(CH_3)_3Si$]$_2$-2,2-(CH_3)$_2$-1,3,2-$N_2SnC_4H_8$	Sn: Org.Comp.19-76/7
–	$(CH_3)_2Sn[\text{-}N(C_3H_7\text{-}i)\text{-}Si(CH_3)_2\text{-}Si(CH_3)_2\text{-}N(C_3H_7\text{-}i)\text{-}]$	Sn: Org.Comp.19-75, 78
$C_{12}H_{32}N_4Si$	$Si(NH\text{-}C_3H_7\text{-}n)_4$	Si: SVol.B4-201
–	$Si(NH\text{-}C_3H_7\text{-}i)_4$	Si: SVol.B4-202
–	$[(CH_3)_2N]_2Si[N(C_2H_5)_2]_2$	Si: SVol.B4-221/2
$C_{12}H_{32}N_4Si_2$	$(CH_3)_2NSiH[\text{-}N(C_4H_9\text{-}t)\text{-}SiHN(CH_3)_2\text{-}N(C_4H_9\text{-}t)\text{-}]$	Si: SVol.B4-245/6
$C_{12}H_{33}InNPSi$	$(CH_3)_3In \cdot (CH_3)_3Si\text{-}N{=}P(C_2H_5)_3$	In: Org.Comp.1-27, 37
$C_{12}H_{33}InNPSn$	$(CH_3)_3In \cdot (CH_3)_3Sn\text{-}N{=}P(C_2H_5)_3$	In: Org.Comp.1-27, 38
$C_{12}H_{33}InSi_3$	$In[CH_2\text{-}Si(CH_3)_3]_3$	In: Org.Comp.1-73, 76/7, 80
$C_{12}H_{33}InSn_3$	$In[CH_2\text{-}Sn(CH_3)_3]_3$	In: Org.Comp.1-73, 77
$C_{12}H_{33}NOsSi_3$	$[(CH_3)_3Si\text{-}CH_2]_3Os(N)$	Os: Org.Comp.A1-26
$C_{12}H_{33}N_3O_4S_2$	$[n\text{-}C_4H_9NH_3]_2[OS(O)\text{-}N(C_4H_9\text{-}n)\text{-}SO_2]$	S: S-N Comp.8-328/9
$C_{12}H_{33}N_4PSn$	$(C_2H_5)_3SnN{=}P(N(CH_3)_2)_3$	Sn: Org.Comp.18-151, 153
$C_{12}H_{33}O_2ReSi_3$	$[(CH_3)_3SiCH_2]_3ReO_2$	Re: Org.Comp.1-3
$C_{12}H_{34}InKSi_3$	$K[HIn(CH_2\text{-}Si(CH_3)_3)_3]$	In: Org.Comp.1-346/7
–	$K[DIn(CH_2\text{-}Si(CH_3)_3)_3]$	In: Org.Comp.1-346/7
$C_{12}H_{34}InNaSi_3$	$Na[HIn(CH_2\text{-}Si(CH_3)_3)_3]$	In: Org.Comp.1-346/7
$C_{12}H_{34}InSi_3^-$	$[HIn(CH_2\text{-}Si(CH_3)_3)_3]^-$	In: Org.Comp.1-346/7

$C_{12}H_{34}InSi_3^-$ ..... $[DIn(CH_2{-}Si(CH_3)_3)_3]^-$ ..... In: Org.Comp.1-346/7
$C_{12}H_{34}In_2N_2$ ..... $(CH_3)_3In \cdot N(CH_3)_2{-}CH_2CH_2{-}N(CH_3)_2 \cdot In(CH_3)_3$ In: Org.Comp.1-49, 50
$C_{12}H_{34}N_6Si$ ..... $SiH(NHN(C_2H_5)_2)_3$ ..... Si: SVol.B4-258
$C_{12}H_{34}N_{22}O_2Pd_3Rh_2$ $[Rh(NH_3)_5(H_2O)]_2[Pd(CN)_4]_3$ ..... Pd: SVol.B2-280
$C_{12}H_{34}O_6OsS_6$ ... $[CH_3{-}S(=O){-}CH_2]_2Os[O{=}S(CH_3)_2]_4 \cdot 2\,(CH_3)_2S{=}O$ Os: Org.Comp.A1-16
$C_{12}H_{35}I_4N_5O_{10}Th_2$ $[ThI_2(H_2O)_3]_2(OH)_2 \cdot [CH_3{-}C(=N{-}O){-}C(CH_3){=}N{-}CH_2CH_2{-}NH{-}CH_2CH_2{-}N{=}C(CH_3){-}C(=N{-}O){-}CH_3]$ ..... Th: SVol.D4-158
$C_{12}H_{35}MnP_4$ ..... $MnH_3((CH_3)_2PCH_2CH_2P(CH_3)_2)_2$ ..... Mn: MVol.D8-78
$C_{12}H_{35}N_9O_{22}Th_2$ .. $[Th(NO_3)_2(H_2O)_3]_2(OH)_2 \cdot [CH_3{-}C(=N{-}O){-}C(CH_3){=}N{-}CH_2CH_2{-}NH{-}CH_2CH_2{-}N{=}C(CH_3){-}C(=N{-}O){-}CH_3]$ ..... Th: SVol.D4-158
$C_{12}H_{36}I_2MnN_6O_2P_2$ $[Mn(O{=}P(N(CH_3)_2)_3)_2I_2]$ ..... Mn: MVol.D8-174/5
$C_{12}H_{36}I_2MnO_6S_6$. $[Mn(O{=}S(CH_3)_2)_6]I_2$ ..... Mn: MVol.D7-96
$C_{12}H_{36}InO_2SbSi_2$ $[Sb(CH_3)_4][(CH_3)_2In(OSi(CH_3)_3)_2]$ ..... In: Org.Comp.1-364
$C_{12}H_{36}In_3P_3$ ..... $[(CH_3)_2In{-}P(CH_3)_2]_3$ ..... In: Org.Comp.1-313
$C_{12}H_{36}In_4K_2S$ .... $K_2[In_4(CH_3)_{12}S]$ ..... In: Org.Comp.1-364/6
$C_{12}H_{36}In_4K_2Se$ ... $K_2[In_4(CH_3)_{12}Se]$ ..... In: Org.Comp.1-365/6
$C_{12}H_{36}In_4S^{2-}$ .... $[In_4(CH_3)_{12}S]^{2-}$ ..... In: Org.Comp.1-364/6
$C_{12}H_{36}In_4Se^{2-}$ ... $[In_4(CH_3)_{12}Se]^{2-}$ ..... In: Org.Comp.1-365/6
$C_{12}H_{36}MnN_2O_{12}S_6$ $[Mn(O{=}S(CH_3)_2)_6](NO_3)_2$ ..... Mn: MVol.D7-94
$C_{12}H_{36}MnN_6O_{16}P_4$ $[Mn(O{=}P(OCH_3)_3)_4(N_3)_2]$ ..... Mn: MVol.D8-160
$C_{12}H_{36}MnN_8O_6P_2$ $[Mn(O{=}P(N(CH_3)_2)_3)_2(NO_2)_2]$ ..... Mn: MVol.D8-174/5
$C_{12}H_{36}MnN_8O_8P_2$ $[Mn(O{=}P(N(CH_3)_2)_3)_2(NO_3)_2]$ ..... Mn: MVol.D8-174/5
$C_{12}H_{36}MnN_9O_9P_3$ $[Mn((O{=}P(N(CH_3)_2)_2NCH_3)_2P(N(CH_3)_2){=}O)(NO_3)_2]$ Mn: MVol.D8-181
$C_{12}H_{36}MnN_{12}S_6^{2+}$ $[Mn(S{=}C(NH_2)NHCH_3)_6]^{2+}$ ..... Mn: MVol.D7-194/5
$C_{12}H_{36}MnO_6S_6^{2+}$ $[Mn(O{=}S(CH_3)_2)_6]^{2+}$ ..... Mn: MVol.D7-94/7
$C_{12}H_{36}MnO_6S_6^{3+}$ $[Mn(O{=}S(CH_3)_2)_6]^{3+}$ ..... Mn: MVol.D7-99/100
$C_{12}H_{36}MnO_6Se_6^{2+}$ $[Mn(OSe(CH_3)_2)_6]^{2+}$ ..... Mn: MVol.D7-243
$C_{12}H_{36}MnO_{13}S_8$ .. $[Mn(O{=}S(CH_3)_2)_6]S_2O_7$ ..... Mn: MVol.D7-96
$C_{12}H_{36}Mn_2N_4O_{18}S_6$ $[Mn(O{=}S(CH_3)_2)_6][Mn(NO_3)_4]$ ..... Mn: MVol.D7-94
$C_{12}H_{36}MoO_{12}S_7^{2+}$ $[MoO_2(CH_3SOCH_3)_6(SO_4)]^{2+}$ ..... Mo: SVol.A2b-288
$C_{12}H_{36}N_2Si_3Sn$ ... $(CH_3)_3SnN(Si(CH_3)_3)N(Si(CH_3)_3)_2$ ..... Sn: Org.Comp.18-121, 123, 125
$C_{12}H_{36}N_3O_{13}P_4Th^+$ $[((CH_3)_3PO)_4Th(NO_3)_3]^+$ ..... Th: SVol.D4-207
$C_{12}H_{36}N_4O_{16}P_4Th$ $Th(NO_3)_4 \cdot 4\,(CH_3)_3PO$ ..... Th: SVol.D4-141
$C_{12}H_{36}N_4O_{18}S_6Th$ $Th(NO_3)_4 \cdot 6\,(CH_3)_2SO$ ..... Th: SVol.D4-160
$C_{12}H_{36}N_4O_{28}P_4Th$ $Th(NO_3)_4 \cdot 4\,(CH_3O)_3P{=}O$ ..... Th: SVol.D4-191/2

$C_{12}H_{36}N_4Si_3Sn$ ... $(CH_3)_3SnN(Si(CH_3)_3)N{=}NN(Si(CH_3)_3)_2$ .... Sn: Org.Comp.18-121, 124, 126
$C_{12}H_{36}N_6Si_2$ ..... $Si_2(N(CH_3)_2)_6$ .... Si: SVol.B4-255
$C_{12}H_{36}N_6Si_3$ ..... $[SiH(NHC_2H_5)N(C_2H_5)]_3$ .... Si: SVol.B4-249
$C_{12}H_{36}N_7NaSi_2$ ... $NaN[Si(N(CH_3)_2)_3]_2$ .... Si: SVol.B4-243
$C_{12}H_{36}N_{10}O_{14}P_2Th$
$Th(NO_3)_4 \cdot 2\ [(CH_3)_2N]_3P{=}O$ .... Th: SVol.D4-142, 207
$C_{12}H_{36}N_{10}O_{16.5}P_3Th$
$Th(NO_3)_4 \cdot 1.5\ [(CH_3)_2N]_2P({=}O){-}O{-}P({=}O)[N(CH_3)_2]_2$
Th: SVol.D4-142
$C_{12}H_{37}N_7Si_2$ ..... $[Si(N(CH_3)_2)_3]_2NH$ .... Si: SVol.B4-242/3
$C_{12}H_{38}MnO_{14}S_8$ .. $[Mn(O{=}S(CH_3)_2)_6](HSO_4)_2$ .... Mn: MVol.D7-96
$C_{12}H_{39}N_9Si_3$ ..... $[Si(N(CH_3)_2)_2NH]_3$ .... Si: SVol.B4-250
$C_{12}K_2N_{10}Pd$ ..... $K_2[Pd(CN)_2(N_2C_3(CN)_2{-}N_2C_3(CN)_2)]$ .... Pd: SVol.B2-286
$C_{12}K_7Mo_{2.5}N_{12}O_2S$
$K_7[Mo_2S(CN)_{12}][MoO_4]_{0.5} \cdot 5\ H_2O$ .... Mo: SVol.B3b-196
$C_{12}MnN_6S_6^{2-}$ .... $[Mn(SC(CN){=}C(CN)S)_3]^{2-}$ .... Mn: MVol.D7-33/8, 48
$C_{12}MnN_6S_6^{3-}$ .... $[Mn(SC(CN){=}C(CN)S)_3]^{3-}$ .... Mn: MVol.D7-33/8
– .... $[Mn(S_2C{=}C(CN)_2)_3]^{3-}$ .... Mn: MVol.D7-31/2
$C_{12}MnN_6S_6^{4-}$ .... $[Mn(SC(CN){=}C(CN)S)_3]^{4-}$ .... Mn: MVol.D7-33/8
$C_{12}Mn_2S_{12}^{2-}$ ..... $[Mn_2(C_4({=}S)_4)_3]^{2-}$ .... Mn: MVol.D7-57
$C_{12}Mo_2N_{12}S^{6-}$ ... $[(CN)_6MoSMo(CN)_6]^{6-}$ .... Mo: SVol.B3b-189, 196
$C_{12}Mo_4N_{12}S_4^{6-}$ ... $[Mo_4S_4(CN)_{12}]^{6-}$ .... Mo: SVol.B3b-189
$C_{12}N_9O_{12}Th^{5-}$ .... $Th[2,4,6{-}(O)_3{-}5{-}(ON){-}1,3{-}N_2C_4]_3^{5-}$ .... Th: SVol.D1-120
$C_{12}N_9Ti$ .... $[((NC)_3C)_3Ti]_n$ .... Ti: Org.Comp.5-324
$C_{12}O_{12}Os_2Re^-$ ... $[(CO)_4Re({-}Os(CO)_4{-}Os(CO)_4{-})]^-$ .... Re: Org.Comp.1-502
$C_{12}O_{12}ReV$ ...... $[Re(CO)_6][V(CO)_6]$ .... Re: Org.Comp.2-218
$C_{12}O_{24}Th^{8-}$ ...... $[Th(C_2O_4)_6]^{8-}$ .... Th: SVol.C7-91
$C_{12.5}H_{27.5}N_{6.5}O_{14.5}Th$
$Th(NO_3)_4 \cdot 2.5\ (CH_3)_2N{-}C({=}O){-}C_2H_5$ .... Th: SVol.D4-134
$C_{13}ClF_3FeH_5LiO_2$ $(C_5H_5)Fe(CO)_2C_6(Cl{-}3)(Li{-}5)F_3{-}2,4,6$ .... Fe: Org.Comp.B13-177, 187
$C_{13}ClF_3FeH_6O_2$ .. $(C_5H_5)Fe(CO)_2C_6H(Cl{-}5)F_3{-}2,4,6$ .... Fe: Org.Comp.B13-176, 186
$C_{13}ClF_3H_{24}N_2O_2S$ $CF_3SCl{-}(4{-}1,4{-}ONC_4H_6{-}3,5{-}(CH_3)_2)_2$ .... S: S-N Comp.8-401, 403
$C_{13}ClF_4FeH_5O_2$ .. $(C_5H_5)Fe(CO)_2C_6F_4Cl{-}3$ .... Fe: Org.Comp.B13-174, 184
$C_{13}ClF_5H_{30}OP_2PtSi$
$[PtCl(CO)(P(C_2H_5)_3)_2][SiF_5]$ .... Si: SVol.B7-272
$C_{13}ClF_6FeH_9O_3$ .. $(C_5H_5)Fe(CO)_2CH[{-}CHCl{-}C(CF_3)_2{-}O{-}CH_2{-}]$ ... Fe: Org.Comp.B14-2, 4
$C_{13}ClF_6FeH_{11}NP$. $[(3{-}Cl{-}2{-}CH_3{-}C_6H_3{-}CN)Fe(C_5H_5)][PF_6]$ ...... Fe: Org.Comp.B19-131
$C_{13}ClF_6FeH_{12}O_2P$ $[(2{-}Cl{-}C_6H_4{-}CH_3)Fe(C_5H_4{-}COOH)][PF_6]$ ..... Fe: Org.Comp.B19-2, 33
– .... $[(3{-}Cl{-}C_6H_4{-}COO{-}CH_3)Fe(C_5H_5)][PF_6]$ .... Fe: Org.Comp.B19-5, 45
– .... $[(4{-}Cl{-}C_6H_4{-}COO{-}CH_3)Fe(C_5H_5)][PF_6]$ .... Fe: Org.Comp.B19-5, 45
$C_{13}ClF_6FeH_{13}NO_2P$
$[(2{-}Cl{-}C_6H_4{-}CH(CH_3){-}NO_2)Fe(C_5H_5)][PF_6]$ ... Fe: Org.Comp.B19-3, 41
$C_{13}ClF_6FeH_{14}P$ ... $[(1,2{-}(CH_3)_2{-}C_6H_3{-}4{-}Cl)Fe(C_5H_5)][PF_6]$ ...... Fe: Org.Comp.B19-100, 129
– .... $[(1,3{-}(CH_3)_2{-}C_6H_3{-}2{-}Cl)Fe(C_5H_5)][PF_6]$ ...... Fe: Org.Comp.B19-99, 128
– .... $[(1,3{-}(CH_3)_2{-}C_6H_3{-}4{-}Cl)Fe(C_5H_5)][PF_6]$ ...... Fe: Org.Comp.B19-100, 129
– .... $[(2{-}Cl{-}C_6H_4{-}C_2H_5)Fe(C_5H_5)][PF_6]$ .... Fe: Org.Comp.B19-35
– .... $[(3{-}Cl{-}C_6H_4{-}CH_3)Fe(C_5H_4{-}CH_3)][PF_6]$ .... Fe: Org.Comp.B19-2, 5/6, 33
– .... $[(4{-}Cl{-}C_6H_4{-}CH_3)Fe(C_5H_4{-}CH_3)][PF_6]$ .... Fe: Org.Comp.B19-2, 5/6, 34

$C_{13}ClF_6FeH_{14}P$ . . . $[(Cl-C_6H_5)Fe(C_5H_4-C_2H_5)][PF_6]$ . . . Fe: Org.Comp.B18-142/6, 246
$C_{13}ClF_6FeH_{15}NP$. $[(3-Cl-2-CH_3-C_6H_3-NH-CH_3)Fe(C_5H_5)][PF_6]$ . . Fe: Org.Comp.B19-100, 131
– . . . $[(2-Cl-C_6H_4-N(CH_3)_2)Fe(C_5H_5)][PF_6]$ . . . Fe: Org.Comp.B19-75
– . . . $[(3-Cl-C_6H_4-N(CH_3)_2)Fe(C_5H_5)][PF_6]$ . . . Fe: Org.Comp.B19-3, 5, 75
– . . . $[(4-Cl-C_6H_4-N(CH_3)_2)Fe(C_5H_5)][PF_6]$ . . . Fe: Org.Comp.B19-3, 5, 75
$C_{13}ClF_6H_{12}N_2O_3OsP$
$[(Cl)Os(CO)(CH_3O-4-NC_5H_3-2-(2-C_5H_3N-$
$4-OCH_3))][PF_6]$ . . . Os: Org.Comp.A1-65, 76
$C_{13}ClF_7H_4N_2S$ . . . $4-Cl-C_6H_4-N{=}S{=}N-C_6F_4-CF_3-4$ . . . S: S-N Comp.7-260
$C_{13}ClF_{10}HN_2$ . . . $C_6F_5-C(Cl){=}N-NH-C_6F_5$ . . . F: PFHOrg.SVol.5-207
$C_{13}ClF_{10}N$ . . . $C_6F_5-N{=}CCl-C_6F_5$ . . . F: PFHOrg.SVol.6-193, 207
$C_{13}ClF_{10}NO$ . . . $(C_6F_5)_2N-CCl{=}O$ . . . F: PFHOrg.SVol.6-227, 238
– . . . $C_6F_5-N{=}CCl-O-C_6F_5$ . . . F: PFHOrg.SVol.6-194, 208
$C_{13}ClF_{12}H_5Mo$ . . . $(C_5H_5)Mo(Cl)(F_3C-C{\equiv}C-CF_3)_2$ . . . Mo: Org.Comp.6-134, 143/4
$C_{13}ClFeH_9N_2$ . . . $(C_5H_5)Fe[C_6H_4-1-Cl-2,6-(CN)_2]$ . . . Fe: Org.Comp.B17-258
$C_{13}ClFeH_9O_2$ . . . $(C_5H_5)Fe(CO)_2C_6H_4Cl-2$ . . . Fe: Org.Comp.B13-171
– . . . $(C_5H_5)Fe(CO)_2C_6H_4Cl-4$ . . . Fe: Org.Comp.B13-171, 182
$C_{13}ClFeH_{11}N^+$ . . . $[(3-Cl-2-CH_3-C_6H_3-CN)Fe(C_5H_5)][PF_6]$ . . . Fe: Org.Comp.B19-131
$C_{13}ClFeH_{12}NO_5S$ $[(C_5H_5)Fe(CO)_2(C_5H_7(C({=}O)NSO_2Cl-3))]$ . . . Fe: Org.Comp.B17-88
$C_{13}ClFeH_{12}O_2^+$ . . $[(2-Cl-C_6H_4-CH_3)Fe(C_5H_4-COOH)][PF_6]$ . . . Fe: Org.Comp.B19-2, 33
– . . . $[(3-Cl-C_6H_4-COO-CH_3)Fe(C_5H_5)][PF_6]$ . . . Fe: Org.Comp.B19-5, 45
– . . . $[(4-Cl-C_6H_4-COO-CH_3)Fe(C_5H_5)][PF_6]$ . . . Fe: Org.Comp.B19-5, 45
$C_{13}ClFeH_{13}NO_2^+$ $[(2-Cl-C_6H_4-CH(CH_3)-NO_2)Fe(C_5H_5)][PF_6]$ . . . Fe: Org.Comp.B19-3, 41
$C_{13}ClFeH_{14}^+$ . . . $[(1,2-(CH_3)_2-C_6H_3-4-Cl)Fe(C_5H_5)][PF_6]$ . . . Fe: Org.Comp.B19-100, 129
– . . . $[(1,3-(CH_3)_2-C_6H_3-2-Cl)Fe(C_5H_5)][PF_6]$ . . . Fe: Org.Comp.B19-99, 128
– . . . $[(1,3-(CH_3)_2-C_6H_3-4-Cl)Fe(C_5H_5)][PF_6]$ . . . Fe: Org.Comp.B19-100, 129
– . . . $[(2-Cl-C_6H_4-C_2H_5)Fe(C_5H_5)][PF_6]$ . . . Fe: Org.Comp.B19-35
– . . . $[(3-Cl-C_6H_4-CH_3)Fe(C_5H_4-CH_3)][PF_6]$ . . . Fe: Org.Comp.B19-2, 5/6, 33
– . . . $[(4-Cl-C_6H_4-CH_3)Fe(C_5H_4-CH_3)][PF_6]$ . . . Fe: Org.Comp.B19-2, 5/6, 34
– . . . $[(Cl-C_6H_5)Fe(C_5H_4-C_2H_5)]^+$ . . . Fe: Org.Comp.B18-142/6, 197/8, 246
$C_{13}ClFeH_{14}NO_5S$ $(C_5H_5)Fe(CO)_2CH[-C(CH_3)_2-C({=}O)-N(SO_2Cl)-CH_2-]$
Fe: Org.Comp.B14-18/9, 24
– . . . $(C_5H_5)Fe(CO)_2[CH_2{=}CH-C(CH_3)_2-C({=}O)-N(SO_2Cl)]$
Fe: Org.Comp.B17-128
$C_{13}ClFeH_{15}$ . . . $(C_5H_5)Fe[1,3-(CH_3)_2-2-Cl-C_6H_4]$ . . . Fe: Org.Comp.B17-294
– . . . $(C_5H_5)Fe[1,6-(CH_3)_2-4-Cl-C_6H_4]$ . . . Fe: Org.Comp.B17-285
– . . . $(C_5H_5)Fe[2,4-(CH_3)_2-3-Cl-C_6H_4]$ . . . Fe: Org.Comp.B17-295
– . . . $(C_5H_5)Fe[2,6-(CH_3)_2-1-Cl-C_6H_4]$ . . . Fe: Org.Comp.B17-285
– . . . $(C_5H_5)Fe[4,6-(CH_3)_2-1-Cl-C_6H_4]$ . . . Fe: Org.Comp.B17-286
$C_{13}ClFeH_{15}N^+$ . . . $[(3-Cl-2-CH_3-C_6H_3-NH-CH_3)Fe(C_5H_5)][PF_6]$ . . Fe: Org.Comp.B19-100, 131
– . . . $[(2-Cl-C_6H_4-N(CH_3)_2)Fe(C_5H_5)][PF_6]$ . . . Fe: Org.Comp.B19-75
– . . . $[(3-Cl-C_6H_4-N(CH_3)_2)Fe(C_5H_5)][PF_6]$ . . . Fe: Org.Comp.B19-3, 5, 75
– . . . $[(4-Cl-C_6H_4-N(CH_3)_2)Fe(C_5H_5)][PF_6]$ . . . Fe: Org.Comp.B19-3, 5, 75
$C_{13}ClFeH_{16}HgN$ . . $(C_5H_5)Fe[C_5H_3(HgCl)-CH_2-N(CH_3)_2]$ . . . Fe: Org.Comp.A10-146/50
$C_{13}ClFeH_{16}N$ . . . $(C_5H_5)Fe[C_5H_3(Cl)-CH_2N(CH_3)_2]$ . . . Fe: Org.Comp.A9-34/5, 46, 73
– . . . $[(CH_3)_2NCH_2-C_5H_4]Fe(C_5H_4-Cl)$ . . . Fe: Org.Comp.A9-11, 17
$C_{13}ClFeH_{16}NO$ . . . $(C_5H_5)Fe(CN-C_6H_{11}-c)(CO)Cl$ . . . Fe: Org.Comp.B15-290, 296
$C_{13}ClFeH_{17}HgSi$ . . $(HgCl-C_5H_4)Fe[C_5H_4-Si(CH_3)_3]$ . . . Fe: Org.Comp.A10-141, 143, 144, 146

$C_{13}ClFeH_{17}O_4S$ . . $[C_5H_5FeC_4S((CH_3)_4-2,3,4,5)][ClO_4]$ . . . . Fe: Org.Comp.B17-214/5
$C_{13}ClFeH_{17}Si$ . . . . $(C_5H_5)Fe[C_5H_3(Cl)-Si(CH_3)_3]$ . . . . Fe: Org.Comp.A9-330, 333
– . . . . $[(CH_3)_3Si-C_5H_4]Fe(C_5H_4-Cl)$ . . . . Fe: Org.Comp.A9-314
$C_{13}ClFe_3H_9O_9S$ . . $(CO)_9(Cl)Fe_3S-C_4H_9$-t . . . . Fe: Org.Comp.C6a-99/100
$C_{13}ClGeH_{16}NO_2S$ $Ge(CH_3)_3-C(S-C_6H_4-NO_2-4)=CCl-CH=CH_2$ . . . Ge: Org.Comp.2-18
– . . . . $Ge(CH_3)_3-C{\equiv}C-CHCl-CH_2S-C_6H_4-NO_2-4$ . . . . Ge: Org.Comp.2-57
$C_{13}ClGeH_{17}$ . . . . . $CH_3CHClCH_2(C_6H_5)Ge(-CH_2CH=CHCH_2-)$ . . . . Ge: Org.Comp.3-266
$C_{13}ClGeH_{17}S$ . . . . $Ge(CH_3)_3-C(S-C_6H_5)=CCl-CH=CH_2$ . . . . Ge: Org.Comp.2-17
– . . . . $Ge(CH_3)_3-C{\equiv}C-CHCl-CH_2S-C_6H_5$ . . . . Ge: Org.Comp.2-57
$C_{13}ClGeH_{17}Se$ . . . $Ge(CH_3)_3-C(Se-C_6H_5)=CCl-CH=CH_2$ . . . . Ge: Org.Comp.2-18
– . . . . $Ge(CH_3)_3-C{\equiv}C-CHCl-CH_2Se-C_6H_5$ . . . . Ge: Org.Comp.2-57
$C_{13}ClGeH_{21}$ . . . . . $Ge(C_2H_5)_3-CH_2C_6H_4Cl-2$ . . . . Ge: Org.Comp.2-117
– . . . . $Ge(C_2H_5)_3-CH_2C_6H_4Cl-3$ . . . . Ge: Org.Comp.2-117
– . . . . $Ge(C_2H_5)_3-CH_2C_6H_4Cl-4$ . . . . Ge: Org.Comp.2-118
$C_{13}ClGeH_{27}O$ . . . . $Ge(C_3H_7-n)_3-CH_2CH_2CH_2-C(=O)Cl$ . . . . Ge: Org.Comp.3-5
– . . . . $Ge(C_3H_7-n)_3-CH_2-CH(CH_3)-C(=O)Cl$ . . . . Ge: Org.Comp.3-5
$C_{13}ClH_8MnNO_2$ . . . $[Mn(-O-C_6H_4-2-CH=N-C_6H_3(Cl-5)-2-O-)]_n$ . . . Mn: MVol.D6-28, 29/30
$C_{13}ClH_8N_2O_3Re$ . . $(CO)_3ReCl[2-(NC_5H_4-2)-NC_5H_4]$ . . . . Re: Org.Comp.1-160/1, 178
– . . . . $(^{13}CO)_3ReCl[2-(NC_5H_4-2)-NC_5H_4]$ . . . . Re: Org.Comp.1-161
$C_{13}ClH_8N_2O_5Re$ . . $(CO)_3ReCl[O=NC_5H_4-2-(2-C_5H_4N=O)]$ . . . . Re: Org.Comp.1-196
$C_{13}ClH_8N_2O_7Re$ . . $(CO)_3Re[NC_5H_4-2-(2-C_5H_4N)]OClO_3$ . . . . Re: Org.Comp.1-156/7
$C_{13}ClH_9MnNO_{2.5}$ $[Mn(-O-C_6H_3(Cl-4)-2-CH=N-C_6H_4-O-2-)(H_2O)_{0.5}]$ Mn: MVol.D6-56/7
$C_{13}ClH_9N_2OS$ . . . . $C_6H_5C(O)N=S=N-C_6H_4-4-Cl$ . . . . S: S-N Comp.7-268/9
$C_{13}ClH_{10}MoN_3O_2$ $[(C_5H_5)Mo(CO)_2(Cl)-2-(1,2,3-N_3C_6H_5)]$ . . . . Mo: Org.Comp.7-56/7, 60
$C_{13}ClH_{10}NOS$ . . . . $(C_6H_5)_2C=N-S(O)Cl$ . . . . S: S-N Comp.8-271
$C_{13}ClH_{10}NO_3STh$ $(OH)_2Th[O-C_6H_3-4-Cl-2-(CH=N-C_6H_4-2-S)] \cdot H_2O$ Th: SVol.D4-133
$C_{13}ClH_{10}N_2O_3Re$ $(CO)_3ReCl(NC_5H_5)_2$ . . . . Re: Org.Comp.1-224/5
– . . . . $(CO)_3ReCl[-1,8-(1,8-N_2C_8H_4(CH_3)_2-2,7)-]$ . . . . Re: Org.Comp.1-163
$C_{13}ClH_{10}N_2O_5Re$ $(CO)_3ReCl(C_5H_5N=O)_2$ . . . . Re: Org.Comp.1-285
$C_{13}ClH_{10}O_5Re$ . . . $(CO)_3ReCl[C_6H_5C(O)CH_2C(O)CH_3]$ . . . . Re: Org.Comp.1-197
$C_{13}ClH_{11}N_2O_2S_3$ . . $4-CH_3-C_6H_4-S(O)_2N=S=NS-C_6H_4-Cl-4$ . . . . . . S: S-N Comp.7-67
$C_{13}ClH_{11}N_2O_4S_3$ . . $4-Cl-C_6H_4-S(O)_2N=S=NS(O)_2-C_6H_4-CH_3-4$ . . . S: S-N Comp.7-100/1
$C_{13}ClH_{11}N_6O_{12}Th$ $Th(NO_3)_4 \cdot NC_5H_4-2-[C(CH_3)=N-C_6H_4-2-Cl]$. Th: SVol.D4-138
– . . . . $Th(NO_3)_4 \cdot NC_5H_4-2-[C(CH_3)=N-C_6H_4-3-Cl]$. Th: SVol.D4-138
$C_{13}ClH_{12}N_2O_2Re$ $(C_5H_5)ReCl(CO)-N=N-C_6H_4-OCH_3-4$ . . . . Re: Org.Comp.3-133/5, 136/7
$C_{13}ClH_{12}N_2O_3Os^+$ $[(Cl)Os(CO)(CH_3O-4-NC_5H_3-2-(2-C_5H_3N-4-OCH_3))][PF_6]$ . . . . Os: Org.Comp.A1-65, 76
$C_{13}ClH_{12}O_2Sb$ . . . $(4-ClC_6H_4)(4-CH_3C_6H_4)Sb(O)OH$ . . . . Sb: Org.Comp.5-224
$C_{13}ClH_{12}O_3Sb$ . . . $(4-ClC_6H_4)(4-CH_3OC_6H_4)Sb(O)OH$ . . . . Sb: Org.Comp.5-223
$C_{13}ClH_{13}MoN_4O_2$ $[(C_5H_5)Mo(CO)_2(1,2-N_2C_3H_4)_2]Cl$ . . . . Mo: Org.Comp.7-283, 285
– . . . . $[(C_5H_5)Mo(CO)_2(1,3-N_2C_3H_4)_2]Cl$ . . . . Mo: Org.Comp.7-283, 285
$C_{13}ClH_{13}N_2O_3Os$ $(CO)Os(NC_5H_5)_2[OC(O)-CH_3]Cl$ . . . . Os: Org.Comp.A1-62
$C_{13}ClH_{13}O_8Sn$ . . . $CH_3SnCl(OC(=CHOC(CH_2OH)=CH)CO)_2$ . . . . . . Sn: Org.Comp.17-144
$C_{13}ClH_{14}OTi$ . . . . . $(C_5H_5)Ti(OC_6H_3(CH_3)_2-2,6)Cl$ . . . . Ti: Org.Comp.5-37
$C_{13}ClH_{16}MoO_5P$ . . $(C_5H_5)Mo(CO)_2(Cl)[P(-OCH_2-)_3CC_2H_5]$ . . . . Mo: Org.Comp.7-56/7, 90, 114
$C_{13}ClH_{16}NOSn$ . . . $(C_2H_5)_2Sn(Cl)OC_9H_6N$ . . . . Sn: Org.Comp.17-94
$C_{13}ClH_{16}N_2O_6Re$ $(CO)_4Re[C(O)CH_3]C(CH_3)=NH-CH_2CH_2N(CH_3)C(O)CH_2Cl$ Re: Org.Comp.1-392

$C_{13}ClH_{16}N_4O_3Re$ $(CO)_3ReCl[3,5\text{-}(CH_3)_2N_2C_3H_2]_2$ . . . . Re: Org.Comp.1-224

$C_{13}ClH_{17}Mn_2N_2Na_2O_9P_2$
$Na_2[Mn_2(O_3P\text{-}CH_2CH_2N{=}CH\text{-}C_6H_2(Cl)(O)$
$\text{-}CH{=}NCH_2CH_2\text{-}PO_3)(OH)] \cdot CH_3OH \cdot H_2O$ Mn: MVol.D6-211, 214

$C_{13}ClH_{17}Mn_2N_2O_9S_2$
$[Mn_2(\text{-}O_3S\text{-}CH_2CH_2\text{-}N{=}CH\text{-}C_6H_2(Cl)(O)$
$\text{-}CH{=}N\text{-}CH_2CH_2\text{-}SO_3\text{-})(OH)] \cdot CH_3OH \cdot H_2O$ Mn: MVol.D6-211, 213/4

$C_{13}ClH_{17}Mo$ . . . . . $(C_5H_5)Mo(Cl)(H_3C\text{-}C{\equiv}C\text{-}CH_3)_2$ . . . . Mo: Org.Comp.6-133, 143

– . . . . $(C_5H_5)Mo(Cl)(H_3C\text{-}C{\equiv}C\text{-}CH_3)_2 \cdot TlOC(CF_3)CHC(CF_3)O$
Mo: Org.Comp.6-142

$C_{13}ClH_{17}MoNO_4P$ $(C_5H_5)Mo(CO)_2(Cl)(1,3,6,2\text{-}O_2NPC_4H_9\text{-}2\text{-}CH{=}CH_2)$
Mo: Org.Comp.7-249, 271

$C_{13}ClH_{17}MoN_4$ . . . $[(C_5H_5)Mo(CNCH_3)_4]Cl$ . . . . Mo: Org.Comp.6-99

$C_{13}ClH_{17}MoS_4$ . . . $(C_5H_5)Mo(Cl)(H_3CS\text{-}C{\equiv}C\text{-}SCH_3)_2$ . . . . Mo: Org.Comp.6-132

$C_{13}ClH_{17}NOSb$ . . . $(CH_3)_3Sb(Cl)OC_9H_5(CH_3)N$ . . . . Sb: Org.Comp.5-10, 14

$C_{13}ClH_{18}MoO_4P$ . . $(C_5H_5)Mo(CO)_2(H)[2\text{-}Cl\text{-}1,3,2\text{-}O_2PC_2(CH_3)_4\text{-}4,4,5,5]$
Mo: Org.Comp.7-56/7, 93

$C_{13}ClH_{18}NOSn$ . . . $(C_2H_5)_2Sn(Cl)OCH(CH_3)C_6H_4CN\text{-}4$ . . . . Sn: Org.Comp.17-94

$C_{13}ClH_{18}NO_4S$ . . . $4\text{-}CH_3OC(O)CH_2\text{-}C_6H_4\text{-}N(CH_2CH_2Cl)\text{-}S(O)O\text{-}C_2H_5$
S: S-N Comp.8-321

$C_{13}ClH_{19}MoN_2O_2$ $(CH_2CHCH_2)Mo(Cl)(CO)_2(CN\text{-}C_3H_7\text{-}i)_2$ . . . . Mo: Org.Comp.5-296/9

$C_{13}ClH_{20}MnN_2O_2$ $MnCl[\text{-}O\text{-}C(CH_3){=}CH\text{-}C(CH_3){=}NCH(CH_3)CH_2$
$N{=}C(CH_3)CH{=}C(CH_3)O\text{-}] \cdot 0.3\ H_2O$ . . . . Mn: MVol.D6-203, 206/7

$C_{13}ClH_{20}MoO_2P$ . . $[(C_5H_5)Mo(CO)_2(P(C_2H_5)_3)(Cl)]$ . . . . Mo: Org.Comp.7-57, 63

$C_{13}ClH_{20}NOSn$ . . . $(C_2H_5)_3SnN(C_6H_4Cl\text{-}4)CHO$ . . . . Sn: Org.Comp.18-137, 139

$C_{13}ClH_{20}N_2O_3Re$ $(CO)_3ReCl[\text{-}N(C_4H_9\text{-}t){=}CH\text{-}CH{=}N(C_4H_9\text{-}t)\text{-}]$ . . . Re: Org.Comp.1-159

$C_{13}ClH_{21}MoNO_4P$ $(C_5H_5)Mo(CO)_2(Cl)[(C_2H_5\text{-}O)_2PN(CH_3)_2]$ . . . . . . Mo: Org.Comp.7-56/7, 91/2

$C_{13}ClH_{21}MoN_2O_2$ $(CH_2CHCH_2)Mo(CO)_2(Cl)(i\text{-}C_3H_7\text{-}N{=}CHCH{=}N\text{-}C_3H_7\text{-}i)$
Mo: Org.Comp.5-262/3, 264

$C_{13}ClH_{21}MoO_2P_2$ $[(C_5H_5)Mo(CO)_2((CH_3)_2P\text{-}CH_2CH_2\text{-}P(CH_3)_2)]Cl$ Mo: Org.Comp.7-250, 272

$C_{13}ClH_{21}N_2O_2S$ . . $4\text{-}(CH_3)_2N\text{-}C_6H_4\text{-}N(CH_2CH_2Cl)\text{-}S(O)O\text{-}C_3H_7\text{-}n$
S: S-N Comp.8-321

$C_{13}ClH_{21}N_2SSn$ . . $(CH_3)_2Sn(Cl)\text{-}N(C_6H_5)\text{-}C({=}S)\text{-}N(C_2H_5)_2$ . . . . Sn: Org.Comp.19-117, 119

$C_{13}ClH_{21}O_4PRe$ . . $(CO)_4Re(Cl)P(C_3H_7\text{-}i)_3$ . . . . Re: Org.Comp.1-440

$C_{13}ClH_{21}O_7PRe$ . . $(CO)_4Re(Cl)P(OC_3H_7\text{-}i)_3$ . . . . Re: Org.Comp.1-448

$C_{13}ClH_{22}N_2O_3Re$ $(CO)_3ReCl(NC_5H_{11})_2$ . . . . Re: Org.Comp.1-224

$C_{13}ClH_{23}MoN_2OS_4$
$(HC{\equiv}C\text{-}CH_2Cl)Mo(O)[S_2C\text{-}N(C_2H_5)_2]_2$ . . . . Mo: Org.Comp.5-139/40

$C_{13}ClH_{23}MoO_3$ . . . $(C_5H_5)Mo(O)(Cl)(O\text{-}C_4H_9\text{-}n)_2$ . . . . Mo: Org.Comp.6-11/2

– . . . . $(C_5H_5)Mo(O)(Cl)(O\text{-}C_4H_9\text{-}i)_2$ . . . . Mo: Org.Comp.6-11/2

$C_{13}ClH_{23}O_4Sn$ . . . $(C_4H_9)_2Sn(Cl)OOCCH{=}CHCOOCH_3$ . . . . Sn: Org.Comp.17-105

$C_{13}ClH_{24}NO_2SSiSn$
$(C_2H_5)_2Sn(Cl)\text{-}N[S({=}O)_2\text{-}C_6H_5]\text{-}Si(CH_3)_3$ . . . . . Sn: Org.Comp.19-117, 120

$C_{13}ClH_{24}NiO_3PSn$ $(CH_3)_2SnCl\text{-}P(C_4H_9\text{-}t)_2Ni(CO)_3$ . . . . Sn: Org.Comp.19-218, 221

$C_{13}ClH_{25}InNSi_2$ . . $C_6H_5\text{-}CH_2\text{-}InCl\text{-}N[Si(CH_3)_3]_2$ . . . . In: Org.Comp.1-299/300

$C_{13}ClH_{25}O_2Sn$ . . . $(C_4H_9)_2Sn(Cl)OC(CH_3){=}CHCOCH_3$ . . . . Sn: Org.Comp.17-103

$C_{13}ClH_{27}OSn$ . . . . $(C_4H_9)_2Sn(Cl)OCH(CH_3)CH_2CH{=}CH_2$ . . . . Sn: Org.Comp.17-100

$C_{13}ClH_{27}O_2Sn$ . . . $(C_4H_9)_2Sn(Cl)OOCC_4H_9\text{-}t$ . . . . Sn: Org.Comp.17-103

$C_{13}ClH_{30}O_5PSn$ . . $C_4H_9SnCl(OCH_3)OP(O)(OC_4H_9)_2$ . . . . Sn: Org.Comp.17-175

$C_{13}ClH_{36}NOsSi_3$ . . $[(CH_3)_3Si\text{-}CH_2]_3Os({=}NCH_3)Cl$ . . . . Os: Org.Comp.A1-27

$C_{13}Cl_2F_3FeH_5O_2$ .. $(C_5H_5)Fe(CO)_2C_6(Cl_2\text{-}3,5)F_3\text{-}2,4,6$ .... Fe: Org.Comp.B13-176, 187
$C_{13}Cl_2F_3H_7N_2S$ ... $4\text{-}CF_3\text{-}C_6H_4\text{-}N{=}S{=}N\text{-}C_6H_3\text{-}Cl_2\text{-}2,4$ .... S: S-N Comp.7-263
$C_{13}Cl_2F_4GeH_{20}O_2$ $Ge(C_2H_5)_3C({=}C(OC_2H_5)OC(CF_2Cl)_2)$ .... Ge: Org.Comp.2-286
$C_{13}Cl_2F_5HN_4$ ..... $NCFCClC(NHC_6F_3(CN)_2)CClCF$ .... F: PFHOrg.SVol.4-150, 160
$C_{13}Cl_2F_9N$ .... $4\text{-}Cl\text{-}C_6F_4\text{-}N{=}CCl\text{-}C_6F_5$ .... F: PFHOrg.SVol.6-193, 207
– .... $C_6F_5\text{-}N{=}CCl\text{-}C_6F_4\text{-}Cl\text{-}4$ .... F: PFHOrg.SVol.6-193, 208
$C_{13}Cl_2FeH_8N_2$ .... $(C_5H_5)Fe[C_6H_3\text{-}1,2\text{-}Cl_2\text{-}1,6\text{-}(CN)_2]$ .... Fe: Org.Comp.B17-258
$C_{13}Cl_2FeH_{10}HgO$ $(HgCl\text{-}C_5H_4)Fe(C_5H_4\text{-}CCl{=}CHCH{=}O)$ .... Fe: Org.Comp.A10-142, 143/4
$C_{13}Cl_2FeH_{11}N$ .... $(C_5H_5)Fe(1,3\text{-}Cl_2\text{-}2\text{-}CH_3\text{-}6\text{-}NC\text{-}C_6H_3)$ .... Fe: Org.Comp.B17-258
– .... $(C_5H_5)Fe(1,5\text{-}Cl_2\text{-}6\text{-}CH_3\text{-}6\text{-}NC\text{-}C_6H_3)$ .... Fe: Org.Comp.B17-258
$C_{13}Cl_2FeH_{15}Hg_2N$ $(ClHg\text{-}C_5H_4)Fe[C_5H_3(HgCl)\text{-}CH_2\text{-}N(CH_3)_2]$ .... Fe: Org.Comp.A10-329
$C_{13}Cl_2H_7N_2O_3Re$ $(CO)_3ReH[4\text{-}Cl\text{-}NC_5H_3\text{-}2\text{-}(2\text{-}C_5H_3N\text{-}4\text{-}Cl)]$ ... Re: Org.Comp.1-174
$C_{13}Cl_2H_9NO_2Th$ .. $ThCl_2[O\text{-}C_6H_4\text{-}3\text{-}(N{=}CH\text{-}C_6H_4\text{-}2\text{-}O)]$ .... Th: SVol.D4-163
$C_{13}Cl_2H_9Ti$ ...... $[(C_{13}H_9)TiCl_2]_n$ .... Ti: Org.Comp.5-36
$C_{13}Cl_2H_{10}MnN_4O_3$ $[Mn(3\text{-}O_2N\text{-}C_6H_4\text{-}CH{=}N\text{-}NHC(O)\text{-}3\text{-}C_5H_4N)Cl_2]$
Mn: MVol.D6-271/2
– .... $[Mn(4\text{-}O_2N\text{-}C_6H_4\text{-}CH{=}N\text{-}NHC(O)\text{-}3\text{-}C_5H_4N)Cl_2]$
Mn: MVol.D6-271/2
$C_{13}Cl_2H_{10}MnN_4S$ $MnCl_2[NC_5H_4\text{-}2\text{-}CH{=}N\text{-}NH\text{-}2\text{-}(1,3\text{-}SNC_7H_4)]$ .. Mn: MVol.D6-244, 246/7
$C_{13}Cl_2H_{10}N_2S$ .... $4\text{-}CH_3\text{-}C_6H_4\text{-}N{=}S{=}N\text{-}C_6H_3\text{-}Cl_2\text{-}2,4$ .... S: S-N Comp.7-263
$C_{13}Cl_2H_{12}MoN_2O$ $(C_5H_5)Mo(N_2C_6H_4CH_3\text{-}4)(CO)Cl_2$ .... Mo: Org.Comp.6-248, 251
$C_{13}Cl_2H_{12}N_4O_4S_5$ $C_6H_5\text{-}S(O)_2N{=}S{=}NSN{=}S{=}NS(O)_2\text{-}C_6H_5 \cdot CH_2Cl_2$
S: S-N Comp.7-69
$C_{13}Cl_2H_{13}MnN_3$ .. $[Mn(6\text{-}CH_3\text{-}NC_5H_3\text{-}2\text{-}CH{=}NCH_2\text{-}2\text{-}C_5H_4N)Cl_2]$ Mn: MVol.D6-80/2
– .... $[Mn(NC_5H_4\text{-}2\text{-}C(CH_3){=}NCH_2\text{-}2\text{-}C_5H_4N)Cl_2]$ ... Mn: MVol.D6-90
$C_{13}Cl_2H_{13}NO_3Sn$ $CH_3OOCCH_2CH_2SnCl_2(OC_9H_6N)$ .... Sn: Org.Comp.17-179
$C_{13}Cl_2H_{13}Sb$ ..... $(C_6H_5)_2(CH_3)SbCl_2$ .... Sb: Org.Comp.5-56
$C_{13}Cl_2H_{13}Ti$ ..... $[(CH_3(C_6H_5)CHC_5H_4)TiCl_2]_n$ .... Ti: Org.Comp.5-36
$C_{13}Cl_2H_{14}MnN_4$ .. $MnCl_2[NC_5H_3(CH_3\text{-}6)\text{-}2\text{-}CH{=}NNH\text{-}2\text{-}C_5H_3N(CH_3\text{-}6)]$
Mn: MVol.D6-244, 246/7
$C_{13}Cl_2H_{14}MnN_4O$ $MnCl_2[CH_3C({=}NOH)C(CH_3){=}N\text{-}NH\text{-}2\text{-}(C_9H_6N\text{-}1)]$
Mn: MVol.D6-252, 253/4
$C_{13}Cl_2H_{15}NOSn$ .. $C_4H_9SnCl_2(OC_9H_6N)$ .... Sn: Org.Comp.17-178
$C_{13}Cl_2H_{16}MoNOP$ $(C_5H_5)Mo(NO)(P(CH_3)_2C_6H_5)Cl_2$ .... Mo: Org.Comp.6-32
$C_{13}Cl_2H_{16}NOSb$ .. $(C_2H_5)_2Sb(Cl_2)OC_9H_6N$ .... Sb: Org.Comp.5-192
$C_{13}Cl_2H_{16}O_2Sn$ ... $C_6H_5SnCl_2(OOCC_6H_{11}\text{-}c)$ .... Sn: Org.Comp.17-181
$C_{13}Cl_2H_{17}N_3OS_2$ .. $4\text{-}CH_3C_6H_4\text{-}NHC(O)CCl_2SN{=}S{=}N\text{-}C_4H_9\text{-}t$ ..... S: S-N Comp.7-25
$C_{13}Cl_2H_{23}MoO_2$ .. $(C_5H_5)Mo(O\text{-}C_4H_9\text{-}n)_2Cl_2$ .... Mo: Org.Comp.6-4, 6/7
– .... $(C_5H_5)Mo(O\text{-}C_4H_9\text{-}i)_2Cl_2$ .... Mo: Org.Comp.6-4, 6/7
$C_{13}Cl_2H_{25}O_2Sb$ ... $(CH_3)_2SbCl_2\text{-}O\text{-}C(C_4H_9\text{-}t){=}CH\text{-}C({=}O)C_4H_9\text{-}t$ .. Sb: Org.Comp.5-190
$C_{13}Cl_2H_{26}NP_2Re$. $(CH_3)Re[P(CH_3)_3]_2({=}N\text{-}C_6H_5)Cl_2$ .... Re: Org.Comp.1-10
$C_{13}Cl_2H_{29}MnO_3P$ $Mn[P(C_3H_7)_3](OC_4H_8)(O_2)Cl_2$ .... Mn: MVol.D8-51
$C_{13}Cl_2H_{38}N_2Si_4Sn$ $[(CH_3)_3Si]_2N\text{-}Sn(Cl)(CH_2Cl)\text{-}N[Si(CH_3)_3]_2$ ..... Sn: Org.Comp.19-149/50
$C_{13}Cl_2H_{39}N_2SbSi_4$ $(CH_3)Sb(Cl_2)[N(Si(CH_3)_3)_2]_2$ .... Sb: Org.Comp.5-274/5, 276
$C_{13}Cl_3F_{10}N$ ...... $(C_6F_5)_2N\text{-}CCl_3$ .... F: PFHOrg.SVol.6-227, 238
$C_{13}Cl_3FeH_{10}NO_4$. $(C_5H_5)Fe(CO)_2CH_2CH{=}CHCONHCOCCl_3$ ...... Fe: Org.Comp.B13-106, 134
$C_{13}Cl_3FeH_{10}O_2{}^+$. $[(C_5H_5)Fe(CO)_2(CH_2{=}CHCH_2C_3Cl_3\text{-}c)]^+$ ...... Fe: Org.Comp.B17-32
$C_{13}Cl_3FeH_{13}O_3$ ... $(C_5H_5)Fe(CO)_2COCHC_2H_5CH_2CCl_3$ .... Fe: Org.Comp.B13-7
$C_{13}Cl_3FeH_{16}NSi$ .. $(C_5H_5)Fe[C_5H_3(SiCl_3)CH_2N(CH_3)_2]$ .... Fe: Org.Comp.A9-332, 336
$C_{13}Cl_3GaH_8O$ .... $GaCl_3[1\text{-}(O{=})C_{13}H_8]$ .... Ga: SVol.D1-49/50

$C_{13}Cl_3GaH_{10}O$ ... $GaCl_3[O{=}C(C_6H_5)_2]$ ... Ga: SVol.D1-39/42
$C_{13}Cl_3GaH_{13}N$ ... $GaCl_3[N(C_6H_5)_2{-}CH_3]$ ... Ga: SVol.D1-233
$C_{13}Cl_3H_{10}NSSb^-$ $[(C_6H_5)_2Sb(Cl_3)NCS]^-$ ... Sb: Org.Comp.5-158/9
$C_{13}Cl_3H_{10}Sb$ ... $(2{-}C_6H_4CH_2C_6H_4{-}2')SbCl_3$ ... Sb: Org.Comp.5-169
$C_{13}Cl_3H_{11}NO_2Sb$ $(4{-}O_2NC_6H_4)(4{-}CH_3C_6H_4)SbCl_3$ ... Sb: Org.Comp.5-224
$C_{13}Cl_3H_{11}N_2O_4S_3$ $C_6H_5{-}S(O)_2{-}NH{-}S(CCl_3){=}N{-}S(O)_2{-}C_6H_5$ ... S: S-N Comp.8-197/8
$C_{13}Cl_3H_{12}Sb$ ... $(C_6H_5)(4{-}CH_3C_6H_4)SbCl_3$ ... Sb: Org.Comp.5-223, 229
$C_{13}Cl_3H_{15}N_2O_2S$ .. $n{-}C_4H_9{-}C(O){-}NH{-}S(CCl_3){=}N{-}C(O){-}C_6H_5$ ... S: S-N Comp.8-199
$C_{13}Cl_3H_{16}O_2Sb$ ... $3{-}CH_3C_6H_4{-}SbCl_3{-}O{-}C(CH_3){=}C(CH_3){-}C({=}O)CH_3$
Sb: Org.Comp.5-270
– ... $4{-}CH_3C_6H_4{-}SbCl_3{-}O{-}C(CH_3){=}C(CH_3){-}C({=}O)CH_3$
Sb: Org.Comp.5-270
$C_{13}Cl_3H_{17}N_2O_3S_2$ $n{-}C_4H_9{-}C(O){-}NH{-}S(CCl_3){=}N{-}S(O)_2{-}C_6H_4{-}4{-}CH_3$
S: S-N Comp.8-196
– ... $i{-}C_4H_9{-}C(O){-}NH{-}S(CCl_3){=}N{-}S(O)_2{-}C_6H_4{-}4{-}CH_3$
S: S-N Comp.8-196
$C_{13}Cl_3H_{23}InSi_2^-$ .. $[(Cl)_3In{-}C(C_6H_5)(Si(CH_3)_3)_2]^-$ ... In: Org.Comp.1-353
$C_{13}Cl_3H_{25}MoN_2O$ $[N(C_2H_5)_4][C_5H_5Mo(NO)Cl_3]$ ... Mo: Org.Comp.6-29
$C_{13}Cl_3H_{25}O_4Sn$ ... $(C_4H_9)_2Sn(OCH(CCl_3)OCH_3)OOCCH_3$ ... Sn: Org.Comp.16-193
$C_{13}Cl_3H_{34}IIn_2N_4$ .. $In(Cl)(I){-}CH_2{-}In(Cl)_2 \cdot 2\ (CH_3)_2N{-}CH_2CH_2{-}N(CH_3)_2$
In: Org.Comp.1-176
$C_{13}Cl_3H_{37}N_2Si_4Sn$ $[(CH_3)_3Si]_2N{-}Sn(Cl)(CHCl_2){-}N[Si(CH_3)_3]_2$ ... Sn: Org.Comp.19-149/50
$C_{13}Cl_4H_{10}LiMoN_3O_2P_3$
$Li[(C_5H_5)Mo(CO)_2(N_3P_3Cl_4(C_6H_5))]$ ... Mo: Org.Comp.7-49
$C_{13}Cl_4H_{10}MoN_3O_2P_3^-$
$[(C_5H_5)Mo(CO)_2(N_3P_3Cl_4(C_6H_5))]^-$ ... Mo: Org.Comp.7-49
$C_{13}Cl_4H_{10}N_2O_4S_3$ $4{-}Cl{-}C_6H_4{-}S(O)_2{-}N{=}S(CCl_3){-}NH{-}S(O)_2{-}C_6H_5$ S: S-N Comp.8-197
– ... $4{-}Cl{-}C_6H_4{-}S(O)_2{-}NH{-}S(CCl_3){=}N{-}S(O)_2{-}C_6H_5$ S: S-N Comp.8-197
$C_{13}Cl_4H_{10}Sb^-$ ... $[(2{-}C_6H_4CH_2C_6H_4{-}2)SbCl_4]^-$ ... Sb: Org.Comp.5-171
$C_{13}Cl_4H_{11}NO_2Sb^-$ $[(4{-}O_2NC_6H_4)(4{-}CH_3C_6H_4)SbCl_4]^-$ ... Sb: Org.Comp.5-227
$C_{13}Cl_4H_{11}OSb$ ... $(4{-}ClC_6H_4)(4{-}CH_3OC_6H_4)SbCl_3$ ... Sb: Org.Comp.5-223
$C_{13}Cl_4H_{11}Sb$ ... $(4{-}ClC_6H_4)(4{-}CH_3C_6H_4)SbCl_3$ ... Sb: Org.Comp.5-224
$C_{13}Cl_4H_{12}Sb^-$ ... $[(C_6H_5)(4{-}CH_3C_6H_4)SbCl_4]^-$ ... Sb: Org.Comp.5-226
$C_{13}Cl_4H_{15}O_4Sb$ ... $(CH_3)Sb({-}OC(CH_3)_2C(CH_3)_2O{-})({-}1{-}OC_6Cl_4O{-}2{-})$
Sb: Org.Comp.5-313
$C_{13}Cl_4H_{17}O_4Sb$ ... $(CH_3)Sb(OC_3H_7{-}i)_2(1{-}OC_6Cl_4O{-}2)$ ... Sb: Org.Comp.5-312
$C_{13}Cl_4H_{20}O_5PReSn$
$(C_5H_5)Re(CO)_2[P(O{-}C_2H_5)_3] \cdot SnCl_4$ ... Re: Org.Comp.3-210/2, 213
$C_{13}Cl_4H_{20}O_5PReTi$
$(C_5H_5)Re(CO)_2[P(O{-}C_2H_5)_3] \cdot TiCl_4$ ... Re: Org.Comp.3-210/2, 213
$C_{13}Cl_4H_{21}Mo$ ... $n{-}C_8H_{17}C_5H_4MoCl_4$ ... Mo: Org.Comp.6-2
$C_{13}Cl_4H_{24}MoP$ ... $(CH_3)_5C_5Mo(P(CH_3)_3)Cl_4$ ... Mo: Org.Comp.6-3
$C_{13}Cl_4H_{25}N_3OPSb$ $(4{-}CH_3C_6H_4)SbCl_4 \cdot OP[N(CH_3)_2]_3$ ... Sb: Org.Comp.5-255, 260
$C_{13}Cl_4H_{34}In_2N_4$ .. $[{-}N(CH_3)_2{-}CH_2CH_2{-}N(CH_3)_2{-}]In(Cl)_2{-}CH_2{-}$
$In(Cl)_2[{-}N(CH_3)_2{-}CH_2CH_2{-}N(CH_3)_2{-}]$ ... In: Org.Comp.1-174/5
$C_{13}Cl_4H_{36}N_2Si_4Sn$ $[(CH_3)_3Si]_2N{-}Sn(Cl)(CCl_3){-}N[Si(CH_3)_3]_2$ ... Sn: Org.Comp.19-149/50
$C_{13}Cl_5H_6N_2O_3Re$ $(CO)_3ReCl(3{,}5{-}Cl_2{-}C_5H_3N)_2$ ... Re: Org.Comp.1-225
$C_{13}Cl_5H_8N_2O_4S_3^-$ $[4{-}Cl{-}C_6H_4{-}S(O)_2{-}NS(CCl_3)N{-}S(O)_2{-}C_6H_4{-}4{-}Cl]^-$
S: S-N Comp.8-198/9

$C_{13}Cl_5H_9N_2O_4S_3$ . . $4\text{-}Cl\text{-}C_6H_4\text{-}S(O)_2\text{-}NH\text{-}S(CCl_3)=N\text{-}S(O)_2\text{-}C_6H_4\text{-}4\text{-}Cl$ S: S-N Comp.8-197
$C_{13}Cl_5H_{10}NOSb^-$ $[(4\text{-}C_6H_5C(O)NHC_6H_4)SbCl_5]^-$ . . . . Sb: Org.Comp.5-254
$C_{13}Cl_5H_{11}N_3O_2Sb$ $[4\text{-}CH_3\text{-}C_6H_4\text{-}N_2][(4\text{-}O_2N\text{-}C_6H_4)SbCl_5]$ . . . . Sb: Org.Comp.5-255
– . . . . $[4\text{-}O_2N\text{-}C_6H_4\text{-}N_2][(4\text{-}CH_3\text{-}C_6H_4)SbCl_5]$ . . . . Sb: Org.Comp.5-256
$C_{13}Cl_5H_{11}OSb^-$ . . . $[(4\text{-}ClC_6H_4)(4\text{-}CH_3OC_6H_4)SbCl_4]^-$ . . . . Sb: Org.Comp.5-227
$C_{13}Cl_5H_{11}Sb^-$ . . . . $[(4\text{-}ClC_6H_4)(4\text{-}CH_3C_6H_4)SbCl_4]^-$ . . . . Sb: Org.Comp.5-227
$C_{13}Cl_5H_{12}N_2OSb$ $[4\text{-}CH_3OC_6H_4N_2][C_6H_5SbCl_5]$ . . . . Sb: Org.Comp.5-247
$C_{13}Cl_5H_{12}N_2Sb$ . . . $[4\text{-}CH_3\text{-}C_6H_4\text{-}N_2][C_6H_5\text{-}SbCl_5]$ . . . . Sb: Org.Comp.5-246/7
– . . . . $[C_6H_5\text{-}N_2][4\text{-}CH_3\text{-}C_6H_4\text{-}SbCl_5]$ . . . . Sb: Org.Comp.5-256
$C_{13}Cl_5H_{13}Mo$ . . . . $(C_5H_5)Mo(Cl)(ClCH_2\text{-}C{\equiv}C\text{-}CH_2Cl)_2$ . . . . Mo: Org.Comp.6-134
$C_{13}Cl_5H_{13}NO_2SSb$ $[NC_5H_6][3\text{-}(HO\text{-}C(=O)\text{-}CH_2\text{-}S)\text{-}C_6H_4\text{-}SbCl_5]$ . . Sb: Org.Comp.5-253
– . . . . $[NC_5H_6][4\text{-}(HO\text{-}C(=O)\text{-}CH_2\text{-}S)\text{-}C_6H_4\text{-}SbCl_5]$ . . Sb: Org.Comp.5-253
$C_{13}Cl_5H_{13}NO_2Sb$ $[NC_5H_6][3\text{-}CH_3O\text{-}C(=O)\text{-}C_6H_4\text{-}SbCl_5]$ . . . . Sb: Org.Comp.5-257
– . . . . $[NC_5H_6][4\text{-}CH_3O\text{-}C(=O)\text{-}C_6H_4\text{-}SbCl_5]$ . . . . Sb: Org.Comp.5-257
– . . . . $[NC_5H_6][4\text{-}(HO\text{-}C(=O)\text{-}CH_2)\text{-}C_6H_4\text{-}SbCl_5]$ . . . . Sb: Org.Comp.5-257
$C_{13}Cl_5H_{13}NO_3Sb$ $[NC_5H_6][4\text{-}(HO\text{-}C(=O)\text{-}CH_2\text{-}O)\text{-}C_6H_4\text{-}SbCl_5]$ . . Sb: Org.Comp.5-252
$C_{13}Cl_5H_{14}N_2OSSb$ $[NC_5H_6][4\text{-}(H_2N\text{-}C(=O)\text{-}CH_2\text{-}S)\text{-}C_6H_4\text{-}SbCl_5]$ Sb: Org.Comp.5-253
$C_{13}Cl_5H_{15}NOSSb$ $[NC_5H_6][4\text{-}(HO\text{-}CH_2CH_2\text{-}S)\text{-}C_6H_4\text{-}SbCl_5]$ . . . . Sb: Org.Comp.5-252
$C_{13}Cl_6GeH_{16}$ . . . . . $(C_2H_5)_2Ge(C_9H_6Cl_6)$ . . . . Ge: Org.Comp.3-321
$C_{13}Cl_9FeH_{10}O_2Sb$ $[(C_5H_5)Fe(CO)_2(CH_2=CHCH_2C_3Cl_3\text{-}c)][SbCl_6]$ . . Fe: Org.Comp.B17-32
$C_{13}CoFeH_{11}O_4S_2$ $(C_5H_5)(CO)Fe(CO)(C(SCH_2)_2CH_2)Co(CO)_2$ . . . . Fe: Org.Comp.B16b-158/9
$C_{13}CoH_{27}NO_3PSn$ $(CH_3)_3Sn\text{-}P(C_4H_9\text{-}t)_2Co(CO)_2(NO)$ . . . . Sn: Org.Comp.19-181, 183/4
$C_{13}CrFeH_{15}N_6O_2S_4$
$[(C_5H_5)Fe(CO)_2(CH_2=CH_2)][Cr(SCN)_4(NH_3)_2]$ . . Fe: Org.Comp.B17-5
$C_{13}CrH_5O_8Re$ . . . . $(CO)_5ReCr(CO)_3C_5H_5$ . . . . Re: Org.Comp.2-188, 204
$C_{13}CrH_8MoN_2O_7$ . . $(C_5H_5)Mo(CO)_2\text{-}N=NCH_3 \cdot Cr(CO)_5$ . . . . Mo: Org.Comp.7-7, 14
$C_{13}CuH_{13}O_8Re$ . . . $[\text{-}OC(CH_3)Re(CO)_4C(CH_3)O\text{-}]Cu[\text{-}OC(CH_3)CHC(CH_3)O\text{-}]$ Re: Org.Comp.2-347
$C_{13}FFeH_9O_2$ . . . . . $(C_5H_5)Fe(CO)_2\text{-}C_6H_4F\text{-}2$ . . . . Fe: Org.Comp.B13-170
– . . . . $(C_5H_5)Fe(CO)_2\text{-}C_6H_4F\text{-}3$ . . . . Fe: Org.Comp.B13-170
– . . . . $(C_5H_5)Fe(CO)_2\text{-}C_6H_4F\text{-}4$ . . . . Fe: Org.Comp.B13-170
$C_{13}FFeH_{18}O_6PS$ . . $[(C_5H_5)Fe(CO)_2C(O)CH(CH_3)P(CH_3)_3][FSO_3]$ . . . Fe: Org.Comp.B14-154, 157
$C_{13}FFeH_{22}O_5PS$ . . $[(C_5H_5)(CO)(P(CH_3)_3)Fe=C(C_2H_5)OCH_3][FSO_3]$ Fe: Org.Comp.B16a-40, 54
$C_{13}FGeH_{15}$ . . . . . . $1\text{-}(CH_3)_3Ge\text{-}C_{10}H_6F\text{-}4$ . . . . Ge: Org.Comp.2-79, 88/9
– . . . . $2\text{-}(CH_3)_3Ge\text{-}C_{10}H_6F\text{-}6$ . . . . Ge: Org.Comp.2-79, 88/9
– . . . . $2\text{-}(CH_3)_3Ge\text{-}C_{10}H_6F\text{-}7$ . . . . Ge: Org.Comp.2-79, 88/9
$C_{13}FGeH_{19}O$ . . . . . $Ge(C_2H_5)_3COC_6H_4F\text{-}4$ . . . . Ge: Org.Comp.2-120
$C_{13}FGeH_{21}$ . . . . . . $Ge(C_2H_5)_3\text{-}CH_2\text{-}C_6H_4\text{-}4\text{-}F$ . . . . Ge: Org.Comp.2-117
– . . . . $Ge(C_2H_5)_3\text{-}C_6H_3\text{-}2\text{-}F\text{-}5\text{-}CH_3$ . . . . Ge: Org.Comp.2-280, 282/3
$C_{13}FH_8N_2O_3Re$ . . . $(CO)_3ReF[NC_5H_4\text{-}2\text{-}(2\text{-}C_5H_4N)]$ . . . . Re: Org.Comp.1-157
$C_{13}FH_9MoN_2O_2$ . . $(C_5H_5)Mo(CO)_2\text{-}N=N\text{-}C_6H_4\text{-}3\text{-}F$ . . . . Mo: Org.Comp.7-6, 17
– . . . . $(C_5H_5)Mo(CO)_2\text{-}N=N\text{-}C_6H_4\text{-}4\text{-}F$ . . . . Mo: Org.Comp.7-6, 17
$C_{13}FH_9N_2OS$ . . . . . $C_6H_5C(O)N=S=N\text{-}C_6H_4\text{-}4\text{-}F$ . . . . S: S-N Comp.7-268/9
$C_{13}FH_9N_2O_2Re^+$ . . $[(C_5H_5)Re(CO)_2\text{-}N=N\text{-}C_6H_4\text{-}4\text{-}F][PF_6]$ . . . . Re: Org.Comp.3-193/7, 203
$C_{13}FH_{25}InN$ . . . . . . $[(CH_3)_3N\text{-}CH_2\text{-}C_6H_5][(CH_3)_3InF]$ . . . . In: Org.Comp.1-349/50
$C_{13}FMnN_3O_{10}Re$. $(CO)_5ReC_3N_3FMn(CO)_5$ . . . . Re: Org.Comp.2-181
$C_{13}F_2H_8N_2O_5PRe$ $(CO)_3Re[NC_5H_4\text{-}2\text{-}(2\text{-}C_5H_4N)]OP(O)F_2$ . . . . Re: Org.Comp.1-156, 176
$C_{13}F_2H_{13}Sb$ . . . . . $(C_6H_5)_2(CH_3)SbF_2$ . . . . Sb: Org.Comp.5-54
$C_{13}F_2H_{20}MoNOP$ $C_5H_5Mo(CO)(P(N(C_2H_5)_2)F_2)C_3H_5$ . . . . Mo: Org.Comp.6-331

$C_{13}F_2H_{31}N_3SSi$	$[(1\text{-}C_4H_8N)_2S\text{-}N(CH_3)_2][Si(CH_3)_3F_2]$	S: S-N Comp.8-230, 242
$C_{13}F_3FeGeH_{15}O_2$	$(C_5H_5)Fe(CO)_2\text{-}C(CF_3)=CH\text{-}Ge(CH_3)_3$	Fe: Org.Comp.B13-92 Ge: Org.Comp.2-9
$C_{13}F_3FeH_7O_2$	$(C_5H_5)Fe(CO)_2\text{-}C_6H_2\text{-}F_3\text{-}2,5,6$	Fe: Org.Comp.B13-165, 176, 186
–	$(C_5H_5)Fe(CO)_2\text{-}C_6H_2\text{-}F_3\text{-}3,4,6$	Fe: Org.Comp.B13-165, 176, 186
–	$(C_5H_5)Fe(CO)_2\text{-}C_6H_2\text{-}F_3\text{-}3,5,6$	Fe: Org.Comp.B13-165, 176
$C_{13}F_3FeH_{10}NO_4S_2$	$[(C_5H_5)Fe(CS)(CO)NC_5H_5][SO_3CF_3]$	Fe: Org.Comp.B15-271
$C_{13}F_3FeH_{12}O^+$	$[(CF_3CH_2\text{-}O\text{-}C_6H_5)Fe(C_5H_5)]^+$	Fe: Org.Comp.B18-142/6, 198/9, 251
$C_{13}F_3FeH_{13}O_3S$	$[(CH_3C_6H_5)Fe(C_5H_5)][CF_3SO_3]$	Fe: Org.Comp.B18-142/6, 197, 201, 206, 269/72
$C_{13}F_3FeH_{13}O_5S$	$[(C_5H_5)Fe(CO)_2=CH\text{-}CH=C(CH_3)_2][CF_3SO_3]$	Fe: Org.Comp.B16a-88, 95
–	$[(C_5H_5)Fe(CO)_2=CH\text{-}CH=CH\text{-}C_2H_5][CF_3SO_3]$	Fe: Org.Comp.B16a-88, 95
–	$[(C_5H_5)Fe(CO)_2=CH\text{-}C_4H_7\text{-}c][CF_3SO_3]$	Fe: Org.Comp.B16a-89
–	$[(C_5H_5)Fe(CO)_2(H_3CCH=C=CHCH_3)][CF_3SO_3]$	Fe: Org.Comp.B17-113/4
$C_{13}F_3FeH_{13}O_6S$	$[(C_5H_5)Fe(CO)_2=C(C_3H_5\text{-}c)\text{-}OCH_3][CF_3SO_3]$	Fe: Org.Comp.B16a-111
$C_{13}F_3FeH_{15}O_2Si$	$(C_5H_5)Fe(CO)_2C(CF_3)=CHSi(CH_3)_3$	Fe: Org.Comp.B13-92
$C_{13}F_3FeH_{15}O_6S$	$[(C_5H_5)Fe(CO)_2=C(C_3H_7\text{-}i)\text{-}OCH_3][CF_3SO_3]$	Fe: Org.Comp.B16a-109, 120
$C_{13}F_3FeH_{16}NO_5S$	$[(C_5H_5)Fe(CO)_2=CH\text{-}N(C_2H_5)_2][CF_3SO_3]$	Fe: Org.Comp.B16a-115/6, 126
–	$[(C_5H_5)Fe(CO)_2=CH\text{-}NHC_4H_9\text{-}t][CF_3SO_3]$	Fe: Org.Comp.B16a-115, 125
$C_{13}F_3FeH_{20}O_5PS$	$[(C_5H_5)(CO)(P(CH_3)_3)Fe=C(CH_3)OCH_3][CF_3SO_3]$	Fe: Org.Comp.B16a-36, 52
$C_{13}F_3FeH_{23}O_9P_2S_2$	$[(C_5H_5)Fe(CS)(P(OCH_3)_3)_2][SO_3CF_3]$	Fe: Org.Comp.B15-264/5
$C_{13}F_3Fe_3H_9NO_8P$	$(CO)_8Fe_3(CN\text{-}CF_3)[P(CH_3)_3]$	Fe: Org.Comp.C6a-76
$C_{13}F_3Fe_3H_9NO_{11}P$	$(CO)_8Fe_3(CN\text{-}CF_3)[P(OCH_3)_3]$	Fe: Org.Comp.C6a-76/7
$C_{13}F_3Fe_3H_{11}NO_{11}P$	$(CO)_8(H)Fe_3(CH=N\text{-}CF_3)[P(OCH_3)_3]$	Fe: Org.Comp.C6a-77/8
$C_{13}F_3Fe_3NO_{11}$	$(CO)_{11}Fe_3(CN\text{-}CF_3)$	Fe: Org.Comp.C6b-140/4
$C_{13}F_3GeH_{19}$	$Ge(C_2H_5)_3\text{-}C_6H_4CF_3\text{-}2$	Ge: Org.Comp.2-278
–	$Ge(C_2H_5)_3\text{-}C_6H_4CF_3\text{-}3$	Ge: Org.Comp.2-278
–	$Ge(C_2H_5)_3\text{-}C_6H_4CF_3\text{-}4$	Ge: Org.Comp.2-278
$C_{13}F_3GeH_{33}O_3SSi_3$	$Ge(CH_3)_3C[Si(CH_3)_3]_2Si(CH_3)_2OSO_2CF_3$	Ge: Org.Comp.1-145
$C_{13}F_3H_8N_3O_2S$	$4\text{-}NO_2\text{-}C_6H_4\text{-}N=S=N\text{-}C_6H_4\text{-}CF_3\text{-}4$	S: S-N Comp.7-262
$C_{13}F_3H_8N_3O_4S_2$	$CF_3S(O)_2\text{-}4\text{-}C_6H_4\text{-}N=S=N\text{-}C_6H_4\text{-}4\text{-}NO_2$	S: S-N Comp.7-261
$C_{13}F_3H_9IN_2ORe$	$(C_5H_5)ReI(CO)\text{-}N=N\text{-}C_6H_4\text{-}CF_3\text{-}2$	Re: Org.Comp.3-133/5
$C_{13}F_3H_9NO_2Re$	$(C_5H_5)Re(CO)(NO)\text{-}C_6H_4\text{-}CF_3\text{-}3$	Re: Org.Comp.3-160, 164, 169
–	$(C_5H_5)Re(CO)(NO)\text{-}C_6H_4\text{-}CF_3\text{-}4$	Re: Org.Comp.3-160, 164, 169
$C_{13}F_3H_9N_2O_2S_2$	$C_6H_5\text{-}S(O)_2N=S=N\text{-}C_6H_4\text{-}4\text{-}CF_3$	S: S-N Comp.7-55/60
$C_{13}F_3H_9N_2O_4S_3$	$C_6H_5\text{-}S(O)_2N=S=N\text{-}C_6H_4\text{-}4\text{-}S(O)_2CF_3$	S: S-N Comp.7-55/60
$C_{13}F_3H_{11}MoO_3$	$(C_5H_5)Mo(CO)_2[\text{-}C(CH_3)=C(CH_3)\text{-}C(CF_3)=O\text{-}]$	Mo: Org.Comp.8-112, 152, 171
$C_{13}F_3H_{13}NO_5Re$	$(CO)_3Re(CNC_4H_9\text{-}t)(CF_3COCHCOCH_3)$	Re: Org.Comp.2-244/5
$C_{13}F_3H_{13}NO_8Re$	$(CO)_4Re[OC(O)CF_3]=C(CH_3)NHCH(CH_3)COOC_2H_5$	Re: Org.Comp.1-380/1
$C_{13}F_3H_{17}MoNO_5PS$	$[(C_5H_5)Mo(CO)_2(NC\text{-}CH_3)(P(CH_3)_3)][O_3S\text{-}CF_3]$	Mo: Org.Comp.7-283, 293

$C_{13}F_3H_{30}N_{11}O_8Th$ $[C(NH_2)_3]_3[Th((OOCCH_2)_2NC_2H_4N(CH_2COO)_2)F_3]$
Th: SVol.C7-124, 125/6
$C_{13}F_3N_2O_5Re$ .... $(CO)_5ReC_6F_3(CN)_2$-3,4 .... Re: Org.Comp.2-143
$C_{13}F_4FeH_5IO_2$ .... $(C_5H_5)Fe(CO)_2C_6F_4I$-2 .... Fe: Org.Comp.B13-174, 185
$C_{13}F_4FeH_5LiO_2$ ... $(C_5H_5)Fe(CO)_2-C_6F_4Li$-2 .... Fe: Org.Comp.B13-175, 186
– .... $(C_5H_5)Fe(CO)_2-C_6F_4Li$-3 .... Fe: Org.Comp.B13-175, 186
– .... $(C_5H_5)Fe(CO)_2-C_6F_4Li$-4 .... Fe: Org.Comp.B13-175, 186
$C_{13}F_4FeH_6O_2$ .... $(C_5H_5)Fe(CO)_2-C_6F_4H$-2 .... Fe: Org.Comp.B13-173, 184
– .... $(C_5H_5)Fe(CO)_2-C_6F_4D$-2 .... Fe: Org.Comp.B13-173
– .... $(C_5H_5)Fe(CO)_2-C_6F_4H$-3 .... Fe: Org.Comp.B13-173, 184
– .... $(C_5H_5)Fe(CO)_2-C_6F_4H$-4 .... Fe: Org.Comp.B13-165, 173, 184
$C_{13}F_4FeH_{10}O_4$ ... $(C_5H_5)Fe(CO)_2CF{=}C(CF_3)COOC_2H_5$ .... Fe: Org.Comp.B13-91, 119
$C_{13}F_4GeH_{15}N$ .... $Ge(C_2H_5)_3C_6F_4CN$-4 .... Ge: Org.Comp.2-279
$C_{13}F_4HO_5Re$ .... $(CO)_5ReC_6F_4C{\equiv}CH$ .... Re: Org.Comp.2-143/4
$C_{13}F_5FeH_5O_2$ .... $(C_5H_5)Fe(CO)_2$-2-[2.2.0]-C_6F_5 .... Fe: Org.Comp.B13-221, 257
– .... $(C_5H_5)Fe(CO)_2-C_6F_5$ .... Fe: Org.Comp.B13-165, 172/3, 183/4
$C_{13}F_5GeH_{15}O$ .... $Ge(C_2H_5)_3COC_6F_5$ .... Ge: Org.Comp.2-120
$C_{13}F_5H_7N_2OS$ .... 4-$CH_3O-C_6H_4-N{=}S{=}N-C_6F_5$ .... S: S-N Comp.7-261
$C_{13}F_5H_7N_2O_2S_2$ .. 4-$CH_3-C_6H_4-S(O)_2N{=}S{=}N-C_6F_5$ .... S: S-N Comp.7-63/4
$C_{13}F_5H_7N_2S$ .... 4-$CH_3-C_6H_4-N{=}S{=}N-C_6F_5$ .... S: S-N Comp.7-263
$C_{13}F_5O_5Re$ .... $(CO)_5ReC{\equiv}CC_6F_5$ .... Re: Org.Comp.2-145
$C_{13}F_6FeH_9NO_6$ ... $[(C_5H_5)Fe(CO)_2(CH_2{=}C{=}NH)][CF_3CO_2] \cdot CF_3CO_2H$
Fe: Org.Comp.B17-125
$C_{13}F_6FeH_{10}O$ .... $(C_5H_5)Fe(CO)-C(CF_3){=}C(CF_3)-CH_2-CH{=}CH_2$ .. Fe: Org.Comp.B17-169/70
– .... $(C_5H_5)Fe[1,3-(CF_3)_2-5-CH_3-OC_5H_2]$ .... Fe: Org.Comp.B17-325
$C_{13}F_6FeH_{10}O_3$ ... $(C_5H_5)Fe(CO)_2CH[-CH_2-C(CF_3)_2-O-CH_2-]$ .... Fe: Org.Comp.B14-2/3, 11
$C_{13}F_6FeH_{11}O_3P$ .. $[((3.3.2)-C_{10}H_{11})Fe(CO)_3][PF_6]$ .... Fe: Org.Comp.B15-256
– .... $[(8-CH_3CH{=}C_8H_7)Fe(CO)_3][PF_6]$ .... Fe: Org.Comp.B15-241/2, 258
$C_{13}F_6FeH_{11}O_4P$ .. $[(1,4-(HOOC)_2-C_6H_4)Fe(C_5H_5)][PF_6]$ .... Fe: Org.Comp.B19-15
– .... $[(8-CH_3C(O)-C_8H_8)Fe(CO)_3][PF_6]$ .... Fe: Org.Comp.B15-201/2, 250/1
– .... $[(HOOC-C_6H_5)Fe(C_5H_4-COOH)][PF_6]$ .... Fe: Org.Comp.B18-142/6, 203
$C_{13}F_6FeH_{12}NO_2P$ $[4-(C_5H_5)Fe(CO)_2CH_2-NC_5H_5][PF_6]$ .... Fe: Org.Comp.B14-152, 156
– .... $[(C_5H_5)Fe(CO)_2(C_5H_7-3-CN)][PF_6]$ .... Fe: Org.Comp.B17-88
$C_{13}F_6FeH_{12}NP$ ... $[(2-CH_3-C_6H_4-CN)Fe(C_5H_5)][PF_6]$ .... Fe: Org.Comp.B19-2, 5, 23
– .... $[(4-CH_3-C_6H_4-CN)Fe(C_5H_5)][PF_6]$ .... Fe: Org.Comp.B19-2, 23
$C_{13}F_6FeH_{12}NaO_3P$ $[(4-CH_3O-C_6H_4-COO)Fe(C_5H_5)] \cdot Na[PF_6]$ ... Fe: Org.Comp.B19-50
$C_{13}F_6FeH_{13}N_2OP$ $[(C_5H_5)Fe(CNCH_3)(CO)NC_5H_5][PF_6]$ .... Fe: Org.Comp.B15-300
$C_{13}F_6FeH_{13}OP$ ... $[(1-OC_8H_8)Fe(C_5H_5)][PF_6]$ .... Fe: Org.Comp.B19-217, 278
– .... $[(CH_3C(=O)C_6H_5)Fe(C_5H_5)][PF_6]$ .... Fe: Org.Comp.B18-142/6, 210
– .... $[(C_6H_6)Fe(C_5H_4-C(O)CH_3)][PF_6]$ .... Fe: Org.Comp.B18-142/6, 151, 162, 186/7
$C_{13}F_6FeH_{13}O_2P$ .. $[(4-CH_3-C_6H_4-COOH)Fe(C_5H_5)][PF_6]$ .... Fe: Org.Comp.B19-23
– .... $[(CH_3-OOC-C_6H_5)Fe(C_5H_5)][PF_6]$ .... Fe: Org.Comp.B18-142/6, 201, 211
– .... $[(HOOC-CH_2-C_6H_5)Fe(C_5H_5)][PF_6]$ .... Fe: Org.Comp.B18-142/6, 199, 210
– .... $[(CH_3-C_6H_5)Fe(C_5H_4-COOH)][PF_6]$ .... Fe: Org.Comp.B18-142/6, 207

$C_{13}F_6FeH_{13}O_2P$ .. $[(C_5H_5)Fe(CO)_2(C_6H_8\text{-}c)][PF_6]$ ... Fe: Org.Comp.B17-89
– ... $[(C_6H_6)Fe(C_5H_4\text{-}COO\text{-}CH_3)][PF_6]$ ... Fe: Org.Comp.B18-142/6, 151, 162
$C_{13}F_6FeH_{13}O_3P$ .. $[(3\text{-}CH_3O\text{-}C_6H_4\text{-}COOH)Fe(C_5H_5)][PF_6]$ ... Fe: Org.Comp.B19-50
– ... $[(4\text{-}CH_3O\text{-}C_6H_4\text{-}COOH)Fe(C_5H_5)][PF_6]$ ... Fe: Org.Comp.B19-50
– ... $[(C_{10}H_{13})Fe(CO)_3][PF_6]$ ... Fe: Org.Comp.B15-124, 176/7
$C_{13}F_6FeH_{13}O_4P$ .. $[(8\text{-}CH_3CH(OH)\text{-}C_8H_8)Fe(CO)_3][PF_6]$ ... Fe: Org.Comp.B15-249
– ... $[(CH_3O\text{-}C_9H_{10})Fe(CO)_3][PF_6]$ ... Fe: Org.Comp.B15-127, 179
$C_{13}F_6FeH_{13}O_5P$ .. $[(2\text{-}CH_3OC(O)CH_2CH_2\text{-}C_6H_6)Fe(CO)_3][PF_6]$ ... Fe: Org.Comp.B15-104, 156
– ... $[(6\text{-}CH_3C(O)OCH_2CH_2\text{-}C_6H_6)Fe(CO)_3][PF_6]$ ... Fe: Org.Comp.B15-108
$C_{13}F_6FeH_{13}P$ .... $[(c\text{-}C_8H_8)Fe(C_5H_5)][PF_6]$ ... Fe: Org.Comp.B19-340, 342, 344/5
$C_{13}F_6FeH_{13}P_2^+$ .. $[(C_5H_5)Fe(CH_2{=}CHC_6H_5)(PF_3)_2]^+$ ... Fe: Org.Comp.B16b-7
$C_{13}F_6FeH_{14}NOP$ .. $[(CH_3\text{-}C({=}N\text{-}OH)\text{-}C_6H_5)Fe(C_5H_5)][PF_6]$ ... Fe: Org.Comp.B18-142/6, 211/2
– ... $[(CH_3\text{-}C({=}O)NH\text{-}C_6H_5)Fe(C_5H_5)][PF_6]$ ... Fe: Org.Comp.B18-142/6, 197, 199, 260/1
– ... $[(C_6H_6)Fe(C_5H_4\text{-}NH\text{-}C({=}O)CH_3)][PF_6]$ ... Fe: Org.Comp.B18-142/6, 167
$C_{13}F_6FeH_{14}NO_2P$ $[(1,3\text{-}(CH_3)_2\text{-}C_6H_3\text{-}2\text{-}NO_2)Fe(C_5H_5)][PF_6]$ ... Fe: Org.Comp.B19-126
$C_{13}F_6FeH_{14}NO_3P$ $[C_5H_5(CO)_2Fe{=}C(\text{-}N(CH_2CH{=}CH_2)CH_2CH_2O\text{-})][PF_6]$ Fe: Org.Comp.B16a-139, 151
$C_{13}F_6FeH_{14}NP$ ... $[(1\text{-}NC_8H_9)Fe(C_5H_5)][PF_6]$ ... Fe: Org.Comp.B19-216, 278
$C_{13}F_6FeH_{15}OP$ ... $[(2\text{-}CH_3\text{-}C_6H_4\text{-}OCH_3)Fe(C_5H_5)][PF_6]$ ... Fe: Org.Comp.B19-1/3, 5, 46
– ... $[(3\text{-}CH_3\text{-}C_6H_4\text{-}OCH_3)Fe(C_5H_5)][PF_6]$ ... Fe: Org.Comp.B19-1/3, 5, 46
– ... $[(4\text{-}CH_3\text{-}C_6H_4\text{-}OCH_3)Fe(C_5H_5)][PF_6]$ ... Fe: Org.Comp.B19-1/3, 5, 47
– ... $[(CH_3CH(OH)\text{-}C_6H_5)Fe(C_5H_5)][PF_6]$ ... Fe: Org.Comp.B18-142/6, 214
– ... $[(C_2H_5\text{-}O\text{-}C_6H_5)Fe(C_5H_5)][PF_6]$ ... Fe: Org.Comp.B18-142/6, 198/9, 250
– ... $[(C_6H_6)Fe(C_5H_4\text{-}CH(OH)CH_3)][PF_6]$ ... Fe: Org.Comp.B18-142/6, 163
– ... $[(C_6H_6)Fe(C_5H_4\text{-}O\text{-}C_2H_5)][PF_6]$ ... Fe: Org.Comp.B18-142/6, 169
$C_{13}F_6FeH_{15}O_2P$ .. $[(1,2\text{-}(CH_3O)_2\text{-}C_6H_4)Fe(C_5H_5)][PF_6]$ ... Fe: Org.Comp.B19-3, 19
– ... $[(1,3\text{-}(CH_3O)_2\text{-}C_6H_4)Fe(C_5H_5)][PF_6]$ ... Fe: Org.Comp.B19-3, 19
– ... $[(1,4\text{-}(CH_3O)_2\text{-}C_6H_4)Fe(C_5H_5)][PF_6]$ ... Fe: Org.Comp.B19-1, 3, 5, 19, 87
– ... $[(C_5H_5)Fe(CO)_2(CH_2{=}CHCH{=}CHC_2H_5)][PF_6]$ ... Fe: Org.Comp.B17-38
– ... $[(C_5H_5)Fe(CO)_2(CH_2{=}CHCH_2CH{=}CHCH_3)][PF_6]$ Fe: Org.Comp.B17-39
– ... $[(C_5H_5)Fe(CO)_2(CH_2{=}CHCH_2CH_2CH{=}CH_2)][PF_6]$ Fe: Org.Comp.B17-39
– ... $[(C_5H_5)Fe(CO)_2(C_5H_7\text{-}3\text{-}CH_3)][PF_6]$ ... Fe: Org.Comp.B17-88
– ... $[(C_5H_5)Fe(CO)_2(C_6H_{10}\text{-}c)][PF_6]$ ... Fe: Org.Comp.B17-90, 104
$C_{13}F_6FeH_{15}O_3P$ .. $[(1\text{-}(i\text{-}C_3H_7)\text{-}4\text{-}CH_3\text{-}C_6H_5)Fe(CO)_3][PF_6]$ ... Fe: Org.Comp.B15-122, 175
– ... $[(6\text{-}(i\text{-}C_3H_7)\text{-}3\text{-}CH_3\text{-}C_6H_5)Fe(CO)_3][PF_6]$ ... Fe: Org.Comp.B15-125, 178/9
– ... $[((CH_3)_5C_5)Fe(CO)_3][PF_6]$ ... Fe: Org.Comp.B15-52/3
$C_{13}F_6FeH_{15}O_4P$ .. $[(1\text{-}CH_3\text{-}4\text{-}(i\text{-}C_3H_7O)\text{-}C_6H_5)Fe(CO)_3][PF_6]$ ... Fe: Org.Comp.B15-121, 173
$C_{13}F_6FeH_{15}O_5P$ .. $[(1\text{-}CH_3OCH_2CH_2\text{-}4\text{-}CH_3O\text{-}C_6H_5)Fe(CO)_3][PF_6]$ Fe: Org.Comp.B15-114, 170
$C_{13}F_6FeH_{15}P$ .... $[(1,2\text{-}(CH_3)_2\text{-}C_6H_4)Fe(C_5H_5)][PF_6]$ ... Fe: Org.Comp.B19-1, 4, 5, 7
– ... $[(1,3\text{-}(CH_3)_2\text{-}C_6H_4)Fe(C_5H_5)][PF_6]$ ... Fe: Org.Comp.B19-1, 4/5, 9
– ... $[(1,4\text{-}(CH_3)_2\text{-}C_6H_4)Fe(C_5H_5)][PF_6]$ ... Fe: Org.Comp.B19-1, 4/5, 11, 81/7

$C_{13}F_6FeH_{15}P$ .... $[(CH_3-C_6H_5)Fe(C_5H_4-CH_3)][PF_6]$ .... Fe: Org.Comp.B18-142/6, 197/8, 207

– .... $[(C_2H_5-C_6H_5)Fe(C_5H_5)][PF_6]$ .... Fe: Org.Comp.B18-142/6, 197, 199, 200, 201, 212/3

– .... $[(C_6H_6)Fe(C_5H_4-C_2H_5)][PF_6]$ .... Fe: Org.Comp.B18-142/6, 151, 162/3, 187

– .... $[(C_6H_6)Fe(C_5H_4-CHDCH_3)][PF_6]$ .... Fe: Org.Comp.B18-187

– .... $[(C_6H_6)Fe(C_5H_4-CD_2CH_3)][PF_6]$ .... Fe: Org.Comp.B18-187

$C_{13}F_6FeH_{15}PS$ ... $[(4-CH_3-C_6H_4-SCH_3)Fe(C_5H_5)][PF_6]$ .... Fe: Org.Comp.B19-1, 51

$C_{13}F_6FeH_{15}Sb$ ... $[(1,4-(CH_3)_2-C_6H_4)Fe(C_5H_5)][SbF_6]$ .... Fe: Org.Comp.B19-1, 5, 11, 81/7

$C_{13}F_6FeH_{16}NO_3PS$
$[C_5H_5(CO)_2Fe=C(SCH_3)(N(CH_2CH_2)_2O)][PF_6]$ .. Fe: Org.Comp.B16a-146

$C_{13}F_6FeH_{16}NP$ ... $[(2,6-(CH_3)_2-C_6H_3-NH_2)Fe(C_5H_5)][PF_6]$ ...... Fe: Org.Comp.B19-99, 124

– .... $[(2-CH_3-C_6H_4-NH_2)Fe(C_5H_4-CH_3)][PF_6]$ ...... Fe: Org.Comp.B19-2, 5/6, 53

– .... $[(3-CH_3-C_6H_4-NH_2)Fe(C_5H_4-CH_3)][PF_6]$ ...... Fe: Org.Comp.B19-2, 5/6, 53

– .... $[(4-CH_3-C_6H_4-NH_2)Fe(C_5H_4-CH_3)][PF_6]$ ...... Fe: Org.Comp.B19-2, 5/6, 54

– .... $[(2-CH_3-C_6H_4-NH-CH_3)Fe(C_5H_5)][PF_6]$ ...... Fe: Org.Comp.B19-2, 54

– .... $[(3-CH_3-C_6H_4-NH-CH_3)Fe(C_5H_5)][PF_6]$ ...... Fe: Org.Comp.B19-2, 54

– .... $[(4-CH_3-C_6H_4-NH-CH_3)Fe(C_5H_5)][PF_6]$ ...... Fe: Org.Comp.B19-2, 55

– .... $[((CH_3)_2N-C_6H_5)Fe(C_5H_5)][PF_6]$ .... Fe: Org.Comp.B18-142/6, 197, 199, 201, 262

– .... $[(CH_3CH(NH_2)-C_6H_5)Fe(C_5H_5)][PF_6]$ .... Fe: Org.Comp.B18-142/6, 214

– .... $[(C_6H_6)Fe(C_5H_4-NH-C_2H_5)][PF_6]$ .... Fe: Org.Comp.B18-142/6, 167

$C_{13}F_6FeH_{16}P_2$ .... $[(CH_3C_6H_5)Fe(PC_4H_2(CH_3)_2-3,4)][PF_6]$ .... Fe: Org.Comp.B18-130

$C_{13}F_6FeH_{17}N_2P$ .. $[(4-NH_2-C_6H_4-N(CH_3)_2)Fe(C_5H_5)][PF_6]$ .... Fe: Org.Comp.B19-79

– .... $[(NH_2-CH_2CH_2-NH-C_6H_5)Fe(C_5H_5)][PF_6]$ ..... Fe: Org.Comp.B18-142/6, 198, 258

$C_{13}F_6FeH_{17}OP$ ... $[(1,3,5-(CH_3)_3-C_6H_3)Fe(CO)(CH_2CHCH_2)][PF_6]$ Fe: Org.Comp.B18-95/7

– .... $[(C_5H_5)Fe(CO)(CH_2=CHC(CH_3)_2CH=CH_2)][PF_6]$ Fe: Org.Comp.B17-187/8

$C_{13}F_6FeH_{17}O_2P$ .. $[(C_5(CH_3)_5)(CO)_2Fe=CH_2][PF_6]$ .... Fe: Org.Comp.B16a-103/4

$C_{13}F_6FeH_{17}O_3P$ .. $[(CH_2(CH)_4-C_5H_{11}-n)Fe(CO)_3][PF_6]$ .... Fe: Org.Comp.B15-26/7

$C_{13}F_6FeH_{17}O_3PSi$ $[(1-CH_3-3-(CH_3)_3Si-C_6H_5)Fe(CO)_3][PF_6]$ ..... Fe: Org.Comp.B15-113

– .... $[(6-CH_3-3-(CH_3)_3Si-C_6H_5)Fe(CO)_3][PF_6]$ ..... Fe: Org.Comp.B15-125

$C_{13}F_6FeH_{17}O_4P$ .. $[(C_5H_5)Fe(CO)_2(CH_2=CHCH_2CH(OCH_3)_2)][PF_6]$ Fe: Org.Comp.B17-33/4

$C_{13}F_6FeH_{17}PS$ ... $[(C_2H_5-C_5H_4)Fe(SC_4H_2-3,4-(CH_3)_2)][PF_6]$ .... Fe: Org.Comp.B17-222

– .... $[(C_5H_5)Fe(SC_4-2,3,4,5-(CH_3)_4)][PF_6]$ .... Fe: Org.Comp.B17-214/5, 233

$C_{13}F_6FeH_{18}O_3P$ .. $[(C_5(CH_3)_5)Fe(CO)_2CH_2OH][PF_6]$ .... Fe: Org.Comp.B14-165

$C_{13}F_6FeH_{19}O_2PSi$ $[((CH_3)_3SiC_5H_4)Fe(CO)_2(CH_2=CHCH_3)][PF_6]$ ... Fe: Org.Comp.B17-132

$C_{13}F_6FeH_{20}N_2O_6PS_3$
$[(C_5H_5)Fe(CO)_2(CH_2=CH(C_2H_5)N(SO_2CH_3)-S$
$-NH-SO_2CH_3)][PF_6]$ .... Fe: Org.Comp.B14-39

$C_{13}F_6FeH_{21}P_2^+$ .. $[(C_5H_5)Fe(CH_2=CHC_6H_{13}-n)(PF_3)_2]^+$ .... Fe: Org.Comp.B16b-6

$C_{13}F_6FeH_{22}O_5P_2$. $[(C_5H_5)Fe(CH_2=CHOC_2H_5)(CO)P(OCH_3)_3][PF_6]$ Fe: Org.Comp.B16b-87, 90

$C_{13}F_6GeH_{18}O_2$ ... $Ge(C_2H_5)_3-C(COOCH_3)=C=C(CF_3)_2$ .... Ge: Org.Comp.2-217/8

– .... $Ge(C_2H_5)_3-C[=C(OCH_3)-C(CF_3)_2-C(=O)-]$ .... Ge: Org.Comp.2-230, 233

$C_{13}F_6GeH_{20}O_2$ ... $Ge(C_2H_5)_3-C[=C(OC_2H_5)-O-C(CF_3)_2-]$ .... Ge: Org.Comp.2-285

$C_{13}F_6GeH_{25}OP$ ... $Ge(C_2H_5)_3-C(CF_3)_2-P(O)(C_2H_5)_2$ .... Ge: Org.Comp.2-129

$C_{13}F_6H_6MoO_4$ .... $[2,3-(CF_3)_2-(2.2.1)-C_7H_6-2,5]Mo(CO)_4$ .... Mo: Org.Comp.5-338, 339

$C_{13}F_6H_8MoO_3S$ ... $(C_5H_5)Mo(CO)[C(O)C(CF_3)=C(CF_3)C(O)SCH_3]$ .. Mo: Org.Comp.6-343/4

$C_{13}F_6H_8MoO_3S$ ... $(C_5H_5)Mo(CO)_2[-C(CF_3)=C(CF_3)-C(=O)-S(CH_3)-]$ Mo:Org.Comp.8-153/4
$C_{13}F_6H_8O_6ReSb$ .. $[(CO)_5ReOC(CH_3)C_6H_5][SbF_6]$ .... Re: Org.Comp.2-157/8
$C_{13}F_6H_{10}NO_5Re$ .. $(CO)_3Re(CNC_4H_9\text{-t})(CF_3COCHCOCF_3)$ .... Re: Org.Comp.2-244/5
$C_{13}F_6H_{11}MoN_2O_2P$
$[(c\text{-}C_7H_7)Mo(CO)_2(N_2C_3H_3)CH][PF_6]$ Mo:Org.Comp.5-235
$C_{13}F_6H_{11}O_5P_2Re$ $[(CO)_5ReP(CH_3)_2C_6H_5][PF_6]$ .... Re: Org.Comp.2-155
$C_{13}F_6H_{12}MoNO_2P$ $[(C_9H_7)Mo(CO)(NO)(CH_2CHCH_2)][PF_6]$ Mo:Org.Comp.6-346
$C_{13}F_6H_{12}MoOS$ ... $(C_5H_5)Mo(CO)(SC_3H_7\text{-i})(F_3C-C{\equiv}C-CF_3)$ Mo:Org.Comp.6-277
$C_{13}F_6H_{13}MoN_4O_2P$
$[(C_5H_5)Mo(CO)_2(1,2\text{-}N_2C_3H_4)_2][PF_6]$ Mo:Org.Comp.7-283, 285
– .... $[(C_5H_5)Mo(CO)_2(1,3\text{-}N_2C_3H_4)_2][PF_6]$ Mo:Org.Comp.7-283, 286
$C_{13}F_6H_{13}MoO_2P$ .. $[(C_5H_5)Mo(CO)_2(C_6H_8)][PF_6]$ Mo:Org.Comp.8-305/6, 313, 317/8
$C_{13}F_6H_{14}N_2OSn$ .. $(CH_3)_3SnN(C_6H_5)CON=C(CF_3)_2$ .... Sn: Org.Comp.18-47, 50
$C_{13}F_6H_{15}MoN_2OP$ $[(C_5H_5)Mo(NO)(C_5H_5N)(CH_2CHCH_2)][PF_6]$ Mo:Org.Comp.6-178
$C_{13}F_6H_{15}MoN_2O_5P$
$[(C_5H_5)Mo(CO)_2(-C(OH)=C(COO-C_2H_5)$
$-N(CH_3)-NH-)][PF_6]$ Mo:Org.Comp.8-140, 169
$C_{13}F_6H_{17}NOReSb$ $[(CH_3C{\equiv}CCH_3)_2Re(O)NC_5H_5][SbF_6]$ .... Re: Org.Comp.2-359, 368/9
$C_{13}F_6H_{18}MoNO_2P$ $[(C_5H_5)Mo(CO)_2((CH_3)_2CCHN(CH_3)_2)][PF_6]$ Mo:Org.Comp.8-185, 186, 195
$C_{13}F_6H_{18}O_4Sn$ ... $(C_4H_9)_2SnOOC(CF_2)_3COO$ .... Sn: Org.Comp.15-322
$C_{13}F_6H_{21}MoO_2P_3$ $[(C_5H_5)Mo(CO)_2((CH_3)_2P-CH_2CH_2-P(CH_3)_2)][PF_6]$ Mo:Org.Comp.7-250, 273
$C_{13}F_6H_{23}MoO_8P_3$ $[(C_5H_5)Mo(CO)_2(P(OCH_3)_3)_2][PF_6]$ Mo:Org.Comp.7-289
$C_{13}F_6H_{29}N_4PPdS$ $[Pd(SCN)((C_2H_5)_2NCH_2CH_2NHCH_2CH_2N(C_2H_5)_2)][PF_6]$ Pd: SVol.B2-307
$C_{13}F_6H_{37}OOsP_5$ .. $[(H)(CO)Os(P(CH_3)_3)_4][PF_6]$ .... Os: Org.Comp.A1-120
$C_{13}F_7GeH_{15}$ ..... $Ge(C_2H_5)_3-C_6F_4CF_3\text{-}3$ .... Ge: Org.Comp.2-278
– .... $Ge(C_2H_5)_3-C_6F_4CF_3\text{-}4$ .... Ge: Org.Comp.2-278
$C_{13}F_7H_9N_2O_2PRe$ $[(C_5H_5)Re(CO)_2-N=N-C_6H_4\text{-}4\text{-}F][PF_6]$ .... Re: Org.Comp.3-193/7, 203
$C_{13}F_7O_5Re$ ...... $(CO)_5ReC_8F_7$ .... Re: Org.Comp.2-142
$C_{13}F_9FeH_5O_2$ .... $(C_5H_5)Fe(CO)_2C[=CF-CF_2-CF_2-CF_2-CF_2-]$.... Fe: Org.Comp.B13-194, 219, 255
$C_{13}F_9FeH_9S$ ..... $(C_5H_5)Fe[C_4HS\text{-}1\text{-}CH_3\text{-}2,4,5\text{-}(CF_3)_3]$ .... Fe: Org.Comp.B17-213
$C_{13}F_9FeH_{12}OP$ ... $[(CF_3CH_2-O-C_6H_5)Fe(C_5H_5)][PF_6]$........... Fe: Org.Comp.B18-142/6, 198/9, 251
$C_{13}F_9HN_2$ .... $4\text{-}NC-C_6F_4-NH-C_6F_5$................... F: PFHOrg.SVol.5-16, 48, 84
$C_{13}F_9H_2NO_2$ ..... $C_6F_5-NH\text{-}2\text{-}C_6F_4-C(=O)OH$ .... F: PFHOrg.SVol.5-16, 47, 85
$C_{13}F_9LiN_2$ .... $C_6F_5CLi(CN)(4\text{-}C_5F_4N)$.................. F: PFHOrg.SVol.4-151, 166
$C_{13}F_9NO$ .... $C_6F_5C_7F_4NO$ .... F: PFHOrg.SVol.4-283, 298
$C_{13}F_9NO_3$ .... $2\text{-}NO_2-C_6F_4-C(=O)-C_6F_5$.................. F: PFHOrg.SVol.5-182, 197
$C_{13}F_9N_2^-$ .... $[N(C_6F_5)-C_6F_4-CN\text{-}4]^-$ .... F: PFHOrg.SVol.5-16, 47
$C_{13}F_9N_2Na$ ...... $Na[NC_5F_4\text{-}4\text{-}C(CN)-C_6F_5]$ .... F: PFHOrg.SVol.4-151, 166
– .... $Na[N(C_6F_5)-C_6F_4-CN\text{-}4]$ .... F: PFHOrg.SVol.5-16, 47
$C_{13}F_{10}HNO$ ...... $C_6F_5-C(=O)NH-C_6F_5$ .... F: PFHOrg.SVol.5-25, 59, 96
$C_{13}F_{10}H_2N_2O$ .... $C_6F_5-C(=O)NH-NH-C_6F_5$.................. F: PFHOrg.SVol.5-207
$C_{13}F_{10}NNaO$..... $Na[N(C_6F_5)-C(=O)-C_6F_5]$ .... F: PFHOrg.SVol.5-25, 59
$C_{13}F_{10}NO^-$ ...... $[N(C_6F_5)-C(=O)-C_6F_5]^-$ .... F: PFHOrg.SVol.5-25, 59
$C_{13}F_{11}FeH_5O$ .... $C_5H_5Fe(CF_2C(CF_3)C=C(CF_3)_2)CO$ .... Fe: Org.Comp.B17-163/4

$C_{13}F_{11}H_6MoO_2S$	$(C_5H_5)Mo(CO)_2(H)[SC(C_2F_5){=}C(CF_3)_2]$, radical	Mo: Org.Comp.7-53/4
$C_{13}F_{11}N$	$C_6F_5{-}N{=}CF{-}C_6F_5$	F: PFHOrg.SVol.6-193, 207
$C_{13}F_{12}HN$	$4{-}CF_3{-}C_6F_4{-}NH{-}C_6F_5$	F: PFHOrg.SVol.5-16, 48, 84
$C_{13}F_{12}H_5IMo$	$C_5H_5Mo(I)(F_3C{-}CC{-}CF_3)_2$	Mo: Org.Comp.6-135, 143/4
$C_{13}F_{12}H_5MoS_4^-$	$[C_5H_5Mo({-}SC(CF_3){=}C(CF_3)S{-})_2]^-$	Mo: Org.Comp.6-5
$C_{13}F_{12}H_6MoOS$	$(C_5H_5)Mo(O)(F_3C{-}C{\equiv}C{-}CF_3)[{-}S{-}C(CF_3){=}CH(CF_3)]$	Mo: Org.Comp.6-103
$C_{13}F_{12}H_6MoS$	$(C_5H_5)Mo(F_3C{-}C{\equiv}C{-}CF_3)[{-}S{-}C(CF_3){=}CH(CF_3)]$	Mo: Org.Comp.6-140
$C_{13}F_{12}H_{17}N_4P_2Re$	$[(C_5H_5)Re(NC{-}CH_3)_4][PF_6]_2$	Re: Org.Comp.3-5
$C_{13}F_{12}N^-$	$[N(C_6F_5){-}C_6F_4{-}CF_3{-}4]^-$	F: PFHOrg.SVol.5-16, 47
$C_{13}F_{12}NNa$	$Na[N(C_6F_5){-}C_6F_4{-}CF_3{-}4]$	F: PFHOrg.SVol.5-16, 47
$C_{13}F_{12}N_2$	$4{-}CF_3{-}C_6F_4{-}N{=}N{-}C_6F_5$	F: PFHOrg.SVol.5-210, 221
$C_{13}F_{15}O_7Re$	$(CO)_5ReOC(O)C_7F_{15}$	Re: Org.Comp.2-23
$C_{13}F_{17}HN_2$	$1,1{-}CF_3(NHC_6F_5){-}5{-}CF_3{-}C_5F_6N$	F: PFHOrg.SVol.4-140
$C_{13}F_{17}H_{13}O_3Sn$	$C_8F_{17}CH_2CH_2Sn(OCH_3)_3$	Sn: Org.Comp.17-55
$C_{13}F_{20}HN$	$3{-}C_5F{-}({=}NH){-}1{-}(CF_3)_4{-}2,2,5,5{-}[CF{=}C(CF_3)_2]{-}4$	F: PFHOrg.SVol.5-28, 67
$C_{13}F_{21}N$	$1,3{-}(i{-}C_3F_7)_2{-}4,6{-}(CF_3)_2{-}[2.2.0]{-}2{-}NC_5F{-}4$	F: PFHOrg.SVol.4-120, 139
–	$2,6{-}(i{-}C_3F_7)_2{-}3,5{-}(CF_3)_2{-}NC_5F$	F: PFHOrg.SVol.4-118/9, 129, 139
$C_{13}F_{22}N_2O_2$	$(CF_3)_2C{=}N{-}C({=}O){-}(CF_2)_5{-}C({=}O){-}N{=}C(CF_3)_2$	F: PFHOrg.SVol.6-191, 204
$C_{13}F_{23}NO_4$	$CF_3{-}CF(CN)[{-}CF_2{-}O{-}CF(CF_3)]_3{-}CF{=}O$	F: PFHOrg.SVol.6-101, 121
$C_{13}F_{24}H_3N_3O$	$C_3F_7{-}O{-}CF(CF_3)C({=}NH){-}N{=}C(NH_2){-}C_6F_{13}$	F: PFHOrg.SVol.5-5
$C_{13}F_{24}H_3N_3O_2$	$C_3F_7{-}C({=}NH){-}N{=}C(NH_2)CF(CF_3){-}O{-}CF_2CF(CF_3){-}O{-}C_3F_7$	F: PFHOrg.SVol.5-5
$C_{13}F_{24}N_2$	$NCF(i{-}C_3F_7)NC(i{-}C_3F_7)CF_2C(i{-}C_3F_7)$	F: PFHOrg.SVol.4-192/3, 206
$C_{13}F_{24}N_4$	$(CF_2CF_2N(CF_3)CF_2N)C_5F_6(NCF_2N(CF_3)CF_2CF_2)$	F: PFHOrg.SVol.4-36/7, 63
$C_{13}F_{26}N_2$	$(C_6F_{13})_2C{=}N_2$	F: PFHOrg.SVol.5-203
$C_{13}F_{29}NO_2$	$n{-}C_3F_7{-}N(CF_2CF_2{-}O{-}C_3F_7{-}n)_2$	F: PFHOrg.SVol.6-226/7
–	$n{-}C_5F_{11}{-}N(CF_2CF_2{-}O{-}C_2F_5)_2$	F: PFHOrg.SVol.6-226/7
$C_{13}FeGeH_{16}O_3$	$Ge(CH_3)_3(CH(CH{=}CHCH{=}CHCH{=}CH))Fe(CO)_3$	Ge: Org.Comp.2-38
$C_{13}FeGeH_{18}$	$Ge(CH_3)_3C_{10}H_9Fe$	Ge: Org.Comp.2-34
$C_{13}FeH_5O_7ReS_2$	$(C_5H_5)Fe(CO)_2C(S)S{-}Re(CO)_5$	Fe: Org.Comp.B13-57, 59/60 Re: Org.Comp.2-176, 182
$C_{13}FeH_5O_{10}Re$	$(CO)_5ReCH{=}C(H)C(O)CH_3{-}Fe(CO)_4$	Re: Org.Comp.2-181, 184
$C_{13}FeH_7O_6Re$	$(C_7H_7)Fe(CO)_3Re(CO)_3$	Re: Org.Comp.2-417/8
$C_{13}FeH_9IO_2$	$(C_5H_5)Fe(CO)_2C_6H_4I{-}2$	Fe: Org.Comp.B13-171/2
$C_{13}FeH_{10}INO$	$(C_5H_5)Fe(CN{-}C_6H_5)(CO)I$	Fe: Org.Comp.B15-293, 297
$C_{13}FeH_{10}MnNO_3S_2$	$(C_5H_5)(CO)_2Fe{=}CS_2Mn(NO)(C_5H_5)$	Fe: Org.Comp.B16a-183, 188
$C_{13}FeH_{10}N_2$	$(C_5H_5)Fe[C_6H_5(CN)_2{-}1,6]$	Fe: Org.Comp.B17-258
$C_{13}FeH_{10}N_2O$	$(C_5H_5)Fe(CO)[C(CN){=}C(CN)CH_2CH{=}CH_2]$	Fe: Org.Comp.B17-170
$C_{13}FeH_{10}NiO_3$	$(C_5H_5)(CO)Fe(CO)_2Ni(C_5H_5)$	Fe: Org.Comp.B16b-142/3
$C_{13}FeH_{10}OSTe$	$(C_5H_5)Fe(CS)(CO)Te{-}C_6H_5$	Fe: Org.Comp.B15-268
$C_{13}FeH_{10}OS_2$	$(C_5H_5)Fe(CS)(CO)S{-}C_6H_5$	Fe: Org.Comp.B15-268/9
$C_{13}FeH_{10}O_2$	$(C_5H_5)Fe(CO)_2(CH_2C{\equiv}CC{\equiv}CCH_3)$	Fe: Org.Comp.B13-155
–	$(C_5H_5)Fe(CO)_2{-}C_6H_5$	Fe: Org.Comp.B13-168/9, 179/82
–	$(C_5H_4D)Fe(CO)_2{-}C_6H_5$	Fe: Org.Comp.B14-53

$C_{13}FeH_{10}O_3$ ..... $(C_9H_7)Fe(CO)_2C(O)CH_3$ .... Fe: Org.Comp.B14-67, 74
– .... $(C_9H_7)Fe(CO)_2{}^{13}C(O)CH_3$ .... Fe: Org.Comp.B14-74
$C_{13}FeH_{10}O_4$ ..... $(C_5H_5)Fe(CO)_2C(=O)-2-C_4H_2O(CH_3)-5$ .... Fe: Org.Comp.B13-8, 10/1, 23, 43
$C_{13}FeH_{10}O_4S$ .... $[C_5H_5Fe(CO)_2-1-C_6H_5{}^--1-SO_2{}^+]$ .... Fe: Org.Comp.B13-167
$C_{13}FeH_{11}N$ ...... $C_5H_5FeC_8H_6N$ .... Fe: Org.Comp.B17-203
$C_{13}FeH_{11}NO$ ..... $NC-C_5H_4FeC_5H_4-COCH_3$ .... Fe: Org.Comp.A9-114, 120/1
$C_{13}FeH_{11}NO^+$ .... $[NC-C_5H_4FeC_5H_4-COCH_3]^+$ .... Fe: Org.Comp.A9-121
$C_{13}FeH_{11}NO_2$ .... $(C_5H_5)Fe(CO)_2CH_2-2-NC_5H_4$ .... Fe: Org.Comp.B13-3, 6
– .... $(C_5H_5)Fe(CO)_2CH_2-3-NC_5H_4$ .... Fe: Org.Comp.B13-3, 6
– .... $(C_5H_5)Fe(CO)_2CH_2-4-NC_5H_4$ .... Fe: Org.Comp.B13-3, 6
– .... $(C_5H_5)Fe(CO)_2-CH[-CH=CHCH_2CH(CN)-]$ .... Fe: Org.Comp.B13-195, 210
– .... $(C_5H_5)Fe[C_5H_3(CN)-C(O)OCH_3]$ .... Fe: Org.Comp.A9-130, 133, 136, 138
– .... $(NC-C_5H_4)Fe[C_5H_4-C(O)OCH_3]$ .... Fe: Org.Comp.A9-114/5, 122
$C_{13}FeH_{11}N_3O_2$ ... $N_3-C(O)-C_5H_4FeC_5H_4-C(O)CH_3$ .... Fe: Org.Comp.A9-145, 147
$C_{13}FeH_{11}O_2{}^+$ .... $[(C_5H_5)Fe(CO)_2(CH_2=C=CH-C\equiv C-CH_3)]^+$ ..... Fe: Org.Comp.B17-112, 117
– .... $[(C_9H_7)Fe(CO)_2(CH_2=CH_2)]^+$ .... Fe: Org.Comp.B17-133, 136
$C_{13}FeH_{11}O_3{}^+$ .... $[(5-CH_3C_9H_8)Fe(CO)_3]^+$ .... Fe: Org.Comp.B15-254
– .... $[(9-CH_3C_9H_8)Fe(CO)_3]^+$ .... Fe: Org.Comp.B15-254
– .... $[(8-CH_3CH=C_8H_7)Fe(CO)_3]^+$ .... Fe: Org.Comp.B15-241/2, 258
– .... $[(C_{10}H_{11})Fe(CO)_3]^+$ .... Fe: Org.Comp.B15-111, 256
– .... $[(c-C_3H_5-C_7H_6)Fe(CO)_3]^+$ .... Fe: Org.Comp.B15-219, 233
$C_{13}FeH_{11}O_4{}^+$ .... $[(1,4-(HOOC)_2-C_6H_4)Fe(C_5H_5)][BF_4]$ .... Fe: Org.Comp.B19-16
– .... $[(1,4-(HOOC)_2-C_6H_4)Fe(C_5H_5)][PF_6]$ .... Fe: Org.Comp.B19-15
– .... $[(8-CH_3C(O)-C_8H_8)Fe(CO)_3]^+$ .... Fe: Org.Comp.B15-201/2, 250/1
– .... $[(HOOC-C_6H_5)Fe(C_5H_4-COOH)]^+$ .... Fe: Org.Comp.B18-142/6, 203
$C_{13}FeH_{12}N$ ...... $[4-CH_3-C_6H_4-CN]Fe(C_5H_5)$ .... Fe: Org.Comp.B19-23
$C_{13}FeH_{12}N^+$ ..... $[(2-CH_3-C_6H_4-CN)Fe(C_5H_5)][PF_6]$ .... Fe: Org.Comp.B19-2, 5, 23
– .... $[(4-CH_3-C_6H_4-CN)Fe(C_5H_5)][PF_6]$ .... Fe: Org.Comp.B19-2, 23
– .... $[(NC-C_5H_4)Fe(C_5H_4-CHCH_3)]^+$ .... Fe: Org.Comp.A9-120
$C_{13}FeH_{12}NO^+$ .... $[NC-C_5H_4FeC_5H_4-C(CH_3)=OH]^+$ .... Fe: Org.Comp.A9-121
$C_{13}FeH_{12}NO_2{}^+$ ... $[(C_5H_5)Fe(CO)_2CH_2-1-NC_5H_5]^+$ .... Fe: Org.Comp.B14-136
– .... $[(C_5H_5)Fe(CO)_2CH_2-2-NC_5H_5]^+$ .... Fe: Org.Comp.B14-151
– .... $[(C_5H_5)Fe(CO)_2CH_2-3-NC_5H_5]^+$ .... Fe: Org.Comp.B14-151, 156
– .... $[(C_5H_5)Fe(CO)_2CH_2-4-NC_5H_5]^+$ .... Fe: Org.Comp.B14-152, 156
– .... $[(C_5H_5)Fe(CO)_2(C_5H_7-3-CN)]^+$ .... Fe: Org.Comp.B17-88
$C_{13}FeH_{12}N_2O_2$ ... $(C_5H_5)Fe(1-O_2N-2-CH_3-6-NC-C_6H_4)$ .... Fe: Org.Comp.B17-289
– .... $(C_5H_5)Fe(1-O_2N-6-CH_3-6-NC-C_6H_4)$ .... Fe: Org.Comp.B17-284
$C_{13}FeH_{12}N_2O_4$ ... $(C_5H_5)Fe(CO)_2C[=CH-C(=O)-N(CH_3)-C(=O)-N(CH_3)-]$ Fe: Org.Comp.B14-26, 30
$C_{13}FeH_{12}O$ ...... $(C_5H_5)Fe(CO)(CH_2C_6H_5)$ .... Fe: Org.Comp.B17-157, 158/9
– .... $(C_5H_5)Fe(CO)(C_7H_7)$ .... Fe: Org.Comp.B17-160/1
– .... $(C_9H_7)Fe(CO)(CH_2CHCH_2)$ .... Fe: Org.Comp.B17-183
$C_{13}FeH_{12}O_2$ ..... $(C_5H_5)Fe(CO)_2-C_5H_4-CH_3$ .... Fe: Org.Comp.B13-214, 251
– .... $[(4.4.1.0^{2,5})-C_{11}H_{12}]Fe(CO)_2$ .... Fe: Org.Comp.B18-18, 33
$C_{13}FeH_{12}O_3$ ..... $(C_5H_5)Fe(CO)_2-C(=O)-CH=CHC(CH_3)=CH_2$ .... Fe: Org.Comp.B13-10, 19
– .... $(C_5H_5)Fe(CO)_2-C(=O)-CH=CHCH=CHCH_3$ ..... Fe: Org.Comp.B13-10, 19
– .... $(C_5H_5)Fe(CO)_2-C[=C(CH_3)-C(=O)-CH_2CH_2-]$ .. Fe: Org.Comp.B13-195, 210

$C_{13}FeH_{12}O_3$ ..... $(C_5H_5)Fe(CO)_2$-C[=CH-C(=O)-$CH_2CH_2CH_2$-] .. Fe: Org.Comp.B13-195, 217
– ..... $(C_9H_7)Fe(CO)_2$-CH_2OCH_3 ..... Fe: Org.Comp.B14-67
– ..... [(4-CH_3O-C_6H_4-COO)Fe(C_5H_5)] · Na[PF_6] ... Fe: Org.Comp.B19-50
$C_{13}FeH_{12}O_4$ ..... [HOC(=O)-C_5H_4]Fe[$C_5H_3(CH_3)$-C(=O)OH] ..... Fe: Org.Comp.A10-266/70, 276
$C_{13}FeH_{12}O_6$ ..... $(C_5H_5)Fe(CO)_2C(COOCH_3)$=$CHCOOCH_3$ ..... Fe: Org.Comp.B13-96, 124
$C_{13}FeH_{13}^+$ ..... [(c-C_8H_8)Fe(C_5H_5)][BF_4] ..... Fe: Org.Comp.B19-340, 342, 345
– ..... [(c-C_8H_8)Fe(C_5H_5)][PF_6] ..... Fe: Org.Comp.B19-340, 342, 344/5
$C_{13}FeH_{13}IN_2O$ .... [$C_5H_5Fe(CNCH_3)(CO)NC_5H_5$]I ..... Fe: Org.Comp.B15-300
$C_{13}FeH_{13}N$ ..... (CH_3-C_5H_4)Fe[C_5H_3(CN)-CH_3] ..... Fe: Org.Comp.A10-294
– ..... (C_5H_5)Fe(C_6H_5-1-CN-2-CH_3) ..... Fe: Org.Comp.B17-276/7
– ..... (C_5H_5)Fe(C_6H_5-1-CN-6-CH_3) ..... Fe: Org.Comp.B17-272
– ..... (C_5H_5)Fe(C_6H_5-2-CN-1-CH_3) ..... Fe: Org.Comp.B17-277
– ..... (C_5H_5)Fe(C_6H_5-2-CN-3-CH_3) ..... Fe: Org.Comp.B17-277
– ..... (C_5H_5)Fe(C_6H_5-3-CN-2-CH_3) ..... Fe: Org.Comp.B17-277
– ..... (C_5H_5)Fe[$C_5H_3(CH_3)$-CH_2CN] ..... Fe: Org.Comp.A9-131, 134
– ..... (C_5H_5)Fe[C_5H_3(CN)-C_2H_5] ..... Fe: Org.Comp.A9-130, 132/3, 136, 138
– ..... (C_6H_6)Fe(C_6H_7-CN) ..... Fe: Org.Comp.B18-103, 105, 107
– ..... $FeC_{10}H_8(CN)(C_2H_5)$ ..... Fe: Org.Comp.A9-140/1
– ..... (NC-C_5H_4)Fe(C_5H_4-C_2H_5) ..... Fe: Org.Comp.A9-113, 120
$C_{13}FeH_{13}NO$ ..... NC-$C_5H_4FeC_5H_4$-CH(CH_3)OH ..... Fe: Org.Comp.A9-113/4, 120
$C_{13}FeH_{13}NO_2$ .... H_2NCO-$C_5H_4FeC_5H_4$-$COCH_3$ ..... Fe: Org.Comp.A9-79, 82
$C_{13}FeH_{13}NO_3$ .... H_2NCO-$C_5H_4FeC_5H_4$-C(O)OCH_3 ..... Fe: Org.Comp.A9-80, 82
$C_{13}FeH_{13}NO_6S$ ... (C_5H_5)Fe$(CO)_2$C[-CH_2-N($COOCH_3$)-S$(=O)_2$-C(CH_3)=] Fe: Org.Comp.B14-34
$C_{13}FeH_{13}N_2O^+$ ... [$C_5H_5Fe(CNCH_3)(CO)NC_5H_5$]$^+$ ..... Fe: Org.Comp.B15-300
$C_{13}FeH_{13}O$ ..... (C_6H_6)Fe[C_5H_4-C(O)CH_3] ..... Fe: Org.Comp.B18-146/8, 162, 186/7
$C_{13}FeH_{13}O^+$ ..... [(1-OC_8H_8)Fe(C_5H_5)][PF_6] ..... Fe: Org.Comp.B19-217, 278
– ..... [(CH_3C(=O)C_6H_5)Fe(C_5H_5)]$^+$ ..... Fe: Org.Comp.B18-142/6, 210
– ..... [(C_5H_5)Fe(CO)(C_7H_8)]$^+$ ..... Fe: Org.Comp.B17-188, 192
– ..... [(C_6H_6)Fe(C_5H_4-C(O)CH_3)]$^+$ ..... Fe: Org.Comp.B18-142/6, 151, 162, 186/7
$C_{13}FeH_{13}O_2$ ..... (CH_3-OOC-C_6H_5)Fe(C_5H_5) ..... Fe: Org.Comp.B18-146/8, 211
– ..... (C_6H_6)Fe(C_5H_4-COO-CH_3) ..... Fe: Org.Comp.B18-146/8, 162
$C_{13}FeH_{13}O_2^+$ .... [(CH_3-C_6H_5)Fe(C_5H_4-COOH)]$^+$ ..... Fe: Org.Comp.B18-142/6, 207
– ..... [(C_5H_5)Fe(4-CH_3-C_6H_4-CO_2H)][BF_4] ..... Fe: Org.Comp.B19-23
– ..... [(C_5H_5)Fe(4-CH_3-C_6H_4-CO_2H)][PF_6] ..... Fe: Org.Comp.B19-23
– ..... [(C_5H_5)Fe$(CO)_2$(C_6H_8-c)]$^+$ ..... Fe: Org.Comp.B17-89/90
– ..... [(C_5H_5)Fe(C_6H_5-CH_2-COOH)]$^+$ ..... Fe: Org.Comp.B18-142/6, 199, 210
– ..... [(C_5H_5)Fe(C_6H_5-C(O)O-CH_3)]$^+$ ..... Fe: Org.Comp.B18-142/6, 200, 201, 211
– ..... [(C_6H_6)Fe(C_5H_4-C(O)O-CH_3)]$^+$ ..... Fe: Org.Comp.B18-142/6, 151, 162
$C_{13}FeH_{13}O_2^-$ .... [(CH_3-C_5H_4)Fe($C_5H_3(CH_3)$-C(O)O)]$^-$ ..... Fe: Org.Comp.A10-274

$C_{13}FeH_{13}O_3^+$	$[(2,8\text{-}(CH_3)_2\text{-}(5.1.0)\text{-}C_8H_7)Fe(CO)_3]^+$	Fe: Org.Comp.B15-216
–	$[(8,8\text{-}(CH_3)_2\text{-}(5.1.0)\text{-}C_8H_7)Fe(CO)_3]^+$	Fe: Org.Comp.B15-215
–	$[(3\text{-}CH_3O\text{-}C_6H_4\text{-}COOH)Fe(C_5H_5)][PF_6]$	Fe: Org.Comp.B19-50
–	$[(4\text{-}CH_3O\text{-}C_6H_4\text{-}COOH)Fe(C_5H_5)][PF_6]$	Fe: Org.Comp.B19-50
–	$[(6\text{-}(CH_3)_2C{=}C_7H_7)Fe(CO)_3]^+$	Fe: Org.Comp.B15-211
–	$[(i\text{-}C_3H_7\text{-}C_7H_6)Fe(CO)_3]^+$	Fe: Org.Comp.B15-219, 233
–	$[(C_{10}H_{13})Fe(CO)_3]^+$	Fe: Org.Comp.B15-124, 176/7
$C_{13}FeH_{13}O_4^+$	$[(2\text{-}HO\text{-}8,8\text{-}(CH_3)_2\text{-}(5.1.0)\text{-}C_8H_6)Fe(CO)_3]^+$	Fe: Org.Comp.B15-221, 233
–	$[(6,6\text{-}(CH_3)_2\text{-}8\text{-}(O{=})\text{-}C_8H_7)Fe(CO)_3]^+$	Fe: Org.Comp.B15-244
–	$[(8\text{-}CH_3CH(OH)\text{-}(3.2.1)\text{-}C_8H_8)Fe(CO)_3]^+$	Fe: Org.Comp.B15-249
–	$[(CH_3O\text{-}C_9H_{10})Fe(CO)_3]^+$	Fe: Org.Comp.B15-127, 179
$C_{13}FeH_{13}O_5^+$	$[(1\text{-}CH_3OC(O)CH(CH_3)\text{-}C_6H_6)Fe(CO)_3]^+$	Fe: Org.Comp.B15-96/7, 100, 136
–	$[(2\text{-}CH_3OC(O)CH_2CH_2\text{-}C_6H_6)Fe(CO)_3]^+$	Fe: Org.Comp.B15-104, 156
–	$[(6\text{-}CH_3C(O)OCH_2CH_2\text{-}C_6H_6)Fe(CO)_3]^+$	Fe: Org.Comp.B15-108
–	$[(1\text{-}CH_3OC(O)CH_2\text{-}C_7H_8)Fe(CO)_3]^+$	Fe: Org.Comp.B15-205, 222
–	$[(2\text{-}C_2H_5\text{-}OC(O)\text{-}C_7H_8)Fe(CO)_3]^+$	Fe: Org.Comp.B15-206
–	$[(1\text{-}CH_3OC(O)\text{-}C_8H_{10})Fe(CO)_3]^+$	Fe: Org.Comp.B15-238
$C_{13}FeH_{13}O_6^+$	$[(1\text{-}CH_3OC(O)CH_2\text{-}4\text{-}CH_3O\text{-}C_6H_5)Fe(CO)_3]^+$	Fe: Org.Comp.B15-115, 171
$C_{13}FeH_{14}$	$(C_6H_6)Fe(C_7H_8)$	Fe: Org.Comp.B18-103, 109
$C_{13}FeH_{14}K_2$	$Fe[C_{10}H_7(K)_2(C_3H_7\text{-}i)]$	Fe: Org.Comp.A10-328
$C_{13}FeH_{14}LiNO$	$(C_5H_5)Fe[C_5H_3(Li)\text{-}C({=}O)NH\text{-}C_2H_5]$	Fe: Org.Comp.A10-122, 126
$C_{13}FeH_{14}Li_2$	$(Li\text{-}C_5H_4)Fe[C_5H_3(Li)\text{-}C_3H_7\text{-}n]$	Fe: Org.Comp.A10-322/3, 325
–	$(Li\text{-}C_5H_4)Fe[C_5H_3(Li)\text{-}C_3H_7\text{-}i]$	Fe: Org.Comp.A10-322, 325, 326
$C_{13}FeH_{14}N^+$	$[(1\text{-}NC_8H_9)Fe(C_5H_5)][PF_6]$	Fe: Org.Comp.B19-216, 278
$C_{13}FeH_{14}NO^+$	$[(CH_3\text{-}C({=}N\text{-}OH)\text{-}C_6H_5)Fe(C_5H_5)]^+$	Fe: Org.Comp.B18-142/6, 211/2
–	$[(CH_3\text{-}C({=}O)NH\text{-}C_6H_5)Fe(C_5H_5)]^+$	Fe: Org.Comp.B18-142/6, 197, 199, 260/1
–	$[(C_6H_6)Fe(C_5H_4\text{-}NH\text{-}C({=}O)CH_3)]^+$	Fe: Org.Comp.B18-142/6, 167/8
$C_{13}FeH_{14}NO_2^+$	$[(1,3\text{-}(CH_3)_2\text{-}C_6H_3\text{-}2\text{-}NO_2)Fe(C_5H_5)][PF_6]$	Fe: Org.Comp.B19-126
$C_{13}FeH_{14}NO_3^+$	$[C_5H_5(CO)_2Fe{=}C(\text{-}N(CH_2CH{=}CH_2)CH_2CH_2O\text{-})]^+$	Fe: Org.Comp.B16a-139, 151
$C_{13}FeH_{14}Na_2$	$Fe[C_{10}H_7(Na)_2(C_3H_7\text{-}i)]$	Fe: Org.Comp.A10-328
$C_{13}FeH_{14}O$	$(CH_3\text{-}C_5H_4)Fe[C_5H_3(CH_3)\text{-}CH{=}O]$	Fe: Org.Comp.A10-238
$C_{13}FeH_{14}O_2$	$(CH_3\text{-}C_5H_4)Fe[C_5H_3(CH_3)\text{-}C({=}O)OH]$	Fe: Org.Comp.A10-266/74
–	$(C_5H_5)Fe(CO)[CH(CH_3)C_5H_5({=}O)]$	Fe: Org.Comp.B17-157
–	$(C_5H_5)Fe(CO)[CH_2C_5H_4({=}O)(CH_3)]$	Fe: Org.Comp.B17-157
–	$(C_5H_5)Fe(CO)_2\text{-}CH{=}CHC(C_2H_5){=}CH_2$	Fe: Org.Comp.B13-98
–	$(C_5H_5)Fe(CO)_2\text{-}CH[\text{-}CH{=}CHCH_2CH(CH_3)\text{-}]$	Fe: Org.Comp.B13-195, 210
–	$(C_5H_5)Fe(CO)_2\text{-}CH[\text{-}CH{=}CH\text{-}CH_2CH_2CH_2\text{-}]$	Fe: Org.Comp.B13-195, 216, 253/4
–	$(C_5H_5)Fe(CO)_2\text{-}CH_2CH{=}CH\text{-}CH_2CH{=}CH_2$	Fe: Org.Comp.B13-108, 135
–	$(C_5H_5)Fe(CO)_2\text{-}CH_2CH_2\text{-}C(CH_3){=}C{=}CH_2$	Fe: Org.Comp.B13-112, 138/9
–	$(C_5H_5)Fe(CO)_2\text{-}CH_2CH_2\text{-}CH{=}C{=}CHCH_3$	Fe: Org.Comp.B13-112
–	$(C_5H_5)Fe(CO)_2\text{-}C{\equiv}C\text{-}C_4H_9\text{-}n$	Fe: Org.Comp.B13-148, 154
–	$(C_5H_5)Fe(C_6H_6\text{-}1\text{-}COOCH_3)$	Fe: Org.Comp.B17-266, 308
–	$(C_5H_5)Fe(C_6H_5\text{-}1\text{-}COOCH_3\text{-}6\text{-}D)$	Fe: Org.Comp.B17-271

$C_{13}FeH_{14}O_2$ ..... $(C_5H_5)Fe(C_6H_6\text{-}2\text{-}COOCH_3)$ ..... Fe: Org.Comp.B17-266, 308
– ..... $(C_5H_5)Fe(C_6H_6\text{-}3\text{-}COOCH_3)$ ..... Fe: Org.Comp.B17-266, 308
– ..... $(C_5H_5)Fe(C_6H_6\text{-}6\text{-}COOCH_3)$ ..... Fe: Org.Comp.B17-266
$C_{13}FeH_{14}O_3$ ..... $(C_5H_5)Fe(CO)[CH_2C(COO\text{-}C_2H_5)C{=}CH_2]$ ..... Fe: Org.Comp.B17-162
– ..... $(C_5H_5)Fe(CO)_2\text{-}C[{=}CH\text{-}C(CH_3)_2\text{-}O\text{-}CH_2\text{-}]$ ..... Fe: Org.Comp.B14-4, 11
– ..... $(C_5H_5)Fe(CO)_2\text{-}C[{=}CH\text{-}CH_2\text{-}O\text{-}C(CH_3)_2\text{-}]$ ..... Fe: Org.Comp.B14-4, 11/2
– ..... $(C_5H_5)Fe(CO)_2\text{-}CH[\text{-}C({=}CHCH_3)\text{-}O\text{-}CH_2CH_2\text{-}]$ Fe: Org.Comp.B14-3
– ..... $(C_5H_5)Fe(CO)_2\text{-}C({=}CH_2)\text{-}C({=}CH_2)\text{-}O\text{-}C_2H_5$ ..... Fe: Org.Comp.B13-98
– ..... $(C_5H_5)Fe(CO)_2\text{-}C({=}O)\text{-}CH_2CH_2C(CH_3){=}CH_2$ ..... Fe: Org.Comp.B13-19
– ..... $(C_5H_5)Fe(CO)_2\text{-}C({=}O)\text{-}CH_2CH_2CH{=}CHCH_3$ ..... Fe: Org.Comp.B13-19, 39
– ..... $(C_5H_5)Fe(CO)_2\text{-}C({=}O)\text{-}CH_2CH_2CH_2\text{-}CH{=}CH_2$ ..... Fe: Org.Comp.B13-19, 39
– ..... $(C_5H_5)Fe(CO)_2\text{-}CH(CHO)CH_2CH_2CH{=}CH_2$ ..... Fe: Org.Comp.B13-111, 138
– ..... $(C_5H_5)Fe(CO)_2\text{-}CH_2\text{-}C{\equiv}C\text{-}C(OH)(CH_3)_2$ ..... Fe: Org.Comp.B13-155
$C_{13}FeH_{14}O_4$ ..... $(CH_3O\text{-}C_5H_4)Fe[C_5H_3(OCH_3)\text{-}C({=}O)OH]$ ..... Fe: Org.Comp.A10-266, 268
– ..... $(C_5H_5)Fe(CO)_2\text{-}CH_2CH{=}CH\text{-}2\text{-}(1,3\text{-}O_2C_3H_5)$ ..... Fe: Org.Comp.B13-103
$C_{13}FeH_{14}O_4S$ ..... $CH_3O\text{-}S(O)_2\text{-}C_5H_4FeC_5H_4\text{-}C(O)CH_3$ ..... Fe: Org.Comp.A9-231
$C_{13}FeH_{14}O_5S$ ..... $CH_3O\text{-}S(O)_2\text{-}C_5H_4FeC_5H_4\text{-}COOCH_3$ ..... Fe: Org.Comp.A9-232
$C_{13}FeH_{15}$ ..... $(C_2H_5\text{-}C_6H_5)Fe(C_5H_5)$ ..... Fe: Org.Comp.B18-146/8, 200/1, 213/4
– ..... $(C_6H_6)Fe(C_5H_4\text{-}C_2H_5)$ ..... Fe: Org.Comp.B18-146/8, 163, 187
– ..... $[1,3\text{-}(CH_3)_2\text{-}C_6H_4]Fe(C_5H_5)$ ..... Fe: Org.Comp.B19-9, 10
– ..... $[1,4\text{-}(CH_3)_2\text{-}C_6H_4]Fe(C_5H_5)$ ..... Fe: Org.Comp.B19-12/3
$C_{13}FeH_{15}^+$ ..... $[(1,2\text{-}(CH_3)_2\text{-}C_6H_4)Fe(C_5H_5)][BF_4]$ ..... Fe: Org.Comp.B19-1, 4, 5, 7/8
– ..... $[(1,2\text{-}(CH_3)_2\text{-}C_6H_4)Fe(C_5H_5)][B_{11}H_{14}]$ ..... Fe: Org.Comp.B19-1, 5, 8
– ..... $[(1,2\text{-}(CH_3)_2\text{-}C_6H_4)Fe(C_5H_5)][Cr(NH_3)_2(SCN)_4]$ Fe: Org.Comp.B19-1, 8
– ..... $[(1,2\text{-}(CH_3)_2\text{-}C_6H_4)Fe(C_5H_5)][I_3]$ ..... Fe: Org.Comp.B19-1, 5, 7
– ..... $[(1,2\text{-}(CH_3)_2\text{-}C_6H_4)Fe(C_5H_5)][PF_6]$ ..... Fe: Org.Comp.B19-1, 4, 5, 7
– ..... $[(1,2\text{-}(CH_3)_2\text{-}C_6H_4)Fe(C_5H_5)]_2[B_{10}H_{10}]$ ..... Fe: Org.Comp.B19-1, 5, 8
– ..... $[(1,2\text{-}(CH_3)_2\text{-}C_6H_4)Fe(C_5H_5)]_2[B_{12}H_{12}]$ ..... Fe: Org.Comp.B19-1, 5, 8
– ..... $[(1,3\text{-}(CH_3)_2\text{-}C_6H_4)Fe(C_5H_5)]^+$ ..... Fe: Org.Comp.B19-9
– ..... $[(1,3\text{-}(CH_3)_2\text{-}C_6H_4)Fe(C_5H_5)][BF_4]$ ..... Fe: Org.Comp.B19-1, 4/5, 9/10
– ..... $[(1,3\text{-}(CH_3)_2\text{-}C_6H_4)Fe(C_5H_5)][Cr(NH_3)_2(SCN)_4]$ Fe: Org.Comp.B19-1, 10
– ..... $[(1,3\text{-}(CH_3)_2\text{-}C_6H_4)Fe(C_5H_5)][I_3]$ ..... Fe: Org.Comp.B19-1, 5, 9
– ..... $[(1,3\text{-}(CH_3)_2\text{-}C_6H_4)Fe(C_5H_5)][PF_6]$ ..... Fe: Org.Comp.B19-1, 4/5, 9
– ..... $[(1,4\text{-}(CH_3)_2\text{-}C_6H_4)Fe(C_5H_5)]^+$ ..... Fe: Org.Comp.B19-11, 87
– ..... $[(1,4\text{-}(CH_3)_2\text{-}C_6H_4)Fe(C_5H_5)][AsF_6]$ ..... Fe: Org.Comp.B19-1, 5, 11, 81/7
– ..... $[(1,4\text{-}(CH_3)_2\text{-}C_6H_4)Fe(C_5H_5)][B(C_6H_5)_4]$ ..... Fe: Org.Comp.B19-1, 5, 12
– ..... $[(1,4\text{-}(CH_3)_2\text{-}C_6H_4)Fe(C_5H_5)][BF_4]$ ..... Fe: Org.Comp.B19-1, 4/5, 11/2, 81/7
– ..... $[(1,4\text{-}(CH_3)_2\text{-}C_6H_4)Fe(C_5H_5)][B_{11}H_{14}]$ ..... Fe: Org.Comp.B19-1, 5, 12
– ..... $[(1,4\text{-}(CH_3)_2\text{-}C_6H_4)Fe(C_5H_5)][Cr(NH_3)_2(SCN)_4]$ Fe: Org.Comp.B19-1, 12
– ..... $[(1,4\text{-}(CH_3)_2\text{-}C_6H_4)Fe(C_5H_5)][I_3]$ ..... Fe: Org.Comp.B19-1, 5, 11
– ..... $[(1,4\text{-}(CH_3)_2\text{-}C_6H_4)Fe(C_5H_5)][O_3S\text{-}CF_3]$ ..... Fe: Org.Comp.B19-1, 5, 12, 81/7
– ..... $[(1,4\text{-}(CH_3)_2\text{-}C_6H_4)Fe(C_5H_5)][PF_6]$ ..... Fe: Org.Comp.B19-1, 4/5, 11, 81/7

$C_{13}FeH_{15}^+$ ...... $[(1,4-(CH_3)_2-C_6H_4)Fe(C_5H_5)][SbF_6]$ .... Fe: Org.Comp.B19-1, 5, 11, 81/7
– .... $[(1,4-(CH_3)_2-C_6H_4)Fe(C_5H_5)]_2[B_{10}H_{10}]$ .... Fe: Org.Comp.B19-1, 5, 12
– .... $[(1,4-(CH_3)_2-C_6H_4)Fe(C_5H_5)]_2[B_{12}H_{12}]$ .... Fe: Org.Comp.B19-1, 5, 12
– .... $[(CH_3-C_5H_4)Fe(C_5H_3(CH_2)-CH_3)]^+$ .... Fe: Org.Comp.A10-223, 225, 232/3
– .... $[(CH_3-C_6H_5)Fe(C_5H_4-CH_3)]^+$ .... Fe: Org.Comp.B18-142/6, 197/8, 207
– .... $[(C_2H_5-C_6H_5)Fe(C_5H_5)]^+$ .... Fe: Org.Comp.B18-142/6, 197, 199, 200, 201, 212/3
– .... $[(C_6H_6)Fe(C_5H_4-C_2H_5)]^+$ .... Fe: Org.Comp.B18-142/6, 151, 162/3, 187
– .... $[(C_6H_6)Fe(C_5H_4-CHDCH_3)]^+$ .... Fe: Org.Comp.B18-187
– .... $[(C_6H_6)Fe(C_5H_4-CD_2CH_3)]^+$ .... Fe: Org.Comp.B18-187
$C_{13}FeH_{15}I_3$ ...... $[(1,2-(CH_3)_2-C_6H_4)Fe(C_5H_5)][I_3]$ .... Fe: Org.Comp.B19-1, 5, 7
– .... $[(1,3-(CH_3)_2-C_6H_4)Fe(C_5H_5)][I_3]$ .... Fe: Org.Comp.B19-1, 5, 9
– .... $[(1,4-(CH_3)_2-C_6H_4)Fe(C_5H_5)][I_3]$ .... Fe: Org.Comp.B19-1, 5, 11
$C_{13}FeH_{15}KO_2S_2$ .. $K[C_5(CH_3)_5Fe(CO)_2CS_2]$ .... Fe: Org.Comp.B14-163/4
$C_{13}FeH_{15}Li$ ...... $(C_5H_5)Fe[C_5H_3(Li)-C_3H_7-i]$ .... Fe: Org.Comp.A10-129
– .... $(Li-C_5H_4)Fe(C_5H_4-C_3H_7-n)$ .... Fe: Org.Comp.A10-113/5
– .... $(Li-C_5H_4)Fe(C_5H_4-C_3H_7-i)$ .... Fe: Org.Comp.A10-113/5
$C_{13}FeH_{15}LiO$ ..... $(C_5H_5)Fe[C_5H_3(Li)-C(CH_3)_2-OH]$ .... Fe: Org.Comp.A10-118, 119, 122
– .... $(C_5H_5)Fe[C_5H_3(Li)-CH(CH_3)-OCH_3]$ .... Fe: Org.Comp.A10-118, 120, 122, 124, 129/30
– .... $(C_5H_5)Fe[C_5H_3(Li)-CH_2-O-C_2H_5]$ .... Fe: Org.Comp.A10-118, 120, 122
– .... $(Li-C_5H_4)Fe[C_5H_4-CH(CH_3)-OCH_3]$ .... Fe: Org.Comp.A10-113, 115/6
$C_{13}FeH_{15}Li_2N$ .... $(Li-C_5H_4)Fe[C_5H_3(Li)-CH_2-N(CH_3)_2]$ .... Fe: Org.Comp.A10-322/5
– .... $(Li-C_5H_4)Fe[C_5H_3(Li)-CH_2-N(CH_3)_2]$ · x $(CH_3)_2N-CH_2CH_2-N(CH_3)_2$ .... Fe: Org.Comp.A10-322
$C_{13}FeH_{15}NO$ ..... $(CH_3-C_5H_4)Fe[C_5H_3(CH_3)-CH=N-OH]$ .... Fe: Org.Comp.A10-294
– .... $(C_5H_5)Fe[C_5H_3(CH_3)-NHC(O)CH_3]$ .... Fe: Org.Comp.A9-95
– .... $(C_5H_5)Fe[C_5H_3(C_2H_5)-C(O)NH_2]$ .... Fe: Org.Comp.A9-90/1, 93, 97/8
– .... $(C_5H_5)Fe[NC_4H-2,4-(CH_3)_2-3-C(O)CH_3]$ ...... Fe: Org.Comp.B17-203
– .... $[H_2NC(O)-C_5H_4]Fe(C_5H_4-C_2H_5)$ .... Fe: Org.Comp.A9-79
$C_{13}FeH_{15}NO^+$ .... $[C_5H_5FeC_5H_3(CONH_2)C_2H_5]^+$ .... Fe: Org.Comp.A9-97/8
$C_{13}FeH_{15}NO_2$ .... $(C_5H_5)Fe(CO)_2CH_2-2-NC_4H_5-5-CH_3$ .... Fe: Org.Comp.B13-1/2, 5
– .... $(C_5H_5)Fe[C_6H_4-1,3-(CH_3)_2-2-NO_2]$ .... Fe: Org.Comp.B17-293
– .... $(C_5H_5)Fe[C_6H_4-2,4-(CH_3)_2-3-NO_2]$ .... Fe: Org.Comp.B17-294
– .... $(C_5H_5)Fe[C_6H_4-2,6-(CH_3)_2-1-NO_2]$ .... Fe: Org.Comp.B17-285
– .... $(NC_6H_6-1-COO-C_2H_5)Fe(C_4H_4)$ .... Fe: Org.Comp.B18-127
– .... $(O_2N-C_5H_4)Fe(C_5H_4-C_3H_7-n)$ .... Fe: Org.Comp.A9-153
$C_{13}FeH_{15}NO_3$ .... $(C_5H_5)Fe(CO)_2C(=O)-1-NC_5H_{10}$ .... Fe: Org.Comp.B13-9, 25
– .... $(C_5H_5)Fe(CO)_2-CH[-C(CH_3)_2-C(=O)-NH-CH_2-]$ Fe: Org.Comp.B14-18/9, 23, 28
$C_{13}FeH_{15}N_3O$ .... $(C_5H_5)Fe[C_5H_3(CH_3)-CH=N-NHC(O)NH_2]$ ..... Fe: Org.Comp.A9-148, 149
– .... $(N_3-CH_2-C_5H_4)Fe(C_5H_4-CH_2-OCH_3)$ .... Fe: Org.Comp.A9-145

$C_{13}FeH_{15}N_3O$ . . . . $[H_2NC(O)NH-N=CH-C_5H_4]Fe(C_5H_4-CH_3)$ . . . . . Fe: Org.Comp.A9-144
$C_{13}FeH_{15}N_3O_2$ . . . $H_2NCONHN=CH-C_5H_4FeC_5H_4-CH_2OH$ . . . . Fe: Org.Comp.A9-144
$C_{13}FeH_{15}Na$ . . . . . $(C_5H_5)Fe[C_5H_3(Na)-C_3H_7-i]$. . . . . Fe: Org.Comp.A10-130/1
– . . . . $(Na-C_5H_4)Fe(C_5H_4-C_3H_7-i)$ . . . . Fe: Org.Comp.A10-130/1
$C_{13}FeH_{15}O$ . . . . . . $[4-CH_3-C_6H_4-OCH_3]Fe(C_5H_5)$ . . . . Fe: Org.Comp.B19-47
$C_{13}FeH_{15}O^+$ . . . . . $[(2-CH_3-C_6H_4-OCH_3)Fe(C_5H_5)][BF_4]$ . . . . Fe: Org.Comp.B19-1, 5, 46
– . . . . $[(2-CH_3-C_6H_4-OCH_3)Fe(C_5H_5)][Cr(NH_3)_2(SCN)_4]$ Fe: Org.Comp.B19-1, 5, 46
– . . . . $[(2-CH_3-C_6H_4-OCH_3)Fe(C_5H_5)][PF_6]$. . . . . Fe: Org.Comp.B19-1/3, 5, 46
– . . . . $[(3-CH_3-C_6H_4-OCH_3)Fe(C_5H_5)][BF_4]$ . . . . Fe: Org.Comp.B19-1, 5, 47
– . . . . $[(3-CH_3-C_6H_4-OCH_3)Fe(C_5H_5)][Cr(NH_3)_2(SCN)_4]$ Fe: Org.Comp.B19-1, 5, 47
– . . . . $[(3-CH_3-C_6H_4-OCH_3)Fe(C_5H_5)][PF_6]$. . . . . Fe: Org.Comp.B19-1/3, 5, 46
– . . . . $[(4-CH_3-C_6H_4-OCH_3)Fe(C_5H_5)][BF_4]$ . . . . Fe: Org.Comp.B19-1/2, 5, 47
– . . . . $[(4-CH_3-C_6H_4-OCH_3)Fe(C_5H_5)][Cr(NH_3)_2(SCN)_4]$ Fe: Org.Comp.B19-1, 5, 47
– . . . . $[(4-CH_3-C_6H_4-OCH_3)Fe(C_5H_5)][PF_6]$. . . . . Fe: Org.Comp.B19-1/3, 5, 47
– . . . . $[(CH_3CH(OH)-C_6H_5)Fe(C_5H_5)]^+$ . . . . Fe: Org.Comp.B18-142/6, 214
– . . . . $[(C_2H_5-O-C_6H_5)Fe(C_5H_5)]^+$ . . . . Fe: Org.Comp.B18-142/6, 197, 198/9, 250
– . . . . $[(C_6H_6)Fe(C_5H_4-CH(OH)CH_3)]^+$ . . . . Fe: Org.Comp.B18-142/6, 163
– . . . . $[(C_6H_6)Fe(C_5H_4-O-C_2H_5)]^+$ . . . . Fe: Org.Comp.B18-142/6, 153, 169
– . . . . $[(C_6H_7)Fe(CO)(C_6H_8)]^+$ . . . . Fe: Org.Comp.B17-244
– . . . . $[(C_7H_9)Fe(CO)(C_4H_3-CH_3)]^+$ . . . . Fe: Org.Comp.B17-244
$C_{13}FeH_{15}OP$ . . . . . $(C_5H_5)Fe[PC_4H-2-C(=O)CH_3-3,4-(CH_3)_2]$ . . . . . Fe: Org.Comp.B17-209
– . . . . $[CH_3C(O)-C_5H_4]Fe[PC_4H_2-3,4-(CH_3)_2]$ . . . . Fe: Org.Comp.B17-218
$C_{13}FeH_{15}O_2^+$ . . . . $[(1,2-(CH_3O)_2-C_6H_4)Fe(C_5H_5)][1,4-((NC)_2C=)_2-C_6H_4]_2 \cdot 8\ [1,4-((NC)_2C=)_2-C_6H_4]$ . . . . Fe: Org.Comp.B19-19
– . . . . $[(1,2-(CH_3O)_2-C_6H_4)Fe(C_5H_5)][PF_6]$ . . . . Fe: Org.Comp.B19-3, 19
– . . . . $[(1,3-(CH_3O)_2-C_6H_4)Fe(C_5H_5)][PF_6]$ . . . . Fe: Org.Comp.B19-3, 19
– . . . . $[(1,4-(CH_3O)_2-C_6H_4)Fe(C_5H_5)][BF_4]$ . . . . Fe: Org.Comp.B19-1, 5, 19, 87
– . . . . $[(1,4-(CH_3O)_2-C_6H_4)Fe(C_5H_5)][Cr(NH_3)_2(SCN)_4]$ Fe: Org.Comp.B19-1, 5, 19
– . . . . $[(1,4-(CH_3O)_2-C_6H_4)Fe(C_5H_5)][PF_6]$ . . . . Fe: Org.Comp.B19-1, 3, 5, 19, 87
– . . . . $[(C_5H_5)Fe(CO)_2(CH_2=CHCH=CHC_2H_5)]^+$ . . . . . . Fe: Org.Comp.B17-38
– . . . . $[(C_5H_5)Fe(CO)_2(CH_2=CHCH_2CH=CHCH_3)]^+$. . . . Fe: Org.Comp.B17-39
– . . . . $[(C_5H_5)Fe(CO)_2(CH_2=CHCH_2CH_2CH=CH_2)]^+$. . . Fe: Org.Comp.B17-39/40
– . . . . $[(C_5H_5)Fe(CO)_2(CH_2=C_5H_8-c)]^+$ . . . . Fe: Org.Comp.B17-70
– . . . . $[(C_5H_5)Fe(CO)_2(C_2H_5-C \equiv C-C_2H_5)]^+$ . . . . Fe: Org.Comp.B17-122/3
– . . . . $[(C_5H_5)Fe(CO)_2(C_5H_7-3-CH_3)]^+$ . . . . Fe: Org.Comp.B17-88
– . . . . $[(C_5H_5)Fe(CO)_2(C_6H_{10}-c)]^+$ . . . . Fe: Org.Comp.B17-90/1, 104
– . . . . $[(C_5H_5)Fe(CO)_2(HC \equiv C-C_4H_9-n)]^+$ . . . . Fe: Org.Comp.B17-121
$C_{13}FeH_{15}O_2S_2^-$. . . $[C_5(CH_3)_5Fe(CO)_2CS_2]^-$ . . . . Fe: Org.Comp.B14-163/4
$C_{13}FeH_{15}O_3^+$ . . . . $[(1-CH_3-4-CH_2=C(CH_3)-C_6H_7)Fe(CO)_3]^+$ . . . . . Fe: Org.Comp.B15-258
– . . . . $[(1-(i-C_3H_7)-4-CH_3-C_6H_5)Fe(CO)_3]^+$ . . . . Fe: Org.Comp.B15-97, 122, 175
– . . . . $[(6-(i-C_3H_7)-3-CH_3-C_6H_5)Fe(CO)_3]^+$ . . . . Fe: Org.Comp.B15-125, 178/9

$C_{13}FeH_{15}O_3^+$	$[(2,4,6,6\text{-}(CH_3)_4\text{-}C_6H_3)Fe(CO)_3]^+$	Fe: Org.Comp.B15-91
–	$[(2,6,6\text{-}(CH_3)_3\text{-}C_7H_6)Fe(CO)_3]^+$	Fe: Org.Comp.B15-221, 233/4
–	$[(2\text{-}(n\text{-}C_4H_9)\text{-}C_6H_6)Fe(CO)_3]^+$	Fe: Org.Comp.B15-104
–	$[(3\text{-}(n\text{-}C_4H_9)\text{-}C_6H_6)Fe(CO)_3]^+$	Fe: Org.Comp.B15-106
–	$[(6\text{-}(t\text{-}C_4H_9)\text{-}C_6H_6)Fe(CO)_3]^+$	Fe: Org.Comp.B15-108
–	$[(CH_2(CH)_4\text{-}CH_2CH_2CH_2\text{-}CH{=}CH_2)Fe(CO)_3]^+$	Fe: Org.Comp.B15-25
–	$[(CH_2CHC(CH_3)CHCH\text{-}CH{=}C(CH_3)_2)Fe(CO)_3]^+$	Fe: Org.Comp.B15-9
–	$[((CH_3)_5C_5)Fe(CO)_3]^+$	Fe: Org.Comp.B15-52/3
–	$[(C_5H_5)Fe(CO)_2(CH_2{=}C{=}CHC(CH_3)_2OH)]^+$	Fe: Org.Comp.B17-112
–	$[(C_5H_5)Fe(CO)_2(CH_2{=}CHCH_2CH_2COCH_3)]^+$	Fe: Org.Comp.B17-35
–	$[(C_5H_5)Fe(CO)_2(OC_4H_4\text{-}2,3\text{-}(CH_3)_2)]^+$	Fe: Org.Comp.B17-94
–	$[(C_5H_5)Fe(CO)_2(OC_4H_4\text{-}2,5\text{-}(CH_3)_2)]^+$	Fe: Org.Comp.B17-94/5
–	$[(C_5H_5)Fe(CO)_2(OC_4H_4\text{-}5,5\text{-}(CH_3)_2)]^+$	Fe: Org.Comp.B17-95
–	$[(C_5H_5)Fe(CO)_2(OC_4H_5\text{-}2\text{-}C_2H_5)]^+$	Fe: Org.Comp.B17-94
$C_{13}FeH_{15}O_3^-$	$[(C_5H_5)Fe({=}C(O)OC_6H_{10}\text{-}2\text{-})CO]^-$	Fe: Org.Comp.B16b-125, 136/7
$C_{13}FeH_{15}O_4^+$	$[(1\text{-}CH_3\text{-}4\text{-}(i\text{-}C_3H_7O)\text{-}C_6H_5)Fe(CO)_3]^+$	Fe: Org.Comp.B15-121, 173
–	$[(1\text{-}HO\text{-}3,4,6,6\text{-}(CH_3)_4\text{-}C_6H_2)Fe(CO)_3]^+$	Fe: Org.Comp.B15-133, 184
–	$[(C_5H_5)Fe(CO)_2(1,4\text{-}O_2C_4H_4\text{-}5,6\text{-}(CH_3)_2)]^+$	Fe: Org.Comp.B17-96, 105/6
–	$[(C_5H_5)Fe(CO)_2(CH_2{=}CHCH_2\text{-}2\text{-}(1,3\text{-}O_2C_3H_5))]^+$	Fe: Org.Comp.B17-29
–	$[(C_5H_5)Fe(CO)_2(CH_3CH{=}CH\text{-}COO\text{-}C_2H_5)]^+$	Fe: Org.Comp.B17-59
$C_{13}FeH_{15}O_4Si^+$	$[(6\text{-}(CH_3)_3SiO\text{-}C_7H_6)Fe(CO)_3]^+$	Fe: Org.Comp.B15-218, 230/1
$C_{13}FeH_{15}O_5^+$	$[1\text{-}CH_3OCH_2CH_2\text{-}4\text{-}CH_3O\text{-}C_6H_5Fe(CO)_3]^+$	Fe: Org.Comp.B15-114, 170
$C_{13}FeH_{15}S$	$(4\text{-}CH_3\text{-}C_6H_4\text{-}SCH_3)Fe(C_5H_5)$	Fe: Org.Comp.B19-51
$C_{13}FeH_{15}S^+$	$[(4\text{-}CH_3\text{-}C_6H_4\text{-}SCH_3)Fe(C_5H_5)][PF_6]$	Fe: Org.Comp.B19-1, 51
–	$[(C_5H_5)Fe(C_5H_3(SCH_3)CHCH_3)]^+$	Fe: Org.Comp.A9-220
$C_{13}FeH_{16}$	$(CH_3\text{-}C_5H_4)Fe[C_5H_3(CH_3)_2]$	Fe: Org.Comp.A10-191, 194, 198, 201, 204, 206
–	$(CH_3\text{-}C_6H_5)Fe(C_6H_8)$	Fe: Org.Comp.B18-103, 111
–	$(C_5H_5)Fe(C_6H_6\text{-}C_2H_5\text{-}1)$	Fe: Org.Comp.B17-264
–	$(C_5H_5)Fe(C_6H_6\text{-}C_2H_5\text{-}2)$	Fe: Org.Comp.B17-264
–	$(C_5H_5)Fe(C_6H_6\text{-}C_2H_5\text{-}3)$	Fe: Org.Comp.B17-264
–	$(C_5H_5)Fe(C_6H_6\text{-}C_2H_5\text{-}6)$	Fe: Org.Comp.B17-261, 307
–	$(C_5H_5)Fe(C_7H_8\text{-}CH_3\text{-}6)$	Fe: Org.Comp.B17-304
–	$(C_5H_5)Fe(C_8H_{11})$	Fe: Org.Comp.B17-305
–	$(C_5H_5)Fe[C_6H_5(CH_3)\text{-}CH_3\text{-}6]$	Fe: Org.Comp.B17-271
–	$(C_5H_5)Fe[C_6H_5(CH_3)_2\text{-}1,2]$	Fe: Org.Comp.B17-276
–	$(C_5H_5)Fe[C_6H_5(CH_3)_2\text{-}1,3]$	Fe: Org.Comp.B17-276
–	$(C_5H_5)Fe[C_6H_5(CH_3)_2\text{-}1,4]$	Fe: Org.Comp.B17-276
–	$(C_5H_5)Fe[C_6H_5(CH_3)_2\text{-}2,3]$	Fe: Org.Comp.B17-276
–	$(C_5H_5)Fe[C_6H_5(CH_3)_2\text{-}2,4]$	Fe: Org.Comp.B17-276
–	$Fe[C_{10}H_7(CH_3)_3]$	Fe: Org.Comp.A10-213/4
$C_{13}FeH_{16}IN$	$(C_5H_5)Fe[C_5H_3(I)\text{-}CH_2N(CH_3)_2]$	Fe: Org.Comp.A9-35, 46
$C_{13}FeH_{16}INO$	$(C_5H_5)Fe(CN\text{-}C_6H_{11}\text{-}c)(CO)I$	Fe: Org.Comp.B15-292, 296/7
$C_{13}FeH_{16}LiN$	$(C_5H_5)Fe[C_5H_3(Li)\text{-}CH_2\text{-}N(CH_3)_2]$	Fe: Org.Comp.A10-118/22, 124
–	$(Li\text{-}C_5H_4)Fe[C_5H_4\text{-}CH_2\text{-}N(CH_3)_2]$	Fe: Org.Comp.A10-113/5
$C_{13}FeH_{16}N$	$[(CH_3)_2N\text{-}C_6H_5]Fe(C_5H_5)$	Fe: Org.Comp.B18-146/8, 262
$C_{13}FeH_{16}N^+$	$[(2,6\text{-}(CH_3)_2\text{-}C_6H_3\text{-}NH_2)Fe(C_5H_5)][PF_6]$	Fe: Org.Comp.B19-99, 124

$C_{13}FeH_{16}N^+$ ..... $[(2\text{-}CH_3\text{-}C_6H_4\text{-}NH_2)Fe(C_5H_4\text{-}CH_3)][PF_6]$ ...... Fe: Org.Comp.B19-2, 5/6, 53
– ..... $[(3\text{-}CH_3\text{-}C_6H_4\text{-}NH_2)Fe(C_5H_4\text{-}CH_3)][PF_6]$ ...... Fe: Org.Comp.B19-2, 5/6, 53
– ..... $[(4\text{-}CH_3\text{-}C_6H_4\text{-}NH_2)Fe(C_5H_4\text{-}CH_3)][PF_6]$ ...... Fe: Org.Comp.B19-2, 5/6, 54
– ..... $[(2\text{-}CH_3\text{-}C_6H_4\text{-}NH\text{-}CH_3)Fe(C_5H_5)][PF_6]$ ...... Fe: Org.Comp.B19-2, 54
– ..... $[(3\text{-}CH_3\text{-}C_6H_4\text{-}NH\text{-}CH_3)Fe(C_5H_5)][PF_6]$ ...... Fe: Org.Comp.B19-2, 54
– ..... $[(4\text{-}CH_3\text{-}C_6H_4\text{-}NH\text{-}CH_3)Fe(C_5H_5)][PF_6]$ ...... Fe: Org.Comp.B19-2, 55
– ..... $[((CH_3)_2N\text{-}C_6H_5)Fe(C_5H_5)]^+$ ...... Fe: Org.Comp.B18-142/6, 197, 199, 201, 262
– ..... $[(CH_3CH(NH_2)\text{-}C_6H_5)Fe(C_5H_5)]^+$ ...... Fe: Org.Comp.B18-142/6, 214
– ..... $[(C_6H_6)Fe(C_5H_4\text{-}NH\text{-}C_2H_5)]^+$ ...... Fe: Org.Comp.B18-142/6, 167
$C_{13}FeH_{16}NO_3S^+$ .. $[C_5H_5(CO)_2Fe{=}C(SCH_3)(N(CH_2CH_2)_2O)]^+$ ..... Fe: Org.Comp.B16a-146
$C_{13}FeH_{16}N_2O_4$ ... $(C_5H_5)Fe(CN\text{-}C_6H_{11}\text{-}c)(CO)NO_3$ ...... Fe: Org.Comp.B15-286
$C_{13}FeH_{16}O$ ...... $(CH_3\text{-}C_5H_4)Fe[C_5H_3(CH_3)\text{-}CH_2OH]$ ...... Fe: Org.Comp.A10-217, 218, 220, 228/9, 231
– ..... $(C_5H_5)Fe(C_6H_5\text{-}1\text{-}CH_3\text{-}2\text{-}OCH_3)$ ...... Fe: Org.Comp.B17-278
– ..... $(C_5H_5)Fe(C_6H_5\text{-}1\text{-}CH_3\text{-}3\text{-}OCH_3)$ ...... Fe: Org.Comp.B17-278
– ..... $(C_5H_5)Fe(C_6H_5\text{-}1\text{-}CH_3\text{-}4\text{-}OCH_3)$ ...... Fe: Org.Comp.B17-278
– ..... $(C_5H_5)Fe(C_6H_5\text{-}2\text{-}CH_3\text{-}1\text{-}OCH_3)$ ...... Fe: Org.Comp.B17-278
– ..... $(C_5H_5)Fe(C_6H_5\text{-}2\text{-}CH_3\text{-}3\text{-}OCH_3)$ ...... Fe: Org.Comp.B17-278
– ..... $(C_5H_5)Fe(C_6H_5\text{-}3\text{-}CH_3\text{-}2\text{-}OCH_3)$ ...... Fe: Org.Comp.B17-278
– ..... $(C_5H_5)Fe(C_6H_5\text{-}4\text{-}CH_3\text{-}1\text{-}OCH_3)$ ...... Fe: Org.Comp.B17-278
– ..... $(C_5H_5)Fe(C_6H_5\text{-}4\text{-}CH_3\text{-}2\text{-}OCH_3)$ ...... Fe: Org.Comp.B17-278
– ..... $[(C_6H_6)Fe(C_5H_5)][O\text{-}C_2H_5]$ ...... Fe: Org.Comp.B18-142/6, 152, 154, 156
– ..... $[H_3CCH(OH)\text{-}C_5H_4]Fe(C_6H_7)$ ...... Fe: Org.Comp.B17-303, 311
– ..... $[H_3CCD(OH)\text{-}C_5H_4]Fe(C_6H_6D\text{-}6)$ ...... Fe: Org.Comp.B17-303, 311/2
$C_{13}FeH_{16}O_2$ ..... $(C_5H_5)Fe(CO)\text{-}C({=}O)\text{-}CH_2C(CH_3)_2CH{=}CH_2$ .... Fe: Org.Comp.B17-170
– ..... $(C_5H_5)Fe(CO)\text{-}C(O\text{-}C_2H_5){=}CHCH_2CH{=}CH_2$ .... Fe: Org.Comp.B17-173
– ..... $(C_5H_5)Fe(CO)_2\text{-}C(C_2H_5){=}CHC_2H_5$ ...... Fe: Org.Comp.B13-96, 124
– ..... $(C_5H_5)Fe(CO)_2\text{-}CH(CH_3)CH_2CH_2CH{=}CH_2$ ..... Fe: Org.Comp.B13-110
– ..... $(C_5H_5)Fe(CO)_2\text{-}CH_2C(CH_3)_2CH{=}CH_2$ ...... Fe: Org.Comp.B13-109, 136/7
– ..... $(C_5H_5)Fe(CO)_2\text{-}CH_2CH{=}C(CH_3)C_2H_5$ ...... Fe: Org.Comp.B13-106, 134
– ..... $(C_5H_5)Fe(CO)_2\text{-}CH_2CH(CH_3)CH_2CH{=}CH_2$ ..... Fe: Org.Comp.B13-110
– ..... $(C_5H_5)Fe(CO)_2\text{-}CH_2CH_2CH_2CH_2\text{-}CH{=}CH_2$ ..... Fe: Org.Comp.B13-110
– ..... $(C_5H_5)Fe(CO)_2\text{-}CH_2CH_2CH_2\text{-}C(CH_3){=}CH_2$ ..... Fe: Org.Comp.B13-110
– ..... $(C_5H_5)Fe(CO)_2\text{-}CH_2CH_2CH_2\text{-}CH{=}CHCH_3$ ..... Fe: Org.Comp.B13-110
– ..... $(C_5H_5)Fe(CO)_2\text{-}C_5H_8\text{-}2\text{-}CH_3$ ...... Fe: Org.Comp.B13-199, 240
– ..... $(C_5H_5)Fe(CO)_2\text{-}C_6H_{11}\text{-}c$ ...... Fe: Org.Comp.B13-196, 215
– ..... $(C_5H_5)Fe[C_6H_5\text{-}1,4\text{-}(OCH_3)_2]$ ...... Fe: Org.Comp.B17-282
– ..... $C_5H_4\text{-}C(CH_3)_2C(CH_3)_2\text{-}Fe(CO)_2$ ...... Fe: Org.Comp.B18-18, 25
– ..... $(HOCH_2\text{-}C_5H_4)Fe[C_5H_3(CH_3)\text{-}CH_2OH]$ ...... Fe: Org.Comp.A10-217, 218, 228, 229, 231
– ..... $(CH_3\text{-}C_5H_4)Fe(CO)_2\text{-}CH_2CH{=}C(CH_3)_2$ ...... Fe: Org.Comp.B14-54, 62
$C_{13}FeH_{16}O_2Si$ .... $(C_5H_5)Fe(CO)_2CH_2C{\equiv}CSi(CH_3)_3$ ...... Fe: Org.Comp.B13-155
$C_{13}FeH_{16}O_3$ ..... $1\text{-}(C_5H_5)Fe(CO)_2\text{-}1\text{-}CH_3O\text{-}2,3\text{-}(CH_3)_2\text{-}C_3H_2$ .. Fe: Org.Comp.B13-197, 234
– ..... $(C_5H_5)Fe(CO)[CH_2CHC(CH_3)COO\text{-}C_2H_5]$ ...... Fe: Org.Comp.B17-150
– ..... $(C_5H_5)Fe(CO)_2\text{-}C({=}O)\text{-}C_5H_{11}\text{-}n$ ...... Fe: Org.Comp.B13-17
– ..... $(C_5H_5)Fe(CO)_2\text{-}C({=}O)\text{-}C_5H_{11}\text{-}i$ ...... Fe: Org.Comp.B13-8, 17, 37
– ..... $(C_5H_5)Fe(CO)_2\text{-}C_6H_{10}\text{-}2\text{-}OH$ ...... Fe: Org.Comp.B13-215, 253
$C_{13}FeH_{16}O_4$ ..... $(C_5H_5)Fe(CO)_2\text{-}C({=}O)\text{-}CH_2CH_2\text{-}CH(OH)\text{-}C_2H_5$ Fe: Org.Comp.B13-18

$C_{13}FeH_{16}O_4$ ..... $(C_5H_5)Fe(CO)_2-CH[-O-CH(CH_3)-CH(CH_3)-O-CH_2-]$ Fe: Org.Comp.B14-8, 14/5
– ..... $(C_5H_5)Fe(CO)_2-CH_2CH=CHCH(OCH_3)_2$ ..... Fe: Org.Comp.B13-105, 133
– ..... $(c-C_7H_9)Fe(CO)_2-C(=O)O-C_3H_7-i$ ..... Fe: Org.Comp.B14-80, 81/2
$C_{13}FeH_{16}O_4S$ ..... $(C_5H_5)Fe(CO)_2-C(=O)-CH(CH_3)CH_2SCH_2CH_2OH$ Fe: Org.Comp.B13-12, 27
– ..... $(C_5H_5)Fe(CO)_2-C(=O)-CH_2CH(CH_3)SCH_2CH_2OH$ Fe: Org.Comp.B13-12, 26
$C_{13}FeH_{16}P^+$ ..... $[(CH_3C_6H_5)Fe(PC_4H_2(CH_3)_2-3,4)]^+$ ..... Fe: Org.Comp.B18-130
$C_{13}FeH_{17}$ ..... $C_5H_5FeC_8H_{12}$, radical ..... Fe: Org.Comp.B17-188, 192
– ..... $C_5H_5FeC_8H_{12} \cdot 0.5\ O(C_2H_5)_2$, radical ..... Fe: Org.Comp.B17-192/4
$C_{13}FeH_{17}^-$ ..... $[C_5H_5Fe(C_8H_{12})]^-$ ..... Fe: Org.Comp.B17-188/90, 194/5
$C_{13}FeH_{17}IN_2O$ ..... $[(C_5H_5)Fe(CNCH_3)(CO)CN-C_4H_9-s]I$ ..... Fe: Org.Comp.B15-334
– ..... $[(C_5H_5)Fe(CNCH_3)(CO)CN-C_4H_9-t]I$ ..... Fe: Org.Comp.B15-334
– ..... $[(C_5H_5)Fe(CN-C_2H_5)(CO)CN-C_3H_7-n]I$ ..... Fe: Org.Comp.B15-336
– ..... $[(C_5H_5)Fe(CN-C_2H_5)(CO)CN-C_3H_7-i]I$ ..... Fe: Org.Comp.B15-336
$C_{13}FeH_{17}ISi$ ..... $(CH_3)_3Si-C_5H_4FeC_5H_4I$ ..... Fe: Org.Comp.A9-315, 319
$C_{13}FeH_{17}K$ ..... $K[C_5H_5Fe(C_8H_{12})]$ ..... Fe: Org.Comp.B17-190
$C_{13}FeH_{17}LiSi$ ..... $(Li-C_5H_4)Fe[C_5H_4-Si(CH_3)_3]$ ..... Fe: Org.Comp.A10-114, 115, 116
$C_{13}FeH_{17}N$ ..... $(C_5H_5)Fe(C_6H_6-N(CH_3)_2-1)$ ..... Fe: Org.Comp.B17-268
– ..... $(C_5H_5)Fe(C_6H_6-N(CH_3)_2-2)$ ..... Fe: Org.Comp.B17-268
– ..... $(C_5H_5)Fe(C_6H_6-N(CH_3)_2-3)$ ..... Fe: Org.Comp.B17-268
– ..... $(C_5H_5)Fe[C_5H_3(CH_3)-CH_2CH_2NH_2]$ ..... Fe: Org.Comp.A9-26
$C_{13}FeH_{17}N^{2+}$ ..... $[(H_3NCH_2-C_5H_4)Fe(C_5H_4-CHCH_3)]^{2+}$ ..... Fe: Org.Comp.A9-4
$C_{13}FeH_{17}NO$ ..... $(C_5H_5)Fe[C_5H_3(OH)-CH_2N(CH_3)_2]$ ..... Fe: Org.Comp.A9-36, 46
– ..... $(H_2NCH_2-C_5H_4)Fe(C_5H_4-CHCH_3-OH)$ ..... Fe: Org.Comp.A9-3, 4
$C_{13}FeH_{17}NO_2$ ..... $(C_5H_5)Fe(CO)(CN-C_4H_9-t)-C(=O)CH_3$ ..... Fe: Org.Comp.B15-306, 310
– ..... $(C_5H_5)Fe(CO)[=C(N(CH_3)_2)-C(CH_3)_2-C(=O)-]$ ..... Fe: Org.Comp.B16b-117, 132/3
– ..... $(C_5H_5)Fe(CO)_2CH_2-1-NC_5H_{10}$ ..... Fe: Org.Comp.B13-3
– ..... $(C_5H_5)Fe(CO)_2CH_2-2-NC_4H_7(CH_3-5)$ ..... Fe: Org.Comp.B13-1/2, 5
– ..... $(C_5H_5)Fe(CO)_2CH_2-2-NC_5H_{10}$ ..... Fe: Org.Comp.B13-1, 3
– ..... $(C_5H_5)Fe(CO)_2-C[N(CH_3)_2]=C(CH_3)_2$ ..... Fe: Org.Comp.B13-89, 118
$C_{13}FeH_{17}N_2^+$ ..... $[(4-NH_2-C_6H_4-N(CH_3)_2)Fe(C_5H_5)][PF_6]$ ..... Fe: Org.Comp.B19-79
– ..... $[(NH_2-CH_2CH_2-NH-C_6H_5)Fe(C_5H_5)]^+$ ..... Fe: Org.Comp.B18-142/6, 198, 258
$C_{13}FeH_{17}N_2O^+$ ..... $[(C_5H_5)Fe(CO)(CNCH_3)(CN-C_4H_9-s)]^+$ ..... Fe: Org.Comp.B15-334
– ..... $[(C_5H_5)Fe(CO)(CNCH_3)(CN-C_4H_9-t)]^+$ ..... Fe: Org.Comp.B15-334
– ..... $[(C_5H_5)Fe(CO)(CN-C_2H_5)(CN-C_3H_7-n)]^+$ ..... Fe: Org.Comp.B15-336
– ..... $[(C_5H_5)Fe(CO)(CN-C_2H_5)(CN-C_3H_7-i)]^+$ ..... Fe: Org.Comp.B15-336
$C_{13}FeH_{17}NaSi$ ..... $(C_5H_5)Fe[C_5H_3(Na)-Si(CH_3)_3]$ ..... Fe: Org.Comp.A10-131
$C_{13}FeH_{17}O^+$ ..... $[(1,3,5-(CH_3)_3-C_6H_3)Fe(CO)(CH_2CHCH_2)]^+$ ..... Fe: Org.Comp.B18-95/7
– ..... $[(C_5H_5)Fe(CO)(CH_2=CHC(CH_3)_2CH=CH_2)]^+$ ..... Fe: Org.Comp.B17-187/8
$C_{13}FeH_{17}O_2^+$ ..... $[(C_5(CH_3)_5)Fe(CO)_2=CH_2]^+$ ..... Fe: Org.Comp.B14-77
Fe: Org.Comp.B16a-103/4
– ..... $[(C_5H_5)Fe(CO)(=C(O-C_2H_5)CH_2CH_2CH=CH_2)]^+$ Fe: Org.Comp.B17-167
– ..... $[(C_5H_5)Fe(CO)_2(CH_2=CH-C_4H_9-n)]^+$ ..... Fe: Org.Comp.B17-13, 47
– ..... $[(C_5H_5)Fe(CO)_2(CH_2=CH-C_4H_9-s)]^+$ ..... Fe: Org.Comp.B17-13

$C_{13}FeH_{17}O_2^+$	$[(C_5H_5)Fe(CO)_2(CH_2{=}CH\text{-}C_4H_9\text{-}t)]^+$	Fe: Org.Comp.B17-12
$C_{13}FeH_{17}O_3^+$	$[(CH_2(CH)_4\text{-}C_5H_{11}\text{-}n)Fe(CO)_3]^+$	Fe: Org.Comp.B15-26/7
–	$[(C_5(CH_3)_5)Fe(CO)_2{=}CH(OH)]^+$	Fe: Org.Comp.B14-75 Fe: Org.Comp.B16a-128
$C_{13}FeH_{17}O_3P$	$(C_5H_5)Fe(CO)_2COC(CH_3){=}P(CH_3)_3$	Fe: Org.Comp.B13-15, 35/6
–	$(C_5H_5)Fe(CO)_2COCH{=}P(CH_3)_2C_2H_5$	Fe: Org.Comp.B13-14, 34
$C_{13}FeH_{17}O_3Si^+$	$[1\text{-}CH_3\text{-}3\text{-}(CH_3)_3Si\text{-}C_6H_5Fe(CO)_3]^+$	Fe: Org.Comp.B15-113
–	$[(6\text{-}CH_3\text{-}3\text{-}(CH_3)_3Si\text{-}C_6H_5)Fe(CO)_3]^+$	Fe: Org.Comp.B15-125
$C_{13}FeH_{17}O_4^+$	$[(C_5H_5)Fe(CO)_2(CH_2{=}CHCH_2CH(OCH_3)_2)]^+$	Fe: Org.Comp.B17-33/4
–	$[(C_5H_5)Fe(CO)_2(C_2H_5\text{-}OCH{=}CHO\text{-}C_2H_5)]^+$	Fe: Org.Comp.B17-69/70, 81
$C_{13}FeH_{17}P$	$(C_5H_5)Fe[C_4P\text{-}2,3,4,5\text{-}(CH_3)_4]$	Fe: Org.Comp.B17-226
$C_{13}FeH_{17}S$	$(C_5H_5)Fe[C_4S\text{-}2,3,4,5\text{-}(CH_3)_4]$	Fe: Org.Comp.B17-213
$C_{13}FeH_{17}S^+$	$[(C_2H_5\text{-}C_5H_4)Fe(SC_4H_2\text{-}3,4\text{-}(CH_3)_2)]^+$	Fe: Org.Comp.B17-222
–	$[(C_5H_5)Fe(SC_4\text{-}2,3,4,5\text{-}(CH_3)_4)]^+$	Fe: Org.Comp.B17-214/5, 233
$C_{13}FeH_{17}S^-$	$[(C_5H_5)Fe(SC_4(CH_3)_4)]^-$	Fe: Org.Comp.B17-233
$C_{13}FeH_{18}$	$(C_5H_5)Fe(C_8H_{13}\text{-}c)$	Fe: Org.Comp.B17-139
$C_{13}FeH_{18}IN$	$[C_5H_5FeC_5H_3(CH_3)\text{-}CH_2NH_2CH_3]I$	Fe: Org.Comp.A9-26
$C_{13}FeH_{18}IO_3P$	$[(C_5H_5)Fe(CO)_2C(O)CH(CH_3)P(CH_3)_3]I$	Fe: Org.Comp.B14-154, 157
$C_{13}FeH_{18}N^+$	$[C_5H_5FeC_5H_3(CH_2NH_2CH_3)CH_3]^+$	Fe: Org.Comp.A9-26
$C_{13}FeH_{18}NO_2P$	$(C_5H_5)Fe(CNCH_3)(CO)C(O)CH{=}P(CH_3)_3$	Fe: Org.Comp.B15-306, 309
$C_{13}FeH_{18}N_2O_6S_3$	$(C_5H_5)Fe(CO)_2\text{-}C(CH_3)[\text{-}CH_2\text{-}N(SO_2\text{-}CH_3)\text{-}S$ $\text{-}N(SO_2\text{-}CH_3)\text{-}CH_2\text{-}]$	Fe: Org.Comp.B14-35
–	$(C_5H_5)Fe(CO)_2\text{-}CH[\text{-}CH_2\text{-}N(SO_2\text{-}CH_3)$ $\text{-}S({=}N\text{-}SO_2\text{-}CH_3)\text{-}CH_2CH_2\text{-}]$	Fe: Org.Comp.B14-34/5, 38
–	$(C_5H_5)Fe(CO)_2\text{-}N(SO_2\text{-}CH_3)$ $\text{-}S(CH_2\text{-}C_3H_5\text{-}c){=}N\text{-}SO_2\text{-}CH_3$	Fe: Org.Comp.B14-38
$C_{13}FeH_{18}O$	$1,3,5\text{-}(CH_3)_3C_6H_4Fe(CH_2CHCH_2)CO$	Fe: Org.Comp.B17-184
$C_{13}FeH_{18}O_2$	$(C_5H_5)Fe(CO)\text{-}CH(O\text{-}C_2H_5)CH_2CH_2CH{=}CH_2$	Fe: Org.Comp.B17-173
–	$[C_5(CH_3)_5]Fe(CO)_2\text{-}CH_3$	Fe: Org.Comp.B14-69, 75/6
$C_{13}FeH_{18}O_2Si$	$(C_5H_5)Fe(CO)_2CH_2CH{=}CHSi(CH_3)_3$	Fe: Org.Comp.B13-103, 132
–	$[(CH_3)_3Si\text{-}C_5H_4]Fe(CO)_2CH_2CH{=}CH_2$	Fe: Org.Comp.B14-59
$C_{13}FeH_{18}O_3$	$[C_5(CH_3)_5]Fe(CO)_2CH_2OH$	Fe: Org.Comp.B14-69, 76/7
$C_{13}FeH_{18}O_3^+$	$[C_5(CH_3)_5Fe(CO)_2CH_2OH]^+$	Fe: Org.Comp.B14-165
$C_{13}FeH_{18}O_3P^+$	$[(C_5H_5)Fe(CO)_2C(O)CH(CH_3)P(CH_3)_3]^+$	Fe: Org.Comp.B14-154, 157
$C_{13}FeH_{19}NO_2^{2+}$	$[(C_5H_5)Fe(CO)_2(CH_2{=}CH(CH_2)_4NH_3)]^{2+}$	Fe: Org.Comp.B17-36
$C_{13}FeH_{19}NO_2Sn$	$(CH_3)_3SnNC\text{-}Fe(CO)_2(C_7H_{10}\text{-}c)$	Sn: Org.Comp.18-128
$C_{13}FeH_{19}O_2Si^+$	$[((CH_3)_3Si\text{-}C_5H_4)Fe(CO)_2(CH_2{=}CHCH_3)]^+$	Fe: Org.Comp.B17-132
–	$[((CH_3)_3Si\text{-}C_5H_3\text{-}CH_3)Fe(CO)_2(CH_2{=}CH_2)]^+$	Fe: Org.Comp.B14-75 Fe: Org.Comp.B17-132/3
$C_{13}FeH_{19}S^+$	$[(C_5H_5)Fe(C_6H_8)S(CH_3)_2]^+$	Fe: Org.Comp.B17-190, 196
$C_{13}FeH_{19}Se^+$	$[(C_5H_5)Fe(C_6H_8)Se(CH_3)_2]^+$	Fe: Org.Comp.B17-190, 196
$C_{13}FeH_{19}Te^+$	$[(C_5H_5)Fe(C_6H_8)Te(CH_3)_2]^+$	Fe: Org.Comp.B17-190, 196
$C_{13}FeH_{20}IN_3O$	$[(C_5H_5)(CH_3\text{-}NC)(CO)Fe{=}C(NH\text{-}CH_3)\text{-}NH\text{-}C_3H_7\text{-}i]I$	Fe: Org.Comp.B16a-164/5, 168
–	$[(C_5H_5)(C_2H_5\text{-}NC)(CO)Fe{=}C(NH\text{-}CH_3)\text{-}NH\text{-}C_2H_5]I$	Fe: Org.Comp.B16a-164
–	$[(C_5H_5)(i\text{-}C_3H_7\text{-}NC)(CO)Fe{=}C(NH\text{-}CH_3)_2]I$	Fe: Org.Comp.B16a-164, 168
$C_{13}FeH_{20}IOPS$	$[(C_5H_5)Fe(CS)(CO)P(C_2H_5)_3]I$	Fe: Org.Comp.B15-271
$C_{13}FeH_{20}NO_2^+$	$[(C_5H_5)Fe(CO)_2CH_2CH(CH_3)N(CH_3)_3]^+$	Fe: Org.Comp.B14-150

$C_{13}FeH_{20}N_2O_6S_3^+$
$[(C_5H_5)Fe(CO)_2(CH_2{=}CH(C_2H_5)N(SO_2CH_3){-}S{-}NH{-}SO_2CH_3)]^+$ . . . Fe: Org.Comp.B14-39
$C_{13}FeH_{20}N_3O^+$ . . . $[(C_5H_5)(CH_3{-}NC)(CO)Fe{=}C(NH{-}CH_3){-}NH{-}C_3H_7{-}i]^+$ Fe: Org.Comp.B16a-164/5, 168
– . . . $[(C_5H_5)(C_2H_5{-}NC)(CO)Fe{=}C(NH{-}CH_3){-}NH{-}C_2H_5]^+$ Fe: Org.Comp.B16a-164
– . . . $[(C_5H_5)(i{-}C_3H_7{-}NC)(CO)Fe{=}C(NH{-}CH_3)_2]^+$ . . . Fe: Org.Comp.B16a-164, 168
$C_{13}FeH_{20}OPS^+$ . . . $[(C_5H_5)Fe(CS)(CO)P(C_2H_5)_3]^+$ . . . Fe: Org.Comp.B15-271
$C_{13}FeH_{20}O_2Si$ . . . $[2{-}(CH_3)_3Si{-}C_5H_3CH_3]Fe(CO)_2C_2H_5$ . . . Fe: Org.Comp.B14-68, 74/5
– . . . $[3{-}(CH_3)_3Si{-}C_5H_3CH_3]Fe(CO)_2C_2H_5$ . . . Fe: Org.Comp.B14-68, 74/5
– . . . $[n{-}C_3H_7Si(CH_3)_2{-}C_5H_4]Fe(CO)_2CH_3$ . . . Fe: Org.Comp.B14-59
$C_{13}FeH_{20}O_4P^+$ . . . $[(C_5H_5)(CO)(P(OCH_3)_3)Fe{=}C{=}C(CH_3)_2]^+$ . . . Fe: Org.Comp.B16a-208, 211
$C_{13}FeH_{20}O_5P^+$ . . . $[(C_5H_5)Fe(CO)_2C(CH_3)_2P(OCH_3)_3]^+$ . . . Fe: Org.Comp.B14-138, 143/4
$C_{13}FeH_{21}INOP$ . . . $[(C_5H_5)Fe(CN{-}C_3H_7{-}i)(CO)P(CH_3)_3]I$ . . . Fe: Org.Comp.B15-302
$C_{13}FeH_{21}LiO_2Si_2$ $Li[(1,3{-}((CH_3)_3Si)_2C_5H_3)Fe(CO)_2]$ . . . Fe: Org.Comp.B14-120
$C_{13}FeH_{21}NOP^+$ . . $[(C_5H_5)Fe(CN{-}C_3H_7{-}i)(CO)P(CH_3)_3]^+$ . . . Fe: Org.Comp.B15-302
$C_{13}FeH_{21}NO_2Sn$ . . $(C_2H_5)_3SnNC{-}Fe(CO)_2C_4H_6$ . . . Sn: Org.Comp.18-156
$C_{13}FeH_{21}O_2^+$ . . . $[(C_5H_5)Fe(CH_2{=}CHCH_3)(CO)O(C_2H_5)_2]^+$ . . . Fe: Org.Comp.B16b-82, 89/90
$C_{13}FeH_{21}O_2P$ . . . $(C_5H_5)Fe(CH_2CHCHCO_2CH_3)P(CH_3)_3$ . . . Fe: Org.Comp.B17-140
$C_{13}FeH_{21}O_2Si_2^-$ . . $[(1,3{-}((CH_3)_3Si)_2C_5H_3)Fe(CO)_2]^-$ . . . Fe: Org.Comp.B14-120
$C_{13}FeH_{21}O_5P$ . . . $(C_5H_5)Fe(CH_2CHCHCO_2CH_3)P(OCH_3)_3$ . . . Fe: Org.Comp.B17-140
$C_{13}FeH_{22}IO_2P$ . . . $[(C_5H_5)(CO)(P(CH_3)_3)Fe{=}C(C_2H_5)OCH_3]I$ . . . Fe: Org.Comp.B16a-40, 54
$C_{13}FeH_{22}O_2P^+$ . . . $[(C_5H_5)(CO)(P(CH_3)_3)Fe{=}C(C_2H_5)OCH_3]^+$ . . . Fe: Org.Comp.B16a-40, 54
$C_{13}FeH_{22}O_2Si_2$ . . . $[C_5H_4Si(CH_3)_2Si(CH_3)_3]Fe(CO)_2CH_3$ . . . Fe: Org.Comp.B14-60
$C_{13}FeH_{22}O_5P^+$ . . . $[(C_5H_5)Fe(CH_2{=}CHOC_2H_5)(CO)P(OCH_3)_3]^+$ . . . Fe: Org.Comp.B16b-87, 90
$C_{13}FeH_{23}P$ . . . $(CH_2CHCHCHCH_2)Fe(CH_2CHCHCH{=}CH_2){-}P(CH_3)_3$ Fe: Org.Comp.B17-178/9
– . . . $(CH_3C_5H_4)Fe(CH_2CHCHCH_3){-}P(CH_3)_3$ . . . Fe: Org.Comp.B17-181
– . . . $(C_5H_5)Fe(CH_2CHCHC_2H_5){-}P(CH_3)_3$ . . . Fe: Org.Comp.B17-142
– . . . $(C_5H_5)Fe(CH_2CH_2CH_2CH{=}CH_2){-}P(CH_3)_3$ . . . Fe: Org.Comp.B17-166
– . . . $(C_5H_5)Fe(CH_3{-}CHCHCH{-}CH_3){-}P(CH_3)_3$ . . . Fe: Org.Comp.B17-142
$C_{13}FeH_{24}IN_2OP$ . . $[(C_5H_5)(CO)(P(CH_3)_3)Fe{=}C(NHCH_3)NHC_2H_5]I$ . . Fe: Org.Comp.B16a-69, 72/3
$C_{13}FeH_{24}N_2OP^+$ . . $[(C_5H_5)(CO)(P(CH_3)_3)Fe{=}C(NHCH_3)NHC_2H_5]^+$ Fe: Org.Comp.B16a-69, 72/3
$C_{13}FeH_{26}O_6P_2$ . . . $(CH_3C_6H_5)Fe[P(OCH_3)_3]_2$ . . . Fe: Org.Comp.B18-2/3, 6, 12
$C_{13}FeH_{26}P_2$ . . . $(CH_3C_6H_5)Fe[P(CH_3)_3]_2$ . . . Fe: Org.Comp.B18-2, 6
– . . . $(C_5H_4{-}CH_2CH_2)Fe[P(CH_3)_3]_2$ . . . Fe: Org.Comp.B18-10
$C_{13}FeH_{27}O_6P_2^+$ . . $[(CH_3C_6H_5)Fe(H)(P(OCH_3)_3)_2]^+$ . . . Fe: Org.Comp.B18-6
– . . . $[(C_5H_5)Fe(CH_2{=}CH_2)(P(OCH_3)_3)_2]^+$ . . . Fe: Org.Comp.B16b-4/5, 8/9
$C_{13}Fe_3Ge_2H_{10}O_9$ $(CO)_9Fe_3(Ge{-}C_2H_5)_2$ . . . Fe: Org.Comp.C6a-282/3
$C_{13}Fe_3HO_{12}Rh$ . . . $Fe_3RhC(CO)_{12}(H)$ . . . Fe: Org.Comp.C6a-305
$C_{13}Fe_3HO_{12}Ru^-$ . . $[Fe_3RuC(CO)_{12}(H)]^-$ . . . Fe: Org.Comp.C6a-304
$C_{13}Fe_3H_3NO_{11}$ . . . $(CO)_{11}Fe_3(CN{-}CH_3)$ . . . Fe: Org.Comp.C6b-140/1
$C_{13}Fe_3H_3NaO_{13}$ . . $Na[(CO)_{11}Fe_3(COO{-}CH_3)]$ . . . Fe: Org.Comp.C6b-157
$C_{13}Fe_3H_3O_{12}^-$ . . . $[(CO)_{10}Fe_3(C{-}O{-}C({=}O){-}CH_3)]^-$ . . . Fe: Org.Comp.C6b-60/1, 63, 66
$C_{13}Fe_3H_3O_{13}^-$ . . . $[(CO)_{11}Fe_3(COO{-}CH_3)]^-$ . . . Fe: Org.Comp.C6b-157
$C_{13}Fe_3H_4O_{11}$ . . . $(CO)_9Fe_3(C{-}H)(C{-}C(O)O{-}CH_3)$ . . . Fe: Org.Comp.C6a-237, 243
$C_{13}Fe_3H_5O_{11}^-$ . . . $[(CO)_{10}Fe_3(C{-}O{-}C_2H_5)]^-$ . . . Fe: Org.Comp.C6b-60/1, 62, 66

$C_{13}Fe_3H_5O_{12}^-$ .... $[(CO)_{10}Fe_3(C\text{-}O\text{-}CH_2\text{-}O\text{-}CH_3)]^-$ .... Fe: Org.Comp.C6b-62/3, 66, 67
$C_{13}Fe_3H_6NO_{10}^-$ .. $[(CO)_{10}(H)Fe_3(CN\text{-}C_2H_5)]^-$ .... Fe: Org.Comp.C6b-21
$C_{13}Fe_3H_6N_2O_9Te_2$ $(CO)_9Fe_3Te_2(1,3\text{-}N_2C_3H_3\text{-}1\text{-}CH_3)$ .... Fe: Org.Comp.C6a-121
$C_{13}Fe_3H_6O_{10}$ .... $(CO)_9Fe_3(CH)(C\text{-}O\text{-}C_2H_5)$ .... Fe: Org.Comp.C6a-234, 235
– .... $(CO)_9Fe_3(C\text{-}CH_3)(C\text{-}OCH_3)$ .... Fe: Org.Comp.C6a-234, 235, 239
$C_{13}Fe_3H_6O_{11}$ .... $(CO)_9Fe_3(C\text{-}O\text{-}CH_3)_2$ .... Fe: Org.Comp.C6a-236, 240/1
– .... $(CO)_{10}(H)Fe_3(C\text{-}O\text{-}C_2H_5)$ .... Fe: Org.Comp.C6b-49, 52, 58
$C_{13}Fe_3H_7NO_9$ .... $(CO)_9Fe_3(NC\text{-}C_3H_7\text{-}n)$ .... Fe: Org.Comp.C6a-246/7
$C_{13}Fe_3H_7NO_{10}$ ... $(CO)_{10}(H)Fe_3(C\text{-}NH\text{-}C_2H_5)$ .... Fe: Org.Comp.C6b-50, 52
– .... $(CO)_{10}(H)Fe_3[C\text{-}N(CH_3)_2]$ .... Fe: Org.Comp.C6b-49/50, 52, 58, 59
$C_{13}Fe_3H_7O_{10}P$ ... $(CO)_{10}Fe_3P\text{-}C_3H_7\text{-}i$ .... Fe: Org.Comp.C6b-27, 28/9, 31, 33/4
$C_{13}Fe_3H_7O_{11}P$ ... $(CO)_{10}Fe_3P(C_3H_7\text{-}i)(=O)$ .... Fe: Org.Comp.C6b-42/3
$C_{13}Fe_3H_8NNaO_9$ .. $Na[(CO)_9Fe_3(N{=}CH\text{-}C_3H_7\text{-}n)]$ .... Fe: Org.Comp.C6a-289/90
– .... $Na[(CO)_9Fe_3(NH{=}C\text{-}C_3H_7\text{-}n)]$ .... Fe: Org.Comp.C6a-287, 288
$C_{13}Fe_3H_8NO_9^-$ ... $[(CO)_9Fe_3(N{=}CH\text{-}C_3H_7\text{-}n)]^-$ .... Fe: Org.Comp.C6a-289/90
– .... $[(CO)_9Fe_3(NH{=}C\text{-}C_3H_7\text{-}n)]^-$ .... Fe: Org.Comp.C6a-286/8
$C_{13}Fe_3H_9IO_9S$.... $(CO)_9(I)Fe_3S\text{-}C_4H_9\text{-}t$ .... Fe: Org.Comp.C6a-100
$C_{13}Fe_3H_9KO_9S$ ... $K[(CO)_9Fe_3S\text{-}C_4H_9\text{-}t]$ .... Fe: Org.Comp.C6a-136/7, 139
$C_{13}Fe_3H_9LiO_9S$... $Li[(CO)_9Fe_3S\text{-}C_4H_9\text{-}t]$ .... Fe: Org.Comp.C6a-136/9
$C_{13}Fe_3H_9NO_9$ .... $(CO)_9(H)Fe_3(N{=}CH\text{-}C_3H_7\text{-}n)$ .... Fe: Org.Comp.C6a-260/1, 263
– .... $(CO)_9(H)Fe_3(NH{=}C\text{-}C_3H_7\text{-}n)$ .... Fe: Org.Comp.C6a-255/7
– .... $(CO)_9(H)Fe_3[N(C_3H_7\text{-}i){=}CH]$ .... Fe: Org.Comp.C6a-255/7, 259
– .... $(CO)_9(D)Fe_3[N(C_3H_7\text{-}i){=}CH]$ .... Fe: Org.Comp.C6a-259
$C_{13}Fe_3H_9NO_9S_2$ .. $(CO)_9Fe_3S_2N\text{-}C_4H_9\text{-}t$ .... Fe: Org.Comp.C6a-218/9
$C_{13}Fe_3H_9NO_{10}Si$.. $(CO)_{10}Fe_3N\text{-}Si(CH_3)_3$ .... Fe: Org.Comp.C6b-28, 31, 32/3
$C_{13}Fe_3H_9NO_{11}$ ... $[NH_3\text{-}C_2H_5][(H)Fe_3(CO)_{11}]$ .... Fe: Org.Comp.C6b-85, 96
$C_{13}Fe_3H_9NaO_9S$ .. $Na[(CO)_9Fe_3S\text{-}C_4H_9\text{-}n]$ .... Fe: Org.Comp.C6a-136
– .... $Na[(CO)_9Fe_3S\text{-}C_4H_9\text{-}s]$ .... Fe: Org.Comp.C6a-136
– .... $Na[(CO)_9Fe_3S\text{-}C_4H_9\text{-}t]$ .... Fe: Org.Comp.C6a-136/9
$C_{13}Fe_3H_9Na_2O_9P$ $Na_2[(CO)_9Fe_3P\text{-}C_4H_9\text{-}t]$ .... Fe: Org.Comp.C6a-220/1
$C_{13}Fe_3H_9O_9P^{2-}$... $[(CO)_9Fe_3P\text{-}C_4H_9\text{-}t]^{2-}$ .... Fe: Org.Comp.C6a-220/2
$C_{13}Fe_3H_9O_9PS$ ... $(CO)_9Fe_3SP\text{-}C_4H_9\text{-}t$ .... Fe: Org.Comp.C6a-202/3, 204
$C_{13}Fe_3H_9O_9PSe$ .. $(CO)_9Fe_3SeP\text{-}C_4H_9\text{-}t$ .... Fe: Org.Comp.C6a-203, 206, 210
$C_{13}Fe_3H_9O_9S^-$ ... $[(CO)_9Fe_3S\text{-}C_4H_9\text{-}n]^-$ .... Fe: Org.Comp.C6a-136
– .... $[(CO)_9Fe_3S\text{-}C_4H_9\text{-}s]^-$ .... Fe: Org.Comp.C6a-136
– .... $[(CO)_9Fe_3S\text{-}C_4H_9\text{-}t]^-$ .... Fe: Org.Comp.C6a-136/40
$C_{13}Fe_3H_9O_{11}PS_2$ $(CO)_8Fe_3S_2[P(\text{-}OCH_2\text{-})_3C\text{-}CH_3]$ .... Fe: Org.Comp.C6a-42, 43, 46, 47
$C_{13}Fe_3H_9O_{11}PSe_2$ $(CO)_8Fe_3Se_2[P(\text{-}OCH_2\text{-})_3C\text{-}CH_3]$ .... Fe: Org.Comp.C6a-42, 44, 46
$C_{13}Fe_3H_9O_{11}PTe_2$ $(CO)_8Fe_3Te_2[P(\text{-}OCH_2\text{-})_3C\text{-}CH_3]$ .... Fe: Org.Comp.C6a-42, 45, 46
$C_{13}Fe_3H_{10}NO_9PS$ $(CO)_9Fe_3SP\text{-}N(C_2H_5)_2$ .... Fe: Org.Comp.C6a-202/3, 205
$C_{13}Fe_3H_{10}NO_{10}PSi$
$(CO)_{10}Fe_3P\text{-}NH\text{-}Si(CH_3)_3$ .... Fe: Org.Comp.C6b-31, 38
$C_{13}Fe_3H_{10}N_2O_9$ .. $(CO)_9Fe_3(C_2H_5\text{-}N{=}N\text{-}C_2H_5)$ .... Fe: Org.Comp.C6a-181, 182/3

$C_{13}Fe_3H_{10}N_2O_9$ .. $(CO)_9Fe_3(N-C_2H_5)_2$ .... Fe: Org.Comp.C6a-147, 148, 150, 152
$C_{13}Fe_3H_{10}NaO_9P$ $Na[(CO)_9(H)Fe_3P-C_4H_9-t]$ .... Fe: Org.Comp.C6a-223, 224, 225
$C_{13}Fe_3H_{10}O_9$ .... $(CO)_9(H)_3Fe_3C-C_3H_7$ .... Fe: Org.Comp.C6a-249, 250, 251
$C_{13}Fe_3H_{10}O_9P^-$ ... $[(CO)_9(H)Fe_3P-C_4H_9-t]^-$ .... Fe: Org.Comp.C6a-223, 224/5
$C_{13}Fe_3H_{10}O_9P_2$ .. $(CO)_9Fe_3(P-CH_3)(P-C_3H_7-i)$ .... Fe: Org.Comp.C6a-156/7, 161, 162, 168/9
– .... $(CO)_9Fe_3(P-C_2H_5)_2$ .... Fe: Org.Comp.C6a-156/7, 158, 162, 163
$C_{13}Fe_3H_{10}O_9S$ ... $(CO)_9(H)Fe_3S-C_4H_9-n$ .... Fe: Org.Comp.C6a-94, 95, 96
– .... $(CO)_9(H)Fe_3S-C_4H_9-s$ .... Fe: Org.Comp.C6a-94, 95, 96
– .... $(CO)_9(H)Fe_3S-C_4H_9-t$ .... Fe: Org.Comp.C6a-94, 95, 96, 97
– .... $(CO)_9(D)Fe_3S-C_4H_9-t$ .... Fe: Org.Comp.C6a-97
$C_{13}Fe_3H_{10}O_9Si_2$ .. $(CO)_9Fe_3(Si-C_2H_5)_2$ .... Fe: Org.Comp.C6a-282
$C_{13}Fe_3H_{11}NO_9$ ... $(CO)_9(H)_2Fe_3N-C_4H_9-n$ .... Fe: Org.Comp.C6a-188, 191
$C_{13}Fe_3H_{11}O_9P$ ... $(CO)_9(H)_2Fe_3P-C_4H_9-t$ .... Fe: Org.Comp.C6a-187/90
$C_{13}Fe_3H_{11}O_{10}PS_4$ $(CO)_7Fe_3S_2[P(OCH_3)_3](1,3-S_2C_3H_2)$ .... Fe: Org.Comp.C6a-27/8
$C_{13}Fe_3H_{12}N_2O_9P_2$ $(CO)_9Fe_3[P-N(CH_3)_2]_2$ .... Fe: Org.Comp.C6a-157, 160, 162
$C_{13}Fe_3H_{12}O_7S_4$ .. $(CO)_7Fe_3(S-CH_2CH_2CH_2-S)_2$ .... Fe: Org.Comp.C6a-27
$C_{13}Fe_3H_{12}O_9P_2$ .. $(CO)_9(H)Fe_3(PH-CH_2-PH-C_3H_7-i)$ .... Fe: Org.Comp.C6a-198/9
– .... $(CO)_9(H)Fe_3(P-H)(CH_3-P-C_3H_7-i)$ .... Fe: Org.Comp.C6a-193
– .... $(CO)_9(H)Fe_3[CH_3-P-CH_2-P(CH_3)_2]$ .... Fe: Org.Comp.C6a-198, 200
$C_{13}Fe_3H_{18}O_{13}P_2S_2$
$(CO)_7Fe_3S_2[P(OCH_3)_3]_2$ .... Fe: Org.Comp.C6a-18, 21
$C_{13}Fe_3H_{18}O_{13}P_2Se_2$
$(CO)_7Fe_3Se_2[P(OCH_3)_3]_2$ .... Fe: Org.Comp.C6a-19, 21
$C_{13}Fe_3O_{12}Rh^-$ ... $[Fe_3RhC(CO)_{12}]^-$ .... Fe: Org.Comp.C6a-306
$C_{13}GaH_{10}NO_3$ .... $Ga(OH)_2(4-O-10-NC_{13}H_8) \cdot 2\ H_2O$ .... Ga: SVol.D1-305, 306
$C_{13}GaH_{13}MoN_2O_2$ $(c-C_7H_7)Mo(CO)_2[-N_2C_3H_3-]Ga-CH_3$ .... Mo: Org.Comp.5-235
$C_{13}GaH_{21}O_5$ ..... $(i-C_3H_7-O)Ga[CH_3-C(O)CHC(O)-CH_3]_2$ .... Ga: SVol.D1-68/9
$C_{13}GaH_{21}O_7$ ..... $Ga(O-C_3H_7-i)[CH_3-C(O)CHC(O)-O-CH_3]_2$ .... Ga: SVol.D1-201
$C_{13}GaH_{22}MoN_3O_3$ $[CH_2C(CH_3)CH_2]Mo(CO)_2[NH_2-CH_2CH_2-O-Ga(CH_3)_2-N_2C_3H_3]$ .... Mo: Org.Comp.5-205/7
$C_{13}GeH_{12}O_3$ ..... $(OC_4H_3-2)_3Ge-CH_3$ .... Ge: Org.Comp.3-145
– .... $(OC_4H_3-3)_3Ge-CH_3$ .... Ge: Org.Comp.3-145
$C_{13}GeH_{12}S_3$ ..... $(SC_4H_3-2)_3Ge-CH_3$ .... Ge: Org.Comp.3-145
$C_{13}GeH_{13}$ .... $GeCH_3(C_6H_5)_2$, radical .... Ge: Org.Comp.3-361, 363
$C_{13}GeH_{14}N_4O_4$ ... $(CH_3)_2Ge(-CH=CHC(=NNHC_6H_3(NO_2)_2-2,4)CH=CH-)$ Ge: Org.Comp.3-308
$C_{13}GeH_{14}O$ ...... $Ge(CH=CH_2)_3COC_6H_5$ .... Ge: Org.Comp.3-50
$C_{13}GeH_{16}$ .... $1-(CH_2=CH-CH_2)-1-C_6H_5-GeC_4H_6$ .... Ge: Org.Comp.3-266
– .... $1-Ge(CH_3)_3-C_{10}H_7$ .... Ge: Org.Comp.2-78, 88
– .... $2-Ge(CH_3)_3-C_{10}H_7$ .... Ge: Org.Comp.2-78/9, 88
$C_{13}GeH_{17}NO_2$ .... $(CH_3)_2Ge(-CH=CHCH(OC(=O)NHC_6H_5)CH_2-)$ .. Ge: Org.Comp.3-263
$C_{13}GeH_{18}$ .... $5-(CH_2=CH)_3Ge-[2.2.1]-C_7H_9$ .... Ge: Org.Comp.3-52
– .... $CH_2=CHCH_2-Ge(C_6H_5)(-CH_2CH_2CH_2CH_2-)$ ... Ge: Org.Comp.3-249

$C_{13}GeH_{18}$ .... CH_3-Ge(C_6H_5)(-$CH_2C(CH_3)$=$C(CH_3)CH_2$-) .... Ge: Org.Comp.3-266, 273
$C_{13}GeH_{19}N$ ...... 2-(t-C_4H_9)-3,3-$(CH_3)_2$-1,3-$NGeC_7H_4$ .... Ge: Org.Comp.3-321/2
$C_{13}GeH_{20}$ .... CH_2=CH-CH_2-Ge$(C_2H_5)_2$-C_6H_5 .... Ge: Org.Comp.3-211
– .... n-C_3H_7-Ge(C_6H_5)(-$CH_2CH_2CH_2CH_2$-) .... Ge: Org.Comp.3-249
– .... n-C_4H_9-Ge(C_6H_5)(-$CH_2CH_2CH_2$-). .... Ge: Org.Comp.3-236
– .... t-C_4H_9-Ge-$C_6H_2(CH_3)_3$-2,4,6 .... Ge: Org.Comp.3-370
$C_{13}GeH_{20}O$ ..... Ge$(C_2H_5)_3COC_6H_5$ .... Ge: Org.Comp.2-120, 122
$C_{13}GeH_{20}O_2$ ..... 4-(2-CH_3-1,3-$O_2C_3H_4$-2)-C_6H_4-Ge$(CH_3)_3$ .... Ge: Org.Comp.2-76
– .... Ge$(C_2H_5)_3$-C_6H_4COOH-4. .... Ge: Org.Comp.2-279
$C_{13}GeH_{21}^-$ ...... [$(CH_3)_2GeC_5H_4C_6H_{11}$]$^-$. .... Ge: Org.Comp.3-307
$C_{13}GeH_{21}Na$ ..... Na[$(CH_3)_2GeC_5H_4$(C_6H_{11}-c)-4] .... Ge: Org.Comp.3-307
$C_{13}GeH_{22}$ .... CH_2=$CHCH_2(CH_3)$Ge[-CH=CHC(C_4H_9-t)=$CHCH_2$-] Ge: Org.Comp.3-306
– .... $(CH_3)_2$Ge[-CH=CHC(C_6H_{11}-c)=$CHCH_2$-]. ..... Ge: Org.Comp.3-302
– .... Ge$(CH_3)_2(C_6H_5)C_5H_{11}$-n .... Ge: Org.Comp.3-203
– .... Ge$(CH_3)_2(C_6H_5)C_5H_{11}$-i. .... Ge: Org.Comp.3-203
– .... Ge$(C_2H_5)_2(C_6H_5)C_3H_7$-n .... Ge: Org.Comp.3-211
– .... Ge$(C_2H_5)_3CH_2C_6H_5$ .... Ge: Org.Comp.2-117, 121
– .... Ge$(C_2H_5)_3$C≡CC≡CC_3H_7-n. .... Ge: Org.Comp.2-242
– .... Ge$(C_2H_5)_3C_6H_4CH_3$-2 .... Ge: Org.Comp.2-277, 282
– .... Ge$(C_2H_5)_3C_6H_4CH_3$-3 .... Ge: Org.Comp.2-277, 282
– .... Ge$(C_2H_5)_3C_6H_4CH_3$-4 .... Ge: Org.Comp.2-278, 282
– .... Ge$(C_2H_5)_3$-2-[2.2.1]-C_7H_7 .... Ge: Org.Comp.2-232
– .... Ge$(CH_3)_3CH(C_2H_5)CH_2$-C_6H_5 .... Ge: Org.Comp.1-182
– .... Ge$(CH_3)_3CH(C_6H_5)C_3H_7$-n. .... Ge: Org.Comp.1-189
– .... Ge$(CH_3)_3C_6H_4(C_4H_9$-t)-3. .... Ge: Org.Comp.2-74, 85
– .... Ge$(CH_3)_3C_6H_4(C_4H_9$-t)-4. .... Ge: Org.Comp.2-74, 85
$C_{13}GeH_{22}^-$ ...... [Ge$(CH_3)_3C_6H_4(C_4H_9$-t)-4]$^-$, radical anion .... Ge: Org.Comp.3-352
$C_{13}GeH_{22}NO$ ..... Ge$(CH_3)_3C_6H_4$-4-N(O)-C_4H_9-t, radical. ...... Ge: Org.Comp.2-73, 84
$C_{13}GeH_{22}N_2O_2$ ... Ge$(C_2H_5)_3$C[=C($COCH_3$)NHN=C($COCH_3$)-] .... Ge: Org.Comp.2-291
– .... Ge$(C_2H_5)_3$C[=NNHC($COCH_3$)=C($COCH_3$)-] .... Ge: Org.Comp.2-291
$C_{13}GeH_{22}N_2O_3$ ... Ge$(C_2H_5)_3$C[=C($COOCH_3$)NHN=C($COCH_3$)-] ... Ge: Org.Comp.2-292
– .... Ge$(C_2H_5)_3$C[=NNHC($COOCH_3$)=C($COCH_3$)-] ... Ge: Org.Comp.2-292
$C_{13}GeH_{22}O$ ...... CH_2=$CHCH_2(CH_3)$Ge[-CH=CHC(OCH_3)(C_3H_7-i)CH=CH-] Ge: Org.Comp.3-306
– .... Ge$(C_2H_5)_3C_6H_4$-OCH_3-2 .... Ge: Org.Comp.2-276
– .... Ge$(C_2H_5)_3C_6H_4$-OCH_3-3 .... Ge: Org.Comp.2-276
– .... Ge$(C_2H_5)_3C_6H_4$-OCH_3-4 .... Ge: Org.Comp.2-276, 280
$C_{13}GeH_{22}O_3$ ..... Ge$(C_2H_5)_3$C≡$CCH_2OOCCH_2COCH_3$ .... Ge: Org.Comp.2-261
$C_{13}GeH_{22}O_4$ ..... Ge$(C_2H_5)_3$CH=$CHCH_2$OOCCH=CHCOOH. ..... Ge: Org.Comp.2-206
$C_{13}GeH_{23}NO$ ..... Ge$(CH_3)_3C_6H_4N(C_4H_9$-t)OH-4 .... Ge: Org.Comp.2-72/3
$C_{13}GeH_{23}P$ ...... Ge$(CH_3)_3C_3H_6P(CH_3)C_6H_5$. .... Ge: Org.Comp.1-181
$C_{13}GeH_{24}$ .... Ge$(CH_3)_3$-1-[3.3.3.$1^{3,7}$]-$C_{10}H_{15}$ .... Ge: Org.Comp.1-199
– .... Ge$(CH_3)_3$-$C_5(CH_3)_5$ .... Ge: Org.Comp.2-32, 47
– .... Ge$(C_2H_5)_3$C(=CHCH=CH_2)C(CH_3)=CH_2 .... Ge: Org.Comp.2-175
– .... Ge$(C_2H_5)_3$C≡CCH=CHC_3H_7 .... Ge: Org.Comp.2-242
– .... Ge$(CH_3)_3$C≡CCH=C(C_2H_5)C_4H_9-n .... Ge: Org.Comp.2-60, 65
$C_{13}GeH_{24}O$ ...... Ge$(C_2H_5)_3$COC≡CC_4H_9-t .... Ge: Org.Comp.2-268
$C_{13}GeH_{24}O_2$ ..... $(C_2H_5)_2$Ge(-CH_2CH_2C(=$CHCOOC_2H_5$)CH_2CH_2-) Ge: Org.Comp.3-291

$C_{13}GeH_{24}O_2$ . . . . . $(C_2H_5)_2Ge[-CH_2-C_3H_2(CH_3)(COO-C_2H_5)-CH_2-]$ Ge: Org.Comp.3-319
– . . . . $Ge(C_2H_5)_3-C{\equiv}C-C[OC(O)CH_3](CH_3)_2$ . . . . Ge: Org.Comp.2-264
$C_{13}GeH_{24}O_4$ . . . . . $(CH_3C(O)O-CH_2CH_2)_2Ge[-(CH_2)_5-]$. . . . . Ge: Org.Comp.3-288
– . . . . $(C_2H_5)_2Ge[-CH_2CH_2-C(CH_2COOH)_2-CH_2CH_2-]$ Ge: Org.Comp.3-292
$C_{13}GeH_{24}Si$ . . . . . $Ge(CH_3)_3CH_2C_6H_4Si(CH_3)_3-4$ . . . . Ge: Org.Comp.1-157, 163
$C_{13}GeH_{24}Sn$ . . . . . $Ge(CH_3)_3CH_2C_6H_4Sn(CH_3)_3-4$ . . . . Ge: Org.Comp.1-157
$C_{13}GeH_{25}NO$. . . . . $Ge(C_2H_5)_3C{\equiv}CCH_2N((CH_2)_2)_2O$ . . . . Ge: Org.Comp.2-261, 270
$C_{13}GeH_{25}NOSi$ . . . $Ge(C_2H_5)_3C{\equiv}CCH(OSi(CH_3)_3)CN$ . . . . Ge: Org.Comp.2-265
$C_{13}GeH_{25}NO_4$. . . . $Ge(CH_3)_3C(COOC_2H_5){=}C(N(CH_3)_2)COOC_2H_5$ . . Ge: Org.Comp.2-11
$C_{13}GeH_{26}$ . . . . $Ge(CH_3)_2[c-C_3H_3(CH_3)_2-1,2]-c-C_3H_2(CH_3)_3-1,2,2$ Ge: Org.Comp.3-198
– . . . . $Ge(CH_3)_3-c-C_6H_8-4-C_4H_9-t$ . . . . Ge: Org.Comp.2-34/6, 47
– . . . . $Ge(C_2H_5)_3CH{=}C{=}CH-C_4H_9-n$. . . . . Ge: Org.Comp.2-179
– . . . . $Ge(C_2H_5)_3C{\equiv}C-C_5H_{11}-n$ . . . . Ge: Org.Comp.2-242
– . . . . $Ge(C_2H_5)_3-(CH_2)_4-CH{=}C{=}CH_2$. . . . . Ge: Org.Comp.2-179
– . . . . $Ge(C_2H_5)_3-(CH_2)_5-C{\equiv}CH$ . . . . Ge: Org.Comp.2-242, 249
$C_{13}GeH_{26}N_2O_2SSn$
$(CH_3)_3SnN(SO_2C_6H_4CH_3-4)NHGe(CH_3)_3$. . . . . . Sn: Org.Comp.18-121/2, 124
$C_{13}GeH_{26}O$ . . . . . . $(C_2H_5)_2Ge[-(CH_2)_4C(=O)(CH_2)_4-]$ . . . . Ge: Org.Comp.3-311, 314
– . . . . $Ge(CH_3)_2(CH(-C(CH_3)_2CH_2-))C(=CH_2)C(CH_3)_2OCH_3$ Ge: Org.Comp.3-198
– . . . . $Ge(C_2H_5)_3C(C(OH)(CH_3)_2){=}CHCH{=}CH_2$ . . . . Ge: Org.Comp.2-216, 219
– . . . . $Ge(C_2H_5)_3C{\equiv}CC(OH)(C_2H_5)_2$ . . . . Ge: Org.Comp.2-267
$C_{13}GeH_{26}O_2$ . . . . . $Ge(C_2H_5)_3CH{=}CHCH_2OCH_2OCH_2CH{=}CH_2$. . . . . Ge: Org.Comp.2-204
– . . . . $Ge(C_2H_5)_3C{\equiv}CCH(OH)O-C_4H_9-n$. . . . . Ge: Org.Comp.2-260
– . . . . $Ge(C_2H_5)_3C{\equiv}CCH(OH)O-C_4H_9-s$ . . . . Ge: Org.Comp.2-260
– . . . . $Ge(C_2H_5)_3C{\equiv}CC(OH)(CH_3)CH_2CH_2OCH_3$. . . . . . Ge: Org.Comp.2-267
$C_{13}GeH_{26}O_3$ . . . . . $Ge(C_2H_5)_3CH_2CH_2CH_2OOCCH_2COCH_3$ . . . . Ge: Org.Comp.2-145
$C_{13}GeH_{26}O_4$ . . . . . $Ge(C_2H_5)_3CH(COOC_2H_5)_2$ . . . . Ge: Org.Comp.2-134, 139
$C_{13}GeH_{26}S$ . . . . . . $Ge(C_2H_5)_3CH{=}C{=}C(CH_3)CH_2-S-C_2H_5$ . . . . Ge: Org.Comp.2-218/9
– . . . . $Ge(C_2H_5)_3CH{=}C{=}CHCH(CH_3)-S-C_2H_5$. . . . . Ge: Org.Comp.2-224
– . . . . $Ge(C_2H_5)_3C{\equiv}CCH(CH_3)CH_2-S-C_2H_5$ . . . . Ge: Org.Comp.2-264, 271
– . . . . $Ge(C_2H_5)_3C{\equiv}CCH_2CH(CH_3)-S-C_2H_5$ . . . . Ge: Org.Comp.2-267
$C_{13}GeH_{27}N$ . . . . . . $Ge(C_2H_5)_3C{\equiv}CCH_2N(C_2H_5)_2$. . . . . Ge: Org.Comp.2-261, 270
$C_{13}GeH_{27}NOSi$ . . . $Ge(C_2H_5)_3C(CH_3){=}C(CN)-O-Si(CH_3)_3$ . . . . Ge: Org.Comp.2-209
$C_{13}GeH_{28}$ . . . . $n-C_6H_{13}Ge(C_4H_9-n)(-CH_2CH_2CH_2-)$. . . . . Ge: Org.Comp.3-235
– . . . . $Ge(CH_3)_3-(CH_2)_4-C_6H_{11}-c$. . . . . Ge: Org.Comp.1-189
– . . . . $Ge(CH_3)_3-(CH_2)_5-C_5H_9-c$ . . . . Ge: Org.Comp.1-192/3
– . . . . $Ge(CH_3)_3-c-C_6H_{10}-4-C_4H_9-t$ . . . . Ge: Org.Comp.1-196
– . . . . $Ge(C_2H_5)_3-(CH_2)_5-CH{=}CH_2$. . . . . Ge: Org.Comp.2-179
– . . . . $Ge(C_3H_7-i)_3CH{=}C(CH_3)_2$ . . . . Ge: Org.Comp.3-14, 15
– . . . . $Ge(C_3H_7-i)_3CH{=}CHC_2H_5$ . . . . Ge: Org.Comp.3-15
– . . . . $Ge(C_3H_7-n)_3CH{=}C(CH_3)_2$. . . . . Ge: Org.Comp.3-10, 11
– . . . . $Ge(C_3H_7-n)_3CH{=}CHC_2H_5$. . . . . Ge: Org.Comp.3-10, 11
$C_{13}GeH_{28}NO_3P$. . . $Ge(C_2H_5)_3CH(CN)CH_2P(O)(OC_2H_5)_2$ . . . . Ge: Org.Comp.2-135
$C_{13}GeH_{28}NP$. . . . . $Ge(C_2H_5)_3CH(CN)CH_2P(C_2H_5)_2$ . . . . Ge: Org.Comp.2-134
$C_{13}GeH_{28}O$ . . . . . . $Ge(C_2H_5)_3C(C_3H_7-i){=}CHO-C_2H_5$ . . . . Ge: Org.Comp.2-193, 200
– . . . . $Ge(C_2H_5)_3CH{=}CHC(OH)(CH_3)C_3H_7-n$ . . . . Ge: Org.Comp.2-225

$C_{13}GeH_{28}O$	$Ge(C_2H_5)_3C(O-C_2H_5)=CH-C_3H_7-i$	Ge: Org.Comp.2-213, 218
–	$Ge(C_3H_7-n)_3CH=CHO-C_2H_5$	Ge: Org.Comp.3-9
$C_{13}GeH_{28}O_2$	$Ge(C_2H_5)_3C(=CH_2)C(OH)(CH_3)-CH_2CH_2-OCH_3$	Ge: Org.Comp.2-191
–	$Ge(C_2H_5)_3C(CH_2OCH_3)=CHC(CH_3)_2OH$	Ge: Org.Comp.2-214
–	$Ge(C_2H_5)_3CH=CHC(CH_3)_2OCH(OH)CH_3$	Ge: Org.Comp.2-212
–	$Ge(C_2H_5)_3CH=CHC(OH)(CH_3)-CH_2CH_2-OCH_3$	Ge: Org.Comp.2-222
–	$Ge(C_3H_7-n)_3CH_2CH(CH_3)-COOH$	Ge: Org.Comp.3-4
–	$Ge(C_3H_7-n)_3CH_2CH_2CH_2-COOH$	Ge: Org.Comp.3-5
–	$Ge(C_3H_7-n)_3CH_2CH_2-C(O)O-CH_3$	Ge: Org.Comp.3-4
–	$Ge(C_3H_7-n)_3CH_2-C(O)O-C_2H_5$	Ge: Org.Comp.3-2
$C_{13}GeH_{28}O_3$	$Ge(C_2H_5)_3CH_2CH_2C(O)OO-C_4H_9-t$	Ge: Org.Comp.2-150, 154
–	$Ge(C_2H_5)_3CH_2CH_2OCH_2CH_2OCH_2-2-OC_2H_3$	Ge: Org.Comp.2-125
$C_{13}GeH_{29}N$	$Ge(C_2H_5)_3C(CH_3)=CHN(C_2H_5)_2$	Ge: Org.Comp.2-193, 200
$C_{13}GeH_{29}O_4P$	$Ge(C_2H_5)_3CH(CHO)-CH_2P(O)(O-C_2H_5)_2$	Ge: Org.Comp.2-134
–	$Ge(CH_3)_3CH[C(O)CH_3]-C(CH_3)_2P(O)(OC_2H_5)_2$	Ge: Org.Comp.1-183
$C_{13}GeH_{30}$	$Ge(C_3H_7-n)_3C_4H_9-n$	Ge: Org.Comp.3-5
–	$Ge(C_4H_9-n)_2(C_2H_5)C_3H_7-n$	Ge: Org.Comp.3-213
–	$Ge(C_4H_9-n)_3CH_3$	Ge: Org.Comp.3-17, 22
–	$Ge(CH_3)_3C_{10}H_{21}-n$	Ge: Org.Comp.1-192/3
$C_{13}GeH_{30}O$	$Ge(C_2H_5)_3CH_2CH(O-C_4H_9-t)CH_3$	Ge: Org.Comp.2-149
–	$Ge(C_2H_5)_3CH_2CH_2CH_2O-C_4H_9-t$	Ge: Org.Comp.2-144
–	$Ge(C_2H_5)_3CH_2CH_2C(OH)(C_2H_5)_2$	Ge: Org.Comp.2-162
–	$Ge(C_2H_5)_3C(OH)(C_3H_7-i)_2$	Ge: Org.Comp.2-148
–	$Ge(C_2H_5)_3-(CH_2)_4-C(OH)(CH_3)_2$	Ge: Org.Comp.2-164
$C_{13}GeH_{30}OSn$	$Ge(CH_3)_3C(Sn(C_2H_5)_3)=CHOC_2H_5$	Ge: Org.Comp.2-5
$C_{13}GeH_{30}O_2Si$	$Ge(C_2H_5)_3C_3H_6COOSi(CH_3)_3$	Ge: Org.Comp.2-160
$C_{13}GeH_{30}O_5P_2$	$Ge(CH_3)_3CH=C(OP(OC_2H_5)_2)P(OC_2H_5)_2$	Ge: Org.Comp.2-4
$C_{13}GeH_{30}S_2Si$	$Ge(C_2H_5)_3C(Si(CH_3)_3)((SCH_2)_2CH_2)$	Ge: Org.Comp.2-289
$C_{13}GeH_{30}Si$	$Ge(C_2H_5)_3CH_2CH_2Si(CH_3)_2CH_2CH=CH_2$	Ge: Org.Comp.2-126
$C_{13}GeH_{30}Si_2$	$(CH_3)_2Ge(-CH_2CH=CHCH_2C(Si(CH_3)_3)_2-)$	Ge: Org.Comp.3-300, 307
$C_{13}GeH_{31}NO_2S$	$Ge(C_2H_5)_3CH(SO_2N(C_2H_5)_2)C_2H_5$	Ge: Org.Comp.2-149
$C_{13}GeH_{31}NO_2Si$	$Ge(C_2H_5)_3CH_2CH_2CH_2NHCOOSi(CH_3)_3$	Ge: Org.Comp.2-146
$C_{13}GeH_{31}O_3P$	$Ge(C_2H_5)_3CH_2CH_2CH_2P(O)(OC_2H_5)_2$	Ge: Org.Comp.2-147
$C_{13}GeH_{32}O_2Si_2$	$Ge(CH_3)_3C(CH_2OSi(CH_3)_3)=CHCH_2OSi(CH_3)_3$	Ge: Org.Comp.2-10
$C_{13}GeH_{32}O_3SSi$	$Ge(C_2H_5)_3CH(SO_2OSi(CH_3)_3)C_3H_7$	Ge: Org.Comp.2-159
$C_{13}GeH_{32}O_6P_2$	$Ge(CH_3)_3CH[P(O)(OC_2H_5)_2]CH_2P(O)(OC_2H_5)_2$	Ge: Org.Comp.1-168
$C_{13}GeH_{32}Si_2$	$Ge(CH_3)_2(CH(Si(CH_3)_3)_2)CH_2C(CH_3)=CH_2$	Ge: Org.Comp.3-195
$C_{13}GeH_{36}N_2SiSn$	$(CH_3)_3SnN(C_4H_9-t)N(Si(CH_3)_3)Ge(CH_3)_3$	Sn: Org.Comp.18-121/2
–	$(CH_3)_3SnN(Ge(CH_3)_3)N(Si(CH_3)_3)C_4H_9-t$	Sn: Org.Comp.18-121, 123
–	$(CH_3)_3SnN(Si(CH_3)_3)N(Ge(CH_3)_3)C_4H_9-t$	Sn: Org.Comp.18-121, 123
$C_{13}GeH_{36}Si_3$	$Ge(CH_3)_3C[Si(CH_3)_3]_3$	Ge: Org.Comp.1-144, 151
$C_{13}H_5MoN_4S_4^-$	$[C_5H_5Mo(-SC(CN)=C(CN)S-)_2]^-$	Mo: Org.Comp.6-6, 7/8
$C_{13}H_5MoO_5^-$	$[(C_6H_5)_3P=N=P(C_6H_5)_3][C_6H_5-CC-Mo(CO)_5]$	Mo: Org.Comp.5-134
$C_{13}H_5MoO_8Re$	$(CO)_5ReMo(CO)_3C_5H_5$	Re: Org.Comp.2-190/1
$C_{13}H_5N_3O_7Re$	$(CO)_5ReNNC(O)N(C_6H_5)CO$, radical	Re: Org.Comp.2-170/1
$C_{13}H_5O_5Re$	$(CO)_5ReC{\equiv}CC_6H_5$	Re: Org.Comp.2-131
$C_{13}H_5O_7Re$	$(CO)_5ReC(O)C(O)C_6H_5$	Re: Org.Comp.2-119
$C_{13}H_5O_8ReW$	$(CO)_5ReW(CO)_3C_5H_5$	Re: Org.Comp.2-192/3, 205
$C_{13}H_6LiO_7Re$	$Li[(CO)_4ReC(=O)CH(C_6H_5)OC(=O)]$	Re: Org.Comp.2-334, 340
$C_{13}H_6MoN_2O_5$	$(CO)_5Mo(CN-N=CH-C_6H_5)$	Mo: Org.Comp.5-10, 11

$C_{13}H_6O_7Re^-$	$[(CO)_4ReC(=O)CH(C_6H_5)OC(=O)]^-$	Re: Org.Comp.2-334, 340
$C_{13}H_7MnNO_9Re$	$(CO)_4Re(CN-C_3H_7-i)Mn(CO)_5$	Re: Org.Comp.2-247
$C_{13}H_7MoNO_5$	$(CO)_5Mo(CN-C_6H_4-CH_3-4)$	Mo: Org.Comp.5-6, 8
$C_{13}H_7MoNO_6$	$(CO)_5Mo(CN-C_6H_4-OCH_3-4)$	Mo: Org.Comp.5-6, 8
$C_{13}H_7N_2O_6STh^+$	$Th[O_3SC_6H_4NNC_6H_3(O)COO]^+$	Th: SVol.D1-122
$C_{13}H_7O_5Re$	$(CO)_4Re[-O=C(C_6H_5)-CH=CH-]$	Re: Org.Comp.1-418
$C_{13}H_7O_6Re$	$(CO)_5ReC(O)CH_2C_6H_5$	Re: Org.Comp.2-118/9
–	$(CO)_5ReC(O)C_6H_4CH_3-2$	Re: Org.Comp.2-122
–	$(CO)_5ReC(O)C_6H_4CH_3-3$	Re: Org.Comp.2-122
–	$(CO)_5ReC(O)C_6H_4CH_3-4$	Re: Org.Comp.2-122
$C_{13}H_7O_7Re$	$(CO)_5ReC(O)C(OH)(H)C_6H_5$	Re: Org.Comp.2-119
$C_{13}H_8IN_2O_3Re$	$(CO)_3ReI[NC_5H_4-2-(2-C_5H_4N)]$	Re: Org.Comp.1-173
$C_{13}H_8IN_2O_5Re$	$(CO)_3ReI[O=NC_5H_4-2-(2-C_5H_4N=O)]$	Re: Org.Comp.1-209
$C_{13}H_8MnN_2O_4$	$[Mn(-O-C_6H_4-2-CH=N-C_6H_3(NO_2-4)-2-O-)]_n$	Mn: MVol.D6-28, 29/30
$C_{13}H_8MnO_{10}Re$	$(CO)_5ReMn(CO)_4C_4H_8O$	Re: Org.Comp.2-198
$C_{13}H_8MoO_5S$	$C_6H_5-S-C(CH_3)=Mo(CO)_5$	Mo: Org.Comp.5-104, 111
$C_{13}H_8MoO_6$	$CH_3O-C(C_6H_5)=Mo(CO)_5$	Mo: Org.Comp.5-100, 101, 103, 109/10
$C_{13}H_8NO_9OsRe$	$(CO)_5ReOs(CO)_3(NCCD_3)C(O)C_2H_5$	Re: Org.Comp.2-202
$C_{13}H_8N_2NaO_3Re$	$Na[(CO)_3Re(NC_5H_4-2-(2-C_5H_4N))]$	Re: Org.Comp.1-175
$C_{13}H_8N_2O_3Re^-$	$[(CO)_3Re(NC_5H_4-2-(2-C_5H_4N))]^-$	Re: Org.Comp.1-175
$C_{13}H_8N_3O_6Re$	fac-$(CO)_3Re[NC_5H_4-2-(2-C_5H_4N)]ONO_2$	Re: Org.Comp.1-155
$C_{13}H_8O_6Re^+$	$[(CO)_5ReOC(CH_3)C_6H_5]^+$	Re: Org.Comp.2-157/8
$C_{13}H_8O_6Re^-$	$[(CO)_4Re(C(O)CH_3)C(O)C_6H_5]^-$	Re: Org.Comp.1-425/9
$C_{13}H_9KMoN_4O_2$	$K[(C_5H_5)Mo(CO)_2(-C_3H_2N_2-C_3H_2N_2-)]$	Mo: Org.Comp.7-157, 173
$C_{13}H_9MnNOS$	$Mn[-O-C_6H_4-2-CH=N-C_6H_4-2-S-]$	Mn: MVol.D6-28
$C_{13}H_9MnNO_2$	$Mn[-O-C_6H_4-2-CH=N-C_6H_4-2-O-]$	Mn: MVol.D6-28/30
–	$Mn[-O-C_6H_4-2-CH=N-C_6H_4-3-O-]$	Mn: MVol.D6-28
–	$Mn[-O-C_6H_4-2-CH=N-C_6H_4-4-O-]$	Mn: MVol.D6-28, 29/30
–	$[Mn(-O-C_6H_4-2-CH=N-C_6H_4-2-O-)]_n$	Mn: MVol.D6-28, 29/30
$C_{13}H_9MnN_2O_6S_2^-$	$Mn[-3-O-C_6H_3(SO_3)-4-CH=NC_6H_4-SO_2NH-4-]^-$	Mn: MVol.D6-50
$C_{13}H_9MnN_4O_5$	$[Mn(-O-2-C_6H_3(NO_2-5)-CH=N-N=C(4-C_5H_4N)O-)(OH)]$	Mn: MVol.D6-280, 284/5
$C_{13}H_9MoN_3O_4$	$(C_5H_5)Mo(CO)_2-N=N-C_6H_4-4-NO_2$	Mo: Org.Comp.7-6, 16
$C_{13}H_9MoN_4O_2^-$	$[(C_5H_5)Mo(CO)_2(-C_3H_2N_2-C_3H_2N_2-)]^-$	Mo: Org.Comp.7-157, 173/4
$C_{13}H_9N_2O_3Re$	fac-$(CO)_3ReH[NC_5H_4-2-(2-C_5H_4N)]$	Re: Org.Comp.1-173
$C_{13}H_9N_3O_3S$	$C_6H_5C(O)N=S=N-C_6H_4-4-NO_2$	S: S-N Comp.7-268/9
$C_{13}H_9O_4Re$	$(CH_2CHCHC_6H_5)Re(CO)_4$	Re: Org.Comp.2-388
$C_{13}H_9O_5Re$	$(CO)_5ReCH_2CH_2C_6H_5$	Re: Org.Comp.2-107/8
$C_{13}H_9O_6Re$	cis-$(CO)_4Re[C(=O)CH_3]=C(OH)C_6H_5$	Re: Org.Comp.1-389
–	fac-$(CO)_3(^{13}CO)Re[C(=O)CH_3]=C(OH)C_6H_5$	Re: Org.Comp.1-428/9
$C_{13}H_{10}IN_2O_3Re$	$(CO)_3ReI(C_5H_5N)_2$	Re: Org.Comp.1-234
$C_{13}H_{10}I_2MnN_4S$	$MnI_2[NC_5H_4-2-CH=N-NH-2-(1,3-SNC_7H_4)]$	Mn: MVol.D6-244, 246/7
$C_{13}H_{10}MnNO^+$	$[Mn(O-C_6H_4-2-CH=N-C_6H_5)]^+$	Mn: MVol.D6-15/6
$C_{13}H_{10}MnNOS^+$	$[Mn(O-C_6H_4-1,3-SNC_7H_6)]^+$	Mn: MVol.D7-237/9
–	$[Mn(O-N(C_6H_5)C(C_6H_5)=S)]^+$	Mn: MVol.D7-215/7
$C_{13}H_{10}MnN_3OS^+$	$Mn[C_6H_5-C(O)NC(S)NH-2-NC_5H_4]^+$	Mn: MVol.D7-198/9
–	$Mn[(O)C_{12}H_8-N=NC(=S)NH_2]^+$	Mn: MVol.D6-346
$C_{13}H_{10}MnN_5O_3^+$	$Mn[O-C(=N-NH-C_6H_5)-N=N-C_6H_4(NO_2-4)]^+$	Mn: MVol.D6-365

$C_{13}H_{10}MnN_5O_3^+$. $Mn[O-C(N=N-C_6H_5)=N-NH-C_6H_4(NO_2-4)]^+$ . . . Mn:MVol.D6-365
$C_{13}H_{10}MnN_6S_2$ . . . $[Mn(NC_5H_4-2-CH=NNH-2-C_5H_4N)(NCS)_2]$ Mn:MVol.D6-244, 246/7
$C_{13}H_{10}MoN_2O_2$ . . . $(C_5H_5)Mo(CO)_2-N=N-C_6H_5$ Mo:Org.Comp.7-6, 15
$C_{13}H_{10}MoN_2O_2S$. . $(C_5H_5)Mo(CO)_2[-S-C(3-NC_5H_4)=NH-]$. Mo:Org.Comp.7-207/10, 225
$C_{13}H_{10}MoN_2O_3$ . . . $(C_5H_5)Mo(CO)_2[-C(=O)-NH-1-NC_5H_4-2-]$ Mo:Org.Comp.8-132
$C_{13}H_{10}MoN_4O_2$ . . . $(C_5H_5)Mo(CO)_2[-C_3H_3N_2-C_3H_2N_2-]$ Mo:Org.Comp.7-157, 175
$C_{13}H_{10}MoO_4$ . . . . . $[(6.1.0)-C_9H_{10}-2,4,6]Mo(CO)_4$ Mo:Org.Comp.5-338, 339, 341/2
– . . . . $[(6.1.0)-C_9H_{10}-2,4,6]Mo(CO)_3(^{13}CO)$ Mo:Org.Comp.5-341/2
$C_{13}H_{10}NO_2Th^{3+}$ . . $Th[C_6H_5C(O)N(C_6H_5)O]^{3+}$ . . . . Th: SVol.D1-113
$C_{13}H_{10}NO_3Re$ . . . . $(C_5H_5)Re(CO)_2(NC_5H_4-4-CH=O)$ . . . . Re: Org.Comp.3-193/7, 199
– . . . . $(C_5H_5)Re(CO)_2(O=N-C_6H_5)$ . . . . Re: Org.Comp.3-187
$C_{13}H_{10}NO_3SSb$ . . . 4,5-$(-N=C(C_6H_5)S-)C_6H_3Sb(O)(OH)_2$. . . . . Sb: Org.Comp.5-307
$C_{13}H_{10}NO_4Sb$ . . . . 4,5-$(-N=C(C_6H_5)O-)C_6H_3Sb(O)(OH)_2$ . . . . Sb: Org.Comp.5-307
$C_{13}H_{10}N_2OS$ . . . . . $C_6H_5C(O)N=S=N-C_6H_5$ . . . . S: S-N Comp.7-268/9
$C_{13}H_{10}N_2OS_2$ . . . . $O=S=NSN=C(C_6H_5)_2$ . . . . S: S-N Comp.6-42/6
$C_{13}H_{10}N_2O_2Re^+$ . . $[(C_5H_5)Re(CO)_2-N=N-C_6H_5][BF_4]$ . . . . Re: Org.Comp.3-193/7, 203
$C_{13}H_{10}N_2O_2S$ . . . . $O=S=NNHC_6H_4C(O)C_6H_5-4$ . . . . S: S-N Comp.6-57/60
$C_{13}H_{10}N_2O_4Re^+$ . . $[(CO)_3Re(NC_5H_4-C_5H_4N)(H_2O)]^+$. . . . . Re: Org.Comp.1-301/2
$C_{13}H_{10}N_3O_6Re$ . . . $O_2NO-Re(CO)_3(NC_5H_5)_2$ . . . . Re: Org.Comp.1-221
– . . . . $O_2NO-Re(CO)_3[-1,8-(1,8-N_2C_8H_4(CH_3)_2-2,7)-]$ Re: Org.Comp.1-156
$C_{13}H_{10}N_4O_4S_3$ . . . 2,4-$(NO_2)_2C_6H_3-SN=S=NS-C_6H_4CH_3-4$ . . . . . . S: S-N Comp.7-40
$C_{13}H_{10}N_8O_{10}STh$ $Th(NO_3)_3[SNC_3H_2-N-N=N_2C_3(CH_3)(C_6H_5)=O]$. . Th: SVol.D4-143
$C_{13}H_{10}O_4Sn$ . . . . . $C_6H_5Sn(OC_6H_4COO-2)OH$ . . . . Sn: Org.Comp.17-77
$C_{13}H_{11}KMoN_2O_3$. $K[(C_5H_5)Mo(CO)_2(CN)-C(=O)-CH_2CH_2CH_2-CN]$ Mo:Org.Comp.8-1, 5
– . . . . $K[(C_5H_5)Mo(CO)_2(CN)-C(=O)-CH_2CH_2CH_2-CN] \cdot n\ OC_4H_8$ Mo:Org.Comp.8-1, 5
$C_{13}H_{11}MnN_3O_3S_2$ $[Mn(-5-S-1,3,4-SN_2C_2-2-N=CH-1-C_{10}H_6-2-O-)(H_2O)_2]_n$ Mn:MVol.D6-72
$C_{13}H_{11}MnN_4O^+$ . . $Mn[O-C(N=N-C_6H_5)=N-NH-C_6H_5]^+$ Mn:MVol.D6-365
$C_{13}H_{11}MnN_4S^+$ . . $Mn[S-C(N=N-C_6H_5)=N-NH-C_6H_5]^+$. Mn:MVol.D6-366
$C_{13}H_{11}MoN_2O_2^+$. $[(c-C_7H_7)Mo(CO)_2(N_2C_3H_3)CH][PF_6]$. Mo:Org.Comp.5-235
$C_{13}H_{11}MoN_2O_3^-$ . . $K[(C_5H_5)Mo(CO)_2(CN)-C(=O)-CH_2CH_2CH_2-CN]$ Mo:Org.Comp.8-1, 5
– . . . . $K[(C_5H_5)Mo(CO)_2(CN)-C(=O)-CH_2CH_2CH_2-CN] \cdot n\ OC_4H_8$ Mo:Org.Comp.8-1, 5
– . . . . $[As(C_6H_5)_4][(C_5H_5)Mo(CO)_2(CN)-C(=O)-CH_2CH_2CH_2-CN]$ Mo:Org.Comp.8-1, 5
$C_{13}H_{11}MoN_3O_2$ . . . $(C_5H_5)Mo(CO)_2[(NC)_2C=C=N(CH_3)_2]$ Mo:Org.Comp.8-185, 190
$C_{13}H_{11}NOS$ . . . . . . $O=S=NCH(C_6H_5)_2$ . . . . S: S-N Comp.6-109
$C_{13}H_{11}NO_2Sn$ . . . . $(C_6H_5)_2Sn(NCO)OH$ . . . . Sn: Org.Comp.17-136, 139
$C_{13}H_{11}NO_3STh$ . . . $(OH)_2Th[O-C_6H_4-2-(CH=N-C_6H_4-2-S)] \cdot H_2O$ Th: SVol.D4-133
$C_{13}H_{11}NO_6Sb_2$ . . . $[(HO)_2(O)Sb]_2C_{13}H_7N$. Sb: Org.Comp.5-316
$C_{13}H_{11}N_2O_2Re$ . . . $(C_5H_5)Re(CO)_2(HN=N-C_6H_5)$ . . . . Re: Org.Comp.3-193/7, 201
$C_{13}H_{11}N_2O_3Sb$ . . . $(4-C_6H_4NHC(O)NHC_6H_4-4')Sb(O)OH$. . . . . Sb: Org.Comp.5-213
$C_{13}H_{11}N_2O_6Sb$ . . . 4-$(3'-HO_2C-4'-HOC_6H_3N=N)C_6H_4Sb(O)(OH)_2$. Sb: Org.Comp.5-291
$C_{13}H_{11}N_3OS_2$ . . . . 2-$(O=S=N-N=)-3-(C_2H_5)C_{11}H_6NS$-3,1. . . . . S: S-N Comp.6-69/71
$C_{13}H_{11}N_3O_2S$ . . . . 3-$NO_2-C_6H_4-N=S=N-C_6H_4-CH_3$-4 . . . . S: S-N Comp.7-262

$C_{13}H_{11}N_3O_2S$	$4\text{-}NO_2\text{-}C_6H_4\text{-}N{=}S{=}N\text{-}C_6H_4\text{-}CH_3\text{-}4$	S: S-N Comp.7-262, 264
$C_{13}H_{11}N_3O_3S$	$4\text{-}CH_3O\text{-}C_6H_4\text{-}N{=}S{=}N\text{-}C_6H_4\text{-}NO_2\text{-}4$	S: S-N Comp.7-260/1, 264
$C_{13}H_{11}N_3O_4S_2$	$4\text{-}CH_3\text{-}C_6H_4\text{-}S(O)_2N{=}S{=}N\text{-}C_6H_4\text{-}NO_2\text{-}4$	S: S-N Comp.7-63
$C_{13}H_{11}N_3O_4S_3$	$4\text{-}CH_3\text{-}C_6H_4\text{-}S(O)_2N{=}S{=}NS\text{-}C_6H_4\text{-}NO_2\text{-}2$	S: S-N Comp.7-67
$C_{13}H_{11}N_5O_{10}Th$	$Th(NO_3)_3[2\text{-}O\text{-}C_6H_4\text{-}CH{=}N\text{-}NH\text{-}C_6H_5]$	Th: SVol.D4-143
$C_{13}H_{11}O_2Sb$	$(2\text{-}C_6H_4CH_2C_6H_4\text{-}2')Sb(O)OH$	Sb: Org.Comp.5-213
$C_{13}H_{11}O_5PRe$	$(CO)_5ReP(CH_3)_2C_6H_5$, radical	Re: Org.Comp.2-171
$C_{13}H_{11}O_5PRe^+$	$[(CO)_5ReP(CH_3)_2C_6H_5]^+$	Re: Org.Comp.2-155
$C_{13}H_{11}O_6Re$	$(CO)_5ReC(O)CH[\text{-}CH{=}C(C_4H_9\text{-}t)\text{-}]$	Re: Org.Comp.2-120
–	$(CO)_5ReC(O)CD[\text{-}CH{=}C(C_4H_9\text{-}t)\text{-}]$	Re: Org.Comp.2-120
$C_{13}H_{12}IMoNO_2$	$C_5H_5Mo(NO)(I)(C(O)C_6H_4CH_3\text{-}4)$	Mo: Org.Comp.6-90, 93
$C_{13}H_{12}IMoNO_3$	$C_5H_5Mo(NO)(I)O_2CC_6H_4CH_3\text{-}4$	Mo: Org.Comp.6-46/7
$C_{13}H_{12}IN_2ORe$	$(C_5H_5)ReI(CO)\text{-}N{=}N\text{-}C_6H_4\text{-}CH_3\text{-}4$	Re: Org.Comp.3-133/5, 136
$C_{13}H_{12}IN_2O_2Re$	$(C_5H_5)ReI(CO)\text{-}N{=}N\text{-}C_6H_4\text{-}OCH_3\text{-}4$	Re: Org.Comp.3-133/5, 137
$C_{13}H_{12}I_2MoN_2O$	$(C_5H_5)Mo(I_2)(CO)\text{-}NN\text{-}C_6H_4\text{-}4\text{-}CH_3$	Mo: Org.Comp.6-248, 251
$C_{13}H_{12}I_2MoN_2O_2$	$C_5H_5Mo(NO)(CNC_6H_4OCH_3\text{-}4)I_2$	Mo: Org.Comp.6-84
$C_{13}H_{12}I_4MoN_2OSn$	$C_5H_5Mo(N_2C_6H_4CH_3\text{-}4)(CO)(SnI_3)I$	Mo: Org.Comp.6-248, 251
$C_{13}H_{12}MnNO_3P$	$Mn[O_2PHCH(NH\text{-}C_6H_5)C_6H_4\text{-}O\text{-}2]$	Mn: MVol.D8-113/4
$C_{13}H_{12}MnNO_5S$	$[Mn(SC_4H_3\text{-}2\text{-}C(O)CH{=}N\text{-}4\text{-}C_6H_4\text{-}COO)(H_2O)_2]$	Mn: MVol.D6-92
$C_{13}H_{12}MnN_2O_2S_2$	$[Mn(\text{-}O\text{-}2\text{-}C_{10}H_6\text{-}1\text{-}CH{=}NN{=}C(SCH_3)S\text{-})(H_2O)]$	Mn: MVol.D6-361/2
$C_{13}H_{12}MnN_2O_4P^-$	$[Mn(O_3P\text{-}CH(NHCH_2\text{-}2\text{-}NC_5H_4)C_6H_4\text{-}O\text{-}2)]^-$	Mn: MVol.D8-125/6
–	$[Mn(O_3P\text{-}CH(NHCH_2\text{-}3\text{-}NC_5H_4)C_6H_4\text{-}O\text{-}2)]^-$	Mn: MVol.D8-125/6
–	$[Mn(O_3P\text{-}CH(NHCH_2\text{-}4\text{-}NC_5H_4)C_6H_4\text{-}O\text{-}2)]^-$	Mn: MVol.D8-125/6
$C_{13}H_{12}MnN_2O_5$	$Mn[C_2H_5OC(O)C(C(O)CH_3){=}N\text{-}N\text{-}2\text{-}C_6H_4COO]$	Mn: MVol.D6-259/60
$C_{13}H_{12}MnN_3O_5P$	$Mn[O_2PH\text{-}CH(NHCH_2\text{-}2\text{-}NC_5H_4)C_6H_3\text{-}O\text{-}2\text{-}NO_2\text{-}5]$	Mn: MVol.D8-113/4
–	$Mn[O_2PH\text{-}CH(NHCH_2\text{-}3\text{-}NC_5H_4)C_6H_3\text{-}O\text{-}2\text{-}NO_2\text{-}5]$	Mn: MVol.D8-113/4
$C_{13}H_{12}MnN_4O_2$	$[Mn(\text{-}O\text{-}C_6H_4\text{-}2\text{-}N{=}CHCH{=}N\text{-}N(C_5H_4N\text{-}2)\text{-})(H_2O)]$	Mn: MVol.D6-268/9
$C_{13}H_{12}MnN_6$	$[Mn(NC_5H_4\text{-}2\text{-}N\text{-}N{=}C(CH_3)\text{-}CH{=}N\text{-}N\text{-}2\text{-}C_5H_4N)]$	Mn: MVol.D6-264/5
$C_{13}H_{12}MnN_6O_3S_3$	$Mn(NH_2\text{-}C_6H_4\text{-}SO_2\text{-}NH\text{-}N_2C_4H_2\text{-}OCH_3)(NCS)_2$	Mn: MVol.D7-116/23
$C_{13}H_{12}MoNO_2^+$	$[C_9H_7Mo(CO)(NO)C_3H_5]^+$	Mo: Org.Comp.6-346
$C_{13}H_{12}MoO_2$	$(C_5H_5)Mo(CO)_2(C_6H_7)$	Mo: Org.Comp.8-296
$C_{13}H_{12}MoO_2S$	$(C_9H_7)Mo(CO)_2[\text{-}CH_2\text{-}S(CH_3)\text{-}]$	Mo: Org.Comp.8-128
$C_{13}H_{12}MoO_3$	$(C_5H_5)Mo(CO)_2[(CH_2)C_5H_5({=}O)]$	Mo: Org.Comp.8-206, 207, 234
–	$(C_5H_5)Mo(CO)_2[(CHD)C_5H_4D({=}O)]$	Mo: Org.Comp.8-206, 207, 234
$C_{13}H_{12}MoO_4$	$(C_5H_5)Mo(CO)_2[(CH_2)OC_5H_5({=}O)]$	Mo: Org.Comp.8-206, 236
$C_{13}H_{12}MoO_6$	$C_2H_5\text{-}O\text{-}C(C_5H_7\text{-}c){=}Mo(CO)_5$	Mo: Org.Comp.5-100, 104
$C_{13}H_{12}NO_2Re$	$(C_5H_5)Re(CO)(NO)\text{-}C_6H_4\text{-}CH_3\text{-}2$	Re: Org.Comp.3-160, 164, 169
–	$(C_5H_5)Re(CO)(NO)\text{-}C_6H_4\text{-}CH_3\text{-}3$	Re: Org.Comp.3-160, 164, 169
–	$(C_5H_5)Re(CO)(NO)\text{-}C_6H_4\text{-}CH_3\text{-}4$	Re: Org.Comp.3-160, 164, 169
–	$(C_5H_5)Re(CO)(NO)\text{-}C_7H_7\text{-}c$	Re: Org.Comp.3-165
–	$(C_5H_5)Re(CO)_2(NC_5H_4\text{-}4\text{-}CH_3)$	Re: Org.Comp.3-193/7, 199
$C_{13}H_{12}NO_3Sb$	$(\text{-}NHC_6H_4CH_2C_6H_3\text{-})Sb(O)(OH)_2$	Sb: Org.Comp.5-308
$C_{13}H_{12}NO_4Sb$	$4\text{-}CH_3\text{-}C_6H_4\text{-}Sb(O)(OH)\text{-}C_6H_4\text{-}4\text{-}NO_2$	Sb: Org.Comp.5-224
–	$4\text{-}C_6H_5C(O)NH\text{-}C_6H_4\text{-}Sb(O)(OH)_2 \cdot H_2O$	Sb: Org.Comp.5-289

$C_{13}H_{12}NO_5Re$	fac-C_5H_5N-1-$Re(CO)_3$[-O-$C(CH_3)$=CH-$C(CH_3)$=O-]	Re: Org.Comp.1-127
$C_{13}H_{12}N_2OS$	O=S=$NN(C_6H_5)(CH_2C_6H_5)$.	S: S-N Comp.6-63/8
$C_{13}H_{12}N_2O_2S_2$ . . .	4-CH_3-C_6H_4-$S(O)_2N$=S=N-C_6H_5	S: S-N Comp.7-62/3
–	C_6H_5-$S(O)_2N$=S=N-C_6H_4-4-CH_3	S: S-N Comp.7-55/60
$C_{13}H_{12}N_2O_3S_2$ . . .	C_6H_5-$S(O)_2N$=S=N-C_6H_4-4-OCH_3.	S: S-N Comp.7-55/60
$C_{13}H_{12}N_2O_4S_2$ . . .	4-NO_2-C_6H_4-NH-S(O)-S-C_6H_4-4-OCH_3	S: S-N Comp.8-332
$C_{13}H_{12}N_2O_4S_3$ . . .	C_6H_5-$S(O)_2N$=S=$NS(O)_2$-C_6H_4-CH_3-4.	S: S-N Comp.7-100/1
$C_{13}H_{12}N_2O_5S_3$ . . .	C_6H_5-$S(O)_2N$=S=$NS(O)_2$-C_6H_4-OCH_3-4	S: S-N Comp.7-100/1
$C_{13}H_{12}N_4O_2PoS$ . .	[PoO(OH)(C_6H_5N=NC(S)$NHNC_6H_5$)].	Po: SVol.1-348
$C_{13}H_{12}N_6O_{12}Th$ . .	$Th(NO_3)_4$ · NC_5H_4-2-(CH=N-C_6H_4-2-CH_3). . .	Th: SVol.D4-140
–	$Th(NO_3)_4$ · NC_5H_4-2-(CH=N-C_6H_4-3-CH_3). . .	Th: SVol.D4-141
–	$Th(NO_3)_4$ · NC_5H_4-2-(CH=N-C_6H_4-4-CH_3). . .	Th: SVol.D4-141
–	$Th(NO_3)_4$ · NC_5H_4-2-[$C(CH_3)$=N-C_6H_5]	Th: SVol.D4-138
$C_{13}H_{12}N_6O_{13}Th$ . .	$Th(NO_3)_4$ · NC_5H_4-2-(CH=N-C_6H_4-2-OCH_3)	Th: SVol.D4-141
–	$Th(NO_3)_4$ · NC_5H_4-2-(CH=N-C_6H_4-4-OCH_3)	Th: SVol.D4-141
$C_{13}H_{12}O_5Re^+$	$[(CO)_5Re(C_8H_{12})]^+$	Re: Org.Comp.2-351, 353/4
$C_{13}H_{12}O_7Sb_2$	$(HO)_2(O)Sb$(4-$C_6H_4C(O)C_6H_4$-4')$Sb(O)(OH)_2$ . .	Sb: Org.Comp.5-315
$C_{13}H_{13}IMoO_4$	[CH_3-C(=O)-C_5H_4]$Mo(CO)_2$(I)=C(-O-$CH_2CH_2CH_2$-)	Mo: Org.Comp.8-17, 27
$C_{13}H_{13}MnN_2O_3P$. .	Mn[O_2PH-CH($NHCH_2$-2-NC_5H_4)C_6H_4-O-2] . . .	Mn: MVol.D8-113
–	Mn[O_2PH-CH($NHCH_2$-3-NC_5H_4)C_6H_4-O-2] . . .	Mn: MVol.D8-113
–	Mn[O_2PH-CH($NHCH_2$-4-NC_5H_4)C_6H_4-O-2] . . .	Mn: MVol.D8-113
$C_{13}H_{13}MnN_2O_4P$. .	[MnH(O_3P-CH($NHCH_2$-2-NC_5H_4)C_6H_4-O-2)] . .	Mn: MVol.D8-125/6
–	[MnH(O_3P-CH($NHCH_2$-3-NC_5H_4)C_6H_4-O-2)] . .	Mn: MVol.D8-125/6
–	[MnH(O_3P-CH($NHCH_2$-4-NC_5H_4)C_6H_4-O-2)] . .	Mn: MVol.D8-125/6
$C_{13}H_{13}MnN_5OS_2$. .	[Mn(C_6H_5CH=CHC(CH_3)=N-NHC(O)NH_2)$(NCS)_2$]	Mn: MVol.D6-328
$C_{13}H_{13}MnN_5S_3$ . . .	[Mn(C_6H_5CH=CHC(CH_3)=N-NHC(=S)NH_2)$(NCS)_2$]	Mn: MVol.D6-342
$C_{13}H_{13}MoN_3O_4$. . .	$(C_5H_5)Mo(CO)_2$[-O-C(=O)-CH(CH_2-4-(1,3-$N_2C_3H_3$))-NH_2-]	Mo: Org.Comp.7-208, 209, 232
$C_{13}H_{13}MoN_4O_2^+$.	$[(C_5H_5)Mo(CO)_2(1,2\text{-}N_2C_3H_4)_2]^+$	Mo: Org.Comp.7-283, 285
–	$[(C_5H_5)Mo(CO)_2(1,3\text{-}N_2C_3H_4)_2]^+$	Mo: Org.Comp.7-283, 285/6
$C_{13}H_{13}MoO_2^+$ . . .	$[(C_5H_5)Mo(CO)_2(C_6H_8)][BF_4]$	Mo: Org.Comp.8-305/6, 313, 317/8
–	$[(C_5H_5)Mo(CO)_2(C_6H_8)][PF_6]$	Mo: Org.Comp.8-305/6, 313, 317/8
$C_{13}H_{13}NO_2S_2$	C_6H_5-NH-S(O)-S-C_6H_4-4-OCH_3	S: S-N Comp.8-331
$C_{13}H_{13}N_2NaO_5S_3$	Na[C_6H_5-$S(O)_2$-NS(OCH_3)N-$S(O)_2$-C_6H_5]	S: S-N Comp.8-203/4, 206
$C_{13}H_{13}N_2ORe$	(C_5H_5)ReH(CO)-N=N-C_6H_4-CH_3-4	Re: Org.Comp.3-133/5, 136
$C_{13}H_{13}N_2O_2Re$ . . .	(C_5H_5)ReH(CO)-N=N-C_6H_4-OCH_3-4.	Re: Org.Comp.3-133/5, 136, 139
$C_{13}H_{13}N_2O_4SSb$ . .	(4-(4'-HOC_6H_4NHC(S)NH)C_6H_4)$Sb(O)(OH)_2$ . . .	Sb: Org.Comp.5-290
$C_{13}H_{13}N_2O_4Sb$ . . .	4-(3'-CH_3-4'-HOC_6H_3N=N)$C_6H_4Sb(O)(OH)_2$. . .	Sb: Org.Comp.5-291
$C_{13}H_{13}N_2O_5S_3^-$. . .	$[C_6H_5\text{-}S(O)_2\text{-}NS(OCH_3)N\text{-}S(O)_2\text{-}C_6H_5]^-$	S: S-N Comp.8-203/4, 206
$C_{13}H_{13}N_2O_5Sb$ . . .	5-CH_3O-2-(4'-HOC_6H_4N=N)$C_6H_3Sb(O)(OH)_2$ . .	Sb: Org.Comp.5-297
$C_{13}H_{13}N_2O_6S_2Sb$	(4-(3'-$HO_3SC_6H_4$NHC(S)NH)C_6H_4)$Sb(O)(OH)_2$.	Sb: Org.Comp.5-290
$C_{13}H_{13}N_7O_{14}Th$ . .	$Th(NO_3)_4$ · 2-(6-NH_2-NC_5H_3-2-N=CH)-6-CH_3O-C_6H_3-OH	Th: SVol.D4-139

$C_{13}H_{13}O_2Sb$	$(C_6H_5)(4\text{-}CH_3C_6H_4)Sb(O)OH$	Sb: Org.Comp.5-223
$C_{13}H_{13}O_2Ti$	$[(CH_3)Ti(OC_6H_5)_2]_n$	Ti: Org.Comp.5-320
$C_{13}H_{13}O_3Sb$	$(2\text{-}C_6H_5CH_2C_6H_4)Sb(O)(OH)_2$	Sb: Org.Comp.5-295
$C_{13}H_{14}IORe$	$(CH_3C{\equiv}CCH_3)(CH_3C{\equiv}CC_6H_5)Re(O)I$	Re: Org.Comp.2-362, 370
$C_{13}H_{14}InNS_2$	$CH_3\text{-}In[1,2\text{-}(S)_2\text{-}C_6H_3\text{-}4\text{-}CH_3] \cdot NC_5H_5$	In: Org.Comp.1-246, 249, 250/1
$C_{13}H_{14}InO_3P$	$[CH_3In(OH)(OP(C_6H_5)_2O)]_4 \cdot 6\ NC_5H_5$	In: Org.Comp.1-238/9
$C_{13}H_{14}MnN_4O_3$	$[Mn(\text{-}O\text{-}C_6H_4\text{-}2\text{-}N{=}CHCH{=}N\text{-}N(C_5H_4N\text{-}2)\text{-})(H_2O)_2]$	Mn: MVol.D6-268/9
$C_{13}H_{14}MoN_2O_5$	$(C_5H_5)Mo(CO)_2[\text{-}C(OH){=}C(COO\text{-}C_2H_5)\text{-}N(CH_3)N\text{-}]$	Mo: Org.Comp.8-139
$C_{13}H_{14}MoO_2$	$(CH_3\text{-}C_5H_4)Mo(CO)_2(C_5H_7)$	Mo: Org.Comp.8-250, 254
–	$(C_5H_5)Mo(CO)(C_5H_6\text{-}c)C(O)CH_3$	Mo: Org.Comp.6-364
–	$(C_5H_5)Mo(CO)(C_5H_5D\text{-}c)C(O)CH_3$	Mo: Org.Comp.6-365
–	$(C_5H_5)Mo(CO)_2(CH_2CHCH\text{-}CH{=}CH\text{-}CH_3)$	Mo: Org.Comp.8-205, 207, 216/7
–	$(C_5H_5)Mo(CO)_2(CH_3\text{-}CHCHCH\text{-}CH{=}CH_2)$	Mo: Org.Comp.8-229
–	$(C_5H_5)Mo(CO)_2(C_6H_9)$	Mo: Org.Comp.8-250, 257, 297/8
–	$(C_5H_5)Mo(CO)_2(C_6H_8\text{-}4\text{-}D)$	Mo: Org.Comp.8-250, 258
–	$(C_5H_5)Mo(CO)_2[(CH_2)C_5H_7]$	Mo: Org.Comp.8-205, 233
$C_{13}H_{14}MoO_3$	$(C_5H_5)Mo(CO)_2[\text{-}C(CH_3){=}C(CH_3)\text{-}C(CH_3){=}O\text{-}]$	Mo: Org.Comp.8-112, 151
–	$[CH_3C({=}O)\text{-}C_5H_4]Mo(CO)_2[CH_2C(CH_3)CH_2]$	Mo: Org.Comp.8-205, 207, 219
–	$[CH_3C({=}O)\text{-}C_5H_4]Mo(CO)_2[CH_2CHCH\text{-}CH_3]$	Mo: Org.Comp.8-205, 207, 212
$C_{13}H_{14}MoO_4$	$(C_5H_5)Mo(CO)_2[CH_2C(CH\text{-}CH_3)\text{-}COO\text{-}CH_3]$	Mo: Org.Comp.8-205, 224
$C_{13}H_{14}NO_2PS$	$CH_3NH\text{-}S(O)\text{-}OP(C_6H_5)_2$	S: S-N Comp.8-302
$C_{13}H_{14}NO_3Sb$	$4\text{-}(4'\text{-}H_2NC_6H_4CH_2)C_6H_4Sb(O)(OH)_2$	Sb: Org.Comp.5-295
$C_{13}H_{14}N_2O_6SSb_2$	$(HO)_2(O)Sb(4\text{-}C_6H_4NHC(S{=})NHC_6H_4\text{-}4')Sb(O)(OH)_2$	Sb: Org.Comp.5-314
$C_{13}H_{14}N_2O_7Sb_2$	$(HO)_2(O)Sb(4\text{-}C_6H_4NHC({=}O)NHC_6H_4\text{-}4')Sb(O)(OH)_2$	Sb: Org.Comp.5-314
$C_{13}H_{14}N_3O_5Re$	cis-$(CO)_4Re[C(O)CH_3]C(CH_3){=}NH\text{-}CH_2CH_2\text{-}4\text{-}(1,3\text{-}N_2C_3H_3)$	Re: Org.Comp.1-395
$C_{13}H_{14}N_4O_3PoS$	$[Po(OH)_3(C_6H_5N{=}NC(S)NHNC_6H_5)]$	Po: SVol.1-348
$C_{13}H_{14}O_3Ti$	$[(C_5H_5)_2Ti(OCH_2C(O)CH_2O)]_n$	Ti: Org.Comp.5-335
$C_{13}H_{14}O_4PRe$	cis-$(CO)_4Re[P(CH_3)_3]C_6H_5$	Re: Org.Comp.1-467/8
$C_{13}H_{14}O_6Sb_2$	$(HO)_2(O)Sb(4\text{-}C_6H_4CH_2C_6H_4\text{-}4')Sb(O)(OH)_2$	Sb: Org.Comp.5-315
$C_{13}H_{14}O_7PRe$	$(CO)_4Re[P(OCH_3)_3]C_6H_5$	Re: Org.Comp.1-468
$C_{13}H_{15}MnNO_4S$	$[Mn(\text{-}O\text{-}C_6H_4\text{-}2\text{-}CH{=}N\text{-}C_6H_4\text{-}2\text{-}S\text{-})(H_2O)_3]$	Mn: MVol.D6-28, 30
$C_{13}H_{15}MnN_9O_2S_2$	$[Mn(2,6\text{-}(H_2N\text{-}C(O)NH\text{-}N{=}C(CH_3))_2C_5H_3N)(NCS)_2]$	Mn: MVol.D6-333, 335
$C_{13}H_{15}MoNO_2S_2$	$(C_5H_5)Mo(CO)_2[\text{-}SC(1\text{-}NC_5H_{10})S\text{-}]$	Mo: Org.Comp.7-189, 197
$C_{13}H_{15}MoNO_3$	$(C_5H_5)Mo(CO)_2N(\text{-}O){=}C(CH_3)CH{=}C(CH_3)_2$	Mo: Org.Comp.7-207, 209, 211
–	$(C_5H_5)Mo(CO)_2N(\text{-}O){=}C(CH_3)CH_2CH_2CH{=}CH_2$	Mo: Org.Comp.7-207, 209, 210
–	$(C_5H_5)Mo(CO)_2N(\text{-}O){=}C_6H_{10}$	Mo: Org.Comp.7-207, 209, 211
–	$(C_5H_5)Mo(CO)_2[CH_2{=}C{=}N(\text{-}CH_2CH_2\text{-})_2O]$	Mo: Org.Comp.8-185, 186, 189
$C_{13}H_{15}MoNO_3S_2$	$(CH_2CHCH_2)Mo(CO)_2(NC_5H_5)\text{-}SC(S)\text{-}O\text{-}C_2H_5$	Mo: Org.Comp.5-282, 283, 284
$C_{13}H_{15}MoNO_6$	$(C_5H_5)Mo(CO)_2[\text{-}O\text{-}C({=}O)\text{-}CH(CH_2CH_2CH_2\text{-}COOH)\text{-}NH_2\text{-}]$	Mo: Org.Comp.7-208, 209, 232
$C_{13}H_{15}MoN_2O^+$	$[C_5H_5Mo(NO)(C_5H_5N)C_3H_5]^+$	Mo: Org.Comp.6-178

$C_{13}H_{15}MoN_2O_5^+$. $[(C_5H_5)Mo(CO)_2(-C(OH)=C(COO-C_2H_5)-N(CH_3)-NH-)][PF_6]$ Mo:Org.Comp.8-140, 169

$C_{13}H_{15}MoO_2^+$ . . . $[(C_5H_5)Mo(CO)_2((CH_2)_2CC(CH_3)_2)][BF_4]$ Mo:Org.Comp.8-305/6, 311

– . . . . $[(C_5H_5)Mo(CO)_2(CH_2=CH-C(CH_3)=CH-CH_3)][BF_4]$ Mo:Org.Comp.8-305/6, 309

– . . . . $[(C_5H_5)Mo(CO)_2(CH_3-CHC(CH_2)CH-CH_3)][BF_4]$ Mo:Org.Comp.8-305/6, 311, 312

$C_{13}H_{15}NO_3S_2$ . . . . $O=S=NSCOOC_6H_{10}C_6H_5-2$ . . . . S: S-N Comp.6-190

$C_{13}H_{15}N_3O_2S_2$ . . . $C_6H_5-S(O)_2N=S=NC_6H_{10}(CN-1)$ . . . . S: S-N Comp.7-54

$C_{13}H_{15}N_3O_9Th$ . . . $Th(NO_3)_2[OOC-CH(C_4H_9-i)-N=CH-C_6H_4-2-O] \cdot 3 H_2O$ Th: SVol.D4-144

– . . . . $Th(NO_3)_2[OOC-CH(C_4H_9-s)-N=CH-C_6H_4-2-O] \cdot 3 H_2O$ Th: SVol.D4-144

$C_{13}H_{16}IMoNO_2$ . . . $(C_5H_5)Mo(CO)_2(I)=C[-N(CH_3)-CH_2CH_2CH_2CH_2-]$ Mo:Org.Comp.8-17, 23/4, 29/30

$C_{13}H_{16}IMoO_5P$ . . . $(C_5H_5)Mo(CO)_2(I)[P(-OCH_2-)_3CC_2H_5]$ Mo:Org.Comp.7-56/7, 90, 114

$C_{13}H_{16}INOSn$ . . . . $(C_2H_5)_2Sn(I)OC_9H_6N$ . . . . Sn: Org.Comp.17-133

$C_{13}H_{16}IN_2O_6Re$ . . $(CO)_4Re[C(O)CH_3]C(CH_3)=NH-CH_2CH_2N(CH_3)C(O)CH_2I$ . . . . Re: Org.Comp.1-392

$C_{13}H_{16}I_2MoNOP$ . . $C_5H_5Mo(NO)(P(CH_3)_2C_6H_5)I_2$ Mo:Org.Comp.6-34

$C_{13}H_{16}InNO$ . . . . . $(C_2H_5)_2In-(1-NC_9H_6-8-O)$ . . . . In: Org.Comp.1-190, 193

$C_{13}H_{16}InN_2^+$ . . . . . $[(CH_3)_2In((NC_5H_4-2)_2CH_2)]^+$ . . . . In: Org.Comp.1-223, 226

$C_{13}H_{16}InN_3O_3$ . . . . $[(CH_3)_2In((NC_5H_4-2)_2CH_2)][NO_3] \cdot H_2O$ . . . . In: Org.Comp.1-223, 226, 227/8

$C_{13}H_{16}MnN_2O_4$ . . . $Mn[-3-O-NC_5H(CH_3-2)(CH_2OH-5)-4-CH=N-CH(C_3H_7-i)COO-]$ Mn:MVol.D6-84

$C_{13}H_{16}MnN_4OS$ . . $Mn(C_5H_{10}N-1-C(O)NNC(S)NHC_6H_5) \cdot 5 H_2O$ . . Mn:MVol.D7-203/4

$C_{13}H_{16}MnN_4O_4$ . . . $[Mn(-O-C_6H_4-2-N=CHCH=N-N(C_5H_4N-2)-)(H_2O)_3]$ Mn:MVol.D6-268/9

$C_{13}H_{16}MoNNaO_2$ $Na[(C_5H_5)Mo(CO)_2=C(-CH_2CH_2CH_2CH_2-N(CH_3)-)]$ Mo:Org.Comp.8-1, 2

$C_{13}H_{16}MoNO_2^-$ . . . $Na[(C_5H_5)Mo(CO)_2=C(-CH_2CH_2CH_2CH_2-N(CH_3)-)]$ Mo:Org.Comp.8-1, 2

$C_{13}H_{16}MoNO_4P$ . . $(C_5H_5)Mo(CO)_2[C_4H_8NO_2P(CH=CH_2)]$ Mo:Org.Comp.7-158, 163

$C_{13}H_{16}MoN_2OS_2$ . . $(C_5H_5)Mo(NO)[S_2C-N(CH_3)_2]-C_5H_5$ Mo:Org.Comp.6-86, 92

$C_{13}H_{16}MoN_2O_2$ . . . $(C_5H_5)Mo(CO)(NO)-CN-C_6H_{11}-c$ Mo:Org.Comp.6-263

$C_{13}H_{16}MoN_2O_2S_2$ $(CH_2CHCH_2)Mo(CO)_2(NC_5H_5)-SC(S)-N(CH_3)_2$ Mo:Org.Comp.5-283, 285

– . . . . $(CH_2CHCH_2)Mo(CO)_2(NC_5H_5)-SC(S)-NH-C_2H_5$ Mo:Org.Comp.5-282, 283, 285, 289

$C_{13}H_{16}MoO_2$ . . . . . $(C_5H_5)Mo(CO)[CH(CH_3)CHCHCH_2C(O)CH_3]$ . . . Mo:Org.Comp.6-342

– . . . . $(C_5H_5)Mo(CO)[CH_2C(CH_3)CHCH_2C(O)CH_3]$ Mo:Org.Comp.6-342

– . . . . $(C_5H_5)Mo(CO)[CH_2CHC(CH_3)CH_2C(O)CH_3]$ Mo:Org.Comp.6-342/3

– . . . . $(C_5H_5)Mo(CO)[CH_2CHCHCH(CH_3)C(O)CH_3]$ . . . Mo:Org.Comp.6-342

– . . . . $(C_5H_5)Mo(CO)_2(C-CH_2-C_4H_9-t)$ Mo:Org.Comp.8-33

– . . . . $(C_5H_5)Mo(CO)_2[CH_2C(C_3H_7-i)CH_2]$ Mo:Org.Comp.8-206, 220

– . . . . $(C_5H_5)Mo(CO)_2[(CH_3)_2CC(CH_3)CH_2]$ Mo:Org.Comp.8-205, 229/30

$C_{13}H_{16}MoO_2S_2$ . . . $(C_5H_5)Mo(CO)_2[-S-C(CH_2-C_4H_9-t)-S-]$ Mo:Org.Comp.7-188, 191, 202/3

$C_{13}H_{16}MoO_2Se_2$ .. $(C_5H_5)Mo(CO)_2[-Se-C(CH_2-C_4H_9-t)-Se-]$ Mo:Org.Comp.7-188, 198
$C_{13}H_{16}NO_3Sb$ .... $(CH_3)_3Sb(OC_9H_6N)O_2CH$ .... Sb: Org.Comp.5-46
$C_{13}H_{16}NO_5Re$ .... $(CO)_3Re(CNC_4H_9-t)(CH_3COCHCOCH_3)$ .... Re: Org.Comp.2-244/5
$C_{13}H_{16}NO_7Re$ .... cis-$(CO)_4Re[C(O)CH_3]C(CH_3)=NH-CH(CH_3)COOC_2H_5$ Re: Org.Comp.1-400, 405/6
$C_{13}H_{16}NO_7ReS$ ... cis-$(CO)_4Re[C(O)CH_3]C(CH_3)=NH-CH(CH_2SH)COOC_2H_5$ Re: Org.Comp.1-397
$C_{13}H_{16}NO_8Re$ .... cis-$(CO)_4Re[C(O)CH_3]C(CH_3)=NH-CH(CH_2OH)COOC_2H_5$ Re: Org.Comp.1-397
$C_{13}H_{16}N_4O_2Ti$ .... $[(C_5H_5)_2Ti(ON=C(NH_2)CH_2C(NH_2)=NO)]_n$...... Ti: Org.Comp.5-343
$C_{13}H_{16}N_4O_7Sb_2$ .. $(HO)_2(O)Sb(4-C_6H_4(NH)_2C(=O)(NH)_2C_6H_4-4')Sb(O)(OH)_2$ Sb: Org.Comp.5-315
$C_{13}H_{16}S_2Ti$ ...... $[(C_5H_5)_2Ti(S(CH_2)_3S)]_n$.................... Ti: Org.Comp.5-347
$C_{13}H_{17}IMo$ ...... $C_5H_5Mo(I)(H_3C-CC-CH_3)_2$................. Mo:Org.Comp.6-133, 143
$C_{13}H_{17}InN_2$...... $(CH_3)_3In \cdot 2-(NC_5H_4-2)-NC_5H_4$ .... In: Org.Comp.1-27, 36
– .... $(CH_3)_3In \cdot 3-(NC_5H_4-3)-NC_5H_4$ .... In: Org.Comp.1-27, 36
– .... $(CH_3)_3In \cdot 4-(NC_5H_4-4)-NC_5H_4$ .... In: Org.Comp.1-27, 36
$C_{13}H_{17}InN_4$...... $CH_3-In[-1-NC_5H_4-2-N(CH_3)-]_2$............. In: Org.Comp.1-302/3
$C_{13}H_{17}MnNO_4P^-$ $[Mn(O_3PCH(NHC_6H_{11}-c)C_6H_4-O-2)]^-$ Mn:MVol.D8-125/6
$C_{13}H_{17}MnNO_7S$ .. $[Mn(-O-2-C_{10}H_6-1-CH=N-CH_2CH_2SO_3-)(H_2O)_3]$ Mn:MVol.D6-70
$C_{13}H_{17}MnN_5O_3S_3$ $Mn(4-NH_2C_6H_4SO_2NH-CONHC_4H_9)(NCS)_2$.... Mn:MVol.D7-116/23
$C_{13}H_{17}MoNO_2$ ... $(C_5H_5)Mo(CO)_2[-1-NC_5H_9-1-CH_3-2-]$ Mo:Org.Comp.8-112, 121/2
– .... $(C_5H_5)Mo(CO)_2[-C(=C(CH_3)_2)-N(CH_3)_2-]$ Mo:Org.Comp.8-112, 122/3
– .... $(C_5H_5)Mo(CO)_2[CH_2=C=N(C_2H_5)_2]$........... Mo:Org.Comp.8-185, 189
– .... $(C_5H_5)Mo(CO)_2[(CH_3)_2C=C=N(CH_3)_2]$ Mo:Org.Comp.8-185, 190
$C_{13}H_{17}MoNO_2S$ .. $(C_5H_5)Mo(CO)_2[-S-C(CH_3)=N(CH(CH_3)C_2H_5)-]$ Mo:Org.Comp.7-209/10, 214
$C_{13}H_{17}MoNO_3$ ... $[C_2H_5-C_5(CH_3)_4]Mo(CO)_2(NO)$.............. Mo:Org.Comp.7-6, 11, 22
$C_{13}H_{17}MoNO_4$ ... $[CH_2C(CH_3)CH_2]Mo(CO)_2[CH_3-C(O)CHC(O)$ $-CH_3](NC-CH_3)$ Mo:Org.Comp.5-283, 288
$C_{13}H_{17}MoNO_4P^+$ $[(C_5H_5)Mo(CO)_2(1,3,6,2-O_2NPC_4H_9-2-CH=CH_2)]^+$ Mo:Org.Comp.7-249, 271, 281
$C_{13}H_{17}MoN_3O_2S_2$ $(C_5H_5)Mo(CO)_2[-S-C(N(CH_3)_2)=N(C(=S)N(CH_3)_2)-]$ Mo:Org.Comp.7-207/10, 226
$C_{13}H_{17}MoN_4{}^+$.... $[C_5H_5Mo(CNCH_3)_4]^+$ Mo:Org.Comp.6-99
$C_{13}H_{17}MoO_3P$.... $(C_5H_5)Mo(CO)_2[P(CH_3)_3]-C(CH_3)=C=O$ Mo:Org.Comp.8-92
$C_{13}H_{17}MoO_4P$.... $(C_5H_5)Mo(CO)_2=[2-(1,3,2-O_2PC_2(CH_3)_4-4,4,5,5)]$ Mo:Org.Comp.7-35/6
$C_{13}H_{17}MoO_4PS$ .. $(C_5H_5)Mo(CO)_2[1,3,2-O_2PC_2(=S)(CH_3)_4]$ Mo:Org.Comp.7-208, 212
$C_{13}H_{17}MoO_5P$.... $(C_5H_5)Mo(CO)_2[P(OCH_3)_3]-CH_2-CCH$ Mo:Org.Comp.8-77, 78, 88
$C_{13}H_{17}NORe^+$ ... $[(CH_3C{\equiv}CCH_3)_2Re(O)NC_5H_5]^+$............... Re: Org.Comp.2-359, 368/9
$C_{13}H_{17}NO_2Sn$.... $(CH_3)_2Sn(OC_2H_5)O-C_9H_6N$................. Sn: Org.Comp.16-173
$C_{13}H_{17}N_2O_5Re$... $(CO)_4Re(NC_4H_9)C(O)-1-NC_4H_8$.............. Re: Org.Comp.1-471
$C_{13}H_{17}N_2O_6Re$... $(CO)_4Re(CNC_4H_9-t)CONH(CH_2)_3OH$ .... Re: Org.Comp.2-247/8
$C_{13}H_{17}N_4O_2Re$... $(C_5H_5)Re(CO)_2(N_4C_6H_{12})$................. Re: Org.Comp.3-193/7, 199
$C_{13}H_{17}N_4Re^{2+}$ ... $[(C_5H_5)Re(NC-CH_3)_4][PF_6]_2$................ Re: Org.Comp.3-5
$C_{13}H_{18}IMoNO$.... $C_5H_5Mo(NO)(I)C_8H_{13}-c$ Mo:Org.Comp.6-176/7, 181/2
$C_{13}H_{18}MnNO_4P$ .. $[MnH(O_3PCH(NHC_6H_{11}-c)C_6H_4-O-2)]$........ Mn:MVol.D8-125/6
$C_{13}H_{18}MnN_3O_2$... $[Mn(-O-C(CH_3)=CH-C(CH_3)=N-C_2H_4$ $-N=C(CH_3)CH=C(CH_3)O-)(CN)]$ Mn:MVol.D6-203, 204/6

$C_{13}H_{18}MnN_3O_2S$. . [Mn(-O-C(CH_3)=CH-C(CH_3)=N-C_2H_4
-N=C(CH_3)CH=C(CH_3)O-)(NCS)] Mn:MVol.D6-203, 204/6
$C_{13}H_{18}MnN_3O_3$. . . [Mn(-O-C(CH_3)=CH-C(CH_3)=N-C_2H_4
-N=C(CH_3)CH=C(CH_3)O-)(NCO)] Mn:MVol.D6-203, 204/6
$C_{13}H_{18}MoNO_2^+$ . . [(C_5H_5)Mo(CO)$_2$((CH_3)$_2$CCHN(CH_3)$_2$)][PF_6] Mo:Org.Comp.8-185, 186, 195
$C_{13}H_{18}MoNO_3P$ . . (C_5H_5)Mo(CO)$_2$=[2-(1,3,2-ONPC_2H_4-3-C_4H_9-t)]
Mo:Org.Comp.7-35/6
$C_{13}H_{18}MoN_2O_4$. . . (C_5H_5)Mo(CO)$_2$[-O-C(=O)-CH(CH_2CH_2-CH_2
CH_2-NH_2)-NH_2-] Mo:Org.Comp.7-208, 209, 232
$C_{13}H_{18}MoOS$ . . . . C_5H_5Mo(CO)(SC_3H_7-i)H_3C-CC-CH_3 . . . . Mo:Org.Comp.6-277
$C_{13}H_{18}MoOS_6$. . . . (CH_3-S-CC-S-CH_3)$_3$Mo(CO) Mo:Org.Comp.5-196
$C_{13}H_{18}MoO_2Si$ . . . (C_5H_5)Mo(CO)$_2$[CH_2C(CH_2)-Si(CH_3)$_3$] Mo:Org.Comp.8-205, 222
– . . . . (C_5H_5)Mo(CO)$_2$[CH_2CHCH-Si(CH_3)$_3$] Mo:Org.Comp.8-205, 207, 218
$C_{13}H_{18}MoO_3$. . . . . (C_5H_5)Mo(CO)$_2$(C_2H_5)-1-OC_4H_8 . . . . Mo:Org.Comp.8-77, 85
$C_{13}H_{18}MoO_3P^-$ . . . [P(CH_3)$_4$][(C_5H_5)Mo(CO)$_2$(CH_3)-C(=O)-CH=P(CH_3)$_3$]
Mo:Org.Comp.8-177
– . . . . [P(CH_3)$_4$][(C_5H_5)Mo(P(CH_3)$_3$)(CO)(-O=C(CH_3)CH=C(O)-)]
Mo:Org.Comp.6-257
$C_{13}H_{18}MoO_6P_2$. . . [OC_4H_2(=O)$_2$-2,5]Mo(CO)$_3$[P(CH_3)$_2$-CH_2CH_2-P(CH_3)$_2$]
Mo:Org.Comp.5-156, 157
$C_{13}H_{18}NO_2Sb$ . . . . (CH_3)$_3$Sb(OCH_3)OC_9H_6N . . . . Sb: Org.Comp.5-46
$C_{13}H_{18}NO_5Re$. . . . (CO)$_4$Re[C(O)C_3H_7-i]C(CH_3)=NH-C_3H_7-n. . . . . Re: Org.Comp.1-402
– . . . . (CO)$_4$Re[C(O)C_3H_7-i]C(C_3H_7-n)=NHCH_3. . . . . . Re: Org.Comp.1-402
$C_{13}H_{18}N_2Ti$. . . . . . [(C_5H_5)$_2$Ti(NH(CH_2)$_3$NH)]$_n$ . . . . Ti: Org.Comp.5-349
$C_{13}H_{18}N_3O_{10}Th^-$ Th[((OOCCH_2)$_2$N)$_2C_2H_4$][CH_3CH(NH_2)COO]$^-$ . . Th: SVol.D1-136
$C_{13}H_{19}I_3NO_5ReSn$ cis-(CO)$_4$Re(SnI_3)C(OC_2H_5)N(C_3H_7-i)$_2$ . . . . Re: Org.Comp.1-379
$C_{13}H_{19}MoNO$ . . . . C_5H_5Mo(NO)((CH_3)$_2$C=CHCH=C(CH_3)$_2$) Mo:Org.Comp.6-187
$C_{13}H_{19}MoN_2O_2^+$. [C_5H_5Mo(NO)$_2$(C_8H_{14}-c)]$^+$ Mo:Org.Comp.6-156
$C_{13}H_{19}MoO_3P$. . . . (C_5H_5)Mo(CO)$_2$[P(CH_3)$_3$]-C(=O)-C_2H_5 . . . . Mo:Org.Comp.8-45, 58
$C_{13}H_{19}MoO_4PS_2$. . (C_5H_5)Mo(CO)$_2$[-S-P(O-C_3H_7-i)$_2$-S-] Mo:Org.Comp.7-189, 200
$C_{13}H_{19}MoO_6P$. . . . (C_5H_5)Mo(CO)$_2$[P(OCH_3)$_3$]-C(=O)-C_2H_5 . . . . . . Mo:Org.Comp.8-45, 58/9
$C_{13}H_{19}MoO_6PSi$ . . (CH_3)$_3$Si-O-C[CH=P(CH_3)$_3$]=Mo(CO)$_5$ Mo:Org.Comp.5-104, 110/1
$C_{13}H_{19}NO_3SSn$ . . . (C_2H_5)$_3$SnN(C(=O)C_6H_4S(=O)$_2$) . . . . Sn: Org.Comp.18-142/3, 148
$C_{13}H_{19}NO_3Sn$ . . . . 2-$CH_3C_6H_4$Sn(OCH_2CH_2)$_3$N. . . . . Sn: Org.Comp.17-69
$C_{13}H_{20}IMoO_2P$ . . . [(C_5H_5)Mo(CO)$_2$(P(C_2H_5)$_3$)(I)] Mo:Org.Comp.7-56/7, 63/4
$C_{13}H_{20}IMoO_5P$ . . . [(C_5H_5)Mo(CO)$_2$(I)(P(O-C_2H_5)$_3$)] Mo:Org.Comp.7-56/7, 86/7
$C_{13}H_{20}I_2MoN_4O_3$ [(-N(CH_3)-CH_2CH_2-N(CH_3)-)C=]$_2$Mo(CO)$_3$(I)$_2$ Mo:Org.Comp.5-122
$C_{13}H_{20}InN$. . . . . . . (C_2H_5)$_3$In · NC-C_6H_5. . . . . In: Org.Comp.1-71
$C_{13}H_{20}InNO$ . . . . . (C_2H_5)$_2$In-N(C_6H_5)-C(O)C_2H_5 . . . . In: Org.Comp.1-267/8
$C_{13}H_{20}MnN_2O_3$. . . Mn[-O-C(CH_3)=CH-C(CH_3)=NCH_2CH(OH)
CH_2N=C(CH_3)CH=C(CH_3)O-] Mn:MVol.D6-203
$C_{13}H_{20}MnN_4O_2S_3$ [Mn(S=C(NH_2)NHC_6H_5)(NCS)$_2$] · 2 C_2H_5OH . . . Mn:MVol.D7-195/6
$C_{13}H_{20}MoN_2OS_2$. . C_5H_5Mo(NO)(S_2CN(C_2H_5)$_2$)CH_2CH=CH_2 . . . . . . Mo:Org.Comp.6-85
$C_{13}H_{20}MoN_2OSn$ C_5H_5Mo(CO)(CNCH_3)$_2$Sn(CH_3)$_3$. Mo:Org.Comp.6-270
$C_{13}H_{20}MoO$. . . . . . C_5H_5Mo(O)(CH_3)(H_3C-CC-C_4H_9-t) Mo:Org.Comp.6-151/2
$C_{13}H_{20}MoO_3P^+$ . . [(C_5H_5)Mo(CO)$_2$(P(CH_3)$_3$)(O=C(CH_3)$_2$)]$^+$ Mo:Org.Comp.7-283, 294
$C_{13}H_{20}NO_5Re$ . . . . [(C_2H_5)$_4$N][(CO)$_5$Re]. . . . . Re: Org.Comp.2-166
$C_{13}H_{20}N_2O_3Sn$ . . . (C_2H_5)$_3$SnN($C_6H_4NO_2$-4)CHO . . . . Sn: Org.Comp.18-137, 139
$C_{13}H_{20}N_2O_6Sn$ . . . (CH_3)$_3$SnN_2C_3($COOCH_3$)$_2$$COOC_2H_5$ . . . . Sn: Org.Comp.18-84, 89
$C_{13}H_{20}N_2Sn$ . . . . . (C_2H_5)$_3$SnN(CH=NC_6H_4-1,2) . . . . Sn: Org.Comp.18-142, 148

$C_{13}H_{20}O_3Sn$	$(C_2H_5)_2Sn(O-C_2H_5)OC(O)-C_6H_5$	Sn: Org.Comp.16-181, 182
–	$n-C_4H_9-Sn(-O-C_6H_4-O-2)O-C_3H_7-i$	Sn: Org.Comp.17-74
$C_{13}H_{20}O_5PRe$	$(C_5H_5)Re(CO)_2[P(O-C_2H_5)_3]$	Re: Org.Comp.3-210, 212
$C_{13}H_{20}PRe$	$(C_5H_5)ReH_4[P(CH_3)_2-C_6H_5]$	Re: Org.Comp.3-6, 7, 9
$C_{13}H_{21}IO_4Sn$	$C_3H_7SnI(OC(CH_3)=CHCOCH_3)_2$	Sn: Org.Comp.17-171
$C_{13}H_{21}IO_6Sn$	$CH_3SnI(OC(CH_3)=CHCOOC_2H_5)_2$	Sn: Org.Comp.17-171
$C_{13}H_{21}InO$	$(C_2H_5)_2In-O-CH(C_2H_5)-C_6H_5$	In: Org.Comp.1-182, 186
$C_{13}H_{21}MoO_2P$	$[(C_5H_5)Mo(CO)_2(H)(P(C_2H_5)_3)]$	Mo: Org.Comp.7-55/7, 63
$C_{13}H_{21}MoO_2P_2^+$	$[(C_5H_5)Mo(CO)_2((CH_3)_2P-CH_2CH_2-P(CH_3)_2)]^+$	Mo: Org.Comp.7-250, 272/3
$C_{13}H_{21}MoO_5P$	$[(C_5H_5)Mo(CO)_2(H)(P(O-C_2H_5)_3)]$	Mo: Org.Comp.7-56/7, 86
$C_{13}H_{21}NOSn$	$(C_2H_5)_3SnN(C_6H_5)CHO$	Sn: Org.Comp.18-136/9
$C_{13}H_{21}NO_2Sn$	$(C_4H_9)_2Sn[-O-(2,3-C_5H_3N)-O-]$	Sn: Org.Comp.15-60
$C_{13}H_{21}N_2O_5Re$	$(CO)_4Re(NH_2C_4H_9-n)C(O)NHC_4H_9-n$	Re: Org.Comp.1-471
$C_{13}H_{21}O_2ReSi$	$(C_5H_5)Re(H)(CO)_2-Si(C_2H_5)_3$	Re: Org.Comp.3-175/6, 178
$C_{13}H_{21}O_4PRe$	$(CO)_4Re[P(C_3H_7-i)_3]$, radical	Re: Org.Comp.1-476/8
$C_{13}H_{21}O_7PRe$	$(CO)_4Re[P(O-C_3H_7-i)_3]$, radical	Re: Org.Comp.1-476/8
$C_{13}H_{21}PSn$	$1-C_6H_5-2,2-(C_2H_5)_2-1,2-PSnC_3H_6$	Sn: Org.Comp.19-202, 206
$C_{13}H_{22}IMoN_2O_4P$	$(C_5H_5)Mo(CO)_2(I)[(C_2H_5-O)_2P-NHN(CH_3)_2]$	Mo: Org.Comp.7-56/7, 91
$C_{13}H_{22}InN$	$(CH_3)_2In-C_6H_4-2-[CH_2-N(C_2H_5)_2]$	In: Org.Comp.1-101, 108, 114/5
–	$(C_2H_5)_2In-C_6H_4-2-[CH_2-N(CH_3)_2]$	In: Org.Comp.1-101, 108/9
$C_{13}H_{22}MoNO_4P$	$(C_5H_5)Mo(CO)_2(H)[(C_2H_5-O)_2PN(CH_3)_2]$	Mo: Org.Comp.7-57, 91
$C_{13}H_{22}MoN_2OS_4$	$(C_2H_5-C{\equiv}C-C_2H_5)Mo(CO)[S_2C-N(CH_3)_2]_2$	Mo: Org.Comp.5-145/8
–	$(n-C_4H_9-C{\equiv}CH)Mo(CO)[S_2C-N(CH_3)_2]_2$	Mo: Org.Comp.5-145/8
–	$(HC{\equiv}CH)Mo(CO)[S_2C-N(C_2H_5)_2]_2$	Mo: Org.Comp.5-145/8
$C_{13}H_{22}MoO_3P^+$	$[(C_5H_5)Mo(CO)_2(P(C_2H_5)_3)(H_2O)]^+$	Mo: Org.Comp.7-283, 294
$C_{13}H_{22}MoO_5P_2$	$[CH_2=CH-C(=O)O-CH_3]Mo(CO)_3$ $[P(CH_3)_2-CH_2CH_2-P(CH_3)_2]$	Mo: Org.Comp.5-156, 157, 160
$C_{13}H_{22}MoS_2$	$C_6H_7Mo(S(CH_3)CH_2CH_2S(CH_3))C_3H_5$	Mo: Org.Comp.6-160
$C_{13}H_{22}N_4O_8Th$	$Th[C_2H_4(N(CH_2COO)_2)_2](H_2NC_3H_6NH_2)$	Th: SVol.D1-138
$C_{13}H_{22}O_3Sn$	$C_6H_5CH_2Sn(OC_2H_5)_3$	Sn: Org.Comp.17-55
$C_{13}H_{22}O_4S_3Sn$	$(C_4H_9)_2SnOOCCH_2SC(S)SCH_2COO$	Sn: Org.Comp.15-324
$C_{13}H_{22}O_4Sn$	$(C_4H_9)_2SnOOCC(CH_2)CH_2COO$	Sn: Org.Comp.15-322
$C_{13}H_{23}IMoN_3O_2P$	$(C_5H_5)Mo(CO)_2(I)[P(N(CH_3)_2)_3]$	Mo: Org.Comp.7-56/7, 83
$C_{13}H_{23}MnNO_2S$	$[Mn(SC_6H_4O-2)(N(C_2H_5)_3)(OHCH_3)]$	Mn: MVol.D7-30/1
$C_{13}H_{23}MoO_2P_2^+$	$[(C_5H_5)Mo(CO)_2(P(CH_3)_3)_2]^+$	Mo: Org.Comp.7-283, 286, 297/8
$C_{13}H_{23}MoO_3PS$	$C_5H_5Mo(P(OCH_3)_3)(SCH_3)H_3C-CC-CH_3$	Mo: Org.Comp.6-104
$C_{13}H_{23}MoO_8P_2^+$	$[(C_5H_5)Mo(CO)_2(P(OCH_3)_3)_2]^+$	Mo: Org.Comp.7-283, 289, 290
$C_{13}H_{23}NOSSn$	$(CH_3)_3SnN(C_3H_7-i)S(=O)C_6H_4CH_3-4$	Sn: Org.Comp.18-58, 62
$C_{13}H_{23}NO_2SSn$	$(C_2H_5)_3SnNHSO_2C_6H_4CH_3-4$	Sn: Org.Comp.18-129/30
$C_{13}H_{23}NO_3SSn$	$(C_2H_5)_3SnNHSO_2C_6H_4OCH_3-4$	Sn: Org.Comp.18-129/30
$C_{13}H_{23}NO_4Sn$	$(C_2H_5)_3SnN(C(=O)CH_2)_2CHCH_2COOH$	Sn: Org.Comp.18-142, 147
$C_{13}H_{23}NSn$	$(C_2H_5)_3SnN(CH_3)C_6H_5$	Sn: Org.Comp.18-137
$C_{13}H_{23}N_3Sn$	$(C_2H_5)_3SnN(C_6H_5)N=N(CH_3)$	Sn: Org.Comp.18-155
$C_{13}H_{23}OOsP_3S_4$	$(CO)Os[-SP(CH_3)_2S-]_2[P(CH_3)_2-C_6H_5]$	Os: Org.Comp.A1-148
$C_{13}H_{23}O_2ReSn_2$	$(C_5H_5)Re(CO)_2[Sn(CH_3)_3]_2$	Re: Org.Comp.3-175/6, 181
$C_{13}H_{23}PSn$	$(CH_3)_3Sn-P(C_6H_5)-C_4H_9-t$	Sn: Org.Comp.19-174
$C_{13}H_{24}IMoN_7O$	$[CH_3-NH-C(NH-C_3H_7-i)=Mo(CN-CH_3)_4(NO)]I$	Mo: Org.Comp.5-118/9
$C_{13}H_{24}MoN_2O_2S_4$	$(HCC-CH_2OH)Mo(O)[S_2C-N(C_2H_5)_2]_2$	Mo: Org.Comp.5-139/40

$C_{13}H_{24}MoN_3O_2P$ . . $(C_5H_5)Mo(CO)_2(H)[P(N(CH_3)_2)_3]$ Mo:Org.Comp.7-56/7, 83
$C_{13}H_{24}MoN_7O^+$ . . $[CH_3\text{-}NH\text{-}C(NH\text{-}C_3H_7\text{-}i)=Mo(CN\text{-}CH_3)_4(NO)]I$. Mo:Org.Comp.5-118
$C_{13}H_{24}N_2OSiSn$ . . $(CH_3)_3SnNHC(C_6H_5)=NOSi(CH_3)_3$ . . . . Sn: Org.Comp.18-15/6, 18
$C_{13}H_{24}N_2O_4SiSn$. $(CH_3)_3SnN_2C_3(COOCH_3)_2Si(CH_3)_3$ . . . . Sn: Org.Comp.18-84, 89
$C_{13}H_{24}OP_2Re$ . . . . $(CO)Re[P(CH_3)_2CH_2CH_2P(CH_3)_2](CH_3)\text{-}C_5H_5$ . . Re: Org.Comp.1-54/5
$C_{13}H_{24}O_4S_2Sn$ . . . $(C_4H_9)_2SnOOCCH_2SCH_2SCH_2COO$ . . . . Sn: Org.Comp.15-323
$C_{13}H_{24}O_6Sn$ . . . . . $C_4H_9Sn(OOCC_2H_5)_3$ . . . . Sn: Org.Comp.17-44
$C_{13}H_{25}I_3MoN_2O$ . . $[N(C_2H_5)_4][C_5H_5Mo(NO)I_3]$ Mo:Org.Comp.6-29
$C_{13}H_{25}InN_2$ . . . . . . $(CH_3)_3In \cdot C_6H_4\text{-}1,4\text{-}[N(CH_3)_2]_2$ . . . . In: Org.Comp.1-27, 34/5
$C_{13}H_{25}InO_2$ . . . . . . $(n\text{-}C_4H_9)_2In[\text{-}OC(CH_3)CHC(CH_3)O\text{-}]$ . . . . In: Org.Comp.1-190, 193, 196
$C_{13}H_{25}MoNO_5P_2S_4$
$C_5H_5Mo(NO)(SP(S)(OC_2H_5)_2)S_2P(OC_2H_5)_2$ Mo:Org.Comp.6-48
$C_{13}H_{25}MoO_8P_2^+$ . . $[(CH_3\text{-}C_5H_4)Mo(CO_2)(P(OCH_3)_3)_2]^+$ Mo:Org.Comp.7-283
$C_{13}H_{25}NO_2S_2Si_2$ . . $((CH_3)_3Si)_2S=N\text{-}S(O)_2\text{-}C_6H_4\text{-}4\text{-}CH_3$ . . . . S: S-N Comp.8-251
$C_{13}H_{25}NO_2Sn$ . . . . $(n\text{-}C_3H_7)_3Sn\text{-}1\text{-}NC_4H_4(=O)_2\text{-}2,5$ . . . . Sn: Org.Comp.18-156/7
– . . . . $(i\text{-}C_3H_7)_3Sn\text{-}1\text{-}NC_4H_4(=O)_2\text{-}2,5$ . . . . Sn: Org.Comp.18-159
$C_{13}H_{25}NO_4Sn$ . . . . $(n\text{-}C_4H_9)_2Sn[\text{-}OC(O)CH_2N(CH_3)CH_2C(O)O\text{-}]$ . . Sn: Org.Comp.15-325
– . . . . $(t\text{-}C_4H_9)_2Sn[\text{-}OC(O)CH_2N(CH_3)CH_2C(O)O\text{-}]$ . . . Sn: Org.Comp.15-376/8
$C_{13}H_{25}N_3O_2S$ . . . . $(1\text{-}C_5H_{10}N)_2S=N\text{-}C(O)OC_2H_5$ . . . . S: S-N Comp.8-207/8, 210
$C_{13}H_{25}N_4O_7PS_2$ . . $(CH_3)_2NC(=O)C(S\text{-}CH_3)=N\text{-}OC(=O)N(CH_3)$
$S(=O)N(CH_3)P(=O)[\text{-}OCH_2C(CH_3)_2CH_2O\text{-}]$ . . . S: S-N Comp.8-356
$C_{13}H_{26}MoOP_2$ . . . . $C_5H_5Mo(P(CH_3)_3)_2(CO)CH_3$ Mo:Org.Comp.6-255
$C_{13}H_{26}MoO_6P_2$ . . . $C_5H_5Mo(P(OCH_3)_3)_2CCH_3$ Mo:Org.Comp.6-69
$C_{13}H_{26}MoO_7P_2$ . . . $[C_5H_5Mo(P(OCH_3)_3)_2(CO)CH_3]$, radical Mo:Org.Comp.6-255
$C_{13}H_{26}MoO_8P_2$ . . . $[CH_2=C(CH_3)\text{-}CH=CH_2]Mo(CO)_2[P(OCH_3)_3]_2$ . . Mo:Org.Comp.5-306/10
$C_{13}H_{26}NO_2P_2Re$ . . $(CO)Re[P(CH_3)_3]_2(NO)(CH_3)\text{-}C_5H_5$ . . . . Re: Org.Comp.1-53/4
$C_{13}H_{26}NO_3P_2Re$ . . $(CO)Re[P(CH_3)_3]_2(NO)(CH_2OH)\text{-}C_5H_5$ . . . . Re: Org.Comp.1-54
$C_{13}H_{26}N_2OSSi_2$ . . $(CH_3)_3Si\text{-}N(CH_3)\text{-}S(O)\text{-}N(C_6H_5)\text{-}Si(CH_3)_3$ . . . . . S: S-N Comp.8-357
$C_{13}H_{26}N_2O_2SSiSn$ $(CH_3)_3SnN(SO_2C_6H_4CH_3\text{-}4)NHSi(CH_3)_3$ . . . . . . Sn: Org.Comp.18-121/2, 124
$C_{13}H_{26}N_2O_2SSn$ . . $(n\text{-}C_4H_9)_2Sn(NCS)\text{-}N(C_2H_5)\text{-}C(=O)O\text{-}CH_3$ . . . . Sn: Org.Comp.19-127
$C_{13}H_{26}N_2O_2Sn$ . . . $(C_2H_5)_3SnN(C(=O)CH(NHC_3H_7\text{-}i)CH_2C(=O))$ . . . Sn: Org.Comp.18-143/4
$C_{13}H_{26}N_3O_2S^+$ . . . $[(1,4\text{-}ONC_4H_8\text{-}4\text{-})_2S\text{-}NC_5H_{10}\text{-}1]^+$ . . . . S: S-N Comp.8-230, 244
$C_{13}H_{26}N_3O_5PS_3$ . . $CH_3S\text{-}C(CH_3)=NOC(=O)N(CH_3)S(=O)$
$N(C_3H_7\text{-}i)P(=S)[\text{-}OCH_2C(CH_3)_2CH_2O\text{-}]$ . . . . . . S: S-N Comp.8-356
– . . . . $CH_3S\text{-}C(CH_3)=NOC(=O)N(CH_3)S(=O)$
$N(C_3H_7\text{-}n)P(=S)[\text{-}OCH_2C(CH_3)_2CH_2O\text{-}]$ . . . . . S: S-N Comp.8-356
– . . . . $CH_3S\text{-}C(CH_3)=NOC(=O)N(CH_3)S(=O)$
$N(C_6H_{11}\text{-}c)P(=S)(OCH_3)_2$ . . . . S: S-N Comp.8-356
$C_{13}H_{26}O_2SSn$ . . . . $(C_4H_9)_2SnOCH_2C(CH_2SCH_2)CH_2O$ . . . . Sn: Org.Comp.15-56
$C_{13}H_{26}O_2Sn$ . . . . . $(C_4H_9)_2SnOC_5H_8O$ . . . . Sn: Org.Comp.15-52
$C_{13}H_{26}O_3Sn$ . . . . . $(C_4H_9)_2SnOCH_2C(CH_2OCH_2)CH_2O$ . . . . Sn: Org.Comp.15-56
$C_{13}H_{26}O_5Sn$ . . . . . $C_4H_9Sn(OOCC_2H_5)_2OC_3H_7\text{-}i$ . . . . Sn: Org.Comp.17-73
$C_{13}H_{26}O_8Sb_2$ . . . . $[(CH_3)_2Sb(O_2CCH_3)_2]_2CH_2$ . . . . Sb: Org.Comp.5-126
$C_{13}H_{27}MoNOSi_2$ . . $C_5H_5Mo(NO)(CH_2Si(CH_3)_3)_2$ Mo:Org.Comp.6-97
$C_{13}H_{27}NO_2OsSi_2$ $[(CH_3)_3SiCH_2]_2Os(O)_2(NC_5H_5)$ . . . . Os: Org.Comp.A1-17, 40/1
$C_{13}H_{27}NO_4Sn$ . . . . $(C_4H_9)_2SnOCH_2C(C_2H_5)(NO_2)CH_2O$ . . . . Sn: Org.Comp.15-56
$C_{13}H_{27}NSn$ . . . . . . $(CH_3)_3SnN(C_6H_{11}\text{-}c)CH=C(CH_3)_2$ . . . . Sn: Org.Comp.18-47, 49, 53
$C_{13}H_{27}N_2NaSn$ . . . $(C_4H_9)_3SnN(Na)CN$ . . . . Sn: Org.Comp.18-200/1
$C_{13}H_{27}N_2Sn^-$ . . . . $[(C_4H_9)_3SnNCN]^-$ . . . . Sn: Org.Comp.18-200/1
$C_{13}H_{27}N_3O_3Sn$ . . . $C_4H_9Sn(ON=C(CH_3)_2)_3$ . . . . Sn: Org.Comp.17-49

$C_{13}H_{27}N_3S_2Sn$ ... $(C_4H_9)_3SnN(N{=}NSC({=}S))$ ... Sn: Org.Comp.18-202, 217

$C_{13}H_{27}N_4O_7PS_2$ .. $(CH_3)_2N{-}C(O){-}C(SCH_3){=}N{-}OC(O){-}N(CH_3)$ $-S(O){-}N(C_2H_5){-}P(O)(OC_2H_5)_2$ ... S: S-N Comp.8-356

$C_{13}H_{27}N_7OOs^{2+}$ .. $[2,6{-}(CH_3)_2{-}NC_5H_3{-}4{-}Os(NH_3)_4(NH_2{-}C({=}O)$ $-4{-}C_5H_4N)]^{2+}$ ... Os: Org.Comp.A1-12

$C_{13}H_{27}O_4Sb$ ... $(CH_3)Sb[-OC(CH_3)_2C(CH_3)_2O-]_2$ ... Sb: Org.Comp.5-265

$C_{13}H_{27}P_2Re$ ... $(C_5H_5)ReH[P(CH_3)_3]_2{-}CH{=}CH_2$ ... Re: Org.Comp.3-40/1, 42

$C_{13}H_{28}N_2O_3Sn$ ... $C_4H_9Sn(ON{=}C(CH_3)_2)_2OC_3H_7{-}i$ ... Sn: Org.Comp.17-73

$C_{13}H_{28}N_2Sn$ ... $(C_4H_9)_3SnNHCN$ ... Sn: Org.Comp.18-160, 162

$C_{13}H_{28}N_3O_4PS$ ... $(1{-}C_5H_{10}N){-}S(NOC_4H_8){=}N{-}P(O)(OC_2H_5)_2$ ... S: S-N Comp.8-207/8, 211

$C_{13}H_{28}N_3O_5PS_3$ .. $CH_3S{-}C(CH_3){=}NOC({=}O)N(CH_3)S({=}O)$ $N(C_4H_9{-}s)P({=}S)(O{-}C_2H_5)_2$ ... S: S-N Comp.8-356

– ... $CH_3S{-}C(CH_3){=}NOC({=}O)N(CH_3)S({=}O)$ $N(C_6H_{13}{-}n)P({=}S)(OCH_3)_2$ ... S: S-N Comp.8-356

$C_{13}H_{28}N_4Sn$ ... $(n{-}C_4H_9)_3SnN(-N{=}NCH{=}N-)$ ... Sn: Org.Comp.18-202, 212, 222

– ... $(n{-}C_4H_9)_3SnN(-N{=}NN{=}CH-)$ ... Sn: Org.Comp.18-202, 211, 222

$C_{13}H_{28}OSn$ ... $(n{-}C_4H_9)_2Sn(H)OCH(CH_3)C(CH_3){=}CH_2$ ... Sn: Org.Comp.17-78

$C_{13}H_{28}O_2Sn$ ... $(n{-}C_4H_9)_2Sn[-OCH(CH_3)CH_2CH(CH_3)O-]$ ... Sn: Org.Comp.15-54

– ... $(n{-}C_4H_9)_2Sn[-O(CH_2)_5O-]$ ... Sn: Org.Comp.15-57

– ... $(n{-}C_4H_9)_2Sn[-OCH_2C(CH_3)_2CH_2O-]$ ... Sn: Org.Comp.15-54

$C_{13}H_{28}O_3SSn$ ... $(n{-}C_4H_9)_2Sn(OC_2H_5)OOCCH_2CH_2SH$ ... Sn: Org.Comp.16-190

$C_{13}H_{28}O_3Sn$ ... $(n{-}C_4H_9)_2Sn(O{-}C_3H_7{-}i)OC(O){-}CH_3$ ... Sn: Org.Comp.16-191, 197

– ... $(n{-}C_4H_9)_2Sn[-OCH_2C(CH_3)(CH_2OH)CH_2O-]$ ... Sn: Org.Comp.15-55

– ... $n{-}C_4H_9{-}Sn(O{-}C_3H_7{-}i)[-OCH(CH_3)CH_2C(CH_3)_2O-]$ Sn: Org.Comp.17-73/4

$C_{13}H_{28}O_4Sn$ ... $n{-}C_4H_9Sn(OC_3H_7{-}i)_2OOCC_2H_5$ ... Sn: Org.Comp.17-72

$C_{13}H_{29}I_2MnO_3P$ .. $Mn[P(C_3H_7)_3](OC_4H_8)(O_2)I_2$ ... Mn: MVol.D8-51

$C_{13}H_{29}NOSn$ ... $(C_2H_5)_3SnN(C_6H_{13})CHO$ ... Sn: Org.Comp.18-136/8

$C_{13}H_{29}NO_2Sn$ ... $(C_2H_5)_3SnN(C_4H_9{-}n)C(O)O{-}C_2H_5$ ... Sn: Org.Comp.18-137/8

– ... $(n{-}C_4H_9)_2Sn(O{-}C_2H_5){-}ON{=}C(CH_3)_2$ ... Sn: Org.Comp.16-190

– ... $(n{-}C_4H_9)_2Sn[-OCH_2CH_2N(CH_3)CH_2CH_2O-]$ ... Sn: Org.Comp.15-72

– ... $(i{-}C_4H_9)_2Sn[-OCH_2CH_2N(CH_3)CH_2CH_2O-]$ ... Sn: Org.Comp.15-370

– ... $(t{-}C_4H_9)_2Sn[-OCH_2CH_2N(CH_3)CH_2CH_2O-]$ ... Sn: Org.Comp.15-374, 376/8

$C_{13}H_{29}NO_3SiSn$ .. $C_4H_9(CH_3)_2SiCH_2Sn(OCH_2CH_2)_3N$ ... Sn: Org.Comp.17-57/8

$C_{13}H_{29}NO_3Sn$ ... $CH_3{-}O{-}Sn(C_4H_9{-}n)_2{-}N(C_2H_5){-}C({=}O)O{-}CH_3$ ... Sn: Org.Comp.19-129, 133

– ... $n{-}C_4H_9{-}Sn(O{-}C_3H_7{-}i)_2{-}ON{=}C(CH_3)_2$ ... Sn: Org.Comp.17-72

$C_{13}H_{29}N_4PdS^+$ ... $[Pd(NCS)((C_2H_5)_2NCH_2CH_2NHCH_2CH_2N(C_2H_5)_2)]^+$ Pd: SVol.B2-307

– ... $[Pd(SCN)((C_2H_5)_2NCH_2CH_2NHCH_2CH_2N(C_2H_5)_2)]^+$ Pd: SVol.B2-307

$C_{13}H_{30}IInN_2Se$ ... $[N(CH_3)_4][(C_4H_9)_2In(SeCN)(I)]$ ... In: Org.Comp.1-363

$C_{13}H_{30}InN$ ... $(i{-}C_3H_7)_2In{-}CH_2CH_2CH_2{-}N(C_2H_5)_2$ ... In: Org.Comp.1-101, 109

– ... $(n{-}C_3H_7)_2In{-}CH_2CH_2CH_2{-}N(C_2H_5)_2$ ... In: Org.Comp.1-101, 109

$C_{13}H_{30}N_2O_2Sn$ ... $(i{-}C_3H_7)_3SnN(C_4H_9{-}t)NO_2$ ... Sn: Org.Comp.18-159

$C_{13}H_{30}N_2O_3Sn$ ... $(C_2H_5)_2N{-}C({=}O)O{-}CH_2CH_2{-}O{-}Sn(CH_3)_2{-}N(C_2H_5)_2$ Sn: Org.Comp.19-129, 130

$C_{13}H_{30}O_3Sn$ ... $CH_3Sn(O{-}C_4H_9{-}i)_3$ ... Sn: Org.Comp.17-14, 16/7

– ... $CH_3Sn(O{-}C_4H_9{-}s)_3$ ... Sn: Org.Comp.17-14, 16/7

$C_{13}H_{30}O_3Sn$	$CH_3Sn(O-C_4H_9-t)_3$	Sn: Org.Comp.17-14, 16/7
–	$n-C_4H_9-Sn(O-C_3H_7-i)_3$	Sn: Org.Comp.17-33, 36/40
–	$n-C_4H_9-Sn(O-C_3H_7-n)_3$	Sn: Org.Comp.17-33, 36/40
$C_{13}H_{30}O_6Sb_2$	$[(C_2H_5)_2SbO]_2(-OO(CH_2)_5OO-)$	Sb: Org.Comp.5-233
$C_{13}H_{30}S_2Sb_2$	$[(C_2H_5)_2SbS]_2(CH_2)_5$	Sb: Org.Comp.5-127
$C_{13}H_{31}NO_2Sn$	$(C_2H_5)_2Sn(O-C_3H_7-i)-OCH_2CH_2-N(C_2H_5)_2$	Sn: Org.Comp.16-181
–	$(n-C_4H_9)_2Sn(O-C_2H_5)-OCH(CH_3)CH_2-NH_2$	Sn: Org.Comp.16-190
–	$(n-C_4H_9)_2Sn(O-C_2H_5)-OCH_2CH_2CH_2-NH_2$	Sn: Org.Comp.16-190
–	$(n-C_4H_9)_2Sn(O-C_2H_5)-OCH_2CH_2-NHCH_3$	Sn: Org.Comp.16-189
$C_{13}H_{31}O_5PSn$	$(n-C_4H_9)_2Sn(OCH_3)OP(O)(OC_2H_5)_2$	Sn: Org.Comp.16-189
$C_{13}H_{32}OOsP_4$	$(CO)Os[(CH_3)_2P-CH_2CH_2-P(CH_3)_2]_2$	Os: Org.Comp.A1-125
$C_{13}H_{32}O_4Si_3Sn$	$(CH_3)[((CH_3)_3Si)_3C]Sn(OOCH)_2$	Sn: Org.Comp.16-209
$C_{13}H_{33}NOSiSn$	$(CH_3)_3SnN(C_4H_9-t)Si(CH_3)_2OC_4H_9-t$	Sn: Org.Comp.18-57, 63
$C_{13}H_{33}N_2PSn$	$(CH_3)_3SnN=P(C_4H_9-t)_2N(CH_3)_2$	Sn: Org.Comp.18-107, 114
$C_{13}H_{33}N_3O_3Sn$	$C_4H_9Sn(OCH_2CH_2NHCH_3)_3$	Sn: Org.Comp.17-34
$C_{13}H_{33}N_3Sn$	$[(C_2H_5)_2N]_3Sn-CH_3$	Sn: Org.Comp.19-111/2
$C_{13}H_{33}OPSiSn$	$(CH_3)_3Si-O-Sn(CH_3)_2-P(C_4H_9-t)_2$	Sn: Org.Comp.19-218, 223
$C_{13}H_{33.5}MnN_{2.5}S_4$	$[N(CH_3)_4]_2[Mn(SCH_2CH_2S)_2] \cdot 0.5\ NCCH_3$	Mn: MVol.D7-38/9
$C_{13}H_{34}N_2SiSn$	$(CH_3)_3SnN(C_4H_9-t)Si(CH_3)_2NHC_4H_9-t$	Sn: Org.Comp.18-57, 63
$C_{13}H_{35}In_3N_4$	$[(CH_3)_2In-N(CH_3)-CH_2CH_2-N(CH_3)-]_2In-CH_3$	In: Org.Comp.1-295/6
$C_{13}H_{35}MoN_2P_3$	$(CH_2=CH_2)_2Mo(N_2)[P(CH_3)_3]_3$	Mo: Org.Comp.5-180, 182
$C_{13}H_{36}In_2N_2$	$(CH_3)_3In \cdot N(CH_3)_2-CH_2CH_2CH_2-N(CH_3)_2 \cdot In(CH_3)_3$	In: Org.Comp.1-49, 50
$C_{13}H_{37}OOsP_4^+$	$[(H)(CO)Os(P(CH_3)_3)_4][PF_6]$	Os: Org.Comp.A1-120
$C_{13}H_{37}O_{13}P_4Re$	$(^{13}CO)ReH[P(OCH_3)_3]_4$	Re: Org.Comp.1-38
$C_{13}H_{39}IN_2Si_4Sn$	$[(CH_3)_3Si]_2N-Sn(I)(CH_3)-N[Si(CH_3)_3]_2$	Sn: Org.Comp.19-149, 152
$C_{13}H_{39}InIrP_3$	$(CH_3)_2In[Ir(CH_3)_2(P(CH_3)_3)_3]$	In: Org.Comp.1-329/32
$C_{13}H_{39}NOP_4Re^+$	$[CH_3Re(P(CH_3)_3)_4(NO)]^+$	Re: Org.Comp.1-9, 10
$C_{13}H_{39}N_3Si_3Sn$	$[(CH_3)_3Si-N(CH_3)]_3Sn-CH_3$	Sn: Org.Comp.19-111/2
$C_{13}H_{39}N_9S_3Sn$	$CH_3-Sn[N(CH_3)-S(=NCH_3)_2-CH_3]_3$	Sn: Org.Comp.19-111, 113
$C_{13}H_{39}P_4Re^+$	$[CH_3Re(P(CH_3)_3)_4]^+$	Re: Org.Comp.1-9/10
$C_{13}H_{43}N_4O_{39}PSnW_{11}$	$[N(CH_3)_3H]_4[CH_3SnO_5W_{11}PO_{34}]$	Sn: Org.Comp.17-16, 23
$C_{13}K_{10}O_{31}Th_2$	$K_{10}Th_2(C_2O_4)_4(CO_3)_5 \cdot 5\ H_2O$	Th: SVol.C7-99, 101
–	$K_{10}Th_2(C_2O_4)_4(CO_3)_5 \cdot 7\ H_2O$	Th: SVol.C7-99, 101
$C_{13}O_{13}Os_3Re$	$ReOs_3(CO)_{13}$	Re: Org.Comp.1-336
$C_{14}ClF_2H_{11}N_2O_3S_2$	$C_6H_5-C(O)-NH-S(CClF_2)=N-S(O)_2-C_6H_5$	S: S-N Comp.8-194
$C_{14}ClF_3FeH_6O_4$	$(C_5H_5)Fe(CO)_2C_6(Cl-3)(COOH-5)F_3-2,4,6$	Fe: Org.Comp.B13-177, 187
$C_{14}ClF_3H_{10}O_2Sn$	$(C_6H_5)_2Sn(Cl)OOCCF_3$	Sn: Org.Comp.17-120, 122
$C_{14}ClF_4H_{25}MoN_2OP_2$	$C_5H_5Mo(CO)(PF_2N(C_2H_5)_2)_2Cl$	Mo: Org.Comp.6-227
$C_{14}ClF_6FeH_{14}OP$	$[(2-Cl-C_6H_4-CH_2-CO-CH_3)Fe(C_5H_5)][PF_6]$	Fe: Org.Comp.B19-3, 6, 35
–	$[(3-Cl-C_6H_4-CH_2-CO-CH_3)Fe(C_5H_5)][PF_6]$	Fe: Org.Comp.B19-3, 6, 35
–	$[(4-Cl-C_6H_4-CH_2-CO-CH_3)Fe(C_5H_5)][PF_6]$	Fe: Org.Comp.B19-3, 6, 36
$C_{14}ClF_6FeH_{14}O_2P$	$[(2-Cl-C_6H_4-CH_2-COO-CH_3)Fe(C_5H_5)][PF_6]$	Fe: Org.Comp.B19-37
$C_{14}ClF_6FeH_{17}NOP$	$[(4-Cl-C_6H_4-NH-CH_2-CH(OH)-CH_3)Fe(C_5H_5)][PF_6]$	Fe: Org.Comp.B19-72
$C_{14}ClF_9N^-$	$[C(C_6F_5)(C_6F_4-Cl-4)(CN)]^-$	F: PFHOrg.SVol.6-108, 127
$C_{14}ClF_9NNa$	$Na[C(C_6F_5)(C_6F_4-Cl-4)(CN)]$	F: PFHOrg.SVol.6-108, 127

$C_{14}ClF_9N_2$	4-$[C_6F_5$-$C(Cl)$=$N]$-C_6F_4-CN	F: PFHOrg.SVol.6-193, 207
$C_{14}ClF_{11}H_2N_2$	4-H_2N-C_6F_4-$C(Cl)$=N-C_6F_4-CF_3-4	F: PFHOrg.SVol.5-10, 41
$C_{14}ClF_{12}N$	4-CF_3-C_6F_4-N=CCl-C_6F_5	F: PFHOrg.SVol.6-193, 208
$C_{14}ClFeH_{11}N_2$	$(C_5H_5)Fe[1$-Cl-6-CH_3-$C_6H_3(CN)_2$-5,6]	Fe: Org.Comp.B17-258
–	$(C_5H_5)Fe[3$-Cl-2-CH_3-$C_6H_3(CN)_2$-1,6]	Fe: Org.Comp.B17-258
$C_{14}ClFeH_{13}O_2$	$[CH_3C(=O)$-$C_5H_4]Fe[C_5H_3(Cl)$-$C(=O)CH_3]$	Fe: Org.Comp.A10-253, 257, 261
$C_{14}ClFeH_{13}O_4$	$[CH_3OC(=O)$-$C_5H_4]Fe[C_5H_3(Cl)$-$C(=O)OCH_3]$	Fe: Org.Comp.A10-277, 278
$C_{14}ClFeH_{13}O_6S$	$[CH_3OC(=O)$-$C_5H_4]Fe[C_5H_3(SO_2Cl)$-$C(=O)OCH_3]$	Fe: Org.Comp.A10-298
$C_{14}ClFeH_{14}O^+$	$[(2$-Cl-C_6H_4-CH_2-CO-$CH_3)Fe(C_5H_5)][PF_6]$	Fe: Org.Comp.B19-3, 6, 35
–	$[(3$-Cl-C_6H_4-CH_2-CO-$CH_3)Fe(C_5H_5)][PF_6]$	Fe: Org.Comp.B19-3, 6, 35
–	$[(4$-Cl-C_6H_4-CH_2-CO-$CH_3)Fe(C_5H_5)][PF_6]$	Fe: Org.Comp.B19-3, 6, 36
$C_{14}ClFeH_{14}O_2^+$	$[(2$-Cl-C_6H_4-CH_2-COO-$CH_3)Fe(C_5H_5)][PF_6]$	Fe: Org.Comp.B19-37
$C_{14}ClFeH_{15}O$	$(C_5H_5)Fe[1$-Cl-6-$CH_3C(O)CH_2$-$C_6H_5]$	Fe: Org.Comp.B17-274
–	$(C_5H_5)Fe[2$-Cl-6-$CH_3C(O)CH_2$-$C_6H_5]$	Fe: Org.Comp.B17-274
$C_{14}ClFeH_{17}NO^+$	$[(4$-Cl-C_6H_4-NH-CH_2-CH(OH)-$CH_3)Fe(C_5H_5)][PF_6]$	Fe: Org.Comp.B19-72
$C_{14}ClFeH_{18}HgN$	$(C_5H_5)Fe[C_5H_3(HgCl)$-$CH(CH_3)$-$N(CH_3)_2]$	Fe: Org.Comp.A10-147, 148, 150/1
–	$(C_5H_5)Fe[C_5H_3(HgCl)$-CH_2CH_2-$N(CH_3)_2]$	Fe: Org.Comp.A10-147, 148
$C_{14}ClFeH_{18}N$	$(C_5H_5)Fe[C_5H_3(Cl)$-$CH_2CH_2N(CH_3)_2]$	Fe: Org.Comp.A9-43
$C_{14}ClFeH_{19}IN$	$[(C_5H_5)Fe(C_5H_3(Cl)$-$CH_2N(CH_3)_3)]I$	Fe: Org.Comp.A9-64, 74
–	$[(Cl$-$C_5H_4)Fe(C_5H_4$-$CH_2N(CH_3)_3)]I$	Fe: Org.Comp.A9-22
$C_{14}ClFeH_{19}N^+$	$[(C_5H_5)Fe(C_5H_3(Cl)$-$CH_2N(CH_3)_3)]^+$	Fe: Org.Comp.A9-64, 74
–	$[(Cl$-$C_5H_4)Fe(C_5H_4$-$CH_2N(CH_3)_3)]^+$	Fe: Org.Comp.A9-22
$C_{14}ClFeH_{19}O_7$	$[(CH_3$-$CHC(CH_3)CHCHCH$-C_4H_9-n$)Fe(CO)_3][ClO_4]$	Fe: Org.Comp.B15-14, 25
$C_{14}ClFeH_{20}NO$	$[C_5H_5FeC_5H_3(CH_2NH(CH_3)_2)CH_2OH]Cl$	Fe: Org.Comp.A9-47
$C_{14}ClFeH_{29}OP_2$	$[C_5H_5(P(CH_3)_3)_2Fe$=$C(CH_3)OCH_3]Cl$	Fe: Org.Comp.B16a-8/9
$C_{14}ClGeH_{19}$	$Ge(C_2H_5)_3C{\equiv}CC_6H_4Cl$-2	Ge: Org.Comp.2-255, 257
–	$Ge(C_2H_5)_3C{\equiv}CC_6H_4Cl$-3	Ge: Org.Comp.2-255, 257
–	$Ge(C_2H_5)_3C{\equiv}CC_6H_4Cl$-4	Ge: Org.Comp.2-255, 257
$C_{14}ClGeH_{19}S$	$Ge(CH_3)_3C(S$-C_6H_4-CH_3-4)=CClCH=CH_2	Ge: Org.Comp.2-18
–	$Ge(CH_3)_3C{\equiv}C$-CHCl-CH_2-S-C_6H_4-CH_3-4	Ge: Org.Comp.2-57
$C_{14}ClGeH_{21}Ti$	$Ge(CH_3)_3CH_2Ti(C_5H_5)_2Cl$	Ge: Org.Comp.1-142
$C_{14}ClGeH_{26}NO_4$	$Ge(C_2H_5)_3(C_6H_4NH(CH_3)_2-4)ClO_4$	Ge: Org.Comp.2-277
$C_{14}ClGeH_{29}$	$Ge(C_4H_9)_3CH$=CHCl	Ge: Org.Comp.3-25
$C_{14}ClGeH_{29}O_2$	$Ge(C_2H_5)_3CH$=$CHCH(OCH(OH)CH_2Cl)C_3H_7$	Ge: Org.Comp.2-225
$C_{14}ClGeH_{31}OSi$	$Ge(C_2H_5)_3CH$=$CHC(CH_3)_2$-O-$Si(CH_3)_2CH_2Cl$	Ge: Org.Comp.2-212
–	$Ge(C_2H_5)_3CH$=$CHCH(C_2H_5)$-O-$Si(CH_3)_2CH_2Cl$	Ge: Org.Comp.2-227
$C_{14}ClH_6LiMoO_{11}$	$Li[(2,5$-$(O=)_2OC_4H_2)_3Mo(CO)_2(Cl)]$	Mo: Org.Comp.5-197/8
$C_{14}ClH_6MoO_{11}^-$	$Li[(2,5$-$(O=)_2OC_4H_2)_3Mo(CO)_2(Cl)]$	Mo: Org.Comp.5-197/8
$C_{14}ClH_7N_4PdS_2$	$Pd(SCN)_2[1,10$-$N_2C_{12}H_7(Cl$-5)]	Pd: SVol.B2-305
$C_{14}ClH_9LiMoN_3O_8$	$Li[(2,5$-$(O=)_2NC_4H_3)_3Mo(CO)_2(Cl)]$	Mo: Org.Comp.5-197/8
$C_{14}ClH_9MnN_2OS$	$Mn(OC(C_6H_5)NC(=S)NC_6H_4Cl$-2) · H_2O	Mn: MVol.D7-198/9
$C_{14}ClH_9MoN_3O_8^-$	$Li[(2,5$-$(O=)_2NC_4H_3)_3Mo(CO)_2(Cl)]$	Mo: Org.Comp.5-197/8
–	$[N(C_2H_5)_4][(2,5$-$(O=)_2NC_4H_3)_3Mo(CO)_2(Cl)]$	Mo: Org.Comp.5-197/8
$C_{14}ClH_{10}MnNOS_6$	$[Mn((S_2C)SC_6H_5)_2(NO)Cl]$	Mn: MVol.D7-210

$C_{14}ClH_{10}MnNO_3S_4$
$[Mn(SC(=S)OC_6H_5)_2(NO)Cl]$ Mn:MVol.D7-130/3
$C_{14}ClH_{10}NO_5Th$ . . $(HO)_2Th[2-(2-O-5-Cl-C_6H_3-CH=N)-C_6H_4-COO] \cdot H_2O$
Th: SVol.D4-158
$C_{14}ClH_{10}N_4O_3Re$ $(CO)_3ReCl[-N(=CH-2-C_5H_4N)-NH-(2,1-C_5H_4N)-]$
Re: Org.Comp.1-160
$C_{14}ClH_{10}O_6Re$ . . . cis-$(CO)_4Re[C(=O)CH_2CH_2Cl]=C(OH)C_6H_5$ . . . . Re: Org.Comp.1-389
$C_{14}ClH_{11}MnN_2O_2$ $[MnCl(O-2-C_6H_4-CH=N-NHC(O)C_6H_5)]$ Mn:MVol.D6-272, 275
$C_{14}ClH_{11}MnN_2O_3$ $MnCl[O-2-C_6H_4-CH=N-NHC(O)-C_6H_4-2-OH]$ Mn:MVol.D6-272, 275
$C_{14}ClH_{11}MoO_2$ . . . $(C_9H_7)Mo(CO)_2(CH_2CClCH_2)$ Mo:Org.Comp.8-205, 210
$C_{14}ClH_{11}N_2O_2S$ . . $C_6H_5-C(O)NH-S(Cl)=N-C(O)-C_6H_5$ . . . . S: S-N Comp.8-184/5
$C_{14}ClH_{12}MnN_3OS_4$
$[Mn(SC(=S)NHC_6H_5)_2(NO)Cl]$ Mn:MVol.D7-136/8
$C_{14}ClH_{12}NOS$ . . . . $4-CH_3-C_6H_4-C(C_6H_5)=N-S(O)Cl$ . . . . S: S-N Comp.8-271/2
$C_{14}ClH_{12}NO_2Sn$ . . $CH_3Sn(Cl)OC_6H_4CH=NC_6H_4O$ . . . . Sn: Org.Comp.17-146
$C_{14}ClH_{13}MnN_2O_5$ $MnCl[O-C(CH_3)=N-N=CH-C_9H_2O(CH_3)(=O)$
$(OCH_3)(OH)] \cdot 2\ H_2O$ Mn:MVol.D6-288, 290/1
$C_{14}ClH_{13}MoN_2O_2$ $[(C_5H_5)Mo(CO)_2(-1-NC_5H_4-2-CH=N(CH_3)-)]Cl$ Mo:Org.Comp.7-249, 250, 252, 277/8
$C_{14}ClH_{13}O_2Sn$ . . . $(C_6H_5)_2Sn(Cl)OOCCH_3$ . . . . Sn: Org.Comp.17-119/20
$C_{14}ClH_{14}In$ . . . . . . $(4-CH_3-C_6H_4)_2InCl$ . . . . In: Org.Comp.1-131
– . . . . $(C_6H_5-CH_2)_2InCl$ . . . . In: Org.Comp.1-118, 120, 126
$C_{14}ClH_{14}InN_2$ . . . . $(CH_3)_2InCl \cdot 1,10-N_2C_{12}H_8$ . . . . In: Org.Comp.1-118, 121
$C_{14}ClH_{14}MnN_3O_4$ $[Mn(O-C_6H_4-2-CH=NNH-2-(1,3-ONC_7H_4))Cl(H_2O)_2]$
Mn:MVol.D6-240, 242
$C_{14}ClH_{14}MoN_2O_5P$
$(C_5H_5)Mo(CO)_2[P(OCH_3)_3]-CCl=C(CN)_2$ Mo:Org.Comp.8-77, 94
$C_{14}ClH_{14}MoO_2P$ . . $[(C_5H_5)Mo(CO)_2(CH_3-PH-C_6H_5)(Cl)]$ Mo:Org.Comp.7-56/7, 79
$C_{14}ClH_{14}NOS$ . . . . $(C_6H_5CH_2)_2N-S(O)Cl$ . . . . S: S-N Comp.8-259, 265
$C_{14}ClH_{14}N_3S_4$ . . . . $[4-CH_3-C_6H_4-S-N=SNS=N-S-C_6H_4-4-CH_3]Cl$ S: S-N Comp.7-42/4
$C_{14}ClH_{15}MoN_2O_2$ $[(C_5H_5)Mo(CO)_2(NH_2-CH(CH_3)-2-NC_5H_4)]Cl$ . . Mo:Org.Comp.7-249, 251
$C_{14}ClH_{15}O_2Ti$ . . . . $[(C_5H_5)_2Ti(Cl)OCOC(CH_3)CH_2]_n$ . . . . Ti: Org.Comp.5-341
$C_{14}ClH_{16}MoNO_2$ . . $(C_5H_5)Mo(CO)_2(Cl)(CN-C_6H_{11}-c)$ Mo:Org.Comp.8-13
$C_{14}ClH_{16}MoO_5P$ . . $[(C_9H_7)Mo(CO)_2(Cl)(P(OCH_3)_3)]$ Mo:Org.Comp.7-56/7, 85
$C_{14}ClH_{16}OTi$ . . . . . $(C_5H_5)Ti(OC_6H_2(CH_3)_3-2,4,6)Cl$ . . . . Ti: Org.Comp.5-37
$C_{14}ClH_{16}PSn$ . . . . $(CH_3)_2SnCl-P(C_6H_5)_2$ . . . . Sn: Org.Comp.19-218, 221
$C_{14}ClH_{17}MoO_5$ . . . $[C_5H_5Mo(CO)(HOCH_2-CC-CH_2OH)_2]Cl$ Mo:Org.Comp.6-318/9
$C_{14}ClH_{17}NO_3Sb$ . . $(CH_3)_3Sb(OC_9H_6N)O_2CCH_2Cl$ . . . . Sb: Org.Comp.5-47
$C_{14}ClH_{18}MoO_5P$ . . $(C_5H_5)Mo(CO)_2(Cl)[P(-OCH_2-)_3CC_3H_7-n]$ Mo:Org.Comp.7-56/7, 90/1, 114
$C_{14}ClH_{18}NO_2Sn$ . . $(C_2H_5)_3SnN(CO)_2C_6H_3Cl$ . . . . Sn: Org.Comp.18-142, 147
$C_{14}ClH_{19}MoNO_4P$ $(C_5H_5)Mo(CO)_2(Cl)(1,3,6,2-O_2NPC_4H_9-2-CH_2CH=CH_2)$
Mo:Org.Comp.7-249, 271
$C_{14}ClH_{19}N_2STi$ . . . $[(C_5H_5)_2TiCl(N=S=N-C_4H_9-t)]$ . . . . S: S-N Comp.7-293/4
$C_{14}ClH_{20}MoO_5P$ . . $(C_9H_{11})Mo(CO)_2(Cl)[P(OCH_3)_3]$ Mo:Org.Comp.7-56/7, 85
$C_{14}ClH_{20}NO_3Sn$ . . $(CH_3)_2Sn(Cl)OOCC_6H_4CON(C_2H_5)_2-2$ . . . . Sn: Org.Comp.17-85, 90
$C_{14}ClH_{20}N_3O_3S$ . . $4-(4-NO_2-2,3,5,6-(CH_3)_4-C_6-N=S(Cl))-1,4-ONC_4H_8$
S: S-N Comp.8-186

$C_{14}ClH_{21}Mn_2N_2Na_2O_{10}P_2$
$Na_2[Mn_2(O_3P-CH(CH_3)N=CH-C_6H_2(Cl)(O)$
$CH=NCH(CH_3)PO_3)(OH)] \cdot 2\ CH_3OH$ Mn:MVol.D6-211, 214

$C_{14}ClH_{21}MoN_2O_5P_2$
$C_5H_5Mo(CO)(P(OC_2H_4)_2N)_2Cl$ Mo:Org.Comp.6-226

$C_{14}ClH_{23}HgOSn$ .. $(C_4H_9)_2Sn(Cl)OHgC_6H_5$ .... Sn: Org.Comp.17-107

$C_{14}ClH_{23}MoN_2O_2$ $[CH_2C(CH_3)CH_2]Mo(Cl)$
$(CO)_2(i-C_3H_7-N=CHCH=N-C_3H_7-i)$ Mo:Org.Comp.5-262/3, 266

$C_{14}ClH_{23}MoO_2PSb$
$(C_5H_5)Mo(CO)_2[P(CH_3)_3]-Sb(Cl)C_4H_9-t$ Mo:Org.Comp.7-119, 140

$C_{14}ClH_{23}OSn$ .... $(C_4H_9)_2Sn(Cl)OC_6H_5$ .... Sn: Org.Comp.17-101

$C_{14}ClH_{27}OSn$ .... $(n-C_4H_9)_2Sn(Cl)OCH(C_2H_5)CH=C=CH_2$ .... Sn: Org.Comp.17-101

– .... $(n-C_4H_9)_2Sn(Cl)OCH(C_2H_5)CH_2C{\equiv}CH$ .... Sn: Org.Comp.17-101

$C_{14}ClH_{28}MnN_3OS_4$
$[Mn(SC(=S)N(C_3H_7-i)_2)_2(NO)Cl]$ Mn:MVol.D7-144

$C_{14}ClH_{28}PSn$ .... $(CH_3)_2SnCl-P(C_6H_{11}-c)_2$ .... Sn: Org.Comp.19-218, 221

$C_{14}ClH_{29}NP_2Re$ .. $(CH_3)_2Re[P(CH_3)_3]_2(=N-C_6H_5)Cl$ .... Re: Org.Comp.1-15

$C_{14}ClH_{29}N_6S_2Sn$ .. $C_6H_5-Sn(Cl)[N(CH_3)-S(=NCH_3)_2-CH_3]_2$ .... Sn: Org.Comp.19-149, 151

$C_{14}ClH_{29}OSn$ .... $(i-C_4H_9)_2Sn(Cl)O-C_6H_{11}-c$ .... Sn: Org.Comp.17-114

– .... $(n-C_4H_9)_2Sn(Cl)-OC(CH_3)_2CH_2CH=CH_2$ ...... Sn: Org.Comp.17-100

– .... $(n-C_4H_9)_2Sn(Cl)-OCH(C_2H_5)CH_2CH=CH_2$ ..... Sn: Org.Comp.17-100

$C_{14}ClH_{29}O_2Sn$ ... $(CH_3)_2Sn(Cl)OOCC_{11}H_{23}$ .... Sn: Org.Comp.17-84

$C_{14}ClH_{31}MoP_2$ ... $i-C_3H_7C_5H_4Mo(P(CH_3)_3)_2ClH_2$ Mo:Org.Comp.6-15

$C_{14}ClH_{32}InN_2$ .... $[(C_2H_5)_2N-CH_2CH_2CH_2]_2InCl$ .... In: Org.Comp.1-122

$C_{14}ClH_{36}N_3O_9P_3Re$
$Re(CNC_4H_9-t)(P(OCH_3)_3)_3(Cl)N_2$ .... Re: Org.Comp.2-231

$C_{14}Cl_2FH_{11}N_2O_3S_2$
$C_6H_5-C(O)-NH-S(CCl_2F)=N-S(O)_2-C_6H_5$ ..... S: S-N Comp.8-194

$C_{14}Cl_2F_2H_{10}O_2Sn$ $(C_6H_5)_2Sn(Cl)OOCCF_2Cl$ .... Sn: Org.Comp.17-120

$C_{14}Cl_2F_8LiN$ ..... $Li[C(C_6F_5)(C_6F_3-2,4-Cl_2)(CN)]$ .... F: PFHOrg.SVol.6-108, 128

$C_{14}Cl_2F_8N^-$ ...... $[C(C_6F_5)(C_6F_3-2,4-Cl_2)(CN)]^-$ .... F: PFHOrg.SVol.6-108, 128

– .... $[C(C_6F_4-Cl-4)_2(CN)]^-$ .... F: PFHOrg.SVol.6-108, 128

$C_{14}Cl_2F_8NNa$ .... $Na[C(C_6F_5)(C_6F_3-2,4-Cl_2)(CN)]$ .... F: PFHOrg.SVol.6-108, 128

– .... $Na[C(C_6F_4-Cl-4)_2(CN)]$ .... F: PFHOrg.SVol.6-108, 128

$C_{14}Cl_2F_{16}GeH_6$ ... $Ge(CH_3)_2(C(-(CF_2)_4CCl=))_2$ .... Ge: Org.Comp.3-154

$C_{14}Cl_2F_{29}N_5O_4$ ... $6-ClN=C_6F_5-[ON(CF_3)_2]_4-1,2,4,5-Cl-3$ .... F: PFHOrg.SVol.5-117, 127

$C_{14}Cl_2FeH_{14}^{2+}$ ... $[(4-Cl-C_6H_4-CH_3)_2Fe][O-C_6H_2-2,4,6-(NO_2)_3]_2$ Fe: Org.Comp.B19-361

$C_{14}Cl_2FeH_{14}O$.... $(C_5H_5)Fe[1,2-Cl_2-6-CH_3C(O)CH_2-C_6H_4]$...... Fe: Org.Comp.B17-287/8

– .... $(C_5H_5)Fe[1,3-Cl_2-6-CH_3C(O)CH_2-C_6H_4]$...... Fe: Org.Comp.B17-288

– .... $(C_5H_5)Fe[1,4-Cl_2-6-CH_3C(O)CH_2-C_6H_4]$...... Fe: Org.Comp.B17-288

– .... $(C_5H_5)Fe[1,5-Cl_2-6-CH_3C(O)CH_2-C_6H_4]$...... Fe: Org.Comp.B17-288

$C_{14}Cl_2FeH_{16}Hg_2$ .. $(ClHg-C_5H_4)Fe[C_5H_3(HgCl)-C_4H_9-t]$ .... Fe: Org.Comp.A10-329

$C_{14}Cl_2FeH_{16}S_2$ ... $Fe[C_5H_4S(CH_2)_2Cl]_2$ .... Fe: Org.Comp.A9-201, 204/5

$C_{14}Cl_2FeH_{19}NPdSe$
$C_5H_5FeC_5H_3(SeCH_3)CH_2N(CH_3)_2 \cdot PdCl_2$ .... Fe: Org.Comp.A9-272

$C_{14}Cl_2FeH_{20}NiP_2$ $Fe[C_5H_4-P(CH_3)_2]_2NiCl_2$ .... Fe: Org.Comp.A10-18, 21, 22/3

$C_{14}Cl_2FeH_{20}P_2Pd$ $Fe[C_5H_4-P(CH_3)_2]_2PdCl_2$ .... Fe: Org.Comp.A10-19, 23/4

$C_{14}Cl_2FeH_{20}P_2Pt$ $Fe[C_5H_4-P(CH_3)_2]_2PtCl_2$ .... Fe: Org.Comp.A10-19, 23, 24

$C_{14}Cl_2FeH_{20}Si_2$ .. $Fe[C_5H_4Si(CH_3)_2Cl]_2$ .... Fe: Org.Comp.A9-295, 307

$C_{14}Cl_2FeH_{22}N_2$ ... $[Fe(C_5H_4-CH(CH_3)NH_3)_2]Cl_2$ ... Fe: Org.Comp.A9-4
$C_{14}Cl_2GaH_8O_2$ ... $GaCl_2[9,10-(O)_2C_{14}H_8]$, radical ... Ga: SVol.D1-90/2
$C_{14}Cl_2GaH_8O_6^+$ .. $[Ga(OC(O)-C_6H_3-5-Cl-2-OH)_2]^+$ ... Ga: SVol.D1-176/80
$C_{14}Cl_2GaH_{17}N_3^+$ $[GaCl_2(6-CH_3-NC_5H_3-2-CH_2-N(CH_3)-CH_2-2-NC_5H_4)][GaCl_4]$ ... Ga: SVol.D1-271/2
$C_{14}Cl_2GaH_{19}NO_4$ $GaCl_2[1,2-(O)_2C_6H-3,5-(C_4H_9-t)_2-6-NO_2]$, radical Ga: SVol.D1-90/2
$C_{14}Cl_2GaH_{20}O_2$ .. $GaCl_2[1,2-(O)_2C_6H_2-3,5-(C_4H_9-t)_2]$, radical ... Ga: SVol.D1-90/2
– ... $GaCl_2[1,2-(O)_2C_6H_2-3,6-(C_4H_9-t)_2]$, radical ... Ga: SVol.D1-90/2
$C_{14}Cl_2GeH_{10}$ ... $Ge(C_6H_5)_2(CCl)_2$ ... Ge: Org.Comp.3-184
$C_{14}Cl_2GeH_{14}$ ... $Ge(CH_3)_2(C_6H_4Cl-4)_2$ ... Ge: Org.Comp.3-156
$C_{14}Cl_2H_8N_2O_2S$ .. $4-Cl-C_6H_4C(O)N=S=NC(O)C_6H_4-4-Cl$ ... S: S-N Comp.7-272/6
$C_{14}Cl_2H_{10}MnN_2O_2S_2$
$[Mn((2-S=)C_7H_5NO-3,1)_2Cl_2]$ ... Mn: MVol.D7-59/60
– ... $[Mn((2-S=)C_7H_5NO-3,1)_2Cl_2] \cdot 2\ H_2O$ ... Mn: MVol.D7-59/60
$C_{14}Cl_2H_{10}MnO_2S_2$ $[Mn(C_6H_5C(=O)SSC(=O)C_6H_5)Cl_2]$ ... Mn: MVol.D7-90/1
$C_{14}Cl_2H_{10}O_6Th$ ... $ThCl_2(2-HO-C_6H_4-COO)_2$ ... Th: SVol.C7-130/1
– ... $Th(OH)_2(2-Cl-C_6H_4-COO)_2$ ... Th: SVol.C7-139/40
$C_{14}Cl_2H_{11}MnN_4P$ $Mn[P(C_6H_5)(CH_3)_2](C_2(CN)_4)Cl_2$ ... Mn: MVol.D8-46/7
$C_{14}Cl_2H_{11}O_3Sb$ ... $(2-ClC_6H_4)_2Sb(O)O_2CCH_3$ ... Sb: Org.Comp.5-213
$C_{14}Cl_2H_{12}MnN_2O_4$ $[Mn(-2-O-C_6H_3(Cl-5)-CH=NN=CH-C_6H_3(Cl-5)-2-O-)(H_2O)_2]$ ... Mn: MVol.D6-261
$C_{14}Cl_2H_{12}MnN_4S$ $MnCl_2[NC_5H_3(CH_3-6)-2-CH=N-NH-2-(1,3-SNC_7H_4)]$ Mn: MVol.D6-244, 246/7
– ... $MnCl_2[NC_5H_4-2-C(CH_3)=N-NH-2-(1,3-SNC_7H_4)]$ Mn: MVol.D6-250, 251/2
$C_{14}Cl_2H_{12}MnN_6$ .. $Mn[NC_5H_3(Cl-4)-2-N-N=C(CH_3)-C(CH_3)=N-N-2-C_5H_3N(Cl-4)]$ ... Mn: MVol.D6-264/5
$C_{14}Cl_2H_{12}N_2O_4Th$ $ThCl_2(2-NH_2-C_6H_4-COO)_2$ ... Th: SVol.C7-137/9
$C_{14}Cl_2H_{13}Sb$ ... $(C_6H_4CH_2C_6H_4)(CH_3)SbCl_2$ ... Sb: Org.Comp.5-76
$C_{14}Cl_2H_{14}MnN_4$ .. $[MnCl_2(NC_5H_4-2-CH=N-C_2H_4-N=CH-2-C_5H_4N)]$ Mn: MVol.D6-189, 190
$C_{14}Cl_2H_{14}MnN_4O_8$ $[Mn(NC_5H_4-2-CH=N-C_2H_4-N=CH-2-C_5H_4N)][ClO_4]_2$ Mn: MVol.D6-189/90
$C_{14}Cl_2H_{14}O_2Po$ ... $(4-CH_3OC_6H_4)_2PoCl_2$ ... Po: SVol.1-334/40
$C_{14}Cl_2H_{14}Po$ ... $(4-CH_3C_6H_4)_2PoCl_2$ ... Po: SVol.1-334/40
$C_{14}Cl_2H_{15}OSb$ ... $(4-CH_3C_6H_4)_2Sb(Cl_2)OH$ ... Sb: Org.Comp.5-186
$C_{14}Cl_2H_{15}Sb$ ... $(C_6H_5)_2(C_2H_5)SbCl_2$ ... Sb: Org.Comp.5-56
$C_{14}Cl_2H_{16}MnN_6$ .. $MnCl_2[NC_5H_4-2-NH-N=C(CH_3)-C(CH_3)=N-NH-2-C_5H_4N]$ Mn: MVol.D6-264, 266/7
$C_{14}Cl_2H_{16}MnN_8S_2$ $MnCl_2[NC_5H_4-2-CH=N-NHC(=S)NH_2]_2$ ... Mn: MVol.D6-347
– ... $MnCl_2[NC_5H_4-3-CH=N-NHC(=S)NH_2]_2$ ... Mn: MVol.D6-347
$C_{14}Cl_2H_{16}NO_3Sb$ $(CH_3)_3Sb(OC_9H_6N)O_2CCHCl_2$ ... Sb: Org.Comp.5-47
$C_{14}Cl_2H_{16}N_2O_4S_2Sn$
$(CH_3)_2Sn[NH-S(=O)_2-C_6H_4-4-Cl]_2$ ... Sn: Org.Comp.19-62
$C_{14}Cl_2H_{18}MnN_2O_{12}S_2$
$[Mn(O=S(C_2H_5)(C_5H_4NO))_2(H_2O)_2][ClO_4]_2$ ... Mn: MVol.D7-106
$C_{14}Cl_2H_{18}MnN_6S_2$ $[Mn(S=C(NHC_6H_5)NHNH_2)_2Cl_2]$ ... Mn: MVol.D7-204/5
– ... $[Mn(S=C(NH_2)NHNHC_6H_5)_2Cl_2]$ ... Mn: MVol.D7-204/5
$C_{14}Cl_2H_{18}N_4O_4S$ .. $(1,4-ONC_4H_8-4)_2S=N-C_6H_2-2,6-Cl_2-4-NO_2$ ... S: S-N Comp.8-207/8, 212

$C_{14}Cl_2H_{18}O_4Sn$ . . . $n-C_4H_9-OC(O)-CH_2CH_2-SnCl_2-O-C_6H_4-CHO-2$ Sn: Org.Comp.17-180

– . . . . $[-Sn(C_4H_9-n)_2-O-C_6(=O)_2(Cl)_2-O-]_n$ . . . . Sn: Org.Comp.15-59

$C_{14}Cl_2H_{19}N_3O_2S$ . . $(1,4-ONC_4H_8-4)_2S=N-C_6H_3-2,4-Cl_2$ . . . . S: S-N Comp.8-207/8, 212

$C_{14}Cl_2H_{20}MnN_4S_2$ $[Mn(1,3-N_2C_4H(=S-2)((CH_3)_3-1,4,6))_2Cl_2]$ . . . . . Mn: MVol.D7-78/9

$C_{14}Cl_2H_{20}MnN_8O_4S_2$ $[Mn(4-NH_2C_6H_4SO_2NHC(NH)NH_2)_2Cl_2(H_2O)_2]$ Mn: MVol.D7-116/23

$C_{14}Cl_2H_{20}N_2S_2Ti_2$ $[(C_5H_5)Ti(SCH_2CH_2NH)Cl]_2$ . . . . Ti: Org.Comp.5-61

$C_{14}Cl_2H_{20}N_2Ti_2$ . . $[(C_5H_5)Ti(NC_2H_5)Cl]_2$ . . . . Ti: Org.Comp.5-54

$C_{14}Cl_2H_{22}MnN_2O_{14}S_2$ $[Mn(O=S(C_2H_5)(C_5H_4NO))_2(H_2O)_2][ClO_4]_2$ . . . . . Mn: MVol.D7-106

$C_{14}Cl_2H_{22}NO_2Sn$ $(C_4H_9)_2Sn(Cl)ON(O)C_6H_4Cl-4$, radical . . . . Sn: Org.Comp.17-106, 110

$C_{14}Cl_2H_{22}N_4Ti_2$ . . $[(C_5H_5)Ti(NN(CH_3)_2)Cl]_2$ . . . . Ti: Org.Comp.5-57

$C_{14}Cl_2H_{22}O_4Sn$ . . . $(C_4H_9)_2Sn(OOCCHCHCl)_2$ . . . . Sn: Org.Comp.15-269

$C_{14}Cl_2H_{23}MnO_2P$ $Mn[P(C_6H_5)(C_4H_9)_2](O_2)Cl_2$ . . . . Mn: MVol.D8-44/5, 53

$C_{14}Cl_2H_{23}MnP$ . . . $Mn[P(C_6H_5)(C_4H_9)_2]Cl_2$ . . . . Mn: MVol.D8-38

$C_{14}Cl_2H_{24}MnN_8O_6S_2$ $[Mn(4-NH_2C_6H_4SO_2NHC(NH)NH_2)_2Cl_2(H_2O)_2]$ Mn: MVol.D7-116/23

$C_{14}Cl_2H_{24}MnN_{10}S_4$ $[Mn(S=C(NH_2)_2)_4(NC_5H_4)_2Cl_2]$ . . . . Mn: MVol.D7-193

– . . . . $[Mn(S=C(NH_2)_2)_4(N_2C_{10}H_8)]Cl_2$ . . . . Mn: MVol.D7-192/3

$C_{14}Cl_2H_{26}MnN_6S_2$ $MnCl_2[c-C_6H_{10}=N-NHC(=S)NH_2]_2$ . . . . Mn: MVol.D6-342/4

$C_{14}Cl_2H_{26}MnN_{10}S_4$ $[Mn(S=C(NH_2)_2)_4(NC_5H_5)_2]Cl_2$ . . . . Mn: MVol.D7-192/3

$C_{14}Cl_2H_{26}O_4Sn$ . . . $(n-C_4H_9)_2Sn[OC(O)-CHClCH_3]_2$ . . . . Sn: Org.Comp.15-256

– . . . . $(n-C_4H_9)_2Sn[OC(O)-CH_2CH_2Cl]_2$ . . . . Sn: Org.Comp.15-256

$C_{14}Cl_2H_{27}NO_3PRe$ $(CO)_2ReCl_2[P(C_4H_9-n)_3](NO)$ . . . . Re: Org.Comp.1-89

$C_{14}Cl_2H_{28}MnN_8O_2$ $MnCl_2[(CH_3)_2C=N-NHC(O)NH-N=C(CH_3)_2]_2 \cdot H_2O$ Mn: MVol.D6-316

$C_{14}Cl_2H_{28}N_2O_2Sn$ $(C_4H_9)_2Sn(ONC(CH_3)CH_2Cl)_2$ . . . . Sn: Org.Comp.15-343

$C_{14}Cl_2H_{29}MoP_2$ . . $i-C_3H_7C_5H_4Mo(P(CH_3)_3)_2Cl_2$ . . . . Mo: Org.Comp.6-15

$C_{14}Cl_2H_{30}O_2P_2Re$ $(CO)_2ReCl_2[P(C_2H_5)_3]_2$ . . . . Re: Org.Comp.1-64

$C_{14}Cl_2H_{31}MnP$ . . . $Mn[P(C_4H_9)_3](CH_2=CH_2)Cl_2$ . . . . Mn: MVol.D8-46/7

$C_{14}Cl_2H_{32}MnP_2$ . . $Mn((i-C_3H_7)_2PCH_2CH_2P(C_3H_7-i)_2)Cl_2$ . . . . Mn: MVol.D8-78

$C_{14}Cl_2H_{32}N_2PtS_2$ $[((n-C_3H_7)_2S)PtCl_2(t-C_4H_9-N=S=N-C_4H_9-t)]$ . . . S: S-N Comp.7-316/8

$C_{14}Cl_2H_{33}N_2PPtS$ $[((C_2H_5)_3P)PtCl_2(t-C_4H_9-N=S=N-C_4H_9-t)]$ . . . . . S: S-N Comp.7-316/8, 321/2

$C_{14}Cl_2H_{33}N_2PtSSb$ $[((C_2H_5)_3Sb)PtCl_2(t-C_4H_9-N=S=N-C_4H_9-t)]$ . . . . S: S-N Comp.7-316/8

$C_{14}Cl_2H_{34}In_2N_2$ . . $[i-C_3H_7-InCl-NH-C_4H_9-t]_2$ . . . . In: Org.Comp.1-298/9

$C_{14}Cl_2H_{34}MnO_6P_2$ $[Mn(CH_3P(O)(OC_3H_7-i)_2)_2Cl_2]$ . . . . Mn: MVol.D8-142/3

$C_{14}Cl_2H_{36}N_2O_{11}Sb_2$ $[((CH_3)_3SbOC(CH_3)N(CH_3)_2)_2O][ClO_4]_2$ . . . . Sb: Org.Comp.5-101

$C_{14}Cl_3F_7LiN$ . . . . . $Li[C(C_6F_5)(C_6F_2-Cl_3-2,4,6)(CN)]$ . . . . F: PFHOrg.SVol.6-108, 128

$C_{14}Cl_3F_7N^-$ . . . . . . $[C(C_6F_5)(C_6F_2-2,4,6-Cl_3)(CN)]^-$ . . . . F: PFHOrg.SVol.6-108, 128

$C_{14}Cl_3F_7NNa$ . . . . $Na[C(C_6F_5)(C_6F_2-Cl_3-2,4,6)(CN)]$ . . . . F: PFHOrg.SVol.6-108, 128

$C_{14}Cl_3GaH_7O_6^+$ . . $[(HO)Ga(2,3-(O)_2-6,7-(HO)_2-9-CCl_3-10-OC_{13}H_4)]^+$ Ga: SVol.D1-134/6

$C_{14}Cl_3GaH_8N_2O_4$ $GaCl_3[2,5-(O=)_2-NC_4H_2-1-(1,3-C_6H_4)-1-NC_4H_2(=O)_2-2,5]$ . . . . Ga: SVol.D1-241/2

$C_{14}Cl_3GaH_{12}O$ . . . $GaCl_3[4-(CH_3-C(=O))-C_6H_4-C_6H_5]$ . . . . Ga: SVol.D1-39/42

$C_{14}Cl_3GaH_{12}O$	$GaCl_3[C_6H_5CH_2-C(=O)-C_6H_5]$	Ga: SVol.D1-39/42
$C_{14}Cl_3GaH_{12}O_2$	$GaCl_3[C_6H_5-C(=O)-C_6H_4-OCH_3-4]$	Ga: SVol.D1-39/42
$C_{14}Cl_3GaH_{20}O_2$	$GaCl_3[1,4-(O=)_2C_6H_2-2,6-(C_4H_9-t)_2]$	Ga: SVol.D1-93/4
$C_{14}Cl_3H_{10}NO_2S$	$3-(2,4,6-Cl_3-C_6H_2-N=)-2,4,3-O_2SC_8H_8$	S: S-N Comp.8-172
$C_{14}Cl_3H_{10}N_3O_5S_2$	$4-NO_2-C_6H_4-C(O)-NH-S(CCl_3)=N-S(O)_2-C_6H_5$	S: S-N Comp.8-197
$C_{14}Cl_3H_{11}N_2O_3S_2$	$C_6H_5-C(O)-N=S(CCl_3)-NH-S(O)_2-C_6H_5$	S: S-N Comp.8-197
–	$C_6H_5-C(O)-NH-S(CCl_3)=N-S(O)_2-C_6H_5$	S: S-N Comp.8-195, 196, 197
$C_{14}Cl_3H_{12}N_2NaO_4S_3$	$Na[4-CH_3-C_6H_4-S(O)_2-NS(CCl_3)N-S(O)_2-C_6H_5]$	S: S-N Comp.8-198/9
$C_{14}Cl_3H_{12}N_2O_4S_3^-$	$[4-CH_3-C_6H_4-S(O)_2-NS(CCl_3)N-S(O)_2-C_6H_5]^-$	S: S-N Comp.8-198/9
$C_{14}Cl_3H_{12}Sb$	$(C_6H_5)_2SbCl_2-CH=CHCl$	Sb: Org.Comp.5-56
–	$[-2-C_6H_4-CH_2CH_2-C_6H_4-2-]SbCl_3$	Sb: Org.Comp.5-169
$C_{14}Cl_3H_{13}NO_3Sb$	$(4-O_2NC_6H_4)(4-C_2H_5OC_6H_4)SbCl_3$	Sb: Org.Comp.5-224
$C_{14}Cl_3H_{13}NSb$	$(C_6H_5)_2SbCl_3 \cdot NCCH_3$	Sb: Org.Comp.5-151
$C_{14}Cl_3H_{13}N_2O_4S_3$	$4-CH_3-C_6H_4-S(O)_2-N=S(CCl_3)-NH-S(O)_2-C_6H_5$	S: S-N Comp.8-198
–	$4-CH_3-C_6H_4-S(O)_2-NH-S(CCl_3)=N-S(O)_2-C_6H_5$	S: S-N Comp.8-197
$C_{14}Cl_3H_{14}OSb$	$4-CH_3-O-C_6H_4-SbCl_3-C_6H_4-4-CH_3$	Sb: Org.Comp.5-224
–	$4-C_2H_5-O-C_6H_4-SbCl_3-C_6H_5$	Sb: Org.Comp.5-223
$C_{14}Cl_3H_{14}Sb$	$2,4-(CH_3)_2C_6H_3-SbCl_3-C_6H_5$	Sb: Org.Comp.5-220
–	$(4-CH_3C_6H_4)_2SbCl_3$	Sb: Org.Comp.5-164, 166
–	$(C_6H_5CH_2)_2SbCl_3$	Sb: Org.Comp.5-141, 145
$C_{14}Cl_3H_{15}NO_3Sb$	$(CH_3)_3Sb(OC_9H_6N)O_2CCCl_3$	Sb: Org.Comp.5-47
$C_{14}Cl_3H_{16}MoOP$	$C_5H_5Mo(CO)(P(CH_3)_2C_6H_5)Cl_3$	Mo: Org.Comp.6-219
$C_{14}Cl_3H_{16}OSSb$	$(C_6H_5)_2SbCl_3 \cdot OS(CH_3)_2$	Sb: Org.Comp.5-153
$C_{14}Cl_3H_{18}NO_2S$	$2-(2,4,6-Cl_3-C_6H_2-N=)-1,3,2-O_2SC_3H_4-6-CH_3-4-C_4H_9-t$	S: S-N Comp.8-167/8
$C_{14}Cl_3H_{18}N_3O_2S$	$(1,4-ONC_4H_8-4)_2S=N-C_6H_2-2,4,6-Cl_3$	S: S-N Comp.8-207/8, 212
$C_{14}Cl_3H_{20}O_3PSSn$	$(C_4H_9)_2SnO_2P(S)OC_6H_2Cl_3-2,4,5$	Sn: Org.Comp.15-359/60
$C_{14}Cl_3H_{22}N_2ORe$	$Re(CNC_6H_{11})_2(Cl_3)O$	Re: Org.Comp.2-259
$C_{14}Cl_3H_{25}InN$	$[N(C_2H_5)_4][C_6H_5-In(Cl)_3]$	In: Org.Comp.1-349/50, 353
$C_{14}Cl_3H_{30}NRe^-$	$[Cl_3Re(NH-C_4H_9-t)(CH_2-C_4H_9-t)\equiv C-C_4H_9-t]^-$	Re: Org.Comp.1-19
$C_{14}Cl_4F_2H_{18}N_2Pd$	$(3-F-5-CH_3-C_6H_3-NH_3)_2[PdCl_4]$	Pd: SVol.B2-123
–	$(4-F-2-CH_3-C_6H_3-NH_3)_2[PdCl_4]$	Pd: SVol.B2-123
–	$(4-F-3-CH_3-C_6H_3-NH_3)_2[PdCl_4]$	Pd: SVol.B2-123
–	$(5-F-2-CH_3-C_6H_3-NH_3)_2[PdCl_4]$	Pd: SVol.B2-123
$C_{14}Cl_4F_6H_{14}N_2Pd$	$(3-CF_3-C_6H_4-NH_3)_2[PdCl_4]$	Pd: SVol.B2-123
–	$(4-CF_3-C_6H_4-NH_3)_2[PdCl_4]$	Pd: SVol.B2-123
$C_{14}Cl_4F_8N_2$	$4-(CCl_2=N)-C_6F_4-C_6F_4-(N=CCl_2)-4$	F: PFHOrg.SVol.6-193/4, 208
$C_{14}Cl_4F_{20}H_3N_5O_2$	$((CF_2Cl)_2(CF_3)_2C_3HN_2O)_2NH$	F: PFHOrg.SVol.4-196, 210
$C_{14}Cl_4Fe_2H_{20}S_2$	$[Fe(C_5H_4S(CH_3)_2)_2][FeCl_4]$	Fe: Org.Comp.A9-197
$C_{14}Cl_4Ga_2H_{12}O_4$	$[Ga(2-OH-C_6H_4-CH=O)_2][GaCl_4]$	Ga: SVol.D1-36
$C_{14}Cl_4H_6N_2O_2S$	$2,3-Cl_2-C_6H_3C(O)N=S=NC(O)C_6H_3-2,3-Cl_2$	S: S-N Comp.7-273, 276
–	$2,4-Cl_2-C_6H_3C(O)N=S=NC(O)C_6H_3-2,4-Cl_2$	S: S-N Comp.7-273, 275/6
$C_{14}Cl_4H_{10}Mo$	$(C_6H_5-CC-C_6H_5)Mo(Cl)_4$	Mo: Org.Comp.5-135/8

$C_{14}Cl_4H_{10}MoN_2$ . . $MoCl_4 \cdot 2\ C_6H_5CN$ . . . . Mo: SVol.B5-323, 374
$C_{14}Cl_4H_{10}N_2O_3S_2$ 4-Cl-C_6H_4-C(O)-NH-S(CCl_3)=N-S(O)$_2$-C_6H_5 . . S: S-N Comp.8-197
– . . . . 4-Cl-C_6H_4-S(O)$_2$-N=S(CCl_3)-NH-C(O)-C_6H_5 . . S: S-N Comp.8-195, 197
$C_{14}Cl_4H_{10}N_2Re^-$ [Re(CNC_6H_5)$_2$$Cl_4$]$^-$ . . . . Re: Org.Comp.2-260, 266
$C_{14}Cl_4H_{10}O_4STi$ . . [(C_5H_5)$_2$Ti($OCOCCl_2SCCl_2COO$)]$_n$ . . . . Ti: Org.Comp.5-339
$C_{14}Cl_4H_{11}N_2NaO_4S_3$
Na[4-CH_3-C_6H_4-S(O)$_2$-NS(CCl_3)N-S(O)$_2$-C_6H_4-4-Cl]
S: S-N Comp.8-198/9
$C_{14}Cl_4H_{11}N_2O_4S_3^-$
[4-CH_3-C_6H_4-S(O)$_2$-NS(CCl_3)N-S(O)$_2$-C_6H_4-4-Cl]$^-$
S: S-N Comp.8-198/9
$C_{14}Cl_4H_{11}O_4Sb$. . . (Cl_2(4-CH_3O)C_6H_2)$_2$Sb(O)OH. . . . . Sb: Org.Comp.5-211
$C_{14}Cl_4H_{12}N_2O_4S_3$ 4-CH_3-C_6H_4-S(O)$_2$-N=S(CCl_3)-NH-S(O)$_2$-C_6H_4-4-Cl
S: S-N Comp.8-198
– . . . . 4-CH_3-C_6H_4-S(O)$_2$-NH-S(CCl_3)=N-S(O)$_2$-C_6H_4-4-Cl
S: S-N Comp.8-197
$C_{14}Cl_4H_{13}NO_3Sb^-$ [(4-$O_2NC_6H_4$)(4-$C_2H_5OC_6H_4$)$SbCl_4$]$^-$ . . . . Sb: Org.Comp.5-227
$C_{14}Cl_4H_{14}N_4Pd$. . . (-C_6H_4NHCH=NH-)$_2$[$PdCl_4$] . . . . Pd: SVol.B2-123
$C_{14}Cl_4H_{14}OSb^-$. . . [(4-CH_3-O-C_6H_4)(4-CH_3-C_6H_4)$SbCl_4$]$^-$ . . . . . . Sb: Org.Comp.5-227
– . . . . [(C_6H_5)(4-C_2H_5-O-C_6H_4)$SbCl_4$]$^-$. . . . . Sb: Org.Comp.5-226
$C_{14}Cl_4H_{14}Sb^-$ . . . . [(4-$CH_3C_6H_4$)$_2$$SbCl_4$]$^-$ . . . . Sb: Org.Comp.5-165
$C_{14}Cl_4H_{16}O_2Ti_2$ . . [(C_5H_5)$TiCl_2$]$_2$(OCH_2CH=$CHCH_2O$). . . . . Ti: Org.Comp.5-52, 53
$C_{14}Cl_4H_{20}N_2Pd$. . . (2-CH_3-C_6H_4-NH_3)$_2$[$PdCl_4$] . . . . Pd: SVol.B2-123
– . . . . (4-CH_3-C_6H_4-NH_3)$_2$[$PdCl_4$] . . . . Pd: SVol.B2-123
– . . . . [2,4-(CH_3)$_2$$NC_5H_4$]$_2$[$PdCl_4$] . . . . Pd: SVol.B2-123
– . . . . [2,5-(CH_3)$_2$$NC_5H_4$]$_2$[$PdCl_4$] . . . . Pd: SVol.B2-123
– . . . . [2,6-(CH_3)$_2$$NC_5H_4$]$_2$[$PdCl_4$] . . . . Pd: SVol.B2-123
– . . . . [3,5-(CH_3)$_2$$NC_5H_4$]$_2$[$PdCl_4$] . . . . Pd: SVol.B2-123
$C_{14}Cl_4H_{24}Mn_2N_8S_4$
[Mn(H_2NC(=S)N=C(CH_3)CH_2C(CH_3)=NC(=S)NH_2)Cl_2]$_2$
Mn: MVol.D6-219
$C_{14}Cl_4H_{25}N_2OOs$ [N(C_2H_5)$_4$][Os(CO)(NC_5H_5)(Cl)$_4$] . . . . Os: Org.Comp.A1-58, 61
$C_{14}Cl_4H_{26}O_4Ti_2$ . . [(C_5H_5)$TiCl_2$(CH_3OH)$_2$]$_2$ . . . . Ti: Org.Comp.5-39
$C_{14}Cl_5H_{10}Re$. . . . . ($C_6H_5C{\equiv}CC_6H_5$)Re(Cl)$_5$ . . . . Re: Org.Comp.2-304, 305
$C_{14}Cl_5H_{10}Re^-$ . . . . [($C_6H_5C{\equiv}CC_6H_5$)Re(Cl)$_5$]$^-$ . . . . Re: Org.Comp.2-304, 306
$C_{14}Cl_5H_{12}N_2O_2Sb$ [2-$CH_3O_2CC_6H_4N_2$][$C_6H_5SbCl_5$] . . . . Sb: Org.Comp.5-246/7
$C_{14}Cl_5H_{12}Sb$ . . . . . (2-CH_3(4-Cl)C_6H_3)$_2$$SbCl_3$ . . . . Sb: Org.Comp.5-164
$C_{14}Cl_5H_{13}N_3O_2S_2Sb$
[C_5H_5NH][SNC_3H_2-NH-S(O)$_2$-C_6H_4-$SbCl_5$] . . . Sb: Org.Comp.5-254
$C_{14}Cl_5H_{13}N_3O_3Sb$ [4-C_2H_5-O-C_6H_4-N_2][(4-O_2N-C_6H_4)$SbCl_5$] . . . Sb: Org.Comp.5-255
– . . . . [4-O_2N-C_6H_4-N_2][(4-C_2H_5-O-C_6H_4)$SbCl_5$] . . . Sb: Org.Comp.5-252
$C_{14}Cl_5H_{14}N_2OSb$ [4-C_2H_5-O-C_6H_4-N_2][C_6H_5-$SbCl_5$]. . . . . Sb: Org.Comp.5-246/7
– . . . . [C_6H_5-N_2][4-C_2H_5-O-C_6H_4-$SbCl_5$]. . . . . Sb: Org.Comp.5-252
$C_{14}Cl_5H_{14}N_2Sb$. . . [2,4-(CH_3)$_2$$C_6H_3$-$N_2$][$C_6H_5$-$SbCl_5$] . . . . Sb: Org.Comp.5-246/7
– . . . . [4-CH_3-C_6H_4-N_2][4-CH_3-C_6H_4-$SbCl_5$]. . . . . . . Sb: Org.Comp.5-256
$C_{14}Cl_5H_{15}NO_2Sb$ [C_5H_5NH][3-C_2H_5-OC(O)-C_6H_4-$SbCl_5$]. . . . . . . Sb: Org.Comp.5-257
– . . . . [C_5H_5NH][4-C_2H_5-OC(O)-C_6H_4-$SbCl_5$]. . . . . . . Sb: Org.Comp.5-257
– . . . . [C_5H_5NH][4-HO_2C-CH_2CH_2-C_6H_4-$SbCl_5$]. . . . . Sb: Org.Comp.5-257
$C_{14}Cl_5H_{15}NO_3Sb$ [C_5H_5NH][3-HO-4-C_2H_5-OC(O)-C_6H_3-$SbCl_5$] Sb: Org.Comp.5-259
– . . . . [C_5H_5NH][4-CH_3-OC(O)-CH_2-O-C_6H_4-$SbCl_5$] Sb: Org.Comp.5-252

$C_{14}Cl_5H_{15}N_3O_3SSb$
$[4\text{-}C_2H_5OC_6H_4N_2][(4\text{-}H_2NSO_2C_6H_4)SbCl_5]$ .... Sb: Org.Comp.5-253
$C_{14}Cl_5H_{15}N_3Sb$... $[4\text{-}(CH_3)_2NC_6H_4N_2][C_6H_5SbCl_5]$ .... Sb: Org.Comp.5-246/7
$C_{14}Cl_6F_{14}N_2P_2$ ... $1,3\text{-}(CF_3\text{-}4\text{-}C_6F_4)_2\text{-}1,3,2,4\text{-}N_2P_2\text{-}2,2,2,4,4,4\text{-}Cl_6$
F: PFHOrg.SVol.6-81, 88
$C_{14}Cl_6FeH_{16}Si_2$ .. $Fe[C_5H_4(CH_2)_2SiCl_3]_2$.................... Fe: Org.Comp.A9-323
$C_{14}Cl_6Fe_2H_{10}O_4Si_2$
$[(C_5H_5)Fe(CO)_3][(C_5H_5)Fe(CO)(SiCl_3)_2]$....... Fe: Org.Comp.B15-38, 47
$C_{14}Cl_6Ga_2H_8N_2O_4$ $Ga_2Cl_6[2,5\text{-}(O{=})_2\text{-}NC_4H_2\text{-}1\text{-}(1,3\text{-}C_6H_4)$
$\text{-}1\text{-}NC_4H_2({=}O)_2\text{-}2,5]$.................... Ga: SVol.D1-241/2
$C_{14}Cl_6Ga_2H_{17}N_3$.. $Ga_2Cl_6(CH_3\text{-}NC_5H_3\text{-}CH_2\text{-}N(CH_3)CH_2\text{-}NC_5H_4)$
$= [GaCl_2(CH_3\text{-}NC_5H_3\text{-}CH_2\text{-}N(CH_3)CH_2$
$\text{-}NC_5H_4)][GaCl_4]$........................ Ga: SVol.D1-271/2
$C_{14}Cl_6H_{14}NOSSb$ $[OSN(CH_2C_6H_5)_2]SbCl_6$ .... S: S-N Comp.6-289, 296
$C_{14}Cl_6H_{26}O_4Sn$... $(C_4H_9)_2Sn(OCH(CCl_3)OCH_3)_2$.............. Sn: Org.Comp.15-5
$C_{14}Cl_7H_{10}O_2Sb$... $(Cl_2(4\text{-}CH_3O)C_6H_2)_2SbCl_3$ .... Sb: Org.Comp.5-165, 167
$C_{14}Cl_8F_{16}H_3N_5O_2$ $((CF_2Cl)_4C_3HN_2O)NH(C_3HN_2O(CF_2Cl)_4)$ ...... F: PFHOrg.SVol.4-196, 210
$C_{14}Cl_8H_{42}Mn_5N_7S_7$
$Mn_5(SCH_2CH_2NH_2)_7Cl_8 \cdot 5\ H_2O$ .... Mn: MVol.D7-28/9
$C_{14}Cl_{12}F_{12}H_3N_5O_2$
$((CF_2Cl)_2(CFCl_2)_2C_3HN_2O)_2NH$ .... F: PFHOrg.SVol.4-196, 210
$C_{14}Cl_{12}H_{15}Mo_6P$.. $[(Mo_6Cl_8)Cl_3(C_2H_5P(C_6H_5)_2)]Cl$ .... Mo: SVol.B5-267
$C_{14}Cl_{12}H_{18}O_{11}S_3Th$
$Th(CCl_3COO)_4 \cdot 3\ (CH_3)_2SO$.............. Th: SVol.C7-59
$C_{14}Cl_{26}H_{42}MnN_7O_7P_7Sb_2$
$[Mn(O{=}PCl_2N(CH_3)_2)_6][SbCl_6]_2 \cdot O{=}PCl_2N(CH_3)_2$
Mn: MVol.D8-187
$C_{14}CoFeH_8Hg_2N_4S_4$
$Fe(C_5H_4\text{-}HgSCN)_2Co(NCS)_2$ .... Fe: Org.Comp.A10-136/7
$C_{14}CoH_{11}N_2O_4Si$ $[H_3SiC_5H_4N(2\text{-}C_5H_4N)\text{-}2][Co(CO)_4]$
$= SiH_3Co(CO)_4 \cdot C_5H_4N(2\text{-}C_5H_4N)\text{-}2$...... Si: SVol.B4-330/1
$C_{14}CoH_{13}N_2O_4Si$ $[SiH_3(NC_5H_5)_2][Co(CO)_4] = SiH_3Co(CO)_4 \cdot 2\ NC_5H_5$
Si: SVol.B4-328/9
$C_{14}CoH_{16}N_2O_6Re$ $(i\text{-}C_3H_7)_2C_2CoH_2N_2Re(CO)_6$ .... Re: Org.Comp.1-337
$C_{14}Co_2GeH_{14}O_6$.. $(CH_3)_2Ge(\text{-}CH{=}C(CH_3)C(CH_3){=}CH\text{-})(CO)Co(CO)_2Co(CO)_3$
Ge: Org.Comp.3-280, 286
$C_{14}Co_2GeH_{18}O_6Si$ $Ge(CH_3)_3C{\equiv}CSi(CH_3)_3 \cdot Co_2(CO)_6$ .... Ge: Org.Comp.2-55
$C_{14}CrFeH_8O_5S$ ... $(C_5H_5)Fe(CO)_2\text{-}2\text{-}SC_4H_3[Cr(CO)_3]$........... Fe: Org.Comp.B14-6, 13/4
$C_{14}CrFeH_8O_7S_2$ .. $C_5H_5(CO)_2Fe{=}C(SCH_3)SCr(CO)_5$ .... Fe: Org.Comp.B16a-180/1
$C_{14}CrFeH_{13}N_6O_5S_4$
$[(1\text{-}HOOCC_6H_6)Fe(CO)_3][Cr(SCN)_4(NH_3)_2]$ .... Fe: Org.Comp.B15-101, 136
– .... $[(2\text{-}HOOCC_6H_6)Fe(CO)_3][Cr(SCN)_4(NH_3)_2]$ .... Fe: Org.Comp.B15-104, 136
– .... $[(3\text{-}HOOCC_6H_6)Fe(CO)_3][Cr(SCN)_4(NH_3)_2]$ .... Fe: Org.Comp.B15-106, 136
$C_{14}CrFeH_{15}N_6O_3S_4$
$[(c\text{-}C_7H_9)Fe(CO)_3][Cr(SCN)_4(NH_3)_2]$ .... Fe: Org.Comp.B15-191/2, 194
$C_{14}CrFe_3HO_{13}^-$... $[CrFe_3(CO)_{13}(CH)]^-$ .... Fe: Org.Comp.C6b-71
$C_{14}CrFe_3HO_{14}^-$... $[Fe_3CrH(CO)_{14}]^-$........................ Fe: Org.Comp.C6b-146, 147
$C_{14}CrFe_3HO_{14}S^-$ $[(CO)_9(H)Fe_3SCr(CO)_5]^-$.................. Fe: Org.Comp.C6a-135
$C_{14}CrFe_3H_2O_{13}$ .. $CrFe_3(CO)_{13}(H)(CH)$ .... Fe: Org.Comp.C6b-71
$C_{14}CrFe_3H_2O_{14}S$ $(CO)_9(H)_2Fe_3SCr(CO)_5$.................... Fe: Org.Comp.C6a-128

$C_{14}CrFe_3O_{13}^{2-}$ ... $[CCrFe_3(CO)_{13}]^{2-}$ .... Fe: Org.Comp.C6b-72
$C_{14}CrFe_3O_{14}^{2-}$ ... $[Fe_3Cr(CO)_{14}]^{2-}$ .... Fe: Org.Comp.C6b-146/9
$C_{14}CrFe_3O_{14}S^{2-}$ $[(CO)_9Fe_3SCr(CO)_5]^{2-}$ .... Fe: Org.Comp.C6a-135
$C_{14}CrH_{19}O_8P_2ReS$ fac-$(CO)_3Re[P(CH_3)_3]_2SH\text{-}Cr(CO)_5$ .... Re: Org.Comp.1-318
– .... mer-trans-$(CO)_3Re[P(CH_3)_3]_2SH\text{-}Cr(CO)_5$ .... Re: Org.Comp.1-318
$C_{14}Cr_2H_{12}N_2O_{10}P_2S$
$[Cr(CO)_5(P(CH_3)_2N{=}S{=}NP(CH_3)_2)Cr(CO)_5]$ .... S: S-N Comp.7-127/8
$C_{14}Cs_3GaH_{10}O_{14}S_2$
$Cs_3[(H_2O)_2Ga(OC(O)\text{-}C_6H_3\text{-}5\text{-}SO_3\text{-}2\text{-}O)_2] \cdot 2\,H_2O$
Ga: SVol.D1-176, 181/2
$C_{14}CuFeH_8Hg_2N_4S_4$
$Fe(C_5H_4\text{-}HgSCN)_2Cu(NCS)_2$ .... Fe: Org.Comp.A10-136/8
$C_{14}FFeH_9NO_2^+$ ... $[(C_5H_5)Fe(CN\text{-}C_6H_4F\text{-}4)(CO)_2]^+$ .... Fe: Org.Comp.B15-315, 318
$C_{14}FFeH_9O_3$ ..... $(C_5H_5)Fe(CO)_2COC_6H_4F\text{-}4$ .... Fe: Org.Comp.B13-10, 23, 42
$C_{14}FFeH_{10}O_2^+$ ... $[C_5H_5(CO)_2Fe{=}CH(C_6H_4F\text{-}4)]^+$ .... Fe: Org.Comp.B16a-84, 90, 97/9
$C_{14}FFeH_{11}O_5S$ ... $[C_5H_5(CO)_2Fe{=}CH(C_6H_5)][FSO_3]$ .... Fe: Org.Comp.B16a-90, 97/9
$C_{14}FFeH_{20}O_6PS$ .. $[(C_5H_5)Fe(CO)_2C(O)C(CH_3)_2P(CH_3)_3][FSO_3]$ ... Fe: Org.Comp.B14-154
– .... $[(C_5H_5)Fe(CO)_2C(OCH_3){=}C(CH_3)P(CH_3)_3][FSO_3]$
Fe: Org.Comp.B14-153
$C_{14}FFeH_{24}O_5PS$ .. $[C_5H_5(CO)(P(CH_3)_3)Fe{=}C(C_3H_7\text{-}i)OCH_3][FSO_3]$ Fe: Org.Comp.B16a-41, 55
$C_{14}FFeH_{29}O_4P_2S$ $[C_5H_5(P(CH_3)_3)_2Fe{=}C(CH_3)OCH_3][FSO_3]$ ...... Fe: Org.Comp.B16a-8/9
$C_{14}FGeH_{17}$ ...... 1-$[Ge(CH_3)_3CH_2]\text{-}C_{10}H_6\text{-}F\text{-}4$ .... Ge: Org.Comp.1-158, 163
– .... 2-$[Ge(CH_3)_3CH_2]\text{-}C_{10}H_6\text{-}F\text{-}6$ .... Ge: Org.Comp.1-158, 163
– .... 2-$[Ge(CH_3)_3CH_2]\text{-}C_{10}H_6\text{-}F\text{-}7$ .... Ge: Org.Comp.1-158, 163
$C_{14}FGeH_{19}$ ...... $Ge(C_2H_5)_3C{\equiv}CC_6H_4F\text{-}4$ .... Ge: Org.Comp.2-255, 257
$C_{14}FH_{11}MoN_2O_2$ .. $(CH_3\text{-}C_5H_4)Mo(CO)_2\text{-}N{=}N\text{-}C_6H_4\text{-}4\text{-}F$ .... Mo: Org.Comp.7-6, 17
$C_{14}FH_{14}In$ .... $(C_6H_5\text{-}CH_2)_2InF$ .... In: Org.Comp.1-116/7
– .... $(C_6H_5\text{-}CH_2)_2InF \cdot n\ NC\text{-}CD_3$ .... In: Org.Comp.1-117
$C_{14}FH_{18}N_3O_{10}Th^-$ $ThF[OOCCH_2N(C_2H_4N(CH_2COO)_2)_2]^-$ .... Th: SVol.D1-141
$C_{14}F_2H_{14}O_2Po$ ... $(4\text{-}CH_3OC_6H_4)_2PoF_2$ .... Po: SVol.1-334/40
$C_{14}F_2H_{14}Po$ ..... $(4\text{-}CH_3C_6H_4)_2PoF_2$ .... Po: SVol.1-334/40
$C_{14}F_3FeH_6NO_2$ ... $(C_5H_5)Fe(CO)_2C_6H(CN\text{-}5)F_3\text{-}3,4,6$ .... Fe: Org.Comp.B13-176, 186
$C_{14}F_3FeH_9O_3$ .... $(C_5H_5)Fe(CO)_2C_6H(OCH_3\text{-}5)F_3\text{-}3,4,6$ .... Fe: Org.Comp.B13-165, 176, 186/7
$C_{14}F_3FeH_{14}O^+$ ... $[(4\text{-}CH_3\text{-}C_6H_4\text{-}O\text{-}CH_2\text{-}CF_3)Fe(C_5H_5)][PF_6]$.... Fe: Org.Comp.B19-2/3, 48
$C_{14}F_3FeH_{15}O_3S$ .. $[(1,4\text{-}(CH_3)_2\text{-}C_6H_4)Fe(C_5H_5)][O_3S\text{-}CF_3]$ ...... Fe: Org.Comp.B19-1, 5, 12, 81/7
$C_{14}F_3FeH_{17}O_5S$ .. $[C_5(CH_3)_5(CO)_2Fe{=}CH_2][CF_3SO_3]$ .... Fe: Org.Comp.B16a-103/4
$C_{14}F_3FeH_{17}O_6S$ .. $[C_5(CH_3)_5(CO)_2Fe{=}CH(OH)][CF_3SO_3]$ .... Fe: Org.Comp.B16a-128
$C_{14}F_3FeH_{20}O_7PS$ $[C_5H_5(CO)(P(OCH_3)_3)Fe{=}C{=}C(CH_3)_2][CF_3SO_3]$ Fe: Org.Comp.B16a-208, 211
$C_{14}F_3FeH_{22}O_7PS$ $[C_5H_5(CO)(P(OCH_3)_3)Fe{=}CH(CH{=}CHC_2H_5)][F_3SO_3]$
Fe: Org.Comp.B16a-19, 26
$C_{14}F_3GeH_{33}O_2Si_3$ $Ge(CH_3)_3C[Si(CH_3)_3]_2Si(CH_3)_2OOCCF_3$....... Ge: Org.Comp.1-145
$C_{14}F_3H_6O_6Re$ .... $(CO)_4Re[\text{-}O{=}C(C_6H_5)\text{-}CH{=}C(CF_3)\text{-}O\text{-}]$ .... Re: Org.Comp.1-356
$C_{14}F_3H_8N_2O_6ReS$ $(CO)_3Re[NC_5H_4\text{-}2\text{-}(2\text{-}C_5H_4N)]OS(O)_2CF_3$..... Re: Org.Comp.1-156, 176, 177
$C_{14}F_3H_9N_2O_2Re^+$ $[(C_5H_5)Re(CO)_2\text{-}N{=}N\text{-}C_6H_4\text{-}2\text{-}CF_3][BF_4]$ ..... Re: Org.Comp.3-193/7, 203, 208
$C_{14}F_3H_9N_2O_3S_2$ .. $C_6H_5C(O)N{=}S{=}N\text{-}C_6H_4\text{-}4\text{-}S(O)_2CF_3$ .... S: S-N Comp.7-268/9
$C_{14}F_3H_{11}MoO_4$ ... $(C_5H_5)Mo(CO)_2[(CH_3)_2(CF_3)OC_4({=}O)]$ .... Mo: Org.Comp.8-256

$C_{14}F_3H_{11}N_3O_2Re$ $(C_5H_5)Re(CO)(N{=}N{-}C_6H_4{-}CF_3{-}2){-}C({=}O)NH_2$ .. Re: Org.Comp.3-133/5
$C_{14}F_3H_{12}NS$ ..... $F_3S{-}6{-}NC_{14}H_{12}$ .... S: S-N Comp.8-389/90
$C_{14}F_3H_{14}Sb$ ..... $(C_6H_4CH_3{-}4)_2SbF_3 \cdot H_2O$ .... Sb: Org.Comp.5-137
$C_{14}F_3H_{16}OSSb$ ... $(C_6H_5)_2SbF_3 \cdot OS(CH_3)_2$ .... Sb: Org.Comp.5-136
$C_{14}F_3H_{23}MoO_5P_2S$
$[(C_5H_5)Mo(CO)_2(P(CH_3)_3)_2][SO_3CF_3]$ .... Mo: Org.Comp.7-286
$C_{14}F_4FeH_5NO_2$ ... $(C_5H_5)Fe(CO)_2C_6F_4CN{-}4$ .... Fe: Org.Comp.B13-174, 185
$C_{14}F_4FeH_6O_4$ .... $(C_5H_5)Fe(CO)_2C_6F_4COOH{-}4$ .... Fe: Org.Comp.B13-175, 185/6
$C_{14}F_4FeH_8O_2$ .... $(C_5H_5)Fe(CO)_2C_6F_4CH_3{-}4$ .... Fe: Org.Comp.B13-165, 173
$C_{14}F_4FeH_8O_2S$ ... $(C_5H_5)Fe(CO)_2C_6F_4{-}4{-}SCH_3$ .... Fe: Org.Comp.B13-175, 185
$C_{14}F_4FeH_8O_3$ .... $(C_5H_5)Fe(CO)_2C_6F_4{-}3{-}OCH_3$ .... Fe: Org.Comp.B13-165, 174, 185
– .... $(C_5H_5)Fe(CO)_2C_6F_4{-}4{-}OCH_3$ .... Fe: Org.Comp.B13-165, 175, 185
$C_{14}F_4H_{14}Sb^-$ ..... $[(C_6H_4CH_3{-}4)_2SbF_4]^-$ .... Sb: Org.Comp.5-137
$C_{14}F_4H_{21}N_3Sn$ ... $(CH_3)_3SnN{=}C(C_6F_4N(CH_3)_2{-}4)N(CH_3)_2$ .... Sn: Org.Comp.18-108, 116
$C_{14}F_5FeH_5O_3$ .... $(C_5H_5)Fe(CO)_2COC_6F_5$ .... Fe: Org.Comp.B13-23, 43
$C_{14}F_5FeH_7O_2$ .... $(C_5H_5)Fe(CO)_2{-}2{-}C_7H_2{-}1,3,4,7,7{-}F_5$ .... Fe: Org.Comp.B13-226
$C_{14}F_5FeH_8NO$ .... $C_5H_5Fe(CNCH_3)(CO)C_6F_5$ .... Fe: Org.Comp.B15-305, 308
$C_{14}F_5GeH_{15}$ ..... $Ge(C_2H_5)_3C{\equiv}CC_6F_5$ .... Ge: Org.Comp.2-255, 258
$C_{14}F_5H_7N_2OS$ .... $4{-}CH_3C(O){-}C_6H_4{-}N{=}S{=}N{-}C_6F_5$ .... S: S-N Comp.7-263
$C_{14}F_6FeGeH_{14}O_2$ $(C_5H_5)Fe(CO)_2C(CF_3){=}C(CF_3)Ge(CH_3)_3$ .... Fe: Org.Comp.B13-94; Ge: Org.Comp.2-11
$C_{14}F_6FeH_8O_7PReS_2$
$[(C_5H_5)(CO)_2Fe{=}C(SCH_3){-}S{-}Re(CO)_5][PF_6]$ ... Fe: Org.Comp.B16a-142, 154; Re: Org.Comp.2-177
$C_{14}F_6FeH_{10}$ ...... $(CF_3{-}C_6H_5)_2Fe$ .... Fe: Org.Comp.B19-348, 358
$C_{14}F_6FeH_{10}NOPS$ $[(C_5H_5)Fe(CN{-}C_6H_5)(CS)CO][PF_6]$ .... Fe: Org.Comp.B15-311
$C_{14}F_6FeH_{10}NO_2P$ $[(C_5H_5)Fe(CN{-}C_6H_5)(CO)_2][PF_6]$ .... Fe: Org.Comp.B15-314/5, 318
$C_{14}F_6FeH_{10}O_3$ ... $(C_5H_5)Fe(CO)_2C[{=}C(CH_3){-}C(CF_3)_2{-}O{-}CH_2{-}]$ .. Fe: Org.Comp.B14-2, 5
$C_{14}F_6FeH_{11}N_2O_2P$ $[C_5H_5(CO)_2Fe{=}C(NH)_2C_6H_4][PF_6]$ .... Fe: Org.Comp.B16a-147
$C_{14}F_6FeH_{11}O_2P$ .. $[(C_5H_5)(CO)_2Fe{=}CH{-}C_6H_5][PF_6]$ .... Fe: Org.Comp.B16a-90, 97/9
– .... $[(C_5H_5)(CO)_2Fe{=}C_7H_6{-}c][PF_6]$ .... Fe: Org.Comp.B16a-93, 101/2
$C_{14}F_6FeH_{11}O_2PS$ $[(C_5H_5)(CO)_2Fe{=}CH{-}S{-}C_6H_5][PF_6]$ .... Fe: Org.Comp.B16a-113, 123/4
$C_{14}F_6FeH_{12}O$ .... $(C_5H_5)Fe(CO)[C(CF_3){=}C(CF_3)CH_2C(CH_3){=}CH_2]$ Fe: Org.Comp.B17-171
– .... $(C_5H_5)Fe(CO)[C(CF_3){=}C(CF_3)CH_2CH{=}CHCH_3]$ .. Fe: Org.Comp.B17-170
$C_{14}F_6FeH_{12}O_3$ ... $(C_5H_5)Fe(CO)_2C(CH_3)[{-}CH_2{-}C(CF_3)_2{-}O{-}CH_2{-}]$ Fe: Org.Comp.B14-2, 4, 11
– .... $(C_5H_5)Fe(CO)_2CH[{-}CH(CH_3){-}C(CF_3)_2{-}O{-}CH_2{-}]$ Fe: Org.Comp.B14-2, 4
$C_{14}F_6FeH_{13}O_3P$ .. $[(8{-}C_2H_5CH{=}C_8H_7)Fe(CO)_3][PF_6]$ .... Fe: Org.Comp.B15-241/2
$C_{14}F_6FeH_{13}P$ .... $[(C_9H_8)Fe(C_5H_5)][PF_6]$ .... Fe: Org.Comp.B19-223
$C_{14}F_6FeH_{14}NO_2P$ $[(C_5H_5)Fe(CO)_2CH_2CH_2{-}NC_5H_5][PF_6]$ .... Fe: Org.Comp.B14-150
$C_{14}F_6FeH_{14}NP$ ... $[(2,6{-}(CH_3)_2{-}C_6H_3{-}CN)Fe(C_5H_5)][PF_6]$ .... Fe: Org.Comp.B19-124
– .... $[(2,6{-}(CH_3)_2{-}C_6H_3{-}NC)Fe(C_5H_5)][PF_6]$ .... Fe: Org.Comp.B19-126
$C_{14}F_6FeH_{14}O_2Si$ .. $(C_5H_5)Fe(CO)_2C(CF_3){=}C(CF_3)Si(CH_3)_3$ .... Fe: Org.Comp.B13-94
$C_{14}F_6FeH_{14}O_2Sn$ $(C_5H_5)Fe(CO)_2C(CF_3){=}C(CF_3)Sn(CH_3)_3$ .... Fe: Org.Comp.B13-94, 122
$C_{14}F_6FeH_{15}N_2P$ .. $[(3{-}CH_3{-}1,2{-}N_2C_8H_7)Fe(C_5H_5)][PF_6]$ .... Fe: Org.Comp.B19-289
– .... $[(NC{-}CH_2CH_2{-}NH{-}C_6H_5)Fe(C_5H_5)][PF_6]$ .... Fe: Org.Comp.B18-142/6, 199, 258

$C_{14}F_6FeH_{15}OP$ ... $[(CH_3C(=O)CH_2-C_6H_5)Fe(C_5H_5)][PF_6]$ .... Fe: Org.Comp.B18-142/6, 198, 214/5
– ... $[(C_6H_6)Fe(C_5H_4-C(=O)C_2H_5)][PF_6]$ .... Fe: Org.Comp.B18-142/6, 151, 163
– ... $[(C_9H_9-5-OH)Fe(C_5H_5)][PF_6]$ .... Fe: Org.Comp.B19-217, 222
$C_{14}F_6FeH_{15}O_2P$ .. $[(4-CH_3-C_6H_4-COO-CH_3)Fe(C_5H_5)][PF_6]$ ..... Fe: Org.Comp.B19-23/4
– ... $[(CH_3-OC(O)-CH_2-C_6H_5)Fe(C_5H_5)][PF_6]$ ...... Fe: Org.Comp.B18-142/6, 199, 215
– ... $[(C_2H_5-OC(O)-C_6H_5)Fe(C_5H_5)][PF_6]$ .... Fe: Org.Comp.B18-142/6, 215
– ... $[(C_6H_6)Fe(C_5H_4-COO-C_2H_5)][PF_6]$ .... Fe: Org.Comp.B18-142/6, 151, 164
$C_{14}F_6FeH_{15}O_3P$ .. $[(3-CH_3O-C_6H_4-COO-CH_3)Fe(C_5H_5)][PF_6]$.... Fe: Org.Comp.B19-50
– ... $[(4-CH_3O-C_6H_4-COO-CH_3)Fe(C_5H_5)][PF_6]$.... Fe: Org.Comp.B19-51
– ... $[(4-C_2H_5-O-C_6H_4-COOH)Fe(C_5H_5)][PF_6]$ ..... Fe: Org.Comp.B19-50
– ... $[(6-C_2H_5C(CH_3)=C_7H_7)Fe(CO)_3][PF_6]$ .... Fe: Org.Comp.B15-211, 226
$C_{14}F_6FeH_{15}O_4P$ .. $[(1-CH_3C(O)CH_2CH(CH_3)-C_6H_6)Fe(CO)_3][PF_6]$ Fe: Org.Comp.B15-100
– ... $[(8-(CH_3)_2C(OH)-C_8H_8)Fe(CO)_3][PF_6]$ .... Fe: Org.Comp.B15-250
– ... $[(CH_3O-C_{10}H_{12})Fe(CO)_3][PF_6]$ .... Fe: Org.Comp.B15-127, 179
$C_{14}F_6FeH_{15}O_5P$ .. $[(1-CH_3C(O)CH_2CH_2-4-CH_3O-C_6H_5)Fe(CO)_3][PF_6]$ Fe: Org.Comp.B15-115, 171
– ... $[(6-CH_3C(O)OCH(CH_3)CH_2-C_6H_6)Fe(CO)_3][PF_6]$ Fe: Org.Comp.B15-108, 159
$C_{14}F_6FeH_{15}O_6P$ .. $[(1-CH_3C(O)OCH_2CH_2-4-CH_3O-C_6H_5)Fe(CO)_3][PF_6]$ Fe: Org.Comp.B15-118, 172
– ... $[(1-CH_3OC(O)CH_2CH_2-4-CH_3O-C_6H_5)Fe(CO)_3][PF_6]$ Fe: Org.Comp.B15-116, 171
$C_{14}F_6FeH_{15}P$ .... $[(C_6H_5-CH_2CH_2CH_2-C_5H_4)Fe][PF_6]$.......... Fe: Org.Comp.B19-410, 412/4
– ... $[(C_9H_{10})Fe(C_5H_5)][PF_6]$ .... Fe: Org.Comp.B19-216, 220/1, 298
$C_{14}F_6FeH_{16}NOP$ .. $[(2-CH_3-C_6H_4-NH-CO-CH_3)Fe(C_5H_5)][PF_6]$ ... Fe: Org.Comp.B19-4, 59
– ... $[(3-CH_3-C_6H_4-NH-CO-CH_3)Fe(C_5H_5)][PF_6]$ ... Fe: Org.Comp.B19-4, 59
– ... $[(4-CH_3-C_6H_4-NH-CO-CH_3)Fe(C_5H_5)][PF_6]$ ... Fe: Org.Comp.B19-1, 4, 59
– ... $[(C_2H_5-C(=O)NH-C_6H_5)Fe(C_5H_5)][PF_6]$ .... Fe: Org.Comp.B18-142/6, 199, 261
$C_{14}F_6FeH_{16}NOPS_2$
$[C_5H_5(CO)(C_5H_5N)Fe=C(SCH_3)_2][PF_6]$........ Fe: Org.Comp.B16a-66
$C_{14}F_6FeH_{16}NO_2P$ $[(CH_3OC(O)-C_5H_4)Fe(C_5H_4-C(CH_3)=NH_2)][PF_6]$ Fe: Org.Comp.A9-103
– ... $[(C_5H_5)Fe(CN-C_6H_{11}-c)(CO)_2][PF_6]$.......... Fe: Org.Comp.B15-314
$C_{14}F_6FeH_{16}NP$ ... $[(1-NC_9H_{11})Fe(C_5H_5)][PF_6]$ .... Fe: Org.Comp.B19-216, 279
$C_{14}F_6FeH_{16}NPS_2$ $[((CH_3S)_2C=N-C_6H_5)Fe(C_5H_5)][PF_6]$ .... Fe: Org.Comp.B18-142/6, 199/200, 264
$C_{14}F_6FeH_{16}N_3P$ .. $[(C_5H_5)Fe(CNCH_3)_2-NC_5H_5][PF_6]$ .... Fe: Org.Comp.B15-327, 328
$C_{14}F_6FeH_{17}OP$ ... $[(2,6-(CH_3)_2-C_6H_3-OCH_3)Fe(C_5H_5)][PF_6]$ ..... Fe: Org.Comp.B19-99, 126
– ... $[(2-CH_3-C_6H_4-O-C_2H_5)Fe(C_5H_5)][PF_6]$....... Fe: Org.Comp.B19-2/3, 47/8
– ... $[(3-CH_3-C_6H_4-O-C_2H_5)Fe(C_5H_5)][PF_6]$....... Fe: Org.Comp.B19-2/3, 48
– ... $[(4-CH_3-C_6H_4-O-C_2H_5)Fe(C_5H_5)][PF_6]$....... Fe: Org.Comp.B19-2/3, 48
$C_{14}F_6FeH_{17}O_2P$ .. $[(1,3-(CH_3O)_2-C_6H_3-2-CH_3)Fe(C_5H_5)][PF_6]$ ... Fe: Org.Comp.B19-129
– ... $[(C_5H_5)Fe(CO)_2(CH_2=CHC(CH_3)_2CH=CH_2)][PF_6]$ Fe: Org.Comp.B17-39

$C_{14}F_6FeH_{17}O_3P$ .. $[(C_5H_5)Fe(CO)_2(H_3COCH{=}CHCH_2CH_2CH{=}CH_2)][PF_6]$ Fe: Org.Comp.B17-66, 78/9

$C_{14}F_6FeH_{17}O_4P$ .. $[(1,2\text{-}(CH_3)_2\text{-}4\text{-}(i\text{-}C_3H_7\text{-}O)\text{-}C_6H_4)Fe(CO)_3][PF_6]$ Fe: Org.Comp.B15-126, 179

– .... $[(1\text{-}C_2H_5\text{-}4\text{-}(i\text{-}C_3H_7\text{-}O)\text{-}C_6H_5)Fe(CO)_3][PF_6]$ .. Fe: Org.Comp.B15-121, 173/4

$C_{14}F_6FeH_{17}O_5P$ .. $[(1\text{-}CH_3OCH(CH_3)CH_2\text{-}4\text{-}CH_3O\text{-}C_6H_5)Fe(CO)_3][PF_6]$ Fe: Org.Comp.B15-115, 170

– .... $[(C_5H_5)Fe(CO)_2(CH_2{=}C(O\text{-}C_2H_5)C(O)O\text{-}C_2H_5)][PF_6]$ Fe: Org.Comp.B17-68, 80

$C_{14}F_6FeH_{17}P$ .... $[(1,2,3\text{-}(CH_3)_3\text{-}C_6H_3)Fe(C_5H_5)][PF_6]$ .... Fe: Org.Comp.B19-99/100, 120/1

– .... $[(1,2,4\text{-}(CH_3)_3\text{-}C_6H_3)Fe(C_5H_5)][PF_6]$ .... Fe: Org.Comp.B19-99, 1OO, 121

– .... $[(1,3,5\text{-}(CH_3)_3\text{-}C_6H_3)Fe(C_5H_5)][PF_6]$ .... Fe: Org.Comp.B19-99/100, 103, 133/6

– .... $[(1,3,5\text{-}(CD_3)_3\text{-}C_6H_3)Fe(C_5H_5)][PF_6]$ .... Fe: Org.Comp.B19-133

– .... $[(1,4\text{-}(CH_3)_2\text{-}C_6H_4)Fe(C_5H_4\text{-}CH_3)][PF_6]$ ...... Fe: Org.Comp.B19-2, 13

– .... $[(2\text{-}CH_3\text{-}C_6H_4\text{-}C_2H_5)Fe(C_5H_5)][PF_6]$ .... Fe: Org.Comp.B19-24

– .... $[(4\text{-}CH_3\text{-}C_6H_4\text{-}C_2H_5)Fe(C_5H_5)][PF_6]$ .... Fe: Org.Comp.B19-1, 24

– .... $[(n\text{-}C_3H_7\text{-}C_6H_5)Fe(C_5H_5)][PF_6]$ .... Fe: Org.Comp.B18-142/6, 197, 215

– .... $[(i\text{-}C_3H_7\text{-}C_6H_5)Fe(C_5H_5)][PF_6]$ .... Fe: Org.Comp.B18-142/6, 197, 199, 216, 273/4

– .... $[(C_6H_6)Fe(C_5H_4\text{-}C_3H_7\text{-}i)][PF_6]$ .... Fe: Org.Comp.B18-142/6, 151, 164, 187

– .... $[(C_6H_6)Fe(C_5H_4\text{-}C_3H_7\text{-}n)][PF_6]$ .... Fe: Org.Comp.B18-142/6, 151, 164, 187

– .... $[(C_6H_6)Fe(C_6H_6\text{-}C_2H_5)][PF_6]$ .... Fe: Org.Comp.B18-142/6, 153, 172

$C_{14}F_6FeH_{17}Sb$ ... $[(1,3,5\text{-}(CH_3)_3\text{-}C_6H_3)Fe(C_5H_5)][SbF_6]$ .... Fe: Org.Comp.B19-103, 133/6

– .... $[(i\text{-}C_3H_7\text{-}C_6H_5)Fe(C_5H_5)][SbF_6]$ .... Fe: Org.Comp.B18-142/6, 197, 216, 273/4

$C_{14}F_6FeH_{18}NOP$ .. $[(3\text{-}CH_3O\text{-}C_6H_4\text{-}N(CH_3)_2)Fe(C_5H_5)][PF_6]$ ..... Fe: Org.Comp.B19-77

– .... $[(4\text{-}CH_3O\text{-}C_6H_4\text{-}N(CH_3)_2)Fe(C_5H_5)][PF_6]$ ..... Fe: Org.Comp.B19-77

– .... $[(CH_3CH(OH)CH_2\text{-}NH\text{-}C_6H_5)Fe(C_5H_5)][PF_6]$ ... Fe: Org.Comp.B18-142/6, 198, 199/200, 258/9

$C_{14}F_6FeH_{18}NO_2PS$
$[C_5H_5(CO)_2Fe{=}C(SCH_3)NC_5H_{10}\text{-}c][PF_6]$ ...... Fe: Org.Comp.B16a-146, 159

$C_{14}F_6FeH_{18}NP$ ... $[(2\text{-}CH_3\text{-}C_6H_4\text{-}N(CH_3)_2)Fe(C_5H_5)][PF_6]$....... Fe: Org.Comp.B19-4, 60

– .... $[(3\text{-}CH_3\text{-}C_6H_4\text{-}N(CH_3)_2)Fe(C_5H_5)][PF_6]$....... Fe: Org.Comp.B19-4, 60

– .... $[(4\text{-}CH_3\text{-}C_6H_4\text{-}N(CH_3)_2)Fe(C_5H_5)][PF_6]$....... Fe: Org.Comp.B19-4, 61

– .... $[(NH_2\text{-}C_6H_4\text{-}4\text{-}C_3H_7\text{-}i)Fe(C_5H_5)][PF_6]$........ Fe: Org.Comp.B19-64

$C_{14}F_6FeH_{18}O_3P_2$. $[(C_6H_6)Fe(C_6H_6\text{-}P({=}O)(OCH_3)_2)][PF_6]$ .... Fe: Org.Comp.B18-142/6, 174

$C_{14}F_6FeH_{18}O_5P_2S$ $[(C_5H_5)Fe(CO)_2CH(SCH_3)P(\text{-}OCH_2\text{-})_3CCH_3][PF_6]$ Fe: Org.Comp.B14-133, 140/1, 144, 145

$C_{14}F_6FeH_{19}N_2P$ .. $[(2\text{-}CH_3\text{-}C_6H_4\text{-}NH\text{-}CH_2CH_2\text{-}NH_2)Fe(C_5H_5)][PF_6]$ Fe: Org.Comp.B19-2, 58

– .... $[(3\text{-}CH_3\text{-}C_6H_4\text{-}NH\text{-}CH_2CH_2\text{-}NH_2)Fe(C_5H_5)][PF_6]$ Fe: Org.Comp.B19-2, 58

$C_{14}F_6FeH_{19}N_2P$ .. $[(4-CH_3-C_6H_4-NH-CH_2CH_2-NH_2)Fe(C_5H_5)][PF_6]$ Fe: Org.Comp.B19-2, 58

$C_{14}F_6FeH_{19}OP$ ... $[(1,3,5-(CH_3)_3-C_6H_3)Fe(CO)(CH_2C(CH_3)CH_2)][PF_6]$ Fe: Org.Comp.B18-95/7

$C_{14}F_6FeH_{19}O_2P$ .. $[((CH_3)_5C_5)Fe(CO)_2(CH_2{=}CH_2)][PF_6]$ .... Fe: Org.Comp.B17-131

$C_{14}F_6FeH_{19}O_3P$ .. $[(CH_2(CH)_4-C_6H_{13}-n)Fe(CO)_3][PF_6]$ .... Fe: Org.Comp.B15-27

– .... $[(C_5(CH_3)_5)(CO)_2Fe{=}CH-OCH_3][PF_6]$ .... Fe: Org.Comp.B16a-129

$C_{14}F_6FeH_{19}O_3PSi$ $[(3-(CH_3)_3Si-6,6-(CH_3)_2-C_6H_4)Fe(CO)_3][PF_6]$ .. Fe: Org.Comp.B15-132, 183

$C_{14}F_6FeH_{19}PS$ ... $[CH_3C_5H_4FeC_4S((CH_3)_4-2,3,4,5)][PF_6]$ .... Fe: Org.Comp.B17-221

$C_{14}F_6FeH_{20}O_4P_2$. $[(C_5H_5)Fe(CH_2{=}CHCH_3)(CO)P(OCH_2)_3CCH_3][PF_6]$ Fe: Org.Comp.B16b-83, 90

$C_{14}F_6FeH_{21}O_2PSi$ $[((CH_3)_3SiC_5H_4)Fe(CO)_2(CH_2{=}C(CH_3)_2)][PF_6]$ .. Fe: Org.Comp.B17-132

$C_{14}F_6FeH_{22}N_3OP$ $[C_5H_5(i-C_3H_7NC)(CO)Fe{=}C(NHCH_3)N(CH_3)_2][PF_6]$ Fe: Org.Comp.B16a-165, 168

$C_{14}F_6FeH_{23}NP_2S$ $[(C_5H_5)Fe(CS)(NCCH_3)P(C_2H_5)_3][PF_6]$ .... Fe: Org.Comp.B15-264

$C_{14}F_6FeH_{24}O_6P_2$. $[C_5H_5(CO)(P(OCH_3)_3)Fe{=}C(CH_2OCH_3)OC_2H_5][PF_6]$ Fe: Org.Comp.B16a-43

$C_{14}F_6H_8N_2O_4PRe$ $[(CO)_4Re(NC_5H_4-C_5H_4N)][PF_6]$ .... Re: Org.Comp.1-475

$C_{14}F_6H_8N_2O_4S_3$ .. $CF_3S(O)_2-4-C_6H_4-N{=}S{=}N-C_6H_4-4-S(O)_2CF_3$ .. S: S-N Comp.7-239

$C_{14}F_6H_{10}IMoNO$ .. $(C_5H_5)Mo(NO)(I)[C_7H_5(CF_3)_2]$ .... Mo: Org.Comp.6-177

$C_{14}F_6H_{10}O_6S_2Sn$ $(C_6H_5)_2Sn[OS(O)_2CF_3]_2$ .... Sn: Org.Comp.16-146

$C_{14}F_6H_{12}Sb^-$ ..... $[(C_6H_3F-3(CH_3-4))_2SbF_4]^-$ .... Sb: Org.Comp.5-138

$C_{14}F_6H_{13}MoN_2O_2P$ $[(C_5H_5)Mo(CO)_2(-1-NC_5H_4-2-CH{=}N(CH_3)-)][PF_6]$ Mo: Org.Comp.7-249, 250, 252

$C_{14}F_6H_{13}MoO_5P$ .. $[(C_5H_5)Mo(CO)(HC{\equiv}C-C(O)O-CH_3)_2][PF_6]$ .... Mo: Org.Comp.6-316, 322

$C_{14}F_6H_{15}MoN_2O_2P$ $[(C_5H_5)Mo(CO)(NO)(H_2C{=}CHCH_2-NC_5H_5)][PF_6]$ Mo: Org.Comp.6-304

– .... $[(C_5H_5)Mo(CO)_2(NH_2-CH(CH_3)-2-NC_5H_4)][PF_6]$ Mo: Org.Comp.7-249, 252, 277

$C_{14}F_6H_{15}MoO_2P$ .. $[(C_5H_5)Mo(CO)_2(C_6H_7-5-CH_3)][PF_6]$ .... Mo: Org.Comp.8-305/6, 313, 317/8

– .... $[(C_5H_5)Mo(CO)_2(C_7H_{10})][PF_6]$ .... Mo: Org.Comp.8-305/6, 314, 318/9

$C_{14}F_6H_{16}MoN_2O_2S_2$ $(C_5H_5)Mo(NO)[S_2C-N(CH_3)_2]-CH[-CH_2-O-C(CF_3)_2-CH_2-]$ .... Mo: Org.Comp.6-86

$C_{14}F_6H_{16}NO_2P_2Re$ $[(C_5H_5)Re(CO)(NO)(P(CH_3)_2-C_6H_5)][PF_6]$ ..... Re: Org.Comp.3-154, 156

$C_{14}F_6H_{17}MoOP$ ... $[(C_5H_5)Mo(CO)(H_3C-C{\equiv}C-CH_3)_2][PF_6]$ .... Mo: Org.Comp.6-318, 323/4

– .... $[(C_5H_5)Mo(CO)(D_3C-C{\equiv}C-CD_3)_2][PF_6]$ .... Mo: Org.Comp.6-318

$C_{14}F_6H_{18}MoNO_2P$ $[C_5H_5Mo(CO)(NO)C_8H_{13}-c][PF_6]$ .... Mo: Org.Comp.6-350, 354

$C_{14}F_6H_{19}NOReSb$ $[(CH_3C{\equiv}CCH_3)_2Re(O)-1-NC_5H_4CH_3-2][SbF_6]$ .. Re: Org.Comp.2-359/60, 369

– .... $[(CH_3C{\equiv}CCH_3)_2Re(O)-1-NC_5H_4CH_3-4][SbF_6]$ .. Re: Org.Comp.2-360, 369

$C_{14}F_6H_{20}MnN_2S_6$ $[Mn(SC({=}S)N(C_2H_5)_2)_2(SC(CF_3){=}C(CF_3)S)]$ .... Mn: MVol.D7-161

$C_{14}F_6H_{20}MoN_2OS_4$ $(CF_3-CC-CF_3)Mo(O)[S_2C-N(C_2H_5)_2]_2$ .... Mo: Org.Comp.5-139/41

$C_{14}F_6H_{29}MoOP_3$ .. $[i-C_3H_7C_5H_4Mo(P(CH_3)_3)_2O][PF_6]$ .... Mo: Org.Comp.6-16

$C_{14}F_7FeH_5O_2$ .... $(C_5H_5)Fe(CO)_2C_6F_4CF_3-4$ .... Fe: Org.Comp.B13-165, 173

$C_{14}F_7FeH_6IO_2$ .... $(I-1-C_9H_6)Fe(CO)_2C_3F_7$ .... Fe: Org.Comp.B14-68, 75

$C_{14}F_7FeH_9NO_2P$	$[(C_5H_5)Fe(CN-C_6H_4F-4)(CO)_2][PF_6]$	Fe: Org.Comp.B15-315, 318
$C_{14}F_7H_7N_2S$	$4-CH_3-C_6H_4-N=S=N-C_6F_4-CF_3-4$	S: S-N Comp.7-263
$C_{14}F_7H_{21}O_4Sn$	$(C_4H_9)_2Sn(OOCCH_3)OOCC_3F_7$	Sn: Org.Comp.16-194
$C_{14}F_8FeH_6O_2$	$(C_5H_5)Fe(CO)_2CH[-C(CF_3)=C(CF_3)-C(=CF_2)-]$	Fe: Org.Comp.B13-198, 239
$C_{14}F_8HN_3$	$(4-NC-C_6F_4)_2NH$	F: PFHOrg.SVol.5-16
$C_{14}F_8H_6N_2O_2S$	$CH_3-O-4-C_6F_4-N=S=N-C_6F_4-4-OCH_3$	S: S-N Comp.7-246, 252/5
–	$CH_3-O-4-C_6F_4-{}^{15}N=S={}^{15}N-C_6F_4-4-OCH_3$	S: S-N Comp.7-246
$C_{14}F_8H_6N_2S$	$CH_3-4-C_6F_4-N=S=N-C_6F_4-4-CH_3$	S: S-N Comp.7-246
$C_{14}F_8H_{13}N_3O_2$	$[(CH_3)_4N][OC(CFCF)_2N]\cdot HOC(CFCF)_2N$	F: PFHOrg.SVol.4-109
$C_{14}F_8N_4$	$4-NC-C_6F_4-N=N-C_6F_4-CN-4$	F: PFHOrg.SVol.5-220
$C_{14}F_9FeH_{14}OP$	$[(4-CH_3-C_6H_4-O-CH_2-CF_3)Fe(C_5H_5)][PF_6]$	Fe: Org.Comp.B19-2/3, 48
$C_{14}F_9H_7MoO_2S$	$C_5H_5Mo(CO)(SCF_3)(c-C_5H_2(=O)(CF_3)_2-2,4)$	Mo: Org.Comp.6-359/60
$C_{14}F_9H_{11}Mo$	$C_5H_5Mo(H_3C-CC-CH_3)(F_3C-CC-CF_3)CF_3$	Mo: Org.Comp.6-157
$C_{14}F_9N_{11}$	$(C_4F_2N_2(N_3))N(C_6F_5)(C_4F_2N_2(N_3))$	F: PFHOrg.SVol.4-189, 203
$C_{14}F_{10}GeH_6$	$Ge(CH_3)_2(C_6F_5)_2$	Ge: Org.Comp.3-156
$C_{14}F_{10}HNO_2$	$[C_6F_5-C(=O)]_2NH$	F: PFHOrg.SVol.5-24, 46
$C_{14}F_{10}KN$	$K[C(C_6F_5)_2CN]$	F: PFHOrg.SVol.6-108, 127
$C_{14}F_{10}LiN$	$Li[C(C_6F_5)_2CN]$	F: PFHOrg.SVol.6-108, 127, 149
$C_{14}F_{10}N^-$	$[C(C_6F_5)_2CN]^-$	F: PFHOrg.SVol.6-108, 127, 149
$C_{14}F_{10}NNa$	$Na[C(C_6F_5)_2CN]$	F: PFHOrg.SVol.6-108, 127, 149
$C_{14}F_{10}N_2O$	$OC(C_6F_5)NNC(C_6F_5)$	F: PFHOrg.SVol.4-74
$C_{14}F_{11}FeH_5O_2$	$(C_5H_5)Fe(CO)_2C(C(CF_3)=CF_2)=C(CF_3)_2$	Fe: Org.Comp.B13-98, 126/8
$C_{14}F_{11}H_5MoO_2$	$(C_5H_5)Mo(CO)_2[CF_2C(CF_3)C=C(CF_3)_2]$	Mo: Org.Comp.8-205, 226/7
$C_{14}F_{12}FeH_{16}P_2$	$[(CH_3-C_6H_5)_2Fe][PF_6]_2$	Fe: Org.Comp.B19-347, 356
$C_{14}F_{12}H_5IMoO$	$C_5H_5Mo(CO)(I)(C_4(CF_3)_4-c)$	Mo: Org.Comp.6-357
$C_{14}F_{12}H_{20}N_4O_4S_4$	$1,4,8,11-(CF_3S(O))_4-1,4,8,11-N_4C_{10}H_{20}$	S: S-N Comp.8-285
$C_{14}F_{12}H_{20}O_2Sn$	$(C_4H_9)_2Sn(OCH(CF_3)_2)_2$	Sn: Org.Comp.15-5, 22
$C_{14}F_{12}H_{21}MoN_7P_2$	$[Mo(CN-CH_3)_7][PF_6]_2$	Mo: Org.Comp.5-83, 84, 87
$C_{14}F_{12}N_2$	$(C_5F_4N)C_4F_4(C_5F_4N)$	F: PFHOrg.SVol.4-120/1, 132
$C_{14}F_{13}H_{13}O_6Sn$	$C_6F_{13}CH_2CH_2Sn(OOCCH_3)_3$	Sn: Org.Comp.17-55
$C_{14}F_{14}HN$	$(4-CF_3-C_6F_4)_2NH$	F: PFHOrg.SVol.5-16
$C_{14}F_{14}H_2N_2O^{2+}$	$[4-CF_3-C_6F_4-N(OH)=NH-C_6F_4-CF_3-4]^{2+}$	F: PFHOrg.SVol.5-202
$C_{14}F_{14}N_2$	$1,2-(NC_5F_4)_2-c-C_4F_6$	F: PFHOrg.SVol.4-120/1, 132
–	$4-CF_3-C_6F_4-N=N-C_6F_4-CF_3-4$	F: PFHOrg.SVol.5-210, 220, 232
–	$C_6F_5-N=CF-N(CF_3)-C_6F_5$	F: PFHOrg.SVol.6-192/3, 207
–	$NC_5F_4-2-CF(CF_3)-CF=CF-2-NC_5F_4$	F: PFHOrg.SVol.4-120/1, 132
$C_{14}F_{14}N_2O$	$4-CF_3-C_6F_4-N(O)=N-C_6F_4-CF_3-4$	F: PFHOrg.SVol.5-202, 213, 223
$C_{14}F_{14}N_4S_3$	$4-CF_3-C_6F_4-N=S=NSN=S=N-C_6F_4-4-CF_3$	S: S-N Comp.7-52
$C_{14}F_{15}H_5Mo$	$C_5H_5Mo(F_3C-CC-CF_3)_2CF_3$	Mo: Org.Comp.6-157, 158
$C_{14}F_{15}H_{14}NSn$	$(CH_3)_3SnNHC(C_7F_{15})=CHCH=CH_2$	Sn: Org.Comp.18-16
$C_{14}F_{17}H_{16}NO_2S$	$(HCF_2CF_2CF_2CF_2CH_2O)_2S(F)-N(C_2H_5)_2$	S: S-N Comp.8-401, 404
$C_{14}F_{18}H_3MoN$	$(CF_3-CC-CF_3)_3Mo(NC-CH_3)$	Mo: Org.Comp.5-196
$C_{14}F_{19}NO_2$	$C_6F_5-N[C(=O)C_3F_7-n]_2$	F: PFHOrg.SVol.6-227, 238, 242
$C_{14}F_{20}H_3N_9O_{10}$	$((CF_2NO_2)_2(CF_3)_2C_3HN_2O)_2NH$	F: PFHOrg.SVol.4-196/7, 210

$C_{14}F_{22}N_2O_4$ ..... $(CF_3)_2C((CF_3)_2C_2NO_2)CFC(CF_3)((CF_3)_2C_2NO_2)$ F: PFHOrg.SVol.4-31, 55

$C_{14}F_{23}N$ .... $(i\text{-}C_3F_7)_3C_5F_2N$ .... F: PFHOrg.SVol.4-119/20, 131, 139

– .... $NC(i\text{-}C_3F_7)CFC(i\text{-}C_3F_7)C(i\text{-}C_3F_7)CF$ .... F: PFHOrg.SVol.4-117/8, 126, 139

– .... $NC(i\text{-}C_3F_7)CFC(i\text{-}C_3F_7)CFC(i\text{-}C_3F_7)$ .... F: PFHOrg.SVol.4-117/8, 126

$C_{14}F_{24}HN_3O_4$ .... $(OC(CF_3)_2OC(CF_3)_2NC)_2NH$ .... F: PFHOrg.SVol.4-176, 181/2

$C_{14}F_{24}H_3N_5O_2$ ... $((CF_3)_4C_3HN_2O)NH(C_3HN_2O(CF_3)_4)$ .... F: PFHOrg.SVol.4-196, 210

$C_{14}F_{24}H_4N_2O_4$ ... $H_2NC(=O)\text{-}(CF_2)_4\text{-}O\text{-}(CF_2)_4\text{-}O\text{-}(CF_2)_4\text{-}C(=O)NH_2$ F: PFHOrg.SVol.5-21, 54

$C_{14}F_{24}H_4N_4O_4$ ... $OC(CF_3)_2OC(CF_3)_2NC[NHC(NC(OH)(CF_3)_2)(NHC(OH)(CF_3)_2)]$ .... F: PFHOrg.SVol.4-176, 181/2

$C_{14}F_{24}N_2O_2$ ..... $NC\text{-}(CF_2)_4\text{-}O\text{-}(CF_2)_4\text{-}O\text{-}(CF_2)_4\text{-}CN$ .... F: PFHOrg.SVol.6-103, 123, 135

$C_{14}F_{24}N_2O_4S_2$ ... $[(OC(CF_3)_2OC(CF_3)_2NC)S]_2$ .... F: PFHOrg.SVol.4-176, 180

$C_{14}F_{25}N$ .... $NC_5F_4\text{-}(C_3F_7\text{-}i)_3$ .... F: PFHOrg.SVol.4-117/8, 127

$C_{14}F_{25}N_3O$ ...... $2,4\text{-}(n\text{-}C_3F_7)_2\text{-}6\text{-}(n\text{-}C_3F_7\text{-}O\text{-}CF_2CF_2)\text{-}1,3,5\text{-}N_3C_3$ F: PFHOrg.SVol.4-228/9

– .... $2,4\text{-}(n\text{-}C_3F_7)_2\text{-}6\text{-}[n\text{-}C_3F_7\text{-}O\text{-}CF(CF_3)]\text{-}1,3,5\text{-}N_3C_3$ F: PFHOrg.SVol.4-228/9

$C_{14}F_{25}N_3O_2$ ..... $2\text{-}CF_3\text{-}4,6\text{-}[n\text{-}C_3F_7\text{-}O\text{-}CF(CF_3)]\text{-}1,3,5\text{-}N_3C_3$ .. F: PFHOrg.SVol.4-229, 242

$C_{14}F_{26}N_2$ .... $c\text{-}C_6F_{11}\text{-}CF_2N{=}N\text{-}CF_2\text{-}C_6F_{11}\text{-}c$ .... F: PFHOrg.SVol.5-208/9, 219

$C_{14}F_{28}N_2O_2$ ..... $(C_4F_8NO)(CF_2)_6(C_4F_8NO)$ .... F: PFHOrg.SVol.4-174

$C_{14}F_{28}N_2O_8S_2$ ... $(C_4F_7NO(OSO_2F))(CF_2)_6(C_4F_7NO(OSO_2F))$ .... F: PFHOrg.SVol.4-174, 182

$C_{14}F_{32}N_2O_3$ ..... $C_2F_5\text{-}O\text{-}CF_2CF_2\text{-}O\text{-}CF_2CF_2\text{-}N(CF_2CF_2\text{-}O\text{-}C_2F_5)\text{-}CF_2CF_2\text{-}N(CF_3)_2$ .... F: PFHOrg.SVol.6-226/7

$C_{14}FeGeH_{16}O_3$ ... $[(CH_3)_3Ge\text{-}(4.2.0)\text{-}C_8H_7]Fe(CO)_3$ .... Ge: Org.Comp.2-41

– .... $[(CH_3)_3Ge\text{-}c\text{-}C_8H_7]Fe(CO)_3$ .... Ge: Org.Comp.2-39

$C_{14}FeGeH_{20}$ ..... $Ge(CH_3)_3CH_2C_5H_4FeC_5H_5$ .... Ge: Org.Comp.1-156, 162

$C_{14}FeH_8Hg_2MnN_4S_4$ $Fe(C_5H_4\text{-}HgSCN)_2Mn(NCS)_2$ .... Fe: Org.Comp.A10-136/7

$C_{14}FeH_8Hg_2N_4NiS_4$ $Fe(C_5H_4\text{-}HgSCN)_2Ni(NCS)_2$ .... Fe: Org.Comp.A10-136/7

$C_{14}FeH_8Hg_2N_4S_4Zn$ $Fe(C_5H_4\text{-}HgSCN)_2Zn(NCS)_2$ .... Fe: Org.Comp.A10-136/7

$C_{14}FeH_8MoO_7S_2$ .. $C_5H_5(CO)_2Fe{=}C(SCH_3)SMo(CO)_5$ .... Fe: Org.Comp.B16a-181

$C_{14}FeH_8O_3$ ...... $2\text{-}C_5H_4\text{-}C_6H_4\text{-}C(=O)\text{-}Fe(CO)_2$ .... Fe: Org.Comp.B18-18, 28

$C_{14}FeH_8O_7ReS_2^+$ $[(C_5H_5)(CO)_2Fe{=}C(SCH_3)S\text{-}Re(CO)_5]^+$ .... Fe: Org.Comp.B16a-142, 154; Re: Org.Comp.2-177

$C_{14}FeH_8O_7S_2W$ ... $(C_5H_5)(CO)_2Fe{=}C(SCH_3)S\text{-}W(CO)_5$ .... Fe: Org.Comp.B16a-181

$C_{14}FeH_9NO_2$ ..... $(C_5H_5)Fe(CO)_2C_6H_4CN\text{-}4$ .... Fe: Org.Comp.B13-172

$C_{14}FeH_9O_3^+$ ..... $[(C_6H_5\text{-}C_5H_4)Fe(CO)_3]^+$ .... Fe: Org.Comp.B15-50

– .... $[(C_{11}H_9)Fe(CO)_3]^+$ .... Fe: Org.Comp.B15-220

$C_{14}FeH_{10}NO_2^+$ ... $[(C_5H_5)Fe(CN\text{-}C_6H_5)(CO)_2]^+$ .... Fe: Org.Comp.B15-314/5, 318

$C_{14}FeH_{10}N_2O_2$ ... $(C_5H_5)Fe(CO)_2C(CN){=}C(CN)CH_2CH{=}CH_2$ .... Fe: Org.Comp.B13-98/9, 128

$C_{14}FeH_{10}O_2$ ..... $C_5H_4\text{-}CH_2\text{-}2\text{-}C_6H_4\text{-}Fe(CO)_2$ .... Fe: Org.Comp.B18-29

$C_{14}FeH_{10}O_2SSe$ .. $(C_5H_5)Fe(CO)_2C(S)SeC_6H_5$ .... Fe: Org.Comp.B13-57

$C_{14}FeH_{10}O_2STe$ .. $(C_5H_5)Fe(CO)_2C(S)TeC_6H_5$ .... Fe: Org.Comp.B13-57/8

$C_{14}FeH_{10}O_2S_2$ ... $(C_5H_5)Fe(CO)_2C(S)SC_6H_5$ .... Fe: Org.Comp.B13-56

$C_{14}FeH_{10}O_3$ ..... $(C_5H_5)Fe(CO)_2$-$C(O)C_6H_5$ ..... Fe: Org.Comp.B13-11, 22, 41/2
– ..... $(C_5H_5)Fe(CO)_2$-C_6H_4-CHO-4 ..... Fe: Org.Comp.B13-172, 183
$C_{14}FeH_{10}O_3^+$ ..... $[(C_5H_5)Fe(CO)_2C(O)C_6H_5]^+$, radical cation ..... Fe: Org.Comp.B13-9
$C_{14}FeH_{10}O_3S$ ..... $(C_5H_5)Fe(CO)_2C(S)OC_6H_5$ ..... Fe: Org.Comp.B13-56/7
$C_{14}FeH_{10}O_4$ ..... $(C_5H_4COOH)Fe(CO)_2C_6H_5$ ..... Fe: Org.Comp.B14-58
$C_{14}FeH_{10}O_4Ru$ ..... cis-$(C_5H_5)(CO)Fe(CO)_2Ru(C_5H_5)(CO)$ ..... Fe: Org.Comp.B16b-144/5
– ..... trans-$(C_5H_5)(CO)Fe(CO)_2Ru(C_5H_5)(CO)$ ..... Fe: Org.Comp.B16b-144/5
$C_{14}FeH_{10}O_7$ ..... C_5H_4-$C[OC(=O)CH_3]=C[OC(=O)CH_3]C(=O)$-$Fe(CO)_2$ Fe: Org.Comp.B18-19, 27
$C_{14}FeH_{11}NO_2^-$ ..... $[NC$-$C_5H_4FeC_5H_4$-$C(O)=C(CH_3)O]^-$ (radical anion) Fe: Org.Comp.A9-121
$C_{14}FeH_{11}N_2O_2^+$ ..... $[C_5H_5(CO)_2Fe=C(NH)_2C_6H_4]^+$ ..... Fe: Org.Comp.B16a-147
$C_{14}FeH_{11}NaO_2$ ..... $Na[(C_6H_5CH_2C_5H_4)Fe(CO)_2]$ ..... Fe: Org.Comp.B14-117
$C_{14}FeH_{11}O_2^+$ ..... $[(C_5H_5)(CO)_2Fe=CH$-$C_6H_5]^+$ ..... Fe: Org.Comp.B16a-83/4, 90, 97/9
– ..... $[C_5H_5(CO)_2Fe=C_7H_6$-c$]^+$ ..... Fe: Org.Comp.B16a-93, 101/2
$C_{14}FeH_{11}O_2^-$ ..... $[(C_6H_5CH_2C_5H_4)Fe(CO)_2]^-$ ..... Fe: Org.Comp.B14-117
$C_{14}FeH_{11}O_2S^+$ ..... $[C_5H_5(CO)_2Fe=CH(SC_6H_5)]^+$ ..... Fe: Org.Comp.B16a-113, 123/4
$C_{14}FeH_{11}O_3^+$ ..... $[(CH_2(CH)_4C_6H_5)Fe(CO)_3]^+$ ..... Fe: Org.Comp.B15-2/3, 12/3, 21
$C_{14}FeH_{11}O_6PW$ ..... $(C_5H_5)Fe(CH_2CHP(CH_3)W(CO)_5)CO$ ..... Fe: Org.Comp.B16b-110
$C_{14}FeH_{12}INO$ ..... $(C_5H_5)Fe(CO)(I)CN$-$CH_2C_6H_5$ ..... Fe: Org.Comp.B15-290, 296
$C_{14}FeH_{12}N_2$ ..... $(C_5H_5)Fe[2$-CH_3-1,6-$(NC)_2$-$C_6H_4]$ ..... Fe: Org.Comp.B17-258, 289
– ..... $(C_5H_5)Fe[6$-CH_3-1,6-$(NC)_2$-$C_6H_4]$ ..... Fe: Org.Comp.B17-258, 283
– ..... $(C_5H_5)Fe[C_5H_3(CH_2CN)_2]$ ..... Fe: Org.Comp.A9-129, 131
– ..... $Fe(C_5H_4$-$CH_2CN)_2$ ..... Fe: Org.Comp.A9-109, 118
$C_{14}FeH_{12}N_2O$ ..... $C_5H_5FeC_6H_4(CN)_2$-1,6$(OCH_3$-2) ..... Fe: Org.Comp.B17-258
$C_{14}FeH_{12}N_2O_4$ ..... $(C_5H_5)Fe(CO)(NO_3)CN$-$CH_2C_6H_5$ ..... Fe: Org.Comp.B15-286
$C_{14}FeH_{12}N_{10}O_4$ ..... $Fe(C_5H_4$-CH_2-2-CN_4-5-$NO_2)_2$ ..... Fe: Org.Comp.A9-177
$C_{14}FeH_{12}O_2$ ..... $C_5H_4CH_3Fe(CO)_2C_6H_5$ ..... Fe: Org.Comp.B14-55, 62
– ..... $(C_5H_4D)Fe(CO)_2$-CH_2-C_6H_5 ..... Fe: Org.Comp.B14-53
– ..... $(C_5H_5)Fe(CO)_2$-C_6H_4-CH_3-2 ..... Fe: Org.Comp.B13-169
– ..... $(C_5H_5)Fe(CO)_2$-C_6H_4-CH_3-3 ..... Fe: Org.Comp.B13-170
– ..... $(C_5H_5)Fe(CO)_2$-C_6H_4-CH_3-4 ..... Fe: Org.Comp.B13-170, 182
– ..... $(C_5H_5)Fe(CO)_2$-C_7H_7-c ..... Fe: Org.Comp.B13-196, 220/1, 256
$C_{14}FeH_{12}O_3$ ..... $(C_5H_5)Fe(CO)_2[C_5H_4$-$C(O)CH_3]$ ..... Fe: Org.Comp.B14-57, 64
– ..... $(C_5H_5)Fe(CO)_2$-C_6H_4-O-CH_3-4 ..... Fe: Org.Comp.B13-172, 183
$C_{14}FeH_{12}O_3Ru$ ..... $(C_5H_5)(CO)Fe(CO)(CH_2)Ru(CO)(C_5H_5)$ ..... Fe: Org.Comp.B16b-152/3
$C_{14}FeH_{13}^+$ ..... $[(C_9H_8)Fe(C_5H_5)][PF_6]$ ..... Fe: Org.Comp.B19-223
$C_{14}FeH_{13}^-$ ..... $[(CH_3$-$C_5H_4)Fe(C_5H_3(CH_3)$-$CC)]^-$ ..... Fe: Org.Comp.A10-200, 203
$C_{14}FeH_{13}IN_2O$ ..... $[(C_9H_7)Fe(CNCH_3)_2CO]I$ ..... Fe: Org.Comp.B15-352
$C_{14}FeH_{13}K$ ..... $K[(CH_3$-$C_5H_4)Fe(C_5H_3(CH_3)$-$CC)]$ ..... Fe: Org.Comp.A10-200, 203
$C_{14}FeH_{13}Li$ ..... $(CH_3$-$C_5H_4)Fe[C_5H_3(CH_3)$-CC-Li$]$ ..... Fe: Org.Comp.A10-322, 323, 324
$C_{14}FeH_{13}N$ ..... $(C_5H_5)Fe[C_5H_3(CH_2CN)CH=CH_2]$ ..... Fe: Org.Comp.A9-131, 134
$C_{14}FeH_{13}NO$ ..... $(C_5H_5)Fe[C_5H_3(CH_2CN)$-$C(O)CH_3]$ ..... Fe: Org.Comp.A9-137
– ..... $(NCCH_2$-$C_5H_4)Fe[C_5H_4$-$C(O)CH_3]$ ..... Fe: Org.Comp.A9-115, 122

$C_{14}FeH_{13}NO$	$(NC-C_5H_4)Fe[C_5H_4-C(O)C_2H_5]$	Fe: Org.Comp.A9-114, 121
$C_{14}FeH_{13}NOS$	$(C_5H_5)Fe(CNCH_3)(CO)S-C_6H_5$	Fe: Org.Comp.B15-288
$C_{14}FeH_{13}NO_2$	$C_5H_5FeC_6H_5CO_2CH_3-1(CN-6)$	Fe: Org.Comp.B17-275
$C_{14}FeH_{13}NS_2$	$(C_5H_5)Fe(CNCH_3)(CS)S-C_6H_5$	Fe: Org.Comp.B15-286
$C_{14}FeH_{13}N_2O^+$	$[(C_9H_7)Fe(CNCH_3)_2CO]^+$	Fe: Org.Comp.B15-352
$C_{14}FeH_{13}N_4O^+$	$[(C_5H_5)Fe(CN-CH_2CN=CH_2)_2CO]^+$	Fe: Org.Comp.B15-338
$C_{14}FeH_{13}Na$	$Na[(CH_3-C_5H_4)Fe(C_5H_3(CH_3)-CC)]$	Fe: Org.Comp.A10-200, 203
$C_{14}FeH_{13}O_2{}^+$	$[(C_5H_5)Fe(CO)_2(C_7H_8)]^+$	Fe: Org.Comp.B17-91, 96, 98
–	$[(C_9H_7)Fe(CO)_2(CH_2=CHCH_3)]^+$	Fe: Org.Comp.B17-133/4, 137
$C_{14}FeH_{13}O_3{}^+$	$[(8-C_2H_5CH=C_8H_7)Fe(CO)_3]^+$	Fe: Org.Comp.B15-241/2
$C_{14}FeH_{13}O_4{}^+$	$[(CH_3O-C_{10}H_{10})Fe(CO)_3]^+$	Fe: Org.Comp.B15-128
$C_{14}FeH_{14}$	$(CH_3-C_5H_4)Fe[C_5H_3(CH_3)-C{\equiv}CH]$	Fe: Org.Comp.A10-191, 192/3, 196, 199/205, 210
–	$(C_6H_6)Fe(C_8H_8)$	Fe: Org.Comp.B18-103, 109
$C_{14}FeH_{14}Hg_2O_4$	$Fe[C_5H_4-HgOC(O)CH_3]_2$	Fe: Org.Comp.A10-141
$C_{14}FeH_{14}N^+$	$[(2,6-(CH_3)_2-C_6H_3-CN)Fe(C_5H_5)][PF_6]$	Fe: Org.Comp.B19-124
–	$[(2,6-(CH_3)_2-C_6H_3-NC)Fe(C_5H_5)][PF_6]$	Fe: Org.Comp.B19-126
$C_{14}FeH_{14}NO_2{}^+$	$[(C_5H_5)Fe(CO)_2CH_2CH_2-NC_5H_5]^+$	Fe: Org.Comp.B14-150
$C_{14}FeH_{14}NO_2S^+$	$[C_5H_5Fe(CO)_2CH(SCH_3)NC_5H_5]^+$	Fe: Org.Comp.B14-141
$C_{14}FeH_{14}N_2O_2$	$(C_5H_5)Fe[1,3-(CH_3)_2-2-NO_2-C_6H_3-6-CN]$	Fe: Org.Comp.B17-299
–	$(C_5H_5)Fe[2,4-(CH_3)_2-3-NO_2-C_6H_3-6-CN]$	Fe: Org.Comp.B17-299
–	$(C_5H_5)Fe[2,6-(CH_3)_2-1-NO_2-C_6H_3-6-CN]$	Fe: Org.Comp.B17-296
$C_{14}FeH_{14}N_8$	$Fe(C_5H_4-CH_2-2-N_4CH)_2$	Fe: Org.Comp.A9-177
–	$(N_4CH-1-CH_2-C_5H_4)Fe(C_5H_4-CH_2-2-N_4CH)$	Fe: Org.Comp.A9-177
$C_{14}FeH_{14}O_2$	$(C_5H_5)Fe(CO)(CH_3-O-CHC_6H_5)$	Fe: Org.Comp.B17-157/8, 159
–	$(C_5H_5)Fe(CO)_2-7-[2.2.1.0^{2,6}]-C_7H_9$	Fe: Org.Comp.B13-222, 257
–	$C_{10}H_{10}-CH_2CH_2-Fe(CO)_2$	Fe: Org.Comp.B18-25, 43
$C_{14}FeH_{14}O_2S_2$	$Fe(C_5H_4-COSCH_3)_2$	Fe: Org.Comp.A9-225/6
$C_{14}FeH_{14}O_3$	$(C_5H_5)Fe(CO)[CH_3-OC(O)C(CH_2)CH-C{\equiv}CCH_3]$	Fe: Org.Comp.B17-151
–	$(O=CH-C_5H_4)Fe[C_5H_3(CH_3)-C(=O)OCH_3]$	Fe: Org.Comp.A10-280, 281
$C_{14}FeH_{14}O_4$	$(C_5H_5)Fe(CO)_2C[=CH-CH(C_3H_7-i)-O-C(=O)-]$	Fe: Org.Comp.B14-5
–	$(C_5H_5)Fe(CO)_2-3-OC_5H_3(=O)-2-(CH_3)_2-6,6$	Fe: Org.Comp.B14-7, 14
–	$(HO_2C-C_5H_4)Fe[C_5H_3(C_2H_5)-CO_2H]$	Fe: Org.Comp.A10-276
$C_{14}FeH_{14}O_6$	$C_5H_5Fe(COCH_2CH(CH=CH_2)CH(CO_2H)_2)CO$	Fe: Org.Comp.B17-171
$C_{14}FeH_{14}O_7S$	$[CH_3OC(=O)-C_5H_4]Fe[C_5H_3(SO_3H)-C(=O)OCH_3]$	Fe: Org.Comp.A10-298
$C_{14}FeH_{15}{}^+$	$[(CH_3-C_5H_4)Fe(C_5H_3(CH_3)-C=CH_2)]^+$	Fe: Org.Comp.A10-192, 202
–	$[(C_6H_5-CH_2CH_2CH_2-C_5H_4)Fe][PF_6]$	Fe: Org.Comp.B19-410, 412/4
–	$[(C_9H_{10})Fe(C_5H_5)][B_{11}H_{14}]$	Fe: Org.Comp.B19-216, 220/1
–	$[(C_9H_{10})Fe(C_5H_5)][PF_6]$	Fe: Org.Comp.B19-216, 220/1, 298
–	$[(C_9H_{10})Fe(C_5H_5)]_2[B_{10}H_{10}]$	Fe: Org.Comp.B19-216, 220/1
–	$[(C_9H_{10})Fe(C_5H_5)]_2[B_{12}H_{12}]$	Fe: Org.Comp.B19-216, 220/1
$C_{14}FeH_{15}N$	$(CH_3-C_5H_4)Fe(CH_3-C_5H_3-CH_2-CN)$	Fe: Org.Comp.A10-294
–	$(C_5H_5)Fe(1-NC_9H_{10})$	Fe: Org.Comp.B19-280
–	$(C_5H_5)Fe(C_2H_5-C_5H_3-CH_2-CN)$	Fe: Org.Comp.A9-131
–	$(C_5H_5)Fe[1,3-(CH_3)_2-C_6H_4-2-CN]$	Fe: Org.Comp.B17-293
–	$(C_5H_5)Fe[2,4-(CH_3)_2-C_6H_4-3-CN]$	Fe: Org.Comp.B17-293
–	$(C_5H_5)Fe[2,6-(CH_3)_2-C_6H_4-1-CN]$	Fe: Org.Comp.B17-284/5
$C_{14}FeH_{15}NO$	$HON=C(CH_3)-C_5H_4FeC_5H_4-CH=CH_2$	Fe: Org.Comp.A9-160

$C_{14}FeH_{15}NO_2$ . . . . $(C_5H_5)Fe[C_5H_3(C(O)CH_3)-NHC(O)CH_3]$ . . . . Fe: Org.Comp.A9-95
– . . . . $[CH_3C(O)NH-C_5H_4]Fe[C_5H_4-C(O)CH_3]$ . . . . Fe: Org.Comp.A9-83/4
– . . . . $[HO-N=C(CH_3)-C_5H_4]Fe[C_5H_4-C(O)CH_3]$ . . . . . Fe: Org.Comp.A9-160, 162
$C_{14}FeH_{15}NO_2^+$ . . . $[(CH_3C(O)NH-C_5H_4)Fe(C_5H_4-C(O)CH_3)]^+$ . . . . . Fe: Org.Comp.A9-84
– . . . . $[(C_5H_5)Fe(C_5H_3(C(O)CH_3)-NHC(O)CH_3)]^+$ . . . . Fe: Org.Comp.A9-95
$C_{14}FeH_{15}NO_2S$ . . . $(C_5H_5)Fe[C_5H_3(CH_3)-SO_2(CH_2)_2CN]$ . . . . Fe: Org.Comp.A9-252
$C_{14}FeH_{15}NO_3$ . . . . $(C_5H_5)Fe[1-NO_2-C_6H_5-6-CH_2C(O)CH_3]$ . . . . . . Fe: Org.Comp.B17-273/4
– . . . . $(C_5H_5)Fe[C_5H_3(C(O)CH_3)-NHC(O)OCH_3]$. . . . . . Fe: Org.Comp.A9-95
– . . . . $[CH_3OC(O)NH-C_5H_4]Fe[C_5H_4-C(O)CH_3]$. . . . . . Fe: Org.Comp.A9-85
– . . . . $[HO-N=C(CH_3)-C_5H_4]Fe[C_5H_4-C(O)OCH_3]$ . . . . Fe: Org.Comp.A9-160
– . . . . $[H_2NC(O)CH_2-C_5H_4]Fe(C_5H_4-CH_2-CO_2H)$ . . . . Fe: Org.Comp.A9-80
$C_{14}FeH_{15}NO_4$ . . . . $(C_5H_5)Fe(CO)_2CH[-O-CH(CH_3)-CH(CH_3)-O-CH(CN)-]$
Fe: Org.Comp.B14-2, 9
$C_{14}FeH_{15}NO_6S$ . . . $[(C_5H_5)Fe(CO)_2(C_5H_7(SO_2NCO_2CH_3-3))]$. . . . . . Fe: Org.Comp.B17-88
$C_{14}FeH_{15}N_2^+$ . . . . $[(3-CH_3-1,2-N_2C_8H_7)Fe(C_5H_5)][PF_6]$. . . . . Fe: Org.Comp.B19-289
– . . . . $[(NC-CH_2CH_2-NH-C_6H_5)Fe(C_5H_5)][PF_6]$ . . . . . . Fe: Org.Comp.B18-142/6, 199, 258
$C_{14}FeH_{15}N_3O_3$ . . . $(C_5H_5)Fe[C_5H_3(CH=NNHCONH_2)C(O)OCH_3]$ . . . Fe: Org.Comp.A9-148
$C_{14}FeH_{15}O^+$ . . . . . $[(CH_3C(=O)CH_2-C_6H_5)Fe(C_5H_5)][PF_6]$ . . . . Fe: Org.Comp.B18-142/6, 198, 214/5
– . . . . $[(CH_3C_6H_5)Fe(C_5H_4-C(=O)CH_3)]^+$ . . . . Fe: Org.Comp.B18-142/6, 197/8, 207
– . . . . $[(C_6H_6)Fe(C_5H_4-C(=O)C_2H_5)][PF_6]$ . . . . Fe: Org.Comp.B18-142/6, 151, 163
– . . . . $[(C_9H_9-5-OH)Fe(C_5H_5)][PF_6]$ . . . . Fe: Org.Comp.B19-217, 222
$C_{14}FeH_{15}O_2^+$ . . . . $[(4-CH_3-C_6H_4-COO-CH_3)Fe(C_5H_5)][PF_6]$ . . . . . Fe: Org.Comp.B19-23/4
– . . . . $[(CH_3-OOC-CH_2-C_6H_5)Fe(C_5H_5)][PF_6]$ . . . . Fe: Org.Comp.B18-142/6, 199, 215
– . . . . $[(C_2H_5-OOC-C_6H_5)Fe(C_5H_5)][PF_6]$ . . . . Fe: Org.Comp.B18-142/6, 215
– . . . . $[(C_5H_5)Fe(CO)_2((2.2.1)-C_7H_{10})][BF_4]$ . . . . Fe: Org.Comp.B17-85/6, 99, 106/7
– . . . . $[(C_5H_5)Fe(CO)_2(CH_2=CHCH=CHCH_2CH=CH_2)][BF_4]$
Fe: Org.Comp.B17-41
– . . . . $[(C_5H_5)Fe(CO)_2(C_6H_7-CH_3)][BF_4]$ . . . . Fe: Org.Comp.B17-85, 91
– . . . . $[(C_5H_5)Fe(CO)_2(c-C_7H_{10})][BF_4]$ . . . . Fe: Org.Comp.B17-85, 91
– . . . . $[(C_5H_5)Fe(CO)_2(c-C_7H_{10})][CF_3-SO_3]$. . . . . Fe: Org.Comp.B17-115, 118
– . . . . $[(C_6H_6)Fe(C_5H_4-COO-C_2H_5)][B(C_6H_5)_4]$ . . . . . . Fe: Org.Comp.B18-142/6, 151, 164
– . . . . $[(C_6H_6)Fe(C_5H_4-COO-C_2H_5)][PF_6]$ . . . . Fe: Org.Comp.B18-142/6, 151, 164
$C_{14}FeH_{15}O_3^+$ . . . . $[(3-CH_3O-C_6H_4-COO-CH_3)Fe(C_5H_5)][PF_6]$. . . . Fe: Org.Comp.B19-50
– . . . . $[(4-CH_3O-C_6H_4-COO-CH_3)Fe(C_5H_5)][PF_6]$. . . . Fe: Org.Comp.B19-51
– . . . . $[(4-C_2H_5O-C_6H_4-COOH)Fe(C_5H_5)][PF_6]$ . . . . . . Fe: Org.Comp.B19-50
– . . . . $[(6-C_2H_5C(CH_3)=C_7H_7)Fe(CO)_3][PF_6]$ . . . . Fe: Org.Comp.B15-204, 211, 226
– . . . . $[((CH_3)_3-(5.1.0)-C_8H_6)Fe(CO)_3]^+$ . . . . Fe: Org.Comp.B15-216/7, 230
– . . . . $[(C_5H_5)Fe(CO)(CH_2=C(COO-CH_3)CH=C=CHCH_3)][BF_4]$
Fe: Org.Comp.B17-188, 191/2
$C_{14}FeH_{15}O_4^+$ . . . . $[(1-CH_3C(O)CH_2CH(CH_3)-C_6H_6)Fe(CO)_3][PF_6]$ Fe: Org.Comp.B15-100
– . . . . $[(8-(CH_3)_2C(OH)-C_8H_8)Fe(CO)_3][PF_6]$. . . . . Fe: Org.Comp.B15-250

$C_{14}FeH_{15}O_4^+$ .... $[(CH_3O-C_{10}H_{12})Fe(CO)_3][BF_4]$ .... Fe: Org.Comp.B15-127, 179
– .... $[(CH_3O-C_{10}H_{12})Fe(CO)_3][PF_6]$ .... Fe: Org.Comp.B15-127, 179
$C_{14}FeH_{15}O_5$ ..... $(C_5H_5)Fe(CO)_2[C(COOCH_3)=CHCH_2C(CH_3)OH]$ Fe: Org.Comp.B13-96, 124
$C_{14}FeH_{15}O_5^+$ .... $[(1-CH_3C(O)CH_2CH_2-4-CH_3O-C_6H_5)Fe(CO)_3][PF_6]$ Fe: Org.Comp.B15-92/4, 115, 171
– .... $[(3-CH_3-6-CH_3C(O)OCH_2CH_2-C_6H_5)Fe(CO)_3][BF_4]$ Fe: Org.Comp.B15-94/5, 125, 178
– .... $[(6-CH_3C(O)OCH(CH_3)CH_2-C_6H_6)Fe(CO)_3][BF_4]$ Fe: Org.Comp.B15-94/5, 108, 159
– .... $[(6-CH_3C(O)OCH(CH_3)CH_2-C_6H_6)Fe(CO)_3][PF_6]$ Fe: Org.Comp.B15-108, 159
$C_{14}FeH_{15}O_6^+$ .... $[(1-CH_3C(O)OCH_2CH_2-4-CH_3O-C_6H_5)Fe(CO)_3][BF_4]$ Fe: Org.Comp.B15-118
– .... $[(1-CH_3C(O)OCH_2CH_2-4-CH_3O-C_6H_5)Fe(CO)_3][PF_6]$ Fe: Org.Comp.B15-118, 172
– .... $[(1-CH_3OC(O)CH_2CH_2-4-CH_3O-C_6H_5)Fe(CO)_3][PF_6]$ Fe: Org.Comp.B15-116, 171
– .... $[(3-CH_3O-6-CH_3C(O)OCH_2CH_2-C_6H_5)Fe(CO)_3][BF_4]$ Fe: Org.Comp.B15-125, 178
$C_{14}FeH_{15}S_2^+$ .... $[(CH_3-SC(=S)CH_2-C_6H_5)Fe(C_5H_5)]^+$ .... Fe: Org.Comp.B18-142/6, 199, 215
$C_{14}FeH_{16}$ .... $(CH_3-C_5H_4)Fe[C_5H_3(CH_3)-CH=CH_2]$ .... Fe: Org.Comp.A10-194, 199, 204, 209
– .... $(CH_3-C_6H_5)Fe[(2.2.1)-C_7H_8]$ .... Fe: Org.Comp.B18-103, 115
– .... $(CH_3-C_6H_5)Fe(c-C_7H_8)$ .... Fe: Org.Comp.B18-103, 115
– .... $(CH_3-C_6H_5)_2Fe$ .... Fe: Org.Comp.B18-111, 131/3
Fe: Org.Comp.B19-348, 357, 377
– .... $(c-C_7H_7)Fe(c-C_7H_9)$ .... Fe: Org.Comp.B17-317, 318
$C_{14}FeH_{16}^{2+}$ ..... $[(CH_3-C_6H_5)_2Fe]Br_2$ .... Fe: Org.Comp.B19-347, 356
– .... $[(CH_3-C_6H_5)_2Fe][B_{12}H_{12}]$ .... Fe: Org.Comp.B19-347, 356
– .... $[(CH_3-C_6H_5)_2Fe][Cr(SCN)_4(NH_3)_2]_2$ .... Fe: Org.Comp.B19-348, 356
– .... $[(CH_3-C_6H_5)_2Fe][I_3]_2$ .... Fe: Org.Comp.B19-347, 356
– .... $[(CH_3-C_6H_5)_2Fe][PF_6]_2$ .... Fe: Org.Comp.B19-347, 356
$C_{14}FeH_{16}I_6$ ...... $[(CH_3-C_6H_5)_2Fe][I_3]_2$ .... Fe: Org.Comp.B19-347, 356
$C_{14}FeH_{16}K_2$ ..... $Fe[C_{10}H_7(K)_2(C_4H_9-t)]$ .... Fe: Org.Comp.A10-328
$C_{14}FeH_{16}LiNO_2$ .. $Li[C_5H_5FeC_5H_3(CH_2N(CH_3)_2)CO_2]$ .... Fe: Org.Comp.A9-51
$C_{14}FeH_{16}Li_2$ ..... $(Li-C_5H_4)Fe[C_5H_3(Li)-C_4H_9-t]$ .... Fe: Org.Comp.A10-322/4
$C_{14}FeH_{16}Li_2O$ .... $[Li-CH(CH_3)-C_5H_4]Fe[C_5H_4-CH(CH_3)-OLi]$ ... Fe: Org.Comp.A10-113/5, 117
$C_{14}FeH_{16}N^+$ ..... $[(1-NC_9H_{11})Fe(C_5H_5)][PF_6]$ .... Fe: Org.Comp.B19-216, 279
$C_{14}FeH_{16}NO$ ..... $(4-CH_3-C_6H_4-NH-CO-CH_3)Fe(C_5H_5)$ .... Fe: Org.Comp.B19-59
$C_{14}FeH_{16}NO^+$.... $[(2-CH_3-C_6H_4-NH-CO-CH_3)Fe(C_5H_5)][PF_6]$ ... Fe: Org.Comp.B19-4, 59
– .... $[(3-CH_3-C_6H_4-NH-CO-CH_3)Fe(C_5H_5)][PF_6]$ ... Fe: Org.Comp.B19-4, 59
– .... $[(4-CH_3-C_6H_4-NH-CO-CH_3)Fe(C_5H_5)][PF_6]$ ... Fe: Org.Comp.B19-1, 4, 59
– .... $[(C_6H_5-NH-CO-C_2H_5)Fe(C_5H_5)][PF_6]$ .... Fe: Org.Comp.B18-142/6, 199, 261
$C_{14}FeH_{16}NOS_2^+$.. $[C_5H_5(CO)(C_5H_5N)Fe=C(SCH_3)_2]^+$ .... Fe: Org.Comp.B16a-66

$C_{14}FeH_{16}NO_2^+$	$[(CH_3OC(O)C_5H_4)Fe(C_5H_4-C(CH_3)=NH_2)][PF_6]$	Fe: Org.Comp.A9-103
–	$[(i-C_3H_7-C_6H_4-4-NO_2)Fe(C_5H_5)][OC(=O)-CF_3]$	Fe: Org.Comp.B19-4, 64, 93
–	$[(C_5H_5)Fe(CN-C_6H_{11}-c)(CO)_2][PF_6]$	Fe: Org.Comp.B15-314
$C_{14}FeH_{16}NO_2^-$	$[C_5H_5FeC_5H_3(CH_2N(CH_3)_2)CO_2]^-$	Fe: Org.Comp.A9-51
$C_{14}FeH_{16}NS_2^+$	$[((CH_3S)_2C=N-C_6H_5)Fe(C_5H_5)]^+$	Fe: Org.Comp.B18-142/6, 199/200, 264
$C_{14}FeH_{16}N_2$	$(C_5H_5)Fe[C_5H_3(CN)CH_2N(CH_3)_2]$	Fe: Org.Comp.A9-130
$C_{14}FeH_{16}N_2O_2$	$Fe[C_5H_4-C(CH_3)=N-OH]_2$	Fe: Org.Comp.A9-160, 161/2
–	$Fe[C_5H_4-C(O)NH-CH_3]_2$	Fe: Org.Comp.A9-77
$C_{14}FeH_{16}N_2O_4$	$Fe(C_5H_4-CH_2CH_2-NO_2)_2$	Fe: Org.Comp.A9-151, 156
–	$Fe[C_5H_4-NH-C(O)OCH_3]_2$	Fe: Org.Comp.A9-85, 86/7
$C_{14}FeH_{16}N_3^+$	$[(C_5H_5)Fe(CNCH_3)_2-NC_5H_5]^+$	Fe: Org.Comp.B15-327, 328
$C_{14}FeH_{16}N_6O_2$	$(C_5H_5)Fe[C_5H_3(CH=NNHCONH_2)_2]$	Fe: Org.Comp.A9-148
$C_{14}FeH_{16}Na_2$	$Fe[C_{10}H_7(Na)_2(C_4H_9-t)]$	Fe: Org.Comp.A10-328
$C_{14}FeH_{16}O$	$(CH_3-C_5H_4)Fe[C_5H_3(CH_3)-C(=O)CH_3]$	Fe: Org.Comp.A10-240, 241, 243, 250, 253/5, 261
–	$(C_5H_5)Fe[C_5H_2(CH_3)_2-C(=O)CH_3]$	Fe: Org.Comp.A10-264
–	$[(CH_3)_2C_5H_3]Fe[C_5H_4-C(=O)CH_3]$	Fe: Org.Comp.A10-241, 250, 253, 254, 261
$C_{14}FeH_{16}O^+$	$[(CH_3-C_5H_4)Fe(C_5H_3(CH_3)-C(=O)CH_3)]^+$	Fe: Org.Comp.A10-261
$C_{14}FeH_{16}O^{2+}$	$[(C_6H_6)Fe(C_5H_4-C(C_2H_5)=OH)]^{2+}$	Fe: Org.Comp.B18-164
$C_{14}FeH_{16}O_2$	$(CH_2C_4H_4-CH_2CH_2-C_4H_4CH_2)Fe(CO)_2$	Fe: Org.Comp.B18-34, 48
–	$(CH_3-C_5H_4)Fe[C_5H_3(CH_3)-C(=O)OCH_3]$	Fe: Org.Comp.A10-277/81
–	$(CH_3-C_5H_4)Fe[C_5H_3(CH_3)-CH_2-CO_2H]$	Fe: Org.Comp.A10-266, 268, 269, 271
–	$(C_5H_5)Fe(CO)_2-1-C_5H_8-3-CH=CH_2$	Fe: Org.Comp.B13-199/200, 240/1
–	$(C_5H_5)Fe(CO)_2-3-C_7H_{11}-c$	Fe: Org.Comp.B13-195, 220, 255
–	$(C_5H_5)Fe(CO)_2-2-[2.2.1]-C_7H_{11}$	Fe: Org.Comp.B13-222, 258
–	$(C_5H_5)Fe(CO)_2-2-[2.2.1]-C_7H_{10}D-2$	Fe: Org.Comp.B13-222, 258
–	$(C_5H_5)Fe(CO)_2-2-[2.2.1]-C_7H_9D_2-5,6$	Fe: Org.Comp.B13-222, 258
–	$(C_5H_5)Fe(CO)_2-7-[4.1.0]-C_7H_{11}$	Fe: Org.Comp.B13-222, 258
–	$(C_5H_5)Fe(CO)_2-CH=CHCH_2CH_2CH_2CH=CH_2$	Fe: Org.Comp.B13-85, 115
–	$[(C_6H_6)Fe(C_5H_4-CH_3)][CH_3-CO_2]$	Fe: Org.Comp.B18-142/6, 151, 154, 161, 185/6
$C_{14}FeH_{16}O_2^{2+}$	$[(CH_3O-C_6H_5)_2Fe][O-C_6H_2-2,4,6-(NO_2)_3]_2$	Fe: Org.Comp.B19-358
$C_{14}FeH_{16}O_3$	$(C_5H_5)Fe(CO)[=C(OCH_3)-C_4H_2(=O)(CH_3)_2-]$	Fe: Org.Comp.B16b-117, 132
–	$(C_5H_5)Fe(CO)[CH_2=CC(CH_2)COO-C_3H_7-i]$	Fe: Org.Comp.B17-162
–	$(C_5H_5)Fe(CO)_2C[=CHCH_2CH_2CH_2CH(OCH_3)-]$	Fe: Org.Comp.B13-216, 254
–	$(C_5H_5)Fe(CO)_2-CH_2CH=CHC(=O)-C_3H_7-i$	Fe: Org.Comp.B13-107, 134
–	$[C_5H_4-C(CH_3)_2C(CH_3)_2C(=O)-]Fe(CO)_2$	Fe: Org.Comp.B18-18, 27
$C_{14}FeH_{16}O_4$	$(CH_3O-C_5H_4)Fe[C_5H_3(OCH_3)-C(=O)OCH_3]$	Fe: Org.Comp.A10-277, 278
–	$(C_5H_5)Fe(CO)_2C(=O)-C_3H_2(OCH_3-1)(CH_3)_2-2,3$	Fe: Org.Comp.B13-20, 39
–	$(C_5H_5)Fe(CO)_2CH=CHC(=CH_2)CH(O-CH_3)_2$	Fe: Org.Comp.B13-98
–	$(C_5H_5)Fe(CO)_2CH_2C(CH_3)=CH-2-(1,3-O_2C_3H_5)$	Fe: Org.Comp.B13-99, 139
–	$(C_5H_5)Fe(CO)_2CH_2CH=CH-2-[1,3-O_2C_3H_4(CH_3-1)]$	Fe: Org.Comp.B13-81, 103
–	$(C_5H_5)Fe(CO)_2C(O)O-C_6H_{11}-c$	Fe: Org.Comp.B13-24, 44/5
$C_{14}FeH_{16}O_4S$	$(C_5H_5)Fe(CO)_2(CH_2=C_6H_9-SO_2)$	Fe: Org.Comp.B17-128

$C_{14}FeH_{16}O_4S$ .... $(C_5H_5)Fe[C_5H_3(CH_3)-S(O)_2CH_2CH_2-COOH]$ ... Fe: Org.Comp.A9-252
$C_{14}FeH_{16}O_5$ ..... $(C_5H_5)Fe(CO)[CH_2C(O-C(O)CH_3)C(CH_3)O-C(O)CH_3]$ Fe: Org.Comp.B17-151
$C_{14}FeH_{17}$ .... $(i-C_3H_7-C_6H_5)Fe(C_5H_5)$ .... Fe: Org.Comp.B18-146/8, 200/1, 217, 274
– .... $[1,3,5-(CH_3)_3-C_6H_3]Fe(C_5H_5)$ .... Fe: Org.Comp.B19-105, 136
$C_{14}FeH_{17}^+$ ...... $[(1,2,3-(CH_3)_3-C_6H_3)Fe(C_5H_5)][BF_4]$ .... Fe: Org.Comp.B19-99, 121
– .... $[(1,2,3-(CH_3)_3-C_6H_3)Fe(C_5H_5)][Cr(NH_3)_2(SCN)_4]$ Fe: Org.Comp.B19-99, 121
– .... $[(1,2,3-(CH_3)_3-C_6H_3)Fe(C_5H_5)][I_3]$ .... Fe: Org.Comp.B19-99, 120
– .... $[(1,2,3-(CH_3)_3-C_6H_3)Fe(C_5H_5)][PF_6]$ .... Fe: Org.Comp.B19-99/100, 120/1
– .... $[(1,2,4-(CH_3)_3-C_6H_3)Fe(C_5H_5)][BF_4]$ .... Fe: Org.Comp.B19-99, 121
– .... $[(1,2,4-(CH_3)_3-C_6H_3)Fe(C_5H_5)][B_{11}H_{14}]$ ...... Fe: Org.Comp.B19-99, 121
– .... $[(1,2,4-(CH_3)_3-C_6H_3)Fe(C_5H_5)][Cr(NH_3)_2(SCN)_4]$ Fe: Org.Comp.B19-99, 121
– .... $[(1,2,4-(CH_3)_3-C_6H_3)Fe(C_5H_5)][I_3]$ .... Fe: Org.Comp.B19-99, 121
– .... $[(1,2,4-(CH_3)_3-C_6H_3)Fe(C_5H_5)][PF_6]$ .... Fe: Org.Comp.B19-99, 100, 121
– .... $[(1,2,4-(CH_3)_3-C_6H_3)Fe(C_5H_5)]_2[B_{10}H_{10}]$ ...... Fe: Org.Comp.B19-99, 121
– .... $[(1,2,4-(CH_3)_3-C_6H_3)Fe(C_5H_5)]_2[B_{12}H_{12}]$ ...... Fe: Org.Comp.B19-99, 121
– .... $[(1,3,5-(CH_3)_3-C_6H_3)Fe(C_5H_5)]^+$ .... Fe: Org.Comp.B19-102
– .... $[(1,3,5-(CH_3)_3-C_6H_3)Fe(C_5H_5)]I$ .... Fe: Org.Comp.B19-100, 101/2, 133/6
– .... $[(1,3,5-(CH_3)_3-C_6H_3)Fe(C_5H_5)][1,4-((NC)_2C{=})_2-C_6H_4]_2$ Fe: Org.Comp.B19-105, 133/6
– .... $[(1,3,5-(CH_3)_3-C_6H_3)Fe(C_5H_5)][1,4-((NC)_2C{=})_2-C_6H_4]_2 \cdot 6\,[1,4-((NC)_2C{=})_2-C_6H_4]$ ..... Fe: Org.Comp.B19-105
– .... $[(1,3,5-(CH_3)_3-C_6H_3)Fe(C_5H_5)][AsF_6]$ .... Fe: Org.Comp.B19-103, 133/6
– .... $[(1,3,5-(CH_3)_3-C_6H_3)Fe(C_5H_5)][B(C_6H_5)_4]$ ..... Fe: Org.Comp.B19-99, 104, 133/6
– .... $[(1,3,5-(CH_3)_3-C_6H_3)Fe(C_5H_5)][BF_4]$ .... Fe: Org.Comp.B19-99, 103/4, 133/6
– .... $[(1,3,5-(CH_3)_3-C_6H_3)Fe(C_5H_5)][B_{11}H_{14}]$ ...... Fe: Org.Comp.B19-99, 104
– .... $[(1,3,5-(CH_3)_3-C_6H_3)Fe(C_5H_5)][Br_3]$ .... Fe: Org.Comp.B19-100, 101/2
– .... $[(1,3,5-(CH_3)_3-C_6H_3)Fe(C_5H_5)][CN]$ .... Fe: Org.Comp.B19-100, 103
– .... $[(1,3,5-(CH_3)_3-C_6H_3)Fe(C_5H_5)][Cr(NH_3)_2(SCN)_4]$ Fe: Org.Comp.B19-99, 104
– .... $[(1,3,5-(CH_3)_3-C_6H_3)Fe(C_5H_5)][I_3]$ .... Fe: Org.Comp.B19-99, 102
– .... $[(1,3,5-(CH_3)_3-C_6H_3)Fe(C_5H_5)][I_x]$ .... Fe: Org.Comp.B19-99, 102
– .... $[(1,3,5-(CH_3)_3-C_6H_3)Fe(C_5H_5)][OC(=O)-CF_3]$ .. Fe: Org.Comp.B19-99, 105, 133/6
– .... $[(1,3,5-(CH_3)_3-C_6H_3)Fe(C_5H_5)][PF_6]$ .... Fe: Org.Comp.B19-99/100, 103, 133/6
– .... $[(1,3,5-(CD_3)_3-C_6H_3)Fe(C_5H_5)][PF_6]$ .... Fe: Org.Comp.B19-133
– .... $[(1,3,5-(CH_3)_3-C_6H_3)Fe(C_5H_5)][SbF_6]$ .... Fe: Org.Comp.B19-103, 133/6
– .... $[(1,3,5-(CH_3)_3-C_6H_3)Fe(C_5H_5)]_2[B_{10}H_{10}]$ ...... Fe: Org.Comp.B19-99, 104
– .... $[(1,3,5-(CH_3)_3-C_6H_3)Fe(C_5H_5)]_2[B_{12}H_{12}]$ ...... Fe: Org.Comp.B19-99, 104
– .... $[(1,3,5-(CH_3)_3-C_6H_3)Fe(C_5H_5)]_2[CoCl_4]$ .... Fe: Org.Comp.B19-105
– .... $[(1,4-(CH_3)_2-C_6H_4)Fe(C_5H_4-CH_3)][BF_4]$ ...... Fe: Org.Comp.B19-2, 13

$C_{14}FeH_{17}^+$ ...... $[(1,4-(CH_3)_2-C_6H_4)Fe(C_5H_4-CH_3)][PF_6]$ ...... Fe: Org.Comp.B19-2, 13
– ...... $[(2-CH_3-C_6H_4-C_2H_5)Fe(C_5H_5)][PF_6]$ ...... Fe: Org.Comp.B19-24
– ...... $[(4-CH_3-C_6H_4-C_2H_5)Fe(C_5H_5)][PF_6]$ ...... Fe: Org.Comp.B19-1, 24
– ...... $[(CH_3-C_5H_4)Fe(C_5H_3(CH_3)-CHCH_3)]^+$ ...... Fe: Org.Comp.A10-223, 225, 227, 232, 233, 235
– ...... $[(CH_3-C_6H_5)Fe(c-C_7H_9)][B(C_6H_5)_4]$ ...... Fe: Org.Comp.B18-209
– ...... $[(CH_3-C_6H_5)Fe(c-C_7H_9)][BF_4]$ ...... Fe: Org.Comp.B18-209
– ...... $[(n-C_3H_7-C_6H_5)Fe(C_5H_5)][B_{11}H_{14}]$ ...... Fe: Org.Comp.B18-197, 215
– ...... $[(n-C_3H_7-C_6H_5)Fe(C_5H_5)][PF_6]$ ...... Fe: Org.Comp.B18-197, 215
– ...... $[(n-C_3H_7-C_6H_5)Fe(C_5H_5)]_2[B_{10}H_{10}]$ ...... Fe: Org.Comp.B18-197, 215
– ...... $[(n-C_3H_7-C_6H_5)Fe(C_5H_5)]_2[B_{12}H_{12}]$ ...... Fe: Org.Comp.B18-197, 215
– ...... $[(i-C_3H_7-C_6H_5)Fe(C_5H_5)][AsF_6]$ ...... Fe: Org.Comp.B18-197, 216, 273/4
– ...... $[(i-C_3H_7-C_6H_5)Fe(C_5H_5)][B_{11}H_{14}]$ ...... Fe: Org.Comp.B18-197, 216
– ...... $[(i-C_3H_7-C_6H_5)Fe(C_5H_5)][CF_3-SO_3]$ ...... Fe: Org.Comp.B18-216, 273/4
– ...... $[(i-C_3H_7-C_6H_5)Fe(C_5H_5)][PF_6]$ ...... Fe: Org.Comp.B18-197, 199, 216, 273/4
– ...... $[(i-C_3H_7-C_6H_5)Fe(C_5H_5)][SbF_6]$ ...... Fe: Org.Comp.B18-197, 216, 273/4
– ...... $[(i-C_3H_7-C_6H_5)Fe(C_5H_5)]_2[1,4-(NC)_2C{=}C_6H_4{=}C(CN)_2]$ Fe: Org.Comp.B18-217
– ...... $[(i-C_3H_7-C_6H_5)Fe(C_5H_5)]_2[B_{10}H_{10}]$ ...... Fe: Org.Comp.B18-197, 216
– ...... $[(i-C_3H_7-C_6H_5)Fe(C_5H_5)]_2[B_{12}H_{12}]$ ...... Fe: Org.Comp.B18-197, 216, 273/5
– ...... $[(C_6H_6)Fe(C_5H_4-C_3H_7-n)][BF_4]$ ...... Fe: Org.Comp.B18-151, 164, 187
– ...... $[(C_6H_6)Fe(C_5H_4-C_3H_7-n)][PF_6]$ ...... Fe: Org.Comp.B18-151, 164, 187
– ...... $[(C_6H_6)Fe(C_5H_4-C_3H_7-i)][BF_4]$ ...... Fe: Org.Comp.B18-151, 164, 187
– ...... $[(C_6H_6)Fe(C_5H_4-C_3H_7-i)][PF_6]$ ...... Fe: Org.Comp.B18-151, 164, 187
– ...... $[(C_6H_6)Fe(C_6H_6-C_2H_5)][PF_6]$ ...... Fe: Org.Comp.B18-153, 172
$C_{14}FeH_{17}I$ ...... $[(1,3,5-(CH_3)_3-C_6H_3)Fe(C_5H_5)]I$ ...... Fe: Org.Comp.B19-100, 101/2, 133/6
$C_{14}FeH_{17}I_3$ ...... $[(1,2,3-(CH_3)_3-C_6H_3)Fe(C_5H_5)][I_3]$ ...... Fe: Org.Comp.B19-99, 120
– ...... $[(1,2,4-(CH_3)_3-C_6H_3)Fe(C_5H_5)][I_3]$ ...... Fe: Org.Comp.B19-99, 121
– ...... $[(1,3,5-(CH_3)_3-C_6H_3)Fe(C_5H_5)][I_3]$ ...... Fe: Org.Comp.B19-99, 102
$C_{14}FeH_{17}I_x$ ...... $[(1,3,5-(CH_3)_3-C_6H_3)Fe(C_5H_5)][I_x]$ ...... Fe: Org.Comp.B19-99, 102
$C_{14}FeH_{17}Li_2N$ .... $(C_5H_5)Fe[C_5H_2(Li)_2-CH_2CH_2-N(CH_3)_2]$ ...... Fe: Org.Comp.A10-327
– ...... $(Li-C_5H_4)Fe[C_5H_3(Li)-CH(CH_3)-N(CH_3)_2]$ ...... Fe: Org.Comp.A10-322/6
– ...... $(Li-C_5H_4)Fe[C_5H_3(Li)-CH(CH_3)-N(CH_3)_2]$ · x $(CH_3)_2N-CH_2CH_2-N(CH_3)_2$ ...... Fe: Org.Comp.A10-322
– ...... $(Li-C_5H_4)Fe[C_5H_3(Li)-CH_2CH_2-N(CH_3)_2]$ ...... Fe: Org.Comp.A10-322/4
$C_{14}FeH_{17}NO$ ..... $(C_5H_5)Fe[C_5H_3(CH{=}O)-CH_2N(CH_3)_2]$ ...... Fe: Org.Comp.A9-38, 49
– ...... $(C_5H_5)Fe[C_5H_3(CH_3)-C(C_2H_5){=}N-OH]$ ...... Fe: Org.Comp.A9-168
– ...... $(C_5H_5)Fe[C_5H_3(C_2H_5)-C(CH_3){=}N-OH]$ ...... Fe: Org.Comp.A9-166, 168
– ...... $(C_5H_5)Fe[C_5H_3(C_3H_7-i)-C(O)NH_2]$ ...... Fe: Org.Comp.A9-91, 93, 98
– ...... $[(CH_3)_2NCH_2-C_5H_4]Fe(C_5H_4-CH{=}O)$ ...... Fe: Org.Comp.A9-12
– ...... $[CH_3C(O)NH-C_5H_4]Fe(C_5H_4-C_2H_5)$ ...... Fe: Org.Comp.A9-83

$C_{14}FeH_{17}NO$	$[HO\text{-}N{=}C(CH_3)\text{-}C_5H_4]Fe(C_5H_4\text{-}C_2H_5)$	Fe: Org.Comp.A9-160
–	$[HO\text{-}N{=}C(C_2H_5)\text{-}C_5H_4]Fe(C_5H_4\text{-}CH_3)$	Fe: Org.Comp.A9-160
–	$[H_2NC(O)\text{-}C_5H_4]Fe(C_5H_4\text{-}C_3H_7\text{-}i)$	Fe: Org.Comp.A9-79, 82
$C_{14}FeH_{17}NO_2$	$(C_5H_5)Fe[C_5H_3(COOH)\text{-}CH_2N(CH_3)_2]$	Fe: Org.Comp.A9-41, 51
–	$(H_2N\text{-}CHCH_3\text{-}C_5H_4)Fe(C_5H_4\text{-}COOCH_3)$	Fe: Org.Comp.A9-3
–	$[C_5(CH_3)_5]Fe(CO)_2\text{-}CH_2CN$	Fe: Org.Comp.B14-70
$C_{14}FeH_{17}NO_2Sn$	$(CH_3)_3SnNC\text{-}Fe(CO)_2(C_8H_8\text{-}c)$	Sn: Org.Comp.18-128
$C_{14}FeH_{17}NO_3$	$(C_5H_5)Fe(CO)_2COCH_2CH(CH_3)NHCH_2CH{=}CH_2$	Fe: Org.Comp.B13-12, 26
–	$(C_5H_5)Fe(CO)_2\text{-}C(O)NH\text{-}C_6H_{11}\text{-}c$	Fe: Org.Comp.B13-25, 46
$C_{14}FeH_{17}NSSi$	$[(CH_3)_3Si\text{-}C_5H_4]Fe(C_5H_4\text{-}SCN)$	Fe: Org.Comp.A9-316, 319
$C_{14}FeH_{17}NSi$	$[(CH_3)_3Si\text{-}C_5H_4]Fe(C_5H_4\text{-}CN)$	Fe: Org.Comp.A9-315
$C_{14}FeH_{17}N_2O_6Re$	$(i\text{-}C_3H_7)_2C_2FeH_3N_2Re(CO)_6$	Re: Org.Comp.1-334/5
$C_{14}FeH_{17}Na$	$(C_5H_5)Fe[C_5H_3(Na)\text{-}C_4H_9\text{-}t]$	Fe: Org.Comp.A10-130/1
–	$(Na\text{-}C_5H_4)Fe(C_5H_4\text{-}C_4H_9\text{-}t)$	Fe: Org.Comp.A10-130/1
$C_{14}FeH_{17}NaO_2$	$Na[(CH_2{=}C(CH_3)C_5(CH_3)_4)Fe(CO)_2]$	Fe: Org.Comp.B14-122
$C_{14}FeH_{17}O^+$	$[(2,6\text{-}(CH_3)_2\text{-}C_6H_3\text{-}OCH_3)Fe(C_5H_5)][BF_4]$	Fe: Org.Comp.B19-99, 126
–	$[(2,6\text{-}(CH_3)_2\text{-}C_6H_3\text{-}OCH_3)Fe(C_5H_5)][Cr(NH_3)_2(SCN)_4]$	Fe: Org.Comp.B19-99, 126
–	$[(2,6\text{-}(CH_3)_2\text{-}C_6H_3\text{-}OCH_3)Fe(C_5H_5)][PF_6]$	Fe: Org.Comp.B19-99, 126
–	$[(2\text{-}CH_3\text{-}C_6H_4\text{-}O\text{-}C_2H_5)Fe(C_5H_5)][PF_6]$	Fe: Org.Comp.B19-2/3, 47/8
–	$[(3\text{-}CH_3\text{-}C_6H_4\text{-}O\text{-}C_2H_5)Fe(C_5H_5)][PF_6]$	Fe: Org.Comp.B19-2/3, 48
–	$[(4\text{-}CH_3\text{-}C_6H_4\text{-}O\text{-}C_2H_5)Fe(C_5H_5)][PF_6]$	Fe: Org.Comp.B19-2/3, 48
–	$[(CH_3\text{-}C_5H_4)Fe(C_5H_3(CH_3)\text{-}C(CH_3){=}OH)]^+$	Fe: Org.Comp.A10-250
–	$[(C_6H_7)Fe(C_7H_{10})(CO)]^+$	Fe: Org.Comp.B17-244
–	$[(C_7H_9)Fe(C_6H_8)(CO)]^+$	Fe: Org.Comp.B17-244
$C_{14}FeH_{17}O_2^+$	$[(1,3\text{-}(CH_3O)_2\text{-}C_6H_3\text{-}2\text{-}CH_3)Fe(C_5H_5)][PF_6]$	Fe: Org.Comp.B19-129
–	$[(CH_3O\text{-}C_6H_6)Fe(CO)(C_6H_8)]^+$	Fe: Org.Comp.B17-244
–	$[(C_5H_5)Fe(CO)_2{=}CH\text{-}(CH_2)_4\text{-}CH{=}CH_2]^+$	Fe: Org.Comp.B16a-88, 95
–	$[(C_5H_5)Fe(CO)_2(CH_2{=}CHC(CH_3)_2CH{=}CH_2)]^+$	Fe: Org.Comp.B17-39
–	$[(C_5H_5)Fe(CO)_2(CH_2{=}CHCH_2CH_2C(CH_3){=}CH_2)]^+$	Fe: Org.Comp.B17-40
–	$[(C_5H_5)Fe(CO)_2(CH_2{=}C_6H_{10}\text{-}c)]^+$	Fe: Org.Comp.B17-72, 82
–	$[(C_5H_5)Fe(CO)_2((CH_3)_2C{=}C{=}C(CH_3)_2)]^+$	Fe: Org.Comp.B17-114, 118
–	$[(C_5H_5)Fe(CO)_2(C_6H_9(CH_3\text{-}1))]^+$	Fe: Org.Comp.B17-91, 104
–	$[(C_5H_5)Fe(CO)_2(C_7H_{12}\text{-}c)]^+$	Fe: Org.Comp.B17-91/2, 104
$C_{14}FeH_{17}O_2^-$	$[(CH_2{=}C(CH_3)C_5(CH_3)_4)Fe(CO)_2]^-$	Fe: Org.Comp.B14-122
$C_{14}FeH_{17}O_3^+$	$[(C_5H_5)Fe(CO)_2(H_3COCH{=}CHCH_2CH_2CH{=}CH_2)]^+$	Fe: Org.Comp.B17-66, 78/9
$C_{14}FeH_{17}O_4^+$	$[(1,2\text{-}(CH_3)_2\text{-}4\text{-}(i\text{-}C_3H_7\text{-}O)\text{-}C_6H_4)Fe(CO)_3]^+$	Fe: Org.Comp.B15-126, 179
–	$[(1\text{-}C_2H_5\text{-}4\text{-}(i\text{-}C_3H_7\text{-}O)\text{-}C_6H_5)Fe(CO)_3]^+$	Fe: Org.Comp.B15-121, 173/4
$C_{14}FeH_{17}O_4Si^+$	$[(2\text{-}CH_3\text{-}6\text{-}(CH_3)_3SiO\text{-}C_7H_5)Fe(CO)_3]^+$	Fe: Org.Comp.B15-231
–	$[(3\text{-}CH_3\text{-}6\text{-}(CH_3)_3SiO\text{-}C_7H_5)Fe(CO)_3]^+$	Fe: Org.Comp.B15-218, 231
–	$[(7\text{-}CH_3\text{-}6\text{-}(CH_3)_3SiO\text{-}C_7H_5)Fe(CO)_3]^+$	Fe: Org.Comp.B15-218, 231
$C_{14}FeH_{17}O_5^+$	$[(1\text{-}CH_3OCH(CH_3)CH_2\text{-}4\text{-}CH_3O\text{-}C_6H_5)Fe(CO)_3]^+$	Fe: Org.Comp.B15-115, 170
–	$[(C_5H_5)Fe(CO)_2(CH_2{=}C(O\text{-}C_2H_5)COO\text{-}C_2H_5)]^+$	Fe: Org.Comp.B17-68, 80
$C_{14}FeH_{18}$	$(CH_3\text{-}C_5H_4)Fe[C_5H_3(CH_3)\text{-}C_2H_5]$	Fe: Org.Comp.A10-191, 194, 201/2, 204
–	$(CH_3\text{-}C_6H_5)Fe(C_6H_7\text{-}CH_3)$	Fe: Org.Comp.B18-103, 111
–	$(C_5H_5)Fe[C_6H_4(CH_3)_3\text{-}1,2,3]$	Fe: Org.Comp.B17-292

$C_{14}FeH_{18}$ .... $(C_5H_5)Fe[C_6H_4(CH_3)_3$-1,2,4]. .... Fe: Org.Comp.B17-292
– .... $(C_5H_5)Fe[C_6H_4(CH_3)_3$-1,3,4]. .... Fe: Org.Comp.B17-293
– .... $(C_5H_5)Fe[C_6H_4(CH_3)_3$-1,3,5]. .... Fe: Org.Comp.B17-293
– .... $(C_5H_5)Fe[C_6H_3D(CH_3)_3$-1,3,5] .... Fe: Org.Comp.B17-296/7
– .... $(C_5H_5)Fe[C_6H_4(CH_3)_3$-1,4,6]. .... Fe: Org.Comp.B17-284
– .... $(C_5H_5)Fe[C_6H_4(CH_3)_3$-2,3,4]. .... Fe: Org.Comp.B17-293
– .... $(C_6H_6)Fe(C_8H_{12})$. .... Fe: Org.Comp.B18-103, 109
– .... $(C_7H_8)Fe(C_7H_{10})$. .... Fe: Org.Comp.B18-103, 127
$C_{14}FeH_{18}IN$. ..... $(C_5H_5)Fe[C_5H_3(I)-CH_2CH_2N(CH_3)_2]$. .... Fe: Org.Comp.A9-43, 58
$C_{14}FeH_{18}LiN$. .... $(C_5H_5)Fe[C_5H_3(Li)-CH(CH_3)-N(CH_3)_2]$ .... Fe: Org.Comp.A10-119/22, 125/6
– .... $(C_5H_5)Fe[C_5H_3(Li)-CH_2CH_2-N(CH_3)_2]$. .... Fe: Org.Comp.A10-118/22
$C_{14}FeH_{18}N^+$ ..... $[(2-CH_3-C_6H_4-N(CH_3)_2)Fe(C_5H_5)][PF_6]$. ...... Fe: Org.Comp.B19-4, 60
– .... $[(3-CH_3-C_6H_4-N(CH_3)_2)Fe(C_5H_5)][PF_6]$. ...... Fe: Org.Comp.B19-4, 60
– .... $[(4-CH_3-C_6H_4-N(CH_3)_2)Fe(C_5H_5)][O-C_6H_2$-2,4,6-$(NO_2)_3]$ Fe: Org.Comp.B19-2, 61
– .... $[(4-CH_3-C_6H_4-N(CH_3)_2)Fe(C_5H_5)][PF_6]$. ...... Fe: Org.Comp.B19-4, 61
– .... $[(NH_2-C_6H_4-4-C_3H_7-i)Fe(C_5H_5)][PF_6]$. .... Fe: Org.Comp.B19-64
$C_{14}FeH_{18}NO^+$.... $[(3-CH_3O-C_6H_4-N(CH_3)_2)Fe(C_5H_5)][PF_6]$ ..... Fe: Org.Comp.B19-77
– .... $[(4-CH_3O-C_6H_4-N(CH_3)_2)Fe(C_5H_5)][PF_6]$ ..... Fe: Org.Comp.B19-77
– .... $[(CH_3CH(OH)CH_2-NH-C_6H_5)Fe(C_5H_5)]^+$ ...... Fe: Org.Comp.B18-142/6, 198, 199/200, 258/9
$C_{14}FeH_{18}NO_2^+$ ... $[(CH_3OC(O)-C_5H_4)Fe(C_5H_4-CH(CH_3)-NH_3)]^+$ .. Fe: Org.Comp.A9-6
– .... $[(C_5H_5)(CO)_2Fe=CH-NH-C_6H_{11}-c]^+$ .... Fe: Org.Comp.B16a-115, 125
– .... $[(C_5H_5)(CO)_2Fe=CD-NH-C_6H_{11}-c]^+$ .... Fe: Org.Comp.B16a-115, 125/6
$C_{14}FeH_{18}NO_2S^+$.. $[(C_5H_5)(CO)_2Fe=C(SCH_3)-NC_5H_{10}]^+$ .... Fe: Org.Comp.B16a-146, 159
$C_{14}FeH_{18}NO_3^+$... $[(C_5H_5)(CO)_2Fe=C(OCH_3)-NC_5H_{10}]^+$. .... Fe: Org.Comp.B16a-138
$C_{14}FeH_{18}N_2O_3$ ... $O_2N-N(CH_3)CH_2-C_5H_4FeC_5H_4-CH_2OCH_3$. .... Fe: Org.Comp.A9-156
$C_{14}FeH_{18}N_4O_4$ ... $Fe[C_5H_4-CH_2N(CH_3)NO_2]_2$. .... Fe: Org.Comp.A9-152
$C_{14}FeH_{18}O$ ...... $(CH_3-C_5H_4)Fe[C_5H_3(CH_3)-CH(CH_3)-OH]$ ..... Fe: Org.Comp.A10-218, 220/1, 228, 229, 231
– .... $(C_5H_5)Fe[C_6H_4(CH_3)_2$-1,3-$OCH_3$-2]. .... Fe: Org.Comp.B17-294
– .... $(C_5H_5)Fe[C_6H_4(CH_3)_2$-2,4-$OCH_3$-3]. .... Fe: Org.Comp.B17-294
– .... $(C_5H_5)Fe[C_6H_6-C(CH_3)_2OH$-1]. .... Fe: Org.Comp.B17-265, 307
– .... $(C_5H_5)Fe[C_6H_5D-C(CH_3)_2OH$-1] .... Fe: Org.Comp.B17-271
– .... $(C_5H_5)Fe(c-C_8H_{13})(CO)$ .... Fe: Org.Comp.B17-161
$C_{14}FeH_{18}OSi$ .... $(C_5H_5)Fe[C_5H_3(CH=O)-Si(CH_3)_3]$. .... Fe: Org.Comp.A9-330, 333/4
– .... $[(CH_3)_3Si-C_5H_4]Fe(C_5H_4-CH=O)$ .... Fe: Org.Comp.A9-315
$C_{14}FeH_{18}O_2$ ..... $(C_5H_5)Fe(CO)_2-C_3H_3(C_2H_5)_2$-2,3. .... Fe: Org.Comp.B13-195/6
– .... $(C_5H_5)Fe(CO)_2-c-C_6H_{10}-CH_3$-1 .... Fe: Org.Comp.B13-215, 252
– .... $(C_5H_5)Fe(CO)_2-c-C_6H_{10}-CH_3$-4 .... Fe: Org.Comp.B13-215, 252/3
– .... $[(CH_3)_6C_6]Fe(CO)_2$ .... Fe: Org.Comp.B18-21
$C_{14}FeH_{18}O_2S_2$ ... $C_5(CH_3)_5Fe(CO)_2C(S)SCH_3$ .... Fe: Org.Comp.B14-72
$C_{14}FeH_{18}O_2Si$. ... $(CH_3)_3Si-C_5H_4FeC_5H_4-C(O)OH$. .... Fe: Org.Comp.A9-315
$C_{14}FeH_{18}O_3$ ..... $(C_5H_5)Fe(CO)[=C(OCH_3)-O-C_6H_{10}-]$. .... Fe: Org.Comp.B16b-126, 137
– .... $(C_5H_5)Fe(CO)[CH_2CHC(CH_3)COO-C_3H_7-i]$ .... Fe: Org.Comp.B17-151
– .... $(C_5H_5)Fe(CO)_2-C(O)CH_2CH_2-C_4H_9-t$ .... Fe: Org.Comp.B13-8, 17
– .... $(C_5H_5)Fe(CO)_2-c-C_6H_{10}-OCH_3$-2 .... Fe: Org.Comp.B13-215, 253
– .... $[C_5(CH_3)_5]Fe(CO)_2-C(O)CH_3$ .... Fe: Org.Comp.B14-71, 79

$C_{14}FeH_{18}O_3$	$[C_5(CH_3)_6]Fe(CO)_3$	Fe: Org.Comp.B14-75/6
$C_{14}FeH_{18}O_3P^+$	$[(C_6H_6)Fe(C_6H_6-P(=O)(OCH_3)_2)]^+$	Fe: Org.Comp.B18-142/6, 174
$C_{14}FeH_{18}O_4$	$(C_5H_5)Fe(CO)_2CH[-O-CH(CH_3)-CH(CH_3)-O-CH(CH_3)-]$	Fe: Org.Comp.B14-2, 8, 14/5
$C_{14}FeH_{18}O_5PS^+$	$[(C_5H_5)Fe(CO)_2CH(SCH_3)P(-OCH_2-)_3CCH_3]^+$	Fe: Org.Comp.B14-133, 140/1, 144, 145
$C_{14}FeH_{18}O_8S_2$	$Fe(C_5H_4SO_3H)_2 \cdot C_4H_8O_2$	Fe: Org.Comp.A9-240
$C_{14}FeH_{19}^-$	$[CH_3C_5H_4Fe(C_8H_{12})]^-$	Fe: Org.Comp.B17-189, 196
$C_{14}FeH_{19}IN^+$	$[(C_5H_5)Fe(C_5H_3(I)-CH_2CH_2NH(CH_3)_2)]^+$	Fe: Org.Comp.A9-58
–	$[(C_5H_5)Fe(C_5H_3(I)-CH_2N(CH_3)_3)]^+$	Fe: Org.Comp.A9-64
$C_{14}FeH_{19}IN_2O$	$[(C_5H_5)Fe(CN-C_3H_7-i)_2CO]I$	Fe: Org.Comp.B15-336
$C_{14}FeH_{19}I_2N$	$[C_5H_5FeC_5H_3(I)-CH_2N(CH_3)_3]I$	Fe: Org.Comp.A9-64
$C_{14}FeH_{19}N$	$(C_2H_5-NH-C_5H_4)Fe(C_5H_4-C_2H_5)$	Fe: Org.Comp.A9-6
–	$(C_5H_5)Fe[C_5H_3(CH_3)-CH_2N(CH_3)_2]$	Fe: Org.Comp.A9-32, 44, 72
–	$(C_5H_5)Fe[C_5H_3(C_2H_5)-NH-C_2H_5]$	Fe: Org.Comp.A9-27, 71
–	$[(CH_3)_2NCH_2-C_5H_4]Fe(C_5H_4-CH_3)$	Fe: Org.Comp.A9-11, 16/7
$C_{14}FeH_{19}NO$	$(C_5H_5)Fe[C_5H_3(CH_2OH)-CH_2N(CH_3)_2]$	Fe: Org.Comp.A9-36, 46/7
–	$(C_5H_5)Fe[C_5H_3(OCH_3)-CH_2N(CH_3)_2]$	Fe: Org.Comp.A9-38, 49, 73
–	$[(CH_3)_2NCH_2-C_5H_4]Fe(C_5H_4-OCH_3)$	Fe: Org.Comp.A9-14
$C_{14}FeH_{19}NO_2$	$(C_5H_5)Fe(CO)_2CH_2CH_2-1-NC_5H_{10}$	Fe: Org.Comp.B13-3, 5/6
$C_{14}FeH_{19}NS$	$(CH_3S-C_5H_4)Fe[C_5H_4-CH_2N(CH_3)_2]$	Fe: Org.Comp.A9-202, 206
–	$(C_5H_5)Fe[C_5H_3(SCH_3)-CH_2N(CH_3)_2]$	Fe: Org.Comp.A9-216/7, 220, 225
$C_{14}FeH_{19}NS_2Si$	$(CH_3)_3Si-C_5H_4FeC_5H_4SCSNH_2$	Fe: Org.Comp.A9-316
$C_{14}FeH_{19}NSe$	$(C_5H_5)Fe[C_5H_3(SeCH_3)CH_2N(CH_3)_2]$	Fe: Org.Comp.A9-272
$C_{14}FeH_{19}N_2^+$	$[(2-CH_3-C_6H_4-NH-CH_2CH_2-NH_2)Fe(C_5H_5)][PF_6]$	Fe: Org.Comp.B19-2, 58
–	$[(3-CH_3-C_6H_4-NH-CH_2CH_2-NH_2)Fe(C_5H_5)][PF_6]$	Fe: Org.Comp.B19-2, 58
–	$[(4-CH_3-C_6H_4-NH-CH_2CH_2-NH_2)Fe(C_5H_5)][PF_6]$	Fe: Org.Comp.B19-2, 58
$C_{14}FeH_{19}N_2O^+$	$[(C_5H_5)Fe(CN-C_3H_7-i)_2CO]^+$	Fe: Org.Comp.B15-336
$C_{14}FeH_{19}N_3Si$	$(C_5H_5)Fe[C_5H_3(Si(CH_3)_3)CH_2N_3]$	Fe: Org.Comp.A9-331, 335
$C_{14}FeH_{19}O^+$	$[(1,3,5-(CH_3)_3-C_6H_3)Fe(CO)(CH_2C(CH_3)CH_2)]^+$	Fe: Org.Comp.B18-95/7
$C_{14}FeH_{19}O_2^+$	$[((CH_3)_5C_5)Fe(CO)_2=CH-CH_3]^+$	Fe: Org.Comp.B16a-104
–	$[((CH_3)_5C_5)Fe(CO)_2(CH_2=CH_2)]^+$	Fe: Org.Comp.B17-131
–	$[(C_5H_5)Fe(CO)_2(CH_2=CH-C_5H_{11}-n)]^+$	Fe: Org.Comp.B17-13/4
–	$[(C_5H_5)Fe(CO)_2(CH_3CH=CH-C_4H_9-n)]^+$	Fe: Org.Comp.B17-58
$C_{14}FeH_{19}O_3^+$	$[(CH_2(CH)_4-C_6H_{13}-n)Fe(CO)_3]^+$	Fe: Org.Comp.B15-27
–	$[(CH_3-CHC(CH_3)CHCHCH-C_4H_9-n)Fe(CO)_3]^+$	Fe: Org.Comp.B15-14, 25, 30
–	$[(C_5(CH_3)_5)Fe(CO)_2=CH(OCH_3)]^+$	Fe: Org.Comp.B16a-129
–	$[(C_5H_5)Fe(CO)_2=C(C_4H_9-t)O-C_2H_5]^+$	Fe: Org.Comp.B16a-111
–	$[(C_5H_5)Fe(CO)_2(C_2H_5CH=C(C_2H_5)OCH_3)]^+$	Fe: Org.Comp.B17-85
$C_{14}FeH_{19}O_3Si^+$	$[3-(CH_3)_3Si-6,6-(CH_3)_2-C_6H_4Fe(CO)_3]^+$	Fe: Org.Comp.B15-132, 183
$C_{14}FeH_{19}O_4^+$	$[(C_5H_5)Fe(CO)_2(CH_2=CHCH(O-C_2H_5)_2)]^+$	Fe: Org.Comp.B17-29
–	$[(C_5H_5)Fe(CO)_2(CH_3CH=CHOCH(CH_3)CH(CH_3)OH)]^+$	Fe: Org.Comp.B17-65, 78
$C_{14}FeH_{19}S^+$	$[CH_3C_5H_4FeC_4S((CH_3)_4-2,3,4,5)]^+$	Fe: Org.Comp.B17-221
$C_{14}FeH_{20}INO$	$[C_5H_5FeC_5H_3(CH_2N(CH_3)_3)OH]I$	Fe: Org.Comp.A9-64
$C_{14}FeH_{20}N^+$	$[(C_5H_5)Fe(1,2,3,4,5-(CH_3)_5NC_4)]^+$	Fe: Org.Comp.B17-225, 235/6

$C_{14}FeH_{20}N^+$	$[(C_5H_5)Fe(CH_3\text{-}C_5H_3\text{-}CH_2NH(CH_3)_2)]^+$	Fe: Org.Comp.A9-44
$C_{14}FeH_{20}NO^+$	$[(C_5H_5)Fe(C_5H_3(CH_2OH)\text{-}CH_2NH(CH_3)_2)]^+$	Fe: Org.Comp.A9-47
–	$[(C_5H_5)Fe(C_5H_3(OCH_3)\text{-}CH_2NH(CH_3)_2)]^+$	Fe: Org.Comp.A9-49
–	$[(C_5H_5)Fe(C_5H_3(OH)\text{-}CH_2N(CH_3)_3)]^+$	Fe: Org.Comp.A9-64
$C_{14}FeH_{20}NP$	$N(CH_3)_2CH_2C_5H_4FeC_4H_2P((CH_3)_2\text{-}3,4)$	Fe: Org.Comp.B17-220
$C_{14}FeH_{20}NS^+$	$[C_5H_5FeC_5H_3(SCH_3)CH_2NH(CH_3)_2]^+$	Fe: Org.Comp.A9-220
$C_{14}FeH_{20}N_2$	$Fe(C_5H_4\text{-}CHCH_3\text{-}NH_2)_2$	Fe: Org.Comp.A9-2, 4
–	$Fe(C_5H_4\text{-}CH_2CH_2\text{-}NH_2)_2$	Fe: Org.Comp.A9-2, 4
–	$Fe[C_5H_4\text{-}N(CH_3)_2]_2$	Fe: Org.Comp.A9-9, 14
$C_{14}FeH_{20}N_2{}^+$	$[Fe(C_5H_4\text{-}N(CH_3)_2)_2]^+$	Fe: Org.Comp.A9-14
$C_{14}FeH_{20}N_2O_2$	$(C_5H_5)Fe(CN\text{-}C_3H_7\text{-}n)(CO)C(O)NH\text{-}C_3H_7\text{-}n$	Fe: Org.Comp.B15-304
–	$(C_5H_5)Fe(CN\text{-}C_3H_7\text{-}i)(CO)C(O)NH\text{-}C_3H_7\text{-}i$	Fe: Org.Comp.B15-304
$C_{14}FeH_{20}N_2O_4$	$(C_5H_5)Fe(CN\text{-}CH_2CH_2CH_2OH)$ $(CO)C(O)NHCH_2CH_2CH_2OH$	Fe: Org.Comp.B15-304
$C_{14}FeH_{20}OSi$	$(C_5H_5)Fe[C_5H_3(CH_2OH)\text{-}Si(CH_3)_3]$	Fe: Org.Comp.A9-330, 333
–	$[(CH_3)_3Si\text{-}C_5H_4]Fe(C_5H_4\text{-}CH_2OH)$	Fe: Org.Comp.A9-315, 319
$C_{14}FeH_{20}O_2$	$C_5(CH_3)_5Fe(CO)_2C_2H_5$	Fe: Org.Comp.B14-69
$C_{14}FeH_{20}O_2Si$	$C_5H_4Si(CH_3)_3Fe(CO)_2CH_2C(CH_3)=CH_2$	Fe: Org.Comp.B14-59
$C_{14}FeH_{20}O_3$	$C_5(CH_3)_5Fe(CO)_2CH_2OCH_3$	Fe: Org.Comp.B14-70, 77
$C_{14}FeH_{20}O_3P^+$	$[(C_5H_5)Fe(CO)_2\text{-}C(O)C(CH_3)_2\text{-}P(CH_3)_3]^+$	Fe: Org.Comp.B14-154
–	$[(C_5H_5)Fe(CO)_2\text{-}C(OCH_3)=C(CH_3)\text{-}P(CH_3)_3]^+$	Fe: Org.Comp.B14-153
$C_{14}FeH_{20}O_4P^+$	$[(C_5H_5)Fe(CH_2=CHCH_3)(CO)P(OCH_2)_3CCH_3]^+$	Fe: Org.Comp.B16b-83, 90
$C_{14}FeH_{20}O_5P^+$	$[C_5H_5(CO)(P(OCH_3)_3)Fe=C(CH_3)OCH_2\text{-}CCH]^+$	Fe: Org.Comp.B16a-38, 53
$C_{14}FeH_{20}P_2$	$[(CH_3)_2P\text{-}C_5H_4]_2Fe$	Fe: Org.Comp.A10-2/3, 5, 11
$C_{14}FeH_{20}S_2{}^+$	$[Fe(C_5H_4S(CH_3)_2)_2]^+$	Fe: Org.Comp.A9-197
$C_{14}FeH_{20}Si$	$(CH_3\text{-}C_5H_4)Fe[C_5H_4\text{-}Si(CH_3)_3]$	Fe: Org.Comp.A9-314
–	$(C_5H_5)Fe[C_5H_3(CH_3)\text{-}Si(CH_3)_3]$	Fe: Org.Comp.A9-329, 332
$C_{14}FeH_{21}NSi$	$(C_5H_5)Fe[C_5H_3(Si(CH_3)_3)CH_2NH_2]$	Fe: Org.Comp.A9-330, 334
$C_{14}FeH_{21}N_2OP$	$C_5H_5Fe(CNCH_3)_2C(O)CH=P(CH_3)_3$	Fe: Org.Comp.B15-330
$C_{14}FeH_{21}O_2Si^+$	$[((CH_3)_3Si\text{-}C_5H_4)Fe(CO)_2(CH_2=C(CH_3)_2)]^+$	Fe: Org.Comp.B17-132
–	$[(n\text{-}C_3H_7\text{-}Si(CH_3)_2\text{-}C_5H_4)Fe(CO)_2(CH_2=CH_2)]^+$	Fe: Org.Comp.B17-132
$C_{14}FeH_{22}$	$(CH_2C(CH_3)CHC(CH_3)CH_2)_2Fe$	Fe: Org.Comp.B17-248, 250/1
–	$(CH_2CHC(CH_3)C(CH_3)CH_2)_2Fe$	Fe: Org.Comp.B17-247, 250
$C_{14}FeH_{22}IN_3O$	$[(C_5H_5)(C_2H_5\text{-}NC)(CO)Fe=C(NH\text{-}C_2H_5)_2]I$	Fe: Org.Comp.B16a-165
–	$[(C_5H_5)(t\text{-}C_4H_9\text{-}NC)(CO)Fe=C(NHCH_3)_2]I$	Fe: Org.Comp.B16a-164, 168
$C_{14}FeH_{22}NO_2{}^+$	$[C_5H_5Fe(CO)_2CH_2N(C_2H_5)_3]^+$	Fe: Org.Comp.B14-136
$C_{14}FeH_{22}N_2{}^{2+}$	$[Fe(C_5H_4\text{-}CH(CH_3)NH_3)_2]^{2+}$	Fe: Org.Comp.A9-4
$C_{14}FeH_{22}N_2O_{11}P_2S$	$Fe(CO)_2(P(OCH_3)_3)_2(O=S=NC_6H_4NO_2\text{-}4)$	S: S-N Comp.6-273/4
$C_{14}FeH_{22}N_3O^+$	$[(C_5H_5)(C_2H_5\text{-}NC)(CO)Fe=C(NH\text{-}C_2H_5)_2]^+$	Fe: Org.Comp.B16a-165
–	$[(C_5H_5)(i\text{-}C_3H_7\text{-}NC)(CO)Fe=C(NHCH_3)\text{-}N(CH_3)_2]^+$	Fe: Org.Comp.B16a-165, 168
–	$[(C_5H_5)(t\text{-}C_4H_9\text{-}NC)(CO)Fe=C(NHCH_3)_2]^+$	Fe: Org.Comp.B16a-164, 168
$C_{14}FeH_{22}O_2Si$	$[n\text{-}C_3H_7\text{-}Si(CH_3)_2\text{-}C_5H_4]Fe(CO)_2\text{-}C_2H_5$	Fe: Org.Comp.B14-60
–	$[C_5H_4\text{-}Si(CH_3)_3]Fe(CO)_2\text{-}C_4H_9\text{-}n$	Fe: Org.Comp.B14-59
$C_{14}FeH_{22}O_2Si_2$	$Fe[C_5H_4Si(CH_3)_2OH]_2$	Fe: Org.Comp.A9-296, 307
$C_{14}FeH_{22}O_4P^+$	$[(C_5H_5)(CO)P(OCH_3)_3Fe(c\text{-}C_4H_5CH_3)]^+$	Fe: Org.Comp.B16b-91/2
–	$[C_5H_5(CO)(P(OCH_3)_3)Fe=CH(CH=C(CH_3)_2)]^+$	Fe: Org.Comp.B16a-17, 20, 26
–	$[C_5H_5(CO)(P(OCH_3)_3)Fe=CH(CH=CHC_2H_5)]^+$	Fe: Org.Comp.B16a-19, 26

$C_{14}FeH_{22}O_5P^+$ . . . $[C_5H_5(CO)(P(OCH_3)_3)Fe{=}C(CH_3)OCH_2CH{=}CH_2]^+$ Fe: Org.Comp.B16a-38, 53

$C_{14}FeH_{22}Si_2$ . . . . . $Fe[C_5H_4{-}SiH(CH_3)_2]_2$ . . . . Fe: Org.Comp.A9-294, 305

$C_{14}FeH_{23}NO_2Sn$ . . $(CH_3)_3SnNC{-}Fe(CO)_2(i{-}C_3H_7CH{=}CHC(CH_3){=}CH_2)$ Sn: Org.Comp.18-128

$C_{14}FeH_{23}NPS^+$ . . . $[(C_5H_5)Fe(CS)(NCCH_3)P(C_2H_5)_3]^+$ . . . . . . Fe: Org.Comp.B15-264

$C_{14}FeH_{24}IO_2P$. . . . $[C_5H_5(CO)(P(CH_3)_3)Fe{=}C(C_3H_7{-}i)OCH_3]I$ . . . . . Fe: Org.Comp.B16a-41, 55

$C_{14}FeH_{24}OP^+$ . . . . $[C_5H_5(CO)(P(C_2H_5)_3)Fe{=}CH(CH_3)]^+$ . . . . . . Fe: Org.Comp.B16a-15, 18, 25

$C_{14}FeH_{24}O_2P^+$ . . . $[C_5H_5(CO)(P(CH_3)_3)Fe{=}C(C_3H_7{-}i)OCH_3]^+$ . . . . . Fe: Org.Comp.B16a-41, 55

– . . . . $[C_5H_5(CO)(P(C_2H_5)_3)Fe{=}C(CH_3)OH]^+$ . . . . . . Fe: Org.Comp.B16a-35, 51

$C_{14}FeH_{24}O_2Si_2$ . . . $(C_5H_4{-}CH_3)Fe(CO)_2{-}CH_2Si(CH_3)_2Si(CH_3)_3$ . . . . Fe: Org.Comp.B14-52/3, 55

– . . . . $[C_5H_4{-}Si(CH_3)_2CH_2Si(CH_3)_3]Fe(CO)_2{-}CH_3$ . . . . Fe: Org.Comp.B14-52/3, 60

$C_{14}FeH_{24}O_5P^+$ . . . $[C_5H_5(CO)(P(OCH_3)_3)Fe{=}C(C_3H_7{-}i)OCH_3]^+$ . . . Fe: Org.Comp.B16a-50

$C_{14}FeH_{24}O_6P^+$ . . . $[C_5H_5(CO)(P(OCH_3)_3)Fe{=}C(CH_2OCH_3)OC_2H_5]^+$ Fe: Org.Comp.B16a-43

$C_{14}FeH_{25}O_2PSi$. . . $[(CH_3)_3PCH_2Si(CH_3)_3][C_5H_5Fe(CO)_2]$ . . . . Fe: Org.Comp.B14-112

$C_{14}FeH_{28}O_6P_2$ . . . $[1,4{-}(CH_3)_2C_6H_4]Fe[P(OCH_3)_3]_2$ . . . . Fe: Org.Comp.B18-2/3, 7, 12

$C_{14}FeH_{29}OP_2{}^+$ . . . $[C_5H_5(P(CH_3)_3)_2Fe{=}C(CH_3)OCH_3]^+$ . . . . Fe: Org.Comp.B16a-8/9

$C_{14}FeH_{29}O_6P_2{}^+$ . . $[(CH_3C_6H_5)Fe(P(OCH_3)_3)_2(CH_3)]^+$ . . . . Fe: Org.Comp.B18-21

– . . . . $[(C_5H_5)Fe(CH_2{=}CHCH_3)(P(OCH_3)_3)_2]^+$ . . . . Fe: Org.Comp.B16b-5/6, 8/9

$C_{14}FeMnO_{14}Re$ . . $(CO)_5ReFe(CO)_4Mn(CO)_5$. . . . . Re: Org.Comp.2-200/1, 210

$C_{14}Fe_2H_{10}O_2S_3$ . . $(C_5H_5)(CO)_2Fe{=}C({-}S{-})_2Fe(CS)(C_5H_5)$ . . . . Fe: Org.Comp.B16a-182, 188

$C_{14}Fe_2H_{10}O_3S_2$ . . $(C_5H_5)(CO)_2Fe{=}C({-}S{-})_2Fe(CO)(C_5H_5)$ . . . . Fe: Org.Comp.B16a-182, 187/8

$C_{14}Fe_3HMnO_{13}$. . . $(H)CFe_3Mn(CO)_{13}$. . . . . Fe: Org.Comp.C6b-75/6

$C_{14}Fe_3HMoO_{14}{}^-$ . . $[Fe_3MoH(CO)_{14}]^-$ . . . . Fe: Org.Comp.C6b-146, 147

$C_{14}Fe_3HO_{13}W^-$ . . . $[HCFe_3W(CO)_{13}]^-$ . . . . Fe: Org.Comp.C6b-74

$C_{14}Fe_3HO_{14}SW^-$ $[(CO)_9(H)Fe_3SW(CO)_5]^-$ . . . . Fe: Org.Comp.C6a-135

$C_{14}Fe_3HO_{14}W^-$ . . . $[Fe_3WH(CO)_{14}]^-$ . . . . Fe: Org.Comp.C6b-146, 147

$C_{14}Fe_3H_2O_{13}W$ . . . $HCFe_3W(CO)_{13}(H)$. . . . . Fe: Org.Comp.C6b-73/4

$C_{14}Fe_3H_5NO_{11}$ . . . $(CO)_{11}Fe_3(CN{-}C_2H_5)$ . . . . Fe: Org.Comp.C6b-142

$C_{14}Fe_3H_5O_{11}{}^-$. . . . $[(CO)_{10}Fe_3(C{-}CH_2{-}C({=}O){-}CH_3)]^-$ . . . . Fe: Org.Comp.C6b-60, 62, 65

$C_{14}Fe_3H_5O_{12}{}^-$. . . . $[(CO)_{10}Fe_3(C{-}CH_2{-}COO{-}CH_3)]^-$ . . . . Fe: Org.Comp.C6b-60, 62, 65

$C_{14}Fe_3H_6N_2O_{10}$ . . $(CO)_{10}Fe_3(CN{-}CH_3)_2$ . . . . Fe: Org.Comp.C6b-13/4

$C_{14}Fe_3H_7O_{12}{}^-$. . . . $[(CO)_{10}(H)Fe_3(CH{-}C(O)O{-}C_2H_5)]^-$. . . . . Fe: Org.Comp.C6b-60

$C_{14}Fe_3H_8NO_{10}{}^-$ . . $[(CO)_{10}(H)Fe_3(CN{-}C_3H_7{-}i)]^-$ . . . . Fe: Org.Comp.C6b-21/2

$C_{14}Fe_3H_8N_2O_9$ . . . $(CO)_9Fe_3[(2.2.1){-}2,3{-}N_2C_5H_8]$ . . . . Fe: Org.Comp.C6a-182, 184/5

$C_{14}Fe_3H_8O_{10}$ . . . . $(CO)_9Fe_3(C{-}CH_3)(C{-}O{-}C_2H_5)$ . . . . Fe: Org.Comp.C6a-234, 235, 238, 240

$C_{14}Fe_3H_9NO_9$ . . . . $(CO)_9Fe_3(CN{-}C_4H_9{-}t)$. . . . . Fe: Org.Comp.C6a-244/6

$C_{14}Fe_3H_9NO_9Te_2$ $(CO)_9Fe_3Te_2(CN{-}C_4H_9{-}t)$. . . . . Fe: Org.Comp.C6a-124

$C_{14}Fe_3H_9NO_{10}$ . . . $(CO)_{10}(H)Fe_3(C{-}NH{-}C_3H_7{-}i)$ . . . . Fe: Org.Comp.C6b-50, 53

– . . . . $(CO)_{10}(H)Fe_3[C{-}N(CH_3){-}C_2H_5]$ . . . . Fe: Org.Comp.C6b-50, 53, 59

$C_{14}Fe_3H_9O_{10}P$ . . . $(CO)_{10}Fe_3P{-}C_4H_9{-}n$. Fe: Org.Comp.C6b-27, 29, 31

– . . . . $(CO)_{10}Fe_3P{-}C_4H_9{-}t$ . . . . Fe: Org.Comp.C6b-27, 29, 31, 34/6

$C_{14}Fe_3H_9O_{11}P$ . . . $(CO)_{10}Fe_3P(C_4H_9{-}t)({=}O)$. . . . . Fe: Org.Comp.C6b-43

$C_{14}Fe_3H_9O_{12}PTe_2$ $(CO)_9Fe_3Te_2[P({-}OCH_2{-})_3C{-}CH_3]$. . . . . Fe: Org.Comp.C6a-121

$C_{14}Fe_3H_9O_{14}P$ . . . $(CO)_{11}Fe_3[P(OCH_3)_3]$ . . . . Fe: Org.Comp.C6b-128, 131/2, 134/7, 139

$C_{14}Fe_3H_9O_{14}P^-$ ... $[(CO)_{11}Fe_3(P(O-CH_3)_3)]^-$, radical anion ...... Fe: Org.Comp.C6b-135, 136
$C_{14}Fe_3H_{10}KNO_9$ .. $K[(CO)_9Fe_3(N(C_4H_9\text{-}t)=CH)]$ .... Fe: Org.Comp.C6a-287, 288, 289
$C_{14}Fe_3H_{10}NO_9^-$ .. $[(CO)_9Fe_3(N(C_4H_9\text{-}t)=CH)]^-$ .... Fe: Org.Comp.C6a-286/9
$C_{14}Fe_3H_{10}NO_{10}P$ $(CO)_{10}Fe_3P-N(C_2H_5)_2$ .... Fe: Org.Comp.C6b-30, 31, 37/8
$C_{14}Fe_3H_{10}N_4O_8S_2$ $(CO)_8Fe_3[1,3-N_2C_3H_5(=S)-2]_2$ .... Fe: Org.Comp.C6a-51/2
$C_{14}Fe_3H_{10}O_9P^-$ ... $[(CO)_9(H)Fe_3(P-C-C_4H_9\text{-}t)]^-$ .... Fe: Org.Comp.C6a-285/6
$C_{14}Fe_3H_{11}NO_9$ ... $(CO)_9(H)Fe_3[N(C_4H_9\text{-}t)=CH]$ .... Fe: Org.Comp.C6a-255/7, 259
$C_{14}Fe_3H_{12}O_9$ .... $(CO)_9(H)_3Fe_3C-C_4H_9$ .... Fe: Org.Comp.C6a-249, 250
$C_{14}Fe_3H_{12}O_9P_2$ .. $(CO)_9Fe_3(P-CH_3)(P-C_4H_9\text{-}t)$ .... Fe: Org.Comp.C6a-156/7, 161/2, 168/9
$C_{14}Fe_3H_{12}O_9S_2$ .. $(CO)_9Fe_3(S-CH_3)(S-C_4H_9\text{-}t)$ .... Fe: Org.Comp.C6a-100/2
$C_{14}Fe_3H_{13}O_{10}P$ .. $(CO)_9(H)Fe_3(i\text{-}C_3H_7-P-O-C_2H_5)$ .... Fe: Org.Comp.C6a-195, 197/8
$C_{14}Fe_3H_{20}NO_{11}PSi$
$(CO)_8(H)_2Fe_3N-Si(CH_3)_3[P(OCH_3)_3]$ .... Fe: Org.Comp.C6a-58
$C_{14}Fe_3MnO_{13}^-$ ... $[CFe_3Mn(CO)_{13}]^-$ .... Fe: Org.Comp.C6b-76
$C_{14}Fe_3MoO_{14}^{2-}$ .. $[Fe_3Mo(CO)_{14}]^{2-}$ .... Fe: Org.Comp.C6b-146/7
$C_{14}Fe_3O_{13}W^{2-}$ ... $[CFe_3W(CO)_{13}]^{2-}$ .... Fe: Org.Comp.C6b-74/5
$C_{14}Fe_3O_{14}SW^{2-}$ .. $[(CO)_9Fe_3SW(CO)_5]^{2-}$ .... Fe: Org.Comp.C6a-135
$C_{14}Fe_3O_{14}S_2W$ ... $(CO)_9Fe_3S[S-W(CO)_5]$ .... Fe: Org.Comp.C6a-301
$C_{14}Fe_3O_{14}W^{2-}$ ... $[Fe_3W(CO)_{14}]^{2-}$ .... Fe: Org.Comp.C6b-146/7, 149
$C_{14}GaH_6N_4O_{14}^+$ .. $[Ga(OC(O)-C_6H_2-3,5-(NO_2)_2-2-OH)_2]^+$ ...... Ga: SVol.D1-176/80
$C_{14}GaH_6O_{10}^{3-}$ ... $[Ga(H(OC(O)-C_6H_2-3,4,5-(O)_3))_2]^{3-}$ .... Ga: SVol.D1-182/3
$C_{14}GaH_6O_{12}S_2^{3-}$ $[Ga(OC(O)-C_6H_3-5-SO_3-2-O)_2]^{3-}$ .... Ga: SVol.D1-176/80
$C_{14}GaH_8I_2O_2$ .... $GaI_2[9,10-(O)_2C_{14}H_8]$, radical .... Ga: SVol.D1-90/2
$C_{14}GaH_8N_2O_8^-$ ... $[Ga(2-(NC_5H_4-2)-NC_5H_4)(OOC-COO)_2]^-$ ..... Ga: SVol.D1-264
$C_{14}GaH_8N_2O_{10}^+$ .. $[Ga(OC(O)-C_6H_3-5-NO_2-2-OH)_2]^+$ .... Ga: SVol.D1-176/80
$C_{14}GaH_8N_6S_4^-$ ... $[2-(NC_5H_5-2)-NC_5H_4][Ga(NCS)_4(2-(NC_5H_4-2)-NC_5H_4)]$ Ga: SVol.D1-269/70
– .... $[Ga(NCS)_2(2-(NC_5H_4-2)-NC_5H_4)_2][Ga(NCS)_4(2-(NC_5H_4-2)-NC_5H_4)]$ .... Ga: SVol.D1-269/70
$C_{14}GaH_8O_7S^+$ ... $[Ga((O)C_{14}H_5(OH)_3-SO_3)]^+$ .... Ga: SVol.D1-35
$C_{14}GaH_8O_8^-$ ..... $[Ga(H(OC(O)-C_6H_3-3,4-(O)_2))_2]^-$ .... Ga: SVol.D1-182/3
– .... $[Ga(H(OC(O)-C_6H_3-3,5-(O)_2))_2]^-$ .... Ga: SVol.D1-182/3
$C_{14}GaH_9O_7S^{2+}$ ... $[Ga((HO)_4C_{14}H_5-SO_3)]^{2+}$ .... Ga: SVol.D1-35
$C_{14}GaH_{10}Na_3O_{14}S_2$
$Na_3[(HO)_2Ga(OC(O)-C_6H_3-5-SO_3-2-OH)_2] \cdot 4\ H_2O$ Ga: SVol.D1-176, 182
$C_{14}GaH_{10}O_6^+$ .... $[Ga(OC(O)-C_6H_4-2-OH)_2]^+$ .... Ga: SVol.D1-176/80
$C_{14}GaH_{10}O_{14}S_2^{3-}$ $Cs_3[(H_2O)_2Ga(OC(O)-C_6H_3-5-SO_3-2-O)_2] \cdot 2\ H_2O$ Ga: SVol.D1-176, 181/2
– .... $Na_3[(HO)_2Ga(OC(O)-C_6H_3-5-SO_3-2-OH)_2] \cdot 4\ H_2O$ Ga: SVol.D1-176, 182
$C_{14}GaH_{11}O_5$ ..... $Ga(OH)[OC(O)-C_6H_5]_2$ .... Ga: SVol.D1-159/60
– .... $Ga(OH)[OC(O)-C_6H_5]_2 \cdot 0.5\ H_2O$ .... Ga: SVol.D1-159/60
$C_{14}GaH_{12}O_4^+$ .... $[Ga(2-OH-C_6H_4-CH=O)_2][GaBr_4]$ .... Ga: SVol.D1-36
– .... $[Ga(2-OH-C_6H_4-CH=O)_2][GaCl_4]$ .... Ga: SVol.D1-36
– .... $[Ga(2-OH-C_6H_4-CH=O)_2][GaI_4]$ .... Ga: SVol.D1-36
$C_{14}GaH_{15}N_3O_4Re$ $(CO)_3Re[-N_2C_3H_3-Ga(CH_3)_2-O(C_6H_4-NH_2-)-]$ Re: Org.Comp.1-118/9

$C_{14}GaH_{16}NO_9$ .... $[Ga(NO_3)(C(=O)-C_6H_2(OH)_2-C(=O)O-(CH_2)_6-O) \cdot H_2O]_n$ Ga: SVol.D1-203
$C_{14}GaH_{16}N_2O_4^+$ .. $[Ga(NC_5H_2-1,2-(CH_3)_2-3,4-(O)_2)_2]^+$ .... Ga: SVol.D1-258
$C_{14}GaH_{19}I_2NO_4$ .. $GaI_2[1,2-(O)_2C_6H-3,5-(C_4H_9-t)_2-6-NO_2]$, radical Ga: SVol.D1-90/2
$C_{14}GaH_{19}O_4$ ..... $(HO)Ga[3,4-(O)_2C_6H_3-C(=O)-C_7H_{15}]$ .... Ga: SVol.D1-39/40
$C_{14}GaH_{19}O_5$ ..... $(HO)Ga[(O)_2C_6H_2(OH)-C(=O)-C_7H_{15}]$ .... Ga: SVol.D1-39/40
$C_{14}GaH_{20}I_2O_2$ .... $GaI_2[1,2-(O)_2C_6H_2-3,5-(C_4H_9-t)_2]$, radical .... Ga: SVol.D1-90/2
– .... $GaI_2[1,2-(O)_2C_6H_2-3,6-(C_4H_9-t)_2]$, radical .... Ga: SVol.D1-90/2
$C_{14}GaH_{20}O_8^-$ .... $[Ga(OC(O)-CH(C_4H_9-n)-C(O)O)_2]^-$ .... Ga: SVol.D1-166/7
$C_{14}GaH_{21}O_5$ ..... $Ga(O-C_3H_7-i)_2[2-O-C_6H_4-C(O)O-CH_3]$ ...... Ga: SVol.D1-202/3
$C_{14}GaH_{24}MoN_3O_3$ $(CH_2CHCH_2)Mo(CO)_2[N(CH_3)_2-CH_2CH_2-O-Ga(CH_3)_2-N_2C_3H_3]$ .... Mo: Org.Comp.5-205/6
– .... $(CH_2CHCH_2)Mo(CO)_2[NH_2-CH_2CH_2-O-Ga(CH_3)_2-N_2C_3H(CH_3)_2]$ .... Mo: Org.Comp.5-205/6
$C_{14}GaH_{26}NO_3$ .... $Ga(NC_5H_5)(O-C_3H_7-i)_3$ .... Ga: SVol.D1-253
$C_{14}GaH_{26}O_4^+$ .... $[Ga(OC(O)-C_6H_{13}-n)_2]^+$ .... Ga: SVol.D1-155
$C_{14}GaH_{42}InN_2P_2Si$ $(CH_3)_3In \cdot (CH_3)_3Ga-N[=P(CH_3)_3]-Si(CH_3)_2-N=P(CH_3)_3$ In: Org.Comp.1-27, 38
$C_{14}Ga_2H_{12}I_4O_4$ ... $[Ga(2-OH-C_6H_4-CH=O)_2][GaI_4]$ .... Ga: SVol.D1-36
$C_{14}GeH_{11}MoNO_5$ $(CO)_5Mo[CN-Ge(CH_3)_2-C_6H_5]$ .... Mo: Org.Comp.5-6, 9, 10/1
$C_{14}GeH_{14}$ .... $(CH_3)_2Ge(-C_6H_4C_6H_4-)$ .... Ge: Org.Comp.3-322, 333
$C_{14}GeH_{14}O$ ...... $(CH_3)_2Ge(-C_6H_4OC_6H_4-)$ .... Ge: Org.Comp.3-325, 336
$C_{14}GeH_{14}OS$ ..... $(CH_3)_2Ge(-C_6H_4S(=O)C_6H_4-)$ .... Ge: Org.Comp.3-326
$C_{14}GeH_{14}O_2$ ..... $Ge(C_6H_5)_2(CH_3)COOH$ .... Ge: Org.Comp.3-214
$C_{14}GeH_{14}S$ ...... $(CH_3)_2Ge(-C_6H_4SC_6H_4-)$ .... Ge: Org.Comp.3-326, 336
$C_{14}GeH_{16}$ .... $Ge(CH_3)_2(C_6H_5)_2$ .... Ge: Org.Comp.3-155/6, 161/2
$C_{14}GeH_{16}O_2$ ..... $Ge(CH_3)_2(C_6H_4OH-4)_2$ .... Ge: Org.Comp.3-156
$C_{14}GeH_{16}S$ ...... $Ge(CH_3)_2(C_4H_3S-2)C(=CH_2)C_6H_5$ .... Ge: Org.Comp.3-204, 207
– .... trans-$Ge(CH_3)_2(C_4H_3S-2)CH=CHC_6H_5$ .... Ge: Org.Comp.3-204, 207
$C_{14}GeH_{17}N_3O_2$ ... $(CH_3)_2Ge(-CH=CHCH(-NNHC(=O)N(C_6H_5)C(=O)-)CH_2-)$ Ge: Org.Comp.3-264, 273
$C_{14}GeH_{18}O$ ...... $(CH_3)_2Ge(-CH=CHC(OCH_3)(C_6H_5)CH=CH-)$ ... Ge: Org.Comp.3-304/5
$C_{14}GeH_{18}O_8$ ..... $(CH_3)_2Ge(-C(COOCH_3)=C(COOCH_3)C(COOCH_3)=C(COOCH_3)-)$ .... Ge: Org.Comp.3-277, 283
$C_{14}GeH_{19}I$ .... $Ge(C_2H_5)_3C{\equiv}CC_6H_4I-4$ .... Ge: Org.Comp.2-255, 257
$C_{14}GeH_{19}NO_2$ .... $(CH_3)_2Ge(-CH=C(CH_3)CH(OC(=O)NHC_6H_5)CH_2-)$ Ge: Org.Comp.3-265
$C_{14}GeH_{20}$ .... $Ge(C_2H_5)_2[C{\equiv}C-C(CH_3)=CH_2]_2$ .... Ge: Org.Comp.3-171
– .... $Ge(C_2H_5)_3-C{\equiv}C-C_6H_5$ .... Ge: Org.Comp.2-236/7, 246
$C_{14}GeH_{20}^-$ ...... $[Ge(C_2H_5)_3-C{\equiv}C-C_6H_5]^-$, radical anion ...... Ge: Org.Comp.3-352
$C_{14}GeH_{20}NP$ ..... $(CH_3)_2Ge(-CH(CN)CH_2P(C_6H_5)(CH_2)_3-)$ ...... Ge: Org.Comp.3-310
$C_{14}GeH_{20}O$ ...... $Ge(C_2H_5)_3C(C_6H_5)=C=O$ .... Ge: Org.Comp.2-199
$C_{14}GeH_{20}O_4$ ..... 5-$(CH_3)_3Ge-[2.2.1]-C_7H_5[C(O)OCH_3]_2$-2,3 .... Ge: Org.Comp.2-40
– .... 7-$(CH_3)_3Ge-[2.2.1]-C_7H_5[C(O)OCH_3]_2$-2,3 .... Ge: Org.Comp.2-41
$C_{14}GeH_{20}Ti$ ...... $(CH_3)_2Ge(-CH_2Ti(C_5H_5)_2CH_2-)$ .... Ge: Org.Comp.3-238, 240/1
$C_{14}GeH_{21}N$ ...... $Ge(C_2H_5)_3CH(C_6H_5)CN$ .... Ge: Org.Comp.2-135
$C_{14}GeH_{22}$ .... $(C_6H_5)(C_4H_9)Ge[-CH_2CH_2CH_2CH_2-]$ .... Ge: Org.Comp.3-249, 252
– .... $Ge(C_2H_5)_3C(C_6H_5)=CH_2$ .... Ge: Org.Comp.2-170

$C_{14}GeH_{22}$ .... $Ge(C_2H_5)_3C(C_6H_5)$=CHD.................. Ge: Org.Comp.2-170
– .... $Ge(C_2H_5)_3CH$=CH-C_6H_5.................. Ge: Org.Comp.2-169, 170, 183
– .... $Ge(C_2H_5)_3CH$=CD-C_6H_5.................. Ge: Org.Comp.2-169
– .... $Ge(C_2H_5)_3CH$=C=C=C=C=$C(CH_3)_2$........... Ge: Org.Comp.2-179
– .... $Ge(C_2H_5)_3C$≡CCH=C=C=$C(CH_3)_2$............ Ge: Org.Comp.2-243
– .... $Ge(C_2H_5)_3C$≡CC≡CCH=$C(CH_3)_2$............. Ge: Org.Comp.2-242/3
– .... $Ge(C_2H_5)_3C_6H_4$-CH=CH_2-4............... Ge: Org.Comp.2-279, 282
$C_{14}GeH_{22}N_4O_4$ ... $Ge(C_2H_5)_3C(=NNHC_6H_3(NO_2)_2$-2,4)$CH_3$....... Ge: Org.Comp.2-139
$C_{14}GeH_{22}O$ ...... $Ge(C_2H_5)_3C(=CH_2)$-O-C_6H_5............... Ge: Org.Comp.2-190
– .... $Ge(C_2H_5)_3CH$=CH-O-C_6H_5................ Ge: Org.Comp.2-188
– .... $Ge(C_2H_5)_3CH_2C(O)$-C_6H_5................. Ge: Org.Comp.2-130, 138
– .... $Ge(C_2H_5)_3C(O)$-$C_6H_4CH_3$-3............... Ge: Org.Comp.2-120
– .... $Ge(CH_3)_3CH_2C(O)$-$C_6H_2(CH_3)_3$-2,4,6........ Ge: Org.Comp.1-174
$C_{14}GeH_{22}S$ ...... $Ge(C_2H_5)_3C(=CH_2)$-S-C_6H_5............... Ge: Org.Comp.2-191
– .... $Ge(C_2H_5)_3CH$=CH-S-C_6H_5................ Ge: Org.Comp.2-188/9
$C_{14}GeH_{23}$ .... $Ge(CH_3)_3C(C_4H_9$-t)C_6H_5, radical............ Ge: Org.Comp.3-349
$C_{14}GeH_{23}NO_2$ .... $(C_2H_5)_2Ge[-CH_2CH_2C(=C(CN)COO-C_2H_5)CH_2CH_2-]$ Ge: Org.Comp.3-291
– .... $Ge(C_2H_5)_3C$≡CCH[-CH_2CH=$CHCH_2CH(NO_2)$-] Ge: Org.Comp.2-254
$C_{14}GeH_{24}$ .... $Ge(CH_3)_2(C_6H_4CH_3$-4)C_5H_{11}-i.............. Ge: Org.Comp.3-203
– .... $Ge(C_2H_5)_3CH$[-CH=$CHCH_2CH$=CHCH=CH-] ... Ge: Org.Comp.2-232, 234
– .... $Ge(C_2H_5)_3CH_2CH_2C_6H_5$.................. Ge: Org.Comp.2-126/7
– .... $Ge(C_2H_5)_3CH_2C_6H_4CH_3$-4................ Ge: Org.Comp.2-118
– .... $Ge(C_2H_5)_3C_6H_4C_2H_5$-4.................. Ge: Org.Comp.2-279
– .... $Ge(CH_3)_3CH(C_6H_5)C_4H_9$-t................ Ge: Org.Comp.1-184
$C_{14}GeH_{24}NP$ ..... $(CH_3)_2Ge(-CH(CH_2NH_2)CH_2P(C_6H_5)(CH_2)_3-)$ .. Ge: Org.Comp.3-310
$C_{14}GeH_{24}O$ ...... CH_2=$CHCH_2(CH_3)Ge(-CH$=$CHC(OCH_3)(C_4H_9$-t)CH=CH-) Ge: Org.Comp.3-306
– .... $(CH_3)_2Ge[-CH$=$CHC(OCH_3)(C_6H_{11}$-c)CH=CH-] Ge: Org.Comp.3-304
– .... $Ge(C_2H_5)_3CH(OCH_3)C_6H_5$................ Ge: Org.Comp.2-118
– .... $Ge(C_2H_5)_3C(OH)(CH_3)C_6H_5$............... Ge: Org.Comp.2-129
– .... $Ge(C_2H_5)_3C_6H_4$-$CH(CH_3)OH$-4.............. Ge: Org.Comp.2-279
– .... $Ge(C_2H_5)_3C_6H_4$-CH_2CH_2OH-4.............. Ge: Org.Comp.2-279
$C_{14}GeH_{24}OSi$ .... $Ge(CH_3)_3C(Si(CH_3)_2OCH_2C_6H_5)$=$CH_2$........ Ge: Org.Comp.2-5
$C_{14}GeH_{24}O_2$ ..... $Ge(CH_3)_2(C$≡$CC(OH)(CH_3)C_2H_5)_2$............ Ge: Org.Comp.3-155
– .... $Ge(C_2H_5)_2(C$≡$CC(CH_3)_2OH)_2$............... Ge: Org.Comp.3-171
– .... $Ge(CH_3)_3C_6H_4$-$CH(O$-$C_2H_5)_2$-4............. Ge: Org.Comp.2-75
$C_{14}GeH_{24}S$ ...... $Ge(C_2H_5)_3CH_2CH_2SC_6H_5$.................. Ge: Org.Comp.2-125
$C_{14}GeH_{25}N$ ...... $Ge(C_2H_5)_3C_6H_4$-$N(CH_3)_2$-2............... Ge: Org.Comp.2-276
– .... $Ge(C_2H_5)_3C_6H_4$-$N(CH_3)_2$-3............... Ge: Org.Comp.2-276
– .... $Ge(C_2H_5)_3C_6H_4$-$N(CH_3)_2$-4............... Ge: Org.Comp.2-277
$C_{14}GeH_{25}NO$ ..... $Ge(C_2H_5)_3C$≡$CC(OCH_2CH_2CN)(CH_3)_2$........ Ge: Org.Comp.2-264
$C_{14}GeH_{25}P$ ...... $Ge(CH_3)_2(C_4H_9)CH_2CH_2PHC_6H_5$............ Ge: Org.Comp.3-197
$C_{14}GeH_{26}$ .... $Ge(C_2H_5)_3C(=CHCH=CH_2)C(CH_3)$=$CHCH_3$..... Ge: Org.Comp.2-175
– .... $Ge(C_2H_5)_3CH_2CH_2CH_2C$≡$CCH_2CH$=$CH_2$...... Ge: Org.Comp.2-244
– .... $Ge(C_2H_5)_3C$≡$CC(C_4H_9$-t)=CH_2............. Ge: Org.Comp.2-239
– .... $Ge(CH_3)_3C$≡CCH=$C(C_3H_7$-n)C_4H_9-n......... Ge: Org.Comp.2-60
$C_{14}GeH_{26}N_2$ ..... $Ge(C_2H_5)_2(CH_2CH_2CH_2CH_2CN)_2$............ Ge: Org.Comp.3-168
$C_{14}GeH_{26}N_2O_4$ ... $Ge(C_2H_5)_3C(=C(CH(OCH_3)_2)NHN=C(COOCH_3))$ Ge: Org.Comp.2-291
$C_{14}GeH_{26}N_4O_6$ ... $[N(CH_3)_4][Ge(CH_3)_3CH_2C_6H_3(NO_2)_3]$......... Ge: Org.Comp.1-161

$C_{14}GeH_{26}O_2$ ..... $(C_2H_5)_2Ge(C_5H_5(CH_3)_2COOC_2H_5)$ .... Ge: Org.Comp.3-320
– .... $(C_2H_5)_2Ge[-(CH_2)_4C(=O)C(=O)(CH_2)_4-]$ .... Ge: Org.Comp.3-312, 315
$C_{14}GeH_{26}S$ ...... $(C_2H_5)_2GeC_2(C(CH_3)_2CH_2)_2S$ .... Ge: Org.Comp.3-317
$C_{14}GeH_{27}N$ ...... $Ge(C_2H_5)_3C{\equiv}CCH_2N((CH_2)_5)$ .... Ge: Org.Comp.2-261, 270
$C_{14}GeH_{27}NOSi$ ... $Ge(C_2H_5)_3C{\equiv}CC(OSi(CH_3)_3)(CH_3)CN$ .... Ge: Org.Comp.2-265
$C_{14}GeH_{27}NO_4$ .... $Ge(C_2H_5)_3C(COOCH_3)=C(N(CH_3)_2)COOCH_3$ ... Ge: Org.Comp.2-209
$C_{14}GeH_{28}$ .... $(C_2H_5)_2Ge[-CH_2C(CH_3)=CH-CH_2CH_2CH_2-CH(CH_3)CH_2-]$
Ge: Org.Comp.3-311, 314
– .... $(n-C_4H_9)_2Ge[-CH_2C(CH_3)=C(CH_3)CH_2-]$ ...... Ge: Org.Comp.3-259
– .... $Ge(CH_3)_2[c-C_3H_2(CH_3)_3]_2$ .... Ge: Org.Comp.3-151
– .... $Ge(C_2H_5)_2[CH_2C(CH_3)=CHCH_3]_2$ .... Ge: Org.Comp.3-170
– .... $Ge(C_2H_5)_3CH=C=CH-C_5H_{11}-n$ .... Ge: Org.Comp.2-181, 185
– .... $Ge(C_2H_5)_3CH[-CH=CHCH_2CH_2CH_2CH_2CH_2-]$ .. Ge: Org.Comp.2-232
– .... $Ge(C_2H_5)_3CH_2CH=CHCH_2CH(CH_3)CH=CH_2$ .... Ge: Org.Comp.2-179
– .... $Ge(C_2H_5)_3CH_2CH=CHCH_2CH_2CH=CHCH_3$ ..... Ge: Org.Comp.2-180
– .... $Ge(C_2H_5)_3CD_2CD=CDCHDCD_2CD=CDCD_3$ .... Ge: Org.Comp.2-180
– .... $Ge(C_2H_5)_3CD_2CD=CDCD_2CD_2CD=CDCHD_2$ .... Ge: Org.Comp.2-180
– .... $Ge(C_2H_5)_3CH_2CH=CHCH_2CH_2CH_2CH=CH_2$ .... Ge: Org.Comp.2-180/1
– .... $Ge(C_2H_5)_3CH_2CH_2CH[-CH_2CH=CHCH_2CH_2-]$ .. Ge: Org.Comp.2-126
– .... $Ge(C_2H_5)_3C{\equiv}C-C_6H_{13}-n$ .... Ge: Org.Comp.2-243/4
– .... $Ge(C_2H_5)_3-(CH_2)_5-C{\equiv}CCH_3$ .... Ge: Org.Comp.2-244
– .... $Ge(C_3H_7-n)_2(CH=CHC_2H_5)_2$ .... Ge: Org.Comp.3-176
– .... $Ge(C_3H_7-i)_2(CH=CHC_2H_5)_2$ .... Ge: Org.Comp.3-177
– .... $Ge(C_3H_7-n)_2[CH=C(CH_3)_2]_2$ .... Ge: Org.Comp.3-176
– .... $Ge(C_3H_7-i)_2[CH=C(CH_3)_2]_2$ .... Ge: Org.Comp.3-177
– .... $Ge(C_4H_9-n)_2(CH=CHCH_3)_2$ .... Ge: Org.Comp.3-179
– .... $Ge(C_4H_9-i)_2(CH=CHCH_3)_2$ .... Ge: Org.Comp.3-180
– .... $Ge(C_4H_9-n)_2(CH_2CH=CH_2)_2$ .... Ge: Org.Comp.3-179, 181
– .... $Ge(C_4H_9-n)_2[C(CH_3)=CH_2]_2$ .... Ge: Org.Comp.3-179
– .... $Ge(C_4H_9-i)_2[C(CH_3)=CH_2]_2$ .... Ge: Org.Comp.3-180
– .... $Ge(C_4H_9-n)_3C{\equiv}CH$ .... Ge: Org.Comp.3-32, 35/6
$C_{14}GeH_{28}O$ ...... $(C_2H_5)_2Ge[-(CH_2)_4C(=O)(CH_2)_5-]$ .... Ge: Org.Comp.3-312, 314
– .... $Ge(CH_3)_2[c-C_3H_3(CH_3)_2]C(=CH_2)C(CH_3)_2-O-C_2H_5$
Ge: Org.Comp.3-198
– .... $Ge(C_2H_5)_3C(=C=CHCH_3)C(CH_3)_2OCH_3$ .... Ge: Org.Comp.2-217
– .... $Ge(C_2H_5)_3C(=CHCH=CH_2)C(CH_3)(OH)C_2H_5$ .... Ge: Org.Comp.2-216, 219
– .... $Ge(C_2H_5)_3C(=CHCH=CH_2)C(CH_3)_2OCH_3$ ...... Ge: Org.Comp.2-216/7, 219
– .... $Ge(C_2H_5)_3C(=CHCH=CH_2)CH(OH)C_3H_7-n$ ..... Ge: Org.Comp.2-216
– .... $Ge(C_2H_5)_3C(CH=CH_2)=CHC(CH_3)_2OCH_3$ ...... Ge: Org.Comp.2-217
– .... $Ge(C_2H_5)_3C{\equiv}CC(OH)(CH_3)C_4H_9-t$ .... Ge: Org.Comp.2-267
$C_{14}GeH_{28}O_2$ ..... $(C_2H_5)_2Ge[-(CH_2)_4-CH(OH)C(=O)-(CH_2)_4-]$ ... Ge: Org.Comp.3-312, 314
– .... $Ge(C_2H_5)_3CH=CHC(CH_3)_2-O-CH_2-2-OC_2H_3$ .. Ge: Org.Comp.2-212
– .... $Ge(C_2H_5)_3CH=CHCH_2O-CH(CH_3)O-CH_2CH=CH_2$
Ge: Org.Comp.2-205
– .... $Ge(C_2H_5)_3C{\equiv}CCH_2-O-CH_2-O-C_4H_9$ .... Ge: Org.Comp.2-260
$C_{14}GeH_{28}O_4$ ..... $Ge(C_2H_5)_2(CH_2CH(CH_3)COOCH_3)_2$ .... Ge: Org.Comp.3-167
– .... $Ge(C_2H_5)_2(CH_2CH_2CH_2CH_2COOH)_2$ .... Ge: Org.Comp.3-168
– .... $Ge(C_2H_5)_2(CH_2CH_2COOCH_3)CH(CH_3)CH_2COO-C_2H_5$
Ge: Org.Comp.3-210

$C_{14}GeH_{28}O_4$ ..... $Ge(C_3H_7\text{-}n)_2(CH_2CH_2COOCH_3)_2$ Ge:Org.Comp.3-176
– .... $Ge(C_4H_9\text{-}n)_2(CH_2COOCH_3)_2$ Ge:Org.Comp.3-178
$C_{14}GeH_{30}$ .... $Ge(C_2H_5)_3C(=CH_2)\text{-}C_6H_{13}\text{-}n$ Ge:Org.Comp.2-170
– .... $Ge(C_2H_5)_3CH{=}CHC_6H_{13}\text{-}n$ Ge:Org.Comp.2-179/80
– .... $Ge(C_4H_9)_3CH{=}CH_2$ Ge:Org.Comp.3-25, 30
$C_{14}GeH_{30}O$ ...... $Ge(C_2H_5)_3C(CH{=}CH_2){=}CH(OH)(CH_3)C_3H_7\text{-}n$ ... Ge:Org.Comp.2-226
– .... $Ge(C_2H_5)_3CH{=}CHC(OH)(CH_3)C_4H_9\text{-}n$ Ge:Org.Comp.2-226
– .... $Ge(C_2H_5)_3CH{=}CHC(OH)(CH_3)C_4H_9\text{-}t$ Ge:Org.Comp.2-222
– .... cis-$Ge(n\text{-}C_3H_7)_3C(C_2H_5){=}CHOCH_3$ Ge:Org.Comp.3-9, 12
– .... cis-$Ge(n\text{-}C_3H_7)_3C(OCH_3){=}CHC_2H_5$ Ge:Org.Comp.3-10, 12
– .... $Ge(C_4H_9\text{-}n)_3C(O)CH_3$ Ge:Org.Comp.3-18
$C_{14}GeH_{30}OSi$ .... $Ge(C_2H_5)_3C(Si(C_2H_5)_3){=}C{=}O$ Ge:Org.Comp.2-198
$C_{14}GeH_{30}OSn$.... $Ge(C_2H_5)_3C(Sn(C_2H_5)_3){=}C{=}O$ Ge:Org.Comp.2-198
$C_{14}GeH_{30}O_2$ ..... $(C_2H_5)_2Ge[\text{-}(CH_2)_4\text{-}CH(OH)CH(OH)\text{-}(CH_2)_4\text{-}]$ .. Ge:Org.Comp.3-312
– .... $Ge(C_2H_5)_3C(C(CH_3)_2OH){=}CHC(CH_3)_2OH$ Ge:Org.Comp.2-213
– .... $Ge(C_2H_5)_3C(CH_2OCH_3){=}CHCH(OH)C_3H_7\text{-}n$ Ge:Org.Comp.2-225
– .... $Ge(C_2H_5)_3CH{=}CHCH_2OCH_2O\text{-}C_4H_9\text{-}n$ Ge:Org.Comp.2-204
– .... $Ge(C_2H_5)_3CH(C_2H_5)\text{-}COO\text{-}C_4H_9\text{-}t$ Ge:Org.Comp.2-149
– .... $Ge(C_3H_7\text{-}n)_3CH_2CH(CH_3)C(O)O\text{-}CH_3$ Ge:Org.Comp.3-4
– .... $Ge(C_3H_7\text{-}n)_3CH_2CH_2C(O)O\text{-}C_2H_5$ Ge:Org.Comp.3-4
– .... $Ge(C_3H_7\text{-}n)_3CH_2CH_2OCH_2\text{-}2\text{-}OC_2H_3$ Ge:Org.Comp.3-2
– .... $Ge(C_3H_7\text{-}n)_3CH_2C(O)O\text{-}C_3H_7\text{-}n$ Ge:Org.Comp.3-2
$C_{14}GeH_{30}O_3$ ..... $Ge(C_2H_5)_3CH_2CH(CH_3)\text{-}C(O)\text{-}OO\text{-}C_4H_9\text{-}t$ Ge:Org.Comp.2-150, 154
– .... $Ge(C_2H_5)_3CH_2CH_2CH_2\text{-}C(O)\text{-}OO\text{-}C_4H_9\text{-}t$ Ge:Org.Comp.2-160, 165/6
– .... $Ge(C_2H_5)_3CH_2CH_2CH_2OCH_2CH[\text{-}CH_2OCH_2CH_2O\text{-}]$
Ge:Org.Comp.2-144
$C_{14}GeH_{30}SSi$ .... $Ge(C_2H_5)_3C(=C=S)Si(C_2H_5)_3$ Ge:Org.Comp.2-199
– .... $Ge(C_2H_5)_3C{\equiv}C\text{-}S\text{-}Si(C_2H_5)_3$ Ge:Org.Comp.2-252
$C_{14}GeH_{30}Si$ ..... $Ge(C_2H_5)_3C{\equiv}CSi(C_2H_5)_3$ Ge:Org.Comp.2-253, 258
$C_{14}GeH_{30}SiSn$ ... $Ge(CH_3)_3C(Sn(CH_3)_3)(CH{=}CHC(Si(CH_3)_3){=}CH)$ Ge:Org.Comp.2-33, 47
$C_{14}GeH_{30}Sn$ ..... $Ge(C_2H_5)_3C{\equiv}CSn(C_2H_5)_3$ Ge:Org.Comp.2-254, 258
$C_{14}GeH_{31}N$ ...... $Ge(C_2H_5)_3C(C_2H_5){=}CH\text{-}N(C_2H_5)_2$ Ge:Org.Comp.2-193, 200
– .... $Ge(C_2H_5)_3CH{=}CHCH_2CH_2\text{-}N(C_2H_5)_2$ Ge:Org.Comp.2-212
– .... $Ge(C_2H_5)_3CH_2CH{=}CHCH_2\text{-}N(C_2H_5)_2$ Ge:Org.Comp.2-214
$C_{14}GeH_{31}NO$..... $Ge(C_2H_5)_3CH(CH_3)CH_2CON(C_2H_5)_2$.......... Ge:Org.Comp.2-151
$C_{14}GeH_{31}NO_3Si$ .. $Ge(C_2H_5)_3CH_2CH_2Si(OCH_2CH_2)_3N$ Ge:Org.Comp.2-137
$C_{14}GeH_{31}N_3Ti$ ... $Ge(CH_3)_3C_5H_4Ti(N(CH_3)_2)_3$ Ge:Org.Comp.2-33
$C_{14}GeH_{32}$ .... $Ge(C_2H_5)_2(C_5H_{11}\text{-}n)_2$ Ge:Org.Comp.3-168
– .... $Ge(C_2H_5)_3C_8H_{17}\text{-}n$ Ge:Org.Comp.2-111
– .... $Ge(C_3H_7\text{-}n)_2(C_4H_9\text{-}n)_2$ Ge:Org.Comp.3-176
– .... $Ge(C_3H_7\text{-}n)_3C_5H_{11}\text{-}n$ Ge:Org.Comp.3-5
– .... $Ge(C_4H_9\text{-}n)_3C_2H_5$ Ge:Org.Comp.3-17, 22/3
$C_{14}GeH_{32}O$ ...... $Ge(C_2H_5)_3CH_2CH(CH_3)C(OH)(C_2H_5)_2$ Ge:Org.Comp.2-162
– .... $Ge(C_2H_5)_3CH_2CH_2CH_2CH_2\text{-}CH(OH)C_3H_7\text{-}n$ ... Ge:Org.Comp.2-165
– .... $Ge(C_4H_9)_3CH_2CH_2OH$ Ge:Org.Comp.3-17
– .... $Ge(C_4H_9)_3CH_2OCH_3$ Ge:Org.Comp.3-17
$C_{14}GeH_{32}O_4$ ..... $Ge(C_2H_5)_3CH_2CH_2OCH_2CH_2OCH_2CH(OH)CH_2OCH_3$
Ge:Org.Comp.2-124
$C_{14}GeH_{32}SSi$ .... cis-$Ge(C_2H_5)_3CH{=}CHSSi(C_2H_5)_3$ Ge:Org.Comp.2-188
– .... trans-$Ge(C_2H_5)_3CH{=}CHSSi(C_2H_5)_3$ Ge:Org.Comp.2-188

$C_{14}GeH_{33}NO_2$.... $Ge(C_2H_5)_3CH_2CH_2CH_2OCH_2CH(OH)CH_2N(CH_3)_2$
Ge: Org.Comp.2-144
$C_{14}GeH_{34}N_2$..... $Ge(C_2H_5)_2(CH_2CH_2CH_2CH_2CH_2NH_2)_2$........ Ge: Org.Comp.3-168
$C_{14}GeH_{36}NPSn$... $(CH_3)_3Sn-P(C_4H_9-t)_2=N-Ge(CH_3)_3$.......... Sn: Org.Comp.19-180/1
$C_{14}GeH_{38}Si_4$..... $Ge(CH(Si(CH_3)_3)_2)_2$.................... Ge: Org.Comp.3-366/7
$C_{14}Ge_2H_{22}MoN_2O_4$
$(CO)_4Mo[CN-CH_2-Ge(CH_3)_3]_2$.............. Mo: Org.Comp.5-25, 26
$C_{14}H_4O_8Th$...... $[Th(2,3,6,7-C_{10}H_4(COO)_4)]_n$................ Th: SVol.C7-149/50
$C_{14}H_5O_7STh^+$.... $Th[1,2,9,10-(O)_4-3-(SO_3)-C_{14}H_5]^+$.......... Th: SVol.D1-117
$C_{14}H_6I_6O_6Th$..... $Th(OH)_2(2,3,5-I_3-C_6H_2COO)_2$............... Th: SVol.C7-139/40
$C_{14}H_6MnNO_9Re$.. $(CO)_5ReMn(CO)_4CNC_4H_6-t$................ Re: Org.Comp.2-199
$C_{14}H_6MnN_2O_8S_4^{2-}$
$[Mn(1,3-SNC_7H_3(O-4)(SO_3-7))_2]^{2-}$.......... Mn: MVol.D7-233/6
$C_{14}H_6MnO_6S^{2-}$... $[Mn(O_2C-C_6H_3(O)-S-C_6H_3(O)-CO_2)]^{2-}$....... Mn: MVol.D7-84/5
$C_{14}H_6MnO_8Re$... $(CO)_5ReCH_2C_5H_4Mn(CO)_3$................ Re: Org.Comp.2-180
$C_{14}H_6N_2O_8Th$.... $Th(NC_5H_3(COO)_2-2,6)_2$.................. Th: SVol.D4-157
–.............. $[Th(NC_5H_3(COO)_2-2,6)_2]_n$................ Th: SVol.C7-153/5
$C_{14}H_6N_2O_{10}Th$... $Th(OO)[(NC_5H_3(COO)_2-2,6)_2] \cdot H_2O$......... Th: SVol.C7-154/5
$C_{14}H_6O_6Re$...... $(CO)_4Re[1,2-(O)_2C_{10}H_6]$, radical............ Re: Org.Comp.1-372
$C_{14}H_6O_8Re$...... $(CO)_4Re[OC_4H_3-2-C(O)C(O)-2-H_3C_4O]$, radical
Re: Org.Comp.1-372
$C_{14}H_7MoO_{11}S^{3-}$.. $[MoO_2(OH)_2(1,2,9,10-(O)_4-C_{14}H_5-3-SO_3)]^{3-}$... Mo: SVol.B3b-156/7
$C_{14}H_7NO_4Th^{2+}$... $Th[1,2,9,10-(O)_4-3-(NH_2)-C_{14}H_5]^{2+}$......... Th: SVol.D1-117
$C_{14}H_7N_5O_2PdS_2$.. $Pd(NCS)_2[1,10-N_2C_{12}H_7(NO_2-5)]$............ Pd: SVol.B2-305
–.............. $Pd(SCN)(NCS)[1,10-N_2C_{12}H_7(NO_2-5)]$........ Pd: SVol.B2-305
–.............. $Pd(SCN)_2[1,10-N_2C_{12}H_7(NO_2-5)]$............ Pd: SVol.B2-304
$C_{14}H_7O_5Re$...... $(CO)_5ReCH_2C{\equiv}CC_6H_5$.................... Re: Org.Comp.2-109, 112/3
$C_{14}H_7O_6Re$...... cis-$(CO)_5ReCH=CHC(O)C_6H_5$.............. Re: Org.Comp.2-128
–.............. trans-$(CO)_5ReCH=CHC(O)C_6H_5$............. Re: Org.Comp.2-128
$C_{14}H_8MnN_2O_2S_2$.. $Mn(1,3-SNC_7H_4-4-O)_2$.................. Mn: MVol.D7-233/6
$C_{14}H_8MnN_2O_6S_2$.. $[Mn(C_7H_4NS(=O)_3)_2(H_2O)_4] \cdot 2\ H_2O$......... Mn: MVol.D7-128/9
$C_{14}H_8MnN_2O_6S_2^{2-}$
$[Mn(O-NH-C(=O)-C_6H_3(-O)-S-S-C_6H_3(-O)$
$-C(=O)NH-O)]^{2-}$...................... Mn: MVol.D7-90/1
$C_{14}H_8MnN_2S_4$.... $Mn(1,3-SNC_7H_4-2-S)_2$.................. Mn: MVol.D7-61/3
$C_{14}H_8MnO_4S_2^{2-}$.. $[Mn(1-S-C_6H_4-2-COO)_2]^{2-}$............... Mn: MVol.D7-50/2
$C_{14}H_8MnO_6S$.... $Mn[OOC-C_6H_3(OH)-S-C_6H_3(OH)-COO]$...... Mn: MVol.D7-84/5
–.............. $Mn[OOC-C_6H_3(OH)-S-C_6H_3(OH)-COO] \cdot 4\ H_2O$
Mn: MVol.D7-86
$C_{14}H_8Mn_2N_2O_6S_2$ $Mn_2((2-O)(1-ONHC(=O))C_6H_3-5-SS-5-C_6H_3$
$(C(=O)NHO-1)(O-2))$.................... Mn: MVol.D7-90/1
$C_{14}H_8MoN_6O_2S_2$.. $Mo(NO)_2(NCS)_2(C_{12}H_8N_2)$................ Mo: SVol.B3b-200
$C_{14}H_8NO_7ReW$... $(CO)_4Re(NCCH_3)W(CO)_3C_5H_5$.............. Re: Org.Comp.1-492
$C_{14}H_8N_2O_4Re$.... $(CO)_4Re[NC_5H_4-2-(2-C_5H_4N)]$, radical....... Re: Org.Comp.1-370
$C_{14}H_8N_2O_4Re^+$... $[(CO)_4Re(NC_5H_4-C_5H_4N)]^+$................ Re: Org.Comp.1-475
$C_{14}H_8N_3O_4Re$.... $(CO)_3Re[NC_5H_4-2-(2-C_5H_4N)]NCO$.......... Re: Org.Comp.1-150
$C_{14}H_8N_4O_4Re$.... $(OC)_2Re(ON_2C_6H_4)_2$, radical............... Re: Org.Comp.1-93
$C_{14}H_8N_4O_6Re$.... $(OC)_2Re(O_2N_2C_6H_4)_2$, radical.............. Re: Org.Comp.1-93
$C_{14}H_8N_4O_6S$..... $4-NO_2-C_6H_4C(O)N=S=NC(O)C_6H_4-4-NO_2$..... S: S-N Comp.7-273, 275/6
$C_{14}H_8N_4O_{16}Th$... $Th(OH)_2[2-OH-3,5-(NO_2)_2C_6H_2COO]_2$....... Th: SVol.C7-134, 135

$C_{14}H_8N_4Pd$ ...... $Pd(CN)_2(1,10\text{-}N_2C_{12}H_8)$ ...... Pd: SVol.B2-286
$C_{14}H_8N_4PdS_2$ .... $Pd(SCN)_2(1,10\text{-}N_2C_{12}H_8)$ ...... Pd: SVol.B2-304
$C_{14}H_8N_6O_2S_4Th$ .. $Th(SCN)_4 \cdot [2\text{-}(1\text{-}O\text{-}NC_5H_4\text{-}2)\text{-}1\text{-}O\text{-}NC_5H_4]$ .. Th: SVol.D4-154, 176
$C_{14}H_8N_6PdS_2$ .... $Pd(SCN)_2(NC\text{-}4\text{-}C_5H_4N)_2$ ...... Pd: SVol.B2-302/3
$C_{14}H_8N_6S_4Si$ ..... $Si(NCS)_4 \cdot C_5H_4N(2\text{-}C_5H_4N)\text{-}2$ ...... Si: SVol.B4-300/1
$C_{14}H_8Na_2O_7Th$ ... $Na_2[ThO(2\text{-}OC_6H_4COO)_2]$ ...... Th: SVol.C7-130/1
$C_{14}H_8O_4Sn$ ...... $Sn(C_6H_4COO\text{-}2)_2$ ...... Sn: Org.Comp.16-221
$C_{14}H_8O_6Th$ ...... $Th(2\text{-}OC_6H_4COO)_2$ ...... Th: SVol.C7-130
$C_{14}H_8O_{10}Th$ ..... $[Th(2,3,6,7\text{-}C_{10}H_4(COO)_4)(H_2O)_2]_n$ ...... Th: SVol.C7-150
$C_{14}H_9MnNO_3$ .... $Mn[\text{-}OC(O)C_6H_3(CH{=}NC_6H_5\text{-}3)\text{-}2\text{-}O\text{-}]$ ...... Mn: MVol.D6-62
$C_{14}H_9MnNO_5Re$ .. $[(C_5H_5)Re(CO)_2\text{-}1\text{-}NC_4H_4]Mn(CO)_3$ ...... Re: Org.Comp.3-193/7, 202, 207/8
$C_{14}H_9MnNO_9Re$ .. $(CO)_4Re(CNC_4H_9\text{-}t)Mn(CO)_5$ ...... Re: Org.Comp.2-248, 250
$C_{14}H_9MnN_2OS^+$ .. $[Mn(1,3\text{-}N_2C_8H_4({=}S\text{-}2)(C_6H_5\text{-}3)({=}O\text{-}4))]^+$ ...... Mn: MVol.D7-75/8
$C_{14}H_9MnN_3O_3S$ .. $Mn(C_6H_5\text{-}C(O)NC(S)N\text{-}C_6H_4\text{-}NO_2\text{-}2) \cdot H_2O$ ... Mn: MVol.D7-198/9
$C_{14}H_9MoNO_5$ .... $(CO)_5Mo(CN\text{-}CHCH_3\text{-}C_6H_5)$ ...... Mo: Org.Comp.5-6, 7
– ...... $(CO)_5Mo[CN\text{-}C_6H_3(CH_3)_2\text{-}2,6]$ ...... Mo: Org.Comp.5-6, 8
$C_{14}H_9NO_5Th$ ..... $(O_2)Th[OOC\text{-}C_6H_4\text{-}2\text{-}(N{=}CH\text{-}C_6H_4\text{-}2\text{-}O)] \cdot H_2O$ Th: SVol.D4-133, 158
$C_{14}H_9NO_7ReS^+$ .. $[(CO)_5Re(CNCH_2SO_2C_6H_4CH_3)]^+$ ...... Re: Org.Comp.2-252, 254
$C_{14}H_9N_2O_5Re$ .... $(CO)_3Re[NC_5H_4\text{-}2\text{-}(2\text{-}C_5H_4N)]OC(O)H$ ...... Re: Org.Comp.1-154, 175/6
– ...... $(^{13}CO)_3Re[NC_5H_4\text{-}2\text{-}(2\text{-}C_5H_4N)]O^{13}C(O)H$ .... Re: Org.Comp.1-154/5
– ...... $C_5H_5N\text{-}1\text{-}Re(CO)_3[\text{-}O\text{-}C(O)\text{-}(2,1\text{-}NC_5H_4)\text{-}]$ ... Re: Org.Comp.1-127
$C_{14}H_9N_2O_6Re$ .... fac-$(CO)_3Re[NC_5H_4\text{-}2\text{-}(2\text{-}C_5H_4N)]OC(O)OH$ ... Re: Org.Comp.1-154
$C_{14}H_9O_5Sb$ ...... $C_6H_4[C({=}O)]_2C_6H_3Sb(O)(OH)_2$ ...... Sb: Org.Comp.5-299
$C_{14}H_9O_6Re$ ...... cis-$(CO)_4Re[C({=}O)CH{=}CH_2]{=}C(OH)C_6H_5$ ...... Re: Org.Comp.1-389
$C_{14}H_9O_6Th^+$ ..... $Th[(2\text{-}HOC_6H_4COO)(2\text{-}(O)C_6H_4COO)]^+$ ...... Th: SVol.D1-110
$C_{14}H_{10}IMoNO_2$ ... $(C_5H_5)Mo(CO)_2(I)(CN\text{-}C_6H_5)$ ...... Mo: Org.Comp.8-9, 13/5
$C_{14}H_{10}IN_4O_3Re$ .. $(CO)_3ReI[\text{-}N({=}CH\text{-}2\text{-}C_5H_4N)\text{-}NH\text{-}(2,1\text{-}C_5H_4N)\text{-}]$ Re: Org.Comp.1-173
$C_{14}H_{10}I_2MnN_2O_2S_2$ $[Mn((2\text{-}S{=})C_7H_5NO\text{-}3,1)_2I_2] \cdot H_2O$ ...... Mn: MVol.D7-59/60
$C_{14}H_{10}I_2O_6Th$ .... $Th(OH)_2(3\text{-}I\text{-}C_6H_4COO)_2$ ...... Th: SVol.C7-139/40
$C_{14}H_{10}K_4O_{18}Th$ .. $K_4[Th(C_2O_4)(HOC(CH_2COO)_2COO)_2] \cdot 3\ H_2O$ Th: SVol.C7-121/3
$C_{14}H_{10}LiO_6Re$ .... $Li[(CO)_4Re(C({=}O)CH_3)C({=}O)CH_2C_6H_5]$ ...... Re: Org.Comp.1-425
$C_{14}H_{10}MnMoNO_7$ $(C_5H_5)Mo(CO)(NO)[CH_2{=}CHCH_2\text{-}Mn(CO)_5]$ .... Mo: Org.Comp.6-306
$C_{14}H_{10}MnNO_3S_4{}^+$ $[Mn(S_2C\text{-}O\text{-}C_6H_5)_2(NO)]^+$ ...... Mn: MVol.D7-130/3
$C_{14}H_{10}MnN_2OS$ .. $Mn(C_6H_5\text{-}C(O)NC(S)N\text{-}C_6H_5) \cdot H_2O$ ...... Mn: MVol.D7-198/9
$C_{14}H_{10}MnN_2O_2$ ... $Mn[\text{-}2\text{-}O\text{-}C_6H_4\text{-}CH{=}NN{=}CH\text{-}C_6H_4\text{-}2\text{-}O\text{-}]$ ...... Mn: MVol.D6-261
$C_{14}H_{10}MnN_2O_4S_5$ $Mn(1,3\text{-}NSC_7H_5({=}S)\text{-}2)_2SO_4$ ...... Mn: MVol.D7-61/3
$C_{14}H_{10}MnN_2O_6S_2$ $MnH_2(O\text{-}NH\text{-}C({=}O)\text{-}C_6H_3(\text{-}O)\text{-}SS\text{-}C_6H_3(\text{-}O)\text{-}C({=}O)\text{-}NH\text{-}O)$ ...... Mn: MVol.D7-90/1
$C_{14}H_{10}MnN_4O_2S_2$ $Mn(2\text{-}NH_2\text{-}1,3\text{-}SNC_7H_3(O)\text{-}4)_2$ ...... Mn: MVol.D7-233/6
$C_{14}H_{10}MnN_5S_2Se_2$ $[Mn(S\text{-}C(SCH_3){=}NN{=}CH\text{-}1\text{-}(C_9H_6N\text{-}2))(NCSe)_2]_n$ Mn: MVol.D6-361, 363
– ...... $[Mn(S\text{-}C(SCH_3){=}NN{=}CH\text{-}2\text{-}(C_9H_6N\text{-}1))(NCSe)_2]_n$ Mn: MVol.D6-361, 363
$C_{14}H_{10}MnN_5S_4$ ... $[Mn(S\text{-}C(SCH_3){=}NN{=}CH\text{-}1\text{-}(C_9H_6N\text{-}2))(NCS)_2]_n$ Mn: MVol.D6-361, 363

$C_{14}H_{10}MnN_5S_4$ ... $[Mn(S-C(SCH_3)=NN=CH-2-(C_9H_6N-1))(NCS)_2]_n$ Mn:MVol.D6-361, 363
$C_{14}H_{10}MnO_2S_4$ ... $Mn(S_2C-O-C_6H_5)_2$ Mn:MVol.D7-130/3
$C_{14}H_{10}MnO_4S_4^{2-}$ $[Mn(SC_4H_4-2-CH=C(-S)-CO_2)_2]^{2-}$ Mn:MVol.D7-50/2
$C_{14}H_{10}MnO_6S_2^{2-}$ $[Mn(OC_4H_4-2-CH=C(-S)-CO_2)_2]^{2-}$ Mn:MVol.D7-50/2
$C_{14}H_{10}MnS_6$ ..... $Mn(S_2C-S-C_6H_5)_2$ Mn:MVol.D7-209/10
$C_{14}H_{10}MoNNaO_2$ $Na[(C_5H_5)Mo(CO)_2(CN-C_6H_5)]$ Mo:Org.Comp.8-1, 3
$C_{14}H_{10}MoNO_2^-$ ... $Na[(C_5H_5)Mo(CO)_2(CN-C_6H_5)]$ Mo:Org.Comp.8-1, 3
$C_{14}H_{10}MoNO_7Re$ $[(CO)_5Re-CH_2CH=CH_2]Mo(NO)(CO)C_5H_5$ ..... Re: Org.Comp.2-179, 183
$C_{14}H_{10}MoN_2O_5$ ... $[-N(CH_3)-C_6H_4-N(CH_3)-]C=Mo(CO)_5$ Mo:Org.Comp.5-100, 106
$C_{14}H_{10}MoN_3OS_2^-$ $[C_5H_5Mo(NO)(-SC(CN)=C(CN)S-)(C_5H_5)]^-$ Mo:Org.Comp.6-88
$C_{14}H_{10}MoN_6O_2S_2$ $Mo(H_2NO)(NO)(NCS)_2(C_{12}H_8N_2)$ Mo:SVol.B3b-200
$C_{14}H_{10}MoNiO_4$ ... $(C_5H_5)Mo(CO)_3Ni(CO)(C_5H_5)$ Mo:Org.Comp.6-378
$C_{14}H_{10}MoO_2S_2$ ... $(C_5H_5)Mo(CO)_2[-S-C(C_6H_5)-S-]$ Mo:Org.Comp.7-191, 203
$C_{14}H_{10}MoO_2S_3$ ... $(C_5H_5)Mo(CO)_2[-S-C(S-C_6H_5)-S-]$ Mo:Org.Comp.7-198
$C_{14}H_{10}MoO_4$ ..... $(C_5H_5)Mo(CO)_2[-O-C(C_6H_5)-O-]$ Mo:Org.Comp.7-188, 190
– ...... $(C_{10}H_{10})Mo(CO)_4$ Mo:Org.Comp.5-344/5
$C_{14}H_{10}MoO_6$ ..... $4-CH_3-C_6H_4-C(OCH_3)=Mo(CO)_5$ Mo:Org.Comp.5-100, 103, 110
– ...... $C_2H_5-O-C(C_6H_5)=Mo(CO)_5$ Mo:Org.Comp.5-100, 104, 110
$C_{14}H_{10}MoO_6^-$ .... $[4-CH_3-C_6H_4-C(OCH_3)=Mo(CO)_5]^-$, radical anion Mo:Org.Comp.5-110
$C_{14}H_{10}N_2OS$ ..... $3-(O=S=N)-2-(C_6H_5)C_8H_5N$ ...... S: S-N Comp.6-170
$C_{14}H_{10}N_2O_2S$ .... $C_6H_5-C(O)-N=S=N-C(O)-C_6H_5$ ...... S: S-N Comp.7-270/7
– ...... $O=S=N-C(C_6H_5)=N-C(O)-C_6H_5$ ...... S: S-N Comp.6-179/80
$C_{14}H_{10}N_2O_{10}Th$ .. $Th(OH)_2(4-NO_2-C_6H_4COO)_2 \cdot H_2O$ ...... Th: SVol.C7-141, 142
$C_{14}H_{10}N_2O_{12}Th$ .. $Th(OH)_2(2-HO-3-NO_2-C_6H_3-COO)_2$ ...... Th: SVol.C7-133, 135
– ...... $Th(OH)_2(2-HO-5-NO_2-C_6H_3-COO)_2$ ...... Th: SVol.C7-133/4
$C_{14}H_{10}N_3O_4Re$ ... $fac-(CO)_3Re[NC_5H_4-2-(2-C_5H_4N)]C(O)NH_2$ ... Re: Org.Comp.1-174
$C_{14}H_{10}N_4OS$ ..... $1,3-N_2C_7H_5-1-S(O)-1-(1,3-N_2C_7H_5)$ ...... S: S-N Comp.8-354, 362
$C_{14}H_{10}N_4O_2PdS_2$ $Pd(SCN)_2[HC(O)-4-C_5H_4N]_2$ ...... Pd: SVol.B2-302/3
$C_{14}H_{10}N_6O_2S_4Si$ .. $Si(NCS)_4 \cdot 2\ O-NC_5H_5$ ...... Si: SVol.B4-300/1
$C_{14}H_{10}N_6S_4Si$ .... $Si(NCS)_4 \cdot 2\ NC_5H_5$ ...... Si: SVol.B4-300/1
$C_{14}H_{10}O_4Th^{2+}$ ... $Th[1,2-(O)_2-3,5,7-c-C_7H_5]_2^{2+}$ ...... Th: SVol.D1-129
$C_{14}H_{10}O_6Re^-$ .... $[(CO)_4Re(C(=O)CH_3)C(=O)CH_2C_6H_5]^-$ ...... Re: Org.Comp.1-425
$C_{14}H_{10}O_7Th$ ..... $ThO(2-HO-C_6H_4-COO)_2$ ...... Th: SVol.C7-130/1
– ...... $ThO(2-HO-C_6H_4-COO)_2 \cdot 3\ H_2O$ ...... Th: SVol.C7-130/1
– ...... $ThO(2-HO-C_6H_4-COO)_2 \cdot 4\ H_2O$ ...... Th: SVol.C7-130/1
$C_{14}H_{10}O_{10}S_2Th$ .. $Th[1,2-(O)_2-5-(O_3S)-3,5,7-c-C_7H_5]_2$ ...... Th: SVol.D1-129
$C_{14}H_{11}I_2MnN_4P$ .. $Mn[P(C_6H_5)(CH_3)_2](C_2(CN)_4)I_2$ Mn:MVol.D8-46/7
$C_{14}H_{11}MnNO_2$ ... $Mn[-O-C_6H_3(CH_3-4)-2-CH=N-C_6H_4-O-2-] \cdot n\ H_2O$ Mn:MVol.D6-56/7, 58
$C_{14}H_{11}MnNO_2S$ .. $Mn[-O-C_6H_3(OCH_3-6)-2-CH=N-C_6H_4-S-2-]_n$ Mn:MVol.D6-56/7
$C_{14}H_{11}MnNO_3$ ... $Mn[-O-2-C_{10}H_6-1-CH=N-CH_2CH_2-C(O)O-]$ .. Mn:MVol.D6-71
– ...... $Mn[-O-C_6H_3(OCH_3-6)-2-CH=N-2-C_6H_4-O-] \cdot n\ H_2O$ Mn:MVol.D6-56/7, 58
$C_{14}H_{11}MnNO_4$ ... $[Mn(-OC(O)C_6H_3(-3-CH=N-C_6H_5)-2-O-)(H_2O)]$ Mn:MVol.D6-62/3
$C_{14}H_{11}MnNO_5S$ .. $Mn[-O_3S-C_6H_4-2-N=CH-C_6H_3(OCH_3-3)-4-O-]$ Mn:MVol.D6-63/4
$C_{14}H_{11}MnN_2O_2S^+$ $[Mn(C_6H_5-C(O)NC(S)N(OH)-C_6H_5)]^+$ Mn:MVol.D7-198

$C_{14}H_{11}MnN_2O_2S^+$	$[Mn(O_2C\text{-}C_6H_4\text{-}2\text{-}NHC(S)NH\text{-}C_6H_5)]^+$	Mn:MVol.D7-197
–	$[Mn(O_2C\text{-}C_6H_4\text{-}3\text{-}NHC(S)NH\text{-}C_6H_5)]^+$	Mn:MVol.D7-197
–	$[Mn(O_2C\text{-}C_6H_4\text{-}4\text{-}NHC(S)NH\text{-}C_6H_5)]^+$	Mn:MVol.D7-197
$C_{14}H_{11}MnN_3O_2S$	$Mn(2\text{-}O\text{-}C_6H_4C(=O)N_2HC(=S)NHC_6H_5)$	Mn:MVol.D7-203
$C_{14}H_{11}MoNO_2S$	$(C_5H_5)Mo(CO)_2[\text{-}S\text{-}C(C_6H_5)=NH\text{-}]$	Mo:Org.Comp.7-207/10, 217
$C_{14}H_{11}MoNO_3$	$(C_5H_5)Mo(CO)_2[\text{-}C(=O)\text{-}CH_2\text{-}2\text{-}NC_5H_4\text{-}1\text{-}]$	Mo:Org.Comp.8-112, 137/8
–	$(C_5H_5)Mo(CO)_2[\text{-}C(O)CH_2\text{-}1\text{-}NC_5H_4\text{-}2\text{-}]$	Mo:Org.Comp.8-112, 132
$C_{14}H_{11}MoNO_5Si$	$(CO)_5Mo[CN\text{-}Si(CH_3)_2\text{-}C_6H_5]$	Mo:Org.Comp.5-6, 9
$C_{14}H_{11}MoN_3O_4$	$(C_5H_5)Mo(CO)(NO)[\text{-}O\text{-}CH_2\text{-}CH=C(CH_2CH(CN)_2)\text{-}C(=O)\text{-}]$	Mo:Org.Comp.6-300
$C_{14}H_{11}MoO^+$	$[(C_9H_7)Mo(CO)(HC{\equiv}CH)_2]^+$	Mo:Org.Comp.6-315
$C_{14}H_{11}NO_5Th$	$(HO)_2Th[2\text{-}(2\text{-}O\text{-}C_6H_4\text{-}CH=N)\text{-}C_6H_4\text{-}COO] \cdot H_2O$	Th: SVol.D4-158
$C_{14}H_{11}N_3O_2S_4$	$2\text{-}[4\text{-}CH_3\text{-}C_6H_4\text{-}S(O)_2N=S=NS]\text{-}1,3\text{-}SNC_7H_4$	S: S-N Comp.7-67/8
$C_{14}H_{11}N_5O_{16}Th$	$Th(NO_3)_4 \cdot HOOC\text{-}C_6H_3\text{-}2\text{-}OH\text{-}4\text{-}(N=CH\text{-}C_6H_4\text{-}2\text{-}OH) \cdot 4\ H_2O$	Th: SVol.D4-139
$C_{14}H_{11}O_6Re$	$cis\text{-}(CO)_4Re[C(=O)CH_3]=C(OH)CH_2C_6H_5$	Re: Org.Comp.1-389
$C_{14}H_{12}I_2MnN_4S$	$MnI_2[NC_5H_4\text{-}2\text{-}C(CH_3)=N\text{-}NH\text{-}2\text{-}(1,3\text{-}SNC_7H_4)]$	Mn:MVol.D6-250, 251/2
$C_{14}H_{12}LiMoNO_3$	$C_5H_5Mo(CO)(NO)=C(C_6H_4CH_3\text{-}4)OLi$	Mo:Org.Comp.6-264, 268
$C_{14}H_{12}LiN_2O_4Re$	$Li[(C_5H_5)Re(CO)(COO)\text{-}N=N\text{-}C_6H_4\text{-}OCH_3\text{-}4] \cdot CH_2Cl_2$	Re: Org.Comp.3-141/2
$C_{14}H_{12}MnNO^+$	$Mn[O\text{-}C_6H_4\text{-}2\text{-}CH=N\text{-}C_6H_4\text{-}CH_3\text{-}2]^+$	Mn:MVol.D6-15/6
–	$Mn[O\text{-}C_6H_4\text{-}2\text{-}CH=N\text{-}C_6H_4\text{-}CH_3\text{-}3]^+$	Mn:MVol.D6-15/6
–	$Mn[O\text{-}C_6H_4\text{-}2\text{-}CH=N\text{-}C_6H_4\text{-}CH_3\text{-}4]^+$	Mn:MVol.D6-15/6
$C_{14}H_{12}MnNOS^+$	$Mn[O\text{-}C_6H_4\text{-}2\text{-}CH=N\text{-}C_6H_4\text{-}SCH_3\text{-}4]^+$	Mn:MVol.D6-15/6
$C_{14}H_{12}MnNO_2^+$	$Mn[O\text{-}C_6H_4\text{-}2\text{-}CH=N\text{-}C_6H_4\text{-}OCH_3\text{-}2]^+$	Mn:MVol.D6-15/6
–	$Mn[O\text{-}C_6H_4\text{-}2\text{-}CH=N\text{-}C_6H_4\text{-}OCH_3\text{-}3]^+$	Mn:MVol.D6-15/6
–	$Mn[O\text{-}C_6H_4\text{-}2\text{-}CH=N\text{-}C_6H_4\text{-}OCH_3\text{-}4]^+$	Mn:MVol.D6-15/6
$C_{14}H_{12}MnN_2O_2S_2$	$Mn(ONHC(C_6H_5)=S)_2 \cdot 0.5\ H_2O$	Mn:MVol.D7-215/7
$C_{14}H_{12}MnN_2O_3$	$[Mn(\text{-}O\text{-}C_6H_4\text{-}2\text{-}CH=NN=CH\text{-}2\text{-}C_6H_4\text{-}O\text{-})(H_2O)]$	Mn:MVol.D6-261
$C_{14}H_{12}MnN_2O_4$	$Mn(O_2N=CHC_6H_5)_2 \cdot 2\ H_2O$	Mn:MVol.D7-20/1
–	$[Mn(\text{-}O\text{-}C_6H_4\text{-}2\text{-}CH=N\text{-}NHC(O)\text{-}2\text{-}C_6H_4\text{-}O\text{-})(H_2O)]$	Mn:MVol.D6-272/3
$C_{14}H_{12}MnN_2O_6S$	$[Mn(\text{-}OC(O)C_6H_3(\text{-}3\text{-}CH=N\text{-}C_6H_4\text{-}SO_2NH_2\text{-}4)\text{-}2\text{-}O\text{-})(H_2O)]$	Mn:MVol.D6-62/3
$C_{14}H_{12}MnN_2S_4$	$Mn(SC(=S)NHC_6H_5)_2$	Mn:MVol.D7-136/8
$C_{14}H_{12}MnN_4S_2$	$Mn(SC(NHC_6H_5)=NN=C(NHC_6H_5)S)$	Mn:MVol.D7-200/1
$C_{14}H_{12}MnN_6S_2$	$[Mn(6\text{-}CH_3\text{-}NC_5H_3\text{-}2\text{-}CH=NNH\text{-}2\text{-}C_5H_4N)(NCS)_2]$	Mn:MVol.D6-244, 246/7
–	$[Mn(NC_5H_4\text{-}2\text{-}CH=NNH\text{-}2\text{-}C_5H_3N\text{-}CH_3\text{-}6)(NCS)_2]$	Mn:MVol.D6-244, 246/7
$C_{14}H_{12}MnO_2P^+$	$[Mn(C_6H_5)_2P(CH_2COO)]^+$	Mn:MVol.D8-81/2
$C_{14}H_{12}MnS_4^-$	$[Mn(1,2\text{-}(S)_2C_6H_3CH_3\text{-}4)_2]^-$	Mn:MVol.D7-33/8, 47
–	$[Mn(1,2\text{-}(S)_2C_6H_3CH_3\text{-}4)_2(NC_5H_5)_n]^-$	Mn:MVol.D7-33/8
–	$[Mn(1,2\text{-}(S)_2C_6H_3CH_3\text{-}4)_2(O=CH\text{-}N(CH_3)_2)_n]^-$	Mn:MVol.D7-33/8
–	$[Mn(1,2\text{-}(S)_2C_6H_3CH_3\text{-}4)_2(O=S(CH_3)_2)_n]^-$	Mn:MVol.D7-33/8
$C_{14}H_{12}MnS_4^{2-}$	$[Mn(1,2\text{-}(S)_2C_6H_3CH_3\text{-}4)_2]^{2-}$	Mn:MVol.D7-36, 39/40
$C_{14}H_{12}Mn_2S_4$	$Mn[Mn(1,2\text{-}(S)_2C_6H_3CH_3\text{-}4)_2] \cdot 4\ H_2O$	Mn:MVol.D7-40

$C_{14}H_{12}MoN_2O_2$ . . . $(C_5H_5)Mo(CO)_2[-NH=C(C_6H_5)-NH-]$ Mo:Org.Comp.7-157, 164
– . . . . $(C_5H_5)Mo(CO)_2-N=N-C_6H_4CH_3-4$ Mo:Org.Comp.7-6, 16, 24
$C_{14}H_{12}MoN_2O_3$ . . . $(C_5H_5)Mo(CO)_2-N=N-C_6H_4-4-OCH_3$ Mo:Org.Comp.7-6, 16/7
$C_{14}H_{12}MoN_2O_5$ . . . $(C_5H_5)Mo(CO)_2[-C(COO-CH_3)=C(CN)-C(OCH_3)=NH-]$
Mo:Org.Comp.8-112, 133
$C_{14}H_{12}MoN_2S_4$ . . . $(HC{\equiv}CH)_2Mo[S_2C-1-NC_4H_4]_2$ Mo:Org.Comp.5-185, 188
$C_{14}H_{12}MoN_4O_2$ . . . $(C_5H_5)Mo(CO)_2[-C_3H_2N_2(CH_3)-C_3N_2H_2-]$ Mo:Org.Comp.7-157, 176
$C_{14}H_{12}MoO_2$ . . . . . $(C_5H_5)Mo(CO)_2(CH_2C_6H_5)$ Mo:Org.Comp.8-205, 206, 232
– . . . . $(C_5H_5)Mo(CO)_2(C_7H_7)$ Mo:Org.Comp.8-250, 293/4, 301/3
– . . . . $(C_9H_7)Mo(CO)_2(CH_2CHCH_2)$ Mo:Org.Comp.8-205, 209/10, 241/2
$C_{14}H_{12}MoO_2S_2$ . . . $(C_5H_5)Mo(CO)_2[-S(CH_3)-1,2-C_6H_4-S-]$ Mo:Org.Comp.7-201
$C_{14}H_{12}MoO_4$ . . . . . $(C_5H_5)Mo(CO)_2[CH_2C(CH_2)-COO-CH_2-C{\equiv}CH]$ Mo:Org.Comp.8-205, 220/1
– . . . . $[(6.2.0)-C_{10}H_{12}-2,4,6]Mo(CO)_4$ Mo:Org.Comp.5-338, 340
$C_{14}H_{12}Mo_2N_2O_{10}P_2S$
$[Mo(CO)_5(P(CH_3)_2N=S=NP(CH_3)_2)Mo(CO)_5]$ . . . S: S-N Comp.7-127/8
$C_{14}H_{12}NO_3Re$ . . . . $(C_5H_5)Re(CO)(NO)-C(=O)-C_6H_4-4-CH_3$ . . . . . . Re: Org.Comp.3-161
– . . . . $(C_5H_5)Re(CO)_2[NC_5H_4-3-C(=O)CH_3]$ . . . . Re: Org.Comp.3-193/7, 199
– . . . . $(C_5H_5)Re(CO)_2[NC_5H_4-4-C(=O)CH_3]$ . . . . Re: Org.Comp.3-193/7, 199
$C_{14}H_{12}NO_4Re$ . . . . $(C_5H_5)Re(CO)(NO)-C(=O)O-CH_2C_6H_5$ . . . . Re: Org.Comp.3-163
$C_{14}H_{12}NO_5Re$ . . . . cis-$(CO)_4Re[C(O)CH_3]C(CH_3)=NH-C_6H_5$ . . . . . . Re: Org.Comp.1-394/5, 404, 405
$C_{14}H_{12}N_2NaO_4Re$ $Na[(C_5H_5)Re(CO)(COO)-N=N-C_6H_4-OCH_3-4]$. Re: Org.Comp.3-142
$C_{14}H_{12}N_2OS$ . . . . . $C_6H_5C(O)N=S=N-C_6H_4-4-CH_3$ . . . . S: S-N Comp.7-268/9
$C_{14}H_{12}N_2O_2Re^+$ . . $[(C_5H_5)Re(CO)_2-N=N-C_6H_4-4-CH_3][BF_4]$ . . . . . Re: Org.Comp.3-193/7, 203
$C_{14}H_{12}N_2O_2S$ . . . . $C_6H_5C(O)N=S=N-C_6H_4-4-OCH_3$ . . . . S: S-N Comp.7-268/9
$C_{14}H_{12}N_2O_3Re^+$ . . $[(C_5H_5)Re(CO)_2-N=N-C_6H_4-2-OCH_3][BF_4]$ . . . . Re: Org.Comp.3-193/7, 203
– . . . . $[(C_5H_5)Re(CO)_2-N=N-C_6H_4-4-OCH_3][BF_4]$ . . . . Re: Org.Comp.3-193/7, 203, 209
$C_{14}H_{12}N_2O_3Re^-$ . . $[(C_5H_5)Re(CO)(COO)-N=N-C_6H_4-CH_3-4]^-$ . . . . Re: Org.Comp.3-141
$C_{14}H_{12}N_2O_4Re^-$ . . $[(C_5H_5)Re(CO)(COO)-N=N-C_6H_4-OCH_3-4]^-$ . . . Re: Org.Comp.3-141
– . . . . $Na[(C_5H_5)Re(CO)(COO)-N=N-C_6H_4-OCH_3-4]$. Re: Org.Comp.3-142
– . . . . $Li[(C_5H_5)Re(CO)(COO)-N=N-C_6H_4-OCH_3-4] \cdot CH_2Cl_2$
Re: Org.Comp.3-141/2
– . . . . $Ca[(C_5H_5)Re(CO)(COO)-N=N$
$-C_6H_4-OCH_3-4]_2 \cdot 2\ CH_2Cl_2$ . . . . Re: Org.Comp.3-142
$C_{14}H_{12}N_2O_4Th^{2+}$ $Th[C_6H_5C(O)NHO]_2{}^{2+}$ . . . . Th: SVol.D1-113
$C_{14}H_{12}N_2O_5S$ . . . . $C_6H_5-C(OCH_3)=N-S(O)O-C_6H_4-4-NO_2$ . . . . S: S-N Comp.8-322
$C_{14}H_{12}N_2O_6Th$ . . . $(O_2)Th(OOC-C_6H_4-2-NH_2)_2$ . . . . Th: SVol.C7-137/9
Th: SVol.D4-133
$C_{14}H_{12}N_3O_3Re$ . . . $(C_5H_5)Re(CO)(N=C=O)-N=N-C_6H_4-OCH_3-4$ . . . Re: Org.Comp.3-133/5, 137
$C_{14}H_{12}N_3O_3Sb$ . . . 3-$[(HO)_2(O)Sb-4-C_6H_4-N=N]-1-NC_8H_6$ . . . . . . Sb: Org.Comp.5-292
$C_{14}H_{12}N_4PdS_2$ . . . $Pd(SCN)(NCS)[2-NC_5H_3(CH_3-4)-2-C_5H_3N(CH_3-4)]$
Pd: SVol.B2-304
$C_{14}H_{12}N_6O_2PdS_2$ $Pd(SCN)_2[NH_2C(O)-3-C_5H_4N]_2$ . . . . Pd: SVol.B2-302/3
– . . . . $Pd(SCN)_2[NH_2C(O)-4-C_5H_4N]_2$ . . . . Pd: SVol.B2-302/3
$C_{14}H_{12}O_3Sn$ . . . . . $(C_6H_5)_2Sn[-OC(O)CH_2O-]$ . . . . Sn: Org.Comp.16-134/5
$C_{14}H_{12}O_4Sn$ . . . . . $(C_6H_5)_2Sn(OOCH)_2$ . . . . Sn: Org.Comp.16-123
$C_{14}H_{12}O_4Ti$ . . . . . . $[-(C_5H_5)_2Ti-OC(O)-CH=CH-C(O)O-]_n$ . . . . Ti: Org.Comp.5-339

$C_{14}H_{12}O_4Ti$	$[-(C_5H_5)_2Ti-O-C_4H_2(=O)_2-O-]_n$	Ti: Org.Comp.5-335
$C_{14}H_{12}O_6ReS^-$	$[H_8C_4O-1-Re(CO)_3(-S-(1,2-C_6H_4)-C(O)-O-)]^-$	Re: Org.Comp.1-137
$C_{14}H_{12}O_6Th$	$Th(OH)_2(C_6H_5COO)_2$	Th: SVol.C7-128/9
$C_{14}H_{12}O_7Th$	$Th(OH)_3(2,2-HOOCC_6H_4C_6H_4COO)$	Th: SVol.C7-146/8
$C_{14}H_{12}O_8Th$	$Th(OH)_2(2-HO-C_6H_4-COO)_2$	Th: SVol.C7-130/1
–	$Th(OH)_2(3-HO-C_6H_4-COO)_2$	Th: SVol.C7-131/3
–	$Th(OH)_2(3-HO-C_6H_4-COO)_2 \cdot x\ H_2O$	Th: SVol.C7-131/3
–	$Th(OH)_2(4-HO-C_6H_4-COO)_2 \cdot x\ H_2O$	Th: SVol.C7-131/3
–	$[ThO(4-HOC_6H_4COO)_2(H_2O)] \cdot 3\ H_2O$	Th: SVol.C7-132
$C_{14}H_{12}O_{10}Th$	$Th(OH)_2[2,4-(HO)_2C_6H_3COO]_2$	Th: SVol.C7-132/3
$C_{14}H_{13}InO_2$	$(C_6H_5)_2InOC(O)-CH_3$	In: Org.Comp.1-196, 199
$C_{14}H_{13}MnNO_4P^-$	$[Mn(O_3PCH(NHCH_2C_6H_5)C_6H_4-O-2)]^-$	Mn: MVol.D8-125/6
$C_{14}H_{13}MnN_4PS_2$	$Mn[(C_6H_5)P(CH_2CH_2CN)_2](NCS)_2$	Mn: MVol.D8-69/70
$C_{14}H_{13}MoNO_2$	$(C_5H_5)Mo(CO)_2(C_6H_8-4-CN)$	Mo: Org.Comp.8-250, 251, 266, 299
–	$(C_5H_5)Mo(CO)_2(C_6H_7-4-D-4-CN)$	Mo: Org.Comp.8-251, 266
$C_{14}H_{13}MoNO_3$	$C_5H_5Mo(CO)(NO)=C(C_6H_5)OCH_3$	Mo: Org.Comp.6-264, 268
$C_{14}H_{13}MoNO_4$	$(C_9H_7)Mo(CO)_2[-O-C(=O)-CH(CH_3)-NH_2-]$	Mo: Org.Comp.7-208, 209, 231
$C_{14}H_{13}MoN_2O_2^+$	$[(C_5H_5)Mo(CO)_2(-1-NC_5H_4-2-CH=N(CH_3)-)]^+$	Mo: Org.Comp.7-249, 250, 252, 277/8
$C_{14}H_{13}MoO_2P$	$(C_5H_5)Mo(CO)_2=P(CH_3)C_6H_5$	Mo: Org.Comp.7-27
$C_{14}H_{13}MoO_2PS$	$(C_5H_5)Mo(CO)_2[-S=P(CH_3)(C_6H_5)-]$	Mo: Org.Comp.7-213
$C_{14}H_{13}MoO_5^+$	$[(C_5H_5)Mo(CO)(HC\equiv C-C(O)O-CH_3)_2]^+$	Mo: Org.Comp.6-316, 322
$C_{14}H_{13}N_2O_2Re$	$(C_5H_5)Re(CO)_2(HN=N-C_6H_4-4-CH_3)$	Re: Org.Comp.3-193/7, 202
$C_{14}H_{13}N_2O_3Re$	$(C_5H_5)Re(CO)(CO_2H)-N=N-C_6H_4-CH_3-4$	Re: Org.Comp.3-133/5, 136
–	$(C_5H_5)Re(CO)_2(HN=N-C_6H_4-4-OCH_3)$	Re: Org.Comp.3-193/7, 202
$C_{14}H_{13}N_2O_4Re$	$(C_5H_5)Re(CO)(CO_2H)-N=N-C_6H_4-OCH_3-4$	Re: Org.Comp.3-133/5, 137, 140
–	$(C_5H_5)Re(CO)(O-CH=O)-N=N-C_6H_4-OCH_3-4$	Re: Org.Comp.3-133/5, 137
$C_{14}H_{13}N_2O_5SSb$	$3-HOC(O)-C_6H_4-NHC(S)NH-4-C_6H_4-Sb(O)(OH)_2$	Sb: Org.Comp.5-290
–	$4-HOC(O)-C_6H_4-NHC(S)NH-4-C_6H_4-Sb(O)(OH)_2$	Sb: Org.Comp.5-290
$C_{14}H_{13}N_2O_9Re$	$(CO)_4Re[C(O)CH_3]C(CH_3)=NH-CH_2C(O)O-1-[NC_4H_4(O)_2-2,5]$	Re: Org.Comp.1-401
$C_{14}H_{13}N_3O_3S$	$C_2H_5-O-4-C_6H_4-N=S=N-C_6H_4-4-NO_2$	S: S-N Comp.7-261
$C_{14}H_{13}N_4NaO_9S_3$	$Na[4-NO_2-C_6H_4-S(O)_2-NS(OC_2H_5)N-S(O)_2-C_6H_4-4-NO_2]$	S: S-N Comp.8-203, 205/6
$C_{14}H_{13}N_4O_9S_3^-$	$[4-NO_2-C_6H_4-S(O)_2-NS(OC_2H_5)N-S(O)_2-C_6H_4-4-NO_2]^-$	S: S-N Comp.8-203, 205/6
$C_{14}H_{13}N_5O_{14}Th$	$Th(NO_3)_4 \cdot HO-C_6H_4-2-(CH=N-C_6H_4-4-OCH_3) \cdot 4\ H_2O$	Th: SVol.D4-139
$C_{14}H_{13}O_3Sb$	$(C_6H_5)_2Sb(O)O_2CCH_3$	Sb: Org.Comp.5-213
$C_{14}H_{13}O_5Sb$	$(HO)_2(O)Sb-C_6H_4-3-C(O)O-CH_2C_6H_5$	Sb: Org.Comp.5-294
–	$(HO)_2(O)Sb-C_6H_4-4-C(O)O-CH_2C_6H_5$	Sb: Org.Comp.5-294
$C_{14}H_{13}O_6ReS$	$H_8C_4O-1-Re(CO)_3[-S(H)-(1,2-C_6H_4)-C(O)-O-]$	Re: Org.Comp.1-134/5
$C_{14}H_{14}IIn$	$(4-CH_3-C_6H_4)_2InI$	In: Org.Comp.1-156, 157
$C_{14}H_{14}IN_2ORe$	$(C_5H_5)ReI(CO)-N=N-C_6H_3(CH_3)_2-2,4$	Re: Org.Comp.3-133/5, 139

$C_{14}H_{14}I_2MnN_2S$ .. $[Mn((6-CH_3)C_5H_3N-2-CH=N-C_6H_4-2-SCH_3)I_2]$ Mn:MVol.D6-82
$C_{14}H_{14}I_2Po$ ...... $(4-CH_3C_6H_4)_2PoI_2$ ...... Po: SVol.1-334/40
$C_{14}H_{14}InN_2^+$ ..... $[(CH_3)_2In(1,10-N_2C_{12}H_8)]^+$ ...... In: Org.Comp.1-223, 226
$C_{14}H_{14}InN_3$ ...... $1-(CH_3)_2In-2-(NC_5H_4-2)-1,3-N_2C_7H_4$ ...... In: Org.Comp.1-276, 279
$C_{14}H_{14}InN_3O_3$ .... $[(CH_3)_2In(1,10-N_2C_{12}H_8)][NO_3]$ ...... In: Org.Comp.1-223, 226
$C_{14}H_{14}KN_2S$ ..... $K[(4-CH_3-C_6H_4-N)_2S]$, radical ...... S: S-N Comp.7-334/6
$C_{14}H_{14}MnNO_3P$ .. $[Mn(O_3PCH(NHCH_2C_6H_5)C_6H_5)]$ Mn:MVol.D8-125/6
$C_{14}H_{14}MnNO_4P$ .. $[MnH(O_3PCH(NHCH_2C_6H_5)C_6H_4-O-2)]$ Mn:MVol.D8-125/6
$C_{14}H_{14}MnN_2O_2S_2$ $[Mn((SC(=S))N(C_2H_4)_2O)NC_9H_6O-8]$ Mn:MVol.D7-168
$C_{14}H_{14}MnN_2O_4$... $[Mn(-O-C_6H_4-2-CH=NN=CH-2-C_6H_4-O-)(H_2O)_2]$ Mn:MVol.D6-261
– ...... $[Mn(-O-C_6H_4-2-N=CHCH=N-2-C_6H_4-O-)(H_2O)_2]$ Mn:MVol.D6-210
$C_{14}H_{14}MnN_2O_4P^-$ $[Mn(O_3PCH(NHCH_2CH_2-2-NC_5H_4)C_6H_4-O-2)]^-$ Mn:MVol.D8-125/6
$C_{14}H_{14}MnN_2O_5$... $Mn[C_2H_5-OC(O)-C(C(O)C_2H_5)=N-N-2-C_6H_4-COO]$ Mn:MVol.D6-259/60
– ...... $Mn[C_2H_5-OC(O)-C(C(O)C_2H_5)=N-N-3-C_6H_4-COO]$ Mn:MVol.D6-259/60
$C_{14}H_{14}MnN_2O_5^{2+}$ $Mn[OC_9H_2(CH_3)(=O)(OCH_3)(OH)CH=N-NH-C(O)CH_3]^{2+}$ Mn:MVol.D6-288/9
$C_{14}H_{14}MnN_4O_2$... $[Mn(-O-C_6H_4-2-N=CH-CCH_3=N-N(C_5H_4N-2)-)(H_2O)]$ Mn:MVol.D6-268/9
– ...... $[Mn(O-C_6H_4-2-CH=NNH_2)_2]$ Mn:MVol.D6-240, 241
$C_{14}H_{14}MnN_6$ ..... $[Mn(NC_5H_3(CH_3-4)-2-N-N=CH-CH=N$ $-N-2-C_5H_3N(CH_3-4))]$ Mn:MVol.D6-264/5
– ...... $[Mn(NC_5H_4-2-N-N=CCH_3-CCH_3=N-N-2-C_5H_4N)]$ Mn:MVol.D6-264/5
$C_{14}H_{14}MnN_6O_2S_3$ $Mn(4-NH_2C_6H_4SO_2NH-2-C_4HN_2-1,3-(CH_3)_2-4,6)(NCS)_2$ Mn:MVol.D7-116/23
$C_{14}H_{14}MnN_6O_4S_3$ $Mn(4-NH_2C_6H_4SO_2NH-4-C_4HN_2-1,3-$ $(OCH_3)_2-2,6)(NCS)_2$ Mn:MVol.D7-116/23
$C_{14}H_{14}MnO_4S_2$... $[Mn(4-CH_3C_6H_4SO_2)_2(H_2O)_2]$. Mn:MVol.D7-109/11
$C_{14}H_{14}MnO_4Se_2$.. $Mn(O_2Se-C_6H_4-CH_3-4)_2$ Mn:MVol.D7-244/6
– ...... $[Mn(O_2Se-C_6H_4-CH_3-4)_2(H_2O)_2]$ · 2 H_2O Mn:MVol.D7-244/6
$C_{14}H_{14}MnO_6S_2$... $Mn(O_3S-C_6H_4-CH_3-4)_2$ Mn:MVol.D7-114/5
– ...... $Mn(O_3S-C_6H_4-CH_3-4)_2$ · 7 H_2O Mn:MVol.D7-114/5
$C_{14}H_{14}MnO_6S_2^{2-}$ $[Mn(OC_4H_4-2-CH_2-CH(-S)-CO_2)_2]^{2-}$ Mn:MVol.D7-50/2
$C_{14}H_{14}MnS_4$ ..... $H_2[Mn(1,2-S_2C_6H_3CH_3-4)_2]$ · 4 H_2O. Mn:MVol.D7-40
$C_{14}H_{14}MoN_2O_2$... $(C_5H_5)Mo(CO)_2[-CH(2-NC_5H_4)-NH(CH_3)-]$ Mo:Org.Comp.8-111, 113/4
$C_{14}H_{14}MoO_3$ ..... $(C_5H_5)Mo(CO)_2[-C(CH_3)=(1,2-C_5H_6-c)=O-]$ ... Mo:Org.Comp.8-113, 153
– ...... $(C_5H_5)Mo(CO)_2[(CH_2)C_5H_4(CH_3)(=O)]$ Mo:Org.Comp.8-206, 207, 234
– ...... $(C_5H_5)Mo(CO)_2[(CH_3-CH)C_5H_5(=O)]$ Mo:Org.Comp.8-206, 207, 234
– ...... $(C_5H_5)Mo(CO)_2[C_5H_6-4-C(=O)CH_3]$. Mo:Org.Comp.8-255
– ...... $(C_5H_5)Mo(CO)_2[C_5H_5-5-D-4-C(=O)CH_3]$. Mo:Org.Comp.8-255
$C_{14}H_{14}MoO_4$ ..... $(C_5H_5)Mo(CO)_2[(CH_2)OC_4H(CH_3)_2(=O)]$. Mo:Org.Comp.8-206, 235
$C_{14}H_{14}MoO_6$ ..... $3-CH_3-C_6H_8-1-C(OCH_3)=Mo(CO)_5$ Mo:Org.Comp.5-103
– ...... $4-CH_3-C_6H_8-1-C(OCH_3)=Mo(CO)_5$ Mo:Org.Comp.5-103
$C_{14}H_{14}NOS^+$ ..... $[OSN(CH_2C_6H_5)_2]^+$ ...... S: S-N Comp.6-289, 296
$C_{14}H_{14}NO_2Re$ .... $(C_5H_5)Re(CO)_2[NC_5H_3(CH_3)_2-3,4]$ ...... Re: Org.Comp.3-193/7, 199

$C_{14}H_{14}NO_5Re$ .... fac-C_6H_5-NH_2-Re(CO)$_3$[-O-C(CH_3)=CH-C(CH_3)=O-] Re: Org.Comp.1-127
$C_{14}H_{14}NO_5Sb$ .... (4-$O_2NC_6H_4$)(4-$C_2H_5OC_6H_4$)Sb(O)OH .... Sb: Org.Comp.5-224
$C_{14}H_{14}N_2O_2S$ .... CH_3O-2-C_6H_4-N=S=N-C_6H_4-2-OCH_3 .... S: S-N Comp.7-238
– .... CH_3O-4-C_6H_4-N=S=N-C_6H_4-4-OCH_3 .... S: S-N Comp.7-238/9
$C_{14}H_{14}N_2O_2S_3$ ... 4-CH_3O-C_6H_4-S-N=S=N-S-C_6H_4-OCH_3-4. ... S: S-N Comp.7-32/7
– .... 4-CH_3-C_6H_4-S(O)$_2$-N=S=N-S-C_6H_4-CH_3-4 ... S: S-N Comp.7-66
$C_{14}H_{14}N_2O_3Re^+$ .. [(C_5H_5)Re(CO)$_2$-NHNH-C_6H_4-4-OCH_3][BF_4] .. Re: Org.Comp.3-193/7, 204
$C_{14}H_{14}N_2O_3S_2$ ... C_6H_5-S(O)$_2$N=S=N-C_6H_4-4-OC_2H_5 .... S: S-N Comp.7-55/60
$C_{14}H_{14}N_2O_4S_3$ ... 4-CH_3-C_6H_4-S(O)$_2$N=S=NS(O)$_2$-C_6H_4-CH_3-4 S: S-N Comp.7-80/94
$C_{14}H_{14}N_2O_6S_3$ ... 4-CH_3O-C_6H_4-S(O)$_2$N=S=NS(O)$_2$-C_6H_4-OCH_3-4 S: S-N Comp.7-95/8
$C_{14}H_{14}N_2O_6Th$ ... Th(OH)$_2$(2-$H_2NC_6H_4COO$)$_2$ .... Th: SVol.C7-137/9
$C_{14}H_{14}N_2O_8Sb_2$ .. (HO)$_2$(O)Sb(4-C_6H_4NHC(O)C(O)NHC_6H_4-4')Sb(O)(OH)$_2$ Sb: Org.Comp.5-315
$C_{14}H_{14}N_2O_{12}Th$ .. [Th(NC_5H_3(COO)$_2$-2,6)$_2$(H_2O)$_4$] .... Th: SVol.C7-154/5
$C_{14}H_{14}N_2S$ ...... CH_3-2-C_6H_4-N=S=N-C_6H_4-2-CH_3 .... S: S-N Comp.7-241
– .... CH_3-3-C_6H_4-N=S=N-C_6H_4-3-CH_3 .... S: S-N Comp.7-242, 252, 253
– .... CH_3-4-C_6H_4-N=S=N-C_6H_4-4-CH_3 .... S: S-N Comp.7-242/3, 248, 251/6
– .... C_6H_5-CH_2-N=S=N-CH_2-C_6H_5 .... S: S-N Comp.7-211
$C_{14}H_{14}N_2S^-$...... [(4-CH_3-C_6H_4-N)$_2$S]$^-$, radical anion .... S: S-N Comp.7-334/6
$C_{14}H_{14}N_2S_3$ ..... (4-$CH_3C_6H_4$-SN=)$_2$S. .... S: S-N Comp.7-32/7
$C_{14}H_{14}N_3O_3Re$ ... (C_5H_5)Re(CO)(N=N-C_6H_4-OCH_3-4)-C(=O)NH_2 Re: Org.Comp.3-133/5, 137
$C_{14}H_{14}N_3S_4^+$ .... [4-CH_3-C_6H_4-S-N=SNS=N-S-C_6H_4-4-CH_3]$^+$ S: S-N Comp.7-42/4
$C_{14}H_{14}N_4O_2S$ .... (CH_3)$_2$N-4-C_6H_4-N=S=N-C_6H_4-2-NO_2 .... S: S-N Comp.7-261
– .... (CH_3)$_2$N-4-C_6H_4-N=S=N-C_6H_4-3-NO_2 .... S: S-N Comp.7-261
– .... (CH_3)$_2$N-4-C_6H_4-N=S=N-C_6H_4-4-NO_2 .... S: S-N Comp.7-261
$C_{14}H_{14}N_4O_2S_2$ ... O=S=NN(C_6H_5)CH_2CH_2N(C_6H_5)N=S=O .... S: S-N Comp.6-74
$C_{14}H_{14}N_4O_4S_2$ ... 4-NO_2-C_6H_4-S(O)$_2$N=S=N-C_6H_4-4-N(CH_3)$_2$ .. S: S-N Comp.7-60/1
$C_{14}H_{14}N_4O_4S_5$ ... 4-CH_3-C_6H_4-S(O)$_2$N=S=NSN=S=NS(O)$_2$-C_6H_4-4-CH_3 S: S-N Comp.7-69/70
$C_{14}H_{14}N_4O_9STh$ .. Th(OOC-CH_2-CN)$_4$ · (CH_3)$_2$SO .... Th: SVol.D4-166
$C_{14}H_{14}N_4PdS_2$ ... Pd(SCN)$_2$(CH_3-2-C_5H_4N)$_2$ .... Pd: SVol.B2-302/3
– .... Pd(SCN)$_2$(CH_3-3-C_5H_4N)$_2$ .... Pd: SVol.B2-302/3
– .... Pd(SCN)$_2$(CH_3-4-C_5H_4N)$_2$ .... Pd: SVol.B2-302/3
$C_{14}H_{14}N_4S_3$ ..... 4-CH_3-C_6H_4-N=S=NSN=S=N-C_6H_4-4-CH_3. ... S: S-N Comp.7-52
$C_{14}H_{14}N_6O_{12}Th$ .. Th(NO_3)$_4$ · NC_5H_4-2-[C(CH_3)=N-C_6H_4-2-CH_3] Th: SVol.D4-138
– .... Th(NO_3)$_4$ · NC_5H_4-2-[C(CH_3)=N-C_6H_4-3-CH_3] Th: SVol.D4-138
– .... Th(NO_3)$_4$ · NC_5H_4-2-[C(CH_3)=N-C_6H_4-4-CH_3] Th: SVol.D4-138
$C_{14}H_{14}N_6O_{13}Th$ .. Th(NO_3)$_4$ · NC_5H_4-2-(CH=N-C_6H_4-4-O-C_2H_5) Th: SVol.D4-141
– .... Th(NO_3)$_4$ · NC_5H_4-2-[C(CH_3)=N-C_6H_4-2-OCH_3] Th: SVol.D4-138
– .... Th(NO_3)$_4$ · NC_5H_4-2-[C(CH_3)=N-C_6H_4-4-OCH_3] Th: SVol.D4-138

$C_{14}H_{14}O_2Po$	$(4\text{-}CH_3OC_6H_4)_2Po$	Po: SVol.1-334/40
$C_{14}H_{14}O_2Sn$	$(C_6H_5)_2Sn[\text{-}OCH_2CH_2O\text{-}]$	Sn: Org.Comp.16-106
$C_{14}H_{14}O_3Sn$	$(C_6H_5)_2Sn(OH)OOCCH_3$	Sn: Org.Comp.16-204
$C_{14}H_{14}O_4STi$	$[(C_5H_5)_2Ti(OCOCH(SH)CH_2COO)]_n$	Ti: Org.Comp.5-339
$C_{14}H_{14}O_6Re$	$(CO)_4Re[C_7H_5(O)_2\text{-}2,3\text{-}(CH_3)_3\text{-}1,7,7]$, radical	Re: Org.Comp.1-372
$C_{14}H_{14}O_6Sb_2$	$(HO)_2(O)Sb(4\text{-}C_6H_4CH{=}CHC_6H_4\text{-}4')Sb(O)(OH)_2$	Sb: Org.Comp.5-315
$C_{14}H_{14}Po$	$(2\text{-}CH_3C_6H_4)_2Po$	Po: SVol.1-334/40
–	$(3\text{-}CH_3C_6H_4)_2Po$	Po: SVol.1-334/40
–	$(4\text{-}CH_3C_6H_4)_2Po$	Po: SVol.1-334/40
–	$(C_6H_5CH_2)_2Po$	Po: SVol.1-334/40
$C_{14}H_{14}Ti$	$[(C_6H_5CH_2)_2Ti]_n$	Ti: Org.Comp.5-323
$C_{14}H_{15}InS$	$(C_6H_5)_2In\text{-}S\text{-}C_2H_5$	In: Org.Comp.1-240, 242
$C_{14}H_{15}MnNO_3P^+$	$[MnH(O_3PCH(NHCH_2C_6H_5)C_6H_5)]^+$	Mn: MVol.D8-125/6
$C_{14}H_{15}MnN_2O_3^+$	$Mn[(CH_3)_2C_6H_4({=}O)_2{=}N\text{-}NH\text{-}C_6H_4\text{-}O]^+$	Mn: MVol.D6-256/7
$C_{14}H_{15}MnN_2O_4P$	$[MnH(O_3PCH(NHCH_2CH_2\text{-}2\text{-}NC_5H_4)C_6H_4\text{-}O\text{-}2)]$	Mn: MVol.D8-125/6
$C_{14}H_{15}MnN_3O_3P$	$Mn(NO)_3[P(C_6H_5)_2(C_2H_5)]$	Mn: MVol.D8-64/7
$C_{14}H_{15}MoNO_2$	$(C_5H_5)Mo(CO)(NO)(c\text{-}C_5H_5\text{-}CH_2CH{=}CH_2\text{-}2)$	Mo: Org.Comp.6-296
$C_{14}H_{15}MoNO_4$	$(C_5H_5)Mo(CO)_2\text{-}C({=}O)\text{-}C({=}CH_2)\text{-}4\text{-}(1,4\text{-}ONC_4H_8)$	Mo: Org.Comp.8-202/3
$C_{14}H_{15}MoNO_6$	$(C_5H_5)Mo(CO)_2[\text{-}O\text{-}C({=}O)\text{-}C(CH_3){=}N(CH_2COO\text{-}C_2H_5)\text{-}]$	Mo: Org.Comp.7-208/9, 238
$C_{14}H_{15}MoN_2O_2^+$	$[(C_5H_5)Mo(CO)(NO)(CH_2{=}CHCH_2\text{-}1\text{-}NC_5H_5)]^+$	Mo: Org.Comp.6-304
–	$[(C_5H_5)Mo(CO)_2(NH_2\text{-}CH(CH_3)\text{-}2\text{-}NC_5H_4)]^+$	Mo: Org.Comp.7-249, 251/2, 277
$C_{14}H_{15}MoN_3O_4$	$[\text{-}N(CH_3)\text{-}CH_2CH_2\text{-}N(CH_3)\text{-}]C{=}Mo(CO)_4(NC_5H_5)$	Mo: Org.Comp.5-97/100
$C_{14}H_{15}MoO_2^+$	$[(C_5H_5)Mo(CO)_2(C_6H_7\text{-}5\text{-}CH_3)][PF_6]$	Mo: Org.Comp.8-305/6, 313, 317/8
–	$[(C_5H_5)Mo(CO)_2(c\text{-}C_7H_{10})][PF_6]$	Mo: Org.Comp.8-305/6, 314, 318/9
$C_{14}H_{15}N_2NaO_5S_3$	$Na[C_6H_5\text{-}S(O)_2\text{-}NS(OC_2H_5)N\text{-}S(O)_2\text{-}C_6H_5]$	S: S-N Comp.8-203/4
$C_{14}H_{15}N_2O_5S_3^-$	$[C_6H_5\text{-}S(O)_2\text{-}NS(OC_2H_5)N\text{-}S(O)_2\text{-}C_6H_5]^-$	S: S-N Comp.8-203/4
$C_{14}H_{15}N_3O_2S_2$	$C_6H_5\text{-}S(O)_2N{=}S{=}N\text{-}C_6H_4\text{-}4\text{-}N(CH_3)_2$	S: S-N Comp.7-55/60
$C_{14}H_{15}N_3O_6S_4$	$C_6H_5\text{-}S(O)_2N{=}S{=}NS(O)_2\text{-}C_6H_4\text{-}S(O)_2N(CH_3)_2\text{-}4$	S: S-N Comp.7-100/1
$C_{14}H_{15}N_4O_3Re$	$(C_5H_5)Re(CO)(N{=}N\text{-}C_6H_4\text{-}OCH_3\text{-}4)\text{-}C({=}O)NHNH_2$	Re: Org.Comp.3-133/5, 138
$C_{14}H_{15}N_8O_8Th^-$	$Th[C_2H_4(N(CH_2COO)_2)_2][(H_2N)HC_4N_5]^-$	Th: SVol.D1-138
$C_{14}H_{15}O_2ReSi$	$(C_5H_5)Re(H)(CO)_2\text{-}SiH_2\text{-}CH_2C_6H_5$	Re: Org.Comp.3-175/6, 178
$C_{14}H_{15}O_2Sb$	$(4\text{-}CH_3C_6H_4)_2Sb(O)OH$	Sb: Org.Comp.5-210
$C_{14}H_{15}O_3Sb$	$4\text{-}CH_3O\text{-}C_6H_4\text{-}Sb(O)(OH)\text{-}C_6H_4\text{-}CH_3\text{-}4$	Sb: Org.Comp.5-224
–	$4\text{-}C_2H_5\text{-}O\text{-}C_6H_4\text{-}Sb(O)(OH)\text{-}C_6H_5$	Sb: Org.Comp.5-223
$C_{14}H_{15}O_5PTh^{2+}$	$Th(OH)[O_2P(OC_6H_4CH_3\text{-}4)_2]^{2+}$	Th: SVol.D1-131
$C_{14}H_{15}O_6ReS$	$(C_2H_5)_2O\text{-}Re(CO)_3[\text{-}S(H)\text{-}(1,2\text{-}C_6H_4)\text{-}C(O)\text{-}O\text{-}]$	Re: Org.Comp.1-135
$C_{14}H_{16}IMnN_4O_3$	$[MnI(O\text{-}C_6H_4\text{-}2\text{-}CH{=}NNH_2)_2(H_2O)]$	Mn: MVol.D6-240, 242/3
$C_{14}H_{16}IMoO_5P$	$[(C_9H_7)Mo(CO)_2(I)(P(OCH_3)_3)]$	Mo: Org.Comp.7-56/7, 86
$C_{14}H_{16}I_2NO_4Re$	$[(CH_3)_3NCH_2C_6H_5][(CO)_4ReI_2]$	Re: Org.Comp.1-343

$C_{14}H_{16}I_3MoOP$ ... $C_5H_5Mo(CO)(P(CH_3)_2C_6H_5)I_3$ ... Mo:Org.Comp.6-219
$C_{14}H_{16}InO_2P$ ... $(CH_3)_2In{-}OP(O)(C_6H_5)_2$ ... In: Org.Comp.1-215
$C_{14}H_{16}InP$ ... $(CH_3)_2In{-}P(C_6H_5)_2$ ... In: Org.Comp.1-312, 314, 317
$C_{14}H_{16}MnN_2OS_2$ ... $[Mn(SC(=S)N(C_2H_5)_2)(NC_9H_6O{-}8)]$ ... Mn:MVol.D7-143/4
$C_{14}H_{16}MnN_2O_6Re$ $(i{-}C_3H_7)_2C_2H_2MnN_2Re(CO)_6$ ... Re: Org.Comp.1-333
$C_{14}H_{16}MnN_2O_6S_2$ $[Mn(SC_4H_3(CH{=}NOH{-}2))_2(OC(CH_3){=}O)_2]$ ... Mn:MVol.D7-222/5
$C_{14}H_{16}MnN_2O_6S_4$ $Mn[2{-}(O_3S{-}CH_2CH_2{-}N{=}CH)C_4H_3S]_2$ ... Mn:MVol.D6-75/7
$C_{14}H_{16}MnN_2O_8S_2$ $Mn[2{-}(O_3S{-}CH_2CH_2{-}N{=}CH)C_4H_3O]_2$ ... Mn:MVol.D6-75/7
$C_{14}H_{16}MnN_2O_{10}S_2$
$[Mn(C_7H_4NS(=O)_3)_2(H_2O)_4] \cdot 2\ H_2O$ ... Mn:MVol.D7-128/9
$C_{14}H_{16}MnN_4O_3$ ... $[Mn({-}O{-}C_6H_4{-}2{-}N{=}CHC(CH_3){=}N{-}N(C_5H_4N{-}2){-})(H_2O)_2]$
Mn:MVol.D6-268/9
$C_{14}H_{16}MnN_6S_6$ ... $Mn(SNC_3H_4{-}C_3H_4NS)_2(NCS)_2$ ... Mn:MVol.D7-230/2
$C_{14}H_{16}MnO_4P_2$ ... $Mn[O_2P(CH_3)(C_6H_5)]_2$ ... Mn:MVol.D8-112
$C_{14}H_{16}MoNO_2P$ ... $C_5H_5Mo(P(CH_3)_2C_6H_5)(NO)CO$ ... Mo:Org.Comp.6-238
$C_{14}H_{16}MoN_4O_4$ ... $[({-}CH{=}CH{-}N(CH_3){-}N(CH_3){-})C{=}]_2Mo(CO)_4$ ... Mo:Org.Comp.5-124, 128
– ... $[({-}N(CH_3){-}CH{=}CH{-}N(CH_3){-})C{=}]_2Mo(CO)_4$ ... Mo:Org.Comp.5-124/5, 127/8, 130/2
– ... $[({-}N(CD_3){-}CH{=}CH{-}N(CD_3){-})C{=}]_2Mo(CO)_4$ ... Mo:Org.Comp.5-130/2
$C_{14}H_{16}MoO_2$ ... $(CH_3{-}C_5H_4)Mo(CO)_2(c{-}C_6H_9)$ ... Mo:Org.Comp.8-250, 257
– ... $(C_5H_5)Mo(CO)[c{-}C_6H_8{-}C(O)CH_3]$ ... Mo:Org.Comp.6-343
– ... $(C_5H_5)Mo(CO)_2(c{-}C_6H_8{-}4{-}CH_3)$ ... Mo:Org.Comp.8-250, 258
– ... $(C_5H_5)Mo(CO)_2(c{-}C_6H_7{-}4{-}D{-}6{-}CH_3)$ ... Mo:Org.Comp.8-258/9
– ... $(C_5H_5)Mo(CO)_2(c{-}C_7H_{11})$ ... Mo:Org.Comp.8-250, 251, 281/2
– ... $(C_5H_5)Mo(CO)_2(c{-}C_7H_{10}{-}4{-}D)$ ... Mo:Org.Comp.8-251, 282
– ... $(C_6H_8)_2Mo(CO)_2$ ... Mo:Org.Comp.5-349, 351, 354/5
$C_{14}H_{16}MoO_4$ ... $(C_5H_5)Mo(CO)_2[CH_2C(CH{-}CH_3){-}COO{-}C_2H_5]$ ... Mo:Org.Comp.8-205, 224
$C_{14}H_{16}NO_2PRe^+$ ... $[(C_5H_5)Re(CO)(NO)(P(CH_3)_2{-}C_6H_5)][PF_6]$ ... Re: Org.Comp.3-154, 156
$C_{14}H_{16}NO_5Re$ ... $(CO)_4Re[{-}C(=NC_5H_{10}){-}C(CH_3)_2C(=O){-}]$ ... Re: Org.Comp.2-331, 335
– ... $(CO)_5Re{-}C(NC_5H_{10}{-}1){=}C(CH_3)_2$ ... Re: Org.Comp.2-127
$C_{14}H_{16}NO_9Re$ ... cis-$(CO)_4Re[C(O)CH_3]C(CH_3){=}NH$
$-CH(COOCH_3)CH_2COOCH_3$ ... Re: Org.Comp.1-398
$C_{14}H_{16}N_2OS$ ... $O{=}S(NCH_3{-}C_6H_5)_2$ ... S: S-N Comp.8-350, 360
$C_{14}H_{16}N_2OSSn$ ... $(C_2H_5)_2Sn(NCS)OC_9H_6N$ ... Sn: Org.Comp.17-137
$C_{14}H_{16}N_2O_3Ti$ ... $[(C_5H_5)_2Ti(NHCH_2C(O)NHCH_2COO)]_n$ ... Ti: Org.Comp.5-351
$C_{14}H_{16}N_2O_5S_3$ ... $C_6H_5{-}S(O)_2{-}NH{-}S(OC_2H_5){=}N{-}S(O)_2{-}C_6H_5$ ... S: S-N Comp.8-201
$C_{14}H_{16}N_2O_{13}Th^{2-}$ $Th[((OOCCH_2)_2N)_2C_2H_4][OOCCH_2CH(OH)COO]^{2-}$
Th: SVol.D1-136
$C_{14}H_{16}N_3O_3Sb$ ... $(4{-}(4'{-}(CH_3)_2NC_6H_4N{=}N)C_6H_4)Sb(O)(OH)_2$ ... Sb: Org.Comp.5-291
$C_{14}H_{16}N_3O_4S_3^-$ ... $[(C_6H_5{-}S(O)_2N)_2SN(CH_3)_2]^-$ ... S: S-N Comp.8-217/8
$C_{14}H_{16}N_4O_2Ti$ ... $[(C_5H_5)_2Ti(ON{=}C(NH_2)CH{=}CHC(NH_2){=}NO)]_n$ ... Ti: Org.Comp.5-343
$C_{14}H_{16}N_4O_8Sb_2$ ... $(HO)_2(O)Sb(4{-}C_6H_4NHC(O{=})(NH)_2$
$C(=O)NHC_6H_4{-}4)Sb(O)(OH)_2$ ... Sb: Org.Comp.5-315
$C_{14}H_{16}N_5O_9Th^-$ ... $Th[C_2H_4(N(CH_2COO)_2)_2][(O)(H_2N)H_2C_4N_2]^-$ ... Th: SVol.D1-138
$C_{14}H_{16}N_8O_8Th$ ... $Th[C_2H_4(N(CH_2COO)_2)_2][(H_2N)HC_4N_5H]$ ... Th: SVol.D1-138
$C_{14}H_{16}O_2Sn$ ... $(C_6H_5)_2Sn(OCH_3)_2$ ... Sn: Org.Comp.16-98/9
– ... $(C_6H_5CH_2)_2Sn(OH)_2$ ... Sn: Org.Comp.16-68
$C_{14}H_{16}O_2Ti$ ... $(CH_3)_2Ti(OC_6H_5)_2$ ... Ti: Org.Comp.5-8

$C_{14}H_{16}O_4S_2Sn$ ... $(C_6H_5)_2Sn(OS(O)CH_3)_2$ ... Sn: Org.Comp.16-145
$C_{14}H_{16}O_4Se_2Sn$ .. $(C_6H_5)_2Sn(OSe(O)CH_3)_2$ ... Sn: Org.Comp.16-152/3
$C_{14}H_{16}O_4Sn$ ..... $(c\text{-}C_5H_5)_2Sn(OOCCH_3)_2$ ... Sn: Org.Comp.16-94, 96
$C_{14}H_{16}O_8Sb_2$ .... $(HO)_2(O)Sb(C_6H_3(OCH_3)C_6H_3(OCH_3))Sb(O)(OH)_2$
Sb: Org.Comp.5-314
$C_{14}H_{17}MoNO_2$ ... $C_5H_5Mo(CO)(NO)(C_8H_{12}\text{-}c)$ ... Mo:Org.Comp.6-304, 311
$C_{14}H_{17}MoNO_3$ ... $(C_5H_5)Mo(CO)_2(CN\text{-}C_4H_9\text{-}t)\text{-}C(=O)\text{-}CH_3$ ... Mo:Org.Comp.8-176, 180/1
– ... $(C_5H_5)Mo(CO)_2[\text{-}C(=O)\text{-}CH(CH=CH_2)\text{-}CH_2\text{-}N(CH_3)_2\text{-}]$
Mo:Org.Comp.8-137
– ... $(C_5H_5)Mo(CO)_2[\text{-}C(=O)\text{-}CH_2\text{-}CH(CH=CH_2)\text{-}N(CH_3)_2\text{-}]$
Mo:Org.Comp.8-137
– ... $(C_5H_5)Mo(CO)_2\text{-}C(=O)\text{-}C(=CH_2)\text{-}N(C_2H_5)_2$ Mo:Org.Comp.8-202/3
$C_{14}H_{17}MoNO_4$ ... $C_5H_5Mo(CO)(NO)CH_2=CHCH_2CH(C(O)CH_3)_2$... Mo:Org.Comp.6-296
$C_{14}H_{17}MoO^+$ .... $[(C_5H_5)Mo(CO)(CH_3\text{-}C\equiv C\text{-}CH_3)_2]^+$ ... Mo:Org.Comp.6-318, 323/5
– ... $[(C_5H_5)Mo(CO)(CD_3\text{-}C\equiv C\text{-}CD_3)_2]^+$ ... Mo:Org.Comp.6-318
$C_{14}H_{17}MoO_5{}^+$ ... $[C_5H_5Mo(CO)(HOCH_2\text{-}CC\text{-}CH_2OH)_2]^+$ ... Mo:Org.Comp.6-318/9
$C_{14}H_{17}MoO_6P$.... $(C_5H_5)Mo(CO)_2[P(\text{-}OCH_2\text{-})_3C\text{-}CH_3]\text{-}C(=O)CH_3$ Mo:Org.Comp.8-45, 54
$C_{14}H_{17}MoO_7P$.... $[OC_4H_2(=O)_2\text{-}2,5]Mo(CO)_4[P(C_2H_5)_3]$ ... Mo:Org.Comp.5-162/3
$C_{14}H_{17}NO_3Sn$ .... $(C_2H_5)_2Sn(OH)O\text{-}C_{10}H_6(NO)$ ... Sn: Org.Comp.16-180, 182
$C_{14}H_{17}N_2O_5Re$ ... $(CO)_4Re(CNC_4H_9\text{-}t)CON(CH_2CH_2)_2$ ... Re: Org.Comp.2-247
$C_{14}H_{17}N_2O_8Re$ ... cis-$(CO)_4Re[C(O)CH_3]C(CH_3)=NH$
$\text{-}CH_2C(O)NHCH_2COOC_2H_5$ ... Re: Org.Comp.1-400
$C_{14}H_{17}N_3O_2S_2$ ... $4\text{-}CH_3\text{-}C_6H_4\text{-}S(O)_2N=S=N\text{-}C_6H_{10}(CN\text{-}1)$...... S: S-N Comp.7-62
$C_{14}H_{17}N_3O_{12}Th^{2-}$ $[Th((OOCCH_2)_2N\text{-}C_2H_4\text{-}N(CH_2COO)_2)(HN(CH_2COO)_2)]^{2-}$
Th: SVol.D1-137
– ... $[Th((OOCCH_2)_2N\text{-}C_2H_4\text{-}N(CH_2COO)_2)$
$(OOCCH_2\text{-}CHNH_2\text{-}COO)]^{2-}$ ... Th: SVol.C7-125
$C_{14}H_{17}N_5O_9Th$ ... $Th[C_2H_4(N(CH_2COO)_2)_2][O(H_2N)H_2C_4N_2H]$ .... Th: SVol.D1-138
$C_{14}H_{17}O_8Sb$ ..... $(C_6H_5)Sb(O_2CCH_3)_4$ ... Sb: Org.Comp.5-266/7
$C_{14}H_{18}IInN_2$ ..... $(C_2H_5)_2InI \cdot 2\text{-}(NC_5H_4\text{-}2)\text{-}NC_5H_4$ ... In: Org.Comp.1-156, 158
$C_{14}H_{18}IMoO_5P$ ... $(C_5H_5)Mo(CO)_2(I)[P(\text{-}OCH_2\text{-})_3CC_3H_7\text{-}n]$ Mo:Org.Comp.7-56/7, 91, 114
$C_{14}H_{18}I_2MoN_2OS$ $C_5H_5Mo(OS(CH_3)_2)(N=NC_6H_4CH_3\text{-}4)I_2$ ... Mo:Org.Comp.6-27
$C_{14}H_{18}InN_2{}^+$..... $[(CH_3)_2In(2\text{-}(4\text{-}CH_3\text{-}NC_5H_3\text{-}2)\text{-}NC_5H_3\text{-}4\text{-}CH_3)]^+$
In: Org.Comp.1-223, 226
$C_{14}H_{18}InN_3O_3$.... $[(CH_3)_2In(2\text{-}(4\text{-}CH_3\text{-}NC_5H_3\text{-}2)\text{-}NC_5H_3\text{-}4\text{-}CH_3)][NO_3]$
In: Org.Comp.1-223, 226
$C_{14}H_{18}In_2$ .... $[CH_3\text{-}CC\text{-}In(CH=CH_2)_2]_2$ ... In: Org.Comp.1-105
$C_{14}H_{18}MnN_2O_4S_2{}^{2+}$
$[Mn(O=S(C_2H_5)(C_5H_4NO))_2(H_2O)_2]^{2+}$ ... Mn:MVol.D7-106
$C_{14}H_{18}MnN_4O_3S_3$ $Mn(4\text{-}CH_3C_6H_4SO_2NH\text{-}CONHC_4H_9)(NCS)_2$.... Mn:MVol.D7-124/5
$C_{14}H_{18}MnN_4O_4$... $[Mn(\text{-}O\text{-}C_6H_4\text{-}2\text{-}N=CHC(CH_3)=N\text{-}N(C_5H_4N\text{-}2)\text{-})(H_2O)_3]$
Mn:MVol.D6-268/9
– ... $[Mn(O\text{-}C_6H_4\text{-}2\text{-}CH=NNH_2)_2(H_2O)_2]$ ... Mn:MVol.D6-240/1
$C_{14}H_{18}MnN_8O_8$... $[Mn(NC_4H_4\text{-}2\text{-}CH=N\text{-}NHC(O)CH_3)_2(NO_3)_2]$.... Mn:MVol.D6-292/4
$C_{14}H_{18}MnO_6S_2$... $[Mn(4\text{-}CH_3C_6H_4SO_2)_2(H_2O)_2]$ ... Mn:MVol.D7-109/11
$C_{14}H_{18}MnO_6Se_2$.. $[Mn(O_2SeC_6H_4CH_3\text{-}4)_2(H_2O)_2] \cdot 2\ H_2O$ Mn:MVol.D7-244/6
$C_{14}H_{18}MoNO_2{}^+$ .. $[C_5H_5Mo(CO)(NO)C_8H_{13}\text{-}c]^+$ ... Mo:Org.Comp.6-350, 354
$C_{14}H_{18}MoNO_4P$ .. $(C_5H_5)Mo(CO)_2[C_4H_8NO_2P(CH_2\text{-}CH=CH_2)]$.... Mo:Org.Comp.7-158, 163
$C_{14}H_{18}MoN_2O_3$... $(C_5H_5)Mo(CO)_2[\text{-}C(=O)\text{-}NH\text{-}(1,1\text{-}C_6H_{10}\text{-}c)\text{-}NH_2\text{-}]$
Mo:Org.Comp.8-140/1

$C_{14}H_{18}MoN_2O_4$ ... $(CO)_4Mo(CN-C_4H_9-t)_2$ Mo:Org.Comp.5-25, 26/7, 29/30
$C_{14}H_{18}MoO_2$ ..... $(C_5H_5)Mo(CO)[CH_2C(CH_3)C(CH_3)-CH_2C(O)CH_3]$ Mo:Org.Comp.6-343
– .... $(C_5H_5)Mo(CO)_2(CH_3-CHCHCH-C_3H_7-i)$ Mo:Org.Comp.8-205, 229
$C_{14}H_{18}Mo_5N_2O_{21}P_2{}^{2-}$
$[(NH_3-C_6H_4-CH_2-PO_3)_2Mo_5O_{15}]^{2-}$ Mo:SVol.B3b-124
$C_{14}H_{18}NO_3Sb$ .... $(CH_3)_3Sb(OC_9H_6N)O_2CCH_3$ .... Sb: Org.Comp.5-47
$C_{14}H_{18}NO_5Re$ .... cis-$(CO)_4Re[C(O)CH_3]C(CH_3){=}NHC_6H_{11}$-c .... Re: Org.Comp.1-391
$C_{14}H_{18}NO_7Re$ .... cis-$(CO)_4Re[C(O)CH_3]C(CH_3){=}NH-CH(C_3H_7-i)COOCH_3$ Re: Org.Comp.1-396
$C_{14}H_{18}N_2O_8Re$ ... $(CO)_4Re[t-C_4H_9-OC(O)N{=}NC(O)-O-C_4H_9-t]$, radical Re: Org.Comp.1-372
$C_{14}H_{18}N_2O_8Th$ ... $Th[c-C_6H_{10}-1,2-(N(CH_2COO)_2)_2]$ .... Th: SVol.C7-124, 126
Th: SVol.D1-93/5, 100
– .... $Th[c-C_6H_{10}-1,2-(N(CH_2COO)_2)_2] \cdot H_2O$ ...... Th: SVol.C7-124, 126
– .... $Th[c-C_6H_{10}-1,2-(N(CH_2COO)_2)_2] \cdot x\ H_2O$ .... Th: SVol.C7-126
$C_{14}H_{18}N_3NaO_{10}Th$ $Na[Th(OOCCH_2N(CH_2CH_2N(CH_2COO)_2)_2)] \cdot 4\ H_2O$ Th: SVol.C7-127
$C_{14}H_{18}N_3O_{10}Th^-$ $Th[OOCCH_2N(C_2H_4N(CH_2COO)_2)_2]^-$ .... Th: SVol.D1-93/4, 96, 101
$C_{14}H_{18}N_4O_2Ti$ .... $[(C_5H_5)_2Ti(ON{=}C(NH_2)(CH_2)_2C(NH_2){=}NO)]_n$.... Ti: Org.Comp.5-343
$C_{14}H_{18}N_6O_2Sn$ ... $(C_4H_9)_2Sn(ONC(CN)_2)_2$.................. Sn: Org.Comp.15-343
$C_{14}H_{18}N_6O_{12}Th$ .. $Th(NO_3)_4 \cdot 2\ [2,4-(CH_3)_2-NC_5H_3]$........... Th: SVol.D4-159
– .... $Th(NO_3)_4 \cdot 2\ [2,6-(CH_3)_2-NC_5H_3]$........... Th: SVol.D4-159
$C_{14}H_{18}N_6O_{16}Th$ .. $Th(NO_3)_4 \cdot 2\ (1-O-NC_5H_5) \cdot CH_3-C({=}O)O-C_2H_5$ Th: SVol.D4-135
$C_{14}H_{18}N_8S_6Si$ .... $(CH_3)_2N_2C_6H_{12}[Si(NCS)_6]$.................. Si: SVol.B4-302/4
$C_{14}H_{18}OPRe$ ..... $(C_5H_5)ReH_2(CO)[P(CH_3)_2-C_6H_5]$ .... Re: Org.Comp.3-144
$C_{14}H_{19}IMoO_3Si$... $[(CH_3)_3Si-C_5H_4]Mo(CO)_2(I){=}C(-O-CH_2CH_2CH_2-)$ Mo:Org.Comp.8-17, 27
$C_{14}H_{19}MnN_2PS_2$ .. $Mn[P(C_6H_5)(C_3H_7)_2](NCS)_2$ Mn:MVol.D8-38
$C_{14}H_{19}MoNO_2$ ... $(C_5H_5)Mo(CO)(NO)(C_8H_{14}-c)$............... Mo:Org.Comp.6-301
– .... $(C_5H_5)Mo(CO)(NO)(C_8H_{13}D-c)$.............. Mo:Org.Comp.6-301
– .... $(C_5H_5)Mo(CO)(NO)[c-C_6H_7-D-1-(CH_3)_2-3,5]$ .. Mo:Org.Comp.6-300
– .... $(C_5H_5)Mo(CO)_2[CH_3-CH{=}C{=}N(C_2H_5)_2]$ Mo:Org.Comp.8-186, 189
$C_{14}H_{19}MoNO_2S_2$.. $(C_5H_5)Mo(CO)_2[S_2C-N(C_3H_7-i)_2]$............ Mo:Org.Comp.7-196, 204, 205
– .... $[(CH_3)_5C_5]Mo(CO)_2(S_2C-NHCH_3)$ Mo:Org.Comp.7-194
$C_{14}H_{19}MoNO_3$ ... $(C_5H_5)Mo(CO)(NO)[CH_3CH{=}CHCH_2C(CH_3)_2-CH{=}O]$ Mo:Org.Comp.6-297, 306/7
– .... $(C_5H_5)Mo(CO)(NO)(c-C_8H_{13}-OH-3)$ Mo:Org.Comp.6-301, 309
– .... $(C_5H_5)Mo(CO)_2[-C({=}O)-CH_2CH_2-NH(C_4H_9-t)-]$ Mo:Org.Comp.8-111, 136
$C_{14}H_{19}MoNO_4P^+$ $[(C_5H_5)Mo(CO)_2(1,3,6,2-O_2NPC_4H_9-2-CH_2CH{=}CH_2)]^+$ Mo:Org.Comp.7-249, 271, 281
$C_{14}H_{19}MoNaO_6$ .. $Na[(C_5H_5)Mo(CO)_2(-CH_2CH_2-O-C({=}O)-)] \cdot CH_3O-CH_2CH_2-OCH_3$ Mo:Org.Comp.8-196
$C_{14}H_{19}MoO_3P$.... $(C_5H_5)Mo(CO)_2[-CH(P(CH_3)_3)-CH{=}C(CH_3)-O-]$ Mo:Org.Comp.8-147
$C_{14}H_{19}MoO_4P$.... $[-Mo(C_5H_5)(CO)_2-CH_2-]P[-O-C(CH_3)_2-C(CH_3)_2-O-]$ Mo:Org.Comp.8-112, 125
$C_{14}H_{19}MoO_5P$.... $(C_5H_5)Mo(CO)_2[P(OCH_3)_3]-CH_2-CC-CH_3$ Mo:Org.Comp.8-77, 78, 89

$C_{14}H_{19}MoO_8P_2ReS$
(CO)$_3$Re[P(CH$_3$)$_3$]$_2$SH-Mo(CO)$_5$ . . . Re: Org.Comp.1-318
$C_{14}H_{19}NORe^+$ . . . [(CH$_3$C≡CCH$_3$)$_2$Re(O)-NC$_5$H$_4$CH$_3$-2]$^+$ . . . Re: Org.Comp.2-359/60, 369
– . . . [(CH$_3$C≡CCH$_3$)$_2$Re(O)-NC$_5$H$_4$CH$_3$-4]$^+$ . . . Re: Org.Comp.2-360, 369
$C_{14}H_{19}NO_2Sn$ . . . (C$_2$H$_5$)$_3$SnN(CO)$_2$C$_6$H$_4$ . . . Sn: Org.Comp.18-142, 147
$C_{14}H_{19}N_2O_8Th^+$ . . Th[H(c-C$_6$H$_{10}$-1,2-(N(CH$_2$COO)$_2$)$_2$)]$^+$ . . . Th: SVol.D1-100
$C_{14}H_{19}N_2O_9Th^-$ . . Th(OH)[c-C$_6$H$_{10}$-1,2-(N(CH$_2$COO)$_2$)$_2$]$^-$ . . . Th: SVol.D1-100
$C_{14}H_{19}N_3O_{10}Th$ . . H[Th(OOCCH$_2$N(CH$_2$CH$_2$N(CH$_2$COO)$_2$)$_2$)] . . . Th: SVol.D1-101
– . . . H[Th(OOCCH$_2$N(CH$_2$CH$_2$N(CH$_2$COO)$_2$)$_2$)] · H$_2$O
Th: SVol.C7-127
$C_{14}H_{19}N_3O_{11}Th^{2-}$ Th[OOCCH$_2$N(C$_2$H$_4$N(CH$_2$COO)$_2$)$_2$](OH)$^{2-}$ . . . Th: SVol.D1-97, 101
$C_{14}H_{19}O_8P_2ReSW$ (CO)$_3$Re[P(CH$_3$)$_3$]$_2$SH-W(CO)$_5$ . . . Re: Org.Comp.1-318, 319
$C_{14}H_{20}In_2N_2^+$ . . . [(CH$_3$)$_2$In-NC$_5$H$_4$=C$_5$H$_4$N-In(CH$_3$)$_2$]$^+$, radical cation
In: Org.Comp.1-294
$C_{14}H_{20}MnN_2S_4$ . . . Mn((SC(=S))C$_6$H$_9$(=NH)-2)$_2$ . . . Mn: MVol.D7-211
$C_{14}H_{20}MoN_2O_2S_2$ C$_5$H$_5$Mo(CO)(NO)CH$_2$=CHCH$_2$SC(S)N(C$_2$H$_5$)$_2$ . . Mo: Org.Comp.6-295
$C_{14}H_{20}MoN_4O_2$ . . . (CO)$_2$Mo(CN-C$_2$H$_5$)$_4$ . . . Mo: Org.Comp.5-50
$C_{14}H_{20}MoN_4O_4$ . . . [(-N(CH$_3$)-CH$_2$CH$_2$-N(CH$_3$)-)C=]$_2$Mo(CO)$_4$ . . . Mo: Org.Comp.5-124/6, 129, 130
$C_{14}H_{20}MoO_2Si_2$ . . [1,1-(CH$_3$)$_2$-SiC$_4$H$_4$]$_2$Mo(CO)$_2$ . . . Mo: Org.Comp.5-351
$C_{14}H_{20}MoO_4P^-$ . . . [C$_5$H$_5$Mo(P(CH$_3$)$_3$)(CO)(-O=C(CH$_2$OCH$_3$)CH=C(O)-)]$^-$
Mo: Org.Comp.6-257
$C_{14}H_{20}MoO_5$ . . . (CH$_2$CHCH$_2$)Mo(CO)$_2$[CH$_3$-C(O)CHC(O)-CH$_3$](OC$_4$H$_8$)
Mo: Org.Comp.5-283, 286
$C_{14}H_{20}MoO_5S_6$ . . . [CH$_3$-OC(=O)-CC-C(=O)O-CH$_3$]Mo(O)(S$_2$C-S-C$_3$H$_7$-i)$_2$
Mo: Org.Comp.5-139/41, 143
$C_{14}H_{20}MoS$ . . . C$_5$H$_5$Mo(SCH$_3$)(H$_3$C-CC-CH$_3$)$_2$ . . . Mo: Org.Comp.6-133/4
$C_{14}H_{20}NO_5Re$ . . . (CO)$_4$Re[C(O)C$_3$H$_7$-i]C(CH$_3$)=NH-C$_4$H$_9$-n . . . Re: Org.Comp.1-402
– . . . (CO)$_4$Re[C(O)C$_3$H$_7$-i]C(CH$_3$)=NH-C$_4$H$_9$-i . . . Re: Org.Comp.1-402/3
$C_{14}H_{20}NO_6Re$ . . . (CO)$_4$Re[C(=O)CH$_3$]=C(OH)CH$_3$ · NH$_2$-C$_6$H$_{11}$-c
Re: Org.Comp.1-390
$C_{14}H_{20}N_2OSn$ . . . (C$_2$H$_5$)$_3$SnN=C=C(CN)CH$_2$C$_4$H$_3$O-2 . . . Sn: Org.Comp.18-152/3
$C_{14}H_{20}N_2O_4Re$ . . . (CO)$_4$Re[t-C$_4$H$_9$-N=CH-CH=N-C$_4$H$_9$-t], radical
Re: Org.Comp.1-370
$C_{14}H_{20}N_2O_8Th$ . . . Th[(OOCCH$_2$)$_2$N-(CH$_2$)$_6$-N(CH$_2$COO)$_2$] . . . Th: SVol.C7-124, 126
Th: SVol.D1-95, 101
– . . . Th[(OOCCH$_2$)$_2$N-(CH$_2$)$_6$-N(CH$_2$COO)$_2$] · 3 H$_2$O
Th: SVol.C7-124, 126
– . . . Th[(OOCCH$_2$)$_2$N-(CH$_2$)$_6$-N(CH$_2$COO)$_2$] · 4.5 H$_2$O
Th: SVol.C7-124, 126
$C_{14}H_{20}N_2O_{10}Th$ . . Th[C$_2$H$_4$(OC$_2$H$_4$N(CH$_2$COO)$_2$)$_2$] . . . Th: SVol.D1-95, 101
$C_{14}H_{20}N_4O_4S$ . . . (1,4-ONC$_4$H$_8$-4)$_2$S=N-C$_6$H$_4$-4-NO$_2$ . . . S: S-N Comp.8-207/8, 212
$C_{14}H_{20}N_4Pb$ . . . Pb(C$_2$H$_5$)$_4$ · (CN)$_2$C=C(CN)$_2$ . . . Pb: Org.Comp.2-164, 270
$C_{14}H_{20}O_4Sn$ . . . (CH$_2$=CH)$_2$Sn[-OC(CH$_3$)CHC(CH$_3$)O-]$_2$ . . . Sn: Org.Comp.16-89/91, 94
– . . . n-C$_4$H$_9$-Sn(C$_6$H$_5$)Sn[OC(O)CH$_3$]$_2$ . . . Sn: Org.Comp.16-212
$C_{14}H_{20}O_6Sn$ . . . c-C$_5$H$_5$Sn(OOCC$_2$H$_5$)$_3$ . . . Sn: Org.Comp.17-60
$C_{14}H_{20}SiTh$ . . . (C$_5$H$_5$)$_2$Th[-CH$_2$-Si(CH$_3$)$_2$-CH$_2$-] . . . Th: SVol.D4-208
$C_{14}H_{21}I_2MoN_7$ . . . [Mo(CN-CH$_3$)$_7$]I$_2$ . . . Mo: Org.Comp.5-83, 84
$C_{14}H_{21}MnN_2O_4$ . . . [Mn(-O-C(CH$_3$)=CH-C(CH$_3$)=N-C$_2$H$_4$
-N=C(CH$_3$)CH=C(CH$_3$)O-)(CH$_3$COO)] · H$_2$O . . Mn: MVol.D6-203, 204, 206/7

$C_{14}H_{21}MnN_2O_4$ ... $[Mn(-O-C(CH_3)=CH-C(CH_3)=N-C_2H_4$ $-N=C(CH_3)CH=C(CH_3)O-)(CH_3COO)]$ ... Mn:MVol.D6-203, 204/6
$C_{14}H_{21}MoNS_2$ ... $(C_5H_5)Mo[S_2C-N(CH_3)_2](HC{\equiv}C-C_4H_9-t)$ ... Mo:Org.Comp.6-108
$C_{14}H_{21}MoN_3O_2S$ .. $(CH_2CHCH_2)Mo(CO)_2(NCS)(i-C_3H_7-N=CHCH=N-C_3H_7-i)$ Mo:Org.Comp.5-263, 264
$C_{14}H_{21}MoN_7^{2+}$ ... $[Mo(CN-CH_3)_7]I_2$ ... Mo:Org.Comp.5-83, 84
– ... $[Mo(CN-CH_3)_7][BF_4]_2$ ... Mo:Org.Comp.5-83, 84, 86/7
– ... $[Mo(CN-CH_3)_7][PF_6]_2$ ... Mo:Org.Comp.5-83, 84, 87
$C_{14}H_{21}MoO_3P$ ... $(C_5H_5)Mo(CO)[P(CH_3)_3][-O=C(CH_3)CH=C(OCH_3)-]$ Mo:Org.Comp.6-257
– ... $(C_5H_5)Mo(CO)_2(CH_3)=C(OCH_3)-CH=P(CH_3)_3$ .. Mo:Org.Comp.8-183
$C_{14}H_{21}MoO_6P$ ... $(C_5H_5)Mo(CO)[P(OCH_3)_3][CH_2C(CH_2)-C(O)OCH_3]$ Mo:Org.Comp.6-333
$C_{14}H_{21}N_2O_8Th^+$ .. $Th[H(n-C_6H_{12}(N(CH_2COO)_2)_2)]^+$ ... Th: SVol.D1-95, 101
$C_{14}H_{21}N_2O_{11}Th^-$ $Th[C_2H_4(OC_2H_4N(CH_2COO)_2)_2](OH)^-$ ... Th: SVol.D1-95/7, 101
$C_{14}H_{21}N_3O_2S$ ... $(1,4-ONC_4H_8-4)_2S=N-C_6H_5$ ... S: S-N Comp.8-207/8, 212
$C_{14}H_{21}N_3O_4S_2$ ... $(1,4-ONC_4H_8-4)_2S=N-S(O)_2-C_6H_5$ ... S: S-N Comp.8-207/8, 213
$C_{14}H_{21}N_3O_8Sn$ ... $(C_4H_9)_2Sn(OH)OC_6H_2(NO_2)_3-2,4,6$ ... Sn: Org.Comp.16-186
$C_{14}H_{21}O_4Sb$ ... $(C_6H_5)Sb[-OCH(CH_3)CH(CH_3)O-]_2$ ... Sb: Org.Comp.5-265
$C_{14}H_{22}MnN_2O_6S_2^{2+}$
$[Mn(O=S(C_2H_5)(C_5H_4NO))_2(H_2O)_2]^{2+}$ ... Mn:MVol.D7-106
$C_{14}H_{22}MnN_2O_8P_2$ $[Mn(O_2P(OCH_3)_2)_2(NC_5H_5)_2]$ ... Mn:MVol.D8-157
$C_{14}H_{22}Mn_2N_4O_{20}P_4^{8-}$
$[Mn_2(NH(CH_2COO)_2)_2(O_3PCH_2)_2NCH_2CH_2N(CH_2PO_3)_2]^{8-}$ Mn:MVol.D8-134/5
$C_{14}H_{22}MoN_2O_4Si_2$ $(CO)_4Mo[CN-CH_2-Si(CH_3)_3]_2$ ... Mo:Org.Comp.5-25, 26
$C_{14}H_{22}NO_5Re$ ... $(C_2H_5)_3N-Re(CO)_3[-O-C(CH_3)=CH-C(CH_3)=O-]$ Re: Org.Comp.1-127
$C_{14}H_{22}O_2Sn$ ... $2,2-(n-C_4H_9)_2-1,3,2-O_2SnC_6H_4$ ... Sn: Org.Comp.15-59, 86/7
– ... $[-Sn(C_4H_9-n)_2-O-(1,4-C_6H_4)-O-]_n$ ... Sn: Org.Comp.15-59
$C_{14}H_{22}O_3Si_2Ti$ ... $[(C_5H_5)_2Ti(O(Si(CH_3)_2O)_2)]_n$ ... Ti: Org.Comp.5-333/4
$C_{14}H_{22}O_8Sn$ ... $(C_4H_9)(CH_2=CH)Sn(OOCCH_2COOCH_3)_2$ ... Sn: Org.Comp.16-211
$C_{14}H_{23}IMoN_2O_2$ .. $[CH_2C(CH_3)CH_2]Mo(I)(CO)_2(i-C_3H_7-N=CHCH=N-C_3H_7-i)$ Mo:Org.Comp.5-263, 266
$C_{14}H_{23}IO_4Sn$ ... $C_4H_9SnI(OC(CH_3)=CHCOCH_3)_2$ ... Sn: Org.Comp.17-171, 172
$C_{14}H_{23}I_2MnO_2PS$ $Mn(I)_2(SO_2)_n-P(C_4H_9-n)_2-C_6H_5$ ... Mn:MVol.D8-46, 57/8
– ... $Mn(I)_2(SO_2)_n-P(C_4H_9-i)_2-C_6H_5$ ... Mn:MVol.D8-46, 57/8
$C_{14}H_{23}I_2MnP$ ... $[MnI_2(P(C_4H_9)_2-C_6H_5)]$ ... Mn:MVol.D8-38
$C_{14}H_{23}InO$ ... $(n-C_4H_9)_2In-O-C_6H_5$ ... In: Org.Comp.1-181, 183, 186
$C_{14}H_{23}InS$ ... $(n-C_4H_9)_2In-S-C_6H_5$ ... In: Org.Comp.1-240, 242, 244
– ... $(i-C_4H_9)_2In-S-C_6H_5$ ... In: Org.Comp.1-240, 242
$C_{14}H_{23}MoO_2PSi_2$ $(C_5H_5)Mo(CO)_2=P=C(Si(CH_3)_3)_2$ ... Mo:Org.Comp.7-27/9
$C_{14}H_{23}NOSn$ ... $(C_2H_5)_3SnN(C_6H_5)COCH_3$ ... Sn: Org.Comp.18-136, 139
$C_{14}H_{23}NO_2Sn$ ... $(C_2H_5)_3SnN(C_6H_5)COOCH_3$ ... Sn: Org.Comp.18-136/7, 139
$C_{14}H_{23}NO_3PReSi$ $(C_5H_5)Re(CO)(NO)-C(=O)-P(C_4H_9-t)-Si(CH_3)_3$ Re: Org.Comp.3-159, 162
– ... $(C_5H_5)Re(CO)(NO)-C(=P-C_4H_9-t)-O-Si(CH_3)_3$ Re: Org.Comp.3-159, 164, 168/9
$C_{14}H_{23}NO_3Sn$ ... $(C_4H_9)_2SnOOCH(C_5H_4N)O$ ... Sn: Org.Comp.15-335, 336
$C_{14}H_{23}N_3O_9Th$ ... $[NH_4][Th(OH)(1,2-((OOCCH_2)_2N)_2C_6H_{10})] \cdot 5\ H_2O$ Th: SVol.C7-124, 126

$C_{14}H_{23}OPSn$	2-(t-C_4H_9)-3-$(CH_3)_3Sn$-1,3-OPC_7H_5	Sn: Org.Comp.19-178
$C_{14}H_{23}O_3P_2Re$ . . .	fac-$(CO)_3Re[P(CH_3)_3]_2$-C_5H_5	Re: Org.Comp.1-273/4, 279
$C_{14}H_{24}HgMoO_7P_2S_2$	$(C_5H_5)Mo(CO)_2[P(OCH_3)_3]$-Hg-SP(=S)(O-$C_2H_5)_2$	Mo: Org.Comp.7-122, 145
$C_{14}H_{24}MnN_3O_2S_4$	$[Mn((SC(=S))NC_5H_{10})_2(NH_2CH_2COO)]$	Mn: MVol.D7-177/8
$C_{14}H_{24}MnN_4S_8Zn$	$MnZn(SC(=S)NHCH_2CH_2CH_2CH_2CH_2NHC(=S)S)_2$	Mn: MVol.D7-180/4
$C_{14}H_{24}MnN_{10}S_4{}^{2+}$	$[Mn(S=C(NH_2)_2)_4(N_2C_{10}H_8)]^{2+}$	Mn: MVol.D7-192/3
$C_{14}H_{24}MoN_2OS_4$ . .	$(CH_3$-C≡CH)Mo(CO)$[S_2C$-$N(C_2H_5)_2]_2$	Mo: Org.Comp.5-145/8
$C_{14}H_{24}MoN_2O_3S_4$	[HC≡C-C(=O)O-CH_3]Mo(O)$[S_2C$-$N(C_2H_5)_2]_2$ . . .	Mo: Org.Comp.5-139/41
$C_{14}H_{24}MoN_2S_4$ . . .	$(HC{\equiv}CH)_2Mo[S_2C$-$N(C_2H_5)_2]_2$	Mo: Org.Comp.5-185, 188
$C_{14}H_{24}MoO_2P_2$ . . .	$(C_5H_5)Mo(CO)_2(CH_3)$-$P(CH_3)_2$-CH_2CH_2-$P(CH_3)_2$	Mo: Org.Comp.8-84
$C_{14}H_{24}N_2O_3SSn$ . .	$(C_2H_5)_3SnN(COCH_3)SO_2C_6H_4NH_2$-2	Sn: Org.Comp.18-142
$C_{14}H_{24}N_2Sn$	2,2-(n-$C_4H_9)_2$-1,3,2-$N_2SnC_6H_6$	Sn: Org.Comp.19-96/8
$C_{14}H_{24}N_4O_2S_4Sn$	[2-(S=)-4-(O=)-1,3-SNC_3H_2-3-NH-$]_2Sn(C_4H_9$-n$)_2$	Sn: Org.Comp.19-86
$C_{14}H_{24}N_4Sn$	[1,3-$N_2C_3H_3$-1-$]_2Sn(C_4H_9$-n$)_2$.	Sn: Org.Comp.19-93/6
$C_{14}H_{24}N_8S_6Si$	$(N(CH_3)_4)_2[Si(NCS)_6]$	Si: SVol.B4-302/4
$C_{14}H_{24}N_8Se_6Si$ . . .	$(N(CH_3)_4)_2[Si(NCSe)_6]$	Si: SVol.B4-307/8
$C_{14}H_{24}N_{10}PdS_2$ . .	$Pd(SCN)_2(N_4C_6H_{12})_2$	Pd: SVol.B2-307
$C_{14}H_{24}O_4Sn$	(n-$C_4H_9)_2Sn[OC(O)$-$CH{=}CH_2]_2$	Sn: Org.Comp.15-268
–	(n-$C_5H_{11})_2Sn$[-OC(O)-CH=CH-C(O)O-]	Sn: Org.Comp.16-17
$C_{14}H_{24}P_2Sn$	1,3-(i-$C_3H_7)_2$-2,2-$(CH_3)_2$-1,3,2-$P_2SnC_6H_4$	Sn: Org.Comp.19-208, 211
$C_{14}H_{25}I_3InN$	$[N(C_2H_5)_4][C_6H_5$-$In(I)_3]$	In: Org.Comp.1-349/50, 356
$C_{14}H_{25}InN_2$	$(CH_3)_2In$-C_6H_3-2,6-$[CH_2$-$N(CH_3)_2]_2$	In: Org.Comp.1-101, 108, 111
$C_{14}H_{25}MnN_2OS_4$ . .	$[Mn((SC(=S))NC_6H_{12})_2OH]$	Mn: MVol.D7-177
$C_{14}H_{25}MoN_2O_3PS_4$	$(C_5H_5)Mo(NO)[S_2P(O$-$C_2H_5)_2][S_2C$-$N(C_2H_5)_2]$	Mo: Org.Comp.6-51
–	$(C_5H_5)Mo(NO)[S_2P(O$-C_3H_7-i$)_2][S_2C$-$N(CH_3)_2]$	Mo: Org.Comp.6-50/1
$C_{14}H_{25}NO_2Sn$	2-$(C_2H_5)_3Sn$-2-$NC_8H_{10}(=O)_2$-1,3	Sn: Org.Comp.18-142, 146
–	(n-$C_4H_9)_2Sn(OH)$-O-C_6H_4-NH_2-2.	Sn: Org.Comp.16-186
$C_{14}H_{25}N_3O_2S_2$ . . .	$((C_2H_5)_2N)_2S$=N-$S(O)_2$-C_6H_5	S: S-N Comp.8-207/8, 210
$C_{14}H_{25}O_3P_2Re$ . . .	$(CO)_3Re[P(CH_3)_3]_2C_5H_7$	Re: Org.Comp.1-272
$C_{14}H_{25}PSn$	$(C_2H_5)_3Sn$-$P(C_6H_5)$-C_2H_5	Sn: Org.Comp.19-190/1
$C_{14}H_{26}I_2MnN_6S_2$ . .	$MnI_2[c$-C_6H_{10}=N-NHC(=S)$NH_2]_2$	Mn: MVol.D6-342/4
$C_{14}H_{26}MnN_2OS$ . .	$[Mn(SC_6H_4O$-2$)(NH(C_2H_5)_2)_2]$	Mn: MVol.D7-30/1
$C_{14}H_{26}MnN_{10}S_4{}^{2+}$	$[Mn(S=C(NH_2)_2)_4(NC_5H_5)_2]^{2+}$	Mn: MVol.D7-192/3
$C_{14}H_{26}MnO_6S_2$ . . .	$[Mn(O=S(CH_3)_2)_2(OC(CH_3)CHC(CH_3)O)_2]$	Mn: MVol.D7-99
$C_{14}H_{26}MoN_2O_3S_4$	$(HOCH_2$-CC-$CH_2OH)Mo(O)[S_2C$-$N(C_2H_5)_2]_2$ . .	Mo: Org.Comp.5-139/41
$C_{14}H_{26}MoO_2P_2$ . . .	$[P(CH_3)_4][(C_5H_5)Mo(CO)_2(P(CH_3)_3)]$	Mo: Org.Comp.7-47/9
$C_{14}H_{26}NOPSn$. . . .	C_6H_5-PH-CH_2CH_2-O-$Sn(CH_3)_2$-$N(C_2H_5)_2$	Sn: Org.Comp.19-129, 130
$C_{14}H_{26}N_2OSiSn$ . .	$(CH_3)_3SnN(C_6H_5)CON(CH_3)Si(CH_3)_3$	Sn: Org.Comp.18-47, 50, 56
$C_{14}H_{26}N_2OSn$	(n-$C_4H_9)_2Sn(OH)$-NH-C_6H_4-2-NH_2.	Sn: Org.Comp.19-129, 132
$C_{14}H_{26}N_2O_3Sn$ . . .	$(C_2H_5)_3SnN(C(=O)CH(N(CH_2CH_2)_2O)CH_2C(=O))$	Sn: Org.Comp.18-143/4
$C_{14}H_{26}N_2SSiSn$. . .	$(CH_3)_3Sn$-$N(C_6H_5)$-$C(=NCH_3)$-S-$Si(CH_3)_3$	Sn: Org.Comp.18-47, 51, 56

$C_{14}H_{26}N_2SSiSn$ . . . $(CH_3)_3Sn-N(C_6H_5)-C(S)-N(CH_3)-Si(CH_3)_3$ . . . . Sn: Org.Comp.18-47, 50, 56
$C_{14}H_{26}N_2SiSn$ . . . . $(CH_3)_3SnN(CH_3)C(C_6H_5)=NSi(CH_3)_3$ . . . . Sn: Org.Comp.18-48, 53
$C_{14}H_{26}N_6Os^{2+}$ . . . $[2,6-(CH_3)_2-NC_5H_3-4-Os(NH_3)_4(NC-C_6H_5)]^{2+}$ Os: Org.Comp.A1-12
$C_{14}H_{26}O_2Sn$ . . . . . $(c-C_6H_{11})_2Sn[-OCH_2CH_2O-]$ . . . . Sn: Org.Comp.16-66
$C_{14}H_{26}O_4SSn$ . . . . $(C_4H_9)_2SnOOC(CH_2)_2S(CH_2)_2COO$ . . . . Sn: Org.Comp.15-324
$C_{14}H_{26}O_4Sn$ . . . . . $(C_4H_9)_2SnOOC(CH_2)_4COO$ . . . . Sn: Org.Comp.15-323
$C_{14}H_{26}O_5Sn$ . . . . . $(n-C_4H_9)_2Sn(OCH_3)-OOC-CH=CH-COO-CH_3$ Sn: Org.Comp.16-189
– . . . . $(n-C_4H_9)_2Sn[-O-C_6H_7O_2(OH)-O-]$ . . . . Sn: Org.Comp.15-68
$C_{14}H_{26}O_6Sn$ . . . . . $2,2-(n-C_4H_9)_2-4,5-[CH_3OC(O)]-1,3,2-O_2SnC_2H_2$
Sn: Org.Comp.15-51
– . . . . $n-C_8H_{17}Sn[OC(O)CH_3]_3$ . . . . Sn: Org.Comp.17-52
$C_{14}H_{27}MnN_2OPS_2$ $[Mn((C_4H_9)_3P=O)(NCS)_2]$ . . . . Mn: MVol.D8-87
$C_{14}H_{27}MnN_2O_2PS_2$
$Mn[P(C_4H_9)_3](O_2)(NCS)_2$ . . . . Mn: MVol.D8-56
$C_{14}H_{27}MnN_2PS_2$ . . $Mn[P(C_4H_9)_3](NCS)_2$ . . . . Mn: MVol.D8-38
$C_{14}H_{27}NO_3Sn$ . . . . $(C_2H_5)_3SnN(C(=O)CH(OC_4H_9)CH_2C(=O))$ . . . . . . Sn: Org.Comp.18-142/4
$C_{14}H_{27}N_3Sn$ . . . . . $(C_4H_9)_3SnN=C=NCN$ . . . . Sn: Org.Comp.18-226, 228
$C_{14}H_{27}N_4O_7PS_2$ . . $(CH_3)_2N-C(O)-C(SCH_3)=N-OC(O)-N(CH_3)$
$-S(O)-N(C_2H_5)-P(O)(-OCH_2C(CH_3)_2CH_2O-)$ . . S: S-N Comp.8-356
$C_{14}H_{27}NiO_3PSn$ . . $(CH_3)_3Sn-P(C_4H_9-t)_2Ni(CO)_3$ . . . . Sn: Org.Comp.19-181, 184/5, 187/8
$C_{14}H_{27}O_5ReSi_4$ . . . $(CO)_5ReSi(Si(CH_3)_3)_3$ . . . . Re: Org.Comp.2-3, 12
$C_{14}H_{27}O_6Sb$ . . . . . $(C_4H_9)_2Sb(O_2CCH_3)_3$ . . . . Sb: Org.Comp.5-181
$C_{14}H_{28}InN$ . . . . $(CH_3)_2In-N(C_6H_{11}-c)_2$ . . . . In: Org.Comp.1-253/4, 256
$C_{14}H_{28}LiN_2O_2Re$. $Li[(t-C_4H_9CH_2)_2ReO_2] \cdot 2\ CH_3CN$ . . . . Re: Org.Comp.1-5, 6
$C_{14}H_{28}MnN_2S_4$ . . . $Mn[S_2C-N(C_3H_7-n)_2]_2$ . . . . Mn: MVol.D7-139/42
– . . . . $Mn[S_2C-N(C_3H_7-i)_2]_2$ . . . . Mn: MVol.D7-139/42
$C_{14}H_{28}MoN_2O_8P_2$ $(CO)_2Mo(CN-C_2H_5)_2[P(OCH_3)_3]_2$ . . . . Mo: Org.Comp.5-21/4
$C_{14}H_{28}N_2OSSi_2$ . . $(CH_3)_3Si-N(C_2H_5)-S(O)-N(C_6H_5)-Si(CH_3)_3$ . . . . S: S-N Comp.8-357
$C_{14}H_{28}N_2O_2Sn$ . . . $1-(C_2H_5)_3Sn-3-(C_2H_5)_2N-NC_4H_3(=O)_2-2,5$ . . . . Sn: Org.Comp.18-143/4
– . . . . $1-(C_2H_5)_3Sn-3-(n-C_4H_9-NH)-NC_4H_3(=O)_2-2,5$ Sn: Org.Comp.18-143/4
$C_{14}H_{28}N_2Si_2Sn$ . . . $1,3-[(CH_3)_3Si]_2-2,2-(CH_3)_2-1,3,2-N_2SnC_6H_4$ . . Sn: Org.Comp.19-76
$C_{14}H_{28}N_3O_5PS_3$ . . $CH_3S-CCH_3=N-OC(O)-NCH_3-S(O)-N(C_4H_9-s)$
$-P(S)(-OCH_2C(CH_3)_2CH_2O-)$ . . . . S: S-N Comp.8-356
– . . . . $CH_3S-CCH_3=N-OC(O)-NCH_3-S(O)-N(C_5H_9-c)$
$-P(S)(OC_2H_5)_2$ . . . . S: S-N Comp.8-356
$C_{14}H_{28}O_2Sn$ . . . . . $2,2-(n-C_4H_9)_2-1,3,2-O_2SnC_6H_{10}$ . . . . Sn: Org.Comp.15-52/3
– . . . . $(-CH_2CH_2CH_2-)C[-CH_2-O-Sn(C_4H_9-n)_2-O-CH_2-]$
Sn: Org.Comp.15-56
– . . . . $(n-C_4H_9)_2Sn(O-CH_2CH=CH_2)_2$ . . . . Sn: Org.Comp.15-29, 31
$C_{14}H_{28}O_3Sn$ . . . . . $(n-C_4H_9)_2Sn(OCH_3)-O-C(CH_3)=CH-C(O)CH_3$ . . Sn: Org.Comp.16-187
– . . . . $[-(CH_2)_5-]C[-O-O-Sn(C_4H_9-n)_2-O-]$ . . . . Sn: Org.Comp.15-335, 336
$C_{14}H_{28}O_4S_2Sn$ . . . $(n-C_4H_9)_2Sn(OOC-CH_2CH_2-SH)_2$ . . . . Sn: Org.Comp.15-259
– . . . . $(n-C_4H_9)_2Sn(OOC-CH_2-SCH_3)_2$ . . . . Sn: Org.Comp.15-258
$C_{14}H_{28}O_4Sn$ . . . . . $(n-C_4H_9)_2Sn(OOC-C_2H_5)_2$ . . . . Sn: Org.Comp.15-127
– . . . . $(n-C_4H_9)_2Sn[-O-C_5H_7O(OCH_3)-O-]$ . . . . Sn: Org.Comp.15-63
– . . . . $[C_2H_5-C(CH_3)_2]_2Sn(OOC-CH_3)_2$ . . . . Sn: Org.Comp.16-17
$C_{14}H_{28}O_5Sn$ . . . . . $(n-C_4H_9)_2Sn(OCH_3)-OOC-CH_2CH_2-COO-CH_3$ Sn: Org.Comp.16-188
– . . . . $(n-C_4H_9)_2Sn[-O-C_4H_4O(CH_2OH)(OCH_3)-O-]$ . . Sn: Org.Comp.15-60
– . . . . $(n-C_4H_9)_2Sn[-O-C_5H_6O(OH)(OCH_3)-O-]$ . . . . . Sn: Org.Comp.15-64

$C_{14}H_{28}O_6Sn$ ..... $(n-C_4H_9)_2Sn(OOC-CHOH-CH_3)_2$ .... Sn: Org.Comp.15-256
$C_{14}H_{28}O_6Th$ ..... $Th(OH)_2(n-C_6H_{13}COO)_2$ .... Th: SVol.C7-73/5
$C_{14}H_{29}InS$ .... $(n-C_4H_9)_2In-S-C_6H_{11}-c$ .... In: Org.Comp.1-240, 242
$C_{14}H_{29}InSi_3$ ..... $In[C_5H_2(Si(CH_3)_3)_3]$ .... In: Org.Comp.1-379, 380/1
$C_{14}H_{29}MoOP_2{}^+$ .. $[i-C_3H_7C_5H_4Mo(P(CH_3)_3)_2O]^+$ .... Mo: Org.Comp.6-16
$C_{14}H_{29}NO_3Sn$ .... $C_2H_5-C(=O)O-Sn(C_4H_9-n)_2-NH-C(=O)-C_2H_5$ .. Sn: Org.Comp.19-129, 133
– .... $n-C_8H_{17}-Sn[-O-CH_2CH_2-]_3N$ .... Sn: Org.Comp.17-52
$C_{14}H_{29}NO_4Sn$ .... $CH_3-C(=O)O-Sn(C_4H_9-n)_2-N(C_2H_5)-C(=O)O-CH_3$ Sn: Org.Comp.19-129, 133
$C_{14}H_{29}NP_2Re^+$ ... $[(CH_3)_2Re(P(CH_3)_3)_2=N-C_6H_5]^+$ .... Re: Org.Comp.1-18/9
$C_{14}H_{29}N_3OSn$ .... $(C_4H_9)_3SnN(CONH_2)CN$ .... Sn: Org.Comp.18-197, 199
$C_{14}H_{29}N_3Sn$ ..... $1-(n-C_4H_9)_3Sn-1,2,4-N_3C_2H_2$ .... Sn: Org.Comp.18-201/2, 210/1, 221/2
– .... $1-(t-C_4H_9)_3Sn-1,2,4-N_3C_2H_2$ .... Sn: Org.Comp.18-232/3
– .... $2-(n-C_4H_9)_3Sn-1,2,3-N_3C_2H_2$ .... Sn: Org.Comp.18-208
$C_{14}H_{29}N_4O_7PS_2$ .. $(CH_3)_2N-C(O)-C(SCH_3)=N-OC(O)-N(CH_3)-S(O)-N(C_3H_7-i)-P(O)(OC_2H_5)_2$ .... S: S-N Comp.8-356
$C_{14}H_{29}N_5PdS_2$ ... $[Pd(SCN)((C_2H_5)_2NCH_2CH_2NHCH_2CH_2N(C_2H_5)_2)][SCN]$ Pd: SVol.B2-307
$C_{14}H_{29}P_2Re$ ..... $(C_5H_5)ReH[P(CH_3)_3]_2-C_3H_5-c$ .... Re: Org.Comp.3-40/1, 43
$C_{14}H_{30}InO_2{}^+$ ..... $[(i-C_3H_7)_2In(OC_4H_8)_2]^+$ .... In: Org.Comp.1-336/7
$C_{14}H_{30}In_2O_4$ ..... $[(CH_3)_2InOC(O)-C_4H_9-t]_2$ .... In: Org.Comp.1-197/8
– .... $[(C_2H_5)_2InOC(O)-C_2H_5]_2$ .... In: Org.Comp.1-202
$C_{14}H_{30}MnO_{12}S_6$ .. $Mn[C(S(O)_2-C_2H_5)_3]_2$ .... Mn: MVol.D7-113/4
– .... $[Mn(H_2O)_6][C(S(O)_2-C_2H_5)_3]_2$ .... Mn: MVol.D7-113/4
$C_{14}H_{30}Mn_2N_3O_9P$ $[Mn_2(O=P(N(CH_3)_2)_3)(CH_3COO)_4]$ .... Mn: MVol.D8-176
$C_{14}H_{30}MoP_2$ ..... $(CH_2=CHCH=CH_2)_2Mo[P(CH_3)_3]_2$ .... Mo: Org.Comp.5-347/8
$C_{14}H_{30}NP_2Re^+$ ... $[(CH_3)_2Re(P(CH_3)_3)_2(NH-C_6H_5)]^+$ .... Re: Org.Comp.1-19
$C_{14}H_{30}N_2O_2Sn$ ... $(C_2H_5)_3Sn-NHC(O)-CH_2CH_2-C(O)NH-C_4H_9-n$ Sn: Org.Comp.18-130
– .... $(n-C_4H_9)_2Sn[O-N=C(CH_3)_2]_2$ .... Sn: Org.Comp.15-343
$C_{14}H_{30}N_2O_6P_2PdS_2$ $Pd(NCS)_2[P(O-C_2H_5)_3]_2$ .... Pd: SVol.B2-341
– .... $Pd(SCN)_2[P(O-C_2H_5)_3]_2$ .... Pd: SVol.B2-341
$C_{14}H_{30}N_2P_2PdS_2$ $Pd(NCS)_2[P(C_2H_5)_3]_2$ .... Pd: SVol.B2-333
$C_{14}H_{30}N_4Sn$ ..... $(C_4H_9)_3SnN(N=NC(CH_3)=N)$ .... Sn: Org.Comp.18-202, 212, 222/3
$C_{14}H_{30}N_6O_{12}Th$ .. $[C(NH_2)_3]_2[Th(CH_3COO)_6]$ .... Th: SVol.C7-51/2
– .... $[C(NH_2)_3]_2[Th(CH_3COO)_6] \cdot 3\ H_2O$ .... Th: SVol.C7-51/2
– .... $[C(NH_2)_3]_2[Th(CH_3COO)_6] \cdot 3.5\ H_2O$ .... Th: SVol.C7-51/2
– .... $[C(NH_2)_3]_2[Th(CH_3COO)_6] \cdot 8\ H_2O$ .... Th: SVol.C7-51/2
$C_{14}H_{30}N_8O_{10}S_4Th_2$ $[Th(SCN)_2(H_2O)_3]_2(OH)_2 \cdot [CH_3C(=NO)-CCH_3=N-CH_2CH_2-N=CCH_3-C(=NO)CH_3]$ Th: SVol.D4-158
$C_{14}H_{30}O_2Sn$ ..... $2,2-(n-C_4H_9)_2-1,3,2-O_2SnC_6H_{12}$ .... Sn: Org.Comp.15-57
– .... $2,2-(n-C_4H_9)_2-4,4,5,5-(CH_3)_4-1,3,2-O_2SnC_2$ .. Sn: Org.Comp.15-50, 84
– .... $2,2-(n-C_4H_9)_2-4,4,6-(CH_3)_3-1,3,2-O_2SnC_3H_3$ .. Sn: Org.Comp.15-54/5
$C_{14}H_{30}O_3SSn$ .... $(C_4H_9)_2Sn(OC_4H_9)OOCCH_2SH$ .... Sn: Org.Comp.16-192
$C_{14}H_{30}O_4Sn$ ..... $(C_4H_9)_2SnOCH(CH_2OCH_3)CH(CH_2OCH_3)O$ .... Sn: Org.Comp.15-50
$C_{14}H_{30}O_5SiSn$ .... $(C_4H_9)_2Sn(OOCCH_3)OSi(CH_3)_2OOCCH_3$ ...... Sn: Org.Comp.16-194
$C_{14}H_{30}O_6S_2Sn$ ... $(C_4H_9)_2SnO_3S(CH_2)_6SO_2O$ .... Sn: Org.Comp.15-346

$C_{14}H_{31}NOSn$ ..... 2,2-(n-C_4H_9)$_2$-3-(t-C_4H_9)-1,3,2-ONSnC_2H_4 ... Sn: Org.Comp.19-135
– ..... (n-C_4H_9)$_3$Sn-N(CH_3)-CH=O ..... Sn: Org.Comp.18-181/2
– ..... (n-C_4H_9)$_3$Sn-NHC(=O)-CH_3 ..... Sn: Org.Comp.18-160/2
$C_{14}H_{31}NO_2Sn$ ..... (C_2H_5)$_3$Sn-N(C_6H_{13}-n)-COO-CH_3 ..... Sn: Org.Comp.18-136/8, 140
– ..... (n-C_4H_9)$_2$Sn(O-C_3H_7-i)-O-N=C(CH_3)$_2$ ..... Sn: Org.Comp.16-191
– ..... (n-C_4H_9)$_2$Sn[-O-CH_2CH_2N(C_2H_5)CH_2CH_2-O-] Sn: Org.Comp.15-72
– ..... (t-C_4H_9)$_2$Sn[-O-CH_2CH_2N(C_2H_5)CH_2CH_2-O-] Sn: Org.Comp.15-374, 376/7
$C_{14}H_{31}NO_3Sn$ ..... (C_4H_9)$_2$SnOCH_2CH_2N(CH_2CH_2OH)CH_2CH_2O ... Sn: Org.Comp.15-72
$C_{14}H_{31}O_3PSn$ ..... (C_4H_9)$_2$SnOP(O)(C_6H_{13})O ..... Sn: Org.Comp.15-359, 360
$C_{14}H_{31}O_3PTh^{2+}$ .. Th(OH)[(C_7H_{15})$_2$PO$_2$]$^{2+}$ ..... Th: SVol.D1-131
$C_{14}H_{31}O_5PTh^{2+}$ .. Th(OH)[O$_2$P(OC_7H_{15})$_2$]$^{2+}$ ..... Th: SVol.D1-131
$C_{14}H_{32}I_2MnP_2$ ..... Mn((i-C_3H_7)$_2$PCH_2CH_2P(C_3H_7-i)$_2$)I_2 · $C_6H_5CH_3$ Mn: MVol.D8-78
$C_{14}H_{32}In_2N_2O_2$ ... [(C_2H_5)$_2$In-N(CH_3)-C(O)CH_3]$_2$ ..... In: Org.Comp.1-267
$C_{14}H_{32}MoO_{11}P_3^{+}$ [(CH_2CHCH_2)Mo(CO)$_2$(P(OCH_3)$_3$)$_3$][BF_4] ..... Mo: Org.Comp.5-200, 201
$C_{14}H_{32}MoP_2$ ..... i-$C_3H_7C_5H_4$Mo(P(CH_3)$_3$)$_2H_3$ ..... Mo: Org.Comp.6-14, 20
$C_{14}H_{32}N_2O_5Sn$ ... 2,2-(n-C_4H_9)$_2$-3-[NH_2-CH_2-C(=O)]-1,3,2-ONSnC_2H_2(=O)-5 · 2 CH_3-OH ..... Sn: Org.Comp.19-138
$C_{14}H_{32}N_2P_2S$ ..... (t-C_4H_9)$_2$PN=S=NP(C_3H_7-i)$_2$ ..... S: S-N Comp.7-110/2
$C_{14}H_{32}O_2Sn$ ..... (n-C_4H_9)$_2$Sn(O-C_3H_7-n)$_2$ ..... Sn: Org.Comp.15-3, 14, 17
– ..... (n-C_4H_9)$_2$Sn(O-C_3H_7-i)$_2$ ..... Sn: Org.Comp.15-3, 17/8
$C_{14}H_{32}O_2Ti_2$ ..... [C_4H_9O(C_2H_5)Ti]$_2$(CH_2CH_2) ..... Ti: Org.Comp.5-25
$C_{14}H_{32}O_3Sn$ ..... C_2H_5-Sn(O-C_4H_9-n)$_3$ ..... Sn: Org.Comp.17-25
– ..... C_2H_5-Sn(O-C_4H_9-i)$_3$ ..... Sn: Org.Comp.17-26
– ..... C_2H_5-Sn(O-C_4H_9-t)$_3$ ..... Sn: Org.Comp.17-26
$C_{14}H_{32}O_4Sn$ ..... (C_4H_9)$_2$Sn(OCH(CH_3)OCH_3)$_2$ ..... Sn: Org.Comp.15-5
$C_{14}H_{32}O_6S_2Sn$ ... (i-C_6H_{13})$_2$Sn(OSO$_2CH_3$)$_2$ ..... Sn: Org.Comp.16-19
$C_{14}H_{32}O_9P_2Sn$ ... CH_3CH=CHCH_2Sn(OP(O)(OC_2H_5)$_2$)$_2$OC_2H_5 ... Sn: Org.Comp.17-75
$C_{14}H_{32}P_3Re$ ..... (C_5H_5)ReH[P(CH_3)$_3$]$_2$-CH_2P(CH_3)$_2$ ..... Re: Org.Comp.3-40/1, 42
– ..... (C_5H_5)ReH[P(CH_3)$_3$]$_2$[CH_2=P(CH_3)$_2$] ..... Re: Org.Comp.3-40/1, 44
– ..... (C_5H_5)Re[P(CH_3)$_3$]$_3$ ..... Re: Org.Comp.3-5
$C_{14}H_{32}S_2Sb_2$ ..... [(C_2H_5)$_2$SbS]$_2$(CH_2)$_6$ ..... Sb: Org.Comp.5-127
$C_{14}H_{33}MoP_3$ ..... C_5H_5Mo(P(CH_3)$_3$)$_3$H ..... Mo: Org.Comp.6-13, 20
$C_{14}H_{33}NO_2Sn$ ..... (n-C_4H_9)$_2$Sn(O-C_2H_5)-O-CH_2CH_2-NH-C_2H_5 .. Sn: Org.Comp.16-189
– ..... (n-C_4H_9)$_2$Sn(O-C_2H_5)-O-N(C_2H_5)$_2$ ..... Sn: Org.Comp.16-190
$C_{14}H_{33}NO_3Sn$ ..... C_4H_9Sn(OC_3H_7-i)$_2$OCH_2CH_2N(CH_3)$_2$ ..... Sn: Org.Comp.17-71
$C_{14}H_{33}NSn$ ..... (C_4H_9)$_3$SnN(CH_3)$_2$ ..... Sn: Org.Comp.18-165/8
$C_{14}H_{33}N_2PSn$ ..... n-C_4H_9-Sn(CH_3)$_2$[-N(C_4H_9-t)-P-N(C_4H_9-t)-] .. Sn: Org.Comp.19-55, 58
$C_{14}H_{33}O_3Sb$ ..... (C_4H_9)$_2$Sb(OC_2H_5)$_3$ ..... Sb: Org.Comp.5-179
$C_{14}H_{34}I_2MnO_6P_2$. [Mn(CH_3P(O)(OC_3H_7-i)$_2$)$_2I_2$] ..... Mn: MVol.D8-142/3
$C_{14}H_{34}InN$ ..... CH_3-In(CH_2-C_4H_9-t)$_2$ · N(CH_3)$_3$ ..... In: Org.Comp.1-105, 114/5
$C_{14}H_{34}In_2N_4$ ..... [1-(CH_3)$_2$In-1,4-$N_2C_4H_8$-4-CH_3]$_2$ ..... In: Org.Comp.1-275
$C_{14}H_{34}MnN_2O_{12}P_2$ [Mn(CH_3P(O)(OC_3H_7-i)$_2$)$_2$(NO_3)$_2$] ..... Mn: MVol.D8-142/3
$C_{14}H_{34}MoP_3^{+}$ ..... [C_5H_5Mo(P(CH_3)$_3$)$_3H_2$]$^{+}$ ..... Mo: Org.Comp.6-13
$C_{14}H_{34}N_2O_2Sn$ ... (n-C_4H_9)$_2$Sn(O-$CH_2CH_2CH_2$-NH_2)$_2$ ..... Sn: Org.Comp.15-7
– ..... (n-C_4H_9)$_2$Sn(O-CH_2CH_2-NHCH_3)$_2$ ..... Sn: Org.Comp.15-6
– ..... (n-C_4H_9)$_2$Sn[O-CH(CH_3)CH_2-NH_2]$_2$ ..... Sn: Org.Comp.15-7
$C_{14}H_{34}N_2S_2Sb_2$ .. [(CH_3)$_2$SbS]$_2$[-(CH_2)$_3$N(CH_3)(CH_2)$_2$N(CH_3)(CH_2)$_3$-] Sb: Org.Comp.5-127

$C_{14}H_{34}N_2Si_2Sn$ . . . $Sn[-CH_2-Si(CH_3)_2-N(C_4H_9-t)-]_2$ . . . . Sn: Org.Comp.19-110
$C_{14}H_{34}N_2Sn$ . . . . . $(i-C_3H_7)_2N-Sn(CH_3)_2-N(C_3H_7-i)_2$ . . . . Sn: Org.Comp.19-66
$C_{14}H_{35}MoOP_3$. . . . $(CH_2=CH_2)_2Mo(CO)[P(CH_3)_3]_3$ Mo:Org.Comp.5-191
$C_{14}H_{35}N_3Sn$ . . . . . $[(C_2H_5)_2N]_3Sn-C_2H_5$. . . . . Sn: Org.Comp.19-111, 113
$C_{14}H_{36}IInN_4^{2+}$ . . . $[InI(CH_2-N(CH_3)_2-CH_2CH_2-N(CH_3)_2)_2]^{2+}$ . . . . . In: Org.Comp.1-368/9
$C_{14}H_{36}I_3InN_4$. . . . . $[InI(CH_2-N(CH_3)_2-CH_2CH_2-N(CH_3)_2)_2]I_2$ . . . . . . In: Org.Comp.1-368/9
$C_{14}H_{36}In_2N_2Si$ . . . $(CH_3)_2In-N(C_4H_9-t)-Si(CH_3)_2-N(C_4H_9-t)-In(CH_3)_2$
In: Org.Comp.1-284, 288
$C_{14}H_{36}MnN_8O_2P_2S_2$
$[Mn(O=P(N(CH_3)_2)_3)_2(NCS)_2]$ Mn:MVol.D8-174/5
$C_{14}H_{36}MnN_8O_2P_2Se_2$
$[Mn(O=P(N(CH_3)_2)_3)_2(NCSe)_2]$ Mn:MVol.D8-174/5
$C_{14}H_{36}NPSiSn$ . . . $(CH_3)_3Sn-P(C_4H_9-t)_2=N-Si(CH_3)_3$ . . . . Sn: Org.Comp.19-180/1
$C_{14}H_{36}N_2O_3Sb_2^{2+}$ $[((CH_3)_3SbOC(CH_3)N(CH_3)_2)_2O]^{2+}$. . . . . Sb: Org.Comp.5-101
$C_{14}H_{36}N_2Si_2Sn$. . . $(CH_3)_2Sn[-N(C_4H_9-t)-Si(CH_3)_2-Si(CH_3)_2-N(C_4H_9-t)-]$
Sn: Org.Comp.19-75, 78
– . . . . $(n-C_4H_9)_2Sn[-N(CH_3)-Si(CH_3)_2-Si(CH_3)_2-N(CH_3)-]$
Sn: Org.Comp.19-96/8
$C_{14}H_{36}N_4Si$. . . . . . $Si(N(CH_3)_2)_3N(C_4H_9)_2$. . . . . Si: SVol.B4-220, 222
$C_{14}H_{36}O_2P_3ReSSn$
$(CO)_2Re[P(CH_3)_3]_3SSn(CH_3)_3$ . . . . Re: Org.Comp.1-97
$C_{14}H_{36}O_2Si_2Sn$. . . $(C_4H_9)_2Sn(OSi(CH_3)_3)_2$. . . . . Sn: Org.Comp.15-352
$C_{14}H_{36}O_6Ti_2$ . . . . . $[(CH_3)Ti(OC_2H_5)_3]_2$. . . . . Ti: Org.Comp.5-6
$C_{14}H_{37}N_2PSiSn$. . . $(CH_3)_3SnNHP(C_4H_9-t)_2=NSi(CH_3)_3$ . . . . Sn: Org.Comp.18-16, 20
$C_{14}H_{38}N_2Si_2Sn$. . . $(CH_3)_3Si-NH-Sn(C_4H_9-t)_2-NH-Si(CH_3)_3$ . . . . . . Sn: Org.Comp.19-98/101
$C_{14}H_{38}O_6P_2Ti_2$ . . . $[(CH_3O)_3Ti(CH_2P(CH_3)_2CH_2)]_2$. . . . . Ti: Org.Comp.5-27/8
$C_{14}H_{40}I_4N_6O_{10}Th_2$ $[ThI_2(H_2O)_3]_2(OH)_2 \cdot [CH_3C(=N-O)C(CH_3)=N$
$CH_2CH_2NHCH_2CH_2NHCH_2CH_2N=C(CH_3)$
$C(=N-O)CH_3]$ . . . . Th: SVol.D4-158
$C_{14}H_{40}InNSi_4$ . . . . $[(CH_3)_3Si-CH_2]_2In-N[Si(CH_3)_3]_2$ . . . . In: Org.Comp.1-253, 258
$C_{14}H_{40}InPSi_4$ . . . . $[(CH_3)_3Si-CH_2]_2In-P[Si(CH_3)_3]_2$ . . . . In: Org.Comp.1-312, 316
$C_{14}H_{40}Li_2N_2Re$. . . $Li_2[(CH_3)_8Re] \cdot (CH_3)_2NCH_2CH_2N(CH_3)_2$ . . . . . Re: Org.Comp.1-2
$C_{14}H_{40}NO_2OsSi_4^-$ $[((CH_3)_3SiCH_2)_2Os(N)(OSi(CH_3)_3)_2]^-$ . . . . Os: Org.Comp.A1-20
$C_{14}H_{40}N_{10}O_{22}Th_2$ $[Th(NO_3)_2(H_2O)_3]_2(OH)_2 \cdot [CH_3C(=N-O)$
$C(CH_3)=NCH_2CH_2NHCH_2CH_2NHCH_2CH_2$
$N=C(CH_3)C(=N-O)CH_3]$ . . . . Th: SVol.D4-158
$C_{14}H_{41}IN_2Si_4Sn$ . . $[(CH_3)_3Si]_2N-Sn(I)(C_2H_5)-N[Si(CH_3)_3]_2$. . . . . . . Sn: Org.Comp.19-149, 152
$C_{14}H_{42}In_2N_2P_2Si$. $(CH_3)_3In \cdot (CH_3)_3P=N-Si(CH_3)_2-N=P(CH_3)_3 \cdot In(CH_3)_3$
In: Org.Comp.1-49, 51
$C_{14}H_{42}MnO_{18}S_6$ . . $[Mn(H_2O)_6](C((O=)_2SC_2H_5)_3)_2$ Mn:MVol.D7-113/4
$C_{14}H_{42}N_4O_{19}S_7Th$ $Th(NO_3)_4 \cdot 7\ (CH_3)_2SO$ . . . . Th: SVol.D4-160
$C_{14}H_{42}OsP_4$ . . . . . $(CH_3)_2Os[P(CH_3)_3]_4$ . . . . Os: Org.Comp.A1-15
$C_{14}O_{28}Th_2^{6-}$ . . . . . $[Th_2(C_2O_4)_7]^{6-}$ . . . . Th: SVol.C7-91